Methods in Enzymology

Volume 341
RIBONUCLEASES
Part A

METHODS IN ENZYMOLOGY

Methods in Enzymology

Volume 341

Ribonucleases

Part A

EDITED BY

Allen W. Nicholson

DEPARTMENT OF BIOLOGICAL SCIENCES
WAYNE STATE UNIVERSITY
DETROIT, MICHIGAN

ACADEMIC PRESS

San Diego London Boston New York Sydney Tokyo Toronto

This book is printed on acid-free paper. ∞

Academic Press
A Harcourt Science and Technology Company
525 B Street, Suite 1900, San Diego, California 92101-4495, USA
http://www.academicpress.com

Academic Press
Harcourt Place, 32 Jamestown Road, London NW1 7BY, UK
http://www.academicpress.com

International Standard Book Number: 0-12-182242-7

PRINTED IN THE UNITED STATES OF AMERICA
01 02 03 04 05 06 07 SB 9 8 7 6 5 4 3 2 1

Table of Contents

Section I. Ribonuclease Classification and Review

Section II. Ribonuclease Assays

Section III. Secreted Ribonucleases

Section IV. Ribonucleases H

Section V. Synthetic Ribonucleases

Section VI. Ribonucleolytic Nucleic Acids

Section VII. Ribonuclease Inhibitors

Section VIII. Nonenzymatic Cleavage of RNA

Contributors to Volume 341

Article numbers are in parentheses following the names of contributors.
Affiliations listed are current.

STEFFEN ABEL (22), *Institut für Biochemie Martin-Luther-Universitat Halle-Wittenberg D-06099 Halle, Germany**

K. RAVI ACHARYA (40), *Department of Biology and Biochemistry, University of Bath, Bath BA2 7AY, United Kingdom*

SIRINART ANANVORANICH (35), *Department of Chemistry and Biochemistry, University of Windsor, Windsor, Ontario N9B 3P4, Canada*

L. ARAVIND (1), *National Center for Biotechnology Information, National Library of Medicine, National Institutes of Health, Bethesda, Maryland 20894*

ARULVATHANI ARUDCHANDRAN (25), *Laboratory of Molecular Genetics, National Institute of Child Health and Human Development, National Institutes of Health, Bethesda, Maryland 20892*

MARK A. BEHLKE (5), *Integrated DNA Technologies Inc., Coralville, Iowa 52241*

NATALIA BELOGLAZOVA (30), *Laboratory of Nucleic Acids Biochemistry, Novosibirsk Institute of Bioorganic Chemistry, Novosibirsk-90, 630090, Russia*

ANTONI BENITO (13), *Laboratori d'Enginyeria de Proteïnes, Departament de Biologia, Universitat de Girona, E-17071 Girona, Spain*

ESTER BOIX (18), *Departament de Bioquímica i Biologia Molecular, Universitat Autònoma de Barcelona, E-08193 Bellaterra, Spain*†

DAVID BROWN (41), *Ambion, Inc., Austin, Texas 78744*

JAMES W. BROWN (4), *Department of Microbiology, North Carolina State University, Raleigh, North Carolina 27695*

JOHN M. BURKE (36), *Department of Microbiology and Molecular Genetics, University of Vermont, Burlington, Vermont 05405*

ALBERT CANALS (13), *Laboratori d'Enginyeria de Proteïnes, Departament de Biologia, Universitat de Girona, E-17071 Girona, Spain*

VINCENT J. CANNISTRARO (11), *Department of Molecular Microbiology, Washington University School of Medicine, St. Louis, Missouri 63110*

CHRISTIAN CAZENAVE (7), *INSERM U 386, Université Victor Segalen, 33076 Bordeaux cédex, France*

SUSANA M. CERRITELLI (25), *Laboratory of Molecular Genetics, National Institute of Child Health and Human Development, National Institutes of Health, Bethesda, Maryland 20892*

RETO CRAMERI (20), *Swiss Institute of Allergy and Asthma Research, CH-7270 Davos, Switzerland*

STANLEY T. CROOKE (27, 31), *Department of Molecular and Structural Biology, Isis Pharmaceuticals Inc., Carlsbad, California 92008*

*Current affiliation: Department of Vegetable Crops, University of California, Davis, California 95616.
†Current affiliation: Department of Biology and Biochemistry, University of Bath, Bath BA2 7AY, United Kingdom.

ROBERT J. CROUCH (25), *Laboratory of Molecular Genetics, National Institute of Child Health and Human Development, National Institutes of Health, Bethesda, Maryland 20892*

JORGE CRUZ-REYES (10), *Department of Biological Chemistry, Johns Hopkins University School of Medicine, Baltimore, Maryland 21205*[*]

CLAUDI M. CUCHILLO (13), *Departament de Bioquímica i Biologia Molecular, Universitat Autònoma de Barcelona, E-08193 Bellaterra, Spain*

GIUSEPPE D'ALESSIO (15), *Dipartimento di Chimica Organica e Biologica, Università di Napoli Federico II, 80134 Napoli, Italy*

JOSE-ANTONIO DAROS (34), *Instituto de Biología Molecular y Celular de Plantas (UPV-CSIC), Universidad Politécnica de Valencia, 46022 Valencia, Spain*

LAURENT DAVID (33), *Département d'Informatique et de Recherche Opérationnelle, Université de Montréal, Montréal, Québec H3C 3J7, Canada*[†]

JULIAN DAVIES (20, 21), *Department of Microbiology and Immunology, University of British Columbia, Vancouver, British Columbia V6T 1Z3 Canada*

MARCOS DE LA PEÑA (34) *Instituto de Biología Molecular y Cellular de Plantas (UPV-CSIC), Universidad Politécnica de Valencia, 46022 Valencia, Spain*

PATRICK DESCHÊNES (35), *Département de Biochimie, Université de Sherbrooke, Sherbrooke, Québec J1H 5N4, Canada*

ALBERTO DI DONATO (15), *Dipartimento di Chimica Organica e Biologica, Università di Napoli Federico II, 80134 Napoli, Italy*

JOSEPH B. DOMACHOWSKE (17), *Department of Pediatrics, Division of Infectious Diseases, State University of New York Upstate Medical University, Syracuse, New York 13210*

SHAD R. EUBANKS (5), *Integrated DNA Technologies Inc., Coralville, Iowa 52241*

KARINE FIOLA (35), *Département de Biochimie, Université de Sherbrooke, Sherbrooke, Québec J1H 5N4, Canada*

RICARDO FLORES (34), *Instituto de Biología Molecular y Celular de Plantas (UPV-CSIC), Universidad Politécnica de Valencia, 46022 Valencia, Spain*

PETER FRANK (26), *Division of Molecular Genetics, Institute of Cancer Research, University of Vienna, 1090 Vienna, Austria*

LUCÍA GARCÍA-ORTEGA (21), *Departamento de Bioquímica y Biología Molecular I, Universidad Complutense de Madrid, E-28040 Madrid, Spain*

JOSÉ G. GAVILANES (21), *Departamento de Bioquímica y Biología Molecular I, Universidad Complutense de Madrid, E-28040 Madrid, Spain*

PATRICK GENDRON (33), *Département d'Informatique et de Recherche Opérationnelle, Université de Montréal, Montréal, Québec H3C 3J7, Canada*

RICHARD GIEGÉ (30), *Institut de Biologie Moléculaire et Cellulaire du CNRS, UPR 9002, F-67084 Strasbourg cédex, France*

THOMAS GREINER-STÖFFELE (9), *Fakultät für Biowissenschaften, Pharmazie, and Psychologie Universität Leipzig, D-04103 Leipzig, Germany*

ULRICH HAHN (9), *Institut für Biochemie, Universität Leipzig, D-04103 Leipzig, Germany*

THOMAS A. HALL (4), *Department of Microbiology, North Carolina State University, Raleigh, North Carolina 27695*

[*]Current affiliation: Department of Biochemistry and Biophysics, Texas A&M University, College Station, Texas 77843.

[†]Current affiliation: AstraZeneca R&D Lund, S-221 87 Lund, Sweden.

KENNETH J. HAMPEL (36), *Department of Microbiology and Molecular Genetics, University of Vermont, Burlington, Vermont 05405*

ROBERT W. HARTLEY (38), *Laboratory of Cellular and Developmental Biology, National Institute of Diabetes and Digestive and Kidney Diseases, National Institutes of Health, Bethesda, Maryland 20892*

CARMEN HERNANDEZ (34), *Instituto de Biología Molecular y Celular de Plantas (UPV-CSIC), Universidad Politécnica de Valencia, 46022 Valencia, Spain*

MASACHIKA IRIE (3), *Department of Microbiology, Hoshi College of Pharmacy, Tokyo 142-8501, Japan*

GERALD F. JOYCE (32), *Departments of Chemistry and Molecular Biology and The Skaggs Institute for Chemical Biology, The Scripps Research Institute, La Jolla, California 92037*

SHIGENORI KANAYA (24), *Department of Material and Life Science, Graduate School of Engineering, Osaka University, Osaka 565-0871, Japan*

RICHARD KAO (20, 21), *HKU-Pasteur Research Center, The University of Hong Kong, Pokfulam, Hong Kong*

BRADLEY R. KELEMEN (5), *Promega Corporation, Madison, Wisconsin 53711*

DAVID KENNELL (11), *Department of Molecular Microbiology, Washington University School of Medicine, St. Louis, Missouri 63110*

RYSZARD KIERZEK (42), *Institute of Bioorganic Chemistry, Polish Academy of Sciences, 61-704 Poznañ, Poland*

KOICHIRO KISHI (6), *Department of Legal Medicine, Gunma University School of Medicine, Gunma 371-8511, Japan*

TONY A. KLINK (5), *Department of Biochemistry, University of Wisconsin—Madison, Madison, Wisconsin 53706*

MARGRET KÖCK (22), *Biozentrum, Martin-Luther-Universität Halle-Wittenberg, D-06099 Halle, Germany*

MAKOTO KOMIYAMA (29), *Research Center for Advanced Science and Technology, University of Tokyo, Tokyo 153-8904, Japan*

EUGENE V. KOONIN (1), *National Center for Biotechnology Information, National Library of Medicine, National Institutes of Health, Bethesda, Maryland 20894*

KERSTIN KORN (9), *GNOTHIS AB, Electrum 212, SE-16440 Kista, Sweden*

AKINORI KUZUYA (29), *Research Center for Advanced Science and Technology, University of Tokyo, Tokyo 153-8904, Japan*

JAVIER LACADENA (21), *Facultad de Biologia, Universidad SEK, 40003 Segovia, Spain*

DOMINIC LAMBERT (33), *Département d'Informatique et de Recherche Opérationnelle, Université de Montréal, Montréal, Québec H3C 3J7, Canada*

MASSIMO LIBONATI (14), *Dipartimento di Scienze Neurologiche e della Visione, Sezione di Chimica Biologica, Università di Verona, Verona 37134, Italy*

WALT F. LIMA (27, 31), *Department of Molecular and Structural Biology, Isis Pharmaceuticals Inc., Carlsbad, California 92008*

STEFAN LOVERIX (19), *Dienst Ultrastructuur, Instituut voor Moleculaire Biologie, Vrije Universiteit Brussel, B-1640 Sint-Genesius-Rode, Belgium*

FRANÇOIS MAJOR (33), *Département d'Informatique et de Recherche Opérationnelle, Université de Montréal, Montréal, Québec H3C 3J7, Canada*

JOSÉ M. MANCHEÑO (21), *Departamento de Bioquímica y Biolgía Molecular I, Universidad Complutense de Madrid, E-28040 Madrid, Spain*

ÁLVARO MARTÍNEZ DEL POZO (20, 21), *Departamento de Bioquimica y Biolgía Molecular I, Universidad Complutense de Madrid, E-28040 Madrid, Spain*

ANTONIO MARTÍNEZ-RUIZ (20, 21), *Centro de Investigaciones Biologicas–CSIC, E-28006 Madrid, Spain*

ALESSANDRO MICHIENZI (37), *Molecular Biology Department, Beckman Research Institute of the City of Hope, Duarte, California 91010*

SOREN NAABY-HANSEN (8), *Ludwig Institute for Cancer Research (RF&UCMS Branch), London WIW 7BS, United Kingdom*

KAZUO NITTA (23), *Cancer Research Institute, Tehoku Pharmaceutical University, Sendai 981-8558, Japan*

M. VICTÒRIA NOGUÉS (13), *Departament de Bioquímica i Biologia Molecular, Facultat de Ciències, Universitat Autònoma de Barcelona, E-08193 Bellaterra, Spain*

KAZUKO OHGI (3), *Department of Microbiology, Hoshi College of Pharmacy, Tokyo 142-8501, Japan*

MERCEDES OÑADERRA (21), *Departamento de Bioquímica y Biología Molecular I, Universidad Complutense de Madrid, E-28040 Madrid, Spain*

JONATHAN OUELLET (35), *Départemement de Biochimie, Université de Sherbrooke, Sherbrooke, Québec J1H 5N4, Canada*

CHIWOOK PARK (5), *Department of Biochemistry, University of Wisconsin—Madison, Madison, Wisconsin 53706*

BRITTAN L. PASLOSKE (41), *Ambion, Inc., Austin, Texas 78744*

JEAN-PIERRE PERREAULT (35), *Départemement de Biochimie, Université de Sherbrooke, Sherbrooke, Québec J1H 5N4, Canada*

RENATA PICCOLI (15), *Dipartmento di Chimica Organica e Biologica, Università di Napoli Federico II, 80134 Napoli, Italy*

ROBERT PINARD (36), *Department of Microbiology and Molecular Genetics, University of Vermont, Burlington, Vermont 05405*

RONALD T. RAINES (5), *Departments of Biochemistry and Chemistry, University of Wisconsin-Madison, Madison, Wisconsin 53706*

MARC RIBÓ (13), *Laboratori d'Enginyeria de Proteïnes, Departament de Biologia, Universitat de Girona, E-17071 Girona, Spain*

JAMES F. RIORDAN (16), *Center for Biochemical and Biophysical Sciences and Medicine, Harvard Medical School, Boston, Massachusetts 02115*

HELENE F. ROSENBERG (17), *Laboratory of Host Defenses, National Institute of Allergy and Infectious Diseases, National Institutes of Health, Bethesda, Maryland 20892*

JOHN J. ROSSI (37), *Molecular Biology Department, Beckman Research Institute of the City of Hope, Duarte, California 91010*

MONICA J. ROTH (28), *Department of Biochemistry, Robert Wood Johnson Medical School, University of Medicine and Dentistry of New Jersey, Piscataway, New Jersey 08854*

LAURA N. RUSCHÉ (10), *Department of Molecular and Cell Biology, University of California, Berkeley, California 94720*

ANIELLO RUSSO (15, 40), *Department of Life Sciences, Second University of Naples, 81100 Caserta, Italy*

A. D. J. SCADDEN (8), *Department of Biochemistry, University of Cambridge, Cambridge CB2 1GA, United Kingdom*

HAROLD A. SCHERAGA (12), *Baker Laboratory of Chemistry and Chemical Biology, Cornell University, Ithaca, New York 14853*

ROBERT SHAPIRO (39, 40), *Center for Biochemical and Biophysical Sciences and Medicine, and Department of Pathology, Harvard Medical School, Boston, Massachusetts 02115*

VLADIMIR SIL'NIKOV (30), *Laboratory of Organic Synthesis, Novosibirsk Institute of Bioorganic Chemistry, Novosibirsk-90, 630090, Russia*

CHRISTINE SMITH SNYDER (28), *Department of Molecular Biology, School of Osteopathic Medicine, University of Medicine and Dentistry of New Jersey, Stratford, New Jersey 08084*

BARBARA SOLLNER-WEBB (10), *Department of Biological Chemistry, Johns Hopkins University School of Medicine, Baltimore, Maryland 21205*

SALVATORE SORRENTINO (14), *Dipartimento di Chimica Biologica, Università di Napoli Federico II, Napoli 80134, Italy*

JAN STEYAERT (19), *Dienst Ultrastructuur, Instituut voor Moleculaire Biologie, Vrije Universiteit Brussel, B-1640 Sint-Genesius-Rode, Belgium*

JUN SUMAOKA (29), *Research Center for Advanced Science and Technology, University of Tokyo, Tokyo 153-8904, Japan*

ROZAMOND Y. SWEENEY (5), *Department of Biochemistry, University of Wisconsin—Madison, Madison, Wisconsin 53706*

HARUO TAKESHITA (6), *Department of Legal Medicine, Gunma University School of Medicine, Gunma 371-8511, Japan*

JEAN-JACQUES TOULMÉ (7), *INSERM U 386, Université Victor Segalen Bordeaux 2 146, 33076 Bordeaux cédex, France*

ANTONIO VERA (34), *División de Genética, Universidad Miguel Hernández, 03550 Alicante, Spain*

MARIA VILANOVA (13), *Laboratori d'Enginyeria de Proteïnes, Departament de Biología, Facultat de Ciències, Universitat de Girona, E-17071 Girona, Spain*

VALENTIN VLASSOV (30), *Laboratory of Nucleic Acids Biochemistry, Novosibirsk Institute of Bioorganic Chemistry, Novosibirsk-90, 630090, Russia*

WILLIAM J. WEDEMEYER (12), *Baker Laboratory of Chemistry and Chemical Biology, Cornell University, Ithaca, New York 14853*

ERVIN WELKER (12), *Baker Laboratory of Chemistry and Chemical Biology, Cornell University, Ithaca, New York 14853*

ULRIKE WINTERSBERGER (26), *Division of Molecular Genetics, Institute of Cancer Research, University of Vienna, 1090 Vienna, Austria*

HONGJIANG WU (27), *Department of Molecular and Structural Biology, Isis Pharmaceuticals Inc., Carlsbad, California 92008*

YOJI YAMAMOTO (29), *Research Center for Advanced Science and Technology, University of Tokyo, Tokyo 153-8904, Japan*

TOSHIHIRO YASUDA (6), *Department of Biology, Fukui Medical University, Fukui 910-1193, Japan*

HIROSHI YOSHIDA (2), *Department of Chemistry, Shimane Medical University, Izumo 693-8501, Japan*

MARINA ZENKOVA (30), *Laboratory of Nucleic Acids Biochemistry, Novosibirsk Institute of Bioorganic Chemistry, Novosibirsk-90, 630090, Russia*

Preface

The cleavage of RNA has remarkably diverse biological consequences. The agents that catalyze this reaction—ribonucleases—have been the objects of intensive study for many years and from many angles. As noted by one of the contributing authors, research on ribonucleases has produced no less than four Nobel prizes. The archetypal member, pancreatic ribonuclease, was the first protein whose covalent structure was determined, as well as the first protein to be chemically synthesized. There has been a major expansion in our knowledge of the structures, mechanisms, and evolutionary relatedness of ribonucleases, as well as an understanding of ribonuclease function in RNA maturation, RNA degradation, gene regulation, and cellular physiology. Engineered ribonucleases, ribozymes, and synthetic ribonucleases are firmly established, and their applications in medicine and biotechnology are underway. In view of the distinguished history and new research developments it was surprising to find that a *Methods in Enzymology* volume on ribonucleases was not on the shelves. The enthusiastic response to the invitation to contribute to such a volume led to over 80 chapters, which now appear in Volumes 341 and 342 of *Methods in Enzymology*. These volumes should be of lasting value in providing techniques and tools for identifying, characterizing, and applying ribonucleases, and will impart some of the current excitement in research on these fascinating enzymes.

ALLEN W. NICHOLSON

METHODS IN ENZYMOLOGY

VOLUME XXXVI. Hormone Action (Part A: Steroid Hormones)
Edited by BERT W. O'MALLEY AND JOEL G. HARDMAN

VOLUME XXXVII. Hormone Action (Part B: Peptide Hormones)
Edited by BERT W. O'MALLEY AND JOEL G. HARDMAN

VOLUME XXXVIII. Hormone Action (Part C: Cyclic Nucleotides)
Edited by JOEL G. HARDMAN AND BERT W. O'MALLEY

VOLUME XXXIX. Hormone Action (Part D: Isolated Cells, Tissues, and Organ Systems)
Edited by JOEL G. HARDMAN AND BERT W. O'MALLEY

VOLUME XL. Hormone Action (Part E: Nuclear Structure and Function)
Edited by BERT W. O'MALLEY AND JOEL G. HARDMAN

VOLUME XLI. Carbohydrate Metabolism (Part B)
Edited by W. A. WOOD

VOLUME XLII. Carbohydrate Metabolism (Part C)
Edited by W. A. WOOD

VOLUME XLIII. Antibiotics
Edited by JOHN H. HASH

VOLUME XLIV. Immobilized Enzymes
Edited by KLAUS MOSBACH

VOLUME XLV. Proteolytic Enzymes (Part B)
Edited by LASZLO LORAND

VOLUME XLVI. Affinity Labeling
Edited by WILLIAM B. JAKOBY AND MEIR WILCHEK

VOLUME XLVII. Enzyme Structure (Part E)
Edited by C. H. W. HIRS AND SERGE N. TIMASHEFF

VOLUME XLVIII. Enzyme Structure (Part F)
Edited by C. H. W. HIRS AND SERGE N. TIMASHEFF

VOLUME XLIX. Enzyme Structure (Part G)
Edited by C. H. W. HIRS AND SERGE N. TIMASHEFF

VOLUME L. Complex Carbohydrates (Part C)
Edited by VICTOR GINSBURG

VOLUME LI. Purine and Pyrimidine Nucleotide Metabolism
Edited by PATRICIA A. HOFFEE AND MARY ELLEN JONES

VOLUME LII. Biomembranes (Part C: Biological Oxidations)
Edited by SIDNEY FLEISCHER AND LESTER PACKER

VOLUME LIII. Biomembranes (Part D: Biological Oxidations)
Edited by SIDNEY FLEISCHER AND LESTER PACKER

VOLUME LIV. Biomembranes (Part E: Biological Oxidations)
Edited by SIDNEY FLEISCHER AND LESTER PACKER

VOLUME 73. Immunochemical Techniques (Part B)
Edited by JOHN J. LANGONE AND HELEN VAN VUNAKIS

VOLUME 74. Immunochemical Techniques (Part C)
Edited by JOHN J. LANGONE AND HELEN VAN VUNAKIS

VOLUME 75. Cumulative Subject Index Volumes XXXI, XXXII, XXXIV–LX
Edited by EDWARD A. DENNIS AND MARTHA G. DENNIS

VOLUME 76. Hemoglobins
Edited by ERALDO ANTONINI, LUIGI ROSSI-BERNARDI, AND EMILIA CHIANCONE

VOLUME 77. Detoxication and Drug Metabolism
Edited by WILLIAM B. JAKOBY

VOLUME 78. Interferons (Part A)
Edited by SIDNEY PESTKA

VOLUME 79. Interferons (Part B)
Edited by SIDNEY PESTKA

VOLUME 80. Proteolytic Enzymes (Part C)
Edited by LASZLO LORAND

VOLUME 81. Biomembranes (Part H: Visual Pigments and Purple Membranes, I)
Edited by LESTER PACKER

VOLUME 82. Structural and Contractile Proteins (Part A: Extracellular Matrix)
Edited by LEON W. CUNNINGHAM AND DIXIE W. FREDERIKSEN

VOLUME 83. Complex Carbohydrates (Part D)
Edited by VICTOR GINSBURG

VOLUME 84. Immunochemical Techniques (Part D: Selected Immunoassays)
Edited by JOHN J. LANGONE AND HELEN VAN VUNAKIS

VOLUME 85. Structural and Contractile Proteins (Part B: The Contractile Apparatus and the Cytoskeleton)
Edited by DIXIE W. FREDERIKSEN AND LEON W. CUNNINGHAM

VOLUME 86. Prostaglandins and Arachidonate Metabolites
Edited by WILLIAM E. M. LANDS AND WILLIAM L. SMITH

VOLUME 87. Enzyme Kinetics and Mechanism (Part C: Intermediates, Stereochemistry, and Rate Studies)
Edited by DANIEL L. PURICH

VOLUME 88. Biomembranes (Part I: Visual Pigments and Purple Membranes, II)
Edited by LESTER PACKER

VOLUME 89. Carbohydrate Metabolism (Part D)
Edited by WILLIS A. WOOD

VOLUME 90. Carbohydrate Metabolism (Part E)
Edited by WILLIS A. WOOD

VOLUME 126. Biomembranes (Part N: Transport in Bacteria, Mitochondria, and Chloroplasts: Protonmotive Force)
Edited by SIDNEY FLEISCHER AND BECCA FLEISCHER

VOLUME 127. Biomembranes (Part O: Protons and Water: Structure and Translocation)
Edited by LESTER PACKER

VOLUME 128. Plasma Lipoproteins (Part A: Preparation, Structure, and Molecular Biology)
Edited by JERE P. SEGREST AND JOHN J. ALBERS

VOLUME 129. Plasma Lipoproteins (Part B: Characterization, Cell Biology, and Metabolism)
Edited by JOHN J. ALBERS AND JERE P. SEGREST

VOLUME 130. Enzyme Structure (Part K)
Edited by C. H. W. HIRS AND SERGE N. TIMASHEFF

VOLUME 131. Enzyme Structure (Part L)
Edited by C. H. W. HIRS AND SERGE N. TIMASHEFF

VOLUME 132. Immunochemical Techniques (Part J: Phagocytosis and Cell-Mediated Cytotoxicity)
Edited by GIOVANNI DI SABATO AND JOHANNES EVERSE

VOLUME 133. Bioluminescence and Chemiluminescence (Part B)
Edited by MARLENE DELUCA AND WILLIAM D. MCELROY

VOLUME 134. Structural and Contractile Proteins (Part C: The Contractile Apparatus and the Cytoskeleton)
Edited by RICHARD B. VALLEE

VOLUME 135. Immobilized Enzymes and Cells (Part B)
Edited by KLAUS MOSBACH

VOLUME 136. Immobilized Enzymes and Cells (Part C)
Edited by KLAUS MOSBACH

VOLUME 137. Immobilized Enzymes and Cells (Part D)
Edited by KLAUS MOSBACH

VOLUME 138. Complex Carbohydrates (Part E)
Edited by VICTOR GINSBURG

VOLUME 139. Cellular Regulators (Part A: Calcium- and Calmodulin-Binding Proteins)
Edited by ANTHONY R. MEANS AND P. MICHAEL CONN

VOLUME 140. Cumulative Subject Index Volumes 102–119, 121–134

VOLUME 141. Cellular Regulators (Part B: Calcium and Lipids)
Edited by P. MICHAEL CONN AND ANTHONY R. MEANS

VOLUME 142. Metabolism of Aromatic Amino Acids and Amines
Edited by SEYMOUR KAUFMAN

VOLUME 143. Sulfur and Sulfur Amino Acids
Edited by WILLIAM B. JAKOBY AND OWEN GRIFFITH

VOLUME 144. Structural and Contractile Proteins (Part D: Extracellular Matrix)
Edited by LEON W. CUNNINGHAM

VOLUME 145. Structural and Contractile Proteins (Part E: Extracellular Matrix)
Edited by LEON W. CUNNINGHAM

VOLUME 146. Peptide Growth Factors (Part A)
Edited by DAVID BARNES AND DAVID A. SIRBASKU

VOLUME 147. Peptide Growth Factors (Part B)
Edited by DAVID BARNES AND DAVID A. SIRBASKU

VOLUME 148. Plant Cell Membranes
Edited by LESTER PACKER AND ROLAND DOUCE

VOLUME 149. Drug and Enzyme Targeting (Part B)
Edited by RALPH GREEN AND KENNETH J. WIDDER

VOLUME 150. Immunochemical Techniques (Part K: *In Vitro* Models of B and T Cell Functions and Lymphoid Cell Receptors)
Edited by GIOVANNI DI SABATO

VOLUME 151. Molecular Genetics of Mammalian Cells
Edited by MICHAEL M. GOTTESMAN

VOLUME 152. Guide to Molecular Cloning Techniques
Edited by SHELBY L. BERGER AND ALAN R. KIMMEL

VOLUME 153. Recombinant DNA (Part D)
Edited by RAY WU AND LAWRENCE GROSSMAN

VOLUME 154. Recombinant DNA (Part E)
Edited by RAY WU AND LAWRENCE GROSSMAN

VOLUME 155. Recombinant DNA (Part F)
Edited by RAY WU

VOLUME 156. Biomembranes (Part P: ATP-Driven Pumps and Related Transport: The Na, K-Pump)
Edited by SIDNEY FLEISCHER AND BECCA FLEISCHER

VOLUME 157. Biomembranes (Part Q: ATP-Driven Pumps and Related Transport: Calcium, Proton, and Potassium Pumps)
Edited by SIDNEY FLEISCHER AND BECCA FLEISCHER

VOLUME 158. Metalloproteins (Part A)
Edited by JAMES F. RIORDAN AND BERT L. VALLEE

VOLUME 159. Initiation and Termination of Cyclic Nucleotide Action
Edited by JACKIE D. CORBIN AND ROGER A. JOHNSON

VOLUME 160. Biomass (Part A: Cellulose and Hemicellulose)
Edited by WILLIS A. WOOD AND SCOTT T. KELLOGG

VOLUME 161. Biomass (Part B: Lignin, Pectin, and Chitin)
Edited by WILLIS A. WOOD AND SCOTT T. KELLOGG

VOLUME 162. Immunochemical Techniques (Part L: Chemotaxis and Inflammation)
Edited by GIOVANNI DI SABATO

VOLUME 163. Immunochemical Techniques (Part M: Chemotaxis and Inflammation)
Edited by GIOVANNI DI SABATO

VOLUME 164. Ribosomes
Edited by HARRY F. NOLLER, JR., AND KIVIE MOLDAVE

VOLUME 165. Microbial Toxins: Tools for Enzymology
Edited by SIDNEY HARSHMAN

VOLUME 166. Branched-Chain Amino Acids
Edited by ROBERT HARRIS AND JOHN R. SOKATCH

VOLUME 167. Cyanobacteria
Edited by LESTER PACKER AND ALEXANDER N. GLAZER

VOLUME 168. Hormone Action (Part K: Neuroendocrine Peptides)
Edited by P. MICHAEL CONN

VOLUME 169. Platelets: Receptors, Adhesion, Secretion (Part A)
Edited by JACEK HAWIGER

VOLUME 170. Nucleosomes
Edited by PAUL M. WASSARMAN AND ROGER D. KORNBERG

VOLUME 171. Biomembranes (Part R: Transport Theory: Cells and Model Membranes)
Edited by SIDNEY FLEISCHER AND BECCA FLEISCHER

VOLUME 172. Biomembranes (Part S: Transport: Membrane Isolation and Characterization)
Edited by SIDNEY FLEISCHER AND BECCA FLEISCHER

VOLUME 173. Biomembranes [Part T: Cellular and Subcellular Transport: Eukaryotic (Nonepithelial) Cells]
Edited by SIDNEY FLEISCHER AND BECCA FLEISCHER

VOLUME 174. Biomembranes [Part U: Cellular and Subcellular Transport: Eukaryotic (Nonepithelial) Cells]
Edited by SIDNEY FLEISCHER AND BECCA FLEISCHER

VOLUME 175. Cumulative Subject Index Volumes 135–139, 141–167

VOLUME 176. Nuclear Magnetic Resonance (Part A: Spectral Techniques and Dynamics)
Edited by NORMAN J. OPPENHEIMER AND THOMAS L. JAMES

VOLUME 177. Nuclear Magnetic Resonance (Part B: Structure and Mechanism)
Edited by NORMAN J. OPPENHEIMER AND THOMAS L. JAMES

VOLUME 178. Antibodies, Antigens, and Molecular Mimicry
Edited by JOHN J. LANGONE

VOLUME 179. Complex Carbohydrates (Part F)
Edited by VICTOR GINSBURG

VOLUME 180. RNA Processing (Part A: General Methods)
Edited by JAMES E. DAHLBERG AND JOHN N. ABELSON

VOLUME 181. RNA Processing (Part B: Specific Methods)
Edited by JAMES E. DAHLBERG AND JOHN N. ABELSON

VOLUME 182. Guide to Protein Purification
Edited by MURRAY P. DEUTSCHER

VOLUME 183. Molecular Evolution: Computer Analysis of Protein and Nucleic Acid Sequences
Edited by RUSSELL F. DOOLITTLE

VOLUME 184. Avidin-Biotin Technology
Edited by MEIR WILCHEK AND EDWARD A. BAYER

VOLUME 185. Gene Expression Technology
Edited by DAVID V. GOEDDEL

VOLUME 186. Oxygen Radicals in Biological Systems (Part B: Oxygen Radicals and Antioxidants)
Edited by LESTER PACKER AND ALEXANDER N. GLAZER

VOLUME 187. Arachidonate Related Lipid Mediators
Edited by ROBERT C. MURPHY AND FRANK A. FITZPATRICK

VOLUME 188. Hydrocarbons and Methylotrophy
Edited by MARY E. LIDSTROM

VOLUME 189. Retinoids (Part A: Molecular and Metabolic Aspects)
Edited by LESTER PACKER

VOLUME 190. Retinoids (Part B: Cell Differentiation and Clinical Applications)
Edited by LESTER PACKER

VOLUME 191. Biomembranes (Part V: Cellular and Subcellular Transport: Epithelial Cells)
Edited by SIDNEY FLEISCHER AND BECCA FLEISCHER

VOLUME 192. Biomembranes (Part W: Cellular and Subcellular Transport: Epithelial Cells)
Edited by SIDNEY FLEISCHER AND BECCA FLEISCHER

VOLUME 193. Mass Spectrometry
Edited by JAMES A. MCCLOSKEY

VOLUME 194. Guide to Yeast Genetics and Molecular Biology
Edited by CHRISTINE GUTHRIE AND GERALD R. FINK

VOLUME 195. Adenylyl Cyclase, G Proteins, and Guanylyl Cyclase
Edited by ROGER A. JOHNSON AND JACKIE D. CORBIN

VOLUME 196. Molecular Motors and the Cytoskeleton
Edited by RICHARD B. VALLEE

VOLUME 197. Phospholipases
Edited by EDWARD A. DENNIS

VOLUME 198. Peptide Growth Factors (Part C)
Edited by DAVID BARNES, J. P. MATHER, AND GORDON H. SATO

VOLUME 199. Cumulative Subject Index Volumes 168–174, 176–194

VOLUME 200. Protein Phosphorylation (Part A: Protein Kinases: Assays, Purification, Antibodies, Functional Analysis, Cloning, and Expression)
Edited by TONY HUNTER AND BARTHOLOMEW M. SEFTON

VOLUME 201. Protein Phosphorylation (Part B: Analysis of Protein Phosphorylation, Protein Kinase Inhibitors, and Protein Phosphatases)
Edited by TONY HUNTER AND BARTHOLOMEW M. SEFTON

VOLUME 202. Molecular Design and Modeling: Concepts and Applications (Part A: Proteins, Peptides, and Enzymes)
Edited by JOHN J. LANGONE

VOLUME 203. Molecular Design and Modeling: Concepts and Applications (Part B: Antibodies and Antigens, Nucleic Acids, Polysaccharides, and Drugs)
Edited by JOHN J. LANGONE

VOLUME 204. Bacterial Genetic Systems
Edited by JEFFREY H. MILLER

VOLUME 205. Metallobiochemistry (Part B: Metallothionein and Related Molecules)
Edited by JAMES F. RIORDAN AND BERT L. VALLEE

VOLUME 206. Cytochrome P450
Edited by MICHAEL R. WATERMAN AND ERIC F. JOHNSON

VOLUME 207. Ion Channels
Edited by BERNARDO RUDY AND LINDA E. IVERSON

VOLUME 208. Protein–DNA Interactions
Edited by ROBERT T. SAUER

VOLUME 209. Phospholipid Biosynthesis
Edited by EDWARD A. DENNIS AND DENNIS E. VANCE

VOLUME 210. Numerical Computer Methods
Edited by LUDWIG BRAND AND MICHAEL L. JOHNSON

VOLUME 211. DNA Structures (Part A: Synthesis and Physical Analysis of DNA)
Edited by DAVID M. J. LILLEY AND JAMES E. DAHLBERG

VOLUME 212. DNA Structures (Part B: Chemical and Electrophoretic Analysis of DNA)
Edited by DAVID M. J. LILLEY AND JAMES E. DAHLBERG

VOLUME 213. Carotenoids (Part A: Chemistry, Separation, Quantitation, and Antioxidation)
Edited by LESTER PACKER

VOLUME 214. Carotenoids (Part B: Metabolism, Genetics, and Biosynthesis)
Edited by LESTER PACKER

VOLUME 215. Platelets: Receptors, Adhesion, Secretion (Part B)
Edited by JACEK J. HAWIGER

VOLUME 216. Recombinant DNA (Part G)
Edited by RAY WU

VOLUME 217. Recombinant DNA (Part H)
Edited by RAY WU

VOLUME 218. Recombinant DNA (Part I)
Edited by RAY WU

VOLUME 219. Reconstitution of Intracellular Transport
Edited by JAMES E. ROTHMAN

VOLUME 220. Membrane Fusion Techniques (Part A)
Edited by NEJAT DÜZGUÜNES

VOLUME 221. Membrane Fusion Techniques (Part B)
Edited by NEJAT DÜZGÜNES

VOLUME 222. Proteolytic Enzymes in Coagulation, Fibrinolysis, and Complement Activation (Part A: Mammalian Blood Coagulation Factors and Inhibitors)
Edited by LASZLO LORAND AND KENNETH G. MANN

VOLUME 223. Proteolytic Enzymes in Coagulation, Fibrinolysis, and Complement Activation (Part B: Complement Activation, Fibrinolysis, and Nonmammalian Blood Coagulation Factors)
Edited by LASZLO LORAND AND KENNETH G. MANN

VOLUME 224. Molecular Evolution: Producing the Biochemical Data
Edited by ELIZABETH ANNE ZIMMER, THOMAS J. WHITE, REBECCA L. CANN, AND ALLAN C. WILSON

VOLUME 225. Guide to Techniques in Mouse Development
Edited by PAUL M. WASSARMAN AND MELVIN L. DEPAMPHILIS

VOLUME 226. Metallobiochemistry (Part C: Spectroscopic and Physical Methods for Probing Metal Ion Environments in Metalloenzymes and Metalloproteins)
Edited by JAMES F. RIORDAN AND BERT L. VALLEE

VOLUME 227. Metallobiochemistry (Part D: Physical and Spectroscopic Methods for Probing Metal Ion Environments in Metalloproteins)
Edited by JAMES F. RIORDAN AND BERT L. VALLEE

VOLUME 228. Aqueous Two-Phase Systems
Edited by HARRY WALTER AND GÖTE JOHANSSON

VOLUME 266. Computer Methods for Macromolecular Sequence Analysis
Edited by RUSSELL F. DOOLITTLE

VOLUME 267. Combinatorial Chemistry
Edited by JOHN N. ABELSON

VOLUME 268. Nitric Oxide (Part A: Sources and Detection of NO; NO Synthase)
Edited by LESTER PACKER

VOLUME 269. Nitric Oxide (Part B: Physiological and Pathological Processes)
Edited by LESTER PACKER

VOLUME 270. High Resolution Separation and Analysis of Biological Macromolecules (Part A: Fundamentals)
Edited by BARRY L. KARGER AND WILLIAM S. HANCOCK

VOLUME 271. High Resolution Separation and Analysis of Biological Macromolecules (Part B: Applications)
Edited by BARRY L. KARGER AND WILLIAM S. HANCOCK

VOLUME 272. Cytochrome P450 (Part B)
Edited by ERIC F. JOHNSON AND MICHAEL R. WATERMAN

VOLUME 273. RNA Polymerase and Associated Factors (Part A)
Edited by SANKAR ADHYA

VOLUME 274. RNA Polymerase and Associated Factors (Part B)
Edited by SANKAR ADHYA

VOLUME 275. Viral Polymerases and Related Proteins
Edited by LAWRENCE C. KUO, DAVID B. OLSEN, AND STEVEN S. CARROLL

VOLUME 276. Macromolecular Crystallography (Part A)
Edited by CHARLES W. CARTER, JR., AND ROBERT M. SWEET

VOLUME 277. Macromolecular Crystallography (Part B)
Edited by CHARLES W. CARTER, JR., AND ROBERT M. SWEET

VOLUME 278. Fluorescence Spectroscopy
Edited by LUDWIG BRAND AND MICHAEL L. JOHNSON

VOLUME 279. Vitamins and Coenzymes (Part I)
Edited by DONALD B. MCCORMICK, JOHN W. SUTTIE, AND CONRAD WAGNER

VOLUME 280. Vitamins and Coenzymes (Part J)
Edited by DONALD B. MCCORMICK, JOHN W. SUTTIE, AND CONRAD WAGNER

VOLUME 281. Vitamins and Coenzymes (Part K)
Edited by DONALD B. MCCORMICK, JOHN W. SUTTIE, AND CONRAD WAGNER

VOLUME 282. Vitamins and Coenzymes (Part L)
Edited by DONALD B. MCCORMICK, JOHN W. SUTTIE, AND CONRAD WAGNER

VOLUME 283. Cell Cycle Control
Edited by WILLIAM G. DUNPHY

VOLUME 284. Lipases (Part A: Biotechnology)
Edited by BYRON RUBIN AND EDWARD A. DENNIS

VOLUME 285. Cumulative Subject Index Volumes 263, 264, 266–284, 286–289

VOLUME 286. Lipases (Part B: Enzyme Characterization and Utilization)
Edited by BYRON RUBIN AND EDWARD A. DENNIS

VOLUME 287. Chemokines
Edited by RICHARD HORUK

VOLUME 288. Chemokine Receptors
Edited by RICHARD HORUK

VOLUME 289. Solid Phase Peptide Synthesis
Edited by GREGG B. FIELDS

VOLUME 290. Molecular Chaperones
Edited by GEORGE H. LORIMER AND THOMAS BALDWIN

VOLUME 291. Caged Compounds
Edited by GERARD MARRIOTT

VOLUME 292. ABC Transporters: Biochemical, Cellular, and Molecular Aspects
Edited by SURESH V. AMBUDKAR AND MICHAEL M. GOTTESMAN

VOLUME 293. Ion Channels (Part B)
Edited by P. MICHAEL CONN

VOLUME 294. Ion Channels (Part C)
Edited by P. MICHAEL CONN

VOLUME 295. Energetics of Biological Macromolecules (Part B)
Edited by GARY K. ACKERS AND MICHAEL L. JOHNSON

VOLUME 296. Neurotransmitter Transporters
Edited by SUSAN G. AMARA

VOLUME 297. Photosynthesis: Molecular Biology of Energy Capture
Edited by LEE MCINTOSH

VOLUME 298. Molecular Motors and the Cytoskeleton (Part B)
Edited by RICHARD B. VALLEE

VOLUME 299. Oxidants and Antioxidants (Part A)
Edited by LESTER PACKER

VOLUME 300. Oxidants and Antioxidants (Part B)
Edited by LESTER PACKER

VOLUME 301. Nitric Oxide: Biological and Antioxidant Activities (Part C)
Edited by LESTER PACKER

VOLUME 302. Green Fluorescent Protein
Edited by P. MICHAEL CONN

VOLUME 303. cDNA Preparation and Display
Edited by SHERMAN M. WEISSMAN

VOLUME 304. Chromatin
Edited by PAUL M. WASSARMAN AND ALAN P. WOLFFE

VOLUME 305. Bioluminescence and Chemiluminescence (Part C)
Edited by THOMAS O. BALDWIN AND MIRIAM M. ZIEGLER

VOLUME 306. Expression of Recombinant Genes in Eukaryotic Systems
Edited by JOSEPH C. GLORIOSO AND MARTIN C. SCHMIDT

VOLUME 307. Confocal Microscopy
Edited by P. MICHAEL CONN

VOLUME 308. Enzyme Kinetics and Mechanism (Part E: Energetics of Enzyme Catalysis)
Edited by DANIEL L. PURICH AND VERN L. SCHRAMM

VOLUME 309. Amyloid, Prions, and Other Protein Aggregates
Edited by RONALD WETZEL

VOLUME 310. Biofilms
Edited by RON J. DOYLE

VOLUME 311. Sphingolipid Metabolism and Cell Signaling (Part A)
Edited by ALFRED H. MERRILL, JR., AND YUSUF A. HANNUN

VOLUME 312. Sphingolipid Metabolism and Cell Signaling (Part B)
Edited by ALFRED H. MERRILL, JR., AND YUSUF A. HANNUN

VOLUME 313. Antisense Technology (Part A: General Methods, Methods of Delivery, and RNA Studies)
Edited by M. IAN PHILLIPS

VOLUME 314. Antisense Technology (Part B: Applications)
Edited by M. IAN PHILLIPS

VOLUME 315. Vertebrate Phototransduction and the Visual Cycle (Part A)
Edited by KRZYSZTOF PALCZEWSKI

VOLUME 316. Vertebrate Phototransduction and the Visual Cycle (Part B)
Edited by KRZYSZTOF PALCZEWSKI

VOLUME 317. RNA–Ligand Interactions (Part A: Structural Biology Methods)
Edited by DANIEL W. CELANDER AND JOHN N. ABELSON

VOLUME 318. RNA–Ligand Interactions (Part B: Molecular Biology Methods)
Edited by DANIEL W. CELANDER AND JOHN N. ABELSON

VOLUME 319. Singlet Oxygen, UV-A, and Ozone
Edited by LESTER PACKER AND HELMUT SIES

VOLUME 320. Cumulative Subject Index Volumes 290–319

VOLUME 321. Numerical Computer Methods (Part C)
Edited by MICHAEL L. JOHNSON AND LUDWIG BRAND

VOLUME 322. Apoptosis
Edited by JOHN C. REED

VOLUME 323. Energetics of Biological Macromolecules (Part C)
Edited by MICHAEL L. JOHNSON AND GARY K. ACKERS

VOLUME 324. Branched-Chain Amino Acids (Part B)
Edited by ROBERT A. HARRIS AND JOHN R. SOKATCH

VOLUME 325. Regulators and Effectors of Small GTPases (Part D: Rho Family)
Edited by W. E. BALCH, CHANNING J. DER, AND ALAN HALL

VOLUME 326. Applications of Chimeric Genes and Hybrid Proteins (Part A: Gene Expression and Protein Purification)
Edited by JEREMY THORNER, SCOTT D. EMR, AND JOHN N. ABELSON

VOLUME 327. Applications of Chimeric Genes and Hybrid Proteins (Part B: Cell Biology and Physiology)
Edited by JEREMY THORNER, SCOTT D. EMR, AND JOHN N. ABELSON

VOLUME 328. Applications of Chimeric Genes and Hybrid Proteins (Part C: Protein-Protein Interactions and Genomics)
Edited by JEREMY THORNER, SCOTT D. EMR, AND JOHN N. ABELSON

VOLUME 329. Regulators and Effectors of Small GTPases (Part E: GTPases Involved in Vesicular Traffic)
Edited by W. E. BALCH, CHANNING J. DER, AND ALAN HALL

VOLUME 330. Hyperthermophilic Enzymes (Part A)
Edited by MICHAEL W. W. ADAMS AND ROBERT M. KELLY

VOLUME 331. Hyperthermophilic Enzymes (Part B)
Edited by MICHAEL W. W. ADAMS AND ROBERT M. KELLY

VOLUME 332. Regulators and Effectors of Small GTPases (Part F: Ras Family I)
Edited by W. E. BALCH, CHANNING J. DER, AND ALAN HALL

VOLUME 333. Regulators and Effectors of Small GTPases (Part G: Ras Family II)
Edited by W. E. BALCH, CHANNING J. DER, AND ALAN HALL

VOLUME 334. Hyperthermophilic Enzymes (Part C)
Edited by MICHAEL W. W. ADAMS AND ROBERT M. KELLY

VOLUME 335. Flavonoids and Other Polyphenols (in preparation)
Edited by LESTER PACKER

VOLUME 336. Microbial Growth in Biofilms (Part A: Developmental and Molecular Biological Aspects)
Edited by RON J. DOYLE

VOLUME 337. Microbial Growth in Biofilms (Part B: Special Environments and Physicochemical Aspects)
Edited by RON J. DOYLE

Section I

Ribonuclease Classification and Review

[1] A Natural Classification of Ribonucleases

By L. ARAVIND and EUGENE V. KOONIN

In this article an evolutionary classification of ribonucleases (RNases) is constructed using database searches with sequence profiles, clustering of proteins by sequence similarity, and comparison of protein structures and domain architectures. The classification includes, as its central concept, protein superfamilies that share detectable sequence conservation. Superfamilies are united into folds with distinct structures and are subdivided into families and subfamilies on the basis of clustering by sequence similarity, domain architectures, and phyletic distribution. Many different folds from all structural classes of proteins have independently evolved RNase activity. Typically, protein superfamilies include both RNases and DNases, and in some cases, also generic hydrolases, for example, phosphatases. Nevertheless, several families of RNases could be traced to the last common ancestor of all extant life forms. Fusion of the catalytic domain with various RNA-binding domains and, less frequently, with other enzymes of nucleic acid metabolism such as helicases is typical of most RNase families. During the construction of this classification, several uncharacterized protein families were predicted to possess RNase activity.

Introduction and Classification Approach

Ribonucleases (RNases) cleave phosphodiester bonds in RNA and are essential both for nonspecific RNA degradation and for numerous forms of RNA processing.[1,2] Consequently, multiple forms of RNases are present in all cellular life forms. Most of the known RNases are protein enzymes, but, in several well-characterized RNases, the catalytic moiety is an RNA molecule. Thus, RNases are among the few enzymes that appear to still retain links to the postulated primeval RNA world, the ancient precellular system of RNA replicators and catalysts.[3] In this article, we concentrate primarily on the protein RNases although the protein components of the ribozyme RNase P are also discussed in some detail.

The classification presented here is based on evolutionary relationships revealed through sequence and structure conservation, rather than functional

[1] S. M. Linn and R. J. Roberts (eds.), "Nucleases (Cold Spring Harbor Monograph Series 14)." Cold Spring Harbor Laboratory Press, Cold Spring Harbor, NY, 1994.

[2] N. C. Mishra, "Molecular Biology of Nucleases." CRC Press, Boca Raton, FL, 1995.

[3] R. F. Gesteland, T. R. Cech, and J. F. Atkins (eds.), "The RNA World," 2nd ed. Cold Spring Harbor Laboratory Press, Cold Spring Harbor, NY, 2000.

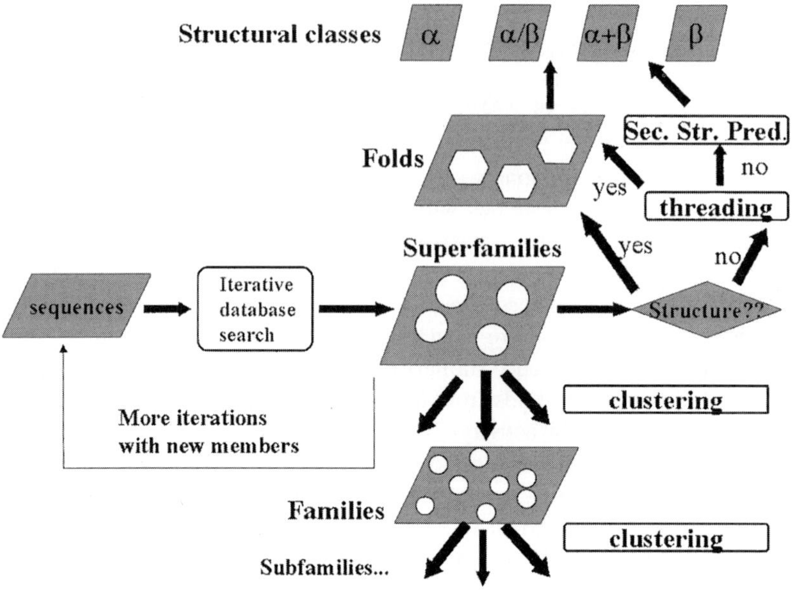

FIG. 1. The protein classification procedure as applied to ribonucleases.

properties, although on most occasions this classification correlates with the biochemical and catalytic properties of the proteins. The starting material for this analysis comprised the protein sequences of all experimentally characterized RNases that were identified through a survey of the relevant literature and retrieved from the Nonredundant (NR) protein database at the National Center for Biotechnology Information (NIH, Bethesda). Additionally, proteins from several generic hydrolase superfamilies, such as the HD,[4] DHH,[5] and metallo-β-lactamase (MBL)[6] superfamilies, that are either known or predicted to possess RNase activity were also included in this classification. The prediction of RNases among these generic hydrolases was based on specific similarity to characterized proteins with RNase activity or on fusions to specific RNA-binding domains such as S1, KH, and N-OB. The catalytic RNase domain was identified in the sequences of the analyzed proteins on the basis of the available biochemical data, crystal structures, and sequence analysis that often helped in separating the catalytic domain from accessory, typically RNA-binding domains.

The methodology of protein classification that was applied here to RNases, but can be used for any class of proteins, is schematically depicted in Fig. 1.

[4] L. Aravind and E. V. Koonin, *Trends Biochem. Sci.* **23,** 469 (1998).
[5] L. Aravind and E. V. Koonin, *Trends Biochem. Sci.* **23,** 17 (1998).
[6] L. Aravind, *In Silico Biol.* **1,** 8 (1998).

Classification of RNases was based primarily on their catalytic domains, with other, accessory domains used as additional characters. The principally employed method was iterative database search using the PSI-BLAST program.[7,8] The searches were initially run against the NR database, starting with the sequences of the catalytic domains of experimentally characterized RNases, usually with an E value of 0.01 used as the cutoff for including sequences into the profile. These searches were run until convergence or until apparent false positives started to appear among the retrieved sequences. Because the outcome of database searches strongly depends on the query sequence,[9] the next round of iterative searches was run with a diverse, representative subset of the sequences retrieved in the first round. Such "superiterative" search was continued until no new sequences could be detected. In complicated cases, additional searches were run against smaller databases comprised of protein sets from completely sequenced genomes. The results were validated by inspection of database search outputs and multiple alignments constructed with the Clustal X program[10] for the presence of known and new sequence motifs. These procedures resulted in the delineation of domain *superfamilies* where a superfamily is defined as an assembly of domains that showed recognizable (however subtle) sequence conservation. At least in the case of enzymes, such conservation almost invariably assumes the form of conserved motifs that center at polar amino acid residues known to be involved or implicated in catalysis.

From the superfamily level, the classification procedure expanded both in the upward direction, toward further unification, and in the downward direction, toward more specific subdivisions. At the higher level, the classification followed the general conventions of the SCOP database.[11,12] The superfamilies whose catalytic domains shared the same structural core and folding pattern, which was typically determined through comparison to structurally characterized representatives, were unified as sharing the same *fold*. Occasionally, structural comparisons may result in identification of distinct patterns of conserved amino acid residues that have not been detected in the sequence analysis phase. In such cases, two or more superfamilies may be merged. At the top level of the classification, folds are grouped, on the basis of their predominant secondary structure, into all α, all β, α/β, and $\alpha + \beta$ structural classes. In the absence of detectable sequence similarity to proteins with known structures, sequence-structure threading was attempted.[13] If

[7] S. F. Altschul, T. L. Madden, A. A. Schäffer, J. Zhang, Z. Zhang, W. Miller, and D. J. Lipman, *Nucleic Acids Res.* **25,** 3389 (1997).

[8] S. F. Altschul and E. V. Koonin, *Trends Biochem. Sci.* **23,** 444 (1998).

[9] L. Aravind and E. V. Koonin, *J. Mol. Biol.* **287,** 1023 (1999).

[10] J. D. Thompson, T. J. Gibson, F. Plewniak, F. Jeanmougin, and D. G. Higgins, *Nucleic Acids Res.* **25,** 4876 (1997).

[11] A. G. Murzin, S. E. Brenner, T. Hubbard, and C. Chothia, *J. Mol. Biol.* **247,** 536 (1995).

[12] L. Lo Conte, B. Ailey, T. J. Hubbard, S. E. Brenner, A. G. Murzin, and C. Chothia, *Nucleic Acids Res.* **28,** 257 (2000).

[13] D. Fischer, *Pac. Symp. Biocomput.,* 119 (2000).

this method did not result in fold recognition either, it was operationally concluded that the given domain belongs to a new fold, and secondary structure prediction was carried out using a multiple alignment as the input,[14] to assign the domain to one of the structural classes.

Within superfamilies, different hierarchical levels of classification were introduced; the exact hierarchy depends on the number of clearly identifiable groups, the two basic levels being *family* and *subfamily*. These groups were initially identified by similarity-based single linkage clustering using the BLASTCLUST program (I. Dondoshansky, Y. I. Wolf, and EVK, unpublished; ftp://ftp.ncbi.nlm.nih.gov/blast/). The clusters defined using BLASTCLUST were used as nuclei to identify robust groups (families or subfamilies) using specific criteria that may be considered shared derived characters (synapomorphies) for each group. These included specific versions of sequence motifs or fusions to particular domains. The latter criterion is especially useful for defining subfamilies and predicting specific functions of enzymes. Additional domains in RNAses were identified by comparing their sequences to previously developed domain-specific profiles using the PSI-BLAST program.[15] Fusion of the catalytic domains to various RNA-binding domains and, less frequently, to other enzymes of nucleic acid metabolism was detected for almost all RNase superfamilies (Fig. 2).

The taxonomic distribution of proteins of each superfamily was determined using the Taxfilt and Taxbreak scripts of the SEALS package.[16] When a clear-cut

[14] B. Rost and C. Sander, *Proteins* **19,** 55 (1994).
[15] S. A. Chervitz, L. Aravind, G. Sherlock *et al.*, *Science* **282,** 2022 (1998).
[16] D. R. Walker and E. V. Koonin, *ISMB* **5,** 333 (1997).

FIG. 2. Domain architectures of ribonucleases. The domain architectures of selected representatives of RNAses are drawn approximately to scale. Only globular domains detectable with statistically significant score are shown. The name of a representative gene/protein is indicated for each domain architecture, along with the source organism name, which is abbreviated in brackets. Species abbreviations: Hs, *Homo sapiens;* Dm, *Drosophila melanogaster;* Ce, *Caenorhabditis elegans;* At, *Arabidopsis thaliana;* Mj, *Methanococcus jannaschii;* Mtu, *Mycobacterium tuberculosis;* Af, *Archaeoglobus fulgidus;* Ph, *Pyrococcus horikoshii;* Ap, *Aeropyrum pernix;* Pa, *Pseudomonas aeruginosa;* Bs, *Bacillus subtilis;* Ec, *Escherichia coli;* Hi, *Haemophilus influenzae;* Ssp, *Synechocystis* sp.; Tma, *Thermotoga maritima;* Dr, *Deinococcus radiodurans;* Ct, *Chlamydia trachomatis.* Domain abbreviations are: TBP, the TBP-like domain in RNase HII family proteins; PGMase, phosphoglycerate mutase; RRM, RNA recognition motif; RN, RNase H N-terminal domain; WD40, WD40 repeat containing β propeller domain; HRDC, helicase–RNase D C-terminal domain; KH, ribonucleoprotein K homology domain; CCCH, CCCH RNA binding domain; Mut7-C, C-terminal of Mut7 domain; S11, S1 like RNA binding domain; Cor, CoRNase domain, C4, DnaJ-like four-cysteine motif; MBL hydrolase, metallo-β-lactamase; PAZ, PIWI, Argonaute, Zwille domain; Dsrbd, double-stranded RNA binding domain; Thermo, thermonuclease domain; ENDO, tRNA endonuclease domain; PIN, PilT N-terminal domain; EGL-N, EGL N-terminal domain; 2C, two cysteine domain in RNase E/G; RecQ-CT, RecQ C-terminal domain. A crossed-out domain indicates an inactive version.

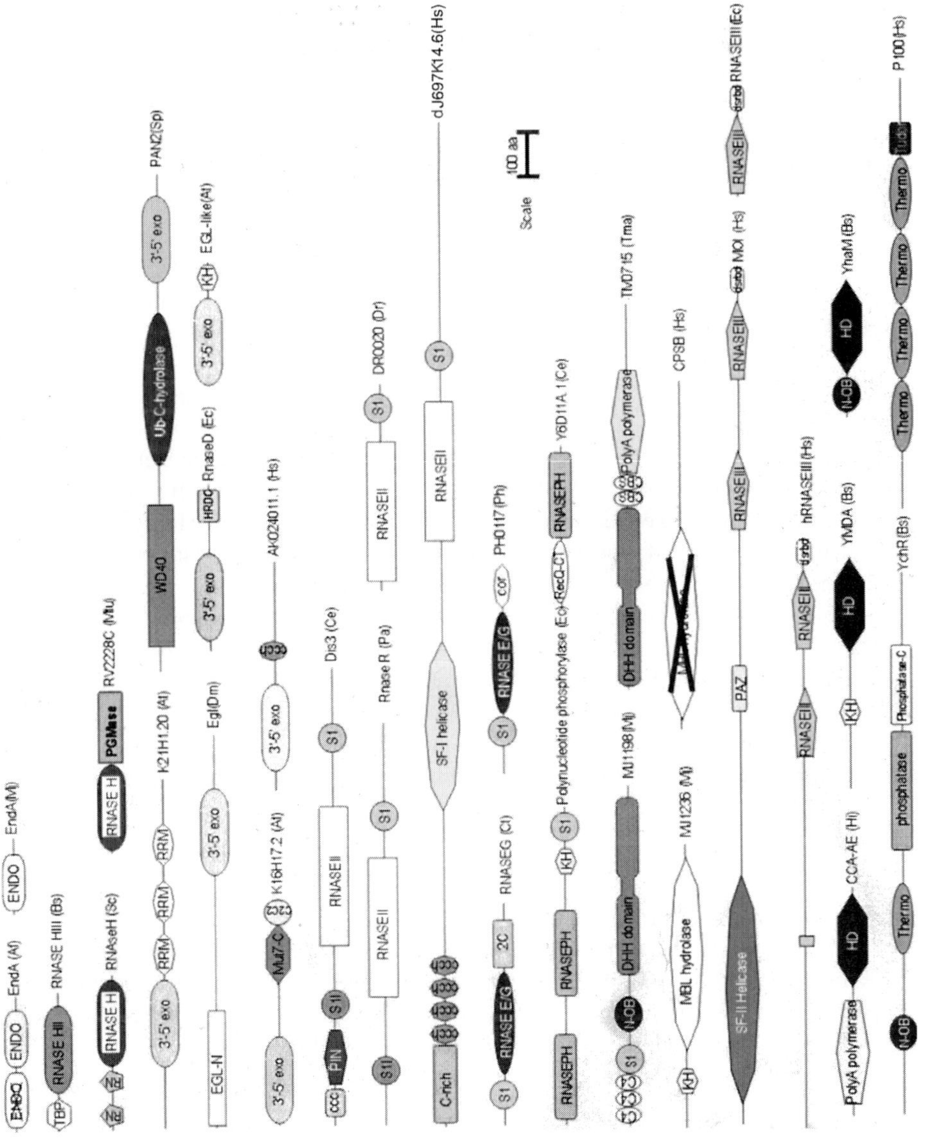

EndA (Af)
EndA(Mj)
RNASE HIII (Bs)
RNAseH (Sc)
RV2228C (Mtu)
K21H1.20 (At)
PAN2(Sp)
Egl(Dm)
RnaseD (Ec)
EGL-like(At)
K16H17.2 (At)
AK024011.1 (Hs)
Dis3 (Ce)
Rnase R (Pa)
DR0020 (Dr)
dJ697K14.6(Hs)
RNASEG (Ct)
PH0117 (Ph)
Scale
100 aa
Polynucleotide phosphorylase (Ec)
Y6D11A.1 (Ce)
MJ1198(Mj)
TM0715 (Tma)
MJ1236 (Mj)
CPSB (Hs)
MOI (Hs)
RNASEIII (Ec)
hRNASEIII (Hs)
CCA-AE (Hi)
YMDA (Bs)
YhaM (Bs)
YchR (Bs)
P100 (Hs)
ENDO
RNASE HII
RNASE H
PGMase
3'-5' exo
RRM
WD40
Ub-C-hydrolase
EGL-N
HRDC
KH
Mut7-C
PIN
S1
RNASEII
C-rich
SF-I helicase
RNASE E/G
2C
RNASEPH
RecQ-CT
N-OB
DHH domain
PolyA polymerase
MBL hydrolase
SF-II Helicase
PAZ
RNASEIII
dsrbd
HD
Thermo
phosphatase
Phosphatase-C

phyletic pattern, for example, grouping of proteins from eukaryotes and archaea to the exclusion of bacterial homologs, was detected, it was employed as an additional criterion in the definition of groups. The lowest level in the classification includes groups of proteins that comprise a set of orthologs (direct evolutionary counterparts)[17] from various organisms or a set of recently duplicated paralogs from a single or a few organisms. The individual groups of orthologs are generally in line with such groups defined in the Clusters of Orthologous Groups of proteins (COGs) database.[18,19] This database was also used to verify the phyletic patterns of the proteins, especially in prokaryotes, and to obtain corrected or unannotated protein sequences from completely sequenced genomes.[20] On several occasions, protein families and subfamilies identified during the construction of this classification, include proteins that are not known and have not been predicted previously to possess RNase activity. In the ensuing discussion, we highlight these groups that may be targets for future experimental studies.

Evolutionary Classification of Ribonucleases

Below we present the classification of RNases based on the results of sequence and structure comparisons (Table I), followed by a brief discussion of the structural, functional, and evolutionary aspects of each group. RNase activity has been independently derived in protein domains of all the basic secondary structure types. In line with what has been observed in the rest of the enzyme universe, the RNases are most frequently derived in the α/β and $\alpha + \beta$ classes of proteins. The former class includes domains that are mainly composed of α-β units following in succession, whereas the latter class includes proteins with isolated α and β elements. These types of structures are likely to combine the flexibility of α helices with the rigidity offered by the β sheets and thus act as better scaffolds for catalytic activities. Nevertheless, there are a few families of predominantly α-helical and β-strand-containing nucleases.

α/β Ribonucleases

Endonuclease Fold. The endonuclease fold includes a large number of DNases such as restriction endonucleases, archaeal Holliday junction resolvases, the RecB-like DNA repair nucleases, Vsr-like nucleases, and the lambda(λ)-type exonucleases.[21] These nucleases have a characteristic set of charged residues in a β-hairpin that coordinates a divalent cation required for catalysis. Despite this

[17] W. M. Fitch, *Systemat. Zool.* **19,** 99 (1970).

[18] R. L. Tatusov, E. V. Koonin, and D. J. Lipman, *Science* **278,** 631 (1997).

[19] R. L. Tatusov, D. A. Natale, I. V. Garkavtsev *et al., Nucleic Acids Res.* **29,** 22 (2001).

[20] D. A. Natale, M. Y. Galperin, R. L. Tatusov, and E. V. Koonin, *Genetica* **108,** 9 (2000).

[21] L. Aravind, K. S. Makarova, and E. V. Koonin, *Nucleic Acids Res.* **28,** 3417 (2000).

TABLE I

CLASSIFICATION OF RIBONUCLEASES

α/β nucleases

Endonuclease fold

This fold contains the restriction Endonucleases, Archaeal Holliday junction resolvases, the lambda exonucleases and several related enzymes that act primarily on DNA. The fold includes only one RNAse superfamily

tRNA-splicing endonuclease superfamily

Family 1: tRNA endonuclease subunit; the Sen2p/Sen34p family

subfamily 1: Archaeal group (archaea only)

subfamily 2:Sen34p group (eukaryotes)

subfamily 2:Sen2p group (plants and fungi only)

Family 2: Sen54p family (eukaryote-specific)

Ribonuclease H/3'-5'exonuclease fold

This very common fold, in addition to numerous nucleases, including the RuvC-like Holliday junction resolvases, is present in a variety of nucleotide-binding and hydrolyzing proteins of the actin-HSP70 class.

Ribonuclease HI superfamily

Family 1: Eukaryotic-type RNAse HI

Domain architecture: 1-2*Rnh-N+RNAseH (Rnh-N - distinct N-terminal domain)

Phyletic distribution: Eukaryotes + retroelements + some bacteria (Thermotoga,Bacillus halodurans)

Family 2: Bacterial-type RNAse HI

Domain architecture: No Rnh-N domain

Phyletic distribution: bacteria+ retroposons + one archaeon (Halobacterium)

Ribonuclease HII superfamily

Family 1: Archaeo-eukaryotic-type RNAse HII

Phyletic distribution: Eukaryotes + archaea

Family 2: Bacterial-type RNAse HII

Subfamily 1:Rnase HII with a TBP-like domain

Domain architecture:TBP+RNASEHII

Phyletic distribution: some bacteria (Aquifex, Chlamydia, Gram-positive bacteria)

Subfamily 1:classic bacterial RNAse HII

Domain architecture: RNAse HII domain only

Phyletic distribution: almost all bacteria

3'-5'exonuclease superfamily

Family 1: Ribonuclease T family

Includes numerous DNAses, for example, the epsilon subunit of bacterial DNA polymerase III, the proofreading exonuclease domain of bacterial DNA polymerase I and the exonuclease of the Werner syndrome helicase. Only the RNAses of this family are considered here.

Phyletic distribution: Proteobacteria only.

Family 2:Oligoribonuclease family

Highly conserved nucleases with a limited phyletic distribution, involved in degradation of oligoribonucleotides

Phyletic distribution:Proteobacteria+Actinomycetes+Crown group eukaryotes

Family 3:KapD family (predicted RNAses)

(Continued)

TABLE I
(Continued)

A small family of potential RNAses
Domain architecture: 3′-5′exo or SAP+3′-5′exo
Phyletic distribution: Some bacteria (Bacillus, Pseudomonas, Vibrio, Deinococcus)+Crown-group eukaryotes (6 members in a small paralog expansion in *C.elegans*)

Family 4:Rnh70p/Rex3/4p/PAN2 family
Large, eukaryote-specific family of nucleases with several subfamilies encompassing diverse activities. Involved in degradation of poly(A) tails in mRNAs; some have RNAseH activity

Subfamily 1: Rex3/4
RNAses involved in processing of 5S, 5.8S, U4, U5,RNAseP RNAs
Phyletic distribution:Crown group eukaryotes

Subfamily 2: Plant-specific, RRM-containing RNases
Domain architecture: 3′-5′exo +1-3 RRM
Phyletic distribution: plants only

Subfamily 3: Rnh70p
Eukaryotic RNAses with RNAse H activity
Phyletic distribution: Crown group eukaryotes

Subfamily 4: PAN2
RNAses involved in degradation of Poly A tails
Domain architecture: WD40+UBC-hydrolase(inactive)+3′-5′exo
Phyletic distribution:Crown group eukaryotes

Family 5:Ribonuclease D family
This family consists of diverse nucleases including DNAses such as the Werner′syndrome nuclease and DNA polymerase proofreading nucleases; Involved in rRNA processing, post-transcriptional gene silencing and other regulatory RNA processing events.

Subfamily 1: Typical RNAse D
Involved in rRNA and tRNA processing.
Domain architecture: 3′-5′exo+HRDC domain
Distribution: Several groups of bacteria (Proteobacteria, Cyanobacteria, Actinomycetes) + Crown group eukaryotes

Subfamily 2: Mut7-like predicted RNAses
Enzymes involved in post-transcriptional gene silencing.
Domain architecture : 3′-5′exo or 3′-5′exo +Mut7-C domain
Phyletic distribution: Plants and animals only

Subfamily 3: predicted RNAses typified by Drosophila CG6744
Domain architecture: 3′-5′ exo + C-terminal extenstion
Phyletic distribution: Plants and animals only

Subfamily 4: EGL-type RNAses (typified by Drosophila Egl)
Family of RNAses with potential role in regulatory RNA processing.
Domain architecture: 3′-5′ exo or distinct N-terminal domain + 3′-5′ exo; plant form:3′-5′ exo +KH domain
Phyletic distribution: Plants, animals and P.falciparum

Family 6:Deadenylating nucleases
The characterized members of this family are involved in Poly(A) degradation

Subfamily 1: DAN (typified by vertebrate DAN enzyme)
Phyletic distribution: Crown group eukaryotes, missing in *S. cerevisiae*

Subfamily 2: Predicted RNAses containing a C-terminal CCCH RNA-binding domain
Domain architecture:3′-5′exo +CCCH
Phyletic distribution: animals only

Ribonuclease II Fold
Ribonuclease II Superfamily

(Continued)

TABLE I
(Continued)

Family 1: Typical RNAse R/II
Involved in mRNA degradation and rRNA processing
Domain architecture: S1/CS-like+Rnase II+S1
Subfamily 1: RNAse R
Phyletic distribution: Bacteria+*Halobacterium*
Subfamily 2: RNAse B
Phyletic distribution:Some Gamma-proteobacteria
Family 2: Eukaryotic DIS3-like exosome subunits
Domain architecture: CR3+PIN+S1/CS-like+Rnase II+S1
Phyletic distribution:Crown group Eukaryotes
Family 3: Eukaryote-specific RNAse II-like proteins (typified by Drosophila CG16940)
Domain architecture: Contain disrupted N- and C-terminal S1-like OB-fold domains
Distribution:Crown group Eukaryotes, missing in *S. cerevisiae*
Family 4: Predicted bacterial RNAses (typified by *Treponema* TP0805)
Domain architecture:RNAse II+S1
Distribution: Some bacteria(Treponema, Synechocystis, Deinococcus, Neisseria)+ Arabidopsis
Family 5: Sts5p-like fungal family
Domain architecture:S1/CS-like+Rnase II+S1
Possibly involved in U5 snRNA processing
Distribution: Fungi
Unassociated members:11137520 - A human protein with a distinct domain architecture:Cys-rich domain+4*CCCH+Helicase+Rnase II+S1; 557724, S.cerevisiae DSS1

Ribonuclease E/G fold
Ribonuclease E/G superfamily
Family 1: Typical RNAse E/G
Domain architecture: S1+Rnase E/G+2C-domain
Subfamily 1:RNAse G
Phyletic distribution: Bacteria
Subfamily 1:RNAse E
Phyletic distribution: Only proteobacteria
Family 2:Archaeal-type RNAse E/G
Domain architecture:S1+Rnase E/G+coRNAse domain
Phyletic distribution:Archaea

Ribonuclease PH fold
A compound fold containing a core S5-like domain with an extra N-terminal β-hairpin, packed against another α+β unit. Enzymes of this family are involved in tRNA processing and also are subunits of the 3'->5' processing complex, the exosome
Ribonuclease PH superfamily
Family 1: polynucleotide phosphorylase
Domain architecture: 2*Rnase PH+KH+S1
Phyletic distribution:Bacteria+Arabidopsis+Drosophila
Family 2:RPR45 (Exosomal subunit)
Phyletic distribution: Archaea+Eukaryotes
Family 3: Ski6p (Exosomal subunit)
Phyletic distribution:Archaea+Eukaryotes
Family 4:RPR46 (Exosomal subunit)
Phyletic distribution:Eukaryotes
Family 5:Bacterial Rnase PH
Phyletic distribution:Bacteria

(Continued)

TABLE I
(Continued)

Unassociated superfamily members: 7503205 from *C.elegans* and 6320092,6321597 from S.cerevisiae

----

Metallo-beta-lactamase fold

Metallo-beta-lactamase superfamily

This superfamily includes diverse hydrolases among which a single family appears to have RNAse activity. Some other members of this superfamily might be still uncharacterized nucleases

Family 1: CSPF-SNM1 family

This family includes predicted DNAses typified by the eukaryotic SNM1 proteins. These are not considered here.

Subfamily 1:Eukaryotic CPSF-type RNase

Phyletic distribution:Archaea+Eukaryotes+ some bacteria (Synechocystis, Pseudomonas, Deinococcus, Vibrio)

Subfamily 2:Archaea specific CPSF-type enzymes

Domain architecture:KH+ MBL

Phyletic distribution:Archaea

Subfamily 3:eukaryote-specific, inactive CPSF proteins

The metal-binding active site is disrupted in these proteins, but they function as subunits of the active RNAses

Phyletic distribution: Eukaryotes

Subfamily 4:Prokaryotic CPSF-like enzymes

Phyletic distribution: Many bacteria + several archaea (Pyrococcus, Methanothermobacter, Halobacterium, Methanococcus) + Arabidopsis

----

Barnase fold

Barnase superfamily

Secreted RNAses of bacteria and fungi; distinct catalytic triad, E-R-H

Family 1: Bacterial barnase

Phyletic distribution: Scattered among bacteria (Deinococcus, Xylella, some Bacillus species, actinomycetes)

Family 2: Fungal barnase

Phyletic distribution: Scattered in fungi(Ustilago, Aspergillus)

Ribonuclease A fold

Ribonuclease A superfamily (pancreatic RNAse)

Secreted nucleases with disulfide bonds, catalytic triad HKH

Phyletic distribution: Only vertebrates

Helix-Grip fold

START domain superfamily

A widespread superfamily of diverse small-molecule-binding and metabolic enzymes and regulatory proteins.

Family 1: Birch allergen family

Only a few members of this family have been reported to possess RNAse activity.

Phyletic distribution:Plants

$\alpha+\beta$ nucleases

DHH fold

DHH superfamily

This superfamily consists of diverse hydrolases including DNAses, potential RNAses and phosphoesterases. Only families with clear

(Continued)

TABLE I
(Continued)

evidence of RNAse activity are considered here

Family 1: Predicted archaeal nucleases
Domain architecture:3*DNAJ-ZnF+S1+N-OB+DHH
Phyletic distribution:Archaea

Family 2:Poly(A) polymerase-associated nucleases
Domain architecture:Poly(A)-Pol+2*CBS+DHH
Phyletic distribution: Scattered in bacteria (*Synechocystis,Aquifex,Thermotoga*)

Ribonuclease T2 Fold

Ribonuclease RNAse T2 Superfamily

Secreted RNAses involved in RNA degradation and self-incompatibility in plants; active site with two conserved histidines

Family 1: RNAse T2
Phyletic distribution: Crown group eukaryotes + some proteobacteria (*E.coli* and *H.influenzae*)

KEM1/RAT1 fold

KEM1/RAT1 superfamily

5'->3'nucleases involved in cytoplasmic RNA degradation and 25S rRNA processing; some members of this family may also have DNAse activity.

Family 1: RAT1p family
Phyletic distribution: Crown group eukaryotes

Family 2: KEM1p family
Phyletic distribution: S.cerevisiae and S.pombe

Unassociated superfamily member: 7494394 from Plasmodium falciparum

----

All α-helical nucleases

Ribonuclease III fold

Ribonuclease III superfamily

Family 1: Typical RNAse III
Involved in processing of RNAs containing extensive double-stranded regions (mainly rRNAs and some mRNAs)

Subfamily 1: Bacterial and eukaryotic RNAse III
Domain architecture: RNAse III+DSRBD
Phyletic distribution: bacteria+Eukaryotes+several Large DNA viruses

Subfamily 2: Animal-specific RNAse III
Domain architecture: N-extension+2*RNASEIII+DSRBD
Phyletic distribution: Animals only

Subfamily 3: Yeast-specific RNAse III
Called "Mitochondrial ribosomal protein L15" in sequence databases
Domain architectures: RNAse III domain only, no DSRBD
Phyletic distribution: *S.cerevisiae* and *S.pombe* only

Family 2: CAF family
Potentially involved in post-transcriptional gene silencing
Domain architecture: RNA-helicase+PAZ+2*RNASEIII+DSRBD
Phyletic distribution: Crown group eukaryotes, missing in S.cerevisiae

Family 3: CG2109 family, RNAse III family typified by *Drosophila* CG2109
Domain architecture: RNAse III+DSRBD
Phyletic distribution: Crown group eukaryotes

(Continued)

TABLE I
(Continued)

Family 4: Predicted RNAses distantly related to RNAse III
Phyletic distribution: *Bacillus, Thermotoga, Synechocystis, Arabidopsis*

HD-phosphoesterase fold

HD-superfamily

This superfamily consists of diverse phosphoesterases and phosphatases, including the cyclic nucleotide phosphodiesterases; most of the members are poorly studied. Only those families for which RNAse activity could be confidently predicted are included.

Family 1: predicted RNAses containing KH domain
Domain architecture: KH+HD
Phyletic distribution: bacteria only

Family 2: predicted RNAses associated with nucleotidyl Transferases
Domain architecture: NTase+HD
Phyletic distribution: bacteria only

Family 3: predicted RNAses with N-OB domain
Domain architecture: N-OB+NTase+HD
Phyletic distribution: Gram-positive bacteria only

Family 4: predicted RNAses
Phyletic distribution: archaea + eukaryotes

All β-strand nucleases

OB-fold proteins

This fold largely includes non-catalytic, nucleic-acid-binding domains; only one superfamily consists of RNAses

Thermonuclease superfamily

Non-specific nucleases with no specific affinities for particular nucleic acids

Family 1: typical thermonucleases
Phyletic distribution:
bacteria+archaea+plants+*S.cerevisiae*+*S.pombe*

Family 2:Predicted RNAse (typified by *B. subtilis* YchR)
Domain architecture: N-OB+Thermo+Phosphatase
Phyletic distribution: *Bacillus* only

Family 3:The eukaryotic 100KD coactivators
Domain architecture: 4*Thermo+Tudor; may be inactive
Phyletic distribution: eukaryotic crown group (Animals+plants+fungi)

Note: Recently, two important studies have been published that require amendments to be made to the classification of RNases presented here. One study resulted in the identification of two new groups of RNases, namely CCR4p and CAF1p that have been identified as subunits of a conserved complex that interacts with the PAN2/3 complex in the degradation of the poly(A) tails [M. Tucker, M. A. Valencia-Sanchez, R. R. Staples, J. Chen, C. L. Denis, R. Parker, *Cell* **104,** 377 (2001)]. CCR4p and its several paralogs, that are predicted to have RNase activity, belong to a distinct family within the exodeoxyribonuclease–sphingomyelinase fold. Thus, these proteins are the first RNase members of this fold that also includes various endo- and exo-DNases, and phosphoesterases. CAF1p and its orthologs define a new subfamily of the deadenylating nuclease family of the 3′-5′ exonuclease superfamily (Family 6). In the second study, it has been shown that proteins of the CAF family of the RNase III superfamily, that combine two RNase III domains with helicase and RNA-binding domains, are RNases required for posttranscriptional gene silencing [E. Bernstein, A. A. Caudy, S. M. Hammond, G. J. Hannon, *Nature* **409,** 363 (2001)], as predicted here and elsewhere [L. Aravind, H. Watanabe, D. J. Lipman, E. V. Koonin, *Proc. Natl. Acad. Sci. USA* **97,** 11319 (2000)].

wealth of DNases, there is a single superfamily of RNAses that have this fold, namely the tRNA-splicing endonucleases that are responsible for the generation of the mature tRNA from its precursor in eukaryotes and archaea through two cleavages at the 3′ and 5′ ends.[22,23] These proteins function as a tetrameric complexes with two catalytic sites.[24] In some archaea, such as *Methanococcus,* the enzyme is a tetramer of a single subunit with a core endonuclease domain. The enzyme from *Archaeoglobus, Halobacterium,* and *Thermoplasma* is a dimer, with each monomer containing an N-terminal inactive and a C-terminal active endonuclease domain. In *Aeropyrum,* the inactive and active versions of the nuclease domain are separately encoded in distinct proteins which probably form a heterodimer that, in turn, further dimerizes. Comparative analysis of eukaryotic endonuclease subunits suggests an early triplication, with the three paralogs typified by the yeast proteins Sen2p, Sen34p, and Sen54p. However, Sen34p is retained only in the fungal and plant lineages and apparently has been lost in animals. Sen2p and Sen34p have been shown to be active,[24] whereas Sen54p appears to be inactive and analogous to the archaeal inactive versions of this nuclease.

RNase H 3′→5′-Exonuclease Fold. This fold contains a vast assemblage of proteins most of which have nucleotide-hydrolyzing or nuclease activities including HSP70-like molecular chaperones, actin, sugar kinases, bacterial Holliday junction resolvases, and a variety of nucleases.[9,21,25] This suggests that polyphosphate/phosphoester bond cleavage activity has been inherited from the common ancestor of all proteins that have this fold. This is supported by the presence of a conserved acidic residue at the end of the first strand present in these domains. Nevertheless, the sequence–structure affinities for the members of this fold suggest that the nuclease activity, and in particular, the RNase activity have evolved independently on several occasions within this fold.

RNase HI and RNase HII Superfamilies. The RNase H and RNase HII superfamilies share all their active site residues and also the specific features of the angle and packing of the C-terminal helix against the central sheet[26,27] and thus can be inferred to have shared a more recent common ancestor than other members of this fold. It is most likely that this common ancestor was already a ribonuclease. RNase HII is typically present in just a single copy in practically all the genomes of organisms sequenced to date, with the archaeal proteins being closer to the eukaryotic than to bacterial forms. This phyletic pattern suggests that this enzyme was inherited vertically from the common ancestor of all extant

[22] H. Li and J. Abelson, *J. Mol. Biol.* **302,** 639 (2000).
[23] J. Abelson, C. R. Trotta, and H. Li, *J. Biol. Chem.* **273,** 12685 (1998).
[24] C. R. Trotta, F. Miao, E. A. Arn, S. W. Stevens, C. K. Ho, R. Rauhut, and J. N. Abelson, *Cell* **89,** 849 (1997).
[25] P. Bork, C. Sander, and A. Valencia, *Proc. Natl. Acad. Sci. USA* **89,** 7290 (1992).
[26] L. Lai, H. Yokota, L. W. Hung, R. Kim, and S. H. Kim, *Structure Fold. Des.* **8,** 897 (2000).
[27] E. R. Goedken and S. Marqusee, *J. Biol. Chem.* **276,** 7266 (2001).

life forms (Last Universal Common Ancestor, or LUCA). Within the framework of the hypothesis that LUCA probably possessed an RNA genome with a DNA replication intermediate,[28,29] the conservation of an ancestral RNase HII that, in LUCA, could catalyze processing of RNA–DNA duplexes is not surprising. A distinct version of RNase HII is present in chlamydiae, *Aquifex,* and gram-positive bacteria, in addition to the regular RNase HII. This version contains a conserved N-terminal domain that is predicted to adopt a TATA-binding protein (TBP)-like fold[30,31] on the basis of sequence similarity and sequence-structure threading (LA, unpublished observations). This is the first homolog of the TBP domain detected in proteins other than archaeal and eukaryotic TBPs. The TBP-like domain probably plays a role in nucleic acid recognition by this form of RNase HII.

The RNase HI superfamily is present in bacteria and eukaryotes. In addition, it is widespread in many retroelements where it processes the DNA–RNA duplexes.[32] The only archaeon that encodes RNase HI is *Halobacterium* that appears to have recently acquired it from bacteria through horizontal gene transfer. The eukaryotic RNase HI contains a distinct N-terminal RNA-binding domain (RNH-N) that has the same fold as ribosomal protein L9[33] and is present as a stand-alone version in caulimoviruses[34] and in a *Borrelia* protein. *B. halodurans* and *Thermotoga* appear to have acquired this form of RNase HI from eukaryotes.

3′→5′-Exonuclease Superfamily. This is one of the largest superfamilies of dedicated nucleases that includes closely related DNases and RNases,[35] suggesting that these proteins have often shifted their catalytic identity in course of evolution. This superfamily clearly is an ancient one inherited from LUCA. However, almost all RNase families in this superfamily are present only in eukaryotes and bacteria, but not in archaea, which suggest a later origin, in some cases accompanied by horizontal gene transfer. We briefly discuss the diverse families of this superfamily emphasizing some new observations.

The RNase T family is seen thus far only in proteobacteria, where it initiates the 3′ exonucleolytic degradation of tRNA.[36] Interestingly, RNase T is closely related to the 3′–5′ DNase domain corresponding to epsilon subunit in *E. coli* of gram-positive DNA polymerase III.[37] This suggests that this gene has been relatively recently acquired by proteobacteria via horizontal transfer from the gram-positive

[28] U. Wintersberger and E. Wintersberger, *Trends Genet.* **3,** 198 (1987).

[29] D. D. Leipe, L. Aravind, and E. V. Koonin, *Nucleic Acids Res.* **27,** 3389 (1999).

[30] D. B. Nikolov and S. K. Burley, *Nat. Struct. Biol.* **1,** 621 (1994).

[31] P. F. Kosa, G. Ghosh, B. S. DeDecker, and P. B. Sigler, *Proc. Natl. Acad. Sci. USA* **94,** 6042 (1997).

[32] S. Kanaya and M. Ikehara, *Subcell. Biochem.* **24,** 377 (1995).

[33] S. P. Evans and M. Bycroft, *J. Mol. Biol.* **291,** 661 (1999).

[34] A. R. Mushegian, H. K. Edskes, and E. V. Koonin, *Nucleic Acids Res.* **22,** 4163 (1994).

[35] M. J. Moser, W. R. Holley, A. Chatterjee, and I. S. Mian, *Nucleic Acids Res.* **25,** 5110 (1997).

[36] Z. Li, S. Pandit, and M. P. Deutscher, *Proc. Natl. Acad. Sci. USA* **95,** 2856 (1998).

[37] E. V. Koonin and M. P. Deutscher, *Nucleic Acids Res.* **21,** 2521 (1993).

lineage followed by exaptation for the RNase function. Oligoribonuclease, which has an important role in the degradation of small RNA fragments, also shows a restricted distribution, predominantly proteobacteria and the eukaryotes.[38,39] The oligoribonuclease gene probably has been acquired horizontally by eukaryotes at an early stage of their evolution, most likely from the ancestor of mitochondria. The KapD family of predicted RNases has a comparable distribution, with a sporadic presence in few unrelated bacteria and a high degree of conservation throughout the eukaryotic crown group. *Caenorhabditis elegans* alone shows multiple paralogous proteins of this family resulting from what appears to be recent, lineage-specific proliferation.

Rnh70p/Rex3/4p/PAN2 family is a large family of RNases that so far have been found only in the eukaryotes, with several distinct subfamilies. The Rex3/4p subfamily is involved in the processing of several small RNAs and is highly conserved throughout eukaryotes; these proteins are subunits of the eukaryotic exosome, a multisubunit complex that consists of several RNases, RNA-binding proteins and helicases and mediates processing of various RNA species.[40,41] The Rnh70p subfamily shows the same pattern of conservation, but its exact function remains unknown. One of the subfamilies of this family is seen only in plants where the RNase domain is fused to one to three RNA-binding RRM domains (Table I). This subfamily could potentially have a role in posttranscriptional regulation through degradation of specific RNAs bound by the RRM domain. The Pan2 subfamily, which is conserved in the eukaryotic crown group, has been implicated in the degradation of poly(A) tails of mRNAs, which contributes to the regulation of mRNA stability.[42] This RNase has an unusual domain architecture, with an N-terminal WD40-like β-propeller domain, followed by an inactive ubiquitin C-terminal hydrolase domain. This may suggest an ancient connection between the RNA and protein degradation systems.

The RNase D family also includes certain DNases such as bacterial DNA polymerase I and the Werner syndrome protein exonuclease domains.[35,43,44] Some distinct subfamilies of this group, including the typical RNase D, contain a C-terminal HRDC domain shared with the RecQ subfamily of DNA helicases.[45] The other notable subfamily of this family includes Mut-7 protein from *C. elegans* that, along

[38] E. V. Koonin, *Curr. Biol.* **7,** R604 (1997).

[39] X. Zhang, L. Zhu, and M. P. Deutscher, *J. Bacteriol.* **180,** 2779 (1998).

[40] A. van Hoof and R. Parker, *Cell* **99,** 347 (1999).

[41] A. van Hoof, P. Lennertz, and R. Parker, *EMBO J.* **19,** 1357 (2000).

[42] R. Boeck, S. Tarun, Jr., M. Rieger, J. A. Deardorff, S. Muller-Auer, and A. B. Sachs, *J. Biol. Chem.* **271,** 432 (1996).

[43] A. R. Mushegian, D. E. Bassett, Jr., M. S. Boguski, P. Bork, and E. V. Koonin, *Proc. Natl. Acad. Sci. USA* **94,** 5831 (1997).

[44] S. Huang, B. Li, M. D. Gray, J. Oshima, A. S. Kamath-Loeb, M. Fry, and L. A. Loeb, *Nat. Genet.* **20,** 114 (1998).

[45] V. Morozov, A. R. Mushegian, E. V. Koonin, and P. Bork, *Trends Biochem. Sci.* **22,** 417 (1997).

with its homologs, has been implicated in posttranscription gene silencing.[46,47] The RNases of this subfamily have so far been seen only in plants and animals. The *C. elegans* and *Arabidopsis* members of this subfamily share a conserved C-terminal module (Mut7-C) that occurs as a stand-alone protein in most archaea and some bacteria such as *Streptomyces* and *Thermotoga*. This module has a bipartite structure with a C-terminal Zn-ribbon and an N-terminal α/β domain resembling the receiver domain of the two-component systems, with a conserved aspartate at the end of the first strand (LA, unpublished observations). The Mut7-C module could be a distinct nucleic-acid-binding domain of these nucleases. Another highly conserved subfamily of the RNase D family is the Egl subfamily, typified by the Egalitarian protein of *Drosophila* that is a part of a mRNA-binding complex required for oocyte specification.[48,49] The conservation of this subfamily throughout eukaryotes suggests that it is part of an ancient RNA processing complex that is likely to participate in the regulated processing of specific mRNAs.

The deadenylating nuclease family is a specific family of $3' \rightarrow 5'$- nucleases seen only in eukaryotes and is distantly related to other members of the superfamily. The vertebrate members of this family have been shown to possess $3'$–$5'$ poly(A) degradation activity.[50,51] A distinct, animal-specific subfamily of this family is characterized by a C-terminal fusion to a CCCH RNA-binding domain.

Ribonuclease II Fold. The RNase II fold so far includes only its namesake superfamily, which has not been structurally characterized. However, secondary structure prediction for the catalytic domain of these enzymes suggests a unique α/β fold (data not shown). The active site of this family has not been described in detail. Sequence comparisons suggest it probably includes conserved motifs found near the N and C termini of the RNase II domain. The N-terminal motif contains a conserved DD dyad followed by an invariant histidine; the C-terminal motif contains two conserved basic residues, typically an RR dyad. The N-terminal histidine is probably the principal catalytic residue, whereas the C-terminal conserved motif could contribute to the interaction with the substrates.

Members of this superfamily are present mainly in eukaryotes and bacteria. Distinct families could be identified on the basis of the domain architectures.

[46] A. Grishok, H. Tabara, and C. C. Mello, *Science* **287,** 2494 (2000).
[47] R. F. Ketting, T. H. Haverkamp, H. G. van Luenen, and R. H. Plasterk, *Cell* **99,** 133 (1999).
[48] W. Deng and H. Lin, *Int. Rev. Cytol.* **203,** 93 (2001).
[49] J. M. Mach and R. Lehmann, *Genes Dev.* **11,** 423 (1997).
[50] E. Dehlin, M. Wormington, C. G. Korner, and E. Wahle, *EMBO J.* **19,** 1079 (2000).
[51] C. G. Korner, M. Wormington, M. Muckenthaler, S. Schneider, E. Dehlin, and E. Wahle, *EMBO J.* **17,** 5427 (1998).

The typical RNase II/R family enzymes contain an OB-fold domain of the S1-like/cold shock domain family at the N terminus, and a bona fide S1 domain following the C terminus of the nuclease domain. The RNase R subfamily is seen in most bacteria and the archaeon *Halobacterium*, which probably acquired it via a relatively recent horizontal transfer from a bacterial source. The RNase B subfamily is represented only in the γ-proteobacteria and apparently has evolved as a result of a recent duplication and divergence from the pan-bacterial RNase R. Another distinct family of RNase II like enzymes, typified by *Treponema* TP0805, was detected in a limited range of bacteria, with a single member in *Arabidopsis*. These proteins lack the N-terminal S1-like/cold shock domain. In *Escherichia coli*, RNase II participates in mRNA degradation, and in particular 3′ poly(A) removal, along with polynucleotide phosphorylase[52,53]; the role of RNase R is poorly understood, but genetic data also suggest a function overlapping that of polynucleotide phosphorylase.[54]

The Dis3-like family and the CG16940-like families are eukaryote-specific RNase II–like families. Dis3 is a subunit of the exosome that is highly conserved in all eukaryotes.[40,55] This family is characterized by the distinct accretion of an N-terminal 3-cysteine domain and a PIN domain[56] to the RNase II core. The PIN domain itself has been proposed to possess RNase activity, based on of a limited sequence similarity to 5′→3′-exonucleases.[57] However, the presence of another, well-characterized RNase domain in the Dis3p protein, together with the pattern of sequence conservation in the PIN domain, make it seem more likely that PIN is an RNA-binding domain.[58]

The STS5p family so far has been detected only in fungi and might be involved in the processing of US snRNAs.[59] An unusual member of this superfamily is the human protein dJ697K14.6, which could not be assigned to any of the families. This protein is a dramatic example of the domain accretion that is characteristic of the vertebrates (Fig. 2). It contains a cysteine-rich domain followed by four CCCH RNA-binding domains, an RNA helicase domain, the RNase II domain and, finally, a C-terminal S1 domain. The RNA helicase domain of this

[52] M. Grunberg-Manago, *Annu. Rev. Genet.* **33,** 193 (1999).

[53] B. K. Mohanty and S. R. Kushner, *Mol. Microbiol.* **36,** 982 (2000).

[54] Z. F. Cheng, Y. Zuo, Z. Li, K. E. Rudd, and M. P. Deutscher, *J. Biol. Chem.* **273,** 14077 (1998).

[55] T. Shiomi, K. Fukushima, N. Suzuki, N. Nakashima, E. Noguchi, and T. Nishimoto, *J. Biochem. (Tokyo)* **123,** 883 (1998).

[56] K. S. Makarova, L. Aravind, M. Y. Galperin, N. V. Grishin, R. L. Tatusov, Y. I. Wolf, and E. V. Koonin, *Genome Res.* **9,** 608 (1999).

[57] P. M. Clissold and C. P. Ponting, *Curr. Biol.* **10,** R888 (2000).

[58] E. V. Koonin, Y. I. Wolf, and L. Aravind, *Genome Res.* **11,** 240 (2001).

[59] B. G. Luukkonen and B. Seraphin, *Nucleic Acids Res.* **27,** 3455 (1999).

protein belongs to the eukaryotic NAM7 family that is implicated in nonsense-mediated mRNA degradation.[60] This suggests that some of the eukaryotic RNases of the RNase II superfamily functionally interact with these helicases in RNA degradation.

RNase E/G Fold. In the absence of an experimentally determined structure, secondary structure prediction shows that the RNase E/G superfamily catalytic domain adopts an α/β structure that is likely to define a distinct fold. In its phyletic distribution, this superfamily is restricted to prokaryotes, with one distinct family seen in archaea and bacteria, respectively. The active site of these enzymes has not been characterized. A comparison of the conserved polar residues between the most divergent members of this family suggests that an N-terminal acidic residue, typically a glutamate, and another glutamate at the extreme C terminus could form the potential active site of this RNAse superfamily (data not shown). The bacterial family has a distinct domain architecture, with an S1-like OB-fold domain and another distinct conserved domain with two cysteines at the N and C termini of the nuclease domain, respectively. The RNase G subfamily is represented in almost all bacteria, whereas RNase E is restricted to the proteobacteria. RNase G and RNase both contribute to 5′-processing of 16S RNA,[61] whereas RNase E, which is a subunit of the RNA degradosome, has been additionally implicated in the processing of 5S RNA and other small RNAs, and in the general degradation of mRNA.[62,63]

The archaeal family of RNase E/G-related enzymes is extremely divergent from the bacterial forms and has a distinct architecture with an N-terminal S1 domain and a conserved C-terminal domain (the coRNase domain) that also occurs as a stand-alone protein in several archaea, bacteria, and the early branching eukaryote *Leishmania*. The coRNase domain consists of approximately 100 residues and cannot be identified with any known fold, but is predicted to form a globular α/β structure. This domain contains a characteristic motif with two conserved aspartates that form a Dx(3)D signature in a predicted loop between a helix and a strand, followed by another motif 14–20 residues downstream that centers around another conserved aspartate (LA, unpublished observations). This pattern is reminiscent of metal-coordinating sites present in hydrolytic enzymes and may define a second, hitherto uncharacterized nuclease domain. It cannot be ruled out, however, that coRNase is a previously undetected RNA-binding domain.

[60] G. Serin, A. Gersappe, J. D. Black, R. Aronoff, and L. E. Maquat, *Mol. Cell. Biol.* **21,** 209 (2001).

[61] Z. Li, S. Pandit, and M. P. Deutscher, *EMBO J.* **18,** 2878 (1999).

[62] P. J. Lopez, I. Marchand, S. A. Joyce, and M. Dreyfus, *Mol. Microbiol.* **33,** 188 (1999).

[63] G. G. Liou, W. N. Jane, S. N. Cohen, N. S. Lin, and S. Lin-Chao, *Proc. Natl. Acad. Sci. USA* **98,** 63 (2001).

$\alpha+\beta$ Ribonucleases

Ribonuclease PH Fold. The RNase PH superfamily[64] defines a compound fold comprised of two distinct substructures.[65] The N-terminal subdomain is clearly derived from the ribosomal protein S5 fold that is also seen in other proteins involved in RNA metabolism and translation, such as bacterial RNase P protein component and the elongation factor G domain IV.[66,67] In the RNase PH fold, a predicted β-hairpin has been added to the S5 core as an N-terminal extension. This subdomain is packed against another subdomain with a βb$\alpha\beta$bαa structure, with the β sheet of this subdomain interacting with the two α helices of the first subdomain. Based on the sequence conservation between the most divergent members of this family, we predict that the active site of these enzymes includes a nearly universal aspartate at the end of the second strand in the second subdomain of the RNase PH domain. A highly conserved arginine located near the N terminus of the first strand in subdomain 1 is in spatial proximity to this aspartate, and is likely to be an important part of the active site. In the same region, also in spatial proximity with the above-mentioned residues, a motif typically containing a RXD signature is conserved in many domains of this superfamily (LA, unpublished observations). These are also likely to be important in stabilizing the active site region. Thus it appears that the RNase PH fold has evolved through elaboration of an S5-like RNA-binding domain into a catalytic domain through extension of a sheet and addition of a subdomain. The members of this superfamily are involved in 3′ processing of RNAs in all the three domains of life and are major components of the exosome in eukaryotes and probably in archaea.[36,40,58]

Polynucleotide phosphorylases comprise one of the principal families of this superfamily, which is present largely in bacteria, although some eukaryotes, such as *Drosophila* and *Arabidopsis,* encode members of this family that probably have been acquired through horizontal gene transfer. This family is characterized by a duplication of the RNase PH domain and fusion to C-terminal KH and S1 RNA-binding domains.[64] The duplication of the RNase PH domain results in a quasi-threefold symmetry in these proteins, with only the C-terminal RNase PH domain possessing nuclease activity.[65] The typical RNase PH that appears to consist of the catalytic domain alone represents another bacteria-specific family. The RPR45p and Ski6p families are conserved in eukaryotes and archaea and include the nuclease subunits of the exosome.[40,58] Eukaryotes additionally have a specific family of exosomal RNase PH: the RPR46 family, that probably evolved through duplication of one of the above archaeoeukaryotic families. There are

[64] I. S. Mian, *Nucleic Acids Res.* **15,** 3187 (1997).

[65] M. F. Symmons, G. H. Jones, and B. F. Luisi, *Structure Fold Des.* **8,** 1215 (2000).

[66] M. Bycroft, S. Grunert, A. G. Murzin, M. Proctor, and D. St. Johnston, *EMBO J.* **14,** 3563 (1995).

[67] A. G. Murzin, *Nat. Struct. Biol.* **2,** 25 (1995).

multiple members of the Ski6p family in *C. elegans,* one of which contains a catalytic RNase PH domain fused to an uncharacterized domain that is bound to the C terminus of the helicase domain in the RecQ subfamily of helicases.[45] The demonstration of the involvement of a RecQ-like helicase in posttranscriptional gene silencing[68] suggests that some of these nucleases might contribute to this process in cooperation with RecQ-like helicases.

Metallo-β-Lactamase Fold. Metallo-β-lactamases are a large family of Zn-dependent hydrolases that contain a characteristic metal binding signature, typically of the form HXHXDH.[6] One family from this superfamily includes CPSF-I that is involved in the endonucleolytic cleavage of pre-mRNAs that precedes the addition of the poly(A) tail.[69] This family additionally includes the DNA repair enzyme SNM1,[70] suggesting that the entire family could be involved in nucleic acid processing functions. The CPSF family consists of four subfamilies. The largest subfamily includes classic CPSF orthologs that are conserved in archaea and eukaryotes and are also seen in certain bacteria. In turn, bacteria have their own, widespread CPSF subfamily that is also represented in several archaea and *Arabidopsis.* The eukaryotes possess a specific subfamily of inactive CPSF paralogs that are known to form subunits of the CPSF protein complex, probably mediating specific protein–protein interactions.[71] The archaea possess a distinct CPSF subfamily in which the metallo-β-lactamase domain is fused to an N-terminal KH domain. In theory, some other, poorly characterized families of the MBL fold also could include nucleases, but there is no direct or circumstantial evidence to predict such activity in any of these proteins.

Barnase Fold. The barnase fold, typified by the barnase superfamily, comprises two known families of secreted RNases from bacteria and fungi, respectively. These proteins have a simple, disulfide bond-stabilized core with four strands with an active site formed by a catalytic triad of E, R, and H residues.[72] Their restricted distribution, which is generally characteristic of secreted proteins, suggests that they evolved rather recently as an adaptation for the utilization of environmental polyribonucleotides.

Ribonuclease A Fold. This is another family of secreted nucleases with a very limited distribution; so far, it has been detected only in the vertebrates. The entire fold is stabilized by disulfide bonds, and the active site of these proteins involves a triad of H, K, and H. These proteins appear to be a recent derivation in the vertebrate lineage in relation to pathogen response and environmental polyribonucleotide degradation.[73]

[68] C. Cogoni and G. Macino, *Science* **286,** 2342 (1999).

[69] J. Zhao, M. Kessler, S. Helmling, J. P. O'Connor, and C. Moore, *Mol. Cell Biol.* **19,** 7733 .

[70] D. Richter, E. Niegemann, and M. Brendel, *Mol. Gen. Genet.* **231,** 194 (1992).

[71] P. J. Preker, M. Ohnacker, L. Minvielle-Sebastia, and W. Keller, *EMBO J.* **16,** 4727 (1997).

[72] G. Schreiber, C. Frisch, and A. R. Fersht, *J. Mol. Biol.* **270,** 111 (1997).

[73] J. J. Beintema and R. G. Kleineidam, *Cell. Mol. Life Sci.* **54,** 825 (1998).

Helix–Grip Fold. A small number of proteins of the START domain superfamily that possess the Helix–Grip fold[74] appear to possess ribonuclease activity.[75] These proteins, ribonucleases 1 and 2 from ginseng and their orthologs, belong to the Birch allergen family of START domains that are only known from the plants. These START domains also bind other ligands, such as cytokinins, which is more in line with the ligand-binding activities described for other members of the START superfamily.[74] The START domain contains a large ligand-binding channel that could potentially serve as the active site for these proteins. These proteins are known to be induced in response to pathogens in plant tissues and could function as part of a pathogen-specific RNA degradation mechanism.[76]

DHH Fold. The DHH superfamily of hydrolases is defined by a characteristic DHH signature, which is predicted to contribute to the active site and includes several metal-dependent hydrolases with phosphatase and nuclease activities.[5] The domain is predicted to adopt a bipartite $\alpha+\beta$ fold with the metal-chelating sites located in the N-terminal part. The best-characterized members of this fold are RecJ, a $5'\rightarrow3'$-exonuclease involved in DNA repair in bacteria[77] and probably archaea[77] and eukaryotic exopolyphosphatase.[78] Two families of DHH-fold proteins are predicted to possess RNase activity on the basis of the accessory domains that are fused to their hydrolase domains. The first of these is found only in the archaea and has a unique architecture with the DHH-domain preceded by an N-terminal zinc finger of the DnaJ variety followed by two RNA-binding OB-fold domains, one of the S1 superfamily and one of the N-OB superfamily.[79] These proteins could be involved in an archaea-specific $5'\rightarrow3'$ RNA degradation function. Anther DHH family with a potential RNase function is suggested by the fusion of the DHH domain to an N-terminal poly(A) polymerase and CBS domains. These could be bifunctional proteins capable of both degrading bacterial mRNAs and lengthening their poly(A) tails.

Ribonuclease T2 Fold. The RNase T2 superfamily is seen in all crown-group eukaryotes and in some γ-proteobacteria, which suggests a relatively recent horizontal transfer from eukaryotes to bacteria. All these nucleases are secreted and contain a core stabilized by disulfide bonds and an active site with two histidines. In most eukaryotes they are likely to be utilized in degradation of extracellular polyribonucleotides, for the acquisition of phosphate or bases. The genes encoding these RNases in plants have undergone multiple duplications and

[74] L. M. Iyer, E. V. Koonin, and L. Aravind, *Proteins,* in press (2001).

[75] A. Bufe, M. D. Spangfort, H. Kahlert, M. Schlaak, and W. M. Becker, *Planta* **199,** 413 (1996).

[76] I. Swoboda, O. Scheiner, D. Kraft, M. Breitenbach, E. Heberle-Bors, and O. Vicente, *Biochim. Biophys. Acta* **1219,** 457 (1994).

[77] V. A. Sutera, E. S. Han, L. A. Rajman, and S. T. Lovett, *J. Bacteriol.* **181,** 6098 (1999).

[78] A. Kornberg, N. N. Rao, and D. Ault-Riche, *Annu. Rev. Biochem.* **68,** 89 (1999).

[79] E. V. Koonin, Y. I. Wolf, and L. Aravind, *Adv. Protein Chem.* **54,** 245 (2000).

appear to be utilized in other roles such as pathogen response and pollen self-incompatibility.[80]

KEM1/RAT1 Fold. The KEM1/RAT1 superfamily is defined by a large composite module that is conserved in at least the crown-group eukaryotes. More divergent versions are seen in the earlier-branching eukaryotes, *Leishmania* and *Plasmodium falciparum*. The members of this superfamily share a very large conserved region and show a high level of sequence conservation with several invariant polar residues, which hampers prediction of the active site through sequence analysis. Comparisons with the plasmodial form suggest that it might be located in the N-terminal part of the catalytic module. This module probably adopts multiple, distinct substructures of the $\alpha+\beta$ type. The characterized members of this superfamily function both in the nucleus and in the cytoplasm, in mRNA maturation and turnover, and in 5'-terminal processing of 23S rRNA.[81,82] The superfamily consists of two families, the RAT1p family that is represented in all crown-group eukaryotes and the KEM1p family that is restricted to fungi.

All α-Helical Ribonucleases

Ribonuclease III Fold. The RNase III superfamily has no representative with an experimentally determined structure, but these proteins are confidently predicted to adopt an all α-fold. The experimentally characterized members of this superfamily hydrolyze double-stranded (ds)RNA or base-paired regions in single-stranded RNAs and are involved in processing of rRNAs and mRNAs.[83] The most widespread family includes the classic RNase III enzymes. One subfamily of this family is present in eukaryotes and bacteria and contains a single RNase III domain fused to a C-terminal dsRNA-binding domain (DSRBD). Another subfamily is restricted to animals and includes proteins with two RNase III domains fused to a DSRBD domain. A yeast-specific family includes proteins that contain the RNase III domain and have been described as mitochondrial ribosomal proteins.[84]

The CAF family of RNase III proteins, typified by the plant protein carpel factory, is found only in eukaryotes and has been implicated in posttranscriptional gene silencing (PTGS).[85,86] These proteins contain two C-terminal RNase III

[80] H. Sassa, T. Nishio, Y. Kowyama, H. Hirano, T. Koba, and H. Ikehashi, *Mol. Gen. Genet.* **250,** 547 (1996).

[81] M. Kenna, A. Stevens, M. McCammon, and M. G. Douglas, *Mol. Cell. Biol.* **13,** 341 (1993).

[82] L. H. Qu, A. Henras, Y. J. Lu, H. Zhou, W. X. Zhou, Y. Q. Zhu, J. Zhao, Y. Henry, M. Caizergues-Ferrer, and J. P. Bachellerie, *Mol. Cell Biol.* **19,** 1144 (1999).

[83] A. W. Nicholson, *FEMS Microbiol. Rev.* **23,** 371 (1999).

[84] L. Grohmann, H. R. Graack, V. Kruft, T. Choli, S. Goldschmidt-Reisin, and M. Kitakawa, *FEBS Lett.* **284,** 51 (1991).

[85] S. E. Jacobsen, M. P. Running, and E. M. Meyerowitz, *Development* **126,** 5231 (1999).

[86] B. L. Bass, *Cell* **101,** 235 (2000).

domains fused to a DSRBD domain and preceded by a RNA helicase and a PAZ domain. The PAZ domain is additionally found in proteins of the argonaute family that are implicated in PTGS.[87] The YLPD family is another group of eukaryote-specific RNase III enzyme, whereas *Bacillus* YazC defines a family of proteins that are only distantly related to the classic RNase III.

HD Fold. Hydrolases of the HD superfamily, so named after a sequence signature predicted to be part of the active site, possess diverse phosphoesterase activities and have been predicted to have an all α-helical fold.[4] The recent solution of the cNMP phosphodiesterase crystal structure showed that these proteins indeed adopt a unique all-α fold with two chelated metal cations.[88] The nucleases that belong to this superfamily have not been experimentally characterized and could only be predicted on the basis of fusion to nucleic-acid-binding domains (Fig. 2). Such fusions, seen in distinct families of HD hydrolases, include a KH domain, an N-OB domain, and a CCA-adding nucleotidyltransferase; the last architecture is analogous to the fusion of the DHH domain with poly(A) polymerase (see above). Another family of HD hydrolases may be predicted to include RNases on the basis of their phyletic pattern, namely conservation in archaea and eukaryotes, that resembles the phyletic patterns of many RNA processing enzymes.

All β-Strand RNases

OB Fold Nucleases. A single superfamily of OB-fold proteins, the thermonuclease superfamily, contains proteins with nuclease activity. In these proteins the active site is situated within the OB-fold domain, suggesting that they have been exapted for nuclease function from an ancestral nucleic acid-binding domain.[89] These proteins are general nucleases that degrade both RNA and DNA. The classic thermonucleases are found in bacteria, archaea, yeasts, and plants. *Bacillus* has a distinct family that includes multidomain proteins which contain, in addition to the OB-fold nuclease, an N-OB-fold domain and a calcineurin-like phosphatase domain (Fig. 2). Eukaryotes possess a distinct family with four thermonuclease domains and a C-terminal Tudor domain.[90] These proteins have been shown to function as transcriptional coactivators, but there is no evidence, so far, that they possess RNAse activity.

Protein Components of Ribonuclease P. RNase P is a ribozyme involved in tRNA processing in both eukaryotes and prokaryotes. Although the catalytic

[87] L. Cerutti, N. Mian, and A. Bateman, *Trends Biochem. Sci.* **25,** 481 (2000).

[88] R. X. Xu, A. M. Hassell, D. Vanderwall, M. H. Lambert, W. D. Holmes, M. A. Luther, W. J. Rocque, M. V. Milburn, Y. Zhao, H. Ke, and R. T. Nolte, *Science* **288,** 1822 (2000).

[89] A. G. Murzin, *EMBO J.* **12,** 861 (1993).

[90] A. Porta, S. Colonna-Romano, I. Callebaut, A. Franco, L. Marzullo, G. S. Kobayashi, and B. Maresca, *Biochem. Biophys. Res. Commun.* **254,** 605 (1999).

moiety of RNase P is an RNA molecule, the enzyme also contains protein subunits.[91,92] The bacterial version of the RNase P protein has no counterparts in eukaryotes and contains an S5-fold domain whose primary function appears to be binding the tRNA substrate and altering its conformation.[93,94] The archaeal and eukaryotic RNase P enzymes are more complex ribonucleoproteins that contain several protein subunits. At least four of these subunits, namely Rrp1p, Rrp2p, Pop4p, and Pop5p, can be traced to the common ancestor of archaea and eukaryotes[58] and probably define the RNase P ribonucleoprotein core. Beyond this core, there seems to be a eukaryote-specific layer of proteins, including POP1, a large protein conserved in all eukaryotes, and P38, an RNase P subunit that is found only in animals and plants and contains an RNA-binding domain of the Pelota superfamily. Several RNase P subunits that have been identified in yeast, including POP3, POP6, POP7, and POP8,[95] have no detectable homologs in other eukaryotes. These might be either peripheral, lineage-specific components of this complex or very fast-evolving proteins.

Functional Systems and Evolutionary History of Ribonucleases

The diverse RNase families that we briefly discussed above participate in a vast variety of polyribonucleotide maturation and degradation processes in various cellular compartments. The originally characterized RNases such as RNase A, RNase T2, and barnase are classical examples of simple, single-domain enzymes. More recent studies have shown, however, that many RNases are organized, through both domain fusion and noncovalent protein–protein interactions, into elaborate macromolecular complexes that mediate concerted RNA degradation and/or processing. Such complexes apparently have evolved independently in bacteria (the degradosome[52,63]) and in the common ancestor of archaea and eukaryotes (the exosome [40,96,97]). The rudimentary version of the exosome that can be traced to the common ancestor of eukaryotes and archaea probably consisted of nucleases of the RNase PH family and several RNA-binding proteins.[58] A recent analysis of predicted operons in archaea suggests that, at least in these organisms, the exosome probably functions in conjunction with the proteasome and might also be

[91] D. N. Frank and N. R. Pace, *Annu. Rev. Biochem.* **67,** 153 (1998).

[92] A. Schon, *FEMS Microbiol. Rev.* **23,** 391 (1999).

[93] S. Niranjanakumari, T. Stams, S. M. Crary, D. W. Christianson, and C. A. Fierke, *Proc. Natl. Acad. Sci. USA* **95,** 15212 (1998).

[94] T. Stams, S. Niranjanakumari, C. A. Fierke, and D. W. Christianson, *Science* **280,** 752 (1998).

[95] J. R. Chamberlain, Y. Lee, W. S. Lane, and D. R. Engelke, *Genes Dev.* **12,** 1678 (1998).

[96] C. Allmang, J. Kufel, G. Chanfreau, P. Mitchell, E. Petfalski, and D. Tollervey, *EMBO J.* **18,** 5399 (1999).

[97] C. J. Decker, *Curr. Biol.* **8,** R238 (1998).

associated with another nucleolytic complex, RNase P.[58] In eukaryotes, many new subunits were added to the exosome complex through both duplication of ancestral subunits and acquisition of bacterial-type nucleases, probably from the endosymbiotic organelles. These subunits include RNases of the RNase II fold that are largely missing in archaea.

Another eukaryotic RNase complex is the CPSF that is involved in mRNA cleavage for poly(A)denylation.[71,98,99] The family of MBL-fold RNAses that are the catalytic subunits of CPSF is nearly universal, indicating that these enzymes were involved in mRNA processing even in LUCA. In contrast, the noncatalytic subunits of the CPSF complex are eukaryotic-specific, which suggests that the complex emerged as a consequence of the regulatory demands placed on the formation of eukaryotic poly(A) tails.

Genetic analysis of the posttranscriptional gene silencing (PTGS) system suggests that it is probably mediated by yet another complex containing a characteristic set of RNases.[46,86] These complexes are thought to contain also an RNA-dependent RNA polymerase, a RecQ-like helicase, and an argonaute-like protein containing a Piwi and a Paz domain.[46,68,87,100] The Mut-7 family of nucleases have been identified as one of the candidate nucleases of this complex.[47] On the basis of domain architecture analysis and phyletic patterns, it is possible to predict that the CAF family of RNase III enzymes and some RNase PH enzymes of the SKI6p family might also interact with the PTGS system.[101] Similarly, our phyletic profile analysis suggests an interaction of the PTGS RNA degradation system with the poly(A) degradation system which involves the DAN family of nucleases.

Some of the RNA processing nucleases such as RNase PH, the CPSF MBL hydrolases, and RNase HII are among the most ancient enzymes that were already present in LUCA. However, the majority of nucleases are present only in one or two domains of life, which suggests substantial secondary elaboration of more specific RNA degradation systems. The large number of nucleases shared by bacteria and eukaryotes seem to indicate massive acquisition of bacterial enzymes by eukaryotes from their bacterial endosymbionts. Conversely, archaea show a relative paucity of enzymes of most nuclease superfamilies. This may mean that some of the archaeal RNases remain to be discovered. There are many potential candidates from the HD, DHH, and MBL superfamilies of hydrolases,

[98] K. S. Dickson, A. Bilger, S. Ballantyne, and M. P. Wickens, *Mol. Cell Biol.* **19,** 5707 (1999).
[99] P. J. Preker, M. Ohnacker, L. Minvielle-Sebastia, and W. Keller, *EMBO J.* **16,** 4727 (1997).
[100] H. Tabara, M. Sarkissian, W. G. Kelly, J. Fleenor, A. Grishok, L. Timmons, A. Fire, and C. C. Mello, *Cell* **99,** 123 (1999).
[101] L. Aravind, H. Watanabe, D. J. Lipman, and E. V. Koonin, *Proc. Natl. Acad. Sci. USA* **97,** 11319 (2000).

although none other than those described above can be currently pinpointed as probable RNases. Biochemical genomics of these hydrolase families may produce significant results in understanding the real complexity of RNA degradation and processing.

Eukaryotes, particularly the multicellular forms, show a proliferation of nucleases from different families. Some of these enzymes additionally show notable domain accretion that is typical of eukaryotes.[102] The 3′→5′-exonuclease superfamily in particular shows a great diversity of forms in eukaryotes, which probably corresponds to the emergence of new posttranscriptional regulatory systems. The secreted, disulfide-bonded RNases resemble other secreted proteins in showing a narrow phyletic distribution and apparently have been independently evolved in terminal branches of life.

[102] E. V. Koonin, L. Aravind, and A. S. Kondrashov, *Cell* **101,** 573 (2000).

[2] The Ribonuclease T1 Family

By HIROSHI YOSHIDA

Introduction

Since its discovery by Sato and Egami (1957)[1] in a commercial enzyme mixture from *Aspergillus oryzae* called Taka-diastase, ribonuclease (RNase) T1 [EC 3.1.27.3] has been the subject of extensive studies. In 1982, it was reviewed in detail by Takahashi and Moore.[2]Since then, explosive progress has been witnessed mainly in two directions: (1) determination of the three-dimensional structure by X-ray crystallography and NMR spectroscopy, and (2) creation of various mutants by protein engineering. Among numerous original and review papers that have been published, only the following two reviews will be cited here because of limited space: Pace *et al.*[3] on the former aspect and Steyaert[4] on the latter. Through these studies, the structure–function relationship of RNase T1 is being understood at the atomic level, and RNase T1 is now among the best-characterized proteins. A number of RNases with sequence similarities to RNase T1 have been isolated from

[1] K. Sato and F. Egami, *J. Biochem. (Tokyo)* **44,** 753 (1957).
[2] K. Takahashi and S. Moore, *in* "Enzymes," 3rd ed., Vol. 15 (P. D. Boyer, ed.) p. 435. Academic Press, New York and London, 1982.
[3] C. N. Pace, U. Heinemann, U. Hahn, and W. Saenger, *Angew. Chem. Int. Ed. Engl.* **30,** 343 (1991).
[4] J. Steyaert, *Eur. J. Biochem.* **247,** 1 (1997).

METHODS IN ENZYMOLOGY, VOL. 341
0076-6879/01 $35.00

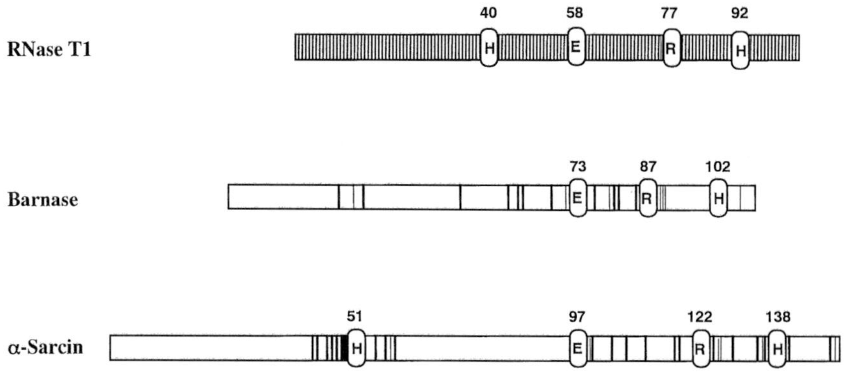

FIG. 1. Schematic representation of sequence similarities among RNase T1, barnase, and α-sarcin. The active site residues are shown with their residue numbers. For barnase and α-sarcin, residues identical to those of RNase T1 are indicated by thick lines and conservative substitutions by thin lines. Matching was made by introducing a minimum number of deletions or insertions, which are not shown in the figure.

various eukaryotic microbes. Structurally, they are small, acidic proteins having approximately 100 amino acid residues. They all retain the active site residues His-40, Glu-58, Arg-77, and His-92 (RNase T1 numbering) with high adjacent sequence homologies. They thus form a protein family called the RNase T1 family, reviewed recently by Irie.[5]

Two protein families related to the RNase T1 family are currently known. One is the family of prokaryotic microbial RNases, exemplified by barnase from *Bacillus amyloliquefaciens*. The other is the fungal cytotoxin family represented by α-sarcin from *Aspergillus giganteus*. Sequence similarities among RNase T1, barnase, and α-sarcin are schematically shown in Fig. 1.

Barnase has the active site residues corresponding to Glu-58, Arg-77, and His-92 of RNase T1, but lacks the one corresponding to His-40.[6] Overall homology between barnase and RNase T1 is low: only marginal to moderate similarity can be observed even around the active site residues. Functionally, however, they are very similar: they are secreted by microbes and degrade RNA endonucleolytically by similar reaction mechanisms. In addition, X-ray studies have revealed their similarities in the three-dimensional structure, leading to a hypothesis that they share the same evolutionary origin.[7] On the other hand, α-sarcin has all four active site residues with relatively high homologies around them. Major differences are

[5] M. Irie, *in* "Ribonucleases: Structures and Functions" (G. D'Alessio and J. F. Riordan, eds.), p. 101. Academic Press, San Diego, 1997.

[6] R. W. Hartley and E. A. Barker, *Nature New Biol.* **235,** 15 (1972).

[7] C. Hill, G. Dodson, U. Heineman, W. Saenger, Y. Mitsui, K. Nakamura, S. Borisov, G. Tischenko, K. Polyakov, and S. Pavlovsky, *Trends Biochem. Sci.* **14,** 1 (1983).

an insertion at the middle of the molecule and an N-terminal extension in α-sarcin. However, they are functionally very different: α-sarcin is a basic protein that exerts cytotoxicity by cleaving ribosomal RNA in a very restricted manner.[8] Therefore, it will be appropriate to postulate a superfamily composed of the RNase T1 family, the barnase family, and the α-sarcin family. Here, only the RNase T1 family will be discussed.

Similarities and Differences among RNase T1 Family Members

Structure

Figure 2a shows the sequence alignment of currently known RNase T1 family members. The evolutionary tree constructed from the alignment (Fig. 2b) coincides fairly well with the taxonomy of organisms of their origins. Note that only RNases U1, U2, and Pol are from species belonging to the phylum *Basidomycota,* while the others are from *Ascomycota.*

The three-dimensional structures of RNases T1,[9] Ms,[10] F1,[11] and U2[12] have been determined at high resolution. All these structures have the same basic architecture (four antiparallel β strands over an α helix), but are fairly diverse in peripheral loop and terminal regions. As a reflection of this fact, a rabbit polyclonal antibody against RNase F1 did not cross-react with RNase T1.[13] Especially noteworthy is the variation in disulfide bond locations, as disulfide bonds tend to be conserved among homologous proteins. Five different disulfide bond patterns are found (Fig. 3) with only one conserved disulfide bond, Cys^{6}–Cys^{103} (RNase T1 numbering).

Function

Functionally, the RNase T1 family of RNases performs endonucleolytic cleavage of RNA in two steps: (1) transphosphorylation to form a 2′,3′-cyclic phosphate intermediate and (2) hydrolysis of the intermediate to a 3′-phosphate (Fig. 4). The first step is reversible and is much faster than the second. The base specificity (B_0 in Fig. 4) is guanine specific with only a few exceptions. Namely, RNase U2 is purine specific and RNase Ms is guanine preferential, acting more or less on the other bases. However, it should be noted that the guanine specificity is not absolute

[8] Y. Endo, P. W. Huber, and I. G. Wool, *J. Biol. Chem.* **258,** 2662 (1983).

[9] R. Arni, U. Heinemann, R. Tokuoka, and W. Saenger, *J. Biol Chem.* **263,** 15358 (1988).

[10] T. Nonaka, Y. Mitsui, M. Irie, and K. Nakamura, *FEBS Lett.* **283,** 207 (1991).

[11] D. G. Vassylyev, K. Katayanagi, K. Ishikawa, M. Tsujimoto-Hirano, M. Danno, A. Pähler, O. Matsumoto, M. Matsushima, H. Yoshida, and K. Morikawa, *J. Mol. Biol.* **230,** 979 (1993).

[12] S. Noguchi, Y. Satow, T. Uchida, C. Sasaki, and T. Matsuzaki, *Biochemistry* **34,** 15583 (1995).

[13] H. Yoshida, H. Hanazawa, and M. Nishioka, unpublished results (1986).

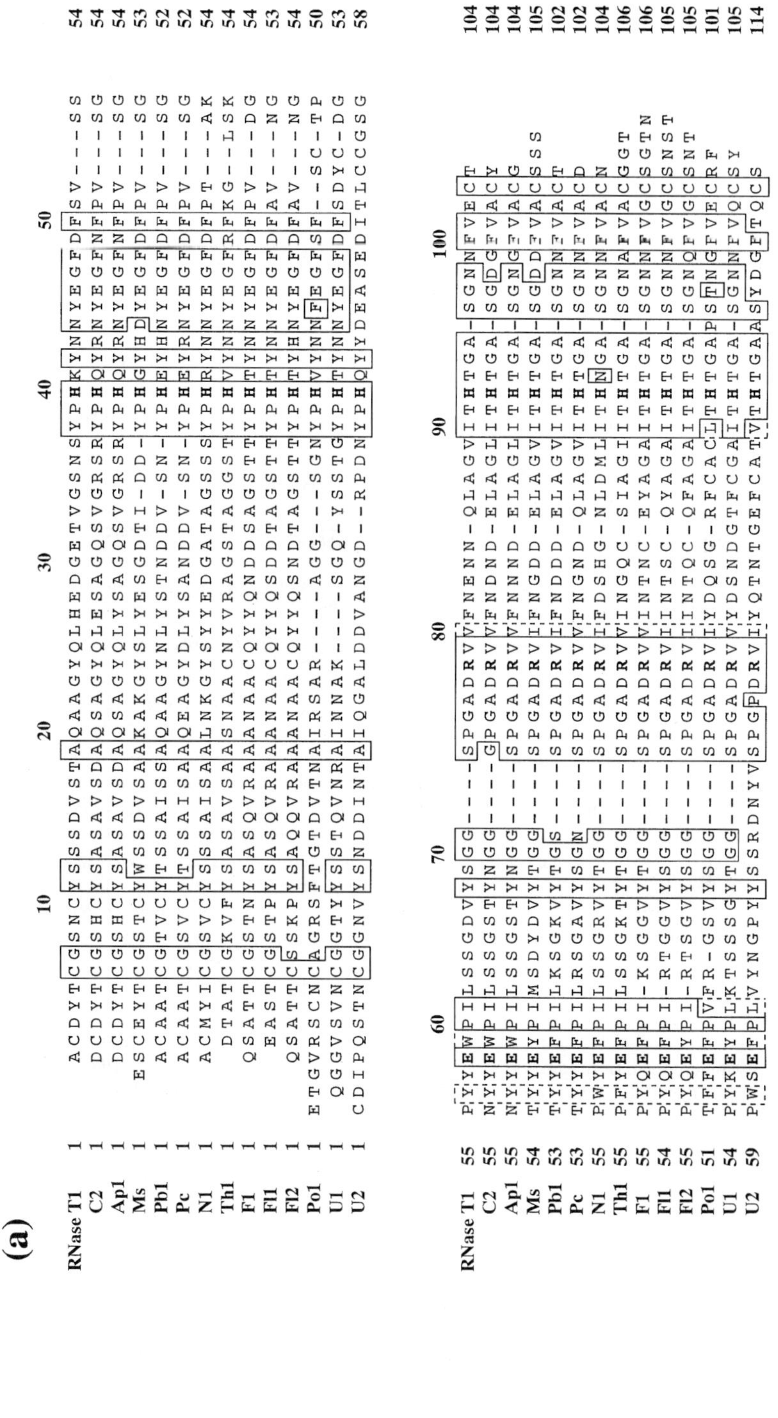

FIG. 2.

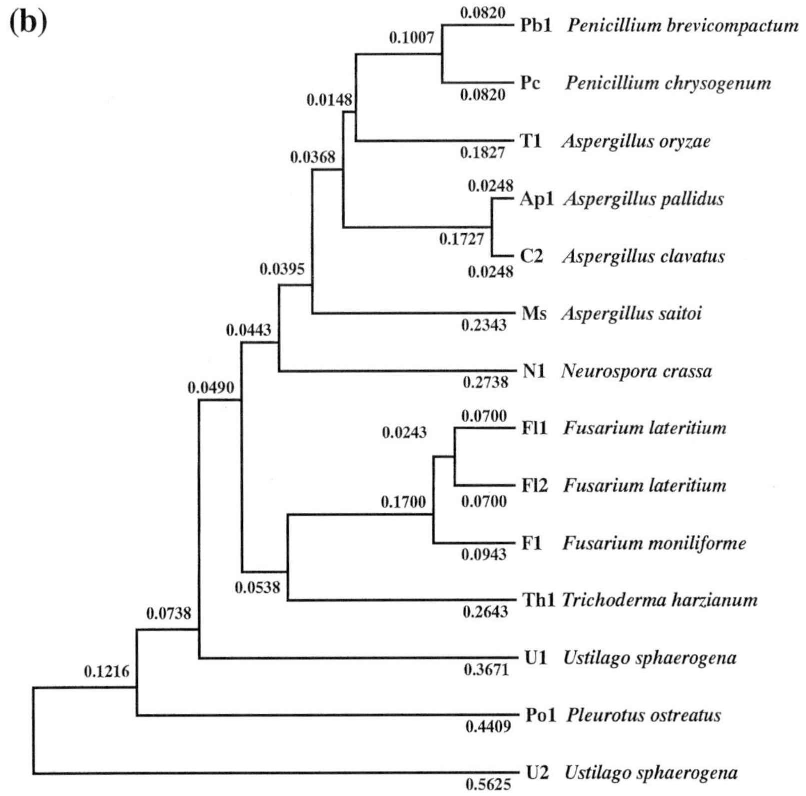

FIG. 2. (a) Sequence alignment of RNase T1 family RNases. Conserved residues with few exceptions are boxed. Conservative substitutions are boxed with dashed lines. The active site residues are in boldface letters. RNase T1 numbering is shown at the top of the matrix. The sequence data were obtained from the data base SWISS-PROT through the integrated database retrieval system GenomeNet and are reproduced without modification. It has been shown that the N-terminal residues of RNases F1 (Q), F11 (E), F12 (Q), Po1 (E), and U1 (Q) are actually pyrrolidone carboxylic acid (pyroglutamic acid). References for the first determination of the complete sequence are as follows. Note that for some RNases revisions were made afterward. RNase T1 [K. Takahashi, *J. Biochem. (Tokyo)* **70,** 945 (1971)]. RNase C2 [S. I. Bezborodova, O. M. Khodova, and V. M. Stepanov, *FEBS Lett.* **159,** 256 (1983)]. RNase Ap1 [S. V. Shlyapnikov, V. A. Kulikov, and G. I. Yakovlev, *FEBS Lett.* **177,** 246 (1984)]. RNase Ms [H. Watanabe, K. Ohgi, and M. Irie, *J. Biochem. (Tokyo)* **91,** 1495 (1982)]. RNase Pb1 [S. V. Shlyapnikov, G. I. Yakovlev, and V. A. Kulikov, *Dokl. Akad. Nauk SSSR* **281,** 226 (1985)]. RNase Pc [S. V. Shlyapnikov, S. I. Bezborodova, V. A. Kulikov, and G. I. Yakovlev, *FEBS Lett.* **196,** 29 (1986)]. RNase N1 [K. Takahashi, *J. Biochem. (Tokyo)* **104,** 375 (1988)]. RNase Th1 [S. I. Bezborodova, E. S. Vasileva-Tonkova, and K. M. Polyakov, *Bioorg. Khim.* **14,** 453 (1988)]. RNase F1 [J. Hirabayashi and H. Yoshida, *Biochem. Int.* **7,** 255 (1983)]. RNase F11 [S. I. Bezborodova, N. K. Chepurnova, and S. V. Shlyapnikov, *Bioorg. Khim.* **14,** 893 (1988)]. RNase F12 [S. V. Shlyapnikov, S. I. Bezborodova, and A. A. Dementiev, *Bioorg. Khim.* **14,** 589 (1988)]. RNase Po1 [H. Nomura, N. Inokuchi, H. Kobayashi, T. Koyama, M. Iwama, K. Ohgi, and M. Irie, *J. Biochem. (Tokyo)* **116,** 26 (1994)]. RNase U1 [K. Takahashi and J. Hashimoto, *J. Biochem. (Tokyo)* **103,** 313 (1988)]. RNase U2 [S. Sato and T. Uchida, *Biochem. J.* **145,** 353 (1975)]. (b) Evolutionary tree of the RNase T1 family constructed from the alignment by the unweighted pair group method using arithmetic averages (UPGMA) of software GENETYX-MAC, Ver. 10 (Software Development Co.). Each number indicates length of the corresponding branch. The species name of the organism from which the corresponding RNase has been isolated is shown.

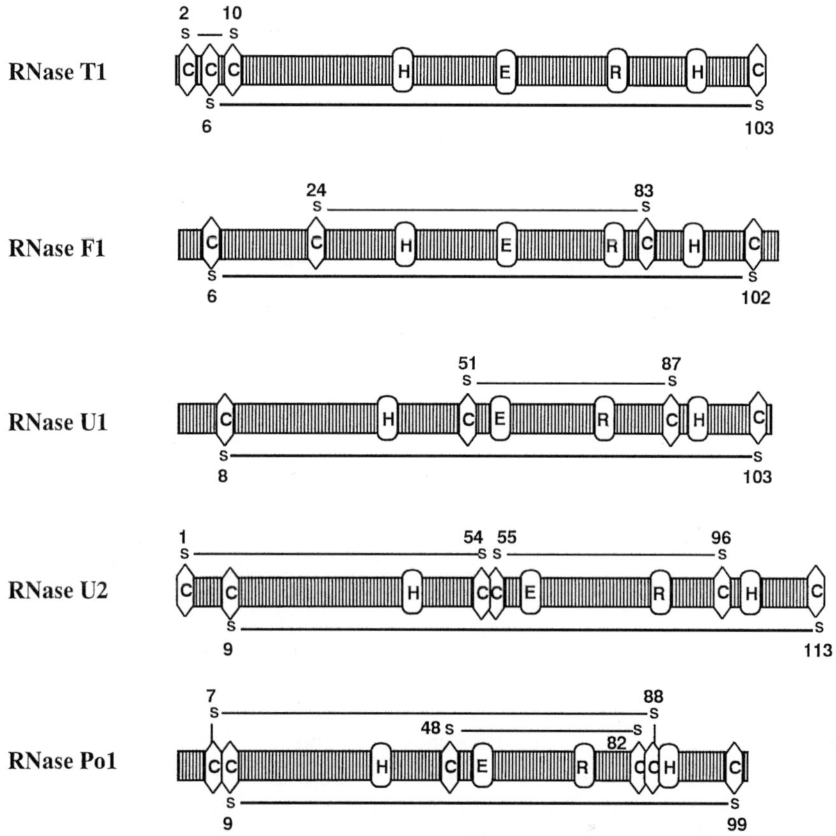

FIG. 3. Five disulfide bond patterns in the RNase T1 family. RNases from *Fusarium* species (F1, F11, and F12) are of RNase F1 type. RNases U1, U2, and Pol are each unique. All the others are of RNase T1 type. The conserved disulfide bond is shown with a thicker line.

but relative. In the enzyme nomenclature, only RNase U2 is classified under the number, EC 3.1.27.4, while all others are under the number, EC 3.1.27.3.

The quartet, His-40, Glu-58, Arg-77, and His-92, were first implicated in the active site of RNase T1 by chemical modification studies (reviewed in Ref. 2). Strict evolutionary conservation of these residues strongly suggests their indispensability. Site-directed mutagenesis studies are also generally in accord with this notion (reviewed in Ref. 4). Although there still remains some uncertainty, the present consensus on the reaction mechanism of RNase T1 is essentially the same as proposed by Takahashi[14] (Fig. 5), although His-92 was not unambiguously

[14] K. Takahashi, *J. Biochem. (Tokyo)* **67,** 833 (1970).

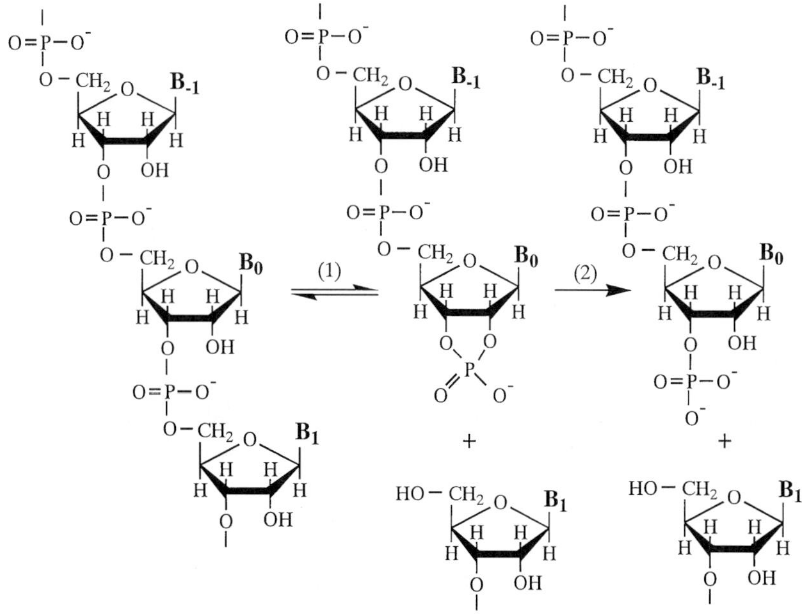

FIG. 4. Reactions catalyzed by RNase T1 family RNases.

assigned at that time. In the first transphosphorylation step (Fig. 5a), Glu-58 in its deprotonated form abstracts the proton of 2′-OH as a general base, facilitating its nucleophilic attack on the adjacent phosphorus. On the other hand, His-92 in its protonated form donates the proton as a general acid to the 5′-alkoxide leaving group. In the second hydrolysis step (Fig. 5b), the roles of general acid and base are presumably reversed between Glu-58 and His-92. However, it should be noted that these two steps do not necessarily proceed consecutively. The first step is much faster than the second step so that the first step would be repeated many times before the second step takes place. This means that there must be a mechanism to restore deprotonated Glu-58 and protonated His-92 after the transphosphorylation. As for His-40 and Arg-77, the following roles have been assumed. His-40 is located near Glu-58 and in some manner strengthens its basicity. Arg-77 will stabilize the pentacovalent phosphate intermediate by interacting with the negative charge on the phosphate oxygens. The precise details of this mechanism, however, await further studies.

The guanine base specificity of RNase T1 is well understood on a structural basis. The base is held by several hydrogen bonds: guanine N(7) and O(6) (bifurcated) accept hydrogen bonds from the peptide N–H group of Asn-43, and

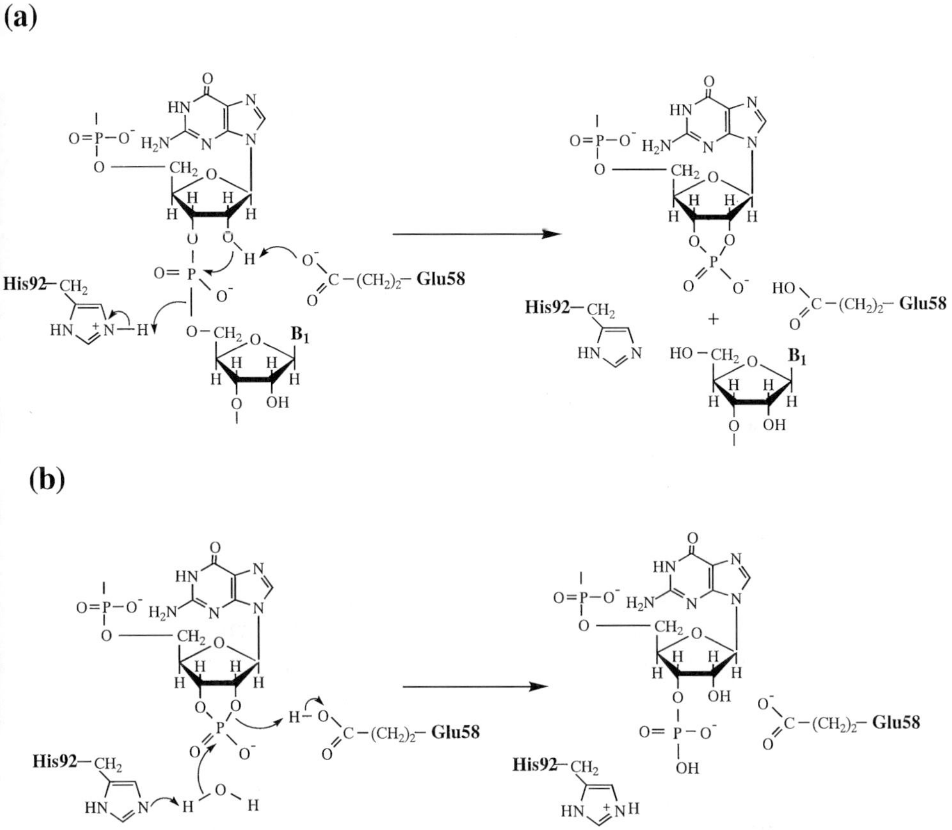

FIG. 5. Proposed reaction mechanism of RNase T1. (a) Transphosphorylation step. (b) Hydrolysis step.

the peptide N–H groups of Asn-44 and Tyr-45, respectively; the guanine N(1)–H and N(2)–H hydrogens bond to the side chain of Glu-46; the second N(2)–H hydrogen bonds to the peptide carbonyl of Asn-98. The base is also sandwiched by Tyr-42 and Tyr-45. In short, the segment from Tyr-42 to Glu-46 is primarily responsible for the guanine binding. Near perfect conservation of this segment among RNase T1 family members except RNase U2 accords with the fact that all but RNase U2 are guanine specific or preferential. However, most of the functional groups responsible for the guanine binding are main-chain peptide bonds. Therefore, the meaning of the conservation of the amino acid residues except Glu-46 is not clear. Presumably, these residues may be important in forming subtle features of the binding site.

RNase Assay Method

Overview

A number of routine RNase assay methods have so far been reported. However, they are based on one of the following three phenomena related to digestion of the RNA substrate: (1) absorbance at wavelengths around 300 nm decreases, (2) acid-soluble nucleotides are produced, and (3) metachromasy of methylene blue caused by interaction with RNA decreases.

A method based on principle 3 was developed by Greiner-Stoeffele *et al.*[15] to overcome drawbacks of principles 2 and 3, which nevertheless have long been used. By any principle, it is impossible to determine RNA cleavage sites quantitatively. As a consequence, definition of enzyme unit is inevitably arbitrary so that direct comparison of enzyme activities among different assay methods is meaningless. A method based on principle 2 delineated by Uchida and Egami[16] has so far been widely used. Here a modified version[17] of the method will be described.

Reagents

Buffer: 0.2 *M* Tris-HCl, pH 7.5
20 m*M* EDTA (pH adjusted to approximately 7)
Substrate solution: 6 mg/ml *Torula* yeast RNA (Calbiochem, La Jolla, CA)
Stop reagent: 0.1 *M* $LaCl_3$/25% (w/v) $HClO_4$
Dilution solution: 0.1% (w/v) gelatin

Procedure

A premix solution is prepared consisting of the buffer, the EDTA solution, and distilled water in a ratio of 5 : 2 : 1 (v/v). To 0.4 ml of the premix is added 0.1 ml of an enzyme solution properly diluted with 0.1% gelatin and the mixture is warmed to 37°. A blank is prepared by replacing the sample with the dilution solution. The reaction is started by addition of 0.25 ml of the substrate solution. The reaction mixture is incubated at 37° for 15 min, then 0.25 ml of the stop solution is added. After storage on ice for 10–30 min, the mixture is centrifuged at 1000*g* for 10 min. A portion (0.2 ml) of the supernatant is diluted with 5.0 ml of distilled water and absorbance at 260 nm (A_{260}) of the solution is measured against the blank, which has been treated in the same way. One unit (U) of the enzyme is defined as that amount which gives an A_{260} of 1 under the specified conditions. Therefore, the enzyme unit in the sample is equal to the A_{260} value obtained. However, care should be taken so that the A_{260} value does not exceed 0.3.

[15] T. Greiner-Stoeffele, M. Grunow, and U. Hahn, *Anal. Biochem.* **240,** 24 (1996).
[16] T. Uchida and F. Egami, *in* "Procedures in Nucleic Acid Research" (G. L. Cantoni and D. R. Davies, eds.), p. 3. Harper and Row, New York and London, 1966.
[17] H. Yoshida and H. Hanazawa, *Biochimie* **71,** 687 (1989).

Comments

Lanthanum(III) ions are added to perchloric acid to enhance RNA precipitation. Uranyl acetate has been a preferred additive for this purpose giving satisfactory results. However, its use requires a cumbersome managerial procedure, because it is now under strict control as a nuclear substance. Lanthanum chloride is slightly less effective as the RNA precipitant than uranyl acetate. In fact, the precipitate formed by the present method is somewhat loose, so care should be taken not to disturb the precipitate when the supernatant is withdrawn. Another demerit of this method, though common with the method of Uchida and Egami,[16] is that the dose–response curve is slightly convex in the entire region of A_{260}. At a lower A_{260} region (<0.3), however, the curve is almost linear and thus can be utilized for determination of the enzyme activity.

Purification

Overview

Currently, production of an enzyme by protein engineering techniques is a common practice. Application of the technique to RNase T1 is already abundant. Here, however, the scope will be limited to purification from natural sources. Most RNase T1 family RNases are small, stable proteins secreted by fungi into culture media. Thus, their separation from other proteins is relatively easy. One can take advantage of their heat stability and small size. Heat treatment at acidic pH denatures most other proteins while leaving the RNases intact. They are sufficiently retarded on Sephadex G-75 gel filtration to be separated from most other proteins. However, major difficulty lies often in elimination of a brown pigment whose chemical nature is not well understood. The pigment is adsorbed to anion exchangers such as DEAE-cellulose and is eluted in a very diffuse manner, tenaciously accompanying the object enzyme. Since the first complete purification of RNase T1 by Takahashi,[18] many improvements have been put forward. However, repeated anion-exchange chromatography has long been the only solution to this problem. Pace *et al.*[19] introduced a cation exchanger, sulfopropyl-Sepharose, which adsorbed RNase T1 at acidic pH but not the pigment.

One possible solution to this problem is the use of specific affinity chromatography. Various affinity adsorbents have been devised (reviewed in ref. 3). However, if the synthesis of the affinity adsorbent requires an elaborate technique, the method will not be convenient to most biochemists. Yoshida *et al.*[20] reported purification of RNase F1 using affinity chromatography as a key step. Although the affinity

[18] K. Takahashi, *J. Biochem. (Tokyo)* **49,** 1 (1961).

[19] C. N. Pace, G. R. Grimsley, and B. J. Barnett, *Anal. Biochem.* **167,** 418 (1987).

[20] H. Yoshida, I. Fukuda, and M. Hashiguchi, *J. Biochem. (Tokyo)* **88,** 1813 (1980).

adsorbent is not commercially available, it can be prepared easily from commercial products. The method is potentially applicable to any RNase T1 family enzyme. In fact, essentially the same method was successfully used for purification of RNase T1 by Kanaya and Uchida.[21] Here, a slightly modified version of the original method will be described.

Materials

Enzyme Source. The starting material is a freeze-dried ethanol precipitate of the culture filtrate of *Fusarium moniliforme*. Growth of a 100 liter scale culture is carried out at 25° for 90 h under aeration in a medium containing 3.0% (w/v) glucose, 0.1% ammonium sulfate, and 0.1% yeast extract (Difco) in 50 m*M* sodium citrate buffer, pH 5.8. The whole culture is filtered and the filtrate mixed with 5 volumes of ethanol. The resulting precipitate is collected by centrifugation, dissolved in 300 ml of distilled water, and freeze-dried. Yield 153 g.

Preparation of the Affinity Adsorbent pG-AH-Sepharose. Aminohexyl (AH)-Sepharose 4B (30 g) is swollen in 600 ml of 0.5 *M* NaCl overnight, washed on a glass filter with distilled water until chloride free, and equilibrated with 0.1 *M* borax. The gel is finally suspended in 100 ml of 0.1 *M* borax. To the suspension is added 3.0 mmol of 5′-GMP (pG) dissolved in 15 ml of 0.21 *M* $NaIO_4$. The suspension is gently stirred for 2 h at room temperature, then 400 mg of $NaBH_4$ is added. After stirring for 1 h, the mixture is left overnight at room temperature. The gel is filtered onto a glass filter and washed with 2 liter of 2 *M* NaCl. pG-AH-Sepharose thus prepared gives 2.3 μmol of inorganic phosphate per ml after mineralization by the method of Fiske and SubbaRow.[22] The reactions involved in the preparation are shown in Fig. 6.

Procedure

All the procedures are carried out at room temperature, except where otherwise stated. The degree of purification is monitored by an index, RNase activity (U) over absorbance unit at 280 nm (A_{280} U), rather than milligrams of protein, because elimination of the nonprotein pigment is a major concern as stated above. One A_{280} unit is defined as that amount of substance present in 1 ml of the solution that gives an A_{280} value of 1.0 at 1-cm optical path.

Step 1: Extraction. The starting material (92.6 g) is suspended in 280 ml of distilled water and centrifuged at 10,000 *g* for 10 min. The precipitate is reextracted with the same volume of distilled water.

Step 2: Batchwise Treatment with DEAE-Cellulose. The combined supernatant is dialyzed in Visking cellophane tubes against three changes (5 liter each) of distilled water. The dialysis is carried out in a cold room to minimize damage of

[21] S. Kanaya and T. Uchida, *J. Biochem. (Tokyo)* **89,** 591 (1981).

[22] C. H. Fiske and Y. SubbaRow, *J. Biol. Chem.* **66,** 375 (1925).

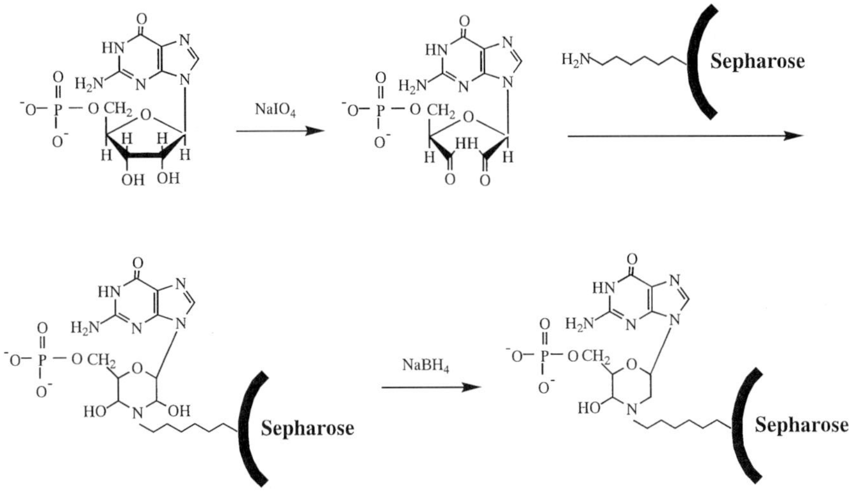

FIG. 6. Reactions involved in the preparation of pG-AH-Sepharose.

cellophane by cellulase. To the dialyzate is added 30 ml of 0.5 *M* Tris–acetate buffer (pH 6.4); then the mixture is diluted to 3 liter with distilled water. Microgranular DEAE-cellulose (Whatman DE 52, 150 g) is added to the solution, stirred, and filtered. After having been washed thoroughly with 5 mM Tris–acetate buffer (pH 6.4), the filter cake is eluted twice with 200 ml of 50 m*M* sodium acetate buffer–0.3 *M* NaCl, pH 4.7 (buffer A).

Step 3: Affinity Chromatography. The eluate is loaded onto a column (1.6 × 12 cm) of pG-AH-Sepharose that has been equilibrated with buffer A. The column is washed with 150 ml of buffer A, then eluted with 200 ml of 1 mM 2′(3′)-GMP in buffer A. The effluent is collected in 10 ml fractions and assayed for RNase activity. Active fractions (the last 80 ml of the wash and the first 170 ml of the GMP eluate) are pooled, concentrated by ultrafiltration through UH-1 membrane (Advantec, Tokyo, Japan), and desalted through a Sephadex G-25 column equilibrated with 10 m*M* NH_4HCO_3.

Step 4: DEAE-Cellulose Column Chromatography. The protein fraction is freeze-dried, dissolved in 15 ml of sodium–potassium phosphate buffer, pH 6.8 (buffer B), and charged onto a column (1.6 × 25 cm) of DE 52 that has been equilibrated with buffer B. Elution is carried out at a flow rate of 40 ml/h by an exponential gradient of increasing ionic strength and decreasing pH. The gradient is made by 800 ml of 0.2 *M* KH_2PO_4–0.15 *M* NaCl in a reservoir and a constant volume mixing chamber that initially contained 300 ml of buffer B. The elution profile is shown in Fig. 7. Fractions comprising a single major A_{280} peak are pooled, concentrated, desalted through a Sephadex G-25 column equilibrated with 10 m*M* NH_4HCO_3, and freeze-dried. Yield 90.4 mg.

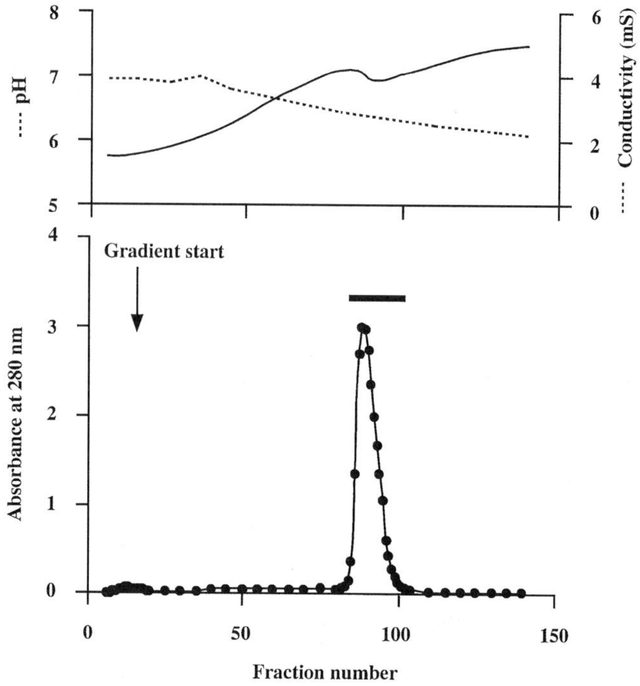

FIG. 7. Elution profile of the DEAE-cellulose column chromatography. The freeze-dried sample from step 3 is dissolved in 15 ml of sodium–potassium phosphate buffer, pH 6.8 (buffer B) and charged onto a column (1.6 × 25 cm) of DE 52 that had been equilibrated with buffer B. Elution was carried out at a flow rate of 40 ml/h by an exponential gradient made by 800 ml of 0.2 *M* KH_2PO_4–0.15 *M* NaCl in a reservoir and a constant-volume mixing chamber that had initially contained 300 ml of buffer B. Fractions of 4.9 ml were collected. Change in pH and ionic strength (conductivity) is shown in the upper panel. Pooled fractions are indicated by a bar.

Comments

RNase F1 was purified almost quantitatively in four steps, the key step being the affinity chromatography (see Table I). Most impurities flowed through the affinity column, whereas RNase F1 was tightly adsorbed to and eluted specifically from the column by a low concentration of 2′(3′)-GMP. In this particular experiment, the amount of RNase F1 contained in the sample exceeded the capacity of the column and a part of the enzyme appeared in the wash. Even so, however, the unadsorbed RNase F1 was sufficiently retarded to be separated from major impurities. Although most of the brown pigment flowed through the column, a part of it was adsorbed so that the whole column became brown after the chromatography. Therefore, the partially purified preparation from this step contained a small amount of the pigment, which was completely eliminated at the next step,

TABLE I
PURIFICATION OF RNASE F1

Purification step	Activity: Total (kU)	Activity: Yield (%)	Protein (A_{280} U)	Specific activity (kU/A_{280} U)	Purification (−fold)
Extraction	4310	100	9840	0.438	1
Batchwise treatment with DEAE-cellulose	4730	110	1520	3.11	7.1
Affinity chromatography	4230	98	121	35.0	79.9
DEAE-cellulose column chromatography	4480	104	102	43.9	100

the DEAE-cellulose column chromatography. The affinity column was reusable several times after being washed successively with 5 column volumes each of 0.1 *M* sodium borate buffer–1 *M* NaCl (pH 9.0), distilled water, and 0.1 *M* sodium acetate buffer–1 *M* NaCl (pH 4.0).

Most enzymes of RNase T1 family are secreted into the fungal culture medium. Only RNase Pol was isolated from a mushroom. Therefore, culture conditions are an important factor for enrichment of an RNase. Although a sufficiently enriched enzyme source of RNase T1 is commercially available (Takadiastase, Sankyo Pharmaceutical Co.), this is an exception. In most cases, purification starts from the culture of a microbe. A search for conditions that favor RNase production may bring a much more enriched source. However, it is not predictable which conditions are suitable for RNase production. For example, Arima *et al.*[23] reported induction of RNase production by *Ustilago sphaerogena* using RNA. However, addition of RNA was inhibitory for production of RNase F1 by *Fusarium moniliforme*.[24] Instead, yeast extract at a very narrow concentration range (0.1–0.3%) was effective for the production, though it was not clear what component(s) in yeast extract caused the induction. The starting material described above was one order of magnitude richer in RNase F1 than that previously reported.[20]

[23] T. Arima, T. Uchida, and F. Egami, *Biochem. J.* **106,** 601 (1968).
[24] H. Yoshida, K. Hoshi, and M. Iizuka, *Bull. Shimane. Med. Univ.* **22,** 27 (1999).

[3] Ribonuclease T2

By MASACHIKA IRIE and KAZUKO OHGI

Classification

RNase T2 was first isolated from a commercial digestive, Taka-diastase (*Aspergillus oryzae*), along with RNase T1, by Sato and Egami.[1] In contrast to RNase T1, which is an exclusively guanylic acid-specific RNase, RNase T2 is a base nonspecific acid RNase. Many fungi and commercial digestives produced from fungi contain similar RNase T2-like RNases. Thus, they were first thought to be typical fungal RNases. The primary structures of RNase T2 and the similar enzyme, RNase Rh from *Rhizopus niveus,* were determined by Kawata *et al.*[2] and Horiuchi *et al.,*[3] respectively, in 1988, and the primary structures of other fungal RNase T2-like enzymes have also been elucidated.[4–6] They contained two unique sequences, CAS1 and CAS 2, which have amino acid residues present at the active site of the enzymes. RNases with similar structures were found in self-incompatibility factors of *Nicotiania alata,* encoded by an allelic S-gene, and it was further recognized that these factors have RNase activity.[7] Thus RNase T2-like enzymes also exist in plants. On the other hand, several similar RNases have been isolated from plants, such as RNase LE from cultured tomato cells (*Lycopersicon esculentum*)[8,9] and from seeds of Cucurbitaceae plants such as bitter gourd,[10] cucumber, and melon, and their primary structures elucidated.[11,12] Since they are not S-gene products, they are called S-like RNases.

[1] K. Sato and F. Egami, *J. Biochem. (Tokyo)* **44,** 753 (1957).

[2] Y. Kawata, F. Sakiyama, and H. Tamaoki, *Eur. J. Biochem.* **176,** 683 (1988).

[3] H. Horiuchi, K. Yanai, M. Takagi, K. Yano, E. Wakabayashi, A. Sanda, S. Mine, K. Ohgi, and M. Irie, *J. Biochem. (Tokyo)* **103,** 408 (1988).

[4] H. Watanabe, A. Naitoh, Y. Suyama, N. Inokuchi, H. Shimada, K. Ohgi, and M. Irie, *J. Biochem. (Tokyo)* **108,** 303 (1990).

[5] Y. Inada, H. Watanabe, K. Ohgi, and M. Irie, *J. Biochem. (Tokyo)* **110,** 896 (1991).

[6] H. Kobayashi, N. Inokchi, T. Koyama, H. Watanabe, M. Iwama, K. Ohgi, and M. Irie, *Biosci. Biotechnol. Biochem.* **56,** 2003 (1992).

[7] B. A. McClure, V. Harring, P. R. Ebert, M. A. Anderson, R. J. Simpson, F. Sakiyama, and A. E. Clake, *Nature* **342,** 955 (1995).

[8] W. Jost, H. Bak, K. Glund, P. Terpstra, and J. J. Beintema, *Eur. J. Biochem.* **198,** 1 (1991).

[9] M. Köck, A. Löffler, S. Abel, and K. Glund, *Plant Mol. Biol.* **27,** 477 (1995).

[10] H. Ide, M. Kimura, M. Arai, and G. Funatsu, *FEBS Lett.* **284,** 161 (1991).

[11] M. A. Rojo, F. J. Arias, R. Igresias, J. M. Ferreras, R. Muñoz, B. Escarmis, F. Sorrano, J. López-Fando, and T. Girbés, *Planta* **194,** 328 (1994).

[12] M. A. Rojo, F. J. Arias, R. Igresias, J. M. Ferrerras, F. Sorrano, Mendez, C. Escamis, and T. Girbes, *Plant Sci.* **103,** 127 (1994).

0076-6879/01 $35.00

Similar RNases are also found in animal tissues and organs, such as bovine and porcine spleen,[13,14] chicken liver,[15] bullfrog liver (*Rana catesbeiana*),[16] squid liver,[17] and oyster.[18] Thus, the broad distribution of RNase T2 in animal organ/tissues has been confirmed. The presence of these RNases in many protozoans such as *Physarum polycephalum*[19] and *Dictyostelium discoidium*[20] has also been demonstrated. In addition to eukaryotes, the presence of RNase T2-like enzymes in the periplasm of gram-negative bacteria, such as *Escherichia coli* and *Aeromonas hydrophila,* has been reported by Meador and Kennell[21] and Favre *et al.,*[22] respectively. Therefore, RNase T2 is widely distributed in living organisms.

Protein Structure

The amino acid sequences of RNase T2 family enzymes are shown in Fig. 1. All share two unique sequences, CAS 1 and CAS 2.[23] Thus, RNase T2 family enzymes can be defined as RNases that contain these two sequences, and that work most effectively at weakly acidic pH (except for *E. coli* RNase I) and are base nonspecific.

RNase T2 family RNases are divided into two groups, based on the number of disulfide bridges. (1) Fungal RNases have five disulfide bridges, C1–C4, C2–C8, C3–C12, C9–C11, and C13–C16, whereas plant, animal, and protozoan RNases have four disulfide bridges, C5–C6, [C7–C10, only for RNase LE], C9–C11, C13–C16, and C14–C15. Only two of these (C9–C11 and C13–C16) are common to both types (Fig. 1).

The three-dimensional structure of only three RNase T2-like enzymes have been elucidated, to date. Three-dimensional structures of RNase Rh (*Rhizopus niveus*),[24] RNase MC1 (*Momordica charantia,* bitter gourd),[25] and RNase LE

[13] A. Kusano, M. Iwama, K. Ohgi, and M. Irie, *Biosci. Biotechnol Biochem.* **62,** 82 (1998).

[14] A. Kusano, M. Iwama, A. Sanda, K. Suwa, E. Nakaizumi, Y. Nakatani, H. Ohkawa, K. Ohgi, and M. Irie, *Acta Biochim. Pol.* **44,** 689 (1997).

[15] K. Hayano, M. Iwama, H. Sakamoto, H. Watanabe, A. Sanda, K. Ohgi, and M. Irie, *J. Biochem. (Tokyo)* **114,** 156 (1993).

[16] N. Inokuchi, H. Kobayashi, M. Miyamoto, T. Koyama, M. Iwama, K. Ohgi, and M. Irie, *Biol. Pharm. Bull.* **20,** 471 (1997).

[17] H. Watanabe, H. Narumi, T. Inaba, K. Ohgi, and M. Irie, *J. Biochem. (Tokyo)* **114,** 800 (1993).

[18] A. Kusano, M. Iwama, K. Ohgi, and M. Irie, *J. Biochem. (Tokyo)* **62,** 87 (1998).

[19] N. Inokuchi, T. Koyama, F. Sawada, and M. Irie, *J. Biochem. (Tokyo)* **113,** 425 (1993).

[20] N. Inokuchi, S. Saitoh, H. Kobayashi, T. Itagaki, T. Koyama, S. Uchiyama, M. Iwama, K. Ohgi, and M. Irie, *J. Biochem. (Tokyo)* **124,** 848 (1998).

[21] J. Meador III and D. Kennell, *Gene* **95,** 1 (1990).

[22] D. Favre, P. K. Ngai, and K. N. Timmis, *J. Bacteriol.* **175,** 3710 (1993).

[23] M. Irie, *Pharmacol. Ther.* **81,** 77 (1998).

[24] Y. Kurihara, Y. Mitsui, K. Ohgi, M. Irie, H. Mizuno, and K. T. Nakamura, *FEBS Lett.* **306,** 189 (1992).

[25] A. Nakagawa, I. Tanaka, R. Sakai, T. Nakashima, G. Funatsu, and M. Kimura, *Biochim. Biophys. Acta* **1433,** 253 (1999).

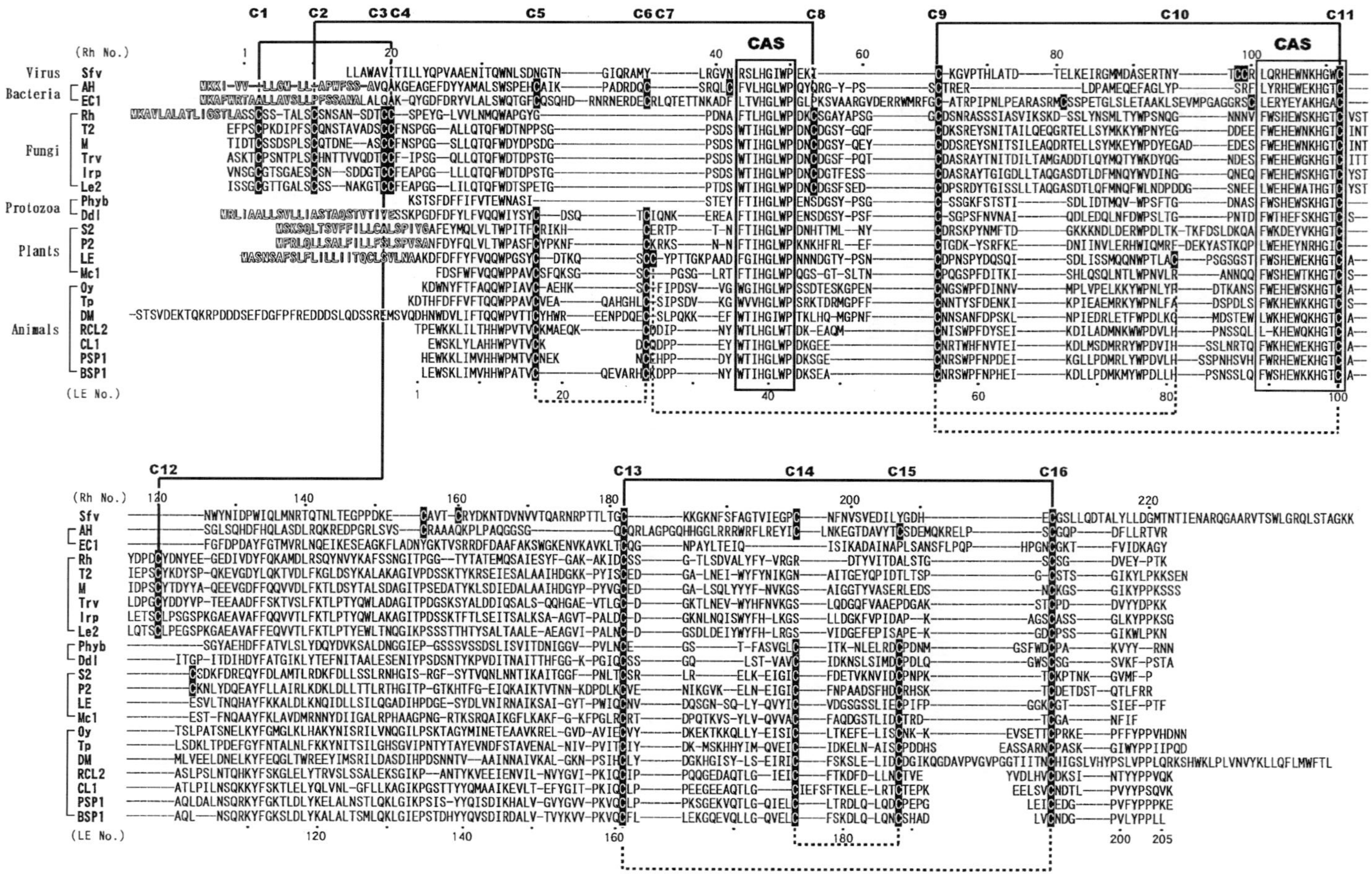

C1
C2
C3 C4
C5
C6 C7
C8
C9
C10
C11
C12
C13
C14
C15
C16
CAS
CAS
(Rh No.)
(LE No.)
Virus
Bacteria
Fungi
Protozoa
Plants
Animals
Sfv
AH
EC1
Rh
T2
M
Trv
Irp
Le2
Phyb
Ddl
S2
P2
LE
Mc1
Oy
Tp
DM
RCL2
CL1
PSP1
BSP1

(*Lycopersicon esculentum*,[26] tomato) have been elucidated by X-ray crystallography and are highly similar. They have ($\alpha + \beta$)-type structures consisting of six α helices and seven β strands (RNase Rh), seven α helices and seven β strands (RNase LE), and six α helices, four 3_{10} helices and eight β strands (RNase MC1). The three-dimensional structure of one of them, (RNase Rh), is shown in Fig. 2 as an example.

Mechanism of Action

RNase T2 endonucleolytically cleaves RNA to form oligonucleotides with a 2′, 3′-cyclic nucleotide at the 3′ end as intermediates, then forms oligonucleotides with 3′-phosphate (EC 3.1.27.1). The mechanism of action has mostly been studied using RNase Rh from *Rhizopus niveus*,[23] RNase M from *A. saitoi*[27] and RNase T2 from *Aspergillus oryzae*,[28] employing chemical modification and the enzymatic properties of several mutants produced by protein engineering.[29,30] The catalytic site consists of three His residues and one glutamic acid residue. In the first step reaction (transphosphorylation reaction), His-109 (RNase Rh numbering) acts as a base catalyst, His-46 acts as an acid catalyst,[29] and His-104 acts as a phosphate binding site.[29] In addition to these groups, Glu-105 and possibly Lys-108 stabilize the pentacovalent intermediate. In the hydrolysis step, His-46 acts as a base catalyst

[26] N. Tanaka, J. Arai, N. Inokuchi, T. Koyama, K. Ohgi, M. Irie, and K. T. Nakamura, *J. Mol. Biol.* **298,** 859 (2000).

[27] M. Harada and M. Irie, *J. Biochem. (Tokyo)* **73,** 705 (1973).

[28] Y. Kawata, F. Sakiyama, F. Hayashi, and Y. Kyogoku, *Eur. J. Biochem.* **187,** 255 (1991).

[29] K. Ohgi, H. Horiuchi, H. Watanabe, M. Iwama, M. Takagi, and M. Irie, *J. Biochem. (Tokyo)* **112,** 132 (1992).

[30] M. Irie, in "Ribonucleases, Structures and Functions" (G. D'Alessio and J. Riordan, eds.), p. 101. Academic Press, New York, 1997.

FIG. 1. Amino acid sequences of RNase T2 family RNases. Sfv, Swine fever virus; AH, *Aeromonas hydrophila;* EC, RNase I from *Escherichia coli;* Rh, RNase from *Rhizopus niveus;* T2, RNase T2 from *Aspergillus oryzae;* M, RNase from *Aspergillus saitoi;* Trv, RNase from *Trichoderma viride;* Irp, RNase from *Irpex lacteus;* Le2, an RNase from *Lentinus edodes;* Phyb, RNase from *Physarum polycephalum;* Dd, RNase from *Dictyostelium discoideum;* S2, an S-gene product from *Nicotiana alata;* P2, an S-gene product of *Petunia inflata;* LE, tomato RNase (*Lycopersicon esculentum*); MC1, bitter gourd RNase (*Momordica charantia*); Oy, RNase from oyster; Tp, squid liver RNase; DM, RNase from *Drosophila melanogaster;* CL1, chicken liver acid RNase; RC, bullfrog liver RNase (*Rana catesbeiana*); PS1, porcine spleen acid RNase: Bsp 1, bovine spleen RNase. Numbers at the top of the matrix, RNase Rh numbering; CASI and CAS2 were conserved active site sequences. C1- C16, location of half-cysteine residues numbered from the N terminus to the C terminus. Half-cysteine residues are expressed in white letters on a black background. Heavy and dotted lines connecting two half-cysteine residues indicate the disulfide bridges observed in fungal RNases and protozoan, plant, and animal RNases.

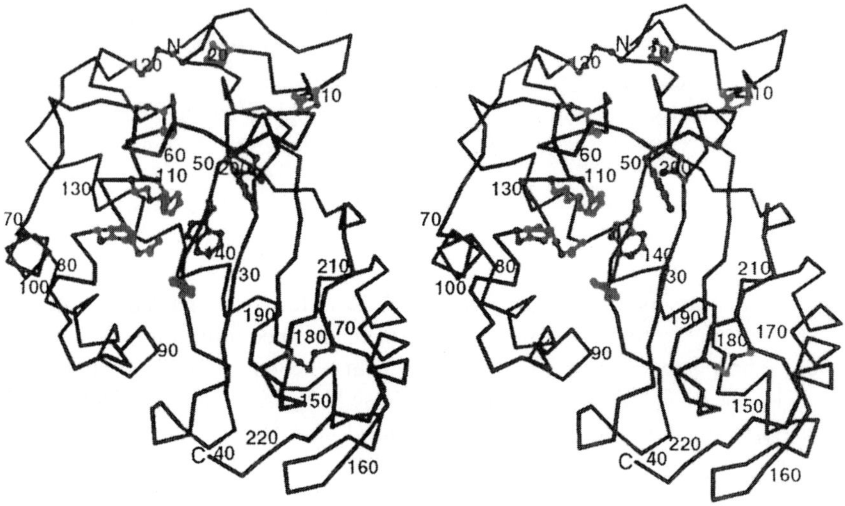

FIG. 2. Three-dimensional structure of RNase Rh. The structure was drawn from 1BOL RCFB data (H. Kurihara and K.T. Nakamura).[24] Disulfide bridges, side chains of amino acid residues involved in catalysis, and B1 base recognition sites are indicated.

and His-109 as an acid catalyst.[31] The enzyme reaction and base specificity are controlled by two base recognition sites (B1 and B2) that are located on the 5′ and 3′ sides of the scissile phosphate bond, respectively.[23,30] In RNase Rh, the B1 site consists of Asp-51, Tyr-57, and Trp-49,[23,30] whereas the B2 site consists of Gln-32, Pro-92, Ser-93, Asn-94, Gln-95, and Phe-101.[23,30,32]

Because of the lack of such studies on other enzymes, detailed clarification of mechanisms of action awaits further research.

Base Specificity

RNase T2 family enzymes are essentially base nonspecific. However, their precise base preferences differ depending on their amino acid sequences. Base specificity was estimated by three methods. The first method measured the rates of release of four nucleotides on digestion of RNA.[30] The fungal RNases released nucleotides in the order of A>G>C, U, but for plant RNases such as RNase LE, the order is G>A = U>C,[30] and for RNase MC1, U≫G>A>C.[33] The second

[31] M. Irie, K. Ohgi, M. Iwama, M. Koizumi, E. Sasayama, K. Harada, Y. Yano, J. Udagawa, and M. Kawasaki, *J. Biochem. (Tokyo)* **121,** 849 (1997).

[32] M. Hamashima, M.S. Thesis, Nagaoka University of Science and Technology (1994).

[33] M. Irie, H. Watanabe, K. Ohgi, Y. Minami, H. Yamada, and G. Funatsu, *Biosci. Biotechnol. Biochem.* **57,** 497 (1993).

method compares the rates of hydrolysis of four homopolynucleotides, poly(A), poly(I), poly(C), and poly(U).[33] The third method measured rates of hydrolysis of 16 dinucleosidephosphates,[34] which indicated the most susceptible nucleotidic bond.[34] For example, RNase MC1 most rapidly hydrolyzes CpU,[33] and RNase from *D. discoideum* hydrolyzes ApG, then GpG.[35] Using this method, we were also able to identify the base preferences of B1 and B2 sites.[33,35]

Physiological Roles

The RNase T2 family of enzymes perform a variety of roles in many organisms. In the following section, we will summarize some of those that have been well studied to date.

Self-Incompatibility Factor

Self-incompatibility is a mechanism that prevents pollen from one flower from fertilizing other flowers of the same plant. Self-incompatibility is often observed in plants belonging to such families as Solanaceae and Rosaceae. In these plants, self-incompatibility is controlled by a single gene locus with a large number of alleles, the S-locus. In self-incompatible plants, when a pollen grain lands on a style expressing the same S-allele, pollen tube growth is stopped or retarded, and there is no delivery of sperm to the ovary.[36] Anderson *et al.* cloned[37] the cDNA encoding the extracellular basic protein associated with S-alleles of *Nicotiana alata*. McClure *et al.*[7] identified the CAS sequences of RNase T2 in these proteins. They also found that these proteins exhibit RNase activity[7] (for a review, see Parry *et al.*[36]). Similar S-associated RNases have been found by several groups in many members of Solanaceae and other plants. The roles of S-RNases in self-incompatibility have not been fully elucidated. The importance of RNase activity in rejecting the pollen from the same plant has been discussed.[38] In a transgenic *Nicotiana* plant, two S-RNases and *E. coli* RNase I were expressed and tested for rejection by these plants of the pollen of self-incompatible plants. Plants expressing S-RNase reject the pollen of a self-incompatible plant, but *E. coli* RNase I expression did not

[34] M. Imazawa, M. Irie, and T. Ukita, *J. Biochem. (Tokyo)* **63,** 649 (1968).

[35] N. Inokuchi, S. Saitoh, H. Kobayashi, T. Itagaki, T. Koyama, S. Uchiyama, and M. Irie, *Biosci. Biotechnol. Biochem.* **63,** 141 (1999).

[36] S. K. Parry, Y. H. Liu, and A. E. Clarke, in "Ribonucleases: Structures and Functions" (G. D'Alessio and J. Riordan, eds.), p. 191. Academic Press, New York, 1997.

[37] M. A. Anderson, E. C. Cornish, S.-L. Mau, E. G. R. Hoggart, A. Atkinson, I. Bönig, B. Grego, R. Simpson, P. J. Roche, J. Haley, J. D. Penshow, H. D. D. Niall, G. W. Tregear, J. P. Coghlaan, R. J. Crawford, and A. E. Clarke, *Nature* **321,** 38 (1986).

[38] S. Huang, H.-S. Lee, B. Karunanandaa, and T.-H. Kao, *Plant Cell* **6,** 1021 (1994).

result in such rejection. Therefore, *E. coli* RNase I cannot serve as a substitute for S-RNases.[39]

Phosphate Remobilization

An increase in S-like RNase levels in response to phosphate deprivation is very often observed in the plant kingdom. This phenomenon was discovered by Nürunberger *et al.*[40] in cultured tomato cells. The induction of RNases parallels increases in phosphatase, thus accelerating the uptake of phosphate from the environment. Similar induction of RNases by phosphate deprivation has been observed in *Arabidopsis thaliana*[41] and RNase NE from *N. alata*.[42] However, not all S-like RNases are necessarily induced by phosphate deprivation.[43]

Defense against Pathogens

S-like RNases are found in plant seeds, especially in cucurbit species. S-like RNases are thought to protect the seeds from pathogens. Bitter gourd (*Momordica charantia*) contains RNase MC1[10]; sponge gourd (*Luffer cylindrica*) seeds contain RNases LC1 and LC2 (K. Nakamura unpublished); cucumber seeds, cusativin[11]; and melon seeds, melosin.[12] All of these have been identified as belonging to the RNase T2 family. Furthermore, all these plants are self-compatible, showing no relation to self-incompatibility. Cusativin located in the coat and cotyledon of cucumber seeds inhibited the translation of cell free protein synthesis.[11] Expression of a double strand specific RNase in transgenic potato reportedly conferred resistance to pathogenic viroids.[44] A similar result was reported for plant acid RNase. A hypersensitive response and aquired resistance are induced by cryptogenin secreted by *Phytopathora* var. *nicotiana*.[45]

Wounding

Ye *et al.*[46] have described induction of acid RNase by wounding in *Zinnia elegans*. The result appears to be rapid death of the cells around the wound with protection of the remaining tissues, thus serving a defense mechanism.[46]

[39] B. Beecher, J. Murfett, and A. MacClure, *Plant Mol. Biol.* **36,** 553 (1998).

[40] T. Nürunberger, S. Abel, W. Jost, and K. Glund, *Plant Physiol.* **92,** 970 (1990).

[41] P. A. Barriola, C. J. Howard, C. B. Taylor, M. M. Verburg, V. D. Jaglan, and P. J. Green, *Plant J.* **6,** 673 (1994).

[42] P. N. Dodds, A. E. Clarke, and E. Newbigin, *Plant Mol. Biol.* **31,** 227 (1996).

[43] P. Bariola and P. J. Green, in "Ribonuclease: Structures and Functions" (G. D'Alessio and J. Riordan, eds.), p. 163. Academic Press, New York, 1997.

[44] T. Sano, A. Nagayana, T. Ogawa, I. Ishida, and Y. Okada, *Nature Biotech.* **51,** 1291 (1997).

[45] E. Galiana, P. Bonnet, S. Conrod, H. Keller, F. Panabieres, M. Poncher, A. Poupet, and P. Ricci, *Plant Physiol.* **115,** 1557 (1997).

[46] Z.-H. Ye and D. L. Droste, *Plant Mol. Biol.* **30,** 697 (1996).

Senescence

During senescence of tomato leaves, two RNase genes, RNase LE and RNase LX, are expressed. Two RNases have been identified in cultured tomato cells.[47] Thus, RNase Lx is related to RNA metabolism in the final stage of senescence.

Localization

Many investigators have reported an acid RNase T2-like RNase in lysosomes of rat liver. deDuve *et al.* suggested that acid RNase is localized in lysosome-like particles in the rat liver.[48] The characteristics of this RNase in partially purified form were reported by Futai *et al.*[49] It has a very sharp pH optimum at acidic pH and is base nonspecific.

The pH–rate profile of RNase activity is very similar to those of bovine spleen and porcine spleen RNase.[50,51] In both bovine and porcine spleen, only one species of acid RNase has been characterized.[50,51] The primary structures of these two RNases elucidated indicated that they are members of the RNase T2 family.[13,14]

In protozoa *Tetrahymena pyriformis*[52] and *D. discoideum,*[53] the presence of RNase T2-like acid RNase in lysosome-like particles was reported. Inokuchi *et al.*[20] isolated only one acid RNase from the latter organism, *D. discoideum,* and identified it as a member of the RNase T2-family enzyme.[20] These data suggested that lysosome acid RNases are very likely to be members of the RNase T2 family.

RNase I of *Escherichia coli* was recognized long ago by Neu and Heppel as a base nonspecific RNase, which located in periplasmic space.[54] Furthermore it is encoded by the *ran* gene. The amino acid sequence indicated that it belongs to the RNase T2 family.[21] A similar RNase was also identified in the gram-negative bacterium Aeromonas hydrophila.[22]

In plants, the situation is more complex. In tomato cells, RNase T2 family enzymes are distributed extracellularly, in vacuoles, and probably in the endoplasmic reticulum (ER).[8,9] One of the tomato RNases, RNase Lx, has a putative C-terminal signal for ER retention, whereas RNase LV 2, located in vacuoles, is very similar but does not have this signal sequence. All of these tomato RNase T2-like enzymes are involved in phosphate remobilization.

RNase T2 Family RNase in Virus

The presence of an RNase T2–like RNase in viral coat protein was reported by Schneider *et al.*[55] The primary structure deduced from the nucleotide sequence confirmed that it belongs to the RNase T2 family.

[47] A. Lers, A. Kalchitsuki, E. Lomantes, S. Burd, and P. J. Green, *Plant Mol. Biol.* **36,** 439 (1998).
[48] C. deDuve, B. C. Pressman, R. Glanetto, P. Wattlaux, and P. Appelman, *Biochem. J.* **60,** 604 (1955).
[49] F. Futai, S. Miyata, and D. Mizuno, *J. Biol. Chem.* **244,** 4951 (1969).
[50] A. Bernardi and G. Bernardi, *Biochim. Biophys. Acta* **129,** 23 (1966).
[51] K. Ohgi, A. Sanda, Y. Takizawa, and M. Irie, *J. Biochem. (Tokyo)* **103,** 267 (1988).

Enzyme Assay

Assay methods using yeast RNA as a substrate were essentially the same as those of RNases belonging to RNase A and RNase T1 families, except for the use of buffer of pH ~5.0. The two methods described below differ only in the use of two different reagents to stop the reaction. Method (2) is about 4–5 times more sensitive than Method (1).

Method (1)

To a substrate solution containing 5 mg RNA in 2 ml of 0.1 *M* sodium acetate buffer (pH 5.0), add a small amount of enzyme solution.

Incubate at 37° for then appropriate time. Add 1.0 ml MacFadyen reagent[56] to stop the reaction, then centrifuge at 3000 rpm for 5 min.

To 0.3 ml of the supernatant add 2 ml water, and measure the absorbancy of the solution at 260 nm.

Method (2)

To a substrate solution containing 5 mg RNA in 2 ml of 0.1 *M* sodium acetate buffer (pH 5.0), add a small amount of enzyme solution,

Incubate at 37° for appropriate time, then add 1.0 ml of ice-cold reagent consisting of 2 m*M* lanthanum nitrate [$La(NO_3)_3$] in 15% perchloric acid at room temperature. Maintain for 10 min at 0°, then centrifuge for 5 min at 3000 rpm.

To 0.3 ml of the supernatant, add distilled water (2 ml) and measure absorbancy at 260 nm.

Enzyme Units

There is no standard definition of enzyme units for the RNase T2 family of enzymes. However, most define one unit as the amount of enzyme that increases absorbancy by 1.0 of acid-soluble nucleotide per 5 min (or an specified time).

Preparation of Fungal T2–like RNase

Most RNase T2-like enzymes are acidic proteins or acidic glycoproteins. Thus, preparation procedures include column chromatography on anion exchangers such as DEAE-cellulose or DEAE-Sephadex, or chromatography on cation exchangers such as phosphocellulose or SP-Sephadex at acidic pH. In earlier reports such as

[52] M. Müller, M. Baudhuin, and deDuve, *Cell Physiol.* **68,** 1165 (1966).

[53] E. Wiener and J. M. Ashworth, *Biochem. J.* **118,** 505 (1970).

[54] H. Neu and L. A. Heppel, *Biochem. Biophys. Res. Commun.* **14,** 109 (1964).

[55] R. Schneider, G. Unger, R. Stark, E. Schneider-Scherzer, and H. J. Thiele, *Science* **261,** 1169 (1993).

[56] M. R. MacDonald, *Methods in Enzymol.* **Vol. II,** p. 427 (1955).

those on the preparation of RNase T2 from Taka-diastase (*A. oryzae*) or of RNase M from molsin (*A. saitoi*),[4] DEAE-cellulose or phosphocellulose was used as the chromatography resin. However, the presence of cellulase in such fungi caused many difficulties, e.g, digestion of resins, resulting in slowing of the chromatographic flow rate, and collapse of the cellulose dialysis tubing in the early steps of purification. Therefore, these purification steps have been replaced by DEAE-Sephadex or SP-Sephadex procedures. In purification of RNase T2-like enzymes it might be necessary to consider that most fungal cells contain RNase T1–like enzymes as well as RNase T2–like enzymes. Thus, the separation of these two enzymes by gel filtration is inevitable. In the following section, purification of a fungal RNase (*Trichoderma viride*) from a commercial enzyme preparation will be described as an example.

Preparation of RNase Trv from Commercial Enzyme Preparation "Cellulase T-AP" Produced from *Trichoderma viride*[5]

Step 1

About 100 g of the crude enzyme, cellulase T-AP (Amano Pharm. Co., Nagoya) is added to ice-cold water (300 ml). The solution is stirred for 30 min, then centrifuged at 10,000 rpm for 10 min. The supernatant is used as a crude extract.

Step 2

The supernatant is diluted to 2.5 liters with ice-cold water, and the pH of the solution adjusted to 7.5 with 0.1 *N* NaOH. The solution is adsorbed on DEAE-Sephadex A-50, equilibrated with 50 m*M* Tris-HCl buffer (pH 7.5), and the resin packed into a column (7.5 × 20 cm). The column is eluted with 50 m*M* Tris-HCl buffer containing 0.5 *M* NaCl. The RNase-active fractions are pooled.

Step 3

The enzymatically active fractions of step 2 are pooled and concentrated to a small volume *in vacuo,* then gel filtrated with Sephadex G-50 (5.5 × 70 cm) preequilibrated with 50 m*M* Tris-HCl buffer (pH 7.5). The enzymatically active fractions are pooled.

Step 4

The enzymatically active fractions from step 3 are adsorbed on a DEAE-Sephadex A-25 column (4 × 20 cm) preequilibrated with 50 m*M* Tris-HCl buffer (pH 7.5). The column is eluted with a linear gradient of NaCl (0–0.5 *M*) in 2 liters of the same buffer. The RNase-active fractions are pooled and desalted as described above (step 3).

Step 5

Three batches of the step 4 enzyme fractions are pooled and used for further purification. The pH of the enzyme solution is adjusted to 6.0 by the addition of acetic acid. The solution is then applied to a DEAE-Sephadex A-25 column (4 × 20 cm) preequilibrated with 50 m*M* of Tris–acetate buffer (pH 6.0). The column is eluted with a linear gradient of NaCl (0–0.35 *M*) in 2 liters of the same buffer. The enzymatically active fractions are pooled and desalted as described in step 3.

Step 6

The pH of the RNase active fraction from step 5 is adjusted to 6.0 and the solution applied to a DEAE-cellulose column (3 × 20 cm) preequilibrated with 50 m*M* sodium acetate buffer (pH 6.0). The column is eluted with a linear gradient of NaCl (0–0.4 *M*) in the same buffer (1 liter). The RNase-active fractions are pooled.

Step 7

The RNase fractions in step 6 are dialyzed against deionized water, and the pH adjusted to 6.0. The solution is applied to a DEAE Toyopearl 650 M column (2.1 × 90 cm) preequilibrated with 50 m*M* sodium acetate buffer (pH 6.0). The column is washed with 1 liter of the same buffer (pH 6.0) containing 0.1 *M* NaCl and a linear gradient of NaCl (0–1 *M*) in the same buffer applied. The RNase-active fractions are pooled and desalted with Sephadex G-50 as described in step 3.

Step 8

The pH of the RNase-active fraction from step 7 is adjusted to 6.0, and the solution is applied to a 2′, 5′-ADP Sepharose 4B column (1 × 15 cm) preequilibrated with 50 m*M* acetate buffer (pH 5.0). The column is eluted with a linear gradient of NaCl (0–0.8 *M* NaCl) in 500 ml of the same buffer. The RNase-active fractions are pooled.

The RNase fraction from Step 8 is concentrated to a small volume and gel-filtered on a Sephadex G-50 column equilibrated with 50 m*M* trimethylamine–acetate buffer (pH 8.0). The RNase-active fractions are pooled. The results are summarized in Table I.

Preparation of Animal T2–like RNase from Oyster

In contrast to fungal RNase T2–like RNases, the preparation of RNase T2–like enzymes from plant and animal sources is not hampered by the presence of cellulase. This simplifies preparation procedures. We will describe the preparation of oyster RNase as an example.

TABLE I
PURIFICATION OF RNASE TRV FROM COMMERCIAL DIGESTIVE CELLULASE T-AP[a]

Step	Total Activity (units)	Specific protein (mg)	Yield activity (units/mg)	Yield (%)
1. Crude extract	70,000	23,700	0.30	100
2. DEAE-Sephadex A-50	59,000	96,000	0.61	78
3. Sephadex G-50	28,000	28,000	1.00	41
4. DEAE-Sephadex A-50	21,000	3500	5.90	31
5. DEAE-Sephadex A-50	18,600	1300	14.3	27
6. DEAE-cellulose	16,100	510	31.6	23
7. DEAE-Toyopearl M650	15,400	63	244	22
8. 2′, 5′-ADP Sepharose 4B	14,000	24	583	20

[a] Produces from *Trichoderma viride*.[5]

Preparation of Oyster (*Crassostrea gigas*) RNase[18]

Step 1

Oysters (15 kg) are homogenized in an equal volume of ice-cold 0.25 *N* H_2SO_4, The homogenate is centrifuged at 10,000 rpm for 10 min at 5° (crude extract).

Step 2

The supernatant is fractionated with ice-cold acetone and the precipitate between 33 and 60% acetone is collected (acetone fractionation). The precipitate is dialyzed against deionized water.

Step 3

The pH of the dialyzate is adjusted to 3.0 with 1 *N* HCl and the solution treated with four batches of wet SP-Sephadex C-50, equilibrated with 20 m*M* sodium citrate buffer (pH 3.0). SP-Sephadex C-50 adsorbed protein is packed on a column (6 × 25 cm). Elution is performed with a linear gradient of the same buffer (1 liter) and 50 m*M* sodium citrate buffer (1 liter). Fractions of 15 ml each are collected. The fractions with RNase activity are pooled. Four batches of RNase fractions are pooled and dialyzed against deionized water.

Step 4

The dialyzate is adsorbed on a DEAE-cellulose column (3.5 × 26 cm) equilibrated with 10 m*M* Na_2HPO_4. The RNase-active fractions not adsorbed to the column are pooled and then dialyzed against deionized water.

Step 5

The RNase-active fraction from step 4 is applied to a CM-Sephadex C-50 column (4 × 26 cm) equilibrated with 20 m*M* sodium acetate buffer (pH 4.5). Elution is performed with a linear gradient of the same buffer (500 ml) and 50 m*M* trisodium citrate containing 50 m*M* sodium chloride, and 15 ml fractions are collected. Fractions with RNase activity are pooled.

Step 6

The RNase fraction from step 5 is dialyzed against deionized water, and the pH adjusted to 5.0 by adding 1 *N* HCl. The enzyme solution is applied to a heparin Sepharose column (1.8 × 30 cm) equilibrated with 20 m*M* sodium acetate buffer (pH 5.0). The elution of RNase is performed with a linear gradient of 20 m*M* sodium acetate buffer (pH 5.0, 350 ml) and Tris–acetate buffer (pH 7.0, 250 ml). Fractions of 3.5 ml each are collected. The fractions with RNase activity are pooled.

Step 7

The RNase fraction from step 6 is again applied to a heparin Sepharose column as described above. The RNase is eluted with linear gradient of NaCl (0–0.5 *M*) in the same buffer (total volume 600 ml). The fractions with RNase activity are pooled.

Step 8

The RNase fraction is dialyzed against deionized water, and the pH of the solution is adjusted to 3.0. The enzyme solution is applied to a CM-Toyopearl column (1.9 × 75 cm) equilibrated with 20 m*M* sodium citrate buffer (pH 3.0). The RNase is eluted with a linear gradient of 20 *M* sodium citrate buffer (pH 3.0, 500 ml) and 50 m*M* trisodium citrate buffer (500 ml). Fractions of 6 ml each are collected.

Step 9

Gel filtration with Sephadex G-75. The RNase fraction from step 8 is dialyzed against deionized water, then concentrated to a small volume *in vacuo* with a rotary evaporator. The concentrated enzyme is applied to a Sephadex G-75 column (1.5 × 170 cm) equilibrated with trimethylamine acetate buffer pH 8.0. Fractions of 2.5 ml each are collected. The fractions with RNase activity and showing a single band on SDS–PAGE are pooled. The results of purification are shown in Table II.

As described above, the RNase T2-like enzymes from animals are more easily purified than those from fungi. However, peptide bond cleavage often occurs in purified preparations at about 20 residues from the N terminus and C terminus. We are not able to ascertain whether the cleavages occur during preparation because of protease action, or have some physiological significance. We know only that the cleavage sites are very similar and independent of species differences, and according to

TABLE II
PURIFICATION OF OYSTER RNASE[a]

Step	Activity (units)	Protein (mg)	Specific activity (units/mg)	Yield (%)
1. Crude extract	54,000	1,780,000	0.03	100
2. Acetone fractionation	28,000	43,000	0.99	52
3. SP-Sephadex C-50	11,000	4640	2.38	20.5
4. DEAE-cellulose	12,200	3150	3.87	22.6
5. CM-Sephadex C 50	14,000	890	15.7	26.1
6. Heparin Sepharose I	12,000	200	60.4	22.2
7. Heparin Sepharose II	11,300	65	175	21.0
8. Sephadex G-75	11,100	40	282	20.7
9. CM-Toyopearl	10,000	21	473	18.6
10. Sephadex G-75	6400	6.6	968	11.9

[a] Starting with 15 kg of oysters.

the three-dimensional structures of RNase Rh, they are located on the surface of the enzyme. Such cleavages are seen in the case of bovine and porcine spleen RNase T2–like RNase, as well as chicken and bullfrog liver RNase T2-like RNases.[13–17]

Inhibitors

Fungal RNases such as RNase T2 from *A. oryzae,* RNase M from *A. saitoi,* and RNase Rh from *R. niveus* are strongly inhibited by Cu^{2+}, Zn^{2+}, and Hg^{2+}, and to an extent similar to those of RNase A and RNase T1 family of RNases. Various mononucleotide, products, or their analogs competively inhibit these enzymes.[57–59] In RNase T2, RNase Rh, and RNase M, inhibitory effects are in the order of 2′-AMP>2′-GMP>3′-AMP>2′-CMP>2′,(3′)-UMP. A potent inhibitor of RNase A, human placental RNase inhibitor, did not inhibit animal acid RNases such as bovine spleen RNase and squid RNase. Heparin inhibited bovine spleen RNase and squid RNases fairly effectively.[23] The presence of intracellular RNase inhibitor has not been demonstrated to date. The absence of RNase inhibitors such as RNase A inhibitor is probably due to the localization of this enzyme in animal cells. The inhibitory effects of heparin and nucleotides made possible the application of heparin Sepharose and 2′, 5′-ADP Sepharose for purification of the RNase T2 family of RNases. Fungal RNase, RNase T2, RNase M, and RNase Rh are irreversibly inhibited by iodoacetate and iodoacetamide at pH 5.0.[27,28] The site of modification was one of the two histidine residues in the catalytic site.

[57] S. Sato, T. Uchida, and F. Egami, *Arch. Biochem. Biophys.* **115,** 48 (1966).
[58] M. Irie, *J. Biochem. (Tokyo)* **65,** 133 (1969).
[59] T. Komiyama and M. Irie, *J. Biochem. (Tokyo)* **71,** 973 (1972).

[4] The Ribonuclease P Family

By THOMAS A. HALL and JAMES W. BROWN

Introduction

Although the vast majority of biological processes are catalyzed by protein enzymes, the synthesis of these proteins is directed by RNA at each step. The information directing the sequence of a nascent protein (mRNA), the synthesis machinery (ribosome), and the molecule that carries individual amino acids and the growing peptide chain (tRNA) are all composed in full or in part of RNA. Transfer RNAs are initially transcribed as precursors (pre-tRNA) that are processed at both their 3′ and 5′ ends.[1] The enzyme responsible for the 5′ processing of pre-tRNA, ribonuclease P (RNase P, EC 3.1.26.5), is itself composed largely of RNA. RNase P cleaves the 5′ leader sequence from all pre-tRNAs during tRNA maturation. In all bacterial, eukaryotic nuclear, and archaeal systems in which the composition is known, RNase P is a ribonucleoprotein complex containing a single RNA and a protein component that varies between phylogenetic domains.

The most remarkable aspect of RNase P is that the RNA is the catalyst,[2] and in contrast to other "ribozymes" (e.g., the group I intron,[3] the hammerhead[4]), the catalytic RNA is not destroyed in the process; it is a true enzyme that processes multiple substrates. All organisms examined for it contain an RNase P activity, and RNase P enzymes have been characterized from representatives of all three major phylogenetic domains (the Bacteria, Eukarya, and Archaea[5]), and from representatives of both chloroplasts and mitochondria. This discussion is intended as a very general overview to the diverse set of characterized RNase P enzymes. Several more specialized reviews have been published discussing RNase P structure and function from a variety of perspectives.[6–10]

[1] M. P. Deutscher, *in* "tRNA: Structure, Biosynthesis, and Function" (D. Söll and U. L. RaBhandary, eds.), p. 51. ASM Press, Washington, D.C., 1995.
[2] C. Guerrier-Takada, K. Gardiner, T. Marsh, N. R. Pace, and S. Altman, *Cell* **35,** 849 (1983).
[3] K. Kruger, P. J. Grabowski, A. J. Zaug, J. Sands, D. E. Gottschling, and T. R. Cech, *Cell* **31,** 147 (1982).
[4] A. C. Forster and R. H. Symons, *Cell* **49,** 211 (1987).
[5] C. R. Woese, O. Kandler, and M. L. Wheelis, *Proc. Natl. Acad. Sci. USA* **87,** 4576 (1990).
[6] A. Schön, *FEMS Microbiol. Rev.* **23,** 391 (1999).
[7] D. N. Frank and N. R. Pace, *Annu. Rev. Biochem.* **67,** 153 (1998).
[8] J. M. Nolan and N. R. Pace, *in* "Nucleic Acids and Molecular Biology," Vol. 10, "Catalytic RNA" (F. Eckstein and D. M. J. Lilley, eds.), p. 109. Springer-Verlag, Berlin, 1996.
[9] N. R. Pace and J. W. Brown, *J. Bacteriol.* **177,** 1919 (1995).
[10] L. A. Kirsebom, *Mol. Microbiol.* **17,** 411 (1995).

0076-6879/01 $35.00

RNase P Subunit Composition

In bacteria, RNase P contains a ca. 350–450 nucleotide (nt) RNA and a single 13–14 kDa protein. Bacterial RNase P RNA alone (in all cases studied) is catalytic *in vitro* without its protein subunit.[2] Eukaryotic and archaeal RNase P enzymes appear to be more complex. Although all eukaryotic nuclear and archaeal RNase P holoenzymes appear to have a single RNA (300–450 nt), buoyant densities of RNase P holoenzymes from eukaryotic nuclei, archaea, mitochondria, and plastids (in Cs_2SO_4 and/or CsCl) suggest greater and more variable protein compositions than their bacterial counterparts (see Table I). In accordance with these observations, ribozyme activity has not been demonstrated in the absence of protein for any isolated eukaryotic, mitochondrial, or plastid RNase P RNA. However, a subset of the euryarchaeal branch of the Archaea has been shown to contain RNase P RNAs capable of specific pre-tRNA cleavage *in vitro* in the absence of protein, although this activity requires extreme ionic conditions.[11]

Nuclear RNase P from *Saccharomyces cerevisiae* has been highly purified and contains nine protein subunits, all of which have been shown to be essential for RNase P activity *in vivo* by genetic analyses[12] (Table I). Eight of these subunits are also shared by RNase MRP, a ribonucleoprotein enzyme found only in eukaryotic nuclei, which contains an RNA related to RNase P and is involved in rRNA processing.[13] Interestingly, none of these proteins contain any common RNA binding motifs, and none have apparent homologs in bacteria. Equally intriguing, even though the full annotations of five archaeal genomes are published, no homologs of bacterial RNase P proteins are obvious on the basis of sequence, and no genome projects have assigned archaeal RNase P protein subunits.[14–18] A putative homolog of the yeast RNase P subunit Pop5p[12] has been identified in the methanogenic archaeon *Methanobacterium thermoautotrophicum* ΔH (hypothetical ORF *MTH687*) by distant sequence similarity. Antiserum against the recombinant archaeal protein immunoprecipitates *M. thermoautotrophicum* ΔH activity and indicates copurification of the natural protein with RNase P activity by Western blot (T. A. Hall and J. W. Brown, 2001, unpublished data). The buoyant densities of archaeal RNase P holoenzymes in Cs_2SO_4 (Table I) suggest a quite variable protein content. The enzyme from *Haloferax volcanii* exhibits a density

[11] J. A. Pannucci, E. S. Haas, T. A. Hall, J. K. Harris, and J. W. Brown, *Proc. Natl. Acad. Sci. USA* **96,** 7803 (1999).

[12] J. R. Chamberlain, Y. Lee, W. S. Lane, and D. R. Engelke, *Genes Dev.* **12,** 1678 (1998).

[13] L. Lindhal and J. M. Zengel, *Mol. Biol. Rep.* **22,** 69 (1996).

[14] C. J. Bult, O. White, G. J. Olsen, *et al.*, *Science* **273,** 1058 (1996).

[15] D. R. Smith, L. A. Doucette-Stamm, C. Deloughery, *et al.*, *J. Bacteriol.* **179,** 7135 (1997).

[16] H. P. Klenk, R. A. Clayton, J-F. Tomb, *et al.*, *Nature* **390,** 364 (1997).

[17] Y. Kawarabayasi, M. Sawada, H. Horikawa, *et al.*, *DNA Res.* **5,** 55 (1998).

[18] Y. Kawarabayasi, Y. Hino, H. Horikawa, *et al.*, *DNA Res.* **6,** 83 (1999).

TABLE I
PROPERTIES OF REPRESENTATIVE RNASE P ENZYMES THAT HAVE BEEN BIOCHEMICALLY CHARACTERIZED

Source	Buoyant density[a]	Sensitive to micrococcal nuclease	Protein subunits	RNA subunit (nt)	RNA alone is catalytic	References
Bacteria						
Escherichia coli	1.71,[b] 1.55	+	14 kDa (*rnpA*, 119 aa)	377 (*rnpB*)	+	*e, f, g, h, i, j, k*
Bacillus subtilis	1.7	+	14 kDa (119 aa)	401	+	*e, i, m*
Eukaryotes						
Schizosaccharomyces pombe	1.40	+	100 kDa	285	–	*n, o, p, q*
Saccharomyces cerevisiae	*n.d.*	+	100.5 kDa (POP1p, 875 aa)	369 (RPR1)	–	*r, s, t, u, v, w*
			32.9 kDa (POP4p, 279 aa)			
			32.2 kDa (RPP1p, 293 aa)			
			22.6 kDa (POP3p, 195 aa)			
			19.6 kDa (POP5p, 173 aa)			
			18.2 kDa (POP6p, 158 aa)			
			16.4 kDa (RPR2p, 144 aa)			
			15.8 kDa (POP7p/RPP2, 140 aa)			
			15.5 kDa (POP8p, 133 aa)			
Xenopus laevis	1.34	+	***	320	–	*x*
Homo sapiens	1.28	+	115 kDa (hPOP1)	340	–	*y, z, aa, bb, cc, dd, ee, ff, gg*
			40 kDa (Rpp40, 302 aa)			
			38 kDa (Rpp38, 283 aa)			
			30 kDa (Rpp30, 268 aa)			
			29 kDa (Rpp29, 220 aa)			
			20 kDa (Rpp20, 140 aa)			
			14 kDa (Rpp14, 124 aa)			
			25 kDa (Rpp25)[c]			

Rattus rattus	1.36	–	***	***	–	*hh*
Dictyostelium discoideum	1.23	+	***	***	–	*ii*
Tetrahymena thermophila	1.42	+	***	***	–	*jj*
Tetrahymena pyriformis	*n.d.*	+	100 kDa[c] 44 kDa[c] 35 kDa[c]			*kk*
Wheat germ	1.34[b]	–	***	***	–	*ll*
Veal heart	1.33	+	***	***	–	*j*
Archaea						
Sulfolobus acidocaldarius	1.27	–	***	315	–	*mm, nn*
Haloferax volcanii	1.61	+	***	435	+	*oo, pp*
Methanobacterium thermoautotrophicum ΔH	1.42	*n.d.*	14.5 kDa (MTH687, 124 aa)	314	+	*qq, rr, ss, tt*
Methanococcus jannaschii	1.39	*n.d.*	***	274	–	*uu, vv*
Organelles						
H. sapiens mitochondria	1.23	??[d]	***	??	–	*ww, xx*
S. cerevisiae mitochondria	1.28		105 kDa (RPM2)	490 (RPM1)	–	*yy, zz, aaa, bbb*
Spinach chloroplast	1.28[b]	–	***	---	*n.a.*	*ccc, ddd*
Cyanophora paradoxa cyanelle	1.29	+	***	***	–	*eee, fff*

n.d., Not determined.
n.a., Not applicable.
***, Element is present, but uncharacterized.
---, Element does not appear to be present.
+, "Yes".
–, "No".
[a] Buoyant density measurements are in Cs_2SO_4 unless noted otherwise.
[b] Measured in CsCl.
[c] Copurifies with RNase P activity, but physical association with catalytic complex not yet shown.
[d] YES, according to (*ww*), NO according to (*xx*).
[e] C. Guerrier-Takada, K. Gardiner, T. Marsh, N. R. Pace, and S. Altman, *Cell* **35,** 849 (1983).
[f] P. Schedl and P. Primakoff, *Proc. Natl. Acad. Sci. USA* **70,** 2091 (1973).

TABLE I (*continued*)

[g] F. G. Hansen, E. B. Hansen, and T. Atlung, *Gene* **38,** 85 (1985).
[h] H. Sakamoto, N. Kimura, F. Nagawa, and Y. Shimura, *Nucleic Acids Res.* **11,** 8237 (1983).
[i] R. E. Reed, M. F. Baer, C. Guerrier-Takada, H. Donis-Keller, and S. Altman, *Cell* **30,** 627 (1982).
[j] E. Akaboshi, C. Guerrier-Takada, and S. Altman, *Biochem. Biophys. Res. Commun.* **96,** 831 (1980).
[k] B. C. Stark, R. Kole, E. J. Bowman, and S. Altman, *Proc. Natl. Acad. Sci. USA* **75,** 3717 (1978).
[l] C. Reich, K. J. Gardiner, G. J. Olsen, B. Pace, T. L. Marsh, and N. R. Pace, *J. Biol. Chem.* **261,** 7888 (1986).
[m] K. Gardiner and N. R. Pace, *J. Biol. Chem.* **255,** 7507 (1980).
[n] G. Krupp, B. Cherayil, D. Frendewey, S. Nishikawa, and D. Söll, *EMBO J.* **5,** 1697 (1986).
[o] L. Kline, S. Nishikawa, and D. Söll, *J. Biol. Chem.* **256,** 5058 (1981).
[p] S. Zimmerly, D. Drainas, L. A. Sylvers, and D. Söll, *Eur. J. Biochem.* **217,** 501 (1993).
[q] S. Zimmerly, V. Gamulin, U. Burkard, and D. Söll, *FEBS Lett.* **271,** 189 (1990).
[r] J. Y. Lee, C. E. Rohlman, L. A. Molony, and D. R. Engelke, *Mol. Cell. Biol.* **11,** 721 (1991).
[s] J. Y. Lee and D. R. Engelke, *Mol. Cell. Biol.* **9,** 2536 (1989).
[t] Z. Lygerou, P. Mitchell, E. Petfalski, B. Seraphin, and D. Tollervey, *Genes Dev.* **8,** 1423 (1994).
[u] B. Dichtl and D. Tollervey, *EMBO J.* **16,** 417 (1997).
[v] S. Chu, J. M. Zengel, and L. Lindahl, *RNA* **3,** 382 (1997).
[w] J. R. Chamberlain, Y. Lee, W. S. Lane, and D. R. Engelke, *Genes Dev.* **12,** 1678 (1998).
[x] M. Doria, G. Carrara, P. Calandra, and G. P. Tocchini-Valentini, *Nucleic Acids Res.*, **19,** 2315 (1991).
[y] M. Bartkiewicz, H. Gold, and S. Altman, *Genes Dev.* **3,** 488 (1989).
[z] Z. Lygerou, H. Pluk, W. J. van Venrooij, and B. Seraphin, *EMBO J.* **15,** 5936 (1996).
[aa] Y. Yuan, E. Tan, and R. Reddy, *Mol. Cell. Biol.* **11,** 5266 (1991).
[bb] P. S. Eder, R. Kekuda, V. Stolc, and S. Altman, *Proc. Natl. Acad. Sci. USA* **94,** 1101 (1997).
[cc] M. J. Mamula, M. Baer, J. Craft, and S. Altman, *Proc. Natl. Acad. Sci. USA* **86,** 8717 (1989).
[dd] N. Jarrous, P. S. Eder, C. Guerrier-Takada, C. Hoog and S. Altman, *RNA* **4,** 407 (1998).
[ee] H. Pluk, H. van Eenennaam, S. A. Rutjes, G. J. M. Pruijn, and W. J. van Venrooij, *RNA* **5,** 512 (1999).
[ff] N. Jarrous, P. S. Eder, D. Wesolowski, and S. Altman, *RNA* **5,** 153 (1999).
[gg] H. van Eenennaam, G. J. Pruijn, and W. J. van Venrooij, *Nucleic Acids Res.* **27,** 2465 (1999).
[hh] G. P. Jayanthi and G. C. Van Tuyle, *Arch. Biochem. Biophys.* **296,** 264 (1992).
[ii] C. Stathopoulos, D. L. Kalpaxis, and D. Drainas, *Eur. J. Biochem.* **228,** 976 (1995).

[jj] H. L. True and D. W. Celander, *J. Biol. Chem.* **271,** 16559 (1996).
[kk] S. Vainauskas, V. Stribinskis, L. Padegimas, and B. Juodka, *Biochimie* **80,** 595 (1998).
[ll] S. Arends and A. Schön, *Eur. J. Biochem.* **244,** 635 (1997).
[mm] S. C. Darr, B. Pace, and N. R. Pace, *J. Biol. Chem.* **265,** 12927 (1990).
[nn] T. E. LaGrandeur, S. C. Darr, E. S. Haas, and N. R. Pace, *J. Bacteriol.* **175,** 5043 (1993).
[oo] N. Lawrence, D. Wesolowski, H. Gold, M. Bartkiewicz, C. Guerrier-Takada, W. H. McClain, and S. Altman, *Cold Spring Harb. Symp. Quant. Biol.* **52,** 233 (1987).
[pp] D. T. Nieuwlandt, E. S. Haas, and C. J. Daniels, *J. Biol. Chem.* **266,** 5689 (1991).
[qq] E. S. Haas, D. W. Armbruster, B. M. Vucson, C. J. Daniels, and J. W. Brown, *Nucleic Acids Res.* **24,** 1252 (1996).
[rr] J. A. Pannucci, E. S. Haas, T. A. Hall, J. K. Harris, and J. W. Brown, *Proc. Natl. Acad. Sci. USA* **96,** 7803 (1999).
[ss] D. R. Smith, L. A. Doucette-Stamm, C. Deloughery, *et al.*, *J. Bacteriol.* **179,** 7135 (1997).
[tt] T. A. Hall and J. W. Brown, 2001, unpublished data.
[uu] C. J. Bult, O. White, G. J. Olsen, *et al.*, *Science* **273,** 1058 (1996).
[vv] A. J. Andrews and J. W. Brown, 2001, unpublished data.
[ww] C. J. Doersen, C. Guerrier-Takada, S. Altman, and G. Attardi, *J. Biol. Chem.* **260,** 5942 (1985).
[xx] W. Rossmanith and R. M. Karwan, *Biochem. Biophys. Res. Commun.* **247,** 234 (1998).
[yy] D. L. Miller and N. C. Martin, *Cell* **34,** 911 (1983).
[zz] M. J. Hollingsworth and N. C. Martin, *Mol. Cell. Biol.* **6,** 1058 (1986).
[aaa] M. J. Morales, C. A. Wise, M. J. Hollingsworth, and N. C. Martin, *Nucleic Acids Res.* **17,** 6865 (1989).
[bbb] Y. L. Dang and N. C. Martin, *J. Biol. Chem.* **268,** 19791 (1993).
[ccc] P. Gegenheimer, *Mol. Biol. Rep.* **22,** 147 (1996).
[ddd] B. C. Thomas, X. Li, and P. Gegenheimer, *RNA* **6,** 545 (2000).
[eee] M. Baum, A. Cordier, and A. Schön, *J. Mol. Biol.* **257,** 43 (1996).
[fff] A. Cordier and A. Schön, *J. Mol. Biol.* **289,** 9 (1999).

in Cs_2SO_4 of 1.61 g/ml (near 1.65 g/ml, the density of pure RNA),[19] whereas that of *Sulfolobus acidocaldarius* is 1.27 g/ml (near the density of pure protein, 1.2 g/ml).[20] *Methanobacterium thermoautotrophicum* ΔH (1.42 g/ml, T. A. Hall and J. W. Brown, 2001, unpublished data) and *Methanococcus jannaschii* (1.39 g/ml, A. J. Andrews and J. W. Brown, 2001, unpublished data) both have buoyant densities intermediate between those of RNA and protein, suggesting that protein subunits in addition to the Pop5p-like protein are likely to be present. Another hypothetical ORF in the *M. thermoautotrophicum* ΔH genome (*MTH688*) has been suggested to be a distant homolog of the human RNase P subunit Rpp38.[21] Incidentally, *MTH687* and *MTH688* are contiguous in the genome, are oriented in the same direction, appear by visual inspection of the sequence to be part of a cotranscribed operon, and have obvious homologs in all published archaeal genomes (although not contiguous in those genomes).

The RNase P enzymes from eukaryotic organelles present an interesting problem. Mitochondria and chloroplasts, although clearly related to their bacterial and cyanobacterial ancestors,[22] have given up much of their genetic information to the nucleus. Nuclear gene products with organellar function must be specifically transported into the appropriate organelle. Fungal and protist mitochondria typically encode an RNase P RNA in their genomes.[23,24] The human mitochondrial genome does not have a gene that encodes a recognizable RNase P RNA[25]. In accordance with this, the RNase P holoenzyme isolated from HeLa cell mitochondria may not have an associated RNA cofactor.[26] Although mitochondria import necessary tRNAs that have been lost from the mitochondrial genome,[27,28] mitochondrial import of an RNA the size of RNase P RNA (~300 nt) has not yet been demonstrated. However, the protein cofactor of yeast RNase P (Rpm2p), which is encoded in the nucleus and transported into the mitochondrion,[29] is nearly the molecular mass of the mitochondrial RNase P RNA subunit. There is conflicting data about the nature of RNase P from human mitochondria. It has been reported that this enzyme contains an essential RNA that is sensitive to micrococcal nuclease,[30] but a more

[19] N. Lawrence, D. Wesolowski, H. Gold, M. Bartkiewicz, C. Guerrier-Takada, W. H. McClain, and S. Altman, *Cold Spring Harb. Symp. Quant. Biol.* **52,** 233 (1987).

[20] S. C. Darr, B. Pace, and N. R. Pace, *J. Biol. Chem.* **265,** 12927 (1990).

[21] K. S. Makarova, L. Aravind, M. Y. Galperin, N. V. Grishin, R. L. Tatusov, Y. I. Wolf, and E. V. Koonin, *Genome Res.* **9,** 608 (1999).

[22] M. W. Gray, *Trends Genet.* **5,** 294 (1989).

[23] M. W. Gray, B. F. Lang, R. Cedergren, *et al., Nucleic Acids Res.* **26,** 865 (1998).

[24] N. C. Martin and B. F. Lang, *Nucl. Acids Symp. Ser.* **36,** 42 (1997).

[25] S. Anderson, A. T. Bankier, B. G. Barrell, *et al., Nature* **290,** 457 (1981).

[26] W. Rossmanith and R. M. Karwan, *Biochem. Biophys. Res. Commun.* **247,** 234 (1998).

[27] C. P. Rusconi and T. R. Cech, *Genes Dev.* **10,** 2870 (1996).

[28] R. P. Martin, J.-M. Schneller, A. J. Stahl, and G. Dirheimer, *Biochemistry* **18,** 4600 (1979).

[29] M. J. Hollingsworth and N. C. Martin, *Mol. Cell. Biol.* **6,** 1058 (1986).

[30] C. J. Doersen, C. Guerrier-Takada, S. Altman, and G. Attardi, *J. Biol. Chem.* **260,** 5942 (1985).

recent study found the enzyme to be insensitive to both micrococcal nuclease and RNase A pretreatments, and to resolve in isopycnic Cs_2SO_4 gradients with bulk protein (1.23 g/ml).[26]

RNase P isolated from spinach chloroplast is insensitive to micrococcal nuclease treatment (100-fold over levels that inactivate *E. coli* RNase P), fractionates with bulk protein in CsCl buoyant density gradients (1.28 g/ml), has an apparent size of ~70 kDa (similar to BSA) in Sephacryl S-200 and S-300 columns, and copurifies with two major proteins of ~30 and ~50 kDa after tRNA-agarose affinity chromatography.[31] Taken together, these observations suggest that spinach chloroplast RNase P does not contain an essential RNA component, but rather is a purely protein-based enzyme. Additionally, the catalytic mechanism employed by spinach chloroplast RNase P may differ from other systems described. In contrast to bacterial, yeast, slime mold, and human RNase P enzymes, spinach chloroplast RNase P does not appear to be severely hampered in reactivity against a phosphorothioate modification at the scissile bond of a model pre-tRNA substrate.[32–35] Like vertebrate mitochondria, plastids of green algae and multicellular plants have not been demonstrated to encode an RNase P RNA in their genomes.[36]

The photosynthetic plastid (cyanelle) from *Cyanophora paradoxa* contains an RNase P with a buoyant density in Cs_2SO_4 similar to those of eukaryotic RNase P enzymes (1.28 g/ml, Table I).[37] The RNA component of this enzyme is very similar in structure to RNase P RNAs from free-living cyanobacteria, which conform to the Type-A bacterial secondary structure (Fig. 1, see below). The buoyant density, however, suggests a greater protein composition than the Bacteria, similar to the eukaryotic nuclear enzyme. Activity is sensitive to nuclease treatment, demonstrating the necessity of the RNA component, but unlike cyanbacterial RNase P RNAs (and all other bacterial representatives), the RNA is not catalytically proficient *in vitro* in the absence of protein.[38] The cyanellar genome, as well as the genomes of other plastids and mitochondria, does not encode a protein with similarity to known RNase P protein subunits.[6,23–25,39] Presumably, these gene products are imported into the organelles. There are three possible scenarios to explain these

[31] P. Gegenheimer, *Mol. Biol. Rep.* **22,** 147 (1996).

[32] J. M. Warnecke, J. P. Fürste, W.-D. Hardt, V. A. Erdmann, and R. K. Hartmann, *Proc. Natl. Acad. Sci. USA* **93,** 8924 (1996).

[33] B. C. Thomas, X. Li, and P. Gegenheimer, *RNA* **6,** 545 (2000).

[34] B. C. Thomas, J. Chamberlain, D. R. Engelke, and P. Gegenheimer, *RNA* **6,** 554 (2000).

[35] T. Pfeiffer, A. Tekos, J. M. Warnecke, D. Drainas, D. R. Engelke, B. Séraphin, and R. K. Hartmann, *J. Mol. Biol.* **298,** 559 (2000).

[36] A. Schön, *Mol. Biol. Rep.* **22,** 139 (1996).

[37] A. Cordier and A. Schön, *J. Mol. Biol.* **289,** 9 (1999).

[38] M. Baum, A. Cordier, and A. Schön, *J. Mol. Biol.* **257,** 43 (1996).

[39] M. Reith and J. Munholland, *Plant Mol. Biol. Report.* **13,** 333 (1995).

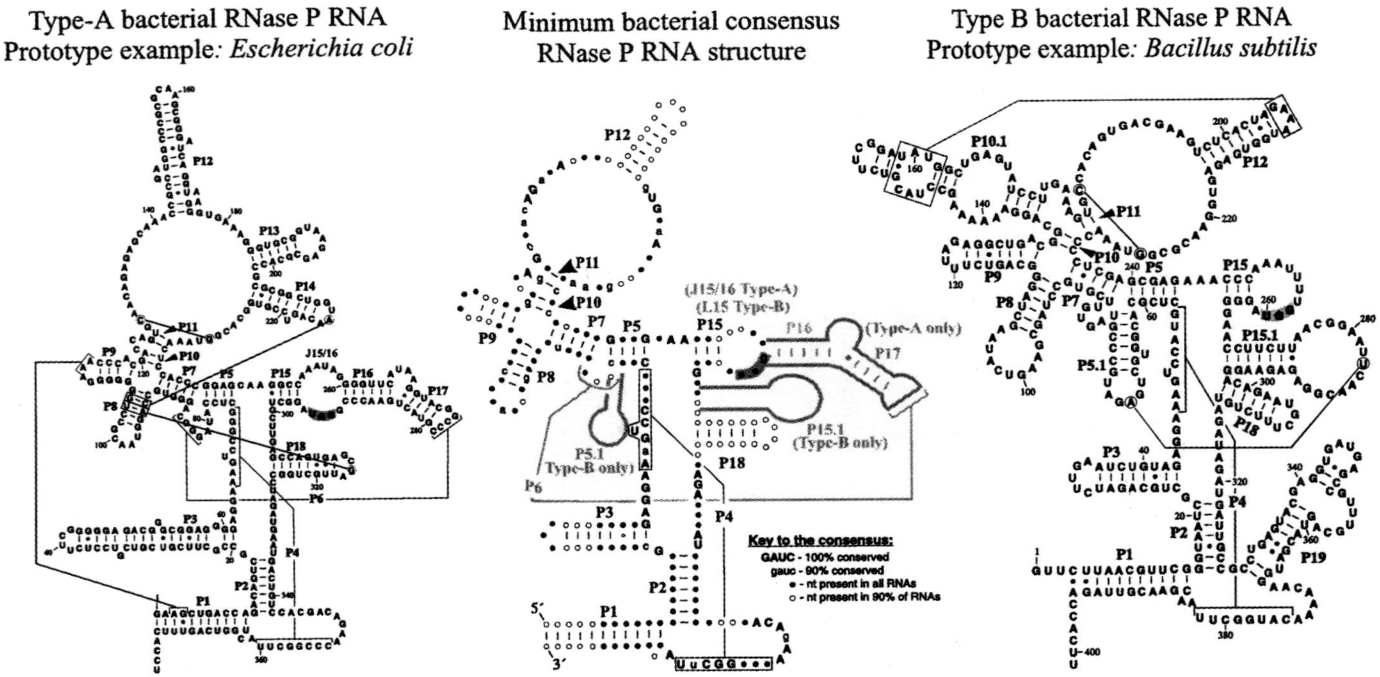

FIG. 1. Secondary structures of examples of the Type-A (left) and Type-B (right) bacterial RNase P RNA. The minimum bacterial consensus structure is shown in the center. Helices are indicated by the designation "P" (*paired*) and are numbered in the 5′ to 3′ direction [E. S. Haas, J. W. Brown, C. Pitulle, and N. R. Pace, *Proc. Natl. Acad. Sci. USA* **91,** 2527 (1994)]. Capital letters indicate residues that are invariant in the domain Bacteria; lowercase letters indicate identities that occur in 90% or more of sequences. Gray lines and helix labels in the minimal consensus structure indicate structural elements that are present in all or nearly all of either the Type-A or Type-B structures, but not both. Helices P4 and P6 are indicated by line segments joining the paired bases. Tertiary interactions are also indicated by line segments connecting long-range base interactions and are supported by data for Type-A [M. E. Harris, J. M. Nolan, A. Malhotra, J. W. Brown, S. C. Harvey, and N. R. Pace, *EMBO J.* **13,** 3953 (1994); M. E. Harris, A. V. Kazantsev, J.-L. Chen, and N. R. Pace, *RNA* **3,** 561 (1997); J.-L. Chen, J. M. Nolan, M. E. Harris, and N. R. Pace, *EMBO J.* **17,** 1515 (1998)] and Type-B [M. A. Tanner and T. R. Cech, *RNA* **1,** 349 (1995); E. S. Haas, A. B. Banta, J. K. Harris, N. R. Pace, and J. W. Brown, *Nucleic Acids Res.* **24,** 4775 (1996)]. The "GGU" motif involved in binding the RCCA tail of bacterial pre-tRNA (see text) is indicated by gray shading in J15/16 (Type-A) or L15 (Type-B). Structures were adapted from images available from the ribonuclease P database [J. W. Brown, *Nucleic Acids Res.* **27,** 314 (1999)].

observations: (1) The original bacterial RNase P protein genes migrated to the nucleus during organellar evolution, (2) the original bacterial protein genes were lost from the organellar genome and existing nuclear RNase P proteins are imported and adapted in the organelles, or (3) the original RNase P protein genes were lost from the organellar genome and non-RNase P protein gene products were adapted for function in the organellar RNase P complex. Elucidation of the complex nature of these interesting examples of this ubiquitous enzyme awaits the full purification of the enzymes and identification of the protein subunit sequences.

RNase P Structure and Function

The majority of understanding about the structure and function of the RNase P ribozyme has come from studies on the bacterial RNAs, most notably those of *E. coli* and *B. subtilis*. The bacterial RNA will be referred to for most of the discussion of the RNA structure and function, and a comparison of common features of the RNAs from the other phylogenetic domains will be made at the end of this section. A detailed account of the bacterial RNA structure and reaction mechanism is beyond the scope of this discussion. Reviews on RNase P enzymes that describe the bacterial, yeast, human, and cyanellar systems in detail are presented elsewhere in this series.[39a]

General RNA Structure

The large size of the RNA subunit of RNase P has hampered structural studies. The three-dimensional structure of neither the RNA nor the holoenzyme has been determined experimentally (e.g., by X-ray crystallography or nuclear magnetic resonance). The most successful method for structural studies of large RNAs has proven to be the comparative analysis of homologous sequences from phylogenetically diverse organisms.[40] The three-dimensional structure of an RNA is determined by the spatial organization of secondary structural elements (double-stranded helices), as well as various tertiary interactions involving coaxial stacking of helices, long-range base pairs, base triplets, base–sugar/phosphate, sugar/phosphate–sugar/phosphate, base-stacking interactions, and pseudo-knots.[41] The evaluation of covarying bases in the sequences of an RNA from diverse organisms can reveal much of the secondary structure of the RNA, as well as shedding light on conserved base–base tertiary interactions.

The determination of RNA structure based on comparative analysis requires a substantial sequence set, ideally one that evenly represents a phylogenetic

[39a] *Methods in Enzymol.* **342,** (2001).

[40] R. R. Gutell, N. Larsen, and C. R. Woese, *Microbiol. Rev.* **58,** 10 (1994).

[41] J. R. Wyatt and I. Tinoco, in "The RNA World" (R. F. Gesteland and J. F. Atkins, eds.), p. 465. Cold Spring Harbor Laboratory Press, Inc., Cold Spring Harbor, NY, 1993.

distribution of the organisms from which the sequences were obtained. Early studies identified RNase P RNA genes based on complementation of mutant bacterial strains deficient in tRNA processing,[42–44] or by isolation of RNA species from biochemical purifications of RNase P holoenzymes or bacterial RNA preparations that contained catalytic RNase P RNA.[45–53] Known RNase P RNAs have been used as probes to isolate homologous sequences from related organisms by hybridization to genomic DNA fragments.[53,54] The ability to sequence entire microbial and organellar genomes has allowed the identification of putative genes directly by similarity to known sequences and potential structure. The majority of RNase P RNA sequences used in defining the structural core of bacterial, archaeal, and eukaryotic RNase P RNA, however, have been obtained relatively recently by using PCR with primers based upon conserved stretches of sequence, generally the "P4" region (see Fig. 1).[55–57] More than 140 bacterial RNase P RNA sequences and secondary structures are available on the RNase P database at http://www.mbio.ncsu.edu/RNaseP/.[58] On the basis of phylogenetic comparative analysis, the secondary structures of bacterial and archaeal RNase P RNAs are well established [55,56,59,60] (Fig. 1), and substantial progress has been made in the determination of a general consensus structure for eukaryotic nuclear RNase P RNAs as well.[53,57,61] Three-dimensional models of the bacterial ribozyme have been

[42] H. Sakano, S. Yamada, T. Ikemura, Y. Shimura, and H. Ozeki, *Nucleic Acids Res.* **1,** 355 (1974).

[43] H. Sakamoto, N. Kimura, and Y. Shimura, *Proc. Natl. Acad. Sci. USA* **80,** 6187 (1983).

[44] D. L. Miller and N. C. Martin, *Cell* **34,** 911 (1983).

[45] R. E. Reed, M. F. Baer, C. Guerrier-Takada, H. Donis-Keller, and S. Altman, *Cell* **30,** 627 (1982).

[46] G. Krupp, B. Cherayil, D. Frendewey, S. Nishikawa, and D. Söll, *EMBO J.* **5,** 1697 (1986).

[47] C. Reich, K. J. Gardiner, G. J. Olsen, B. Pace, T. L. Marsh, and N. R. Pace, *J. Biol. Chem.* **261,** 7888 (1986).

[48] M. Bartkiewicz, H. Gold, and S. Altman, *Genes Dev.* **3,** 488 (1989).

[49] T. E. LaGrandeur, S. C. Darr, E. S. Haas, and N. R. Pace, *J. Bacteriol.* **175,** 5043 (1993).

[50] J. Y. Lee and D. R. Engelke, *Mol. Cell. Biol.* **9,** 2536 (1989).

[51] M. Doria, G. Carrara, P. Calandra, and G. P. Tocchini-Valentini, *Nucleic Acids Res.* **19,** 2315 (1991).

[52] P. S. Eder, A. Srinivasan, M. C. Fishman, and S. Altman, *J. Biol. Chem.* **271,** 21031 (1996).

[53] J. W. Brown, E. S. Haas, B. D. James, D. A. Hunt, J. S. Liu, and N. R. Pace, *J. Bacteriol.* **173,** 3855 (1991).

[54] A. J. Tranguch and D. R. Engelke, *J. Biol. Chem.* **268,** 14045 (1993).

[55] J. W. Brown, J. M. Nolan, E. S. Haas, M. A. Rubio, F. Major, and N. R. Pace, *Proc. Natl. Acad. Sci. USA* **93,** 3001 (1996).

[56] E. S. Haas, D. W. Armbruster, B. M. Vucson, C. J. Daniels, and J. W. Brown, *Nucleic Acids Res.* **24,** 1252 (1996).

[57] C. Pitulle, M. Garcia-Paris, K. R. Zamudio, and N. R. Pace, *Nucleic Acids Res.* **26,** 3333 (1998).

[58] J. W. Brown, *Nucleic Acids Res.* **27,** 314 (1999).

[59] E. S. Haas, A. B. Banta, J. K. Harris, N. R. Pace, and J. W. Brown, *Nucleic Acids Res.* **24,** 4775 (1996).

[60] E. S. Haas and J. W. Brown, *Nucleic Acids Res.* **26,** 4093 (1998).

[61] J.-L. Chen and N. R. Pace, *RNA* **3,** 557 (1997).

proposed and refined on the basis of comparative data utilizing spatial constraints obtained by biochemical experiments.[62–66]

Bacterial RNase P RNAs are subdivided into two distinct structural groups, type A and type B. The type-A group represents the sequences from the majority of bacterial groups, whereas Type B is represented only by the low G+C gram-positive bacteria (Fig. 1). The refinement of the bacterial RNase P RNA secondary structure to base-pair resolution allowed a standard nomenclature to be adopted specifying the general helices present in all prokaryotic lineages,[67] designated P1 through P20, and based on the Type-A prototype (see Fig. 1). Homologous helices (in the strict sense of the word, meaning derived from a common ancestor) are given the same designation in differing structures, and helices not present in the Type-A structure are given intermediate numbers. For example helices P5.1, P10.1, and P15.1 do not exist in the Type-A group, while P13, P14, P16 and P17 do not exist in the Type-B lineage (as *homologous* regions, though presumably *analogous* structures fulfill the same structural/functional role in each group). The two structural groups share a common core of secondary structure represented by the hypothetical minimal consensus shown in Fig. 1.

The conservation of the minimal core structure throughout bacterial phylogeny suggests that this region is likely to represent the minimal set of structural elements required for a functional molecule in the bacterial cell. Accessory structures, generally present as helices that exist in some or most lineages but are absent in others, probably serve to stabilize the overall global structure of an individual RNA against the tendency for repulsions of the sugar–phosphate backbone to disrupt the three-dimensional packing of helices. Helices not present in the minimal core structure tend to be "dispensible" for *in vitro* activity, meaning that deletion constructs which remove individual helices retain *in vitro* ribozyme activity, although the ribozyme then generally requires elevated ionic conditions and certain deletions raise the K_m of the reaction by orders of magnitude.[67–69]

In accordance with these observations, an RNA molecule tailored into a secondary structure very similar to the minimal consensus RNase P RNA structure (micro-P, based on the sequence of the *Mycoplasma fermentans* RNase P RNA) exhibits specific pre-tRNA cleaving activity *in vitro,* but requires very high ionic conditions (2.5–3 *M* monovalent, 300 m*M* divalent salts), has a lower temperature

[62] M. E. Harris, J. M. Nolan, A. Malhotra, J. W. Brown, S. C. Harvey, and N. R. Pace, *EMBO J.* **13,** 3953 (1994).

[63] M. E. Harris, A. V. Kazantsev, J.-L. Chen, and N. R. Pace, *RNA* **3,** 561 (1997).

[64] J.-L. Chen, J. M. Nolan, M. E. Harris, and N. R. Pace, *EMBO J.* **17,** 1515 (1998).

[65] E. Westhof and S. Altman, *Proc. Natl. Acad. Sci. USA* **91,** 5133 (1994).

[66] C. Massire, L. Jaeger, and E. Westhof, *J. Mol. Biol.* **279,** 773 (1998).

[67] E. S. Haas, J. W. Brown, C. Pitulle, and N. R. Pace, *Proc. Natl. Acad. Sci. USA* **91,** 2527 (1994).

[68] D. S. Waugh, C. J. Green, and N. R. Pace, *Science* **244,** 1569 (1989).

[69] S. C. Darr, K. Zito, D. Smith, and N. R. Pace, *Biochemistry* **31,** 328 (1992).

optimum than the native molecule, and has reduced substrate affinity (increased K_m).[70]

Across the full bacterial lineage of known RNase P RNAs, only 40 nt are absolutely invariant in identity,[60] however, statistical examination of the variation/conservation[71] at each position in an alignment of 145 bacterial RNase P sequences has revealed a generally high level of overall sequence conservation throughout the minimum consensus core[60] (Fig. 2). The most conserved sequences encompass and surround the regions that make up the P4 helix (Fig. 2). Tertiary models place the pre-tRNA scissile bond adjacent to the P4 region of RNase P RNA,[63,64,66] and sulfur substitution of phosphate oxygens in the P4 region has directly implicated this region in the pre-tRNA cleavage reaction.[72] Several intramolecular tertiary contacts proposed on the basis of comparative analysis (see Fig. 1) and chemical cross-linking have the effect of folding distal extensions of the RNase P RNA molecule around the P4 helix, the combined effect of which creates a pocket for the substrate that positions the scissile bond of the pre-tRNA at the proposed active site of the ribozyme.[59,62–64,66,73]

On the basis of Fe(II)–EDTA protection studies, it has been proposed that RNase P RNA is composed of two semiautonomous "folding" domains: the "specificity domain" (P7 and everything distal to P7 in the *E. coli* and *B. subtilis* structures shown in Fig. 1) and the "catalytic domain" (the rest of the molecule).[74,75] Substrate specificity can be altered, without changing the basic cleavage reaction, by altering the specificity domain of *B. subtilis* RNase P RNA.[76] An *E. coli* RNase P RNA lacking this entire domain is not catalytic by itself, but can be reconstituted

[70] R. W. Siegel, A. B. Banta, E. S. Haas, J. W. Brown, and N. R. Pace, *RNA* **2,** 452 (1996).

[71] T. D. Schneider and R. M. Stephens, *Nucleic Acids Res.* **18,** 6097 (1990).

[72] M. E. Harris and N. R. Pace, *RNA* **1,** 210 (1995).

[73] M. A. Tanner and T. R. Cech, *RNA* **1,** 349 (1995).

[74] T. Pan, *Biochemistry* **34,** 902 (1995).

[75] A. Loria and T. Pan, *RNA* **2,** 551 (1996).

[76] E. M. Mobley and T. Pan, *Nucleic Acids Res.* **27,** 4298 (1999).

FIG. 2. Conservation of sequence in bacterial RNase P RNA (represented here by the Type-A group). Nucleotides are colored according to region as indicated in the figure. The level of conservation through an alignment of 145 Type-A bacterial RNase P RNA sequences is indicated by the diameters of circles representing individual nucleotides. Variability was assessed by entropy (Hx) at each alignment column [E. S. Haas and J. W. Brown, *Nucleic Acids Res.* **26,** 4093 (1998) and references therein]. A tertiary model (right) is shown with the pre-tRNA substrate bound in the presumed active site groove containing P4 [C. Massire, L. Jaeger, and E. Westhof, *J. Mol. Biol.* **279,** 773 (1998)]. Each ball in the model represents a single nucleotide, and nucleotides are colored by region and scaled to portray sequence conservation as in the secondary structure image. The apparent high level of conservation in the distal region of P3 is due to an extended P3 helix in the Proteobacteria, which phylogenetically biases the level of identity in this region (most Type-A RNase P RNAs lack these residues).

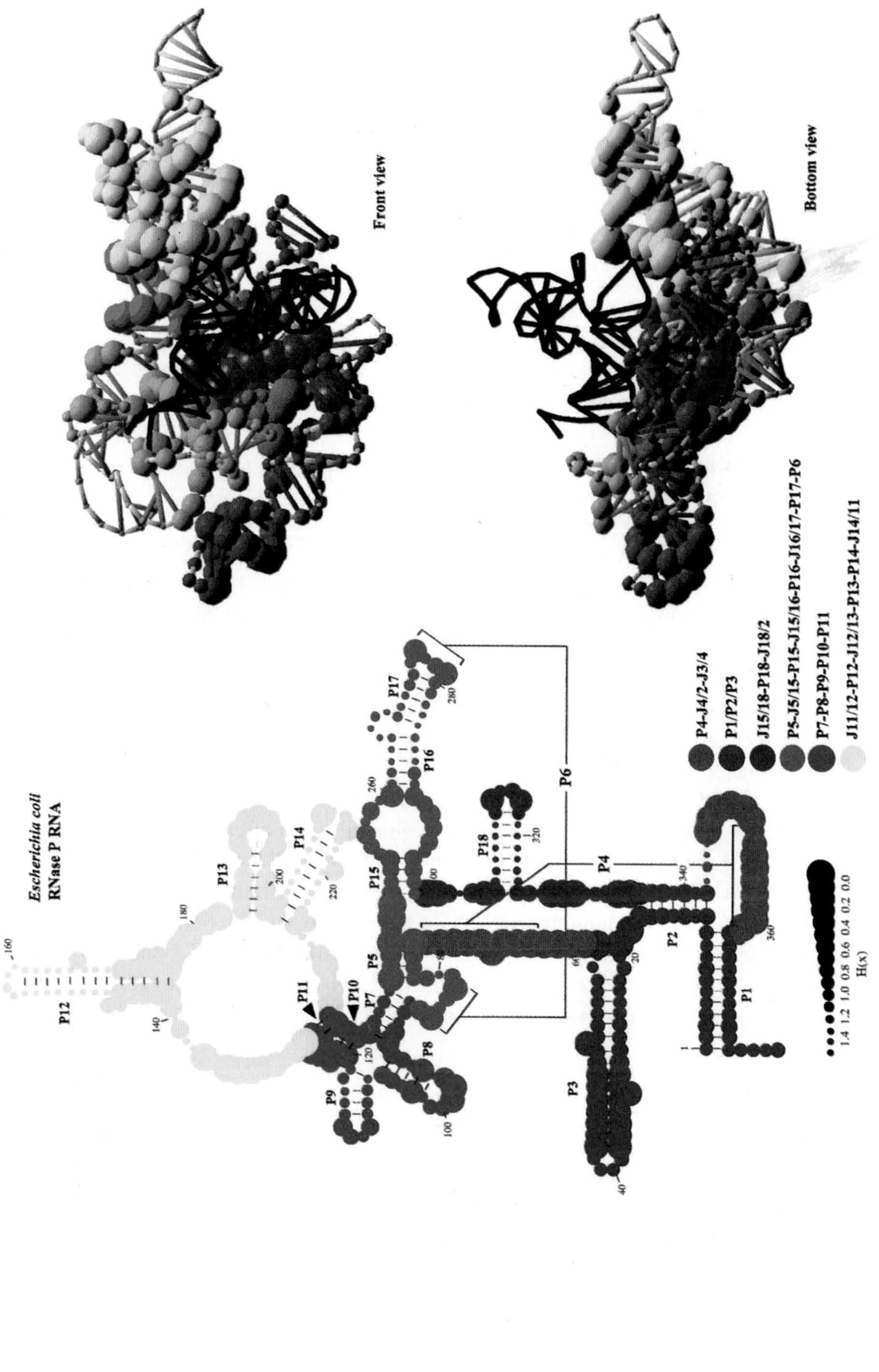

Escherichia coli
RNase P RNA
P1
P2
P3
P4
P5
P6
P7
P8
P9
P10
P11
P12
P13
P14
P15
P16
P17
P18
1.4 1.2 1.0 0.8 0.6 0.4 0.2 0.0
H(x)
P4-J4/2-J3/4
P1/P2/P3
J15/18-P18-J18/2
P5-J5/15-P15-J15/16-P16-J16/17-P17-P6
P7-P8-P9-P10-P11
J11/12-P12-J12/13-P13-P14-J14/11
Front view
Bottom view

in vitro with the recombinant *E. coli* RNase P protein subunit,[77] in contrast to micro-P, the activity of which is not improved by the presence of the protein subunit.[70] This suggests that this specificity domain is able to facilitate substrate binding in the ribozyme reaction, but is dispensible for catalysis, and that the bacterial RNase P protein can compensate to aid in substrate binding in its absence. Also, the protein component is able to specifically bind the RNA subunit through contacts not present in the P7- J14/11-P10 region.

RNase P RNA Reaction: An RNA Enzyme

RNase P RNA from bacteria represents the only known case of a naturally occurring RNA that acts as a true enzyme, but only in bacteria and a subset of the archaea is there a direct demonstration that the RNA is catalytic. Although it is possible that protein side chains could participate in the cleavage reaction, all known RNase P RNAs contain the conserved P4 region that is implicated in the reaction mechanism, and it is assumed likely that the RNA serves as the catalytic unit in nonbacterial RNase P enzymes as well.

RNase P is a phosphodiesterase that catalyzes the hydrolysis of the phosphate backbone of pre- tRNA at the 5′ leader to leave a mature 5′ terminus. The reaction is completely dependent on divalent metals, and Mg^{2+} is the preferred ion. Although the RNA is the enzyme, chemical groups in the RNA itself are not known to be involved directly in the attack on the scissile bond. The reaction mechanism remains unproven, but there is experimental evidence that it involves a hydroxyl attack on the scissile phosphate mediated by coordination of water by a magnesium ion that is itself coordinated by a phosphate at the pre-tRNA cleavage site, and also by oxygens in the RNase P RNA itself.[32,78–80] Coordination of the 2′-OH on the 5′ side of the cleavage site is likely, as deoxy substitution at this site raises the K_m 3400-fold for the *E. coli* reaction.[79,81] The general mode of catalysis is suspected to be analogous to the two metal ion scenario for generalized phosphoryl transfer reactions, which has been implicated in the cleavage reaction catalyzed by group I introns.[82,83] Figure 3 summarizes the general cleavage reaction and diagrams a possible transition state of cleavage at the scissile phosphate bond.[7] A Hill coefficient of 3.2 was measured for the *E. coli* RNA with Mg^{2+}, suggesting that three magnesiums bind cooperatively for maximal activity.[81]

[77] C. J. Green, R. Rivera-León, and B. S. Vold, *Nucleic Acids Res.* **24,** 1497 (1996).

[78] Y. Chen, X. Li, and P. Gegenheimer, *Biochemistry* **36,** 2425 (1997).

[79] R. G. Kleineidam, C. Pitulle, B. Sproat, and G. Krupp, *Nucleic Acids Res.* **21,** 1097 (1993).

[80] J. Kufel and L. A. Kirsebom, *J. Mol. Biol.* **263,** 685 (1996).

[81] D. Smith and N. R. Pace, *Biochemistry* **32,** 5273 (1993).

[82] T. A. Steitz and J. A. Steitz, *Proc. Natl. Acad. Sci. USA* **90,** 6498 (1993).

[83] T. R. Cech, *in* "The RNA World" (R. F. Gesteland and J. F. Atkins, eds.), p. 239. Cold Spring Harbor Laboratory Press, Inc., Cold Spring Harbor, NY, 1993.

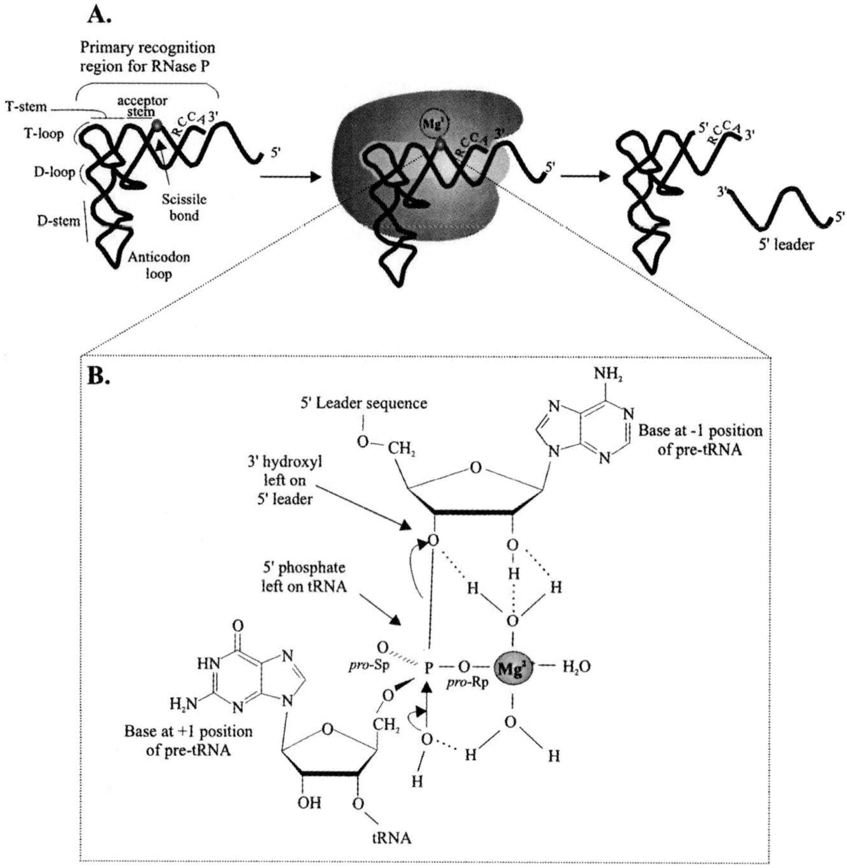

FIG. 3. The RNase P cleavage reaction. RNase P cleaves the 5′ leader sequence from pre-tRNA during tRNA biosynthesis. (*A*) Substrate (pre-tRNA) is bound in a cleft formed by the three-dimensional structure of the RNase P RNA/protein complex. The primary determinants of substrate recognition are the T-stem, T-loop, and acceptor stems of pre-tRNA (three-dimensional structure) and, in the case of most bacterial RNase P enzymes, the 3′ "RCCA" tail of pre-tRNA (base-pairing interaction with J15/16 of Type-A or L15 of Type B RNAs; see Fig. 1). The scissile bond is indicated by a sphere. (***B***) Illustration of a possible transition state configuration during the cleavage reaction. Magnesium is required in the RNase P cleavage reaction, and the reaction rate is pH dependent [D. Smith and N. R. Pace, *Biochemistry* **32,** 5273 (1993)]. Experimental data support the coordination of a magnesium ion to the *pro-Rp* oxygen of pre-tRNA at the 5′ side of the cleavage site [J. M. Warnecke, J. P. Fürste, W.-D. Hardt, V. A. Edermann, and R. K. Hartmann, *Proc. Natl. Acad. Sci. USA* **93,** 8924 (1996); Y. Chen, X. Li, and P. Gegenheimer, *Biochemistry* **36,** 2425 (1997)].

All cells and organelles appear to use RNase P to process all varieties of pre-tRNA. Eighty-six tRNA-encoding genes are present in the *E. coli* K-12 genome.[84] RNase P must recognize the cell's full complement of tRNA precursors, regardless of sequence variation in those substrates. Accordingly, the primary sequence of the substate appears to have very little influence on substrate recognition by RNase P; rather, the enzyme recognizes and binds primarily to the three- dimensional surface presented by the tRNA T-stem, T-loop, and acceptor stem (Fig. 3). Specific base-pairing interactions have been implicated only between a conserved "GGU" motif located in the region joining P15 and P16 (J15/16, see the Type-A structure in Fig. 1) and the "RCC" of the "RCCA" ("R" represents the tRNA "discriminator" base) at the 3′ end of the pre-tRNA acceptor stem. Nucleotides in this region of the RNase P RNA/pre-tRNA complex are protected from chemical modification,[85] and alteration or deletion of nucleotides from the RCCA tail results in an alteration in cleavage site selection[86,87] and an increase in K_m.[88,89] The same interaction occurs in the Type-B RNA in L15 (Fig. 1), which presumably occupies the same general position in space as the Type-A L15/16. The "GGU" motif is not present in members of the Crenarchaea, some euryarchaeal varieties, the *C. paradoxa* cyanelle, and eukaryotic RNase P RNAs. These varieties of RNase P RNA also tend not to exhibit catalytic proficiency *in vitro*. However, cyanobacterial RNase P RNAs and those of the Chlamydiae also lack the "GGU" motif, and these RNAs exhibit catalytic activity *in vitro* in the absence of protein comparable to other bacterial RNase P RNAs. The RNase P RNA from the *C. paradoxa* cyanelle is very similar in secondary structure to those of the cyanobacteria,[6] yet the cyanobacterial RNase P RNAs are catalytic *in vitro* in the absence of protein while the cyanelle RNase P RNA is not. The protein subunit largely alleviates the dependence on the CCA tail seen with bacterial RNase P RNAs, but puromycin, a CCA-tRNA analog, inhibits both the holoenzyme reaction and the ribozyme reaction.[90] These observations suggest that the interaction of the RCCA tail with the J15/16 GGU motif, though important, is certainly not the only specific contact made between RNase P and its substrate.

The T-stem and T-loop domains have been implicated in specific binding of the substrate to the enzyme RNA. Photoactivated cross-linking experiments and mutational studies have suggested contacts between the T-stem of pre-tRNA and L8 and P9 within the "cruciform" of RNase P RNA[91–93] (the cruciform corresponds to

[84] F. R. Blattner, G. Plunkett III, C. A. Bloch, N. T. Perna, and V. Burland, *Science* **277,** 1453 (1997).
[85] T. E. LaGrandeur, A. Hüttenhofer, H. F. Noller, and N. R. Pace, *EMBO J.* **13,** 3945 (1994).
[86] A. Tallsjö, J. Kufel, and L. A. Kirsebom, *RNA* **2,** 299 (1996).
[87] S. G. Svärd, U. Kagardt, and L. A. Kirsebom, *RNA* **2,** 463 (1996).
[88] B.-K. Oh and N. R. Pace, *Nucleic Acids Res.* **22,** 4087 (1994).
[89] W.-D. Hardt, J. Schlegl, V. A. Erdmann, and R. K. Hartmann, *J. Mol. Biol.* **247,** 161 (1995).
[90] A. Vioque, *FEBS Lett.* **246,** 137 (1989).
[91] J. M. Nolan, D. H. Burke, and N. R. Pace, *Science* **261,** 762 (1993).

the structure formed by helices P7, P8, P9, P10, and P11 of RNase P RNA; Fig. 1). It has been proposed that the T-stem and loop, along with the acceptor stem, act to position the tRNA physically within the catalytic groove of RNase P such that the cleavage site is in proximity with the coordinated catalytic magnesium.[94]

Role of the Protein

The role of the protein component of RNase P is not completely understood. It was originally suggested that the RNase P protein shields electrostatic repulsions of the RNA backbone, because the requirement for the protein subunit in the bacterial RNase P reaction is alleviated by elevated ionic strength.[95] However, electrostatic shielding is not the only influence of the protein subunit. For example: (1) The absence of a CCA tail (used in recognition by the RNA alone) does not have a large effect on the holoenzyme reaction,[96] (2) the substrate range of the holoenzyme (also cleaves 4.5S RNA and the tmRNA precursor) is greater than the RNA alone,[2] and (3) mature tRNA is a stronger inhibitor of the RNA alone reaction than the holoenzyme reaction.[95] If electrostatic shielding was the predominant function of the protein, the above changes would be expected to exert the same *relative* effects on the RNA alone as the holoenzyme reaction.

Kinetic, thermodynamic, and cross-linking studies with *in vitro*–reconstituted *Bacillus subtilis* RNase P holenzyme suggest that the bacterial protein subunit has a more direct role than previously thought. Kinetic and substrate affinity experiments have demonstrated that the protein subunit, besides slightly enhancing the rate of product release (<10-fold) and chemical cleavage, creates a 10^4-fold increase in affinity for pre-tRNA substrate over mature tRNA product, which helps explain the enhancement in reaction rate seen with the holoenzyme over the ribozyme reaction.[97] Deletion mutations of the pre-tRNA leader sequence, base modification interferences, and direct cross-linking of 5′ leader sequence nucleotides to the *B. subtilis* RNase P protein subunit suggest a model where the enhancement in substrate affinity seen in the holoenzyme is due to direct binding of the protein subunit to the 5′ leader of the substrate.[98–100] This accords nicely with the observations that substrates such as 4.5S RNA that do not have a T-stem and loop, but do have a 5′ single-stranded extension from a helix that can serve as a cleavage substrate,

[92] T. Pan, A. Loria, and K. Zhong, *Proc. Natl. Acad. Sci. USA* **92,** 12510 (1995).
[93] A. Loria and T. Pan, *Biochemistry* **36,** 6317 (1997).
[94] L. A. Kirsebom and A. Vioque, *Mol. Biol. Rep.* **22,** 99 (1996).
[95] C. Reich, G. J. Olsen, B. Pace, and N. R. Pace, *Science* **239,** 178 (1988).
[96] C. Guerrier-Takada, W. H. McClain, and S. Altman, *Cell* **38,** 219 (1984).
[97] J. C. Kurz, S. Niranjanakumari, and C. A. Fierke, *Biochemistry* **37,** 2393 (1998).
[98] S. M. Crary, S. Niranjanakumari, and C. A. Fierke, *Biochemistry* **37,** 9409 (1998).
[99] A. Loria, S. Niranjanakumari, C. A. Fierke, and T. Pan, *Biochemistry* **37,** 15466 (1998).
[100] S. Niranjanakumari, T. Stams, S. M. Crary, D. W. Christianson, and C. A. Fierke, *Proc. Natl. Acad. Sci. USA* **95,** 15212 (1998).

are substrates of the holoenzyme, but not the RNA alone.[2] The structure of the *B. subtilis* RNase P protein has been determined by X-ray crystallography.[101] The protein is composed of an "α–β sandwich," with the general topology $\alpha\beta\beta\beta\alpha\beta\alpha$, and contains a rare $\beta\alpha\beta$ crossover. The general structure is strikingly similar to the small subunit ribosomal protein S5 and to elongation factor G. These observations have led to the hypothesis that the RNase P protein (and possibly the enzyme in general) may have direct evolutionary links to the translational apparatus,[101] an attractive idea when one wonders how a system as complex as the information processing core of even the simplest cell may have evolved from simpler components.

Universal Features of RNase P

With the two possible exceptions of plant chloroplast and vertebrate mitochondrial RNase P enzymes, all known examples of RNase P contain biologically essential RNA and protein components, and all catalyze the same reaction. The disparate physical characteristics of holoenzymes from the three different domains of life, however, may lead one to wonder if the RNase P complexes from different domains are true homologs or if they are examples of convergent evolution independently evolved to fulfill similar needs as already diverged branches of life developed more complicated transcription/translation mechanisms. Because RNase P from the eukarya and the archaea tend to be refractory to biochemical study compared to the bacterial enzyme, progress on elucidating the RNA structures, and certainly the holoenzyme composition and structure, has been much slower. Sufficient sequence data is now in hand, however, for construction of detailed secondary structures of archaeal RNase P RNAs.[56,58] Lower resolution secondary structure models have been proposed for fungal and vertebrate RNase P RNAs.

Comparative analysis of archaeal RNase P RNAs has revealed an overall secondary structure quite similar to the Type-A bacterial consensus (see Fig. 4). The most obvious difference between the archaeal RNase P RNA structures and the bacterial consensus is the absence of P18. Much more striking than the differences, however, are the similarities. The core sequence and structure, including both pseudo-knots, is maintained in global secondary structure. A survey of a part of vertebrate diversity[57] reveals a structural core with many key elements in common with both the bacterial and archaeal RNAs (Fig. 4). All three consensus core structures contain a P4 region that exists in a similar context in each structure; the P4 helix is surrounded by helical elements that, based upon their overall similarity in placement, spacing, and shape, are likely to occupy similar positions in three-dimensional space. Figure 5 shows a phylogeny of RNase P RNAs across the three domains that emphasizes their overall similarity in gross anatomy.

[101] T. Stams, S. Niranjanakumari, C. A. Fierke, and D. W. Christianson, *Science* **280,** 752 (1998).

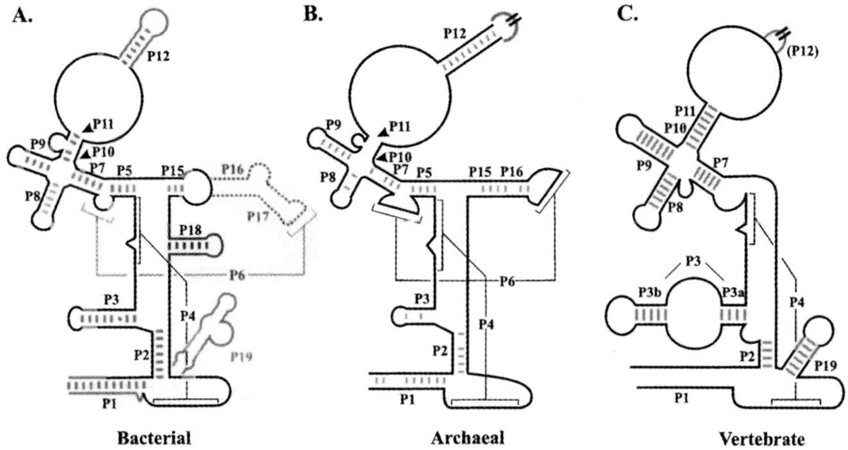

FIG. 4. Consensus RNase P secondary structures from Bacteria (*A*), Archaea (*B*), and Eukarya (*C*) (represented here only by vertebrates). Segments of the structures that exist in 100% of the structures are drawn with a solid black line. Segments that exist in 90% or more (bacterial [J. W. Brown, *Nucleic Acids Res.* **27,** 314 (1999)]) or 80% or more (vertebrate [C. Pitulle, M. Garcia-Paris, K. R. Zamudio, and N. R. Pace, *Nucleic Acids Res.* **26,** 3333 (1998)]) of available structures are drawn with a gray line. The region above P12 in the Archaea and vertebrates is not well defined, and the two slanted lines through a light gray loop at the end of P12 indicate an arbitrary truncation of the structures at this point. For simplicity, the bacterial consensus (*A*) excludes Type-B specific structures (P5.1, P10.1, p15.1; see Fig. 1). The archaeal consensus (*B*) excludes the *Methanococcus* spp. and *Archaeoglobus fulgidus*, which lack P16, P6, and P8.

The RNA and protein subunits from many bacteria are interchangeable both *in vitro* and *in vivo.*[102] Interestingly, a functional reconstitution of enzymatic activity was achieved with the RNase P RNA from the *C. paradoxa* cyanelle and the protein subunit from a cyanobacterial RNase P (*Synechocystis* sp. PCC 6803), but not with the RNase P protein subunit from *E. coli,*[103] suggesting that the cyanellar RNase P, which exhibits a eukaryotic nuclear-like density in Cs_2SO_4, may have retained a protein of similar structure and function to the cyanobacterial ancestor from which it was derived. The *B. subtilis* RNase P protein subunit is able to combine with the RNase P RNA from at least two methanogenic archaea, *Methanobacterium thermoautotrophicum* and *M. formicicum,* to form a catalytic complex at relatively low ionic conditions (100 m*M* Ammonium acetate, 25 m*M* $MgCl_2$).[11] These observations suggest that, at least over the phylogenetic boundaries between the Bacteria and Archaea, and between the cyanobacteria and a primitive plastid, the structural surfaces presented by the presumed homologous

[102] D. S. Waugh and N. R. Pace, *J. Bacteriol.* **172,** 6316 (1990).
[103] A. Pascual and A. Vioque, *FEBS Lett* **442,** 7 (1999).

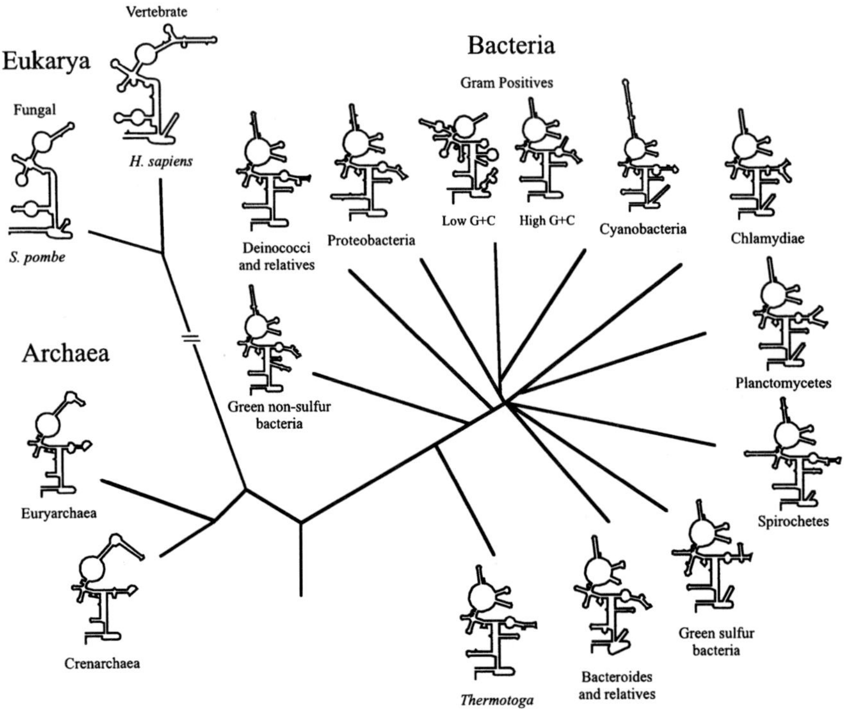

FIG. 5. A phylogenetic view of representative RNase P RNA structures. Secondary structures were obtained from the Ribonuclease P database [J. W. Brown, *Nucleic Acids Res.* **27,** 314 (1999)]. Phylogenetic groups are represented by the following organisms, clockwise from lower left to lower right: *Sulfolobus acidocaldarius, Methanobacterium thermoautotrophicum* ΔH, *Schizosaccharomyces pombe, Homo sapiens, Chloroflexus auranticus, Thermus aquaticus, Escherichia coli, Bacillus subtilis, Streptomyces bikiniensis, Anabaena* PCC7120, *Chlamydia trachomatis, Planctomyces maris, Treponema pallidum, Chlorobium limicola, Bacteroides thetaiotaomicron, Thermotoga maritima.*

RNA molecules are similar enough that they can interact with the same protein to form a catalytic complex, even between phylogenetic domains.

Because a majority of archaeal RNase P RNAs, and all nuclear RNase P RNAs known, are unable to cleave substrate *in vitro* without protein subunits, there has as yet been no definitive proof that it is truly the RNA in these cases that is catalytic. However, all RNase P enzymes so far studied require magnesium ions for catalysis, and recent studies[34,35] have demonstrated that eukaryotic nuclear RNase P holoenzymes are sensitive to phosphorothioate modification at the pre-tRNA cleavage site in the same manner as bacterial RNase P, evidence that the RNase P RNAs from these two domains of life are likely to act through a similar (or identical) catalytic mechanism. In light of the overall structural similarity of known archaeal, bacterial, and eukaryal RNase P RNAs, the dramatically exceptional case is that of the spinach chloroplast RNase P. If it can be demonstrated that viable RNase P function *can* be provided biologically by a protein enzyme (it is not excluded on the basis of any natural principle), then why do we see that the vast majority of all RNase P enzymes from all major phylogenetic domains contain an RNA with an immediately recognizable structural core that encompasses the catalytic domain of the molecule? Why has RNase P maintained itself as an RNA in all walks of life, when nearly every other known true enzyme function has developed as a protein-based system?

Acknowledgments

We thank Elizabeth Haas for careful reading of this manuscript and for useful suggestions. Thanks also to Andy Andrews for unpublished data and to Tim Dean for helpful discussions. Work in our laboratory is supported by NIH Grant GM52894.

Section II

Ribonuclease Assays

[5] Fast, Facile, Hypersensitive Assays for Ribonucleolytic Activity

By CHIWOOK PARK, BRADLEY R. KELEMEN, TONY A. KLINK, ROZAMOND Y. SWEENEY, MARK A. BEHLKE, SHAD R. EUBANKS, and RONALD T. RAINES

Assays for Ribonucleolytic Activity

A sensitive and convenient assay for catalytic activity is an essential tool for studying catalysis. Because of the seminal role of bovine pancreatic ribonuclease A (RNase A; EC 3.1.27.5) in the history of enzymology,[1] assays for ribonucleolytic activity have existed for decades. The recent discovery of ribonucleases[2] and ribozymes[3] with important biological functions, but low ribonucleolytic activity, has spurred the development of assays with high sensitivity.

Early work on the kinetics of catalysis by RNase A used substrates that were either ill-defined heterogeneous strands of RNA (e.g., yeast RNA in the "Kunitz" assay[4]) or nucleoside 2′,3′-cyclic phosphodiesters,[5] which are the products rather than the substrates of the germinal transphosphorylation reaction.[6,7] One appropriate application of assays using RNA polymers is for the detection of ribonucleolytic activity in a complex mixture. For example, the release of methylene blue from yeast RNA provides a sensitive assay at 688 nm, a wavelength of light not absorbed by most biomolecules.[8] Alternatively, zymogram assays can detect as little as 1 pg (0.1 fmol) of impure RNase A. In a zymogram assay, a polymeric substrate is incorporated into a gel, and cleavage is visualized by staining for intact polymers after electrophoresis[9–11] or isoelectric focusing.[12] A zymogram blot is also effective.[13]

[1] R. T. Raines, *Chem. Rev.* **98,** 1045 (1998).

[2] G. D'Alessio and J. F. Riordan, eds., "Ribonucleases: Structures and Functions." Academic Press New York, 1997.

[3] R. F. Gesteland, T. Cech, and J. F. Atkins, eds., "The RNA World." Cold Spring Harbor Laboratory, Cold Spring Harbor, NY, 1999.

[4] M. Kunitz, *J. Biol. Chem.* **164,** 563 (1946).

[5] E. M. Crook, A. P. Mathias, and B. R. Rabin, *Biochem. J.* **74,** 234 (1960).

[6] C. M. Cuchillo, X. Parés, A. Guasch, T. Barman, F. Travers, and M. V. Nogués, *FEBS Lett.* **333,** 207 (1993).

[7] J. E. Thompson, F. D. Venegas, and R. T. Raines, *Biochemistry* **33,** 7408 (1994).

[8] T. Greiner-Stoeffele, M. Grunow, and U. Hahn, *Anal. Biochem.* **240,** 24 (1996).

[9] A. Blank, R. H. Sugiyama, and C. A. Dekker, *Anal. Biochem.* **120,** 267 (1982).

[10] J.-S. Kim and R. T. Raines, *Protein Sci.* **2,** 348 (1993).

[11] J. Bravo, E. Fernández, M. Ribó, R. de Llorens, and C. M. Cuchillo, *Anal. Biochem.* **219,** 82 (1994).

[12] T. Yasuda, D. Nadano, E. Tenjo, H. Takeshita, and K. Kishi, *Anal. Biochem.* **206,** 172 (1992).

[13] D. Nadano, T. Yasuda, K. Sawazaki, H. Takeshita, and K. Kishi, *Anal. Biochem.* **212,** 111 (1993).

0076-6879/01 $35.00

Answering questions about enzymatic catalysis with chemical rigor requires the use of a well-defined substrate. Homopolymeric substrates such as poly(uridylic acid) [poly(U)] and poly(cytidylic acid) [poly(C)] are now readily available. Further, the advent of modern phosphoramidite chemistry enables the facile synthesis of any di-, tri-, or tetranucleotide substrate. Because ribonucleases do not catalyze DNA cleavage, the synthesis of RNA/DNA chimeras extends further the horizon of possible analyses.[14,15]

Chromogenic substrates facilitate assays of ribonucleolytic activity. Uridine 3′-(5-bromo-4-chloroindol-3-yl)phosphate (U-3′-BCIP) is a substrate for RNase A.[16,17] The 5-bromo-4-chloroindol-3-ol product dimerizes rapidly in air to form a blue pigment. This substrate is analogous to 5-bromo-4-chloroindol-3-yl-galactose (X-Gal), a common substrate for β-galactosidase. Other chromogenic substrates rely on the production of yellow phenolates from the cleavage of a uridine 3′-arylphosphates.[18–20]

Fluorogenic substrates for fast, facile, and hypersensitive assays of ribonucleolytic activity are now known.[21–23] Fluorogenic substrates provides the basis for extremely sensitive assays for ribonucleolytic activity. 5′-*O*-[4-(2,4-Dinitrophenylamino)butyl]phosphoryluridylyl-(3′→5′)-2′-deoxyadenosine 3′-{*N*-[(2-aminobenzoyl)aminoprop-3-yl]phosphate} consists of a fluorophore (*o*-aminobenzoic acid) linked via Up(dA) to a quencher (2,4-dinitroaniline).[21] Cleavage of the phosphodiester bond in the Up(dA) linker results in a 60-fold increase in fluorescence, enabling the detection of a 50 f*M* concentration of RNase A. Substrates for even more sensitive assays have been reported and are available commercially. We describe herein the properties of these new fluorogenic substrates and associated protocols for assaying ribonucleolytic activity. We also describe applications for these substrates.

Design and Synthesis of Fluorogenic Substrates

The fluorogenic substrates were designed as RNA/DNA chimeras (Table I). Each substrate contains only one nucleotide with a ribose moiety, which defines a single cleavable site in the substrate. Such chimeric substrates are superior to polymeric or oligomeric RNA substrates for enzymological analysis because the

[14] L. A. Jenkins, J. K. Bashkin, and M. E. Autry, *J. Am. Chem. Soc.* **118,** 6822 (1996).

[15] B. R. Kelemen and R. T. Raines, *Biochemistry* **38,** 5302 (1999).

[16] P. L. Wolf, J. P. Horwitz, J. Freisler, J. Vazquez, and E. Von der Muehll, *Experientia* **24,** 1290 (1968).

[17] M. R. Witmer, C. M. Falcomer, M. P. Weiner, M. S. Kay, T. P. Begley, B. Ganem, and H. A. Scheraga, *Nucleic Acids Res.* **19,** 1 (1991).

[18] A. M. Davis, A. C. Regan, and A. Williams, *Biochemistry* **27,** 9042 (1988).

[19] J. E. Thompson and R. T. Raines, *J. Am. Chem. Soc.* **116,** 5467 (1994).

[20] S. B. delCardayré, M. Ribó, E. M. Yokel, D. J. Quirk, W. J. Rutter, and R. T. Raines, *Prot. Eng.* **8,** 261 (1995).

TABLE I

OPTIMIZATION OF FLUOROGENIC SUBSTRATES FOR CLEAVAGE BY RIBONUCLEASE A[a]

Substrates	k_{cat}/K_m^b ($10^7\ M^{-1}s^{-1}$)	F_{max}/F_0 [c]	Sensitivity[d] ($10^8\ M^{-1}s^{-1}$)
6-FAM~rUdA~6-TAMRA	2.5 ± 0.3	15 ± 2	3.8 ± 0.7
6-FAM~dArU$(dA)_2$~6-TAMRA	3.6 ± 0.4	180 ± 10	65 ± 8
6-FAM~$(dA)_2$rU$(dA)_3$~6-TAMRA	4.7 ± 0.6	26 ± 3	12 ± 2
6-FAM~$(dA)_3$rU$(dA)_4$~6-TAMRA	4.8 ± 0.5	62 ± 2	30 ± 3

[a] Data are from Ref. 23.

[b] Values of k_{cat}/K_m were determined in 0.10 *M* MES–NaOH (pH 6.0) containing NaCl (0.10 *M*).

[c] F_{max}/F_0 is the ratio of fluorescence intensity of the product to that of the substrate.

[d] Sensitivity (*S*) is defined in Eq. (1).

unique cleavable site provides for a homogeneous substrate. The sequences of the substrates were optimized for catalysis by RNase A. RNase A requires a pyrimidine residue 5′ to the scissile bond and prefers a purine residue 3′ to the scissile bond. The sole ribonucleoside residue has a uracil base. To minimize nonproductive binding and product inhibition, all of the deoxyribonucleotide residues are purines.

The assay of ribonucleolytic activity with these fluorogenic substrates is based on the fluorescence resonance energy transfer (FRET).[24] The substrates have a 5′ 6-carboxyfluorescein (6-FAM) label and a 3′ 6-carboxytetramethylrhodamine (6-TAMRA) label (Fig. 1). In an intact substrate, the fluorescence emission of the 6-FAM moiety at 515 nm is quenched by the proximal 6-TAMRA moiety (Figs. 1 and 2). Weak emission of the intact substrate is observed at 577 nm as a result of FRET from 6-FAM to 6-TAMRA. On substrate cleavage, the observed fluorescence emission of 6-FAM at 515 nm increases dramatically (Figs. 1 and 2). The net increase in fluorescence on substrate cleavage depends on the efficiency of FRET, and thus on the length of the substrate. A series of substrates with an incremental change in length substrate is defined as the product of F_{max}/F_0, the total increase in fluorescence intensity on substrate cleavage, and k_{cat}/K_M, the specificity constant of the enzyme, as in Eq. (1):

$$S = (F_{max}/F_0)(k_{cat}/K_m) \tag{1}$$

[21] O. Zelenko, U. Neumann, W. Brill, U. Pieles, H. E. Moser, and J. Hofsteenge, *Nucleic Acids Res.* **22,** 2731 (1994).

[22] D. A. James and G. A. Woolley, *Anal. Biochem.* **264,** 26 (1998).

[23] B. R. Kelemen, T. A. Klink, M. A. Behlke, S. R. Eubanks, P. A. Leland, and R. T. Raines, *Nucleic Acids Res.* **27,** 3696 (1999).

[24] J. R. Lakowicz, "Principles of Fluorescence Spectroscopy." Plenum, New York, 1999.

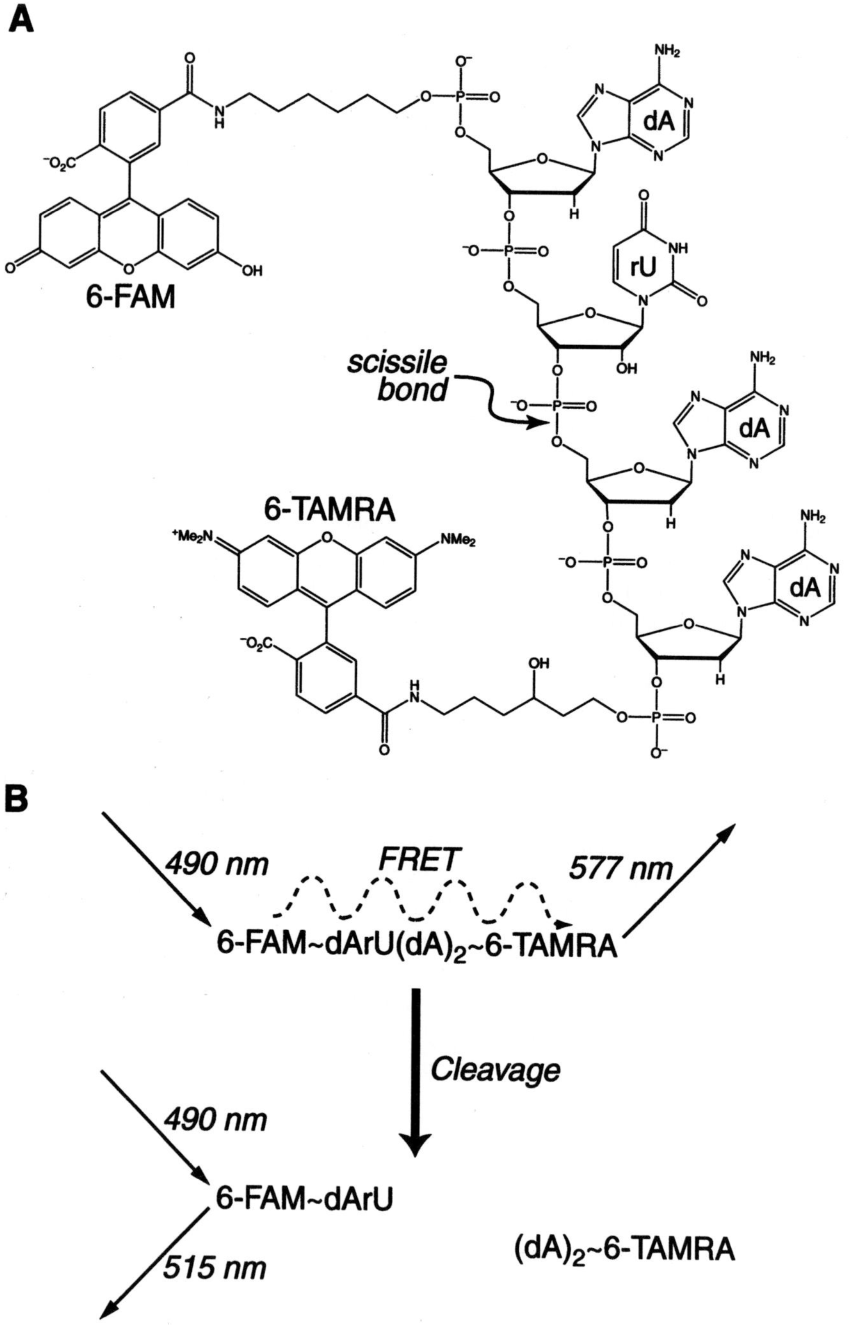
A
6-FAM
dA
rU
scissile bond
6-TAMRA
dA
dA
B
490 nm
FRET
577 nm
6-FAM~dArU(dA)2~6-TAMRA
Cleavage
490 nm
6-FAM~dArU
515 nm
(dA)2~6-TAMRA

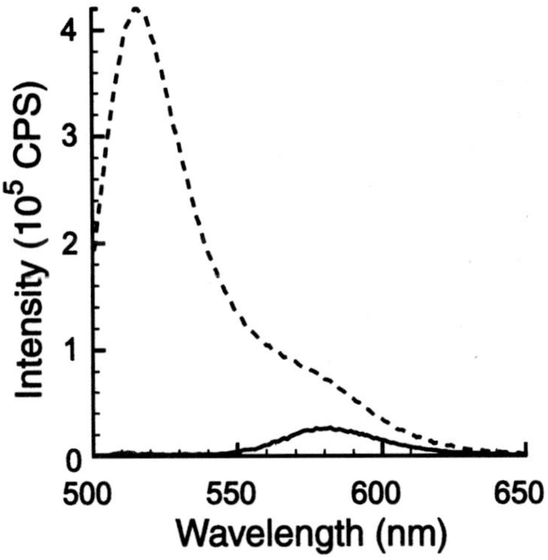

FIG. 2. Emission spectra of 6-FAM–dArU$(dA)_2$–6-TAMRA (solid line) and its cleavage product (dashed line) on excitation at 490 nm.[23] "CPS" refers to photon counts per second.

In Eq. (1), F_0 is the initial intensity before the reaction is initiated and F_{max} is the final fluorescence intensity after the reaction reaches completion. Thus, the value of S reports on the rate at which the fluorescence intensity changes at low concentrations of substrate. The value of S is unique for a particular enzyme and substrate and provides a useful means to compare assays of enzymatic activity.

The substrates can be synthesized by standard phosphoramidite chemistry on a conventional DNA/RNA synthesizer. [The most sensitive substrate for RNase A is available commercially from Integrated DNA Technologies (Coralville, IA; *http://www.idtdna.com*).] To minimize the background fluorescence from by-products, synthetic substrates are purified by high-performance liquid chromatography (HPLC). The concentration of the substrates can be estimated spectrophotometrically at 260 nm with extinction coefficients calculated from their sequences.[25]

[25] G. H. Beaven, E. R. Holiday, and E. A. Johnson, *in* "The Nucleic Acids" (E. Chargraff and J. N. Davidson, eds.), p. 493. Academic Press, New York, 1995.

FIG. 1. (A) Chemical structure of 6-FAM–dArU$(dA)_2$–6-TAMRA, the fluorogenic substrate for ribonucleases. The arrow indicates the scissile bond for ribonucleolytic cleavage. (B) Basis for the fluorescence assay of ribonucleolytic activity. The 6-FAM moiety is excited at 490 nm. In an intact substrate, the fluorescence emission of 6-FAM at 515 nm is quenched efficiently by the proximal 6-TAMRA moiety. Weak emission of the intact substrate is observed at 577 nm as a result of fluorescence energy transfer (FRET) from 6-FAM to 6-TAMRA. On substrate cleavage, the observed fluorescence emission of 6-FAM at 515 nm increases dramatically.

Preparation of Reagents

Pellets of synthetic substrates are dissolved in ribonuclease-free water[26] to a final concentration near 10 μM, dispensed into small aliquots, and stored at $-20°$. Fluorometric cuvettes (1.0 cm path length; 3.5 ml volume) are presoaked in nitric acid (10%, v/v) overnight to remove any residual ribonucleolytic activity. Quartz and glass cuvettes can be used interchangeably without any observable difference in assay results. Buffer [e.g., 0.10 *M* MES–NaOH (pH 6.0) containing NaCl (0.10 *M*)] is prepared with ribonuclease-free water and chemicals. Solutions should be handled with great care to minimize background ribonucleolytic activity from contamination with human ribonucleases.[27] Enzyme is diluted to a suitable concentration (e.g., 0.10–50 n*M* for RNase A) with the assay buffer.

Assay Procedures and Data Analysis

A typical assay for ribonucleolytic activity follows. Buffer (2.0 ml) is placed in a quartz or glass cuvette. A substrate (5.0 μl of a 10 μM solution) is added to the buffer with stirring to give a final concentration near 25 n*M*. The fluorescence emission is monitored at 515 nm, on excitation at 490 nm. The change in fluorescence intensity is monitored initially to check for unwanted ribonuclease contamination. The average fluorescence intensity at this point is designated as F_0, the initial fluorescence intensity. The reaction is initiated by adding enzyme (10 μl of a stock solution) with stirring.

The substrate concentration used in this assay is likely to be much lower than K_m. Thus, k_{cat}/K_m can be determined directly without applying Michaelis–Menten kinetics. Either an *exponential rise analysis* or *an initial velocity analysis* is used to calculate k_{cat}/K_m from the fluorescence change (Fig. 3). The two analyses typically produce equivalent results. If the reaction is fast enough to reach completion, V/K ($= (k_{cat}/K_m)[E]$) for the reaction is determined by an exponential analysis, fitting the observed fluorescence intensity to Eq. (2):

$$F = F_0 + (F_{max} - F_0)(1 - e^{-(V/K)t}) \qquad (2)$$

In Eq. (2), F is the observed fluorescence intensity. Values of F_{max}, F_0, and V/K are determined by nonlinear regression analysis.

If a reaction is too slow to achieve completion, V/K is determined by an initial velocity analysis with Eq. (3):

$$V/K = \frac{(\Delta F/\Delta t)}{F_{max} - F_0} \qquad (3)$$

[26] Y. H. Huang, P. Leblanc, V. Apostolou, B. Stewart, and R. B. Moreland, *Biotechniques* **19,** 656 (1995).

[27] R. Poulson, *in* "The Ribonucleic Acids" (P. R. Stewart and D. S. Letham, eds.), p. 333. Springer-Verlag, New York (1977).

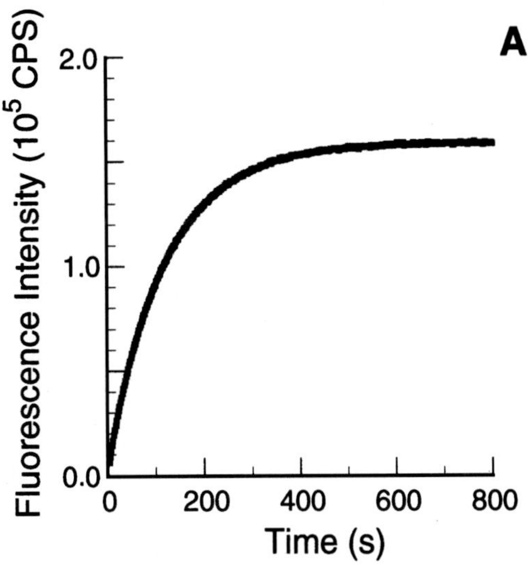

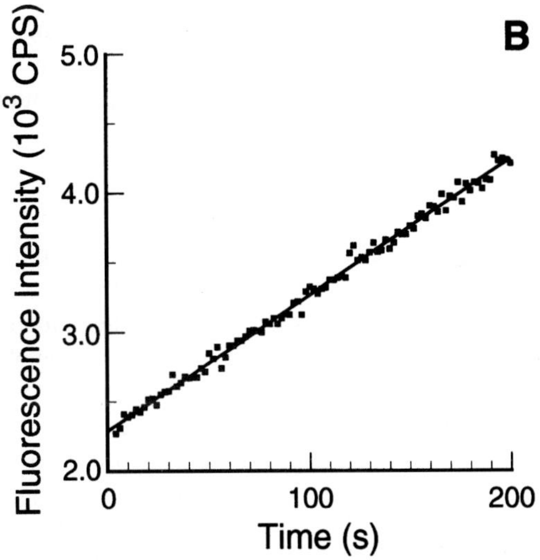

FIG. 3. Comparison of (A) exponential rise analysis and (B) initial velocity analysis of the cleavage of 6-FAM–dArU(dA)$_2$–6-TAMRA by ribonuclease A. Reactions were performed in 0.10 *M* MES–NaOH (pH 6.0) containing NaCl (0.10 *M*), 6-FAM–dArU(dA)$_2$–6-TAMRA (20 n*M*), and ribonuclease A [(A) 0.25 n*M*; (B) 2.5 p*M*]. "CPS" refers to photon counts per second.

In Eq. (3), $\Delta F/\Delta t$ is the slope determined from linear regression analysis of the fluorescence intensity change for an initial linear region. After the initial velocity is determined, F_{max} is measured by adding enough enzyme to complete the reaction. Because the fluorescence of the product can be sensitive to a change in pH or salt concentration, the enzyme used to complete the reaction should be in the assay buffer. The fluorescence intensity often decreases slowly after reaching a maximum, presumably because of the subsequent hydrolysis of the 2′,3′-cyclic phosphodiester product. The value of F_{max} can, however, be obtained from the observed maximum without introducing significant error. The value of k_{cat}/K_m can be calculated by dividing V/K by the concentration of enzyme in the reaction. Knowing the substrate concentration is not necessary for the determination of k_{cat}/K_m in either analysis.

Applications

Ribonuclease Contamination

Contaminating ribonucleolytic activity is the bane of everyone who works with RNA.[27,28] The fluorogenic substrates provide a most sensitive means to detect such contamination. For example, the activity arising from a 10 f*M* solution of human pancreatic ribonuclease, which is the human homolog of RNase A and a common laboratory contaminant, can be detected easily with a spectrofluorometric assay in 50 m*M* Bis–Tris-HCl buffer (pH 6.0).

Steady-State Kinetic Parameters

The hypersensitivity of assays based on fluorogenic substrates makes it possible to assay ribonucleolytic activity with low concentrations of substrates. Under these conditions, it is likely that $[S] \ll K_m$, so that the value of k_{cat}/K_m can be determined directly as described above. Two criteria can be applied to ensure that $[S] \ll K_m$. First, V/K will not change upon doubling the substrate concentration. If [S] is not far enough below K_m, then V/K will decrease. Second, fitting the fluorescence change data to Eq. (2) should not yield significant residual errors when the exponential rise analysis is used. Deviation of the data from Eq. (2) indicates that [S] is not far enough below K_m.

The hypersensitive assays also enable a quantitative description of catalysis by poor catalysts of RNA cleavage. For example, angiogenin is an RNase A homolog that promotes neovascularization. The k_{cat}/K_m values determined with RNase A and angiogenin are listed in Table II. Though k_{cat}/K_m values of angiogenin are 10^5-fold lower than those of RNase A, these values are determined without any

[28] S. L. Berger, *Methods Enzymol.* **152,** 227 (1987).

TABLE II
VALUES OF k_{cat}/K_m FOR CLEAVAGE OF FLUOROGENIC SUBSTRATES BY RIBONUCLEASE A AND ANGIOGENIN

	k_{cat}/K_m (10^7 $M^{-1}s^{-1}$)[a]	
Substrate	RNase A	Angiogenin
6-FAM–dArU(dA)$_2$–6-TAMRA	3.6 ± 0.4	0.000033 ± 0.000004
6-FAM–dArC(dA)$_2$–6-TAMRA	6.6 ± 0.6	0.000054 ± 0.000004

[a] Values of k_{cat}/K_m (± SE) were determined in 0.10 *M* MES–NaOH (pH 6.0) containing NaCl (0.10 *M*). Data are from Ref. 23.

complication. The assay conditions used to determine the ribonucleolytic activity of angiogenin were identical to those with RNase A, except that the concentration of angiogenin used in the reaction was higher (0.10–0.50 μM).

Substrate Specificity

Fluorogenic substrates that differ only in the nucleobase of the sole ribonucleotide residue can be used to determine the substrate specificity of a ribonuclease variant. For example, the T45G variant of RNase A catalyzes the efficient cleavage of poly(adenylic acid), whereas RNase A does not.[29] This change in substrate specificity is also manifested in the cleavage of fluorogenic substrates. Specifically, T45G RNase A cleaves 6-FAM–dArA(dA)$_2$–6-TAMRA 10^2-fold faster than does the wild-type enzyme (Table III). This finding is consistent with the observation that T45G RNase A is significantly more effective than the wild-type enzyme at cleaving heterogeneous RNA to completion.[30]

RNase T_1 from *Aspergillus oryzae* is not homologous to RNase A. This microbial ribonuclease is known to cleave preferentially the P–$O^{5'}$ bond after a guanosine residue.[31] As expected, a fluorogenic substrate that contains a guanosine ribonucleotide is cleaved at least 10^4-fold more quickly than are those that contain uridine, cytosine, or adenosine (Table III).

Inhibitor K_i Values

Fluorogenic substrates provide a simple method to determine the K_i value for competitive inhibition of ribonucleolytic activity. This method has been used to

[29] S. B. delCardayré and R. T. Raines, *Biochemistry* **33,** 6031 (1994).
[30] S. B. delCardayré and R. T. Raines, *Anal. Biochem.* **225,** 176 (1995).
[31] K. Takahashi, T. Uchida, and F. Egami, *Adv. Biophys.* **1,** 53 (1970).

TABLE III
VALUES OF k_{cat}/K_m FOR CLEAVAGE OF 6-FAM–dArXdAdA–6-TAMRA BY WILD-TYPE RIBONUCLEASE A, ITS T45G VARIANT, AND RIBONUCLEASE T_1 FROM *Aspergillus oryzae*

	k_{cat}/K_m ($10^6\ M^{-1}s^{-1}$)[a]		
X	Wild-type RNase A[b]	T45G RNase A[b]	RNase T_1
U	36 ± 3	0.75 ± 0.08	<0.000001[c]
C	66 ± 6	3.3 ± 0.3	<0.000001[c]
A	0.000018 ± 0.000002	0.0023 ± 0.0002	0.00012 ± 0.00001
G	<0.0000001[c]	<0.0000005[c]	1.3 ± 0.1

[a] Values of k_{cat}/K_m (±SE) were determined in 0.10 *M* MES–NaOH (pH 6.0) containing NaCl (0.10 *M*).
[b] Data from Ref. 37.
[c] The activity is below the detection limit.

determine K_i values for inhibition of RNase A by substrate analogs (Fig. 4)[23] and the ribonuclease inhibitor protein (RI).[32,33]

Assays are performed with stirring in 2.00 ml of 0.10 *M* MES–NaOH buffer (pH 6.0) containing NaCl (0.10 *M*), substrate (60 n*M*), and enzyme (5–500 p*M*) [plus dithiothreitol (5 m*M*) for RI K_i determination]. The inhibitor stock solution should be prepared in the same assay buffer. The value of $\Delta F/\Delta t$ is measured for 5 min after enzyme is added. Next, an aliquot of inhibitor is added, and $\Delta F/\Delta t$ of the newly established linear region determined, now in the presence of inhibitor. The concentration of inhibitor in the assay is doubled repeatedly in 5-min intervals until $\Delta F/\Delta t$ decreases to less than 10% of the initial uninhibited value. If the volume increase caused by adding the inhibitor is significant, the concentration of the enzyme and the substrate should be corrected by considering the dilution. For optimal results, the inhibitor concentrations in the reaction should be planned to span the range from $<K_i/10$ to $>10\ K_i$. (It is convenient to prepare a stock solution of inhibitor with a concentration 100-fold greater than K_i.) To obtain a properly weighted set of data for statistical analysis, the inhibitor concentration in the reaction should be increased exponentially by increasing the volume of each inhibitor addition. To ensure that the entire inhibition assay occurs during steady-state conditions, excess ribonuclease is then added to the reaction mixture so as to cleave all substrate. This addition is done to obtain the final fluorescence intensity and to confirm that less than 10% of the substrate is cleaved prior to completion

[32] L. E. Bretscher, R. L. Abel, and R. T. Raines, *J. Biol. Chem.* **275,** 9893 (2000).
[33] T. A. Klink and R. T. Raines, *J. Biol. Chem.* **275,** 17463 (2000).

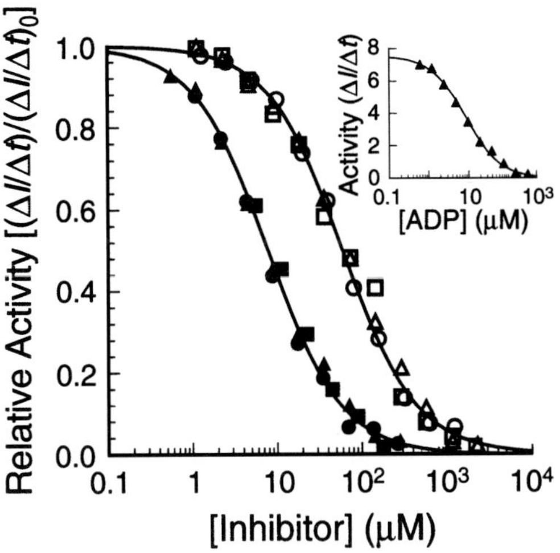

FIG. 4. Competitive inhibition of ribonuclease A catalysis by uridine 3′-phosphate (open symbols) and adenosine 5′-diphosphate (closed symbols). Relative activity is the ratio of the slope determined at each concentration of inhibitor to the slope determined without any inhibitor. Triangle, square, and circle symbols represent three different sets of data determined independently. Absolute activity in one set of data. Data were fitted to Eq. (4).[23]

of the inhibition assay. The value for K_i is determined by non-linear least squares regression analysis of $\Delta F/\Delta t$ fitted to Eq. (4):

$$\Delta F/\Delta t = (\Delta F/\Delta t)_0 \left(\frac{K_i}{[I] + K_i} \right) \tag{4}$$

In Eq. (4), $(\Delta F/\Delta t)_0$ is the initial slope prior to the addition of inhibitor.

Determination of K_i values can be a useful tool for detecting a contaminating ribonuclease. Traditionally, a protein is considered to be pure if it constitutes >95% of the protein visible after polyacrylamide gel electrophoresis performed in the presence of sodium dodecyl sulfate (SDS–PAGE). Analysis by SDS–PAGE is insufficient, however, if the contaminating enzyme is a more efficient catalyst than the test enzyme. For example, consider the inadvertent contamination of angiogenin by RNase A. RNase A is $>10^5$-fold better catalyst of RNA cleavage than is angiogenin (Table II). If RNase A is a 0.01% contaminant in an angiogenin preparation, then the determined K_i value will be similar to that for RNase A rather than the actual value of angiogenin because >90% of the ribonucleolytic activity is from the contaminant. The K_i value, along with SDS–PAGE and zymogram assays, can be used to show that angiogenin is effectively free of ribonucleolytic activity from contaminating RNase A.[23]

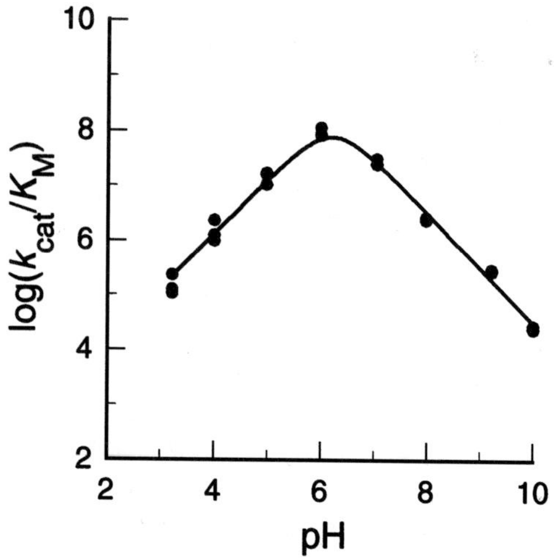

FIG. 5. pH–Rate profile for catalysis of 6-FAM–$(dA)_2rU(dA)_3$–6-TAMRA cleavage by ribonuclease A. Values of k_{cat}/K_m were determined in 0.10 *M* buffer containing NaCl (0.10 *M*) and 6-FAM–$dArU(dA)_2$–DABCYL (5–80 n*M*). Buffers were sodium citrate (pH 3.22), sodium succinate (pH 3.84 and 4.97), MES–NaOH (pH 5.99), MOPS–NaOH (pH 7.04), Tris-HCl (pH 7.97), CHES–NaOH (pH 9.14), and CAPS–NaOH (pH 10.0). Data were fitted to Eq. (5).[37]

pH–Rate Profiles

pH–Rate profiles have been used often and with great success to reveal the role of acidic and basic functional groups in catalysis.[34–36] To obviate artifacts from changing pH, the substrate used in the determination of a pH–rate profile should be stable within the pH range used for the experiments. A 6-TAMRA label is not optimal for pH–rate profiles because of its chemical instability at acidic pH. The pH–rate profile for catalysis by wild-type RNase A, has, however, been determined with 6-FAM–$dArU(dA)_2$–4-DABCYL because the 4-((4-(dimethylamino)phenyl)azo)benzoic acid (DABCYL) label is a stable quencher throughout the pH range (Fig. 5). Also, the fluorescence intensity of fluorescein is sensitive to pH because the anionic form of the molecule has greater fluorescence than does the neutral form ($pK_a \sim 7.5$). As a consequence, a higher concentration of the substrate is required for assaying activity at acidic pH (40–80 n*M*) than at basic pH (5–10 n*M*). Acid dissociation constants were determined to be 6.0 and 6.4

[34] J. R. Knowles, *CRC Crit. Rev. Biochem.* **4,** 165 (1976).
[35] W. W. Cleland, *Methods Enzymol.* **87,** 390 (1982).
[36] K. Brocklehurst, *Prot. Eng.* **7,** 291 (1994).

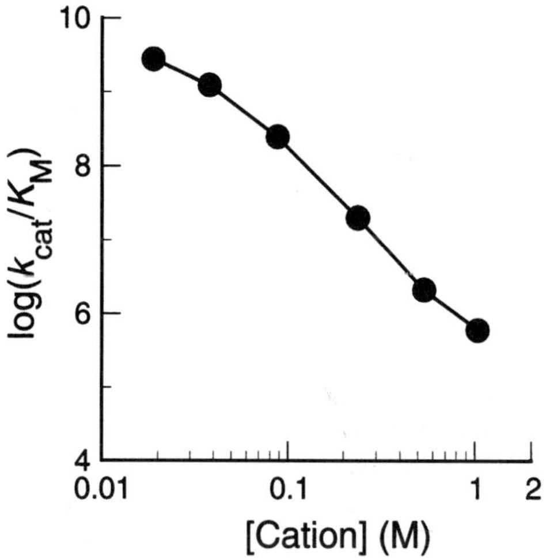

FIG. 6. Salt–rate profile for catalysis of 6-FAM–$(dA)_2rU(dA)_3$–6-TAMRA cleavage by ribonuclease A. Values of k_{cat}/K_m were determined in 25 m*M* or 50 m*M* Bis–Tris-HCl (pH 6.0) containing NaCl (0–1.0 *M*) and 6-FAM–$(dA)_2rU(dA)_3$–6-TAMRA (20 n*M*). [Cation] is the sum of the concentration of Bis–Tris cation and Na^+.[40]

by fitting the pH–rate profile in Fig. 5 to Eq. (5) by nonlinear regression analysis.[37]

$$k_{cat}/K_m = \frac{(k_{cat}/K_m)_0}{1 + 10^{(pH - pK_1)} + 10^{(pK_2 - pH)}} \quad (5)$$

Salt–Rate Profiles

Though less often used than pH–rate profiles, salt–rate profiles can provide valuable insights into the role of Coulombic interactions in catalysis. Such interactions are known to be of great importance in the interaction of ribonucleases with their substrates.[38–40] The determination of k_{cat}/K_m values for ribonucleases at low salt concentration is problematic because K_m decreases significantly at low salt. Moreover, kinetic analyses could be complicated because of product or substrate inhibition at low salt.[41] The fluorogenic substrates can eliminate this pitfall because k_{cat}/K_m can be determined with a low concentration of substrates.

The salt–rate profile for catalysis by RNase A has been determined with 6-FAM–$(dA)_2rU(dA)_3$–6-TAMRA in 25 m*M* or 50 m*M* Bis–Tris-HCl buffer

[37] B. R. Kelemen, Ph.D. Thesis, University of Wisconsin—Madison, 1999.

[38] D. E. Jensen and P. H. von Hippel, *J. Biol. Chem.* **251,** 7198 (1976).

[39] B. M. Fisher, J.-H. Ha, and R. T. Raines, *Biochemistry* **37,** 12121 (1998).

[40] C. Park and R. T. Raines, *FEBS Lett.* **468,** 199 (2000).

[41] S. R. Dickman and B. Ring, *J. Biol. Chem.* **231,** 741 (1958).

(pH 6.0) containing NaCl (0–1.0 *M*) (Fig. 6).[40] The k_{cat}/K_m of RNase A is $> 10^9$ $M^{-1}s^{-1}$ at low salt and decreases monotonically as salt concentration increases. Special care in preparing buffer solution should be taken for assays at low salt to avoid possible inhibition by low-level contaminants in common buffers. Because enhanced Coulombic interactions at low salt decrease K_i of inhibitory contaminants significantly, more than 10^4-fold inhibition was observed in certain types of buffers at low salt concentration.[40]

S·Tag Fusion System

The N-terminal 15 amino acid residues of RNase A (S·Tag) are an effective carrier in protein fusion systems.[10,42] This pentadecapeptide will bind to the C-terminal 104 amino acid residues of RNase A to produce an enzyme with high ribonucleolytic activity. [Reagents to use the S·Tag Fusion System are available commercially from Novagen (Madison, WI; *http://www.novagen.com*).] The hypersensitive assays described herein enable the detection of extremely low levels of proteins with a fused S·Tag. Moreover, because assays based on fluorescence spectroscopy are easy to automate, the fluorogenic substrates facilitate the high-throughput screening of proteins having an S·Tag.

[42] R. T. Raines, M. McCormick, T. R. Van Oosbree, and R. C. Mierendorf, *Methods Enzymol.* **326,** 362 (2000).

[6] Activity Staining for Detection of Ribonucleases Using Dried Agarose Film Overlay Method after Isoelectric Focusing

By TOSHIHIRO YASUDA, HARUO TAKESHITA, and KOICHIRO KISHI

Introduction

In human systems, two major types of ribonucleases (RNases), pancreatic and nonpancreatic, previously designated as secretory and nonsecretory, respectively, have been shown to be widely distributed in various tissues and body fluids.[1] These RNases are present in multiple forms with regard to their molecular masses and to their isoelectric points (p*I* values).[2–10] In order to examine and characterize the biological and clinical significance of their multiplicity, a technique combining

[1] S. Sorrentino and M. Libonati, *FEBS Lett.* **404,** 1 (1997).

METHODS IN ENZYMOLOGY, VOL. 341
0076-6879/01 $35.00

polyacrylamide gel electrophoresis (PAGE) and activity staining[3,11–15] has been developed. It is generally recognized that PAGE is relatively insensitive to differences in the surface charge density of a molecule, whereas isoelectric focusing is insensitive to differences in molecular size, shape, and conformation. Although both pancreatic- and nonpancreatic-type RNases purified from human urine migrate as only a single band on PAGE in the presence of sodium dodecyl sulfate, isoelectric focusing (IEF) electrophoresis in a thin layer of polyacrylamide (IEF–PAGE) followed by immunostaining can resolve the enzyme into several bands with different p*I* values.[4,16] Since these results suggested that differences in surface net charge contribute significantly to the microheterogeneity of the RNases, IEF might be useful for clarifying the details of RNase multiplicity. Warshaw *et al.*[17] and Kottel *et al.*[18] measured RNase activity in each slice of the focused gel, whereas Cranston *et al.*[2] and Hishiki *et al.*[19] assayed the activities in fractions eluted from an IEF column. However, the resolution and sensitivity of their methods were insufficient for analyzing the microheterogeneity of the enzyme. Because the IEF had been expected to provide excellent separation of RNase isoforms, the development of a more sensitive zymogram method was considered desirable for characterization and microheterogeneity analysis of RNases, corresponding to the high resolution of IEF.

In this chapter, we present a new *in situ* zymogram method for detecting RNase activity, suitable for IEF, using the dried agarose film overlay (DAFO) method.

---

[2] J. W. Cranston, F. Perini, E. R. Crisp, and C. V. Hixson, *Biochim. Biophys. Acta* **616,** 239 (1980).

[3] A. Blank and C. A. Dekker, *Biochemistry* **20,** 2261 (1981).

[4] T. Yasuda, W. Sato, and K. Kishi, *Biochim. Biophys. Acta* **965,** 185 (1988).

[5] T. Yasuda, K. Mizuta, W. Sato, and K. Kishi, *Eur. J. Biochem.* **191,** 523 (1990).

[6] T. Yasuda, K. Mizuta, and K. Kishi, *Arch. Biochem. Biophys.* **279,** 130 (1990).

[7] K. Mizuta, S. Awazu, T. Yasuda, and K. Kishi, *Arch. Biochem. Biophys.* **281,** 144 (1990).

[8] T. Yasuda, D. Nadano, H. Takeshita, and K. Kishi, *Biochem. J.* **296,** 617 (1993).

[9] M. Ribó, J. J. Beintema, M. Osset, E. Fernández, J. Bravo, R. de Llorens, and C. M. Cuchillo, *Biol. Chem. Hoppe-Seyler* **375,** 357 (1994).

[10] D. Nadano, T. Yasuda, H. Takeshita, K. Sawazaki, and K. Kishi, *Prot. Expres. Purif.* **7,** 167 (1996).

[11] A. L. Rosenthal and S. A. Lacks, *Anal. Biochem.* **80,** 76 (1977).

[12] J. M. Thomas and M. E. Hodes, *Clin. Chim. Acta* **111,** 185 (1981).

[13] B. Allinquant, C. Musenger, J. Reboul, J. J. Hauw, and E. Schuller, *Neurochem. Res.* **12,** 1067 (1987).

[14] Y. Yen and P. S. Baenziger, *Biochem. Genet.* **31,** 133 (1993).

[15] J. Bravo, E. Fernández, M. Ribó, R. de Llorens, and C. M. Cuchillo, *Anal. Biochem.* **219,** 82 (1994).

[16] T. Yasuda, D. Nadano, Y. Tanaka, and K. Kishi, *Biochim. Biophys. Acta* **1121,** 331 (1992).

[17] A. L. Warshaw, K.-H. Lee, W. C. Wood, and A. M. Cohen, *Am. J. Surg.* **139,** 27 (1980).

[18] R. H. Kottel, S. O. Hoch, R. G. Parsons, and J. A. Hoch, *Br. J. Cancer* **38,** 280 (1978).

[19] S. Hishiki, S. Sakaguchi, and T. Kanno, *Clin. Chim. Acta* **136,** 155 (1984).

Zymogram Method for Detection of RNase Activities after Isoelectric Focusing by Means of Dried Agarose Film Overlay Method

Previously, another type of endonuclease, deoxyribonuclease (DNase), was able to be detected by ordinary electrophoresis as bands by prior incorporation of substrate, high molecular weight DNA, and ethidium bromide into the gel.[11] However, with zymogram methods, detailed microheterogeneity analysis (such as genetic surveys) of DNases has been hampered by the low sensitivity and the poor band resolution produced by ordinary electrophoresis, similar to the RNase analysis. We have devised a new *in situ* zymogram method, the dried agarose film overlay (DAFO) method, for DNase activity, the principle of which is that ethidium bromide produces fluorescence with unhydrolyzed large DNA molecules, but not with DNA degraded by DNase action.[20–22] We have applied this principle to *in situ* detection of RNase activity on IEF–PAGE gel, and have developed the DAFO method based on the fact that ethidium bromide fluoresces only with unhydrolyzed RNA and not with RNA hydrolyzed by RNases.[23]

Materials and Analytical Methods

Acrylamide, *N,N'*-methylenebisacrylamide (Bis), *N,N,N',N'*-tetramethylethylenediamine (Temed), and agarose GP–36, which is characterized by a sulfate content of less than 0.5% (w/v), a gel strength of more than 900 g/cm^2 (1%, w/v), a gel point of about 36°C, and electroendosmosis of 0.10–0.15, can be obtained from Nacalai Tesque (Kyoto, Japan); ammonium persulfate and urea for electrophoresis and ethidium bromide are from Bio-Rad (Richmond, CA), and Roche Diagnostics GmbH (Mannheim, Germany), respectively. RNAs from baker's yeast, *Torula* yeast (Type VI), and calf liver (Type IV); flavine mononucleotide (FMN); acridine orange; acridine yellow; bisbenzimide; propidium iodide; and quinacrine mustard are obtained from Sigma Chemical Co. (St. Louis, MO). Each sample of RNA is dissolved in distilled water, dialyzed against distilled water after adjustment of the pH to 7.0 with NaOH to remove low molecular weight RNA fractions from the commercial products, and then lyophilized. Highly polymerized yeast RNA (Calbiochem, San Diego, CA) can also be used without dialysis. Polyester Agafix MSL, the supporting sheet for agarose electrophoresis, can be obtained from Wako Pure Chemical Industries Ltd. (Osaka, Japan).

[20] T. Yasuda, K. Mizuta, Y. Ikehara, and K. Kishi, *Anal. Biochem.* **183,** 84 (1989).

[21] K. Kishi, T. Yasuda, Y. Ikehara, K. Sawazaki, W. Sato, and R. Iida, *Am. J. Hum. Genet.* **47,** 121 (1990).

[22] T. Yasuda, D. Nadano, S. Awazu, and K. Kishi, *Biochim. Biophys. Acta* **1119,** 185 (1992).

[23] T. Yasuda, D. Nadano, E. Tenjo, H. Takeshita, and K. Kishi, *Anal. Biochem.* **206,** 172 (1992).

The standard assay for RNase is performed according to the procedure described by Uchida and Egami,[24] with some modifications. Yeast RNA solution (1.2%, w/v, 50 μl) is added to 150 μl of reaction mixture consisting of 50 m*M* Tris-HCl buffer, pH 7.5, 5 m*M* EDTA, and enzyme. The reaction is performed at 37° and terminated by adding 50 μl of 25% (v/v) perchloric acid containing 0.75% (w/v) uranyl acetate. The reaction tubes are chilled in an ice bath for 15 min and then centrifuged (1600*g*, 5 min, room temperature). A 100 μl of aliquot of the supernatant is diluted with 2.5 ml of distilled water, and the absorbance of this solution at 260 nm is measured. One unit of RNase activity is defined as the increase by 1.0 in the absorbance at 260 nm per 15 min.

Procedure for IEF in Thin Layer of Polyacrylamide Gel

Stock solution A contains 7.76% acrylamide (w/v), 0.22% Bis (w/v), and 33.3% urea (w/v), dissolved in distilled water, from which contaminant acrylic acid is removed by AG 501-X8 resin (Bio-Rad); *stock solution B* contains 0.002% FMN (w/v) in distilled water. These stock solutions are stored at 4°. Sheets of polyacrylamide gel measuring 0.5 (thickness) × 90 (width) × 120 (length) mm used for IEF–PAGE are prepared as follows: 4.4 ml of stock solution A, 280 μl of Pharmalyte 3–10 or 2.5–5 (Amersham Pharmacia Biotech, Uppsala, Sweden), and 480 μl of 0.5% ammonium persulfate (w/v) dissolved in stock solution B are mixed. In addition, sheets of polyacrylamide gel for IEF–PAGE without urea are prepared using the following materials. *Monomer solution* is prepared from 19.4% acrylamide (w/v) and 0.6% Bis (w/v) in distilled water and treated with AG 501-X8 resin. *Sucrose–glycerol solution* consists of 20% sucrose (w/v) and 10% glycerol (v/v). Monomer solution, 1.4 ml, 2.3 ml of sucrose–glycerol solution, 1 ml of distilled water, 280 μl of Pharmalyte 3–10, 5 μl of Temed, and 40 μl of 1.2% freshly prepared ammonium persulfate (w/v) are mixed. After thorough deaeration, the mixture is poured into a gel mold and polymerized. Wicks are formed from strips of filter paper (3MM, Whatman, Clifton, NJ) and soaked in the electrode solution: 0.04 *M* aspartic acid at the anode and 1.0 *M* NaOH at the cathode for Pharmalyte 3–10 gel, and 0.1 *M* H_2SO_4 at the anode and 0.1 *M* NaOH at the cathode for Pharmalyte 2.5–5 gel. The position of sample application on the gel is known to be very important for good focusing and high resolution of most proteins. Therefore, the position where the samples are applied to the gel should be examined in advance for each of the RNases of interest or kinds of carrier ampholyte incorporated into the gel; 2 μl of the sample solution is applied to the gel at a distance of 3.0 cm and 2.0 cm from the cathode wick for Pharmalyte 3–10 and Pharmalyte 2.5–5 gels, respectively, using an IEF/SDS sample application strip (Amersham Pharmacia Biotech).

[24] T. Uchida and F. Egami, *in* "Procedures in Nucleic Acid Research" (G. L. Cantoni and D. R. Davis, eds.), pp. 46–55. Harper & Row, New York/London, 1966.

In preliminary experiments, we determined the optimum IEF–PAGE conditions for equilibrium focusing. Three kinds of gel were examined: Pharmalyte 3–10 with and without urea and Pharmalyte 2.5–5 with urea. These were run using a Multiphor apparatus (Amersham Pharmacia Biotech) at 5 W under cooling at 15° for different electrophoresis times: 4, 6, and 12 h, respectively. After the IEF–PAGE run, RNase activities were detected *in situ* by the zymogram method described below, and the pH gradient in the gel was determined from the gel pieces simultaneously. For each of the gels, both the migration position of each RNase component and the pH gradient formed in the gel remained almost unchanged, irrespective of the time required for electrophoresis. These results indicated that 4 h of electrophoresis is sufficient for equilibrium focusing of the RNase components.

Procedure for Detection of RNase Activity on Gel by DAFO Zymogram Method

The reaction mixture consists of 0.05 mg of ethidium bromide and 0.8 mg of *Torula* yeast RNA per 1 ml of 0.1 *M* Tris-HCl buffer, pH 7.5, containing 20 m*M* EDTA. The thin agarose film is prepared as follows: To the reaction mixture warmed at 55°, an equal volume of 2% agarose GP-36 (w/v) in distilled water, melted in advance in a microwave oven, is added, mixed, and poured immediately onto a horizontal Agafix sheet (warmed to 55°) to a 2 mm thickness. Another type of supporting sheet for agarose gel electrophoresis, such as a GelBond film (Amersham Pharmacia Biotech), could also be used in the same manner. About 5 ml of the mixed solution is sufficient to prepare an agarose film measuring 5.0 × 7.5 cm. After solidification at room temperature, the agarose gel is dried completely in an incubator at about 55° to produce a dried agarose film, which is stored in a dark place prior to use. After the IEF–PAGE run, the dried agarose film, which is previously cut with scissors to an appropriate size, is placed carefully on top of the gel in full contact, without any air bubbles. For this method, the step involving washing off the carrier ampholyte is omitted, because the ampholyte has been found not to hinder RNase action. The focused gel is incubated in contact with the film at 37° in the moisture chamber, and the progress of RNase action is monitored under ultraviolet (UV) light (312 nm). After incubation for optimal development, the film is removed from the gel. Normally, dark bands corresponding to the RNases are observed clearly on the surface of the focused gel within 5–15 min. The bands are recorded photographically, if necessary.

In Situ Detection of RNase Activity Following IEF–PAGE

In preliminary experiments, we surveyed the best matching of RNA as a substrate and fluorescent dye for fluorogenic *in situ* detection of RNase: Three kinds of RNA, derived from *Torula* yeast, baker's yeast, or calf liver RNA, and six dyes (acridine orange, acridine yellow, bisbenzimide, ethidium bromide, propidium iodide, and quinacrine mustard), all of which are available commercially, were examined.

Ethidium bromide had several advantages: low fluorescence derived from free dye in the absence of RNA, an excitation wavelength of the dye–RNA complex convenient for a standard transilluminator, and an emission wavelength appropriate for observation. Calf liver RNA provided less intense fluorescence than other RNAs under experimental conditions. Therefore, since a combination of *Torula* yeast RNA and ethidium bromide as a substrate and fluorescent dye, respectively, gave best results, we selected this set to be used for the new zymogram method.

Although IEF–PAGE has been employed previously to demonstrate the multiplicity of RNases,[16,17] activity has been detected by scraping off the IEF–PAGE gel, eluting the enzyme, and assaying the resulting fractions. This approach may be rapidly superseded by our direct detection of individual RNase components, or isoenzymes, *in situ* in the IEF–PAGE gel. Bovine pancreatic RNases A and B and the purified human urinary pancreatic-type RNase were applied to this zymogram method following the Pharmalyte 3–10 gel without urea (Fig. 1a). Multiple forms

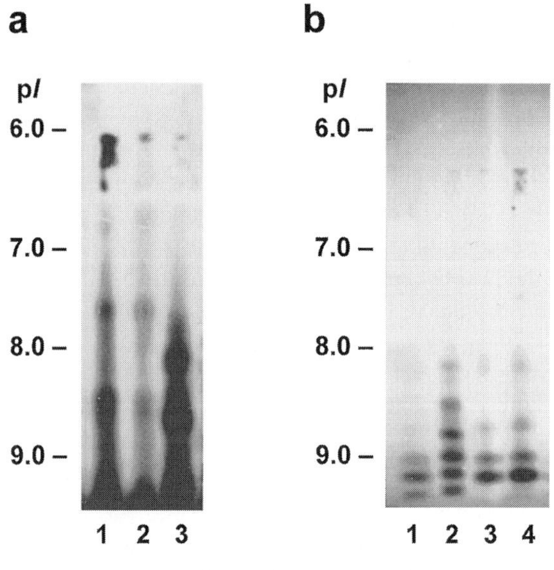

FIG. 1. Isoelectric focusing patterns of purified mammalian RNases revealed by the DAFO zymogram method. IEF–PAGE was performed using polyacrylamide gel with Pharmalyte 3–10 as the carrier ampholyte, (a) without addition of urea, and (b) with addition of urea to a final concentration of 4.8 *M*. (a) Two microliters of each purified enzyme (about 4 ng), bovine pancreatic RNase A (lane 1), and RNase B (lane 2) and human urinary pancreatic type RNase (lane 3), were applied to the gel. (b) Two microliters of each purified enzyme (about 1 ng), human urinary non-pancreatic-type RNase (lane 1), human urinary pancreatic type RNase (lane 2), and bovine pancreatic RNase A (lane 3) and RNase B (lane 4), were applied to the gel at a distance of 3.0 cm from the cathode wick. The zymogram method is described in the text. Addition of urea to the gel improves the separation and sharpness of the bands. Anode is at the top.

of these RNases are observed, but band resolution is poor, perhaps because the p*I*s of these RNases are too high for equilibrium focusing.

Since addition of urea to the IEF–PAGE gel resulted in wide separation of the genetically polymorphic glycoprotein, GP43, which is composed of several isoproteins with high p*I*s of around 9,[25] urea is incorporated into the IEF–PAGE gel. Inclusion of urea at a final concentration of 4.8 *M* improved the separation and sharpness of bands and provided the best results (Fig. 1b). Within 5–15 min of incubation, sharp dark bands corresponding to RNase activity on a fluorescent background of ethidium bromide bound to the unhydrolyzed RNA appear under UV light. Human urinary pancreatic-type RNase is resolved into four major and several minor components, whereas urinary non-pancreatic-type RNase is found to be composed of three components with similar activity. The zymogram patterns of the former are almost identical to its antigenic patterns detected by immunoblotting with anti-human pancreatic-type RNase antibody under the same electrophoresis conditions.[16] Bovine pancreatic RNases A and B, which are glycosylated and nonglycosylated forms, respectively, are both separated into one major and several minor bands migrating more anodally, and their patterns are similar to each other. These findings are consistent with the suggestion by Berman *et al.*[26] that the carbohydrate attachment does not affect the polypeptide structure of the enzyme. None of the human DNases I and II, phosphodiesterases I and II, or alkaline and acid phosphatases exhibited any band on the gel under our experimental conditions. Therefore, our DAFO detection method is specific for RNase activity.

Quantities as small as about 0.1 ng RNase A and 0.25 ng human pancreatic-type RNase, corresponding to 1.0×10^{-4} and 0.6×10^{-4} units, respectively, can be detected by the DAFO zymogram method within 15 min of incubation. If the incubation time is extended, then the sensitivity becomes higher. The minimum detection limit of the enzyme in a urea gel is about one-tenth of that in a normal gel without urea, proving that the presence of urea neither denatures the enzymes during electrophoresis nor interferes with their activity during incubation.

Detection of Microbial RNases and Analysis of RNase Activity in Crude Extracts using DAFO Method

In contrast to most mammalian RNases with relatively high p*I* values and pH optima around the neutral range,[27] several microbial RNases exhibit low p*I* values and/or optimal pH values far from the neutral pH range.[28] The DAFO zymogram method is employed to detect two typical microbial RNases, RNases

[25] K. Mizuta, T. Yasuda, and K. Kishi, *Biochem. Genet.* **27,** 731 (1989).

[26] E. Berman, D. E. Walters, and A. Allerhand, *J. Biol. Chem.* **256,** 3853 (1981).

[27] H. Sierakowska and D. Shugar, *Prog. Nucleic Acid Res. Mol. Biol.* **20,** 59 (1977).

[28] T. Uchida and F. Egami, *in* "The Enzymes" (P. D. Boyer, ed.), 3rd Ed., Vol. 4, pp. 205–250. Academic Press, New York, 1971.

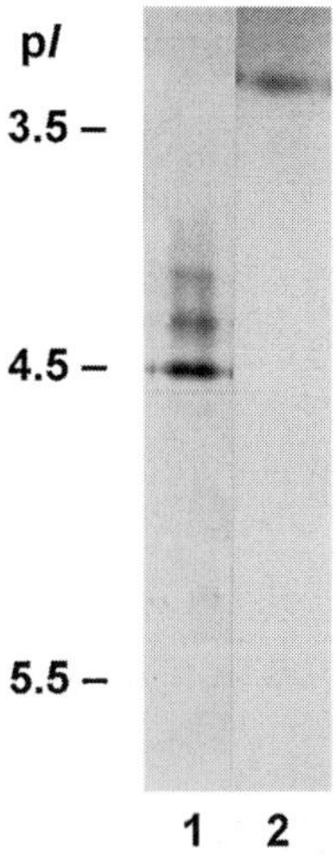

FIG. 2. IEF–PAGE patterns of RNase T_1 and T_2 detected in one gel by the DAFO zymogram. The RNase samples were obtained commercially from Sigma and used without further purification. IEF–PAGE was performed using Pharmalyte 2.5–5 and 4.8 *M* urea. Two microliters of the enzyme (about 10 ng and 100 ng of RNases T_1 and T_2, respectively) were applied to the gel at a distance of 2.0 cm from the cathode wick. A film containing 50 m*M* Tris-HCl buffer, pH 7.5 (lane 2) and one containing 50 m*M* sodium acetate buffer, pH 4.7 (lane 1) were used for detection of RNases T_1 and T_2, respectively. The DAFO zymogram method is described in the text. Lane 1, RNase T2; lane 2, RNase T1. Anode is at the top.

T_1 with a low p*I* (3.8), and T_2 with a low optimal pH of activity (pH 4.5), from *Aspergillus oryzae*. The carrier ampholyte is replaced with Pharmalyte 2.5–5, and the reaction buffer (Tris-HCl, pH 7.5) of the DAFO film is replaced with 50 m*M* acetate buffer, pH 4.7, for detection of RNase T_2. As shown in Fig. 2, these RNases, whose minimum detection units are as small as those of RNase A or human RNases, have distinct zymogram patterns. RNase T_1 shows only one major band on the gel, corresponding to the low pH region, whereas the latter is resolved into three bands. Our method thus has a great advantage over others in that several RNases with different pH optima can be detected easily and simultaneously on only one IEF–PAGE gel. This involves preparation of a detection-film strip, which contains a reaction buffer suitable for each RNase of interest, and then placing the strip on the region of the gel to which the enzyme has migrated. This application confirms a general use for the DAFO zymogram method for detection of RNase activity derived from various origins directly on the IEF–PAGE gel.

When each of the crude extracts prepared separately from bovine pancreas by three different extraction methods,[29] is separated by the pH 3–10 gel with urea,

[29] J. L. Weickmann, M. Elson, and D. G. Glitz, *Biochemistry* **20,** 1272 (1981).

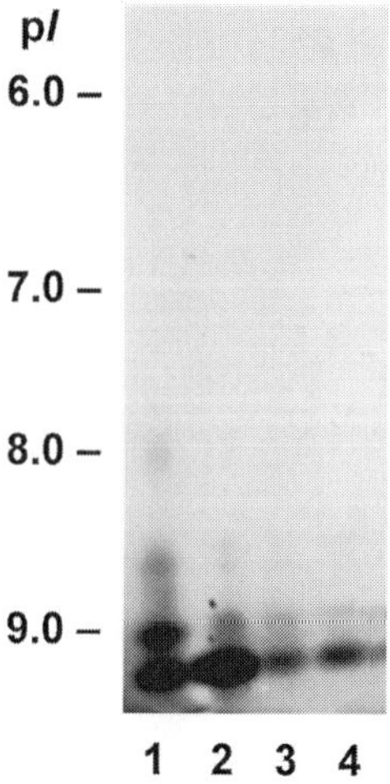

FIG. 3. IEF–PAGE patterns of bovine pancreatic RNases present in crude extracts revealed by the DAFO zymogram method. Five-microliter aliquots of each crude extract prepared from bovine pancreas containing about 0.05 unit of RNase activity were applied to the gel. The conditions of electrophoresis and the zymogram method are described in the text. Other materials present in the crude extracts produced little interference with the observation of RNase activity in the gel. Lane 1, authentic RNase A; lane 2, RNase in acetone precipitates of sulfuric acid extract; lane 3, extract with 0.1 *M* Tris-HCl buffer, pH 7.5; lane 4, extract with 0.25 *M* sulfuric acid.

followed by visualization using the DAFO method, the zymogram patterns of the extracts are very similar to one another, and the major band of each exhibits a very similar p*I* to authentic RNase A (Fig. 3). Less than about 0.5 μg wet weight of pancreas, corresponding to 1×10^{-4} units of activity, is needed to obtain good zymogram patterns. Furthermore, no other material coexisting with RNase in the crude extracts interferes with the observation of RNase activity in the gel. Thus, the DAFO zymogram method can be applied not only to purified enzymes but also to RNases in crude extracts containing other proteins.

Comments

Previously, Tournut *et al.*[30] employed the substrate films method, followed by toluidine blue staining, to detect RNase isoforms in human pancreatic juice subjected to IEF–PAGE. However, bands corresponding to RNase activity during development, accompanied by RNase action, could not appear until the final staining step, and it was difficult to terminate the reaction on the gel after an appropriate incubation time, resulting in broader or eventually fused bands. Our DAFO method, on the other hand, is convenient because the progress of the RNase action

[30] R. Tournut, B. J. Allan, and T. T. White, *Clin. Chim. Acta* **88,** 345 (1978).

can be monitored continuously during incubation. Furthermore, with our zymogram method, termination of the RNase action on the gel at the desired time for the best pattern simply requires removal of the film from the gel, because it seems likely that RNase action progresses at the contact surface of the film. Although it is generally recognized that activity staining requires a relatively large amount of the enzyme, the DAFO method has a higher sensitivity for purified human pancreatic-type RNase than immunological detection.[16] The high sensitivity and resolution of a combined technique of IEF–PAGE and the DAFO zymogram method are well suited for analysis of the various forms of RNases in small quantities, permitting the pinpointing of genetic variants and/or clinically abnormal RNases present in body fluids and tissues.

Specific Detection of Pancreatic-Type Ribonucleases Based on Polycytidylic Acid/Ethidium Bromide Fluorescence Following IEF

Although currently only five distinct proteins with RNase activity encoded by different genes have been identified in the human system, RNases present in human body fluids and/or tissues have been grouped into two broad classes, usually designated as pancreatic- and non-pancreatic-type RNases.[1] In biochemical, genetic, and clinical surveys, specific determination and characterization of each kind of RNase activity would facilitate precise and reliable evaluation of RNase activity present in human tissues and body fluids. Immunological methods, such as an immunoassay and immunoblotting using a specific antibody against a particular enzyme, may be suitable for this purpose; however, such antibodies are scarce and difficult to obtain. With regard to their catalytic properties, pancreatic-type RNase degrades poly(C) markedly faster than RNA and poly(U), under weakly alkaline conditions,[31] in contrast to non-pancreatic-type RNase: Human pancreatic-type RNase purified from kidney cleaves poly(C) about 1000 times faster than non-pancreatic-type RNase present in the same tissue.[7] Hence, since the remaining types of RNase are even less potent than the non-pancreatic type, only the pancreatic-type RNase, of the known human RNases, degrades poly(C) at an appreciable rate.

The zymogram method preferred for pancreatic-type RNase following IEF–PAGE, which is based on polycytidylic acid/ethidium bromide fluorescence, have been developed.[32]

Detection Procedure for Pancreatic-Type RNases by Modified DAFO Zymogram Method Following IEF–PAGE

Poly(C) (sodium salt, $s^0_{20,w} = 8.5–9.4$) is obtained from Seikagaku Kogyo (Tokyo, Japan): Poly(C) is dissolved in distilled water at 3 mg/ml, dispensed into

[31] S. Sorrentino and M. Libonati, *Arch. Biochem. Biophys.* **312,** 340 (1994).

[32] D. Nadano, T. Yasuda, K. Sawazaki, H. Takeshita, and K. Kishi, *Anal. Biochem.* **212,** 111 (1993).

aliquots, and stored at -20° until use. The reaction mixture is prepared as follows: 1.5 ml of reaction buffer (0.2 *M* Tris-HCl buffer, pH 7.5, containing 40 m*M* EDTA), 1.0 ml of poly(C) solution, 15 μl of 1% ethidium bromide (w/v), and 0.5 ml distilled water are placed in a test tube and maintained at 55° for about 5 min (the addition of ethidium bromide at this step is necessary to obtain a high sensitivity). To the reaction mixture, an equal volume of 2% molten agarose GP-36 (w/v) is added and the mixture is immediately poured onto a horizontal Agafix MSL sheet or GelBond film warmed at 55° to about 2 mm thick. After solidification at room temperature, the agarose gel is dried completely at 55° to make a thin film and stored at room temperature in a dark place until use. The final volume (6 ml) is sufficient to prepare a thin agarose film measuring 7.5×7.5 cm.

After the IEF–PAGE is run as described above, the dried film sheet is placed carefully on the top of the focused gel, in full contact. The step that washes off the carrier ampholyte from the focused gel is omitted because the ampholyte does not hinder RNase action against poly(C) substrate, as described above. The gel with the film sheet is incubated for 5–30 min (routinely 15 min) at 37° in a moisture chamber. Under weakly alkaline reaction conditions, the complex formed between ethidium bromide and unhydrolyzed poly(C) produces a weak fluorescence, so that the degradation of poly(C) by RNase action is difficult to detect, in contrast to the ordinary DAFO technique using RNA as a substrate. The conformation of poly(C) and the interaction between ethidium bromide and poly(C) are known to change under various conditions. The film sheet is removed from the gel and soaked in 20 m*M* sodium acetate buffer, pH 5.0, which contains 20 μg/ml ethidium bromide, and is chilled on ice for about 15 min without shaking. Shaking the film sheet in the acidic buffer should be avoided in order to minimize any loss of fluorescent complex from the sheet. Dark bands corresponding to pancreatic-type RNase activity appear on a fluorescent background under UV (312 nm) light. The step of immersing the film sheet after RNase digestion in a cooled acid solution (pH 5.0) improves the intensity of the fluorescence derived from the complex. If the film sheet is dried at room temperature, it will be possible to preserve the zymogram patterns for at least 3 months.

Pancreatic-Type RNase-Specific Detection using Modified DAFO Zymogram Method Following IEF–PAGE

A combined technique of IEF–PAGE and zymogram detection by the modified DAFO method using poly(C) as a substrate can be employed to analyze mammalian pancreatic-type RNases. The zymogram patterns of human urinary pancreatic-type RNase thus obtained are shown in Fig. 4a. On the film sheet, less than 3×10^{-4} units of purified enzyme, corresponding to 0.3 ng, exhibits a distinct zymogram pattern. In contrast to the original DAFO method, the band patterns of the enzyme are not observed on the surface of the focused gel, even after immersion of the

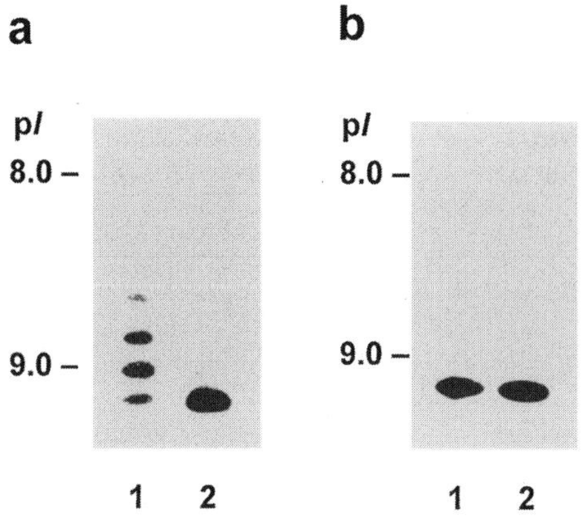

FIG. 4. IEF–PAGE patterns of (a) human and (b) bovine RNases revealed by the modified DAFO method using poly(C) as a substrate. IEF–PAGE was performed using Pharmalyte 3–10 as the carrier ampholyte with addition of urea to a final concentration of 4.8 *M*. The substrate incorporated into the DAFO film was poly(C). The conditions of IEF–PAGE and the modified DAFO method are described in the text. (a) Human pancreatic-type RNase (2×10^{-3} unit, lanes 1 and 2) purified from urine was applied to the gel. Non-pancreatic-type RNase with less preference for poly(C) could not be detected under these conditions, indicating high specificity of the modified DAFO method for pancreatic-type RNase. Prior to electrophoresis, 2×10^{-3} unit of purified human pancreatic-type RNase was treated with 1×10^{-2} unit of sialidase at 20° for 6 h, and the enzyme (lane 1) without and (lane 2) with sialidase treatment was analyzed. (b) Bovine RNase A (2×10^{-3} unit, lane 1) and crude extract from bovine pancreas (2×10^{-3} unit, lane 2) were examined. The purified enzyme gave the same band patterns as those of the crude extract. Anode is at the top.

gel in the cooled acidic buffer. The zymogram patterns of the urinary pancreatic-type RNase are almost identical to those visualized by the original DAFO method using RNA as a substrate, and to the antigenic pattern revealed by immunoblotting with specific antibody under the same IEF–PAGE conditions. Sialidase treatment of the urinary pancreatic type enzyme, which contains five sialic acid residues per molecule,[33] diminishes the anodal bands and induces a parallel increase in the staining intensity of the higher p*I* band (Fig. 4a). These findings indicate that the multiple forms of the enzyme are primarily due to variation in sialic acid content. Thus, the simple and practical combination of the IEF–PAGE and DAFO methods allows us to show the contribution of sialic acid residues to the multiplicity of human pancreatic-type RNase. Human urinary non-pancreatic RNase (up to 5×10^{-2} units, corresponding to 100 ng) and phosphodiesterases I and II (up to 4 μg) cannot be detected as a distinct band on the film sheet. Also, bovine pancreatic RNase A is easily detectable in the same manner: Figure 4b shows the zymogram patterns of the purified enzyme and crude extracts of bovine pancreas.

[33] K. Mizuta, T. Yasuda, Y. Ikehara, W. Sato, and K. Kishi, *Int. J. Legal Med.* **103,** 315 (1990).

RNase activity in the crude extract exhibits the same band patterns as that of the purified enzyme. Therefore, it is obvious that this modified DAFO method may be applicable to detection of pancreatic-type RNases in crude extracts as well as in the purified form after IEF–PAGE.

Comments

At first, we developed the DAFO zymogram method using yeast RNA as a substrate for detection of various RNases following IEF–PAGE, as described above. However, since RNA has been incorporated into the gel, both pancreatic- and non-pancreatic-type RNases have been observed simultaneously, irrespective of their different band patterns. For specific detection and visualization of pancreatic-type RNase alone, it is necessary for the DAFO film to include poly(C) as a substrate and to be soaked in a cooled acidic solution after incubation with the focused gels. Furthermore, the activity of types of RNase distributed in most human tissues and body fluids[32,34] can be differentiated by the modified DAFO method because of its high sensitivity.

pH Gradient Electrophoresis of Basic Ribonucleases in Sealed Slab Polyacrylamide Gels Followed by DAFO Zymogram Detection

Mammalian RNases, including pancreatic- and non-pancreatic-type enzymes, are characterized by high p*I*, which has been considered to be an important property of these RNases, perhaps influencing their catalytic properties and biological functions.[35] However, their high p*I* properties generally prevent a finer resolution of the enzyme on the IEF, mainly owing to methodological problems such as the cathodal shift phenomenon, which has a troublesome effect on analysis of proteins with high p*I* value, such as mammalian RNases. Although the carrier ampholytes, such as Ampholine 9–11 and Pharmalyte 8–10.5, aimed at analysis of basic proteins, are available commercially, this phenomenon seems to be inherent in IEF using carrier ampholytes. Accordingly, in order to address this problem, we have devised a method of pH gradient electrophoresis in a sealed slab polyacrylamide gel to analyze mammalian basic RNases.

Procedure for pH Gradient Electrophoresis in Sealed Slab Polyacrylamide Gels

Stock solution A is composed of 28.8% acrylamide (w/v) and 1.2% Bis (w/v), dissolved in distilled water, from which contaminant acrylic acid has been removed

[34] J. Futami, Y. Tsushima, Y. Murato, H. Tada, J. Sasaki, M. Seno, and H. Yamada, *DNA Cell Biol.* **16,** 413 (1997).

[35] M. Libonati and S. Sorrentino, *Mol. Cell. Biochem.* **117,** 139 (1992).

by AG 501-X8 resin; *stock solutions B* and *C* contain 50% glycerol (v/v) and 40 mg/ml FMN (w/v), respectively, in distilled water. These stock solutions are stored at 4°. A 7.5% T polyacrylamide gel measuring 1.0 (thickness) × 87 (width) × 78 (length) mm is prepared as follows: The mixture comprises 2.5 ml of solution A, 3.0 ml of solution B, 1.8 ml of solution C and 2.0 ml of distilled water. Next, N_2 gas is passed through the mixture, in order to purge other gases, particularly CO_2, from the mixture. Then, 70 mg L-arginine and 670 μl of Pharmalyte 8–10.5 are added to the mixture. After thorough deaeration, 35 μl of freshly prepared 0.1 g/ml ammonium persulfate is added, and the mixture is poured into the slab gel mold, which is composed of two glass plates and a 1.0 mm thick spacer (Nihon Eido, Tokyo, Japan). Sample wells are formed by insertion of a well comb into the top of the polyacrylamide gel, and the gel is polymerized for at least 45 min under a fluorescent lamp. The comb and spacer are then removed, and the gel plate is placed in a vertical slab gel cell (Nihon Eido). The electrode solutions are 9.5 mg/ml HEPES dissolved in distilled water, deaerated prior to use, for the upper anode reservoir and 0.2 *M* NaOH for the lower cathode reservoir. Sample solutions are mixed with an equal volume of 25% glycerol (v/v) solution, containing 20 m*M* HEPES. The gels are run at room temperature at 150 V for 1 h followed by 300 V for 3 h (total 1050 Vh), without prefocusing.

pH Gradient Electrophoresis of Basic RNases and Proteins in Sealed Slab Polyacrylamide Gels

When analyzed by pH gradient electrophoresis in sealed slab polyacrylamide gels using Pharmalyte 8–10.5, bovine pancreatic RNase A migrates toward the cathode and is detected at a different position from human non-pancreatic-type RNase purified from spleen, known to be one of the most basic RNases (Fig. 5a). However, non-pancreatic-type RNase and lysozyme accumulate near and migrate into the most cathodal region and cannot be separated. Even substitution of carrier ampholyte by Ampholine 9–11 gives similar band patterns. In order to improve the situation, L-arginine with a p*I* of 10.8 is added to the gel mixture. The presence of arginine at a concentration of 7.0 mg/ml allows basic proteins with p*I* values above 8, including human non-pancreatic-type RNase and lysozyme, to be clearly resolved (Fig. 5b). As shown in Fig. 6, the pH gradient formed during electrophoresis, especially in the region of higher pH, is improved by addition of arginine. We previously found 2000 Vh to be necessary to reach equilibrium focusing in the sealed gel electrophoresis system.[37] However, prolonged electrophoresis with more than 1500 Vh causes the bands to migrate toward the cathode, resulting in distorted and fuzzy band patterns, probably owing to the gradual shift of arginine and carrier ampholyte to the cathode. So, the sealed slab gel should be run

[36] R. W. Blakesley and J. A. Boezi, *Anal. Biochem.* **82,** 580 (1977).
[37] D. Nadano, T. Yasuda, H. Takeshita, and K. Kishi, *Anal. Biochem.* **227,** 210 (1995).

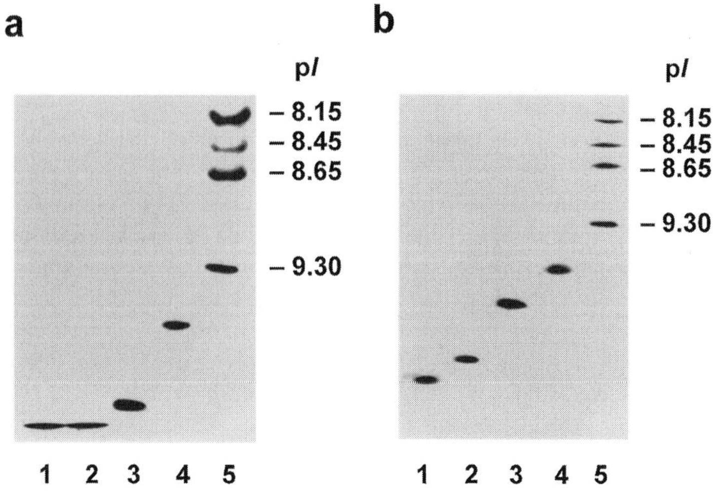

FIG. 5. pH Gradient electrophoresis of basic RNases and proteins followed by protein staining. Electrophoresis was carried out in a sealed slab polyacrylamide gel in the presence of Pharmalyte 8–10.5 (a) without and (b) with L-arginine. Each sample (about 2 μg) was electrophoresed and visualized by protein staining. For protein staining, the gel after electrophoresis was directly immersed in the Serva Blue G solution in 0.12 g/ml trichloroacetic acid, prepared according to the method of Blakesley and Boezi,[36] and kept at 50° for 2 h, or at 37° overnight. Subsequently, the gel was rinsed in distilled water for color intensification. This staining method does not require fixation or the time-consuming destaining step, and is as sensitive as protein staining using Coomassie Brilliant Blue after the pH gradient electrophoresis. The conditions of electrophoresis are described in the text. Lane 1, chicken egg white lysozyme; lane 2, purified human spleen non-pancreatic-type RNase; lane 3, bovine heart cytochrome *c*; lane 4, bovine pancreatic RNase A; lane 5, p*I* calibration kit. Anode is at the top.

with less than 1500 Vh in this system. Furthermore, in order to reduce the cathodal shift phenomenon, 0.2 M NaOH is indispensable for the cathodal solution: Neither 10 m*M* ethylenediamine, 2% (v/v) *N,N,N′,N′*-tetramethylethylenediamine, nor 30 m*M* NaOH is suitable. A 7.5% T polyacrylamide gel is used in this system in order to achieve higher mechanical stability of the polyacrylamide gel. The band patterns of several proteins analyzed in this are almost same as those of a 5% T gel under the same electrophoresis conditions. Therefore, the molecular sieving effect is considered insignificant under the conditions described above.

Pharmalyte 8–10.5 and Ampholine 9–11 are employed for analysis of basic proteins to generate an expanded basic pH gradient. However, carbon dioxide (CO_2) has a detrimental influence upon basic pH gradient, because atmospheric CO_2 is readily absorbed and retained by the mixture of the basic carrier ampholyte. Usually, analytical IEF–PAGE is performed on a horizontal flatbed system with continuous exposure of the gel surface to the open air. In an attempt to minimize such disturbance, IEF–PAGE is performed in a conventional vertical slab gel so that

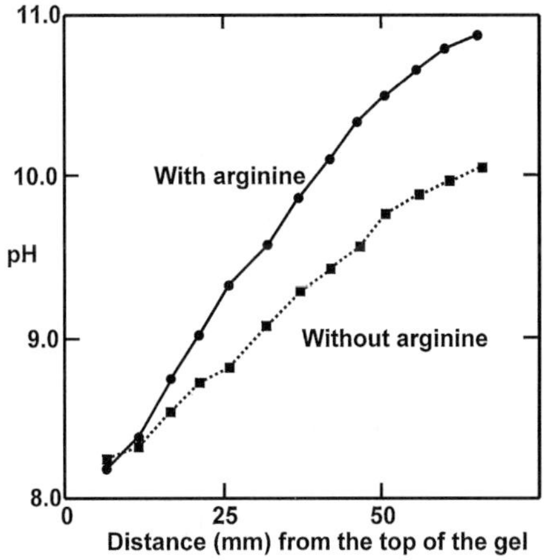

FIG. 6. pH Gradients formed in sealed slab polyacrylamide gels. Electrophoresis using Pharmalyte 8–10.5 with (●) and without (■) L-arginine was performed under the conditions described in the text. The generated pH gradient was determined from the gel piece, excised 5 mm from one side of the slab gel using a razor blade, transferred to polypropylene tubes, and soaked in 0.4 ml distilled water, which was made CO_2-free by extensive bubbling with N_2 gas. After the soaked gel pieces were crushed with the aid of a polypropylene stick, the pH of the resulting solution was measured. The abscissa represents the distance from the bottom of the sample well.

all six sides of the gel are enclosed by glass plates or electrode solutions, and any adverse effect of the atmospheric components on electrophoresis is reduced, resulting in sealing of the gel from the open air during the electrophoretic run. Originally, we had developed this IEF–PAGE method using a sealed slab gel as a continuous dithiothreitol (DTT)-supplying system for investigating sulfhydryl-dependent inhibitory activity of human RNase inhibitor in tissues.[37] The good focused patterns of basic RNases and other proteins may be attributed to the combined use of arginine incorporated in the gel as well as to the sealed gel system. The immobilized pH gradient technique was reported to be more suitable for determining the precise p*I* of a particular basic RNase.[38] However, this approach requires more expensive equipment and cumbersome techniques and has not yet been applied to RNase activity staining. The electrophoresis apparatus and power supply used in this system have been used routinely for ordinary SDS–PAGE on our laboratory.

Detection of RNase Activity by DAFO Zymogram Method after Sealed Slab Polyacrylamide Gel Electrophoresis

Dried agarose film sheets using *Torula* yeast RNA or poly(C) as a substrate are prepared as described above. After the electrophoretic run, the gel plate is

[38] E. C. Coronel, B. W. Little, and J. A. Alhadeff, *Biochem. J.* **296,** 553 (1993).

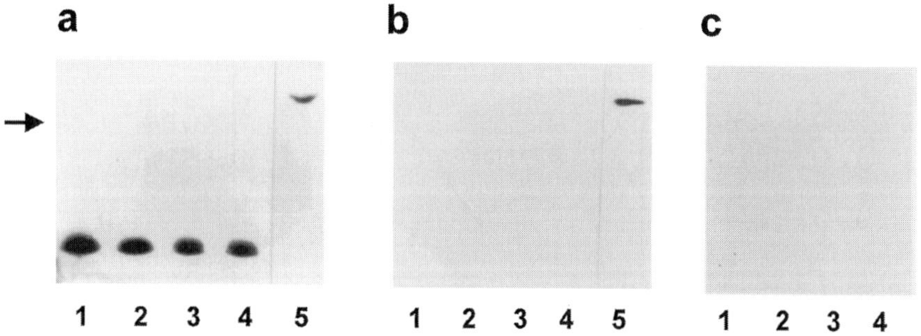

FIG. 7. pH Gradient electrophoresis in sealed slab polyacrylamide gels followed by detection (a, b) and inhibition (c) of RNase activity using the DAFO zymogram method. The zymogram patterns were obtained using a DAFO sheet, containing (a, c) yeast RNA and (b) poly(C) as a substrate. About 0.01 unit of RNase activity was applied to each sample well. Electrophoresis, activity staining, and the inhibition test using RNase inhibitor are described in the text. Lane 1, purified human spleen RNase; lane 2, crude extract of human spleen; lane 3, human granulocyte lysate; lane 4, human mononuclear leukocyte lysate; lane 5, human platelet lysate. Arrow indicates the migration position of bovine pancreatic RNase A used as a marker enzyme. Anode is at the top.

detached from the vertical gel cell, and one of the two glass plates is removed from the gel. The DAFO sheet is carefully overlaid on the open surface of the gel and then incubated at 37° routinely for 15 min. In order to terminate the progress of the RNase action, the DAFO sheet is then removed from the slab gel. When the DAFO sheet containing yeast RNA as a substrate is used, dark bands corresponding to RNase activity are observed on a fluorescent background in both the film and the polyacrylamide gel under UV (312 nm) light. On the other hand, the removed DAFO sheet containing poly(C) is soaked in 20 m*M* sodium acetate buffer, pH 5.0, which contains 20 μg/ml ethidium bromide, and is chilled on ice for 20–30 min without shaking. Dark bands corresponding to poly(C)-preferring RNase activity appear in the film under UV light.

Less than 0.5 ng of the non-pancreatic-type RNase purified from human spleen, corresponding to 2×10^{-3} units of activity, can be detected within 15 min of incubation (Fig. 7a). Although longer incubation allows sensitivity to become even higher, the band resolution tends to be poor, probably because of the high diffusion rate of the spleen RNase, having a lower molecular mass (about 17 kDa).[5] Therefore, longer incubation of the DAFO sheet in contact with the gel should be avoided. Also, replacement of Pharmalyte 8.5–10 by Ampholine 9–11 elevates the detection limit of RNase activity by a factor of 4, irrespective of the similar zymogram pattern, because of a relatively strong inhibitory effect of the Ampholine on RNase activity in the gel. In the case that the RNase activity staining is planned to follow IEF–PAGE, carrier ampholytes must be chosen carefully so as to obtain the highest sensitivity.

The crude extract of human spleen exhibits the same zymogram pattern as that of the purified enzyme (Fig. 7a), indicating that activity staining after sealed slab gel electrophoresis may be applicable to the specific detection of basic RNases in heterogeneous biological materials. As an application, cell lysates fractionated from human peripheral blood are subjected to electric focusing followed by the DAFO zymogram method. Since human platelets have been shown to contain an excess of placental-type RNase inhibitor over that of RNase,[39] the platelet lysates should be treated with 1 m*M* *p*-hydroxymercuribenzoate at 37° for 10 min to release active RNase from inhibitor before sample loading. As shown in Fig. 7, the zymogram pattern for platelets is different from those for granulocytes and mononuclear leukocytes. With poly(C) as a substrate in the DAFO film, the major band of platelet RNase is clearly visible within 15 min of incubation time, whereas RNases from spleen, granulocytes, and mononuclear leukocytes were scarcely observed even after 90 min. Based on these resulting zymogram patterns, the major basic RNase in mononuclear leukocytes and granulocytes appears to be non-pancreatic-type RNase, whereas that in platelets is pancreatic-type RNase. These findings indicate that a combined technique of pH gradient electrophoresis in sealed slab polyacrylamide gel and the DAFO zymogram method would be useful in the survey and characterization of basic RNases distributed in mammalian tissues.

Inhibition Test of RNase Activity by Human Placental-Type RNase Inhibitor on Sealed Slab Polyacrylamide Gel

For a classification of RNases it is necessary to examine the inhibitors and enhancers of enzyme activity as well as substrate specificity. Although small, heat-stable materials such as divalent cations can be added during preparation of the DAFO sheet, another technique will be indispensable for proteins. Human placental-type RNase inhibitor is a heat-labile 50 kDa protein that inhibits the activity of basic RNases, including angiogenin.[40] The inhibition of RNase activity by RNase inhibitor in the gel could be examined by a DAFO method with some modifications.

After the electrophoretic run in a sealed slab polyacrylamide gel as described above, the gel is removed from the gel slab mold, shaken gently for 10 min at room temperature in 0.1 *M* Tris-HCl buffer, pH 7.5, which contains 15% glycerol (v/v), 3 m*M* DTT, and 1 m*M* EDTA, and placed on a thin glass plate. Any remaining Tris-HCl buffer is removed from the top surface of the gel by brief blotting with a Kimwipe (Jujo-Kimberly, Tokyo, Japan). Using a micropipet, purified human

[39] D. Nadano, T. Yasuda, H. Takeshita, and K. Kishi, *Int. J. Biochem. Cell Biol.* **27,** 971 (1995).
[40] D. Nadano, T. Yasuda, H. Takeshita, K. Uchide, and K. Kishi, *Arch. Biochem. Biophys.* **312,** 421 (1994).

placental RNase inhibitor, dissolved in 20 mM Tris-HCl buffer, pH 7.5, containing 15% glycerol (v/v), 2 mM DTT, and 1 mM EDTA, is loaded as small drops onto many points on the agarose film side of the DAFO sheet (up to 500 ng per 1 cm^2 of the sheet), for uniform distribution of RNase inhibitor. Less than 5 μl inhibitor solution per 1 cm^2 of the DAFO sheet should be applied, because the excess solution spreads out from the DAFO sheet when the sheet is overlaid on the gel. The agarose film side of the DAFO sheet is then placed carefully on top of the gel slab in full contact, avoiding air bubbles. The assembly is incubated at 37° for 15 min, and visualization is performed as described above.

The inhibitor solution forms a layer <0.05 mm thick when the sheet is placed on the gel and is absorbed immediately into the dried agarose film and polyacrylamide gel, inhibiting the RNase activity. On application of 150 ng RNase inhibitor per 1 cm^2 of the DAFO sheet, none of the RNase activity (about 0.01 unit per samples) derived from purified non-pancreatic-type RNase and cell lysates from granulocytes, mononuclear leukocytes, and platelets can be visualized (Fig. 7c). In control experiments with buffer without inhibitor protein, the band patterns are similar to those shown in Fig. 7a.

This type of inhibitor is known to be widely distributed in other mammalian cells and to show tight 1:1 binding to pancreatic- and non-pancreatic-type RNases on a molar basis.[40] It has been reported that RNase inhibitor may have a different inhibitory effect on different RNases, some basic RNases being completely resistant to inhibition.[41] Therefore, this inhibition test of RNase activity by RNase inhibitor protein in the gel appears useful for characterization of RNases present in heterogeneous biological samples. Furthermore, the methodology of the inhibition test based on the DAFO zymogram method is potentially also applicable to other inhibitory proteins, such as actin, which inhibits bovine seminal RNase activity.

Acknowledgments

We are grateful to Dr. Daita Nadano, Molecular Oncology Laboratory, RIKEN, for his continuous contribution to this study. This work was supported in part by Grants-in-Aid from the Ministry of Education, Science, Sports and Culture of Japan (09357004, 12307011 to K. K., 12357003 to TY and 12770216 to H.T.).

[41] D. L. Newton, S. Walbridge, S. M. Mikulski, W. Ardelt, K. Shogen, S. J. Ackerman, S. M. Rybak, and R. J. Youle, *J. Neurosci.* **14,** 538 (1994).

[7] Gel Renaturation Assay for Ribonucleases

By CHRISTIAN CAZENAVE and JEAN-JACQUES TOULMÉ

Introduction

In situ detection of enzymatic activities on electrophoretic gels combines the resolving power of electrophoresis with the sensitivity and the specificity of enzymatic testing. This technique is popular as it allows the simultaneous identification and analysis of several isozymes even in crude cell homogenates, complex biological fluids, or subcellular fractions, opening the way to phylogenetic studies as well as compartmentalization or physiological studies. It can be useful for monitoring a purification procedure or for checking the expression of a recombinant enzyme, for instance. Also, the comparison of the mobility of active polypeptides prior to and after treatment with enzymes that eliminate posttranslational modifications can give access to the detection of such modifications. Published procedures for a large number of enzymes mostly use electrophoretic conditions that preserve the native conformation of the enzyme.[1,2] However, separation of proteins by sodium dodecyl sulfate–polyacrylamide gel electrophoresis (SDS–PAGE) can also be used if a suitable procedure for renaturing the enzyme in the gel can be tailored.[3]

In situ detection of enzymatic activity by SDS–PAGE is frequently referred to as an activity gel, or alternatively as a renaturation gel assay: electrophoresis is carried out in denaturing conditions on a gel in which the macromolecular substrate has been embedded. At the end of the migration, the proteins are renatured *in situ* and the gel incubated under conditions that allow the enzyme to act on the substrate. Therefore, such a procedure can be used as long as (i) the substrate can be immobilized in the gel, (ii) the SDS denatured enzyme can be renatured, and (iii) the enzymatic reaction can be visualized. This technique has been applied to many different ribonucleases but has been, from the beginning, particularly popular among research groups studying ribonucleases H. We give below a representative, rather than exhaustive, list of ribonuclease activities that have been visualized using activity gels. This includes RNases H from various sources [*Escherichia coli* RNase H,[4,5] yeast RNase H,[6,7] eukaryotic RNases H,[8,9] RNase H associated with murine

[1] G. M. Rothe, "Electrophoresis of Enzymes: Laboratory Methods." Springer-Verlag, Berlin and Heidelberg, 1994.

[2] T. Yasuda, H. Takeshita, and K. Kishi, *Methods in Enzymol.* **341,** [7] 2001 (this volume).

[3] A. Spanos and U. Hubscher, *Methods in Enzymol.* Vol. 91, p. 263 (1983).

[4] P. L. Carl, L. Bloom, and R. J. Crouch, *J. Bacteriol.* **144,** 28 (1980).

[5] S. Kanaya and R. J. Crouch, *J. Biol. Chem.* **258,** 1276 (1983).

[6] J. Huet, J. M. Buhler, A. Sentenac, and P. Fromageot, *J. Biol. Chem.* **252,** 8848 (1977).

[7] F. Iborra, J. Huet, B. Breant, A. Sentenac, and P. Fromageot, *J. Biol. Chem.* **254,** 10920 (1979).

0076-6879/01 $35.00

sarcoma and leukemia virus (MSV–MuLV) reverse transcriptase,[10] RNase H associated with human immunodeficiency virus (HIV) reverse transcriptase,[11,12] and RNase H activity associated with hepadnaviruses[13]], as well as RNases from different species and tissues (RNases from human body fluids,[14–16] RNase from *Xenopus* liver involved in the selective destabilization of albumin mRNA,[17] RNase from *Xenopus* oocyte,[18] RNase from pollen,[19] nucleases of tobacco leaves,[20] RNases from developing tomato fruit,[21] and RNases of *Dictyostelium discoideum*[22]).

Advantages and Limitations of Activity Gels Applied to Ribonucleases

In contrast to other electrophoretic assays, the gel renaturation assay offers a key advantage in permitting the determination of the molecular mass of the active polypeptide(s) by comparison of their electrophoretic mobility with those of protein standards of known molecular mass. This has been done to study the glycosylation pattern of several human RNases.[16] For RNases, the advantages of SDS–PAGE over native gels have been listed by Blank and Dekker[14]: prevention of aggregation problems observed, for example, with serum RNases, disruption of noncovalent complexes of RNases, including suspected RNase inhibitors, anodal migration of both acidic and basic RNases, determination of molecular masses, examination of catalytic properties, and possible assessment of the influence of various physical and chemical treatments, without separation of RNases from one another or from the bulk of contaminating proteins. Moreover, denaturing conditions prevent interactions between the protein and the embedded substrate so that electrophoretic mobilities are not modified. Attempts performed with native gels may lead to alterations in the electrophoretic mobility of the protein, which can be prevented by adding polyamines such as spermine in the gel.[23] Indeed, the

[8] G. Cathala, J. Rech, J. Huet, and P. Jeanteur, *J. Biol. Chem.* **254,** 7353 (1979).
[9] Y. W. Rong and P. L. Carl, *Biochemistry* **29,** 383 (1990).
[10] M. Rucheton, M. N. Lelay, and P. Jeanteur, *Virology* **97,** 221 (1979).
[11] M. C. Starnes and Y. C. Cheng, *J. Biol. Chem.* **264,** 7073 (1989).
[12] R. T. D'Aquila and W. C. Summers, *J. Acquir. Immune Defic. Syndr.* **2,** 579 (1989).
[13] S. M. Oberhaus and J. E. Newbold, *Methods in Enzymol.* **275,** 328 (1996).
[14] A. Blank and C. A. Dekker, *Biochemistry* **20,** 2261 (1981).
[15] R. H. Sugiyama, A. Blank, and C. A. Dekker, *Biochemistry* **20,** 2268 (1981).
[16] G. L. Schieven, A. Blank, and C. A. Dekker, *Biochemistry* **21,** 5148 (1982).
[17] R. E. Dompenciel, V. R. Garnepudi, and D. R. Schoenberg, *J. Biol. Chem.* **270,** 6108 (1995).
[18] C. W. Seidel and L. J. Peck, *Nucleic Acids Res.* **22,** 1456 (1994).
[19] A. Bufe, M. D. Spangfort, H. Kahlert, M. Schlaak, and W. M. Becker, *Planta* **199,** 413 (1996).
[20] L. C. van Loon, *FEBS Lett.* **51,** 266 (1975).
[21] T. A. McKeon, M. L. Lyman, and G. Prestamo, *Arch. Biochem. Biophys.* **290,** 303 (1991).
[22] S. Uchiyama, K. Isobe, and S. Nagai, *Comp. Biochem. Physiol. [B]* **102,** 343 (1992).
[23] T. P. Karpetsky, G. E. Davies, K. K. Shriver, and C. C. Levy, *Biochem. J.* **189,** 277 (1980).

observation that nucleic acids embedded in polyacrylamide gels alter the mobility of interacting proteins has been exploited for the development of PACE (polyacrylamide gel coelectrophoresis) technology.[24] However, the denaturing conditions used will dissociate an oligomeric protein into its individual components so that detection will be restricted to the catalytic subunit only, provided that interactions with other subunits are not necessary for activity. One of the major limitations is the ease with which a given polypeptide will recover its native conformation and consequently its activity. This appears to be rather unpredictable and various renaturation procedures should generally be empirically tested until one is found satisfactory. However, some general guidelines to be described in the next paragraph have been gained from the numerous assays that have been reported since the initial work of Rosenthal and Lacks.[25]

Interestingly, nucleic acids can be conveniently embedded in the resolving gel prior to polymerization, thus providing a substrate for nucleases. Even rather short polymers such as a 50 base pair oligonucleotide have been used successfully.[26] In these conditions the substrate will stay in the gel during electrophoresis without interfering with the migration of the polypeptides as long as the denaturing conditions are maintained. After renaturation of the protein and subsequent incubation of the gel in an appropriate buffer, the nucleic acid is digested, and the breakdown fragments diffuse out of the gel. Therefore, the absence of the nucleic acid reveals the placement of the active polypeptide. If unlabeled nucleic acid is used, the gel can be stained with ethidium bromide, or in the case of RNA, with toluidine blue; in both cases the active bands will appear as a clear area on a uniformly stained gel. If a radiolabeled substrate is used, autoradiography will identify active bands as white or light gray bands on a dark background. The advantages of using a radioactive substrate are (i) labeling with high specific activity increases the level of detection by allowing the use of substrate concentrations well below the K_m, and (ii) it allows the experimenter to check by autoradiography the same gel as the enzymatic reaction proceeds: the gel can be returned back to the buffered solution for prolonged incubation to reveal lower RNase activity bands. Such flexibility is not found with conventional staining procedures, because of the unknown effect of the stains on the nuclease reaction. An interesting alternative to radiolabeled substrates is the use of fluorescent-labeled substrates, which offers the advantages of a long half-life compared to ^{32}P-labeled substrates, as well as a short analysis time (about 15 min) compared to autoradiographic exposures (several hours or overnight), and do not necessitate the precautions linked to the handling of radioactive compounds.[27]

[24] C. D. Cilley and J. R. Williamson, *RNA* **3,** 57 (1997).
[25] A. L. Rosenthal and S. A. Lacks, *Anal. Biochem.* **80,** 76 (1977).
[26] A. K. Sheaffer, S. P. Weinheimer, and D. J. Tenney, *J. Gen. Virol.* **78,** 2953 (1997).
[27] L. Y. Han, W. P. Ma, and R. J. Crouch, *Biotechniques* **23,** 920 (1997).

General Guidelines for *in Situ* Recovery of Enzymatic Activity

As mentioned above, the critical and limiting step in all activity gel procedures is the renaturation of the protein. The importance of avoiding or eliminating SDS impurities for good recovery of activity has been recognized early on. Comparison of recovered activities obtained with SDS from different sources has implicated impurities such as hexadecyl sulfate and tetradecyl sulfate,[28] but more importantly didodecyl ether, didodecyl sulfate, and long-chain hydrocarbons.[29] Apart from selecting carefully a pure SDS reagent (e.g., from BDH), one can add 25% 2-propanol in the initial washes of the gel after completion of electrophoresis in order to remove SDS. In early times this was accomplished simply by several changes of water, or of the buffer solution appropriate for the enzyme activity so that a unique solution could be used throughout. Later on it was found that the use of aqueous 2-propanol greatly improves the reproducibility and the sensitivity of activity gels by efficiently removing the impurities present in SDS preparations.[30] In addition to SDS impurities, other chemicals used for preparing the gel, acrylamide, bisacrylamide, Temed, ammonium persulfate, and by-products of the polymerization reaction have the potential to inactivate proteins. This is why some investigators include bovine serum albumin (BSA) in the gel mixture to distribute a protective protein throughout the gel. For that purpose, fibrinogen (0.1 mg/ml) is preferable to BSA, as it has been observed that the smaller albumin molecules migrate out of the upper portion of the gel during electrophoresis.[31] However these additive proteins should be free of contaminating nucleases. Alternatively, according to the original observation of Hager and Burgess,[32] one can include in the sample a protein that, migrating faster than the protein to be renatured, will sweep away damaging agents. Han *et al.*[27] reported that, in the case of pure *E. coli* RNase HI, the detection level was decreased from 5 ng down to 0.5–2 ng on addition of lysozyme. This explains why crude extracts often provide better signals as they contain proteins migrating faster than protein(s) under study, which act as protective shields.

A last point to be examined is the denaturation procedure prior to loading the sample on the gel. Standard procedures involve boiling the proteins for 3 min in the sample buffer containing 2% SDS and 5% 2-mercaptoethanol (reducing conditions). However, for some proteins this harsh treatment does not allow proper recovery of activity. A milder treatment can be substituted, for example an incubation at 37° for 3–5 min in the sample buffer.[33] At one extreme, if no incubation is

[28] S. A. Lacks, S. S. Springhorn, and A. L. Rosenthal, *Anal. Biochem.* **100,** 357 (1979).

[29] A. Blank, C. Dekker, G. Schieven, R. Sugiyama, and M. Thelen, *Nucleic Acids Symp. Ser.* 203 (1981).

[30] A. Blank, R. H. Sugiyama, and C. A. Dekker, *Anal. Biochem.* **120,** 267 (1982).

[31] A. Blank, J. R. Silber, M. P. Thelen, and C. A. Dekker, *Anal. Biochem.* **135,** 423 (1983).

[32] D. A. Hager and R. R. Burgess, *Anal. Biochem.* **109,** 76 (1980).

[33] U. Bertazzoni, A. I. Seovassi, M. Mezzina, A. Sarasin, E. Franchi, and R. Izzo, *Trends in Genetics* **2,** 67 (1986).

performed and the reducing agent is omitted from the sample buffer, one can speak of native SDS–PAGE for conditions where proteins migrate according to approximate molecular weight but are not irreversibly denatured.[34] Finally, it is worth saying that in one study, the sample incubation conditions were of little importance, but the gel running temperature was found to be critical, the nuclease activity being permanently inactivated in preparations run at a temperature above 10°.[35]

Preparing, Casting (Pouring), and Running an Activity Gel

We currently run the SDS-discontinuous system based on the method of Laemmli[36] using a 13% acrylamide resolving gel ($13 \times 8 \times 0.1$ cm) and a 5% acrylamide stacking gel 1 cm long. Readers are free to adopt other acrylamide concentrations better suited to the molecular masses of their enzyme under study, or use other gel dimensions (a miniature activity gel has been described[33]).

Nature and Concentration of Substrates

Nonradioactive substrates such as total or ribosomal RNA from different sources, (*Escherichia coli*,[25] wheat germ,[21,14] yeast *Torula*[37]) can be prepared in the laboratory with standard procedures (e.g., phenol extraction), but homopolymers are obtained from commercial suppliers. They are used at final concentrations ranging from 25 μg/ml ribosomal RNA up to 2 mg/ml *Torula* yeast RNA (this high concentration compensates for the abundant presence of short oligomers that diffuse out of the gel during its processing). Alternatively, synthetic homopolymers have been used at concentrations ranging from 0.5 μg/ml to 0.3 mg/ml: poly(U)[37], poly(rA)-poly(dT)[12], poly(rC)[12].

Radiolabeled Substrate

Radiolabeled RNA a few hundred nucleotides long can be conveniently obtained by standard run-off transcription of any DNA template consisting of a linearized plasmid containing a gene under the control of a bacteriophage promoter (T3, T7, or SP6). Below is described the run-off transcription of a gene cloned downstream of the SP6 promoter in a pGEM (Promega, Madison, WI) vector:

To a pellet containing 8 to 10 μg linearized plasmid, add, in a microtube:

10 μl 10$\times$ buffer (0.4 *M* Tris-HCl pH 7.9; 0.1 *M* NaCl; 0.06 *M* $MgCl_2$; 0.02 *M* spermidine)
2.4 μl ATP 10 m*M*
2.4 μl GTP 50 m*M*

[34] H. N. Dodd and J. M. Pemberton, *J. Bacteriol.* **178,** 3926 (1996).
[35] F. C. Minion, K. J. Jarvill-Taylor, D. E. Billings, and E. Tigges, *J. Bacteriol.* **175,** 7842 (1993).
[36] U. K. Laemmli, *Nature* **227,** 680 (1970).
[37] A. Bufe, G. Schramm, M. B. Keown, M. Schlaak, and W. M. Becker, *FEBS Lett.* **363,** 6 (1995).

2.4 μl CTP 50 m*M*
2.4 μl UTP 50 m*M*
4 μl HPRI (human placenta ribonuclease inhibitor, Amersham Pharmacia Biotech, Piscataway, NJ, 100 units/μl)
2 μl SP6 RNA polymerase from Roche, Indianapolis, IN (20 units/μl)
5 μl [α-^{32}P]ATP (3000 Ci/mmol; 10 μCi/μl)
89.4 μl H_2O

After gentle mixing, incubate for 2 hours at 37°.

Add 10 μl DNase I–RNase free RQI from Promega (1U/μl) and incubate further for 10 min at 37°.

Extract with phenol, then with chloroform/isoamyl alcohol (24 : 1) and precipitate with 0.5 volume of 7.5 *M* ammonium acetate and 2.5 volumes of ethanol at −20°.

Centrifuge at 15,000*g* in a microfuge for 30 min at 4°, carefully remove the supernatant, and dry the pellet.

The pellet is resuspended in 70 μl of STE buffer (20 m*M* Tris-HCl pH 7.5; 100 m*M* Nacl; 1 m*M* EDTA) and final purification is achieved by use of a Nuc Trap push column (Stratagene, La Jolla, CA). After humidification of the column with 70 μl STE, the radiolabeled RNA is applied, followed by two additional applications of 70 μl STE, which are employed to rinse the microtube first. Then, 10 μl of the final pooled (210 μl) resulting eluate is quantitated by Cerenkov counting in a liquid scintillation counter.

A detailed description of a similar procedure using T3 RNA polymerase is described by Dompenciel *et al.*[17]

Synthesis of ^{32}P-Radiolabeled RNA–DNA Hybrid

This substrate is used for the detection of RNaseH activity and is routinely obtained by *E. coli* RNA polymerase transcription of single-stranded DNA (M13 or ϕX174 ssDNA or heat-denatured DNA from calf thymus) or homopolymeric DNA such as poly (dT) or poly (dC). The recipe for the synthesis of [^{32}P] poly(rA)-poly(dT), as described by Sarngadharan *et al.*[38] is given below. Synthesis of poly(rG)-poly(dC) can be obtained similarly but using poly(dC) as template and using cold and labeled GTP.[6,11] The detailed synthesis of an RNA–DNA hybrid using heat-denatured calf thymus DNA has been described elsewhere.[39]

The recipe is given for a 3 ml transcription volume.

Prepare the following mix in a 15 ml conical propylene tube:

1 ml poly(dT) (5 A_{260} units dissolved in sterile deionized water)
1.623 ml H_2O

[38] M. G. Sarngadharan, J. P. Leis, and R. C. Gallo, *J. Biol. Chem.* **250,** 365 (1975).
[39] C. Cazenave, P. Frank, and W. Büsen, *Biochimie* **75,** 113 (1993).

150 μl glycerol
150 μl 1 *M* Tris (pH 8.0)
16 μl 1 *M* $MgCl_2$
30 μl 0.1 *M* $MnCl_2$
6 μl 2 M Dithiothreitol (DTT)
9 μl 10 m*M* ATP
10 μl [α-^{32}P]ATP (3000 Ci/mmol; 10 μCi/μl; ICN, Costa Mesa, CA)
6 μl RNA polymerase (10 units/μl; Pharmacia, Piscataway, NJ)

Vortex gently a few seconds, then incubate 30 min at 37°.

Stop by addition of 30 μl 0.5 *M* EDTA.

Add 3 ml phenol, vortex vigorously for 1 min, centrifuge (2000 rpm, 5 min), and carefully remove the aqueous phase.

Reextract the phenol phase with 3 ml 50 m*M* Tris pH8.0, vortex, centrifuge, and carefully remove the aqueous phase, as above.

Pool the two aqueous phases and add 6 ml chloroform/isoamyl alcohol (24 : 1). Vortex, centrifuge (or let sediment a few minutes on the bench), and then carefully remove the upper aqueous phase.

To this aqueous phase add ethanol up to 35% and load the resulting solution on a 2-ml column of cellulose (CF11 Whatman, Clifton, NJ, or fibrous medium from Sigma, St. Louis, MO) equilibrated in a 35 : 65 mixture of ethanol and buffer A (50 m*M* Tris-HCl pH 7.5/0.1 *M* NaCl/1 m*M* EDTA). The column is washed extensively with about 50 ml of this mixture until all unbound radioactivity is removed. The nucleic acid polymer is then eluted in a small volume of buffer A, by collecting 10 fractions of 500 μl in Eppendorf tubes. Fractions with high activity (generally tubes 3 to 7) are pooled together and an aliquot is counted by liquid scintillation (a 10 μl aliquot should contain about 50,000 to 150,000 cpm). The hybrid is stored at −20° until use. A detailed description of the synthesis and use of a fluorescent substrate for ribonuclease H has been given by Han *et al.*[27]

Gel Preparation

For 20 ml of a 13% acrylamide resolving gel, mix

8.77 ml 30% (w/v) acrylamide/ 0.8% (w/v) bisacrylamide
2.5 ml 3 *M* Tris-HCl pH 8.3
0.2 ml 10% (w/v) SDS
0.1 ml 10% (w/v) ammonium persulfate (APS)
8.5 ml H_2O
20 to 200 μl radiolabeled substrate (3×10^5 to 2×10^6 cpm)
33 μl TEMED

Pour the gel, cover it with a thin layer of isopropanol, and let polymerize (about 20 min at room temperature). During that time, prepare the stacking gel:

1.7 ml Acrylamide/bisacrylamide (30/0.8)
2.5 ml 0.5 *M* Tris (pH 6.8)
0.1 ml 10% (w/v) SDS
70 μl 10% (w/v) APS
5.7 ml H_2O
7 μl TEMED, when ready to pour this gel (that is, after polymerization of the resolving gel and removal of the 2-propanol layer)

Pour the gel, place the comb, and let polymerize (30 to 45 min).

After the polymerization is complete, remove the comb carefully, cover with reservoir buffer (0.025 *M* Tris, 0.192 *M* glycine, 0.1% SDS), and immediately rinse the slots by gently flushing this buffer with a syringe to remove any unpolymerized acrylamide.

Sample Preparation

Mix one volume of sample with one volume of 2× sample buffer (1.25 ml 0.5 *M* Tris-HCl, pH 6.8/1.875 ml 20% SDS/0.635 ml 2 *M* DTT/2.5 ml glycerol/6 mg bromphenol blue). Vortex, heat for 3 min at 98° (although we have found this treatment not to be detrimental for subsequent renaturation of the enzymes under study, it appears that for some enzymes it is preferable to avoid heating, or to use lower temperatures), centrifuge a few seconds, and load samples on the gel, together with molecular weight markers if needed.

Electrophoresis

When both the lower and upper buffer reservoirs are filled with reservoir buffer, connect the tank to the power supply and turn power on. For a 8 to 9 cm height resolving gel (1 mm thick) an overnight run at 40 to 45 V (constant voltage) is satisfactory. Faster (3 to 4 hr) runs are obtained by operating at 30 mA constant current.

Electrophoresis is normally stopped when the bromphenol blue dye reaches the bottom of the gel and begins to enter the lower reservoir.

Remove glass plates and spacers, then place the gel in a cuvette suitable for incubations in 200 ml of renaturation and activity buffers for a medium sized gel, as described above. Wear gloves any time a gel is manipulated.

Processing of Gel for Renaturation, and *in Situ* Detection of Ribonuclease Activity

Renaturation and in Situ Digestion of Nucleic Acid Substrates: Designing Buffers

The gel is first incubated in a buffered solution containing 25% 2-propanol, to efficiently remove detergent. After washing with 2-propanol-free buffer the gel is incubated in a buffer suitable for both renaturation of the protein and detection of enzymatic activity. For RNases H, suitable buffers generally contain Tris-HCl (10 to 50 m*M*, pH 8.0), NaCl (20 m*M*,[7] 50 m*M*,[6] 75 m*M*,[4] or 100 m*M*[9]), $MgCl_2$ (10 mM[6,4,9] or 20 m*M*[7]), 2-mercaptoethanol (1 m*M*[4] or 10 m*M*[5] and/or DTT[9]). Although the overall composition is similar for all enzymes, precise concentrations (especially of NaCl and $MgCl_2$) have to be determined for each enzyme. Variations can be introduced in this general procedure, to increase the sensitivity of the activity gel, most likely by improving the renaturation yield. Assays concerning cellular RNases H provide a good illustration: Rong and Carl[9] added 10% glycerol to buffer, and either guanidine hydrochloride or nonidet P-40 (NP-40) detergent. The gel was treated with buffer containing Gdn-HCl 6 *M* for 2 hr, followed by a 20-hr incubation in the same buffer but without guanidine. In an alternative procedure, the gel was first incubated for 2 hr in buffer without NP-40 followed by 20 hr in the same buffer containing 2.5% NP-40. The procedure using NP-40 has been found to be more efficient, at least for the RNase HI from calf thymus.[9] We have routinely used it in our own laboratory and found it to be effective for the *E. coli* enzyme and for RNase HII from wheat germ extracts[39] and *Xenopus* oocytes,[40] but also from beef heart mitochondrial extracts and HeLa cells. We adopted a slight modification of this procedure by substituting 0.5 m*M* $MnCl_2$ for the 10 m*M* $MgCl_2$ generally used.[41] Even though the eukaryotic RNase HII is far less active with manganese than magnesium in liquid assays, we have found that this substitution greatly increased the sensitivity of the activity gel, very likely because the renaturation is favored by using manganese instead of magnesium. Although the final result reflects the combination of both renaturation and activity, it appears that the limiting step is the renaturation process. One should note, however that some chemicals known to be inhibitors in liquid assays can also act as inhibitors in the renaturation gel assay: for example, dextran and *N*-ethylmaleimide (NEM), known to be inhibitors of eukaryotic RNase HII in liquid assays, were also inhibitory in the gel assay when added to the renaturation/activity buffer.[8]

Substituting manganese for magnesium turned out to be effective for most enzymes: this allowed the detection of RNase H activities undetected in earlier gels. Not only could we visualize *E. coli* RNase HII, which has very low activity in liquid assays, but we could also detect RNase H activities associated with the exo III and DNA pol I polypeptides from *E. coli*.[41] An example showing the parallel between the requirement of a particular divalent cation for both liquid assays and activity gels is provided by the calcium-dependent ribonuclease X from *Xenopus laevis*.[18] It is the only ribonuclease detected in *Xenopus* crude oocyte homogenates using the standard procedure (which will be described below) when 1 m*M* calcium chloride is present in the renaturation SRB buffer.[42] An enhanced detection of the

[40] C. Cazenave, P. Frank, J. J. Toulmé, and W. Busen, *J. Biol. Chem.* **269,** 25185 (1994).

[41] P. Frank, C. Cazenave, S. Albert, and J. J. Toulmé, *Biochem. Biophys. Res. Commun.* **196,** 1552 (1993).

RNases of *Arabidopsis thaliana* has been obtained by adding 2 μM $ZnCl_2$ to the buffer.[43]

Incubation in Activity Buffer

Incubation is generally performed for 15 to 20 hr, or longer (up to 40 hr[6]) either at room temperature[9,41] or at higher temperatures (37°,[6,8] 42°[4]). When long incubations (several days) are necessary, an inhibitor of bacterial growth (0.01% sodium azide) should be included in the solution.

Detection of Active Bands

Radioactive Substrates. The gel is fixed in an acidic solution (methanol/acetic acid/water 5 : 1 : 5[6]) or 10% trichloroacetic acid (TCA)/1% sodium pyrophosphate[9]), dried, and autoradiographed. Alternatively, the gel can be fixed first in methanol/acetic acid/water (5 : 1 : 5) for 30 min, then stained with 0.1% Coomassie Blue dissolved in the same solution, followed by destaining in methanol/acetic acid/water (30/7.5/62.5). The stained gel is then dried and autoradiographed. If the main purpose of the staining is to visualize molecular weight marker proteins, this step can be omitted and time saved if prestained molecular weight markers are used. Activity gels involving unlabeled substrates can be stained for 30 min with 1 or 2 μg/ml ethidium bromide added to the incubation buffer, then placed on a long wavelength UV light box and photographed. Alternatively, for staining with toluidine blue: rinse the gel for 10–40 min with 10 mM Tris HCl pH 7.4, then stain for 5 min in 0.2% (w/v) toluidine blue O (Sigma) in the same buffer, and destain in the same buffer until the desired degree of destaining is reached (usually 30–60 min). After a final rinse in 10% (v/v) glycerol–0.01 M Tris-HCl (pH 7.0), the gel is dried on cellophane.

Example: Detection of RNase H from HeLa Cells

We next describe a procedure appropriate for detecting RNase H activities that we have applied routinely in our laboratory (Fig. 1). All incubations are performed at room temperature with gentle agitation.

Soak the gel (13% polyacrylamide) in 3 changes of 200 ml 2-propanol-containing buffer (25% 2-propanol; 50 mM Tris pH 7.5; 1 mM 2-mercaptoethanol; 0.1 mM EDTA) for 20 min each.

Wash the gel in 2 changes of rinsing buffer (10 mM Tris-HCl pH 7.5; 5 mM 2-mercaptoethanol), 15 min each.

[42] C. Cazenave, *in* "Ribonucleases" (H. R. J. Crouch and J. J. Toulmé, eds.), p. 101, John Libbey, Paris, 1998.

[43] Y. Yen and P. J. Green, *Plant Physiol.* **97,** 1487 (1991).

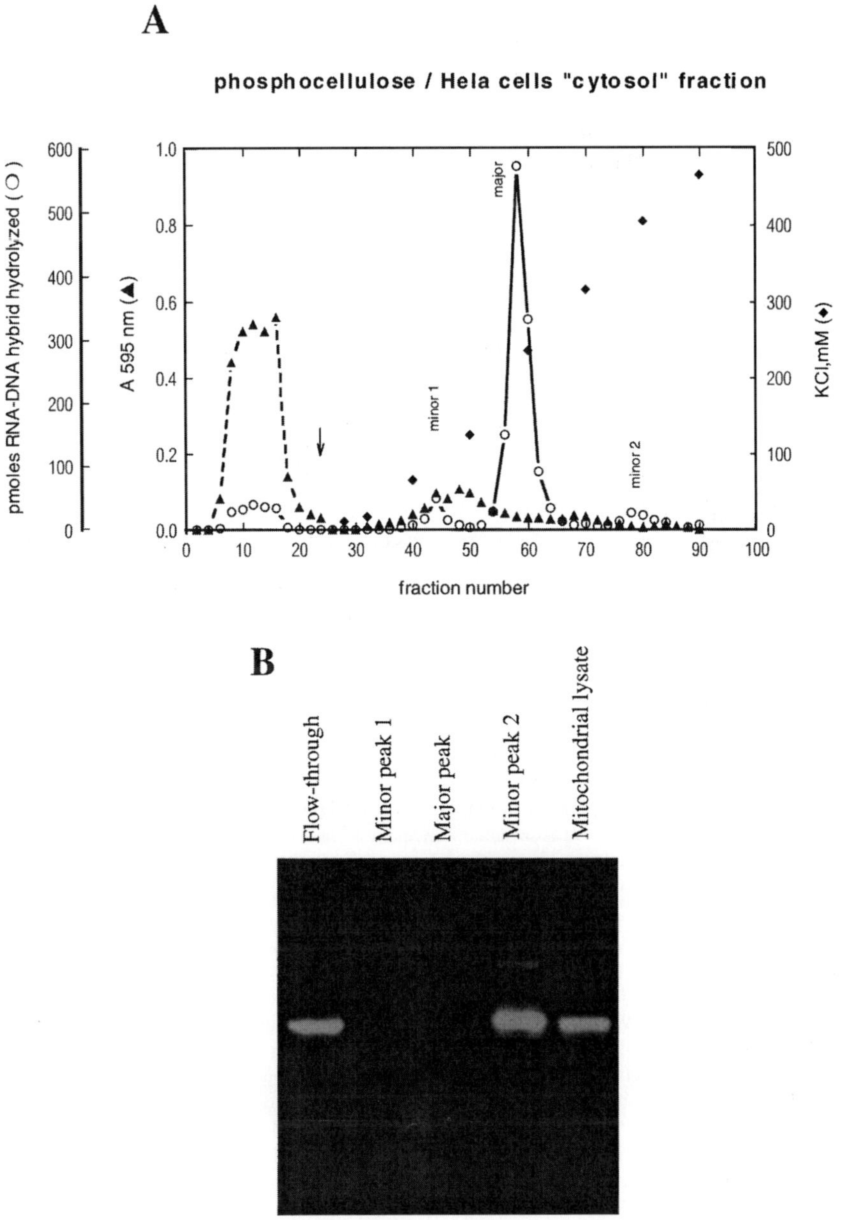
A
phosphocellulose / Hela cells "cytosol" fraction
pmoles RNA-DNA hybrid hydrolyzed (O)
A 595 nm (▲)
KCl,mM (◆)
fraction number
major
minor 1
minor 2
B
Flow-through
Minor peak 1
Major peak
Minor peak 2
Mitochondrial lysate

Immerse the gel for 2 hr in SRB buffer (50 m*M* Tris-HCl pH 8.0/0.5 m*M* $MnCl_2$/100 m*M* NaCl/10% glycerol/1 m*M* 2-mercaptoethanol and 1 m*M* DTT).

Then incubate the gel for 20 hr in 3 changes of SRB plus 2.5% NP-40.

Fix the gel and eliminate the last traces of acid soluble radioactivity by 4 washes of 5% (w/v) TCA/1% (w/v) PP_i, 15 min each.

The gel is left in this solution for 2 to 20 hr, then dried and autoradiographed.

Interpretation of Results

In interpretation of autoradiographs, polypeptides with RNase activity will appear as light gray to white bands on a dark background. It should be noted that the relative intensity will depend not only on the amount of enzyme that has been loaded on the gel, but also on the renaturation yield of the protein and on its relative activity in the renaturation/activity buffer. The relative activity for the same polypeptide can be compared from one lane to another (for example, when checking for the content of several fractions eluted from a chromatography column

FIG. 1. Partial purification and activity gel analysis of RNase H from HeLa cells. (A) Fractionation of ribonucleases H from a HeLa cell cytosolic extract on a phosphocellulose column. Fractionation of HeLa cell cytosolic extract: 9 ml of cytosolic fraction was ultracentrifuged for 1 hr at 50,000*g* and the supernatant dialyzed overnight against buffer P (50 m*M* Tris-HCl pH 7.9 at 4°C; 10 m*M* KCl; 1.5 m*M* 2-mercaptoethanol). The dialyzate (8 ml at 12.5 mg protein per ml) was diluted 5 times in buffer P and loaded on a 35 ml phosphocellulose column (P11 Whatman) equilibrated with buffer P. After extensive washing with buffer, the adsorbed proteins were eluted with a linear (10 m*M* → 500 m*M*) KCl gradient in buffer P (start of the gradient is indicated by an arrow). RNase H activity measured by liquid assay (○) using [^{3}H]RNA–DNA hybrid as substrate[39] and protein content measured with the Bradford assay (▲)[46] have been normalized to volumes of 10 μl (panel A). The KCl concentration was determined by conductimetry (◆). (B) Activities present in the various peaks were pooled and concentrated, and aliquots containing identical RNase H activities were loaded on a 13% acrylamide activity gel (panel B), together with a mitochondrial extract of HeLa cells. The substrate used was [^{32}P]poly(rA)-poly(dT) and the gel was processed as described in the text. Biochemical parameters and hydrolysis profiles on hybrids of defined length and sequence indicate that activities present in minor peak 1 and major peak belong to the class I ribonuclease H, whereas activities present in minor peak 2 and in mitochondrial lysates belong to the class II. Activities present in the flow-through represent a mixture of both classes. This result indicates that presence of a 32 kDa band can be used as an additional criteria for the presence of a class II ribonuclease H. It illustrates the possible absence of correlation between activities tested in liquid assays and in activity gels, as the main activity detected in the liquid assay (major peak) is not detected by the renaturation gel assay. The faint band of 43 kDa detected in minor peak 2 is observed, although not reproducibly, in crude extracts, and cannot be assigned to one of the major RNases H of the cell detected in liquid assays. Nuclei, mitochondria, and cytosolic fractions from HeLa cells, prepared by differential centrifugation, were obtained from Computer Cell Culture Center (Mons, Belgium). Mitochondrial extract was obtained from lysis of mitochondria with Triton X-100 and ultacentrifugation, followed by dialysis of the supernatant against 20% glycerol/20 m*M* potassium phosphate pH7.5/5 m*M* 2-mercaptoethanol.

during a purification procedure[44]), provided that the amounts of enzyme loaded are below the level of saturation of the response of the activity assay. On the other hand, comparing activities between different polypeptides can be misleading. For example, the eukaryotic RNase HI that represents the major activity assessed in liquid assays is not visualized on activity gels, unless very large amounts of the enzyme are loaded,[9] whereas the eukaryotic RNase HII, which is the minor activity found in liquid assays, is easily detected. This point is illustrated in Fig. 1. This might be linked either to the mass of the enzyme, as shorter proteins are expected to renature more easily than larger ones, or to the oligomeric nature of RNase HI compared to the monomeric form of RNase HII. Moreover, denaturation by SDS can dissociate the active polypeptide from other components susceptible to a decrease or even complete inhibition of activity in the liquid assay. A good example is provided by the case of the ribonuclease HII from *Xenopus* oocytes.[40]

However, one might also detect simultaneously bands darker than the background. These correspond to proteins bound to the embedded nucleic acid, preventing partial diffusion out of the gel during the incubation–renaturation step. The prevalence of this phenomenon seems to depend on the type and length of nucleic acid chains. There is probably a mixture of fully renatured genuine nucleic acid–binding proteins and of denatured, or only partially renatured, polypeptides that retain nonspecific nucleic acid binding through elecrostatic interactions. As the pattern of these nucleic acid–binding proteins can vary with the nucleic acid polymer used,[6] these behaviors could be of interest for addressing the question of the selectivity of the interactions.

The same general considerations apply also to stained gels. For the latter, one should keep in mind a possible artifact: it has been observed that some proteins, such as histones, can bind to the nucleic acid and prevent staining by ethidium bromide, leading to false positive bands.[45]

In conclusion, activity gels are easy to perform, give access to the molecular mass of RNase polypeptides without the need for extensive prior purification, and can provide valuable information not attainable by liquid assays. Additional information relative to the specificity of the enzyme toward its substrate can be obtained by running activity gels with different radiolabeled nucleic acid polymers.[6,8,42] We have described the classical one-dimensional electrophoresis procedure that has been until now most exclusively used with activity gels. However, the principle of the technique could be applied to two-dimensional electrophoresis, which is becoming increasingly popular with the burgeoning growth of proteomic studies.

[44] P. Frank, S. Albert, C. Cazenave, and J. J. Toulmé, *Nucleic Acids Res.* **22,** 5247 (1994).

[45] A. L. Rosenthal and S. A. Lacks, *J. Biol. Chem.* **253,** 8674 (1978).

[46] M. M. Bradford, *Anal. Biochem.* **72,** 248 (1976).

[8] Analysis of Ribonucleases following Gel Electrophoresis

By A. D. J. SCADDEN and SOREM NAABY-HANSEN

Zymograms

Zymography describes the experimental technique in which an enzyme activity is analyzed *in situ* following electrophoresis. It allows characterization of important physical properties of the enzyme within crude cell extracts or partially purified fractions. This can assist subsequent purification. Zymograms (also commonly known as "in-gel assays" or "activity gels") have been used to analyze a wide variety of mammalian, plant, and microbial enzymes, including histone acetyltransferase,[1] matrix metalloproteases,[2] lipase,[3] endoglucanase,[4] phytase,[5] poly(ADP-ribose) glycohydrolase,[6] elastase,[7] DNase I,[8] α-L-arabinofuranosidase,[9] and an assortment of ribonucleases.[10–15] Zymograms have certain advantages over more conventional enzyme assays. Primarily, they allow enzyme activities to be attributed to polypeptides with defined physical characteristics, such as molecular weight or isoelectric point. In addition, it is possible using an in-gel assay to study heterogeneity of enzyme isoforms, multiplicity of enzymes (for example, in biological samples), and posttranslational modification of a particular enzyme (for example, glycosylation) and to simultaneously analyze an enzyme activity contained in various protein fractions.

A number of ribonucleases have been studied using different types of zymograms. These include various overlay zymograms,[13,14] positive zymograms,[10]

[1] T. E. Spencer, G. Jenster, M. M. Burcin, C. D. Allis, J. Zhou, C. A. Mizzen, N. J. McKenna, S. A. Onate, S. Y. Tsal, M-J. Tsal, and B. W. O'Malley, *Nature* **389,** 194 (1997).

[2] T. M. Leber and F. R. Balkwill, *Anal. Biochem.* **249,** 24 (1997).

[3] R. P. Yadav, R. K. Saxena, R. Gupta, and W. S. Davidson, *Biotechniques* **24,** 754 (1998).

[4] B. Perito, E. Hanhart, T. Irdani, M. Iqbal, A. J. McCarthy, and G. M. Mastromei, *Gene* **148,** 119 (1994).

[5] H. D. Bae, L. J. Yanke, K. J. Cheng, and L. B. Selinger, *J. Microbiol. Meth.* **39,** 17 (1999).

[6] G. Brochu, G. M. Shah, and G. G. Poirier, *Anal. Biochem.* **218,** 265 (1994).

[7] C. Gardi and G. Lungarella, *Anal. Biochem.* **140,** 472 (1984).

[8] A. L. Rosenthal and S. A. Lacks, *Anal. Biochem.* **80,** 76 (1977).

[9] M. T. Fernandez-Espinar, J. L. Pena, F. Pinaga, and S. Valles, *FEMS Microbiol. Lett.* **115,** 107 (1994).

[10] R. C. Karn, M. Crisp, E. A. Yount, and M. E. Hodes, *Anal. Biochem.* **96,** 464 (1979).

[11] J. M. Thomas and M. E. Hodes, *Anal. Biochem.* **113,** 343 (1981).

[12] A. Blank, R. H. Sugiyama, and C. A. Dekker, *Anal. Biochem.* **120,** 267 (1982).

[13] T. Yasuda, D. Nadano, E. Tenjo, H. Takeshita, and K. Kishi, *Anal. Biochem.* **206,** 172 (1992).

[14] C. Paech, T. Christianson, and K-H. Maurer, *Anal. Biochem.* **208,** 249 (1993).

[15] J. Bravo, E. Fernandez, M. Ribo, R. de Llorens, and C. M. Cuchillo, *Anal. Biochem.* **219,** 82 (1994).

METHODS IN ENZYMOLOGY, VOL. 341
0076-6879/01 $35.00

and negative zymograms.[11,12,15] An example of an overlay zymogram is the dried agarose film overlay (DAFO) technique described by Yasuda *et al.*[13] In this method, isoelectric focusing using thin polyacrylamide gels was used to separate ribonucleases. After incubation of the focused gel with an overlaid dried agarose film containing substrate RNA, ethidium bromide, and a suitable buffer, dark bands corresponding to ribonuclease activity appeared on a fluorescent background under UV light. Positive zymograms are techniques in which ribonucleases are detected by specific staining at the site of ribonuclease activity. For example, in the method described by Karn *et al.*,[10] ribonucleases are separated on an agarose gel that contains the small substrate UpA [uridylyl(3′→5′)-adenosine]. UpA is hydrolyzed by ribonucleases and the adenosine is subsequently deaminated (to inosine) by adenosine deaminase. The inosine produced is linked by a series of enzymatic reactions to formation of a blue tetrazolium salt, which thus indicates the site to which the ribonuclease has migrated. Conversely, in negative zymogram techniques, clear areas on a background of stained RNA represent ribonuclease activity. For example, in the method of Bravo *et al.*,[15] ribonucleases are separated by SDS–PAGE using gels that contain homoribopolymer substrates. Following a series of incubation steps, ribonuclease activity was detected either by direct visualization (by UV shadowing) or by staining the undigested RNA with either toluidine blue or silver nitrate. In this case the ribonuclease activity corresponds to clear bands on the stained gel.

We describe here a sensitive method for performing a negative-staining ribonuclease zymogram utilizing either one- or two-dimensional polyacrylamide gel electrophoresis. In this method radiolabeled RNA (produced by *in vitro* transcription) is polymerized into an SDS–polyacrylamide gel and, following electrophoresis, ribonuclease activity is detected by a short incubation in a suitable buffer and subsequent autoradiography. Ribonuclease activity results in the production of clear zones in a dark background as digested radiolabeled RNA has diffused out of the gel. Radiolabeled RNA has previously been used in negative-staining zymograms where RNA : RNA or DNA : RNA duplexes were incorporated into the gel for analysis of RNase H or RNase III.[16,17] Use of radiolabeled RNA substrates has enabled the use of less RNA in the gel (0.5 μg/ml) than has been used previously to analyze single-strand specific ribonucleases (for example, 0.3 mg/ml[15]). A relatively small amount of labeled RNA is easy to visualize by autoradiography, whereas unlabeled RNA must be in sufficient quantity to be visible by UV shadowing or toluidine blue staining, for example. The use of radiolabeled RNA substrates should therefore result in a more sensitive assay that would be advantageous for assaying a ribonuclease that is in low abundance, has

[16] J. S. Smith and M. J. Roth, *J. Virol.* **67,** 4037 (1993).

[17] H. Ben-Artzi, E. Zeelon, S. F. J. Le-Grice, M. Gorecki, and A. Panet, *Nucleic Acids Res.* **20,** 5115 (1992).

weak activity, or is renatured poorly in the gel. Moreover, the use of *in vitro* transcribed labeled RNAs permits specific RNA substrates rather than homoribopolymers [poly(U) or poly(C)] or tRNA substrates to be used for in-gel assays. Furthermore, RNAs can be synthesized *in vitro* that contain modified nucleotides that may be important for ribonuclease activity [for example, the ribonuclease p29 is assayed using inosine-containing RNA (see below)]. The potential limitation of *in vitro* transcription in producing sufficient RNA for an in-gel assay is overcome by the fact that less RNA is required for this assay. The use of specific substrates will enable a wide range of ribonucleases to be analysed by zymography.

We have previously characterized a ribonuclease activity present in various cell extracts, which we referred to as "I-RNase" because of its ability to preferentially degrade RNA that contained inosine in place of guanosine.[18] We purified an I-RNase activity ~80,000-fold from pig brain extracts. Using one- and two-dimensional zymograms and elution of RNase activity from SDS gel slices, we were able to attribute the I-RNase activity to a polypeptide of 29 kDa (referred to here as p29). Subsequently, we have found that the preference for inosine-containing RNA is not unique to p29, but is shared by a number of common single-strand specific RNases, and probably reflects the greater accessibility of the RNA substrate when intramolecular G-C base pairs are replaced by the weaker I-C pairs. Nevertheless, the analysis of p29 illustrates how zymograms, and elution of enzyme activities from SDS gels, can assist in the identification, characterization, and purification of RNases.

Here, we outline the procedures that we have used for sensitive one- and two-dimensional zymograms and for recovery of RNase activity from SDS gel slices.

One-Dimensional Zymograms

Materials

SDS–PAGE apparatus: We use gels of approximately 17 cm × 12 cm × 0.8 mm. Glass gel plates are baked at 200° for 4–6 h prior to assembly and Teflon spacers and combs are washed thoroughly with 2% SDS before use.

Polyacrylamide gel mixes: Ensure acrylamide is of high quality and deionised before use (this is important for recovery of enzyme activity after electrophoresis):

Resolving gel: 12.5%: 12.5% acrylamide (A), 0.1% bisacrylamide (B); **15%:** 15% A, 0.087% **B; 20%:** 20% A, 0.066% B. 375 m*M* Tris-HCl, pH 8.8; 50 μl 10% (v/v) ammonium persulfate and 5 μl TEMED are added per 10 ml gel mix.

[18] A. D. J. Scadden and C. W. J. Smith, *EMBO J.* **16,** 2140 (1997).

Stacking gel: 5% acrylamide, 0.13% bisacrylamide, 125 m*M* Tris-HCl, pH 6.8; 50 μl 10% (v/v) ammonium persulfate and 5 μl TEMED are added per 5 ml gel mix.

2× protein sample buffer: 4% SDS, 160 m*M* Tris-HCl, pH 6.8, 20% (v/v) glycerol, 0.05% bromphenol blue. A reducing agent such as 2-mercaptoethanol [10% (v/v)] may be added (see note 1).

Prestained protein molecular weight markers (NEB; broad range)

Protein molecular weight markers: Our standard molecular weight markers contain the following components: β-galactosidase (116 kDa), phosphorylase *b* (97 kDa), bovine serum albumin (BSA) (68.5 kDa), catalase (57.5 kDa), glutamate dehydrogenase (55.5 kDa), creatine kinase (43 kDa), glyceraldehyde-3-phosphate dehydrogenase (36 kDa), carbonic anhydrase (29 kDa), and RNase A (13.7 kDa). Molecular weight markers are prepared in the presence of a reducing agent such as 2-mercaptoethanol.

Substrate RNA is transcribed from a suitable plasmid using T3, T7, or SP6 RNA polymerase. RNAs are internally labeled using any [α-^{32}P]NTP (3000 Ci/mmol, 10 Ci/μl; Amersham); we normally use [α-^{32}P]UTP. RNA substrates are labeled to a specific activity of approximately 1×10^6 cpm/pmol. This is achieved by using 10 μCi [α-^{32}P]UTP and 0.25 m*M* unlabeled UTP in a 10 μl transcription reaction. The final RNA concentration used for the in-gel assays is 0.5 μg/ml (this is equivalent to 5 pmol/ml for a 300 nt RNA substrate, as used to assay p29). Inosine-containing RNA (I-RNA), generated by substituting GTP with ITP in the transcription reaction, was generally used to assay p29. Other modifications to the RNA may be made if necessary for assaying particular ribonucleases.

Glass dish (approximately 14 cm × 20 cm; baked at 200° for 4–6 h). A glass plate larger than the dish should be used as a lid to shield radioactivity.

Assay buffer as appropriate for assaying enzyme of interest (20 m*M* HEPES–KOH, pH 7.9, 5 m*M* $MgCl_2$ for p29).

Assay buffer plus 20% (v/v) 2-propanol.

Rocking platform or shaker.

Perspex (Plexiglas) screen (at least 1 cm thick) to shield beta emission. All steps preceding and following electrophoresis should be performed behind this screen.

Benchkote (squares of approximately 50 cm × 50 cm). All manipulations of the gel should be performed on squares of Benchkote. Any radioactive contamination is thus contained and may be disposed of immediately.

Geiger–Müller radioactivity monitor. All work areas should be closely monitored for radioactivity before and after use.

Fuji Film, standard autoradiography, and/or phosphorimaging apparatus.

Method

(i) Prepare the polyacrylamide gel mixes as described above, with particular care taken to ensure RNase-free conditions. An appropriate percentage acrylamide gel (12.5–20%) is used for zymograms depending on the molecular weight of the protein of interest. It is important to ensure that the spacers and comb used are of uniform thickness (0.8 mm). As the assay relies on diffusion of digested RNA from the gel, the thickness of the gel will affect the degradation activity observed.

(ii) Add the substrate RNA to the resolving polyacrylamide gel mix immediately prior to casting the gel (before the addition of TEMED and ammonium persulfate) to give a final concentration of 0.5 μg/ml. The RNA does not appear to migrate significantly in the gel during electrophoresis; therefore a uniform concentration of RNA throughout the gel is maintained.

(iii) Ensure that the resolving RNA gel is completely polymerized before casting the stacking gel (the resolving gel may be cast 12 hr before use). This minimizes the amount of residual free acrylamide that may modify and inactivate proteins during electrophoresis and affect subsequent detection of enzyme activity.

(iv) Prepare the protein samples for electrophoresis by adding an appropriate volume of 2× SDS sample buffer (± 2-mercaptoethanol; see note 1 below). When the zymogram is initially performed, it is useful to load a range of protein concentrations on the gel, as it is difficult to predict how much enzyme is required to observe activity on a zymogram. The amount of activity observed on a zymogram is largely dependent on the amount of renaturation of the enzyme in the gel, which is unpredictable.

(v) Protein molecular weight markers (both prestained and standard) are also loaded onto the gel. Prestained protein markers are necessary in order to later assign the ribonuclease activity to a protein of a particular molecular weight (when gels are not stained after the assay). Unstained protein markers are useful if the gels are to be stained before the gel is dried and exposed to film or phosphorimager screen. Molecular weight markers that include RNase A may serve as a control for ribonuclease activity (see note 2 below).

(vi) Heat the protein samples at 95° for 2 min prior to electrophoresis. It is important to test the effects of heating on the ability to subsequently recover ribonuclease activity. Some enzyme activities may be irrecoverable after heating (see *note* 1 below).

(vii) Electrophoresis of the proteins is carried out at a constant current of 25 mA until the bromphenol blue reaches the bottom of the gel (approximately 100 min for a 15% gel).

(viii) Following electrophoresis, transfer the gel to a baked glass dish that contains approximately 150 ml of an appropriate assay buffer plus 20% (v/v) 2-propanol. Incubation in 2-propanol is necessary to remove SDS from the gel prior to the ribonuclease assay.[12] Incubate the gel with gentle agitation on a rocking platform for 15 min at room temperature. A glass plate should be placed over the dish during incubations to minimize exposure to radioactivity, and all manipulations should be performed behind a Perspex screen. It is important to note that some buffers used for incubations of the gel may become radioactive. At the completion of the wash step the buffer is removed using a 50 ml syringe. This wash step should be repeated once.

(ix) The gel is next subjected to two incubations (15 min each) in approximately 150 ml of an appropriate assay buffer at room temperature. During these incubations the isopropanol is washed out of the gel and renaturation of the proteins occurs. If the ribonuclease is active at room temperature, it may be possible to detect activity following these wash steps (and step x may be omitted).

(x) If necessary, incubate the gel in fresh assay buffer under conditions equivalent to those used for the standard ribonuclease assay (in solution). The amount of time necessary for detection of activity is dependent on the amount of active enzyme present in the gel after renaturation. As this quantity is not possible to predict, it is necessary to optimize the incubation time. Indeed, it is possible to monitor the progress of the assay by exposing the wet gel to film at regular time intervals. As the amount of enzyme or the assay time is increased, the activity band is likely to become more diffuse. It is thus important to keep the assay times as short as possible.

(xi) The zymograms may be stained with either Coomassie blue or silver nitrate, using standard methods (we usually omit this step). In any staining method used the gel should be incubated for the shortest possible time to avoid unnecessary degradation or diffusion of the RNA from the gel. It is important to note that when activity gels are stained with silver nitrate, the RNA may also become a little stained. This is consistent with the previous use of silver nitrate as a negative stain for ribonuclease zymograms.[15]

(xii) Place the gel on 4–5 sheets of Whatman 3MM paper (to prevent contamination of the gel drier), dry under vacuum, and expose to phosphorimager screen or film. It may be necessary to get several different exposures of the activity gel to see the degradation of the RNA most clearly. It is possible to see activity with a relatively short exposure (30–60 min), but a longer exposure will generally give a better contrast between intact and digested RNA (see note 3 below).

Notes

Note 1: Ideally, proteins are separated on polyacrylamide gels under conditions where they are fully denatured; i.e., where they are both treated with a reducing agent and boiled prior to electrophoresis. However, it is unlikely that all enzymes can be treated in this manner and subsequently renatured for recovery of activity. For example, the ribonuclease p29 is inactive on a zymogram following treatment with a reducing agent, although it is insensitive to treatment with a reducing agent when assayed in a conventional (solution phase) assay. In contrast, recovery of p29 activity on a zymogram is unaffected by boiling the sample prior to electrophoresis (without reducing agent). It is therefore important to test the activity of the ribonuclease fraction in the presence or absence of a reducing agent. The sample should also be tested $\pm$ boiling prior to electrophoresis.

If the protein sample cannot be either reduced or boiled, it is important to note that the enzyme may only be partially denatured prior to electrophoresis, and that the mobility on the zymogram may therefore not reflect the true subunit size. Any subsequent gels that rely on information provided by the zymogram (for example, for peptide sequencing) should be performed under equivalent conditions.

Note 2: In the assay with p29, the protein markers used included RNase A (approximately 250 ng/lane), the presence of which provided a useful internal control for ribonuclease activity. However, as the protein markers were prepared in the presence of a reducing agent the RNase A activity is largely inhibited. The amount of RNase A used thus gave a weak signal compared with the activity of interest (see Fig. 1A).

Alternatively, RNase A alone may be used as a control for ribonuclease activity. In this case, a titration of RNase A (in the absence of reducing agent) should be loaded onto a zymogram in order to determine the optimal concentration to give an appropriate signal.

Note 3: As outlined above, it is necessary to get several different exposures of the zymogram to get an optimal picture. Exposure times will depend on the specific activity of the RNA used for the zymogram. Although exposure of the film is faster at $-80°$, the bands will be sharper if the gel is exposed at room temperature. Figure 2 shows two exposures of the same zymogram, one at room temperature (60 min) and the other at $-80°$ (45 min). Although the exposure times are different, it is easy to see that the exposure at room temperature gives sharper bands and that the greater contrast allows all of the bands to be seen more clearly. It is best to start with a relatively short exposure at room temperature (where all of the bands are seen but the contrast is less) and then increase the exposure time to optimize the contrast.

One-Dimensional Zymograms using p29 and RNase A

Various zymograms were performed to assay p29 activity, according to the method outlined above. Samples of p29 were normally electrophoresed on 15% SDS polyacrylamide gels under nonreducing conditions. Following steps i–ix, p29 activity was analyzed by incubating the gels in 20 m*M* HEPES–KOH, pH 7.9, 5 m*M* $MgCl_2$ at room temperature for an additional 0–60 min. Figure 1A shows a titration of p29 (approximately 1500 fmol–3.5 fmol) where the gel was incubated in assay buffer for 30 min. At low enzyme concentrations a single band of

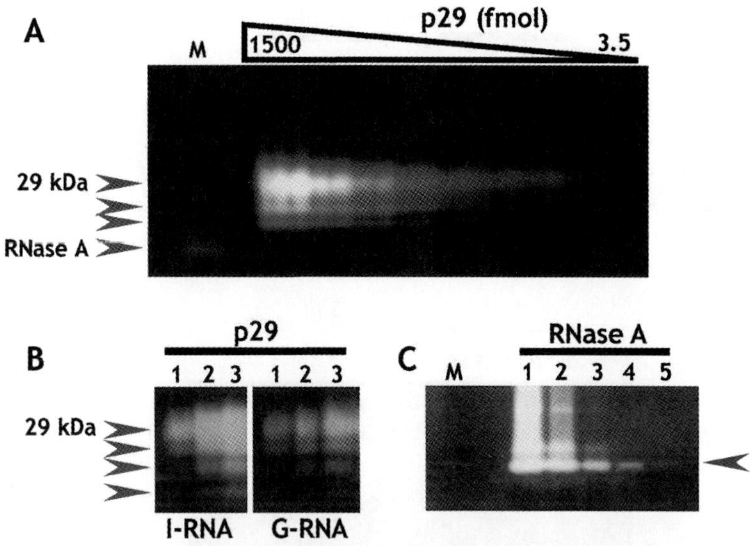

FIG. 1. Analysis of ribonuclease activity using 1D zymograms. (A) A titration of p29 (approximately 1500–3.5 fmol) was loaded onto the I-RNA-containing gel and the in-gel assay was performed at room temperature for 30 min (in 20 m*M* HEPES–KOH, 5 m*M* $MgCl_2$). At low enzyme concentrations a single band of 29 kDa was observed. At the highest enzyme concentration, 3 or 4 bands were detected. It is likely that the additional bands represent degradation products of p29 that retain ribonuclease activity. Lane M contains protein molecular weight markers (including 250 ng RNase A, + reducing agent). (B) Duplicate zymograms were performed that contained either G- or I-RNA as substrate. The amount of p29 loaded onto these gels was equivalent, although the amount of protein was too low to be quantitated. Lanes 1, 2, and 3 in each panel contain 0.5, 1, and 2 ml of p29, respectively. More degradation was observed with the I-RNA substrate than with the G-RNA substrate (compare panels). This is consistent with the accelerated degradation of I-RNA in conventional liquid phase assays. (C) A titration of RNase A (without reducing agent) was loaded onto a G-RNA gel (lanes 1, 2, 3, 4, and 5 contain 100 ng, 10 ng, 1 ng, 100 pg, and 10 pg RNase A, respectively). The gel was exposed to film immediately after step ix of the method (no further incubation in 20 m*M* HEPES–KOH, 5 m*M* $MgCl_2$ following the wash steps). Although multiple bands were seen at the highest enzyme concentration, a single band was observed with either 100 or 10 pg RNase A. Lane M contains protein molecular weight markers (including 250 ng RNase A, + reducing agent).

approximately 29 kDa was observed. At the highest enzyme concentration, 3 or 4 bands were detected. It is likely that the additional bands represent degradation products of p29 that retain ribonuclease activity. The lane labeled M contains protein molecular weight markers, including 250 ng RNase A (with reducing agent). A faint band corresponding to RNase A activity was observed.

As p29 had been shown to degrade inosine-containing RNAs (I-RNA) at a faster rate than guanosine-containing RNAs (G-RNA), duplicate zymograms were performed that contained either G- or I-RNA (Fig. 1B). In this particular experiment

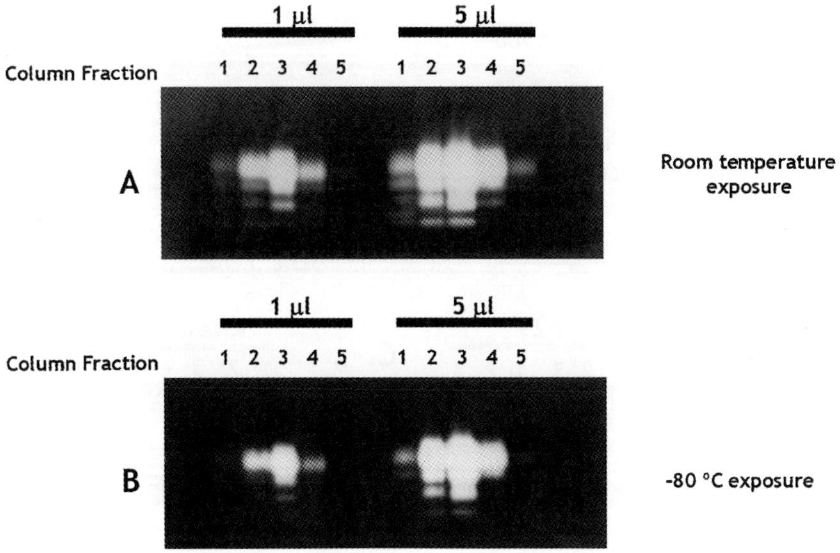

FIG. 2. Exposure of zymograms to film. Samples of octyl-Sepharose column fractions 1–5 (1 and 5 ml of each) were loaded onto a 15% polyacrylamide gel and an in-gel assay was performed for 60 min at room temperature. The amount of protein loaded onto the gel was insufficient for silver staining. (A and B) Results of the assay following exposure of the gel under different conditions. (A) Film that has been exposed to the assay gel for 60 min at room temperature; (B) film that has been exposed to the gel for 45 min at −80°C. Although the exposure times are different, it is easy to see that the exposure at room temperature gives sharper bands and that the contrast is better to see all bands.

it was impossible to quantitate the amount of protein loaded onto the gel, as there was so little it could not be detected by silver staining. Nevertheless, equivalent amounts of p29 were loaded onto the gels and the in-gel assays performed at room temperature for 60 min. The higher rate of degradation of I-RNA seen in solution phase assays was also clearly demonstrated using zymograms.

In addition to p29, RNase A was assayed using a G-RNA zymogram (Fig. 1C). For this analysis a titration of RNase A (100 ng–10 pg; without reducing agent) was loaded onto a 20% polyacrylamide gel and assayed using the same buffer conditions as used to assay p29 (20 m*M* HEPES–KOH, 5 m*M* $MgCl_2$). Following steps i–ix (above) the gel was dried and exposed to film with no further incubation in assay buffer. At the highest concentration of RNase A (lane 1; 100 ng) multiple bands were observed and the substrate RNA was heavily degraded. Nevertheless, the predominant band corresponded to a molecular weight of approximately 14 kDa. At lower enzyme concentrations (lanes 4 or 5; 100 or 10 pg, respectively) the ribonuclease activity was restricted to a discrete band of approximately 14 kDa. Protein molecular weight markers, including RNase A (250 ng; with reducing agent) were also loaded onto the gel (lane M). A very faint band corresponding to RNase A activity was observed.

Two-Dimensional Zymograms

As an extension to the basic method used to monitor ribonuclease activity in SDS polyacrylamide gels, we have also detected enzyme activity following 2D gel electrophoresis.

Although the ribonuclease p29 was purified approximately 80,000-fold from a pig brain extract, peptide sequencing indicated that it comigrated on SDS–PAGE with a more abundant protein (Erp29[19]). It was therefore impossible to obtain peptide data corresponding to p29 following the separation of proteins on a one-dimensional polyacrylamide gel. We thus attempted to separate p29 from other proteins using a two-dimensional gel system. In order to identify the spot on the 2D gel corresponding to p29 for subsequent peptide analysis, an in-gel assay was performed following the second dimension. The method used for the 2D zymogram is outlined below.

Materials

These materials are in addition to those outlined above for one-dimensional gels.

Isoelectric focusing (IEF) performed in a Multiphor II apparatus (Amersham).

SDS–PAGE is performed in a PROTEAN II xi Multicell apparatus (Bio-Rad, Hercules, CA), employing $18.5 \times 19 \times 0.15$ cm uniform polyacrylamide gels ($T=$ 12%).

Pharmacia Immobiline DryStrips, 18 cm, pH 3-10 NL

Ampholines, pH 3-10 (Pharmacia)

IEF sample buffer: 8 *M* urea, 2 *M* thiourea, 4% CHAPS, 0.005% (w/v) bromphenol blue

Reequilibration buffer: 50 m*M* Tris-HCl, pH 6.8, 2% SDS, 6 *M* urea, and 30% (v/v) glycerol

Agarose overlay: 0.5% (v/v) agarose (in running buffer) with a trace of bromphenol blue

Glass dish (25 cm $\times$ 35 cm; baked at 200° for 4–6 h)

Method

The 2D zymogram is carried out after the activity assay has been optimized using a 1D zymogram. Similar precautions and conditions should thus be applied to the 2D assay as determined for the 1D assay.

[19] J. Demmer, C. M. Zhou, and M. J. Hubbard, *FEBS Lett.* **402,** 145 (1997).

(i) Add the protein sample to IEF sample buffer to give a total volume of 350 μl. As for the 1D zymograms (above), samples should be prepared in the presence or absence of a reducing agent as appropriate (see note 1 above). Ampholines (8 μl of pH 3–10) should also be added to the protein sample.

(ii) Reswell the dehydrated Immobiline DryStrips with the protein sample (350 μl) overnight under mineral oil cover (to prevent evaporation).

(iii) Proteins are separated by isoelectric focusing using the following conditions: 300 V, 2 hr, gradient to 3500 V, 3 hr, and finally 3500 V for 18 hr. Focusing was performed under a mineral oil cover; 2D protein markers (Bio-Rad) may be included for subsequent determination of the isoelectric point and molecular weight corresponding to ribonuclease activity (as described for 1D zymograms).

(iv) Following 1D electrophoresis the strips are equilibrated for 2×10 min in reequilibration buffer, before being placed on top of the SDS–PAGE gel and embedded in the agarose overlay.

(v) As described above for the 1D zymogram, the polyacrylamide gel (12% acrylamide) used for the second dimension contained radiolabeled substrate RNA (0.5 μg/ml). Following transfer of the first dimension strip gel to the SDS–polyacrylamide gel, electrophoresis in the second dimension should be carried out using standard conditions (40 mA for 5–5.5 hr).

(vi) At the completion of electrophoresis, carry out the in-gel ribonuclease assay as described above (steps viii–xii). Although gels of different thickness are used for the 1D and 2D zymograms, the conditions of the ribonuclease assay determined for the 1D assay are likely to be similar for the 2D zymogram. It is important to note, however, that proteins separated in two dimensions have been denatured using both urea and SDS, whereas those separated in one dimension have only been treated with SDS. The renaturation efficiency of the enzyme may thus differ following fractionation in two dimensions compared with one.

2D Zymogram using p29

In an assay with p29, approximately 5–15 μg *total* protein (and approximately 1–3 pmol p29) was added to the IEF sample buffer. As for the 1D zymograms (above) the reducing agent was omitted from the IEF sample buffer.

Although the method was successful in that ribonuclease activity was recovered after 2D electrophoresis, it indicated that p29 separated on the 2D gel as at least 13 isoelectric variants (Fig. 3). It is possible that the multiple spots may be the result of various posttranslational modifications such as glycosylation, phosphorylation, or the addition of lipid anchors. However, none of the spots corresponding to ribonuclease activity show any shift in mobility that would typically accompany such modifications. Further investigations are necessary in order to determine the basis of the isoelectric heterogeneity of p29.

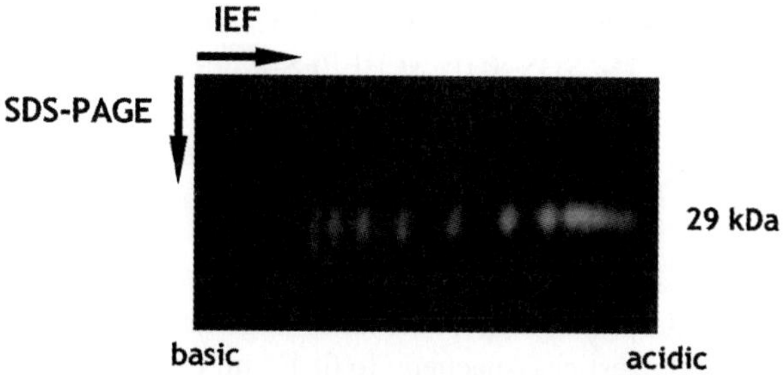

FIG. 3. Analysis of p29 activity using a 2D zymogram. p29 activity was analyzed on a 2D zymogram where approximately 1–3 pmol p29 were loaded onto the gel. The in-gel assay was performed for 45 min at room temperature following isoelectric focusing and SDS–PAGE. Although p29 activity was evident, at least 13 isoforms of p29 were observed across a broad pH range.

If the method had been successful in identifying a single spot corresponding to p29, a duplicate 2D gel (without RNA) would have been run, the spot corresponding to p29 excised, and the peptide data obtained. It was hoped that the presence of RNA in the gel matrix in the second dimension would not have significantly altered the electrophoretic mobility of the proteins. Staining of one-dimensional gels $\pm$RNA with silver nitrate suggested that the presence of RNA did not significantly affect protein migration.

Elution of Proteins from SDS–Polyacrylamide Gels

An alternative method for determining the approximate molecular weight of ribonucleases is elution of proteins from SDS–polyacrylamide gels followed by conventional activity assays in solution. The following method is based on that described by Hager and Burgess.[20]

Materials

Prestained molecular weight protein markers (NEB; Broad range)

β-Lactoglobulin (25 μg/μl), acetone (−20°), pyronine Y (0.5 mg/ml final concentration (in 1 × SDS sample buffer); prepared from 10 mg/ml stock (dissolved in ethanol).

Thioglycolic acid (final concentration 2 mM).

SDS–PAGE apparatus (we use gels of approximately 17 cm × 12 cm × 0.8 mm), polyacrylamide gel mixes (as above). Glass gel plates are baked at 200° for 4–6 hr prior to assembly, and Teflon spacers and combs are washed thoroughly with 2% SDS before use.

[20] D. A. Hager and R. R. Burgess, *Anal. Biochem.* **109,** 76 (1980).

2× protein sample buffer (as above; ± 2-mercaptoethanol).

Elution buffer: 0.1% SDS, 0.05 *M* HEPES–KOH, pH 7.9, 0.1 m*M* EDTA, 0.1 mg/ml BSA, 0.15 *M* NaCl.

Dilution buffer: 0.05 *M* HEPES–KOH, pH 7.9, 0.1 m*M* EDTA, 0.1 mg/ml BSA, 0.15 *M* NaCl, 20% glycerol.

Dilution buffer +6 *M* guanidine hydrochloride.

Razor blades or scalpel.

1.5 and 2 ml hydrophobic Eppendorf tubes (Anachem).

Plastic disposable pestles (Anachem; to fit 1.5 ml Eppendorf tubes).

Rotating mixer (suitable for Eppendorf tubes).

Method

(i). Cast the resolving gel (10–20%) 12 hr prior to use to ensure complete polymerization of the acrylamide.

(ii). Prepare the ribonuclease sample (see notes 4 and 5, below) for electrophoresis by adding an equal volume of 2× protein sample buffer. It is recommended that the protein be mixed with sample buffer in the absence or presence of 2-mercaptoethanol as appropriate (see note 1, above).

(iii). Heat the protein samples at 95° for 2 min prior to electrophoresis. Note that some ribonucleases may be denatured irreversibly by boiling (see note 1, above).

(iv). Load the ribonuclease samples onto the gel, with as much distance between samples as possible to avoid contamination. This is achieved by leaving several empty wells between samples. Pyronine Y may be loaded into wells adjacent to those containing the samples to enable subsequent location of the lane(s) of interest. Prestained markers may also be loaded onto the gel in a manner that will aid later identification of lanes.

(v). Electrophoresis of the proteins is carried out at a constant current of 25 mA until the bromphenol blue reaches the bottom of the gel (approximately 100 min for 15% gel). At the completion of electrophoresis the gel should be photographed immediately, alongside a ruler for subsequent identification of gel slices.

(vi). Excise the lane(s) of interest and cut into 5 mm slices using a clean razor blade or scalpel. Transfer the gel slices to 1.5 ml hydrophobic Eppendorf tubes, and soak each slice in 1 ml water for 5 min. Repeat this wash step once.

(vii). Add 400 μl elution buffer to each Eppendorf tube and then crush the gel slice with a plastic pestle. Incubate the crushed gel slices at room temperature for approximately 12 hr on a rotating mixer (rotating slowly).

(viii). Pellet the crushed gel slices by centrifugation at top speed for 1–2 min in a microcentrifuge. Transfer the supernatants (approximately 250–300 μl) to 2 ml hydrophobic tubes. Precipitate the proteins by adding 1.3 ml cold acetone (−20°) and then incubating the tubes in a dry ice/ethanol bath for 30 min. Recover the precipitated proteins by centrifugation for 10 min at top speed in a microcentrifuge.

It is important to note that the pellets are quite small. Dry the pellets by inverting the tubes to drain.

(ix). Dissolve the protein pellets in 20 μl dilution buffer containing 6 M guanidine hydrochloride. Incubate the tubes at room temperature for 20 min, and then dilute the guanidine hydrochloride 50-fold by adding dilution buffer (final volume 1 ml). Incubate the proteins at room temperature overnight to allow renaturation.

(x). Assay the eluted proteins for ribonuclease activity using a conventional solution phase assay. Note that the amount of guanidine remaining (50 mM) may be sufficient to interfere with the assay. It may be therefore be necessary to dialyze the protein samples prior to analysis. Alternatively, different amounts of enzyme may be tested; if guanidine interferes with the assay, higher enzyme activity may be seen when less of the fraction is assayed.

Notes

Note 4: Sufficient enzyme should be loaded onto the gel to enable detection of the activity following dilution (final volume 1 ml). As it is impossible to predict how much active protein will be recovered from the gel, it is advisable to load more protein than normally necessary for detection of activity.

Note 5: In the case of p29, β-lactoglobulin (25 μg) was also added to the ribonuclease sample. As β-lactoglobulin (18.4 kDa) is smaller than p29, it migrates faster on the gel and therefore reacts with any impurities in the gel that may otherwise react with the protein of interest. Alternatively, thioglycolic acid (2 mM final concentration) may be added to the upper electrode buffer. This mobile thiol will run ahead of the protein and scavenge any free radicals that may affect subsequent recovery of enzyme activity.

Note 6: If sufficient protein is initially loaded onto the gel, the eluted proteins may be visualized by staining with Coomassie blue or silver nitrate following electrophoresis on a second SDS–polyacrylamide gel. However, it is important to note that the amount of BSA added (in elution buffers) may interfere with visualization of the protein(s) of interest.

Recovery of p29 following SDS–PAGE

A sample of p29 was electrophoresed on a 15% SDS–polyacrylamide gel and proteins were then eluted according to the method outlined above. Following elution of the proteins from gel slices (11 slices in total), the eluates were assayed for p29 activity (Fig. 4).

To assay for p29 activity, 2 μl of each gel slice eluate was incubated with 50 fmol ^{32}P-labeled I-RNA or G-RNA substrates for 0 or 30 min (in solution). Assays were stopped by the addition of 1 ml 10% TCA, which precipitated any RNA that was not completely degraded. Following incubation on ice, the

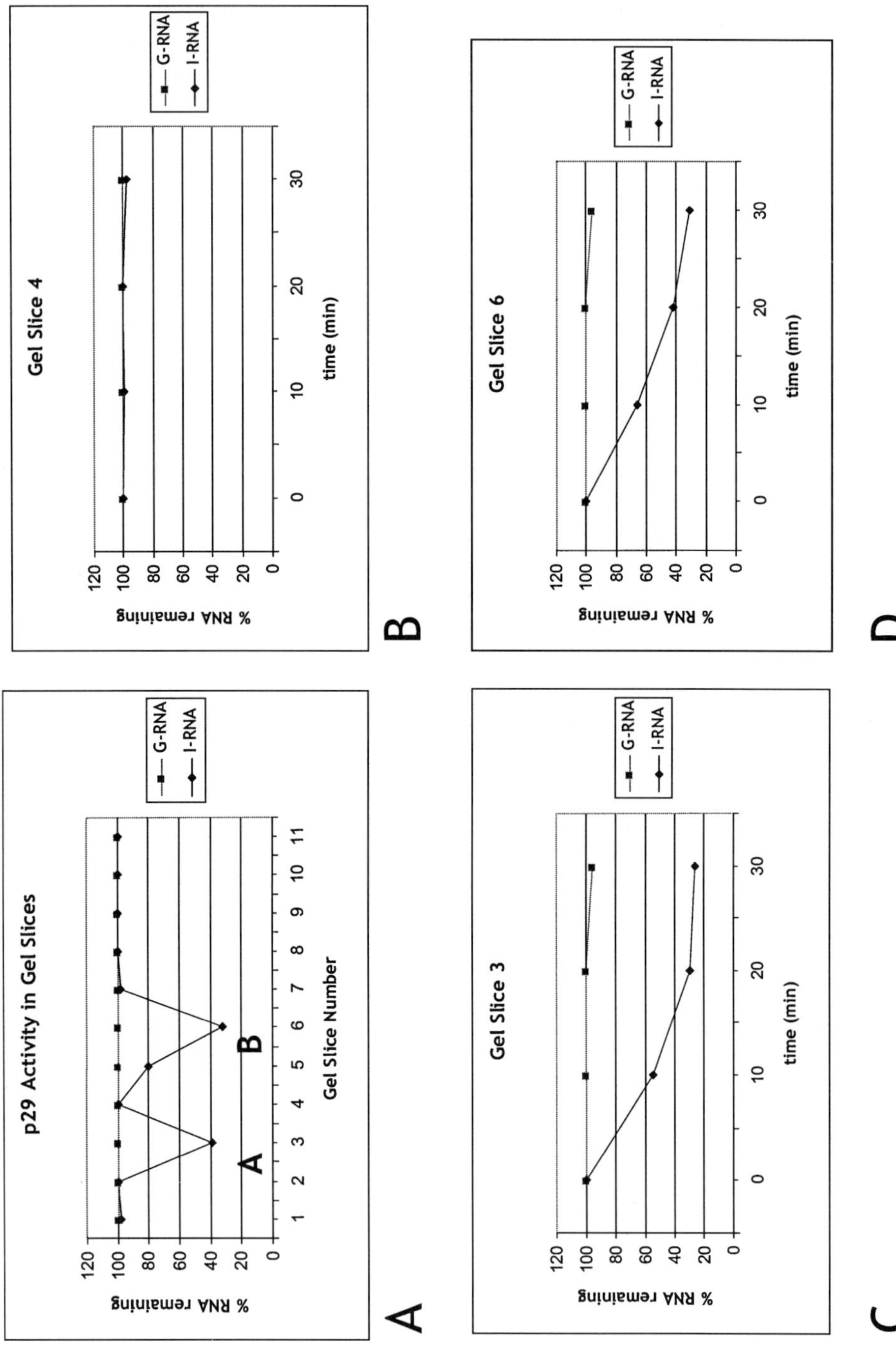
p29 Activity in Gel Slices
% RNA remaining
Gel Slice Number
G-RNA
I-RNA
A
B
Gel Slice 4
time (min)
Gel Slice 3
Gel Slice 6
C
D

TCA-precipitated RNA was filtered onto glass fiber circles under vacuum and the amount of RNA remaining measured by scintillation counting. This assay showed that ribonuclease activity was restricted to 2–3 gel slices (corresponding to two peaks of activity) and that the amount of nonspecific degradation in other gel slices was negligible (Fig. 4A). When the gel slices were correlated with the molecular weight markers in the gel photograph (step v), it was possible to determine the approximate molecular weight of the proteins with ribonuclease activity. Activity peak B (gel slice 6 mainly) corresponds to a molecular weight of approximately 27–34 kDa. This is consistent with the apparent molecular weight of the ribonuclease determined using zymography. Activity peak A (gel slice 3) is likely to contain fragments of p29 that retain ribonuclease activity and that give rise to additional bands on zymograms at high concentrations of p29 (Fig. 1).

Further assays were carried out using eluted proteins from gel fractions 3, 4 and 6, where both I- and G-RNAs were incubated with 2 μl of the fraction for a time course of 0–30 min. The results of this analysis are shown in Figs. 4B–4D. Gel slices 3 and 6 contained ribonuclease activity that degrades I-RNA at a faster rate than the equivalent G-RNA, whereas both I- and G-RNAs incubated with gel slice 4 remain undegraded. We estimate that approximately 10% of the enzyme activity loaded onto the gel was recovered using the outlined method. The observed efficiency of recovery of ribonuclease activity was consistent with that described for other enzymes.[20]

Acknowledgments

Chris Smith is gratefully acknowledged for many helpful discussions and also for critically reading the manuscript. This work was supported by the Wellcome Trust (Grant 052241).

FIG. 4. Elution of p29 from an SDS polyacylamide gel. A fraction containing p29 was electrophoresed using SDS–PAGE. Following electrophoresis, the gel was cut into 11 slices and the proteins were eluted and tested for ribonuclease activity. Two peaks of ribonuclease activity were observed, as seen in (A). Most activity was found in gel slices 3 and 6. (B–D) Assays where the eluted proteins are assayed by incubation with I- or G-RNA over a time course of 0–30 min. In gel slices 3 and 6 I-RNA is degraded at a faster rate than an equivalent G-RNA substrate, whereas degradation of either I- or G-RNA by proteins eluted from gel slice 4 is negligible.

[9] Ribonuclease Assays Utilizing Toluidine Blue Indicator Plates, Methylene Blue, or Fluorescence Correlation Spectroscopy

By KERSTIN KORN, THOMAS GREINER-STÖFFELE, and ULRICH HAHN

Introduction

To investigate ribonuclease (RNase)-mediated cleavage of its presumed natural substrates—long RNA molecules—a variety of different assays have been described.

An assay first reported by Kunitz[1] and modified by Oshima *et al.*[2] monitors the change in absorbance of RNA solutions at 298.5 nm on digestion. This has the disadvantage that proteins present in the sample absorb at the same wavelength, leading to insensitivity at high protein concentrations. This frequently occurs in the course of an RNase purification procedure. Furthermore, this assay is not practical at RNA concentrations below $0.2\ \mathrm{mg \cdot ml^{-1}}$. The assays initially described by Anfinsen *et al.*[3] and modified by Uchida and Egami[4] or by Corbishley *et al.*,[5] which consist of (i) a 15 min incubation followed by the addition of perchloric acid containing uranyl acetate or lanthanum nitrate, (ii) centrifugation, (iii) dilution, and (iv) absorbance measurements, are very time-consuming and/or labor-intensive. The "zymogram methods" (Yasuda *et al.*,[6] Nadano *et al.*,[7] Bravo *et al.*[8]) that have a sensitivity down to the picogram range of RNases require much more time and are not useful for detailed kinetic studies.

Here we present first a method for the rapid detection of microbial colonies that (over)produce ribonucleases, using toluidine blue O indicator plates. This test also allows a qualitative judgment of the nuclease content of more or less impure sample solutions (e.g., in the course of RNase purification). Furthermore, we present two assays to study the degradation of larger RNA substrate molecules using either methylene blue or fluorescence correlation spectroscopy (FCS).

[1] M. Kunitz, *J. Biol. Chem.* **164,** 563 (1946).
[2] T. Oshima, N. Uenishi, and K. Imahori, *Anal. Biochem.* **71,** 632 (1976).
[3] C. B. Anfinsen, R. R. Redfield, W. L. Choate, J. Page, and W. R. Carroll, *J. Biol. Chem.* **207,** 201 (1954).
[4] T. Uchida and F. Egami, *in* "Proceedings in Nucleic Acids Research" (G. L. Cantoni and D. R. Davis, eds.), p. 46. Harper & Row, New York, 1966.
[5] T. P. Corbishley, P. J. Johnson, and R. Williams, *in* "Methods of Enzymatic Analysis" (H. U. Bergmeyer, J. Bergmeyer, and M. Grassl, eds.), Vol. 4, p. 134. Verlag Chemie, Weinheim, 1984.
[6] T. Yasuda, D. Nadano, E. Tenjo, H. Takeshita, and K. Kishi, *Anal. Biochem.* **206,** 172 (1992).
[7] D. Nadano, T. Yasuda, K. Sawazaki, H. Takeshita, and K. Kishi, *Anal. Biochem.* **212,** 111 (1993).
[8] J. Bravo, E. Fernandez, M. Ribo, R. de Llorens, and C. M. Cuchillo, *Anal. Biochem.* **219,** 82 (1994).

0076-6879/01 $35.00

Toluidine Blue O Indicator Plates

Principle

The dye toluidine blue O (TB) shows an absorption maximum at 625 nm. As a cationic compound it interacts with different polyanions, leading to complexes of different absorption maxima (metachromatic effect). Depending on the kind of polyanion and the concentration of each partner a bathochromic or hypsochromic shift in absorption occurs. If TB is added to molten bacterial growth media consisting of nutrient containing agar that contains sulfate esters it forms a red complex after solidification of the agar. If high molecular weight RNA (another polyanion) is additionally present in TB agar plates, the plate color is blue. This is due to the fact that TB forms a blue complex with the RNA. These solid bacterial growth media are ideal to serve as ribonuclease indicator plates: Colonies grown on these plates that (over)produce and/or secrete an RNase form red or pink halos, in contrast to controls that do not produce significant amounts of these enzymes (Fig. 1).[9,10] The smaller (oligo-)nucleotides resulting from RNase-mediated digestion are not capable for forming the blue complex. A similar assay has been described for DNase test agar systems.[11,12]

We have used these TB indicator plates to identify RNase T1 overproducing and secreting *Escherichia coli* clones[13] or M13 phages presenting RNase A or T1 on their surfaces.[14,15] The plates are also convenient for the detection of RNase T1 variants obtained after oligonucleotide-directed specific or random mutagenesis.[16,17]

Furthermore, holes punched into the solid medium with the aid of a pipette tip and filled with aliquots of column chromatography fractions show red halos after short incubation at 37°, if RNases are present.

Method

1. Autoclave usual growth medium (e.g., LB medium) containing 1.5% common agar and TB (Merck, Darmstadt) at a final concentration of 75 mg liter^{-1} in an appropriate flask. Do not forget to add a stirring bar before autoclaving.

[9] R. Quaas, Ph.D. Thesis, Free University of Berlin (1989).

[10] R. Quaas, O. Landt, H.-P. Grunert, M. Beineke, and U. Hahn, *Nucl. Acids Res.* **17,** 3318 (1989).

[11] C. D. Jeffries, D. F. Holtmann, and D. G. Guse, *J. Bacteriol.* **73,** 590 (1957).

[12] M. M. Streitfield, M. Hoffmann, and H. M. Janklow, *J. Bacteriol.* **84,** 77 (1962).

[13] R. Quaas, Y. McKeown, P. Stanssens, R. Frank, H. Blöcker, and U. Hahn, *Eur. J. Biochem.* **173,** 617 (1988).

[14] B. Hubner, K. Korn, H.-H. Förster, and U. Hahn, *Nucleosides Nucleotides* **16,** 727 (1997).

[15] K. Korn, H.-H. Förster, and U. Hahn, *Biol. Chem.* **381,** 179 (2000).

[16] O. Landt, J. Thölke, H.-P. Grunert, W. Saenger, and U. Hahn, *Biol. Chem.* **378,** 553 (1997).

[17] B. Hubner, M. Haensler, and U. Hahn, *Biochemistry* **38,** 1371 (1999).

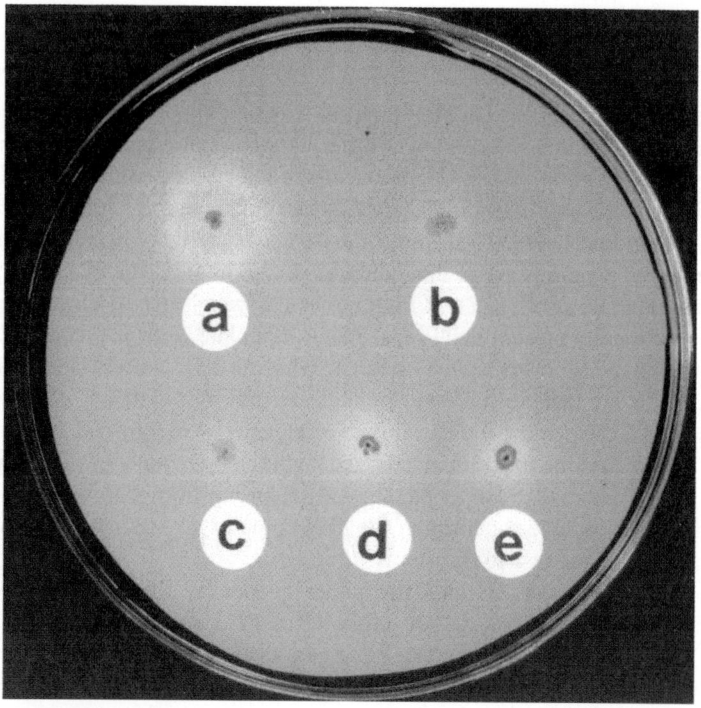

FIG. 1. RNase indicator plate containing LB agar, RNA, and toluidine blue O. The colonies consist of *Escherichia coli* DH5α cells overproducing and secreting RNase T1 wild type (a) or variants thereof with lowered activities (c–e). As a control untransformed *E. coli* (b) was inoculated. From O. Landt, J. Thölke, H.-P. Grunert, W. Saenger, and U. Hahn, *Biol. Chem.* **378,** 553 (1997).

2. Let the medium cool down to 50–60° (this is the temperature at which you can touch the flask without getting burned).
3. Add 2 g of yeast RNA (Boehringer Mannheim) per liter, and the appropriate antibiotic.
4. Stir carefully to avoid air bubbles.
5. Pour plates and handle them as usual. The solution may have a yellow color at higher temperatures that will turn blue during solidification.

Methylene Blue Assay

Principle

Dye molecules such as methylene blue (an acridine derivative) are known to interact with oligonucleotides by intercalation into the nucleic acids. This

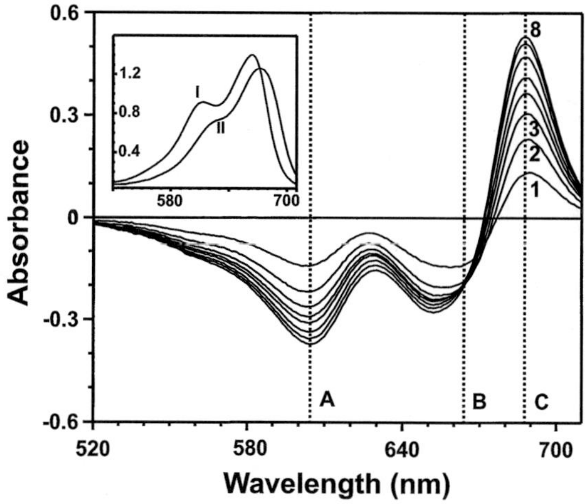

FIG. 2. Difference spectra of methylene blue in complex with various amounts of RNA. The spectrophotometer was calibrated against methylene blue buffer without RNA. RNA concentrations (mg RNA · ml^{-1}) were 0.1 (spectrum 1), 0.2 (2), 0.3 (3), 0.4 (4), 0.5 (5), 0.6 (6), 0.7 (7), and 0.8 (8), respectively. Range A, methylene blue dimer band; B, methylene blue monomer band; C, band of methylene blue intercalated into RNA. The inset shows spectra of methylene blue buffer without RNA (I) and with 0.8 mg RNA · ml^{-1} (II) against buffer. From T. Greiner-Stoeffele, M. Grunow, and U. Hahn, *Anal. Biochem.* **240,** 24 (1996).

intercalation does not interfere with hydrogen bonding between base pairs of the two complementary strands.[18]

The addition of RNA to methylene blue causes a shift of the absorbance maximum of the dye monomer from 653 to 688 nm (see Fig. 2).[19] The shift indicates intercalation of methylene blue molecules into the RNA. The change in absorbance at 688 nm with increasing RNA concentration corresponds to a hyperbolic function (Fig. 3).[20] For a given methylene blue concentration one can generate a calibration curve up to 1500 μg RNA ml^{-1}. However, the accuracy is reduced above 1000 $\mu g \cdot ml^{-1}$ because of the relatively small changes in absorbance with increasing RNA concentrations.

Enzymatic digestion as well as alkaline hydrolysis of the RNA methylene blue complex causes a decrease in absorbance at 688 nm.

Method

1. Dissolve 100 mg yeast RNA in 10 ml MOPS buffer (0.1 *M* MOPS-HCl, pH 7.5; 2 m*M* EDTA).

[18] J. Pritchard, A. Blake, and A. R. Peacocke, *Nature* **212,** 1360 (1966).
[19] J. Ruprecht, Ph.D. Thesis, University of Freiburg, Freiburg (1981).
[20] T. Greiner-Stoeffele, M. Grunow, and U. Hahn, *Anal. Biochem.* **240,** 24 (1996).

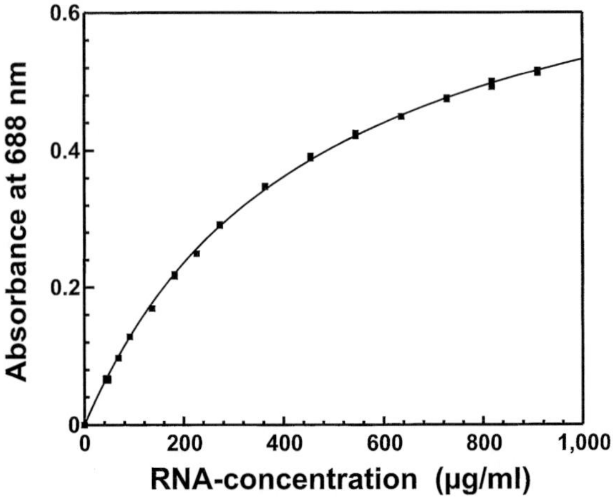

FIG. 3. Calibration curve for the determination of RNA concentrations. From T. Greiner-Stoeffele, M. Grunow, and U. Hahn, *Anal. Biochem.* **240,** 24 (1996).

2. Prepare methylene blue buffer by dissolving 1 mg methylene blue in 100 ml MOPS buffer. Because of the light sensitivity of the methylene blue, the buffer must be stored in the dark.
3. Adjust the absorbance at 688 nm of methylene blue buffer to 0.5 ± 0.01 using MOPS buffer.
4. Mix 2.5–100 μl of RNA solution with methylene blue buffer to a final volume of 1 ml in cuvettes with a light path of 10 mm.
5. Preincubate the sample at 25° for 10 min in the dark.
6. Add RNase solution immediately before measurement. The enzyme should be applied in the smallest volume possible to reduce dilution effects.
7. Record the hydrolysis for 1–30 min (less than 10% of the initial RNA amount should be degraded).

Different RNA preparations vary significantly in their compositions because of the differing size distribution of the RNA molecules. Since samples that vary in purity or isolation procedure lead to different results in the assay, we recommend the use of an identical batch or at least the same lot of RNA substrate in all related studies. Otherwise the comparison of different enzymes will not provide reproducible results.

We determined the range in which the absorbance change is proportional to the enzyme concentration by varying the RNase T1 concentration, and keeping the substrate concentration at 0.8 mg · RNA ml^{-1}. The reaction was started by the addition of 10 μl RNase solution to the preincubated substrate solution. The absorbance change was recorded immediately between 1 and 3 min (Fig. 4B). A linear relationship could be demonstrated between the rate of absorbance change

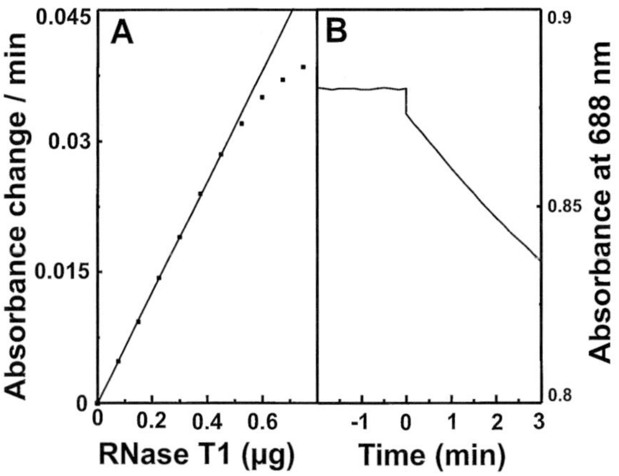

FIG. 4. Relationship between the rate of absorbance change and the amounts of RNase T1. (A) The reactions were initiated by adding the indicated amount of RNase T1 enzyme to 1 ml of RNA solution (0.8 mg ml^{-1}) in methylene blue buffer. The absorbance change was recorded at 688 nm. (B) Time course of absorbance change of RNA–methylene blue mixture at 688 nm upon treatment with RNase T1. Enzyme (0.15 μg) was added to 1 ml RNA solution (0.8 mg $\cdot$ ml^{-1}) in methylene blue buffer at time 0. From T. Greiner-Stoeffele, M. Grunow, and U. Hahn, *Anal. Biochem.* **240,** 24 (1996).

and the amount of RNase T1, up to an absorbance change of 0.025 min^{-1}. This corresponds to an enzyme concentration of 0.4 μg ml^{-1}, as shown in Fig. 4A. It is important to ensure that the "no enzyme blank" does not exceed 0.0005 min^{-1}.

For this assay one unit of enzyme activity is defined as the amount of RNase that leads to an absorbance change of 0.01 min^{-1} at 25° at a substrate concentration of 0.8 mg RNA ml^{-1} methylene blue buffer. This corresponds to 60 units according to Uchida and Egam.[4]

The complete enzymatic digestion by RNase T1 leads to a final absorbance corresponding to about 20% of the initial RNA concentration, whereas alkaline hydrolysis yields a considerably lower value. For the digestion of the RNA methylene blue complex with RNase A and Nuclease S1 similar final values are obtained. This phenomenon is probably caused by the fact that a mixture of enzyme cleavage products (mono-, di-, and oligonucleotides) and undigestible RNA molecules remain after the incubation.

FCS Assay

Principle

Fluorescence correlation spectroscopy (FCS) is based on the analysis of thermal fluctuations of fluorescent particles excited by a laser. Using the principle of confocal microscopy the focused laser beam creates an open illuminated sample volume as small as a bacterial cell (fl).[21–23] Fluorophores or molecules that are labeled with fluorescent dyes are excited during passage through the sample volume and a burst of photons is generated. The intensity fluctuations of the emitted photons of these molecules are statistically analyzed via an autocorrelation function to get information on the molecular processes that induce the fluctuations. These fluctuations can be caused by translational and rotational diffusion and conformational transitions, as well as by molecular processes such as photochemical reactions. $G'(t)$ describes the normalized autocorrelation function of the signal $I(t)$ that is fluctuating around its mean value $\langle I \rangle$ by $\delta I(t)$:[24,25]

$$G'(t) = \langle I \rangle^2 + \langle \delta I(t)\, \delta I(t+\tau) \rangle / \langle I \rangle^2 \tag{1}$$

The number (N) and the diffusion times (τ) of fluorescent molecules as well as the relation of different fluorescent species (y) that are present in the measuring volume can be calculated following the translational diffusion and using a corresponding fitting model for the analysis of the data. Since the diffusion time depends on the radius, and thus on the molecular weight of the molecules ($\tau \sim \sqrt[3]{M_r}$), a sufficient weight difference of the species is a basic requirement for reliable measurements. Binding, accumulation, and cleavage processes can be easily followed by FCS based on the relation of the molecular weight and τ of the molecules. In the following part experiments for the determination of RNase activity via FCS are described.

Method

Substrate Preparation. Since the FCS technique is based on the analysis of intensity fluctuations of fluorescent particles, only fluorophores or target molecules that are labeled with specific fluorescent dyes can be detected. Previous experiments have shown that the commonly used derivatives of the fluorescent dye rhodamine tend to interact with guanine residues.[26,27] Therefore we have chosen a gapped heteroduplex as an RNase substrate, which provides the opportunity to introduce biotin as a second label.[28] The heteroduplex (Fig. 5) consists of a riboligonucleotide (rOligo, 3′-GUG CGA AUA AGG GAU CCG UUC UGA GAA GCC GAC UA-5′, 35-mer) hybridized to two smaller deoxyribooligonucleotides (dOligo) that are complementary to the corresponding ends of the rOligo. One dOligo is labeled with rhodamine B isothiocyanate (Rho-dOligo, 5′-CAC GCT

[21] R. Rigler and J. Widengren, *BioScience* **3,** 180 (1990).

[22] R. Rigler, Ü. Mets, J. Widengren, and P. Kask, *Eur. Biophys. J.* **22,** 169 (1993).

[23] M. Eigen and R. Rigler, *Proc. Natl. Acad. Sci. U.S.A.* **91,** 5740 (1994).

[24] M. Ehrenberg and R. Rigler, *Chem. Phys.* **4,** 390 (1974).

[25] E. L. Elson and D. Magde, *Biopolymers* **13,** 1 (1974).

[26] L. Edman, Ü. Mets, and R. Rigler, *Proc. Natl. Acad. Sci. U.S.A.* **93,** 6710 (1996).

[27] J. Widengren, J. Dapprich, and R. Rigler, *Chem. Physics.* **216,** 417 (1997).

[28] K. Korn, S. Wennmalm, H.-H. Förster, U. Hahn, and R. Rigler, *Biol. Chem.* **381,** 259 (2000).

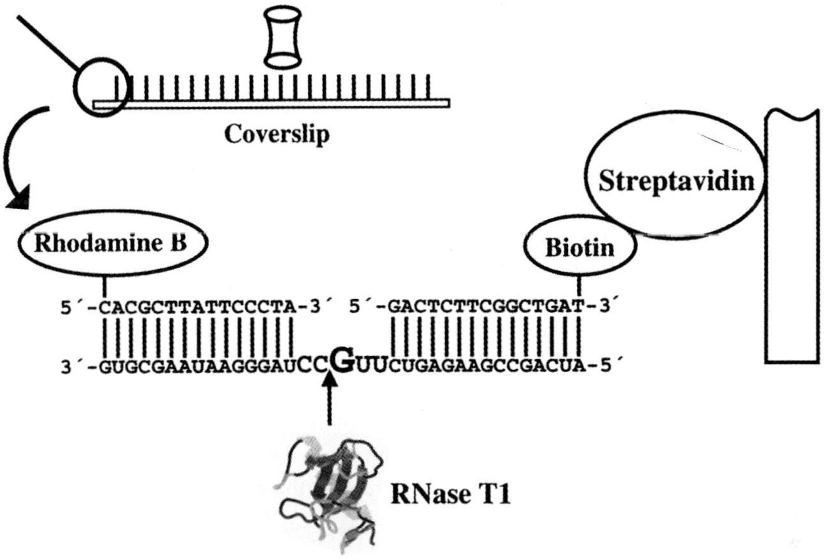

FIG. 5. Substrate for the determination of RNase T1 activity via FCS. A gapped heteroduplex of a ribooligonucleotide and two smaller complementary deoxyribooligonucleotides was used. One dOligo was 5′-labeled with rhodamine B (Rho-dOligo). The other one was 3′-labeled with biotin (Bio- dOligo). The RNase T1 cleavage site is marked by an arrow. The sample volume within the reaction solution droplet is shown above the coverslip. For further details see text. From K. Korn, S. Wennmalm, H.-H. Förster, U. Hahn, and R. Rigler, *Biol. Chem.* **381,** 259 (2000).

TAT TCC CTA-3′, 15-mer), and the other with biotin (Bio-dOligo, 5′-GAC TCT TCG GCT GAT-3′, 15-mer). The relatively stiff conformation of the RNA/DNA hybrid probably protects the single stranded ribooligonucleotide part in the middle (five bases including one guanine residue) from the interacting rhodamine and also prevents the formation of secondary structures by the RNA.

For the FCS measurements the hybridization is carried out with a total amount of 200 pmol of each oligonucleotide in ultrapure water or in a corresponding buffer. Alternatively, the amount of the unlabeled oligonucleotides can be an excess of the amount of the dye-labeled oligonucleotide to make sure that no surplus of the dye-labeled single stranded molecules is left. The sample solutions (50 μl) are denatured at 80° for 10 s and cooled down slowly to room temperature (within 2 hr).

FCS Measurements. For rhodamine B a helium/neon (488 nm) or an argon ion laser (514 nm) should be used. Simple coverslips, capillaries, or different chambers (Nalge Nunc International, Wiesbaden, Germany or Whatman Polyfiltronics, Springfield, UK) are suitable as sample holders. Measurements are carried out with sample droplets having a volume of 20–100 μl. The time of the measurements can vary between 10 and 120 sec depending on the fluorescence intensity obtained.

Initially the diffusion time of the free unlabeled fluorophore (τ_{free}) should be determined for each specific setup. Depending on the size of the illuminated sample volume (differing from device to device as well as from laser to laser) small diffusion times (in the range of 0.25–0.6 ms) are obtained. For the measurements of the hybridization products, the samples have to be diluted with buffer until a fluorescence intensity of 20–100 kHz is observed. Measurements of pure buffer are strongly recommended. Different reagent lots of different manufacturers should be tested beforehand. Since a background fluorescence of buffer ingredients is often observed, the fluorescence intensity of the sample solutions should be at least 10-fold higher compared to background. Measurements should be carried out step by step, starting with the Rho-dOligo, followed by the addition of the rOligo to yield finally the complete substrate by adding Bio-dOligo. The autocorrelation functions are obtained from the different hybridization experiments (Rho-dOligo/rOligo; Rho-dOligo/rOligo/Bio-dOligo). Based on the data for the free dye, a model assuming the existence of two molecular species with different sizes and diffusion constants is used for evaluating the autocorrelation function[29]:

$$G(t) = 1 + \frac{1}{N}\left[\frac{1-y}{(1+t/\tau_{free})\sqrt{1+\left(\frac{\omega_0}{z_0}\right)^2 t/\tau_{free}}} + \frac{y}{(1+t/\tau_{bound})\sqrt{1+\left(\frac{\omega_0}{z_0}\right)^2 t/\tau_{bound}}}\right] \tag{2}$$

N is the average number of fluorescent molecules per volume element, $\tau_{free} = \omega_0{}^2/4D_{dye}$, $\tau_{bound} = \omega_0{}^2/4D_{dOligo}$, ω_0 is the radius of the volume element, $2z_0$ is its length, and y is the fraction of Rho-dOligo and its hybridization products, respectively. D_{dye} is the translational diffusion constant of the free dye, D_{dOligo} is the translational diffusion constant of the Rho-dOligo and its hybridization products, respectively. Table I shows the results of different hybridization experiments with the substrate mentioned above using a custom-made FCS setup.[29] The hybridization of the rOligo with the dOligos can be followed by an increase in the diffusion time.

For cleavage experiments RNase T1 is prepared as described[30] and added to a final concentration of 50 n*M*. After an incubation time of 15 min the measurements are carried out in MES buffer (100 m*M* MES/NaOH, 100 m*M* NaCl, 2 m*M* EDTA, pH 6.0). A pronounced difference in diffusion time between the substrate and the leaving group is essential for the quality of the FCS results. Because of the

[29] M. Kinjo and R. Rigler, *Nucl. Acids Res.* **23,** 1795 (1995).
[30] O. Landt, M. Zirpel-Giesebrecht, A. Milde, and U. Hahn, *J. Biotechnol.* **24,** 189 (1992).

TABLE I
DIFFUSION TIMES OF RHODAMINE B-LABELED DEOXYRIBONUCLEOTIDE (dOligo) AND HETEREODUPLEXES[a]

Hybridization experiment		Diffusion time τ (msec)
Rho-dOligo		0.8 ± 0.1
Rho-dOligo/rOligo		1.2 ± 0.2
Rho-dOligo/rOligo/Bio-dOligo		1.4 ± 0.2
Rho-dOligo/rOligo	+ RNase T1	1.0 ± 0.3
Rho-dOligo/rOligo/Bio-dOligo	+ RNase T1	1.0 ± 0.2

[a] Before and after the addition of RNase T1 to a final concentration of 50 n*M*. The measurements were carried out with a custom-made FCS setup, partly manufactured by Spindler and Hoyer (Göttingen, Germany). The fluorescent dye-labeled substrate was excited at 514.5 nm. A model assuming the existence of two molecular species with different sizes and diffusion constants [M. Kinjo and R. Rigler, *Nucl. Acids Res.* **23,** 1795 (1995)] for the evaluation of the autocorrelation function was used. From K. Korn, S. Wennmalm, H.-H. Förster, U. Hahn, and R. Rigler, *Biol. Chem.* **381,** 259 (2000).

mass difference between the uncut (44 kDa) and the cut substrate (22 kDa), the assumed difference in diffusion times should be about 20%. After the cleavage of both hybridization products (Rho-dOligo/rOligo; Rho-dOligo/rOligo/Bio-dOligo) in solution the diffusion times decreases to 1 ms (Table I). This corresponds to the expected difference between the gapped heteroduplex and the cut substrate. The system can be easily applied for a fast screening of a few samples for RNase activity. In the case of kinetic studies or for screening huge amounts of different samples, larger substrate molecules should be available. This could be achieved by using longer RNA molecules or by binding the gapped heteroduplex substrate to other big particles. Alternatively, the substrates can be immobilized directly via the biotin label to the surface of streptavidin coated coverslips.

Immobilization of Substrates to Coverslips. For the immobilization procedure it is helpful to fix the coverslips in a container made of Teflon. This should fit into a beaker to allow the solutions to be stirred.[31]

1. Clean coverslips (160–180 μm thick) by boiling in 5% HNO_3 (v/v) for about 1 hr.
2. Dry at 50° for 1–2 hr.
3. Incubate in 10% 3-glycidyloxypropyltrimethylsilane in toluene (v/v) for 24 hr.

[31] S. Wennmalm, L. Edman, and R. Rigler, *Proc. Natl. Acad. Sci. U.S.A.* **94,** 10641 (1997).

4. Remove the coverslips from the Teflon container and dry them at room temperature for 30 min.

5. Dry at 120° for 6 hr.

6. While cooling the coverslips down to room temperature, prepare a streptavidin solution (0.75 $\mu g \cdot ml^{-1}$) in carbonate buffer (0.2 *M* $NaHCO_3/Na_2CO_3$, pH 9.0).

7. Apply droplets of 10 μl of the streptavidin solution to the coverslips (4–6 per plate) and incubate at room temperature in closed petri dishes under water-saturated atmosphere for 2–3 days.

8. Mark the positions of the droplets with a pen.

9. Wash the coverslips twice using carbonate buffer.

10. Fix the coverslips in the Teflon container and incubate them in Tris/HCl buffer (0.5 M, pH 8.6) for 12 hr.

11. Store in TE buffer (10 m*M* Tris/HCl, pH 7.6, 2 m*M* EDTA) at 4° for up to 2 weeks.

12. Incubate droplets of the hybridization products at the streptavidin coated spots for 1 hr.

13. Remove unbound substrates by washing several times with MES buffer.

14. Leave 20 μl MES buffer at each substrate spot.

FCS Measurements of Immobilized Substrate. The focus of the measuring volume is adjusted at approximately 200 μm above the substrate layer (Fig. 5). The background of the MES buffer in the presence of the bound substrate has to be recorded first without addition of RNase. Because of the distance of the measuring volume from the bound substrate, the background intensity is comparable to measurements in buffer over plain coverslips. Although no fluorescence-labeled molecule should be present in solution at this stage, autocorrelation functions are often obtained. This results from buffer components or from substrates leaving the coverslip. Because RNase is diluted in TE buffer, an additional blank has to be monitored after the addition of 1 μl of TE buffer.

The hydrolysis of the gapped heteroduplex substrate by RNases is monitored online by recording the change in intensity. On hydrolyzation, the rhodamine B–labeled part of the substrates is released into solution, and subsequently passes the measuring volume. After the addition of high amounts of enzyme a fast transient increase in intensity leading to a maximum is observed. This is followed immediately by a fast decrease indicating that the hydrolytic cleavage rate is comparable to the rate at which the products leave the observation element by diffusion. Because of the high enzyme concentration nearly all substrate molecules are cleaved rapidly. Because of the small distance (200 μm) of the measuring volume to the bound substrate, the burst of fluorescing reaction products first passes the region of the measuring volume before the products diffuse into the whole droplet. After the reaction is followed for 1–10 min, a new measurement is started to yield

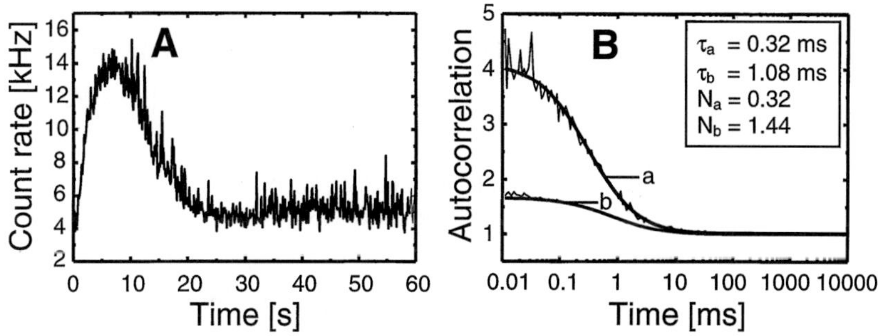

FIG. 6. FCS measurement after the addition of RNase T1 to a final concentration of 50 n*M*. (A) Time course of the product appearance in the sample volume monitored by following the change in fluorescence intensity. RNase T1 was added to a droplet of 20 μl on a coverslip containing immobilized gapped heteroduplex substrate. (B) Comparison of the two autocorrelation functions before the addition of RNase T1 (a) and after the reaction (b). Measured (thin line) and calculated (thick line) autocorrelation curves are shown. τ, Diffusion time; *N*, number of molecules. From K. Korn, S. Wennmalm, H.-H. Förster, U. Hahn, and R. Rigler, *Biol. Chem.* **381,** 259 (2000).

the autocorrelation function together with the number of molecules statistically present in the measuring volume and the diffusion time. Figure 6 shows a measurement after the addition of RNase T1 to a final concentration of 50 n*M* (A) and also autocorrelation functions that were obtained before and after the addition of the enzyme (B). The number of molecules increases to 1.4 and a diffusion time of 1.1 ms corresponds to the data of the hydrolyzed substrates in solution (Table I). Measurements with lower enzyme concentrations lead to a lower increase of intensity, since the substrate molecules are hydrolyzed consecutively. Thus we succeeded to detect the activity of about 600 RNase T1 molecules (about 10^{-21} mol) in a droplet of 20 μl.[28]

The described method allows the fast and sensitive determination of the hydrolytic activity of RNases and variants with low activity. This method permits the screening of a large collection of different RNase variants. Apart from the analysis of the cleavage of RNA by RNase T1 or RNase A, this approach can also be used to study other hydrolyzing enzymes.

Acknowledgments

We thank Hans-Heinrich Förster and Peter Weber for critical reading of the manuscript. This work was supported by the Deutsche Forschungsgemeinschaft and the Fonds der Chemischen Industrie.

[10] Ribonuclease Activities of Trypanosome RNA Editing Complex Directed to Cleave Specifically at a Chosen Site

By BARBARA SOLLNER-WEBB, LAURA N. RUSCHÉ, and JORGE CRUZ-REYES

Nuclease activities that have unique specificities are critical components of the RNA editing complex which catalyzes U deletions and U insertions in trypanosomes and related organisms. These include endonuclease activities that cleave RNA precisely at the first phosphodiester bond 5′ of a duplex region and an exonuclease specific for unpaired U residues. Our recent data suggest that the endonucleases can be directed to cleave specifically at virtually any desired phosphodiester bond in a chosen RNA, merely by synthesizing an appropriate guiding oligonucleotide. Thus, in addition to being important in the field of RNA editing, these nuclease could prove useful to researchers in manipulating RNA.

Background

Trypanosome RNA editing is an amazing form of RNA processing in which U residues are inserted into and deleted from primary mitochondrial transcripts to generate mature protein-encoding sequences[1] (reviewed in Refs. 2–8). It can be very extensive, with U residues inserted at hundreds of discrete sites in a single mRNA to contribute over half the mature protein-encoding length. This RNA editing involves endonuclease and exonuclease activities with specificities virtually unique among RNases. The endonuclease activities cleave at Y-branch RNA junctions precisely between the (5′) single-stranded and (3′) double-stranded residues created by pairing of a *trans*-acting guide RNA. These cleavage reactions for U deletion and U insertion pathways are optimized under notably different conditions, depending on whether the first single stranded residue at the Y-branch is a U or another nucleotide[9]; both generate 3′-OH and 5′-P termini.[10] The

[1] R. Benne, J. Van Den Burg, J. Brakenhoff, P. Sloof, J. Van Boom, and M. Tromp, *Cell* **46,** 819 (1986).

[2] G. Arts and R. Benne, *Biochim. Biophys. Acta.* **1007,** 39 (1996).

[3] B. Sollner-Webb, *Science* **273,** 1182 (1996).

[4] H. Smith, J. Gott, and M. Hanson, *RNA* **3,** 1105 (1997).

[5] S. Hajduk, *Trends Microbiol.* **5,** 1 (1997).

[6] K. Stuart, T. Allen, S. Heidmann, and S. Seiwert, *Microbiol. Molec. Biol. Rev.* **61,** 105 (1997).

[7] J. Alfonzo, O. Thiemann, and L. Simpson, *Nucleic Acids Res.* **25,** 3751 (1997).

[8] M. Kable, S. Heidmann, and K. Stuart, *TIBS* **22,** 162 (1997).

[9] J. Cruz-Reyes, L. Rusché, K. Piller, and B. Sollner-Webb, *Mol. Cell* **1,** 401 (1998).

0076-6879/01 $35.00

exonuclease is U-specific, degrades from a 3′-OH end to generate UMP, and is dependent on abutting single stranded character.[11,12] These nucleases and the other activities needed to catalyze RNA editing cycles are associated in a complex that can be readily purified from *Trypanosoma brucei* mitochondria,[11] and the enzymes are active on RNAs of natural or completely artificial sequence.[10] U editing occurs in all examined kinetoplastid protozoa (see Ref. 13 *and reference therein*), but only from *T. brucei* have the editing cycles been efficiently reproduced *in vitro*,[16,18,21] the nuclease cleavage reactions characterized,[9,10] or the enzymatic complex purified.[12]

Trypanosome RNA editing occurs through repeated cycles of U insertion and U deletion, each involving three sequential reactions[14–17] (Fig. 1). First the mRNA is cleaved by one of the above-noted endonuclease activities. Next U residues are added or removed at the newly created 3′ end by the 3′-U-exonuclease or a terminal-U-transferase. The mRNA then is rejoined by a RNA ligase. This editing is directed by short separately encoded guide RNAs (gRNAs) that are complementary to segments of edited mRNA; they consequently mismatch with the preedited mRNA at each site to be edited and these mismatches specify the U alterations.[14] The initial gRNA corresponds to the 3′ end of the editing domain and several resides beyond, so its 5′ region can base pair to the pre-mRNA, creating an "anchor duplex" (see Fig. 1). This pairing structure directs an endonuclease to cleave at its upstream border, beginning the first editing cycle. Following each editing cycle, the anchor duplex "zips up" to the next mismatch to direct cleavage at this next editing site. The central portion of a typical gRNA directs 5–10 adjoining editing cycles; its 3′ portion is an oligo-U that provides a universal tether to help retain the cleaved upstream pre-edited mRNA that is virtually poly-purine (Fig. 1). Overlapping similarly organized gRNAs can then sequentially anchor, so editing progresses 3′ to 5′ along the mRNA. At U deletion sites, the single stranded mRNA residue abutting the anchor duplex is a U, which will be deleted, and the opposing gRNA residue is a pyrimidine, which does not base pair with the U (Fig. 1; mRNA : gRNA base pairing involves G : U as well as Watson–Crick interactions). At U insertion sites, the single-stranded mRNA residue abutting the anchor duplex is a non-U (generally a purine), and the opposing nonpairing gRNA residue is a purine, to guide the U insertion. Natural gRNAs are presumably optimized for their aggregate

[10] K. J. Piller, L. N. Rusché, J. Cruz-Reyes, and B. Sollner-Webb, *RNA* **3,** 279 (1997).

[11] L. Rusché, J. Cruz-Reyes, K. Piller, and B. Sollner-Webb, *EMBO J.* **16,** 4069 (1997).

[12] J. Cruz-Reyes, A. Zhelonkina, and B. Sollner-Webb, unpublished observations.

[13] D. Blom, A. deHaan, M. van den Berg, P. Sloof, M. Jirku, J. Lucas, and R. Benne, *Nucleic Acids Res.* **26,** 1205 (1998).

[14] B. Blum, N. Bakalara, and L. Simpson, *Cell* **60,** 189 (1990).

[15] S. Seiwert, S. Heidmann, and K. Stuart, *Cell* **84,** 831 (1996).

[16] M. Kable, S. Seiwert, S. Heidmann, and K. Stuart, *Science* **273,** 1189 (1996).

[17] J. Cruz-Reyes and B. Sollner-Webb, *Proc. Natl. Acad. Sci. U.S.A.* **93,** 8901 (1996).

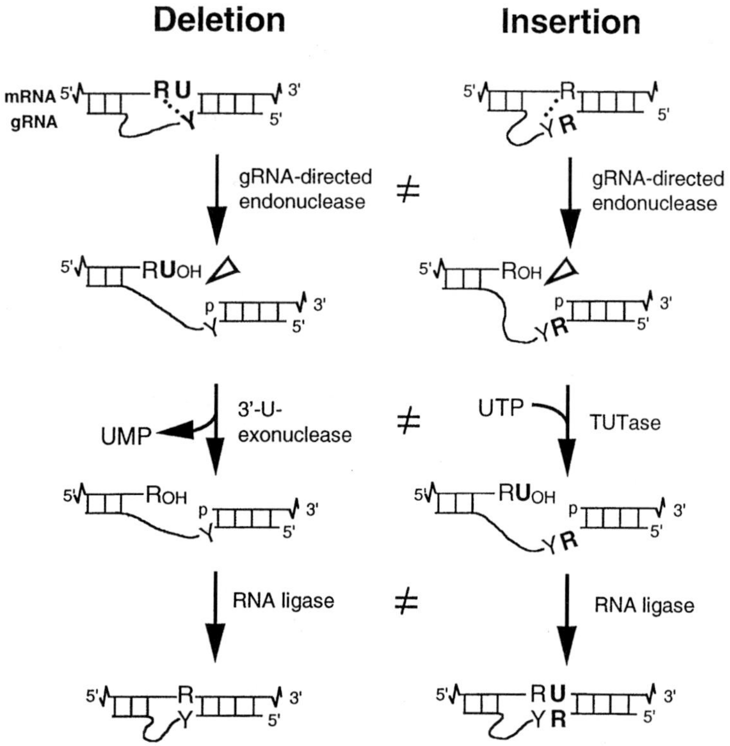

FIG. 1. Current understanding of U-deletion and U-insertion. U deletion and U insertion involve gRNA-directed endonuclease cleavage of the pre-mRNA (shown by an arrowhead), next a 3′-U-exonuclease or terminal-U-transferase acting on the upstream fragment, and then RNA ligase rejoining the mRNA.[15–17] These two forms of editing use distinct cleavage activities[9] and primarily distinct ligase activities,[25,25a] and the 3′-U-exo is not a reverse TUTase reaction.[11,17] Both forms of editing are catalyzed by a complex with seven major polypeptides.[11,22] gRNAs have three main portions[14]: a 5′ region to anchor the gRNA to the mRNA just downstream of the editing site and direct the cleavage, a central region to guide the U deletions and insertions at mismatches in base pairing with the pre-mRNA, and a 3′ oligo (U) that could tether the very purine-rich upstream pre-mRNA. R, Purine(s); Y, pyrimidine(s). Dotted lines indicate potential base pairs.

in vivo function, but we have found that simple artificial gRNAs can be designed that direct individual cycles of *in vitro* U deletion or U insertion much more efficiently.[12,18] Efficient and precise endonuclease cleavage can be directed using only a downstream pairing oligonucleotide, which is effectively an anchor region without the central guiding or 3′ tether regions of a normal gRNA.[12,15]

Description of many methods used in the *in vitro* study of RNA editing was presented in two previous articles in *Methods: A Companion to Methods*

[18] J. Cruz-Reyes, A. Zhelonkina, L. Rusché, and B. Sollner-Webb, *Mol. Cell Biol.* **21,** 884 (2001).

in Enzymology.[19,20] Since their preparation, the field has progressed to include purifying the RNA editing activities, understanding properties of the RNase activities it contains, and designing artificial gRNAs to optimize editing cycles and direct cleavage at desired locations. The present article will focus on these advances. However, to enable purification of these RNases without referencing other articles, previously described methods for isolating trypanosome mitochondria and assaying RNA ligase are also described.

Trypanosoma brucei RNA Editing Complex

The ability to catalyze individual cycles of U deletion and U insertion *in vitro* using mitochondrial extracts[16,21] enabled experimental study of the mechanism of RNA editing. One key finding was the cofractionation of the above-noted endonuclease and exonuclease activities, terminal-U-transferase, two RNA ligases, and the catalysis of both U deletion and U insertion editing cycles.[11,17,22] Q-Sepharose and DNA cellulose chromatography has enabled purification of an editing complex consisting of seven major polypeptides from *T. brucei*[11,23] (Fig. 2; protocols given below). All these activities and polypeptides evidently constitute a single complex since they fractionate together on these and other examined chromatographic matrices, on velocity centrifugation, and on nondenaturing gel electrophoresis in the absence or presence of bound RNA.[11,24] The Q-Sepharose and DNA cellulose purified complex represents $\sim 10^{-4}$ of the protein in the starting mitochondrial extract, with $\geq$500 fold purification of the quantified activities[11,23,24] (Table I and data not shown). The resultant seven major polypeptides (Fig. 2) appear approximately equimolar by silver or amido black staining, and their genes have now been cloned.[23,24] The ~55 kDa and ~50 kDa protein species constitute distinct RNA ligases[11,23]; one functions primarily in U deletion and the other in U insertion.[25,25a] However, the other proteins remain to be assigned specific enzymatic functions. This article describes purification of the enzymatic editing complex; presents properties, gRNA optimization, and potential utility of the endonuclease activities; summarizes aspects of the editing

[19] R. Sabatini, B. Adler, S. Madison-Antenucci, M. McManus, and S. Hajduk, "Methods: A Companion to Methods in Enzymology," Vol. 15, pp. 15–26. Academic Press, San Diego, 1998.

[20] K. Stuart, M. Kable, T. Allen, and S. Lawson, "Methods: A Companion to Methods in Enzymology," Vol. 15, pp. 3–14. Academic Press, San Diego, 1998.

[21] S. Seiwert and K. Stuart, *Science* **266,** 114 (1994).

[22] J. Cruz-Reyes, L. Rusché, and B. Sollner-Webb, *Nucleic Acids Res.* **26,** 3634 (1998).

[23] L. Rusché, C. Huang, K. Piller, M. Heeman, E. Wirtz, and B. Sollner-Webb, *Mol. Cell Biol.* **21,** 979 (2001).

[24] L. Rusché, K. Piller, M. Heeman, C. Huang, S. O'Hearn, and B. Sollner-Webb, unpublished observations.

[25] J. Cruz-Reyes, C. Huang, A. Zhelonkina, and B. Sollner-Webb, submitted.

[25a] C. Huang, J. Cruz-Reyes, A. Zhelonkina, and B. Sollner-Webb, *EMBO J.* (in press), 2001.

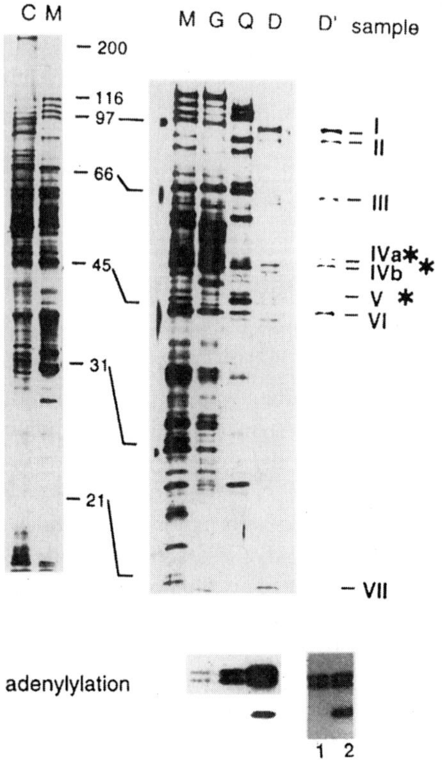

FIG. 2. Purification of the *T. brucei* RNA editing activities. Protein silver stain (*upper*) and autoradiogram (*lower*) of an SDS gel containing in the indicated lanes: (C), Whole cell lysate, 10^5 cell equivalents; (M), mitochondrial extract, 10^6 cell equivalents; (G), standard glycerol gradient fraction ~20S fraction, 5 μl; (Q), Q-Sepharose fraction, 5 μl; (D) and (D′), DNA-cellulose fraction from two different preparations, 2.5 μl (preparation 2) and 30 μl (preparation 3, more dilute). Fractions were adenylylated with [α-^{32}P]ATP prior to electrophoresis. Major bands are identified by roman numerals; asterisks indicate the position of the adenylylated species. Bands IV and V are RNA ligases[11] and that bands IVa and IVb are due to different allelic variants present in some but not all *T. brucei* strains.[23] The bottom right panel shows an autoradiogram of a Q-Sepharose fraction that was adenylylated without prior treatment (lane 1) or after deadenylylation with pyrophosphate followed by its inactivation with pyrophosphatase; this and Fig. 3C demonstrate that band V RNA ligase becomes adenylylated during purification. (Data from Ref. 11.)

exonuclease; and recounts relevant assays. A related article elsewhere in this series[26] reminds the reader that precise electrophoretic sizing of RNA in these and other assays requires marker ladders with the same terminal P or OH character as the experimental RNA.[27]

[26] B. Sollner-Webb, J. Cruz-Reyes, and L. Rusché, *Methods Enzymol.* **345,** [31] 2001.

[27] J. Cruz-Reyes, K. J. Piller, L. N. Rusché, M. Mukherjee, and B. Sollner-Webb, *Biochemistry* **37,** 6059 (1998).

TABLE I
PURIFICATION OF RNA EDITING COMPLEX[a]

Step	Total protein recovered	Activity recovered	Relative specific activity
T. brucei cell	~10×	ND	ND
Mitochondrial extract	≡1	≡1	≡1
Q-Sepharose, fraction 15, 16	1/300	1/6	50
Q-Sepharose, fraction 14–20		1/3	
DNA-cellulose, fraction 19, 20[b]	1/9,000	1/18	500
DNA-cellulose, fraction 18–22[b]		1/12	

[a] Throughout a purification of the RNA editing complex, the indicated fractions were assayed for RNA ligase activity (determined as in Fig. 2, using samples that had been titrated such that activity varied linearly with sample amount) and for total protein (concentration estimated by Bradford assays or silver staining). Values are reported relative to the mitochondrial extract (2–4 μg protein/10^7 cells). ND, Not determined. (Data from Ref. 11.)
[b] Normalized for loading the entire Q-Sepharose peak.

Preparation of T. brucei Mitochondrial Extract

We recommend following the protocol for preparation of *T. brucei* mitochondrial extract, basically as presented earlier in *Methods* by Sabatini *et al.*[19] *(see also Ref. 28),* but with addition of protease inhibitors. Generally 4–16 liter cultures of the procyclic form of *Trypanosoma brucei* (strain TREU667), grown in liquid at 27° to late-log phase (~1–1.5 × 10^{10} cells/liter), are used. The extract is prepared at 0–4° using filter-sterilized solutions. Cells are collected by centrifugation (Sorvall GS-3 rotor, 6000 rpm, 10 min), washed in 5–10% that volume of 0.15 *M* NaCl, 20 m*M* $NaHPO_4$ (pH 7.9), 20 m*M* glucose, removing as much medium as possible, and then suspended using a Dounce homogenizer (B pestle, five strokes), using 8 ml per 10^{10} cells, in 1 m*M* EDTA, 1 m*M* Tris-HCl (pH 8) freshly supplemented to 100 μg/ml Pefabloc, 50 μg/ml antipain, and 10 μg/ml E-64 (Sigma, St. Louis, MO). (The protease inhibitors can be stored as frozen stocks, Pefabloc and antipain at 400× in H_2O and E-64 at 2000× in 50% ethanol.) Cells are lysed in batches by rapid passage through a 26-gauge needle on a 10 ml syringe using maximal hand pressure, and the lysate is brought to 250 m*M* sucrose (adding 0.14 ml 60% sucrose/ml lysate). After centrifugation (Sorvall SS-34 rotor, 11,500 rpm, 10 min), the supernatant is removed with a pipette, and the pellet is suspended in 1/6 the previous lysate volume of 250 m*M* sucrose, 20 m*M* Tris-HCl (pH 8.0), 2 m*M* $MgCl_2$. It is supplemented to 5 m*M* $MgCl_2$ and 0.3 m*M* $CaCl_2$, treated with 9 μg/ml RNase-free DNase I (Life Technologies, Inc.) for 1 hr at 4°, then mixed with an equal volume of STE (250 m*M* sucrose, 20 m*M* Tris-HCl pH 8.0, 2 m*M* EDTA), and recovered by similar centrifugation. The pellet is suspended with a Dounce homogenizer in 1–1.25 ml 50% Percoll per 10^{10} original

[28] M. E. Harris, D. R. Moore, and S. L. Hajduk, *J. Biol. Chem.* **265,** 11368 (1990).

cells. Using a long sterile Pasteur pipette, it is layered under linear 20–35% Percoll gradients, 4 ml per 32 ml gradient in an SW28 centrifuge tube (Beckman). (Percoll solutions are in STE, prepared using sterile water and 2× STE.) Following centrifugation (SW28 rotor, 24,000 rpm, 40 min), the material between the upper and lower bands (~20 ml) is collected from above (using an 18-gauge needle on a 30 ml syringe), diluted at least 3-fold with STE, and sedimented (Sorvall SS-34 rotor, 16,500 rpm, 15 min). The soft pellet, removed with a pipette, is ~3 times washed with STE and resedimented, until the pellet is solid. (Mitochondrial vesicles can be examined microscopically using ethidium bromide[19]; appreciable amounts of the editing complex can also be in the upper and lower Percoll band regions, which can be similarly recovered.[29]) The mitochondrial vesicles are then suspended, at 0.4 ml per original 10^{10} cells, in MRB buffer (25 m*M* Tris-HCl, pH 8.0, 60 m*M* KCl, 10 m*M* magnesium acetate, 1 m*M* EDTA, and 5% glycerol) supplemented to 5 m*M* dithiothreitol (DTT), 1 mg/ml Pefabloc, 50 μg/ml antipain, 10 μg/ml E-64 and 0.5% Triton X-100. The supernatant obtained after 5 min sedimentation in a microfuge is combined with the supernatant obtained from re-extracting the pellets in 1/5 volume of this buffer, aliquoted, and stored at $-70°$. This extract, at 2×10^{10} cell equivalents/ml ($\leq$7 mg protein/ml), is stable for long periods.

All the activities of the editing complex can be assayed using this mitochondrial extract, usually 0.5–1 μl per 20 μl reaction; with active gRNAs the endonuclease can be scored using only 1/40 of a microliter of extract[25,29] (see below). Nonetheless, we find the most quantitative and convenient assays to compare different fractions and their subsequent purification are for the RNA ligase of the editing complex (see below).

Purification of T. brucei RNA Editing Complex

The editing complex is purified from mitochondrial extract by sequential chromatography on Q-Sepharose and DNA cellulose (see Rusché *et al.*[11,23]). These columns were selected because they bind the editing complex under conditions where the vast majority of the extract proteins flow through. The fractionation is at 4° in buffer P [25 m*M* Tris-HCl, pH 8.0, 10 m*M* $MgCl_2$, 1 m*M* EDTA, 5 m*M* DTT, and 10% (v/v) glycerol] supplemented with the amounts of KCl indicated in parentheses (in millimolar). The columns are prepared in disposable syringes, ~2–4 times as tall as wide, and are sized according to the amount of extract being fractionated (see below). Preparation of editing complex from ~4 liters of initial cells uses sequential gradient elutions from Q-Sepharose and DNA cellulose and typically yields purifications as shown in Fig. 2 and summarized in Table I, generating $\geq$1 ml of purified editing complex, of which 0.5–2 μl are generally used for most reactions. However, when very large amounts of the editing activities are

[29] C. Huang and B. Sollner-Webb, unpublished observations.

desired, it can be more convenient to elute the initial Q-Sepharose column in steps, followed by gradient elution from DNA cellulose. This yields preparations with only a few contaminating polypeptides visible by silver staining, which can then be further purified by gradient elution from as second, small Q-Sepharose column. The fractions are stored at -70° and are stable for years.

For the initial Q-Sepharose chromatography, use $\frac{1}{3}$ to 1 ml of column per 10^{10} cells worth of mitochondrial extract. The column is prewashed with $\geq$1 column volume of buffer P(350) and then with $\geq$3 column volumes of buffer P(100). The extract (above) is brought to 100 m*M* KCl and loaded directly onto the column at a flow rate adjusted to be ~0.25 cm vertical column height/min. When loading relatively smaller columns, the flow-through can be repassed to ensure complete binding of the editing activities. After washing the column with ~4 column volumes buffer P(100), sufficient to rinse through the unbound material (>90% of the loaded protein), bound protein is eluted with an ~8 column volume linear gradient, 100 to 350 m*M* KCl in buffer P, collecting fractions of 0.3–0.5 column volume (Fig. 3A). The KCl concentration of each fraction is determined by measuring conductivity; the editing complex peaks at 170–200 m*M* KCl. As noted above, we find most convenient to assay the fractions for the ligase of the editing complex, either its RNA joining (Fig. 3B) or adenylylation (Fig. 3C) activities. When assaying for endonuclease (Fig. 3D), be aware that another endonuclease of the extract elutes in the 250–350 m*M* range. It is not specific for Y branch structures but is selective within single-stranded regions[10,11]; in the case of cytochrome *b* pre-mRNA, its favored cleavage is adjacent to the editing site (Fig. 3D) but is not dependent on addition of gRNA. The fractions of the editing complex contain minimal detectable contaminating other endonuclease activity. The 3′-U- exo can also be assayed in these fractions (Fig. 3E; assay validated in Fig. 3F), as can full round U deletion,[11] although a contaminating other component may inhibit assay of the full round U insertion reaction (data not shown).

DNA-cellulose chromatography is used to further purify the peak fractions pooled from the initial Q-Sepharose column. The DNA-cellulose column is at least 1/5 the size of the previous Q-Sepharose column ($\geq$0.1 ml of column volume per 10^{10} cells worth of initial mitochondrial extract), or appropriately scaled down if only part of the Q-Sepharose product is loaded. The DNA cellulose column is prewashed in 2 column volumes of buffer P(350) and then 3 column volumes of buffer P(30). The Q-Sepharose fractions are either diluted or dialyzed to 30 m*M* KCl and then loaded as described above. After washing the column with 3–4 column volumes of buffer P(30), material is eluted with an ~8 column volume linear gradient, 30 to 350 m*M* KCl in buffer P, collecting fractions of 0.3–0.5 column volume (Fig. 4A). Again, the KCl concentration of each fraction is determined by measuring conductivity; the editing complex peaks at 85–120 m*M* KCl. We generally first assay the fractions for the ligase (Fig. 4B), but assays for gRNA-directed endonuclease (Fig. 4C) or any of the other activities of the

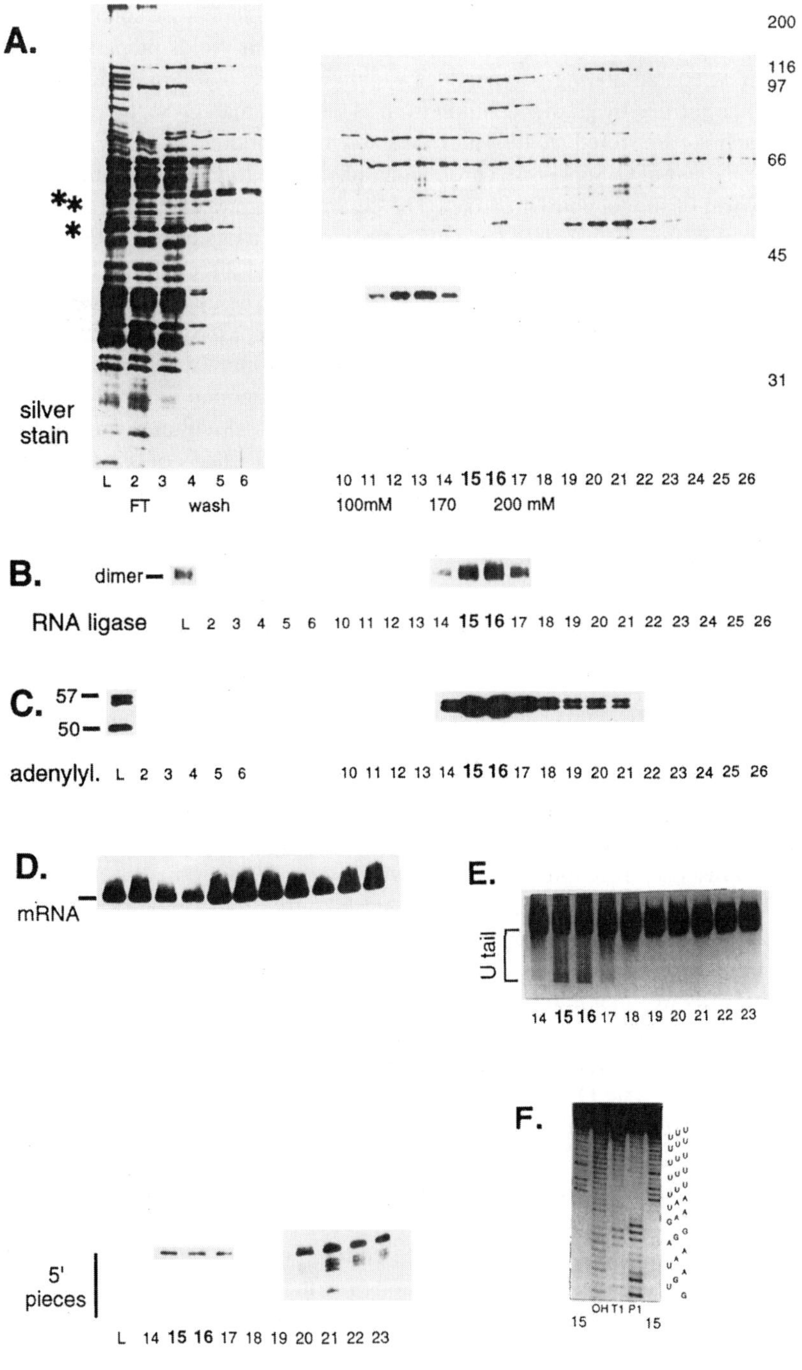
A.
200
116
97
66
45
31
silver
stain
L 2 3 4 5 6
FT wash
10 11 12 13 14 15 16 17 18 19 20 21 22 23 24 25 26
100mM 170 200 mM
B.
dimer
RNA ligase L 2 3 4 5 6 10 11 12 13 14 15 16 17 18 19 20 21 22 23 24 25 26
C.
57
50
adenylyl. L 2 3 4 5 6 10 11 12 13 14 15 16 17 18 19 20 21 22 23 24 25 26
D.
mRNA
5'
pieces
L 14 15 16 17 18 19 20 21 22 23
E.
U tail
14 15 16 17 18 19 20 21 22 23
F.
OH T1 P1
15 15

editing complex are similarly informative. The fractions are aliquoted and frozen directly at −70°.

When purifying from larger volumes of starting extract using an initial step elution from Q-Sepharose, the column is prepared and loaded as described above. After washing with 1 column volume buffer P(100) and 2 column volumes of buffer P(130), the editing complex is largely eluted with 2 column volumes of buffer P(210) and trailing activity is eluted with 2 column volumes of buffer P(310). Fractions are collected, assayed, and subjected to DNA cellulose chromatography as described above. Then a final gradient elution from Q-Sepharose can be performed, using a column prepared and run as described above but ~1/10 the size used for that amount of starting extract. Peak fractions pooled from the previous DNA cellulose chromatography are ~100–110 m*M* KCl and can be loaded directly. Resulting fractions are aliquoted and frozen at −70°.

gRNA-Directed Endonuclease

The *in vitro* systems most commonly used to study trypanosome editing focus on ATPase subunit 6 (A6) RNA and have been particularly informative since they faithfully reproduce the full round U deletion cycle at the natural first editing site (ES1) or the full round U insertion cycle at the natural second editing site (ES2). These A6 systems generally utilize a 3′-end-labeled synthetic pre-mRNA containing the 3′ end of the editing domain and an unlabeled cognate gRNA,

FIG. 3. Enrichment by Q-Sepharose chromatography. (**A**) Protein silver stain of 2 μl of the numbered fractions from Q-Sepharose chromatography (Q4) or 0.5 μl of the loaded mitochondrial extract (L), resolved on 10% SDS–PAGE. Fractions were adenylylated prior to electrophoresis. Protein molecular weight markers (Life Technologies, Inc.) are indicated. FT indicates flow through, and approximate KCl concentrations are shown. The asterisks indicate the approximate position of the adenylylatable polypeptides, which do not coincide with major bands detectable in the extract or fractions from this first column. (**B**) RNA ligase assays using 2 μl of each numbered fraction or 0.5 μl of the column load (L) and ~0.7 pmol substrate RNA per reaction. The dimer ligation product is shown. In this and the following figure, the fractions with peak ligase activity are indicated in bold. (**C**) Autoradiogram of the gel in (A). Approximate kilodaltons of the adenylated species are shown. (**D**) Assay for gRNA-directed endonuclease cleavage at editing site 1 of Cyb pre-mRNA (a U insertion site) using 5′ labeled pre-mRNA, unlabeled gRNA, and 2 μl of the indicated Q-Sepharose fraction or 0.5 μl of the column load. The fragment generated maximally in fractions 15 and 16 represents accurate gRNA-directed cleavage at editing site 1; the fragments generated maximally in fractions 20–23 arise from a different nuclease that cleaves this mRNA at adjoining sites independent of gRNA addition. (**E**) 3′-U- exonulease assay using a 5′ labeled gRNA bearing 16 U residues on its 3′ end (CYb gRNA-558) and 1.5 μl of the indicated Q-Sepharose fractions. The exonuclease stops after removing the final U residue. (**F**) Verification of the 3′-U-exonuclease assay of (E), showing cleavage by fraction 15 and sizing markers from cleavage by alkali (which cuts after any nucleotide leaving a 2′,3′-cyclic-P), T1 RNase (which cuts after G, leaving a 2′,3′-cyclic-P), or nuclease P1 (which here cuts preferentially after G, leaving a 3′ OH) (see discussion for Fig. 5B and Ref. 26, 27). (Data from Ref. 11.)

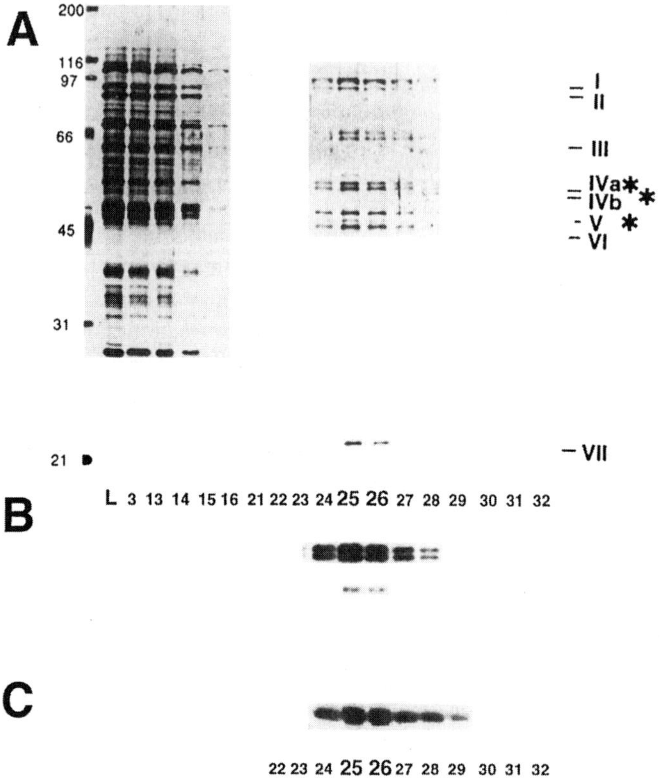

FIG. 4. Purification by subsequent DNA-cellulose chromatography. (**A**) Protein silver stain of 2 μl of the numbered fractions from DNA cellulose chromatography (D7) or 0.5 μl of the loaded pooled peak Q-Sepharose fractions (L), resolved in 10% SDS–PAGE. Fractions were adenylylated prior to electrophoresis. Protein molecular weight markers (Life Technologies, Inc.) are indicated. The asterisks indicate the adenylylatable polypeptides, bands IVa, IVb, and V, which do not represent major bands detectable in the loaded peak Q-Sepharose elution. (**B**) Autoradiograph of the gel in part A. (Data from Ref. 11.) (**C**) Assay for gRNA-directed endonuclease cleavage at ES1 of A6 pre-mRNA (a U deletion site) using 3′-labeled pre-mRNA, unlabeled gRNA, and 2 μl of the indicated fraction. The region of the gel representing cleavage product is shown.

either natural or pseudonatural sequences[15–18] (e.g., Fig. 5A), which are further described in the following section. However, the specific endonuclease cleavage can also be studied using cytochrome *b* pre-mRNA and cognate gRNAs,[10,11] using completely heterologous RNAs that simply form a Y-branch structure,[10] or even using only a short complementary oligonucleotide to direct cleavage of a substrate RNA, as described in a following section. These cleavages, both of the U-deletional type (Fig. 5B, left) and U-insertional type (Fig. 5B, right), are very selective and precise.

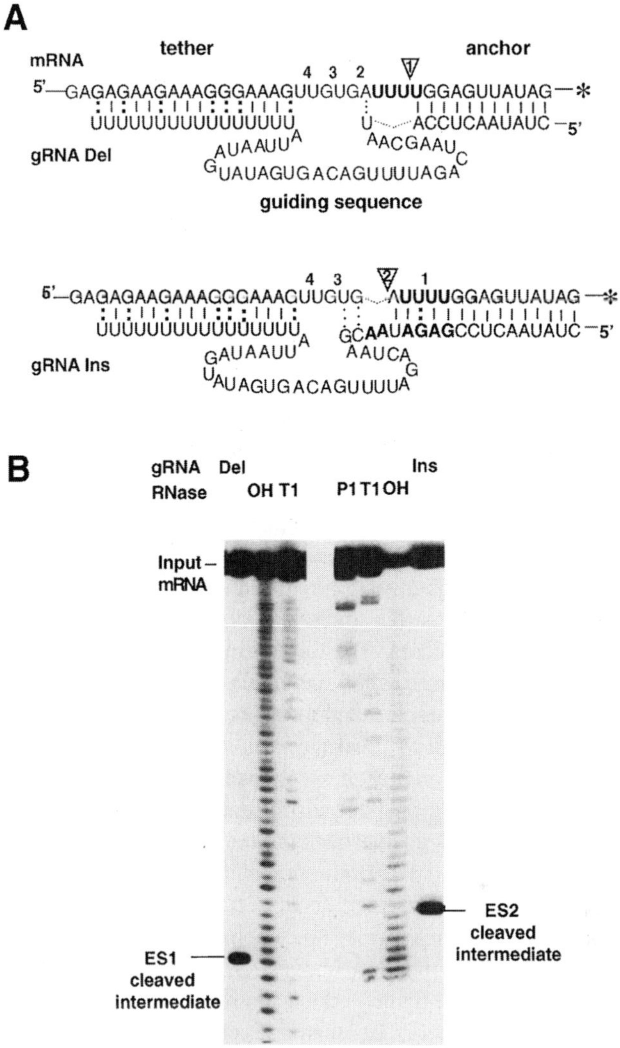

FIG. 5. Specific cleavage of the U deletional and U insertional type. (**A**) A6 pre-mRNA m[0,4] (above) and pseudonatural gRNAs (below), g[2,1] (also called ΔG)[15] to direct deletion of three U's at ES1 or g[2,4][22] to direct insertion of two U's at ES2. (**B**) Cleavage of 3'-end-labeled A6 pre-mRNA using gRNAs directing U deletional cleavage at ES1 (left; g[2,1]+4C)[18] or U insertional cleavage at ES2 (right; g5'Δ-15).[29] The size markers represent the same input mRNA treated with alkali (cleaving every nucleotide), RNase T1 (cleaving after G residues) or nuclease P1 (also cleaving after G residues; 10 units enzyme in 20 m*M* sodium phosphate or sodium citrate, 1 m*M* $ZnCl_2$, 6.5 M urea).[26,27] The molecules of the P1 ladder contain 5'-P termini, like those generated by the editing endonucleases, and unlike the 5'-OH termini generated by RNase T1 and alkali (in this particular gel the migration correction for 5'-P versus 5'-OH termini is ~1 nt).[27] Using 3' pCp-labeled mRNA, the cleavage product is a single size species, unaffected by action of terminal-U-transferase (that can act on 3'-OH ends if UTP is provided) or 3'-U-exonuclease (that can act on the upstream fragment at U deletion sites).

Given the apparent similarity between the two types of editing (Fig. 1), it might seem that U insertion and U deletion would utilize a common endonuclease to cleave the pre-mRNA.[5,6] Surprisingly, the gRNA-directed endonuclease cleavage reactions for U deletion and U insertion have notably different biochemical properties,[9] indicating they involve different activities. Foremost, cleavage at U deletion sites requires adenosine nucleotides (Fig. 6A), whereas cleavage at U insertion sites is inhibited by their presence (Fig. 6B). ADP and AMP-CP are most active, and ATP is also effective, but other ribo- or deoxyribonucleotides have little effect.[9] However, both forms of cleavage are about half-maximally active at ~0.3 m*M* adenosine nucleotide, the concentration estimated to be present in trypanosome mitochondria. The cleavage reactions of U deletion and U insertion also exhibit different magnesium profiles (although 10 m*M* provides near maximal activity for both), different pyrophosphate effects (see below), and different sensitivities to various structural perturbations of the gRNA.[30] Thus, gRNA-directed endonuclease cleavages at U deletion and U insertion sites are differently optimized.

As indicated above, the gRNA-directed endonuclease cleavages are directed by base pairing, to occur just upstream of the anchor duplex. Both endonuclease activities require minimally a (5′) single-strand–(3′) double-strand junction,[30] at which bond the cleavage occurs. The (adenosine nucleotide inhibited) U insertional kind of cleavage is used when the single-stranded mRNA residue immediately upstream from the duplex region is not a U; the (adenosine nucleotide requiring) U deletional kind of cleavage is used when there are U residues at this position.[9] Merely exchanging these characteristic proximal single-stranded mRNA and gRNA residues that distinguish U deletion from U insertion sites (see Fig. 1) interconverts the adenosine nucleotide specificity of the cleavage[9] (Fig. 6C,D; compare with Fig. 6A,B). Both endonuclease activities are also substantially affected by RNA features more distal to the cleavage site, and their effects differ for the two kinds of cleavage[18,30] (see below).

Reactions catalyzed by the editing complex using normal gRNAs generally accumulate cleaved mRNA, ligated edited product, and in certain instances also ligated gRNA–mRNA chimeras. To quantitatively assess cleavage by the editing endonuclease in these reactions, one cannot merely sum up the amounts of these unligated and ligated products because ligation can also occur prior to the U addition/removal; this recreates the input mRNA and causes underestimation of endonuclease activity. To score the endonuclease it therefore is prudent to inactivate the ligase activities. This can be accomplished by treatment with pyrophosphate (PP_i), to decharge the ligase molecules that became activated (adenylylated) during purification of the editing complex[11] (see Fig. 2, bottom right). The ligases then remain largely uncharged during the cleavage reaction, even of the U deletional

[30] J. Cruz-Reyes, A. Zhelonkina, and B. Sollner-Webb, in preparation.

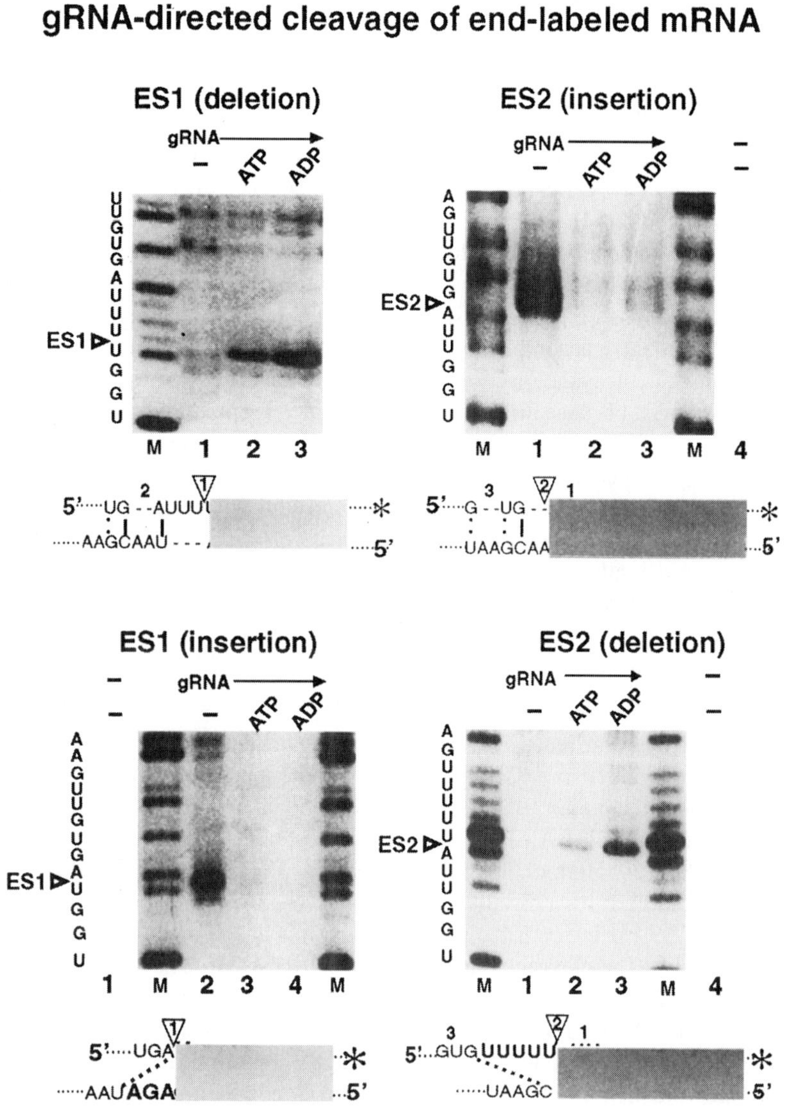

FIG. 6. Adenosine nucleotide effects on cleavage at natural and interconverted U deletion and U insertion sites. *Upper*: Cleavage of A6 pre-mRNA at ES1 (left; U deletional) and ES2 (right; U insertional) using the RNAs diagrammed below and shown in Fig. 5A. *Lower*: Cleavage of analogous RNAs where the sequence just upstream from the anchor duplex has been changed by adding/removing U residues in the mRNA and purines in the gRNA to interconvert the most obvious features of U deletional and U insertional sites. (Data using Q-Sepharose enriched editing complex are contained in Ref. 9.)

kind as long as ATP-free AMP-CP or ADP is used to activate the endonuclease.[9] PP_i inactivation of ligase is also efficient in crude extracts to score the nucleases.[17]

PP_i Pretreatment of Purified Editing Complex

We have found that the major ligase that functions in U insertion[25] (band V) requires a higher PP_i concentration to discharge than does the ligase that functions in U deletion[25] (band IV), but too high levels of PP_i can also inhibit the endonuclease reactions, especially the U deletional kind. With most of our editing complex preparations, 0.25–0.5 m*M* freshly made PP_i is optimal to assess U deletional cleavage and 1–2 m*M* is optimal to assess U insertional cleavage. This PP_i treatment can be conveniently accomplished by preincubating for 5 min on ice in an otherwise complete cleavage reaction (see below) before addition of the mRNA and gRNA. However, because different editing complex preparations can show somewhat different PP_i optima, the PP_i amount should best be titrated for deletional and insertional reactions in each protein preparation. If incomplete deadenylylation is indicated by incomplete suppression of full round editing, one can instead incubate 1 μl of purified editing complex with 1 μl of 20 m*M* freshly made PP_i in MRB for 5 min on ice, allowing deadenylylation at 10 m*M* PP_i, and then dilute it into a 20 μl cleavage reaction (see below), reducing the final PP_i concentration to 1 m*M*.[25,25a]

Endonuclease Cleavage Reactions and Their Analysis

The 20 μl cleavage reactions use 0.5–1.5 μl of purified editing complex in MRB buffer (above) with 10 m*M* KCl, the chosen PP_i amount, 25–100 ng/μl hexokinase [or bovine serum albumin (BSA)] and for U deletional cleavage 3 m*M* ADP or the nonhydrolyzable analog AMP-CP (>99% pure; Fluka). The reactions also contain the pre-mRNA (20–50 fmol, 3′-end-labeled with [^{32}P]pCp to $\sim 10^3$ cpm/fmol), and gRNA ($\sim$1 pmol), further discussed below, which had been preannealed at 37° for 10 min and cooled to room temperature for 5 min in 10 m*M* Tris-HCl (pH 8), 0.1 m*M* EDTA. The cleavage reactions are for 1 hr at 26°. They are then phenol extracted, ethanol precipitated in the presence of 1 μg of glycogen, and analyzed by electrophoresis in a 9% polyacrylamide/8 *M* urea gel in 1× TBE (100 m*M* Tris-HCl, 80 m*M* boric acid, 1 m*M* EDTA, pH 8.3), run warm ($\geq$30 V/cm).

To size the cleavage products to the nucleotide, it is important to use RNA sequencing ladders prepared from the same labeled mRNA and to take account of their terminal moieties,[27] as discussed in an accompanying article.[26] The editing nucleases generate 5′-P and 3′-OH termini,[10,11] which causes offset migration relative to otherwise identical RNAs bearing 5′-OH and 2′,3′- cyclic-P termini, as present on standard sequencing ladders prepared using RNase T1, RNase A, and hydroxide.[27] This is especially critical when sizing 5′-end-labeled RNA, since otherwise identical RNAs bearing different 3′ terminal moieties can migrate offset by anywhere from ~0.25 to >2 nts, but it is also relevant when sizing 3′-end-labeled RNA, since otherwise identical RNAs bearing different 5′-terminal moieties can migrate offset by ~0 to >1 nts.[26,27] Figure 5B shows sequencing standards prepared using nuclease P1, which generates 5′-P (and 3′-OH) termini and at pH 9 cleaves preferentially after G residues[26,27] to align ladders generated using RNase T1 and alkali (which cleave, after G residues and at every residue, respectively).

gRNAs for Cleavage at Natural A6 Editing Sites

With natural or pseudonatural A6 gRNAs and mRNA, at most a few percent of input mRNA becomes cleaved or edited in the *in vitro* reaction with either crude mitochondrial extract or glycerol gradient fractions thereof (e.g., Ref. 15–17) and little more with the purified editing complex.[9,22] We have recently developed more efficient gRNAs, starting with ones for A6 U deletion. These experiments[18,30] examined the basic features common to all natural gRNAs: the 5′ anchor region, central guiding region, 3′ oligo(U) tether, and potential ligation bridge (see Fig. 1 and 5A,B). For this U deletional cleavage, the most critical RNA feature is a previously unappreciated one—having several mRNA residues just upstream of the editing site and several opposing gRNA residues truly single stranded.[12,18] The natural guiding region is actually inhibitory for this U deletion. The potential ligation bridge is fully dispensable and the tether does not need any U residues. Relative to the natural gRNA, minimal gRNAs with these features enhance *in vitro* U deletion more than 100-fold and enhance U deletional cleavage an order of magnitude.[12,18] Using such active artificial gRNAs, the PP_i containing endonuclease reaction can generate 20 times more specifically cleaved mRNA than the remaining input mRNA (Fig. 7), and the full round reaction can generate twice as much specifically edited mRNA as remaining input mRNA.[18] Minimal gRNA features that optimize U insertion are turning out to be somewhat different than for U deletion, although we find that U insertional cleavage is also favored by single strandedness upstream of the editing site and can be markedly enhanced by artificial gRNAs with appropriate features.[30]

RNA Preparation

The mRNA and gRNAs are transcribed by T7 RNA polymerase from templates prepared by PCR. The PCR reactions are incubated at 94° for 5 min; *Taq* DNA polymerase (Life Technologies, Inc.) is added at 65°, followed by 30 cycles of 94° (45 sec), 41° (45 sec), and 72° (45 sec), following the manufacturer's recommendations. We identify mRNAs and gRNAs by m or g followed by the number of U residues (mRNAs) or guiding purines (gRNAs) present at ES2 and ES1. Thus, pre-edited wt mRNA with no U's at ES2 and four U's at ES1 is m[0,4], and gRNA with two guiding purines at ES2 and one at ES1 is g[2,1]. For editing using natural-like gRNAs, we generally use a common pre-mRNA (m[0,4]) and

FIG. 7. Efficient cleavage with an optimized gRNA. U deletional cleavage reaction as in Fig. 5B, but using the efficient artificial gRNA D33′.[18] The relevant parts of the gel, representing the remaining input mRNA and the cleaved product, are shown.

gRNAs g[2,1] or g[2,4] to direct deletion of three U residues at ES1 or insertion of two U residues at ES2, respectively.[9,22]

The template for synthesis of pre-edited A6 wt mRNA (72 nt, m[0,4]) was PCR amplified from A6-TAG DNA that contains the relevant sequence (5′ GGAAAGG-TTA GGGGGAGGAG AGAAGAAAGG GAAAGTTGTG ATTTTGGAGT TAT-AGAATAC TTACCTGGCA TC 3′) using primers T7A6 short (5′ GTAATACGAC TCACTATAGG AAAGGTTAGG G 3′) and A6RT (5′ GATGCCAGGT AAGTAT-TCTA TAACT 3′) (see Ref. 15). The residues encoding the Us at ES1 are shown double underlined; the portions of the template DNA in common with the primers are shown by dotted underlining; the upstream primer adds a T7 promoter sequence, shown by single underlining.

Templates for gRNAs were originally made[15] by double PCR amplification from gA6[14]Δ16G plasmid DNA that contains the relevant sequence (5′ GGATA-TACTA TAACTCCATA ACGAATCAGA TTTTGACAGT GATATGATA ATTA-TTTTTT TTTTTTTTTT AAA 3′) using the upstream primer 5′-GATATACTAT

AACTCCATAA CGAAT-3′ for g[2,1] (also called ΔG)[15] or 5′-GATATACTAT AACTCCGAGA TAACGAAT-3′ for g[2,4] and a downstream primer for plasmid sequences (5′ ATTAACCCTC ACTAAAG 3′; called T3).[15] The ES1 portion is double underlined; the residues in common with this upstream primer and a subsequent downstream primer are shown by dotted underlining. A second PCR reaction using primer T7gA6wt (5′ GTAATACGAC TCACTATAGG ATATACTAT 3′) and the same downstream primer adds the T7 RNA polymerase promoter (single underlining). This DNA was then *Dra*I cleaved just beyond the T_{16} and gel isolated for transcription.[15] More recently we instead amplify using the upstream primer T7gA6wt that adds the T7 RNA polymerase promoter and a downstream primer (5′ AAAAAAAAAA AAAAAATAAT TATCATAT 3′) that directly generates the 3′ end of the transcription template. Constructions of templates for various active artificial gRNAs are described in Ref. 18. The template for D33′ (Fig. 7) could be synthesized simply from the overlapping upstream primer T7gA6wt and downstream primer (5′ GGGAAAGTTG TAGGGTGGAG TTATAGTATA TCC 3′) without added template. This D33′ gRNA contains an artificial tether, no guiding regions, and a 5′/anchor region (whose complement is shown in double underline) in common with g[2,1].

RNAs are transcribed from these templates using [^{3}H]CTP and T7 RNA polymerase (Life Technologies, Inc.) following the manufacturer's recommendations. They are purified on 6 or 8% polyacrylamide–8 *M* urea gels (identified by a neighboring lane containing an aliquot of the same RNA preparation that had been ^{32}P internally labeled) and quantitated by scintillation counting. The mRNA is then 3′ labeled by ligation to 5′-[^{32}P]pCp using T4 RNA ligase (Life Technologies, Inc.) following the manufacturer's instructions and again gel purified. A6 mRNA forms two bands (*n* and *n*+1 nt), the shorter of which was used; the A6 gRNA transcripts have 3′ U tracts of ~12–16 nts. As noted above, the gRNA-directed nucleases can alternatively be assayed using cytochrome *b* mRNA and gRNAs, whose sequences and production have been described.[10]

Design of Minimal Anchor "gRNA" to Cleave Heterologous RNAs

Determination of the minimal gRNA features required for cleavage[12,18] suggests that the editing endonucleases could allow convenient and precise cleavage at virtually any desired phosphodiester bond in an RNA of interest. Importantly, both editing endonuclease activities can be directed to cleave specifically when using a synthetic RNA oligonucleotide that contains only sequences complementary to the substrate RNA just downstream from the cleavage site; this is effectively an anchor region without the usual guiding region or tether of a natural gRNA. Thus, these nucleases do not require a full Y-branch structure for recognition, but are guided simply by a (5′) single-stranded–(3′) double-stranded junction. Consequently, synthesizing a short downstream complementary RNA is sufficient to direct cleavage at a desired site. Experimental analysis to date at a limited number of different positions has confirmed very selective and active endonuclease cleavage using such a downstream pairing RNA oligonucleotide and the purified editing complex.[12] The guiding complementary oligonucleotides used to direct this cleavage opposite their paired 3′ terminus have been relatively short, ~14 nts.

This oligonucleotide-directed cleavage reaction exhibits the same adenosine nucleotide specificity established using complete gRNAs (see above). It functions best with ~3 mM ADP (or close analog) when the substrate RNA has multiple U residues just upstream of the duplex region; otherwise it is inhibited by millimolar concentrations of these nucleotides. In designing such complementary short guiding RNAs, we have noted two kinds of sequences that reduce cleavage efficiency, both of which make good biological sense. First, cleavage is highly inhibited when the guiding base paired oligonucleotide ends in a U residue or contains a sizable internal oligo (U) tract. This structure likely mimics the natural gRNAs′ 3′ oligo (U) tether, and any cleavage it directed would destroy the mRNA during editing. Second, cleavage appears inefficient when the substrate mRNA has C residues just upstream of the duplex region; this probably reflects the strong selection against a C residue at this position in natural pre-mRNAs. However, all attempts to target other sequence using a short base pairing downstream RNAs have yielded successful and selective cleavage by endonuclease activity of the editing complex.

U-Specific Exonuclease

The *T. brucei* RNA editing complex also contains the 3′-U-exonuclease that removes the U residues in U deletion (see Fig. 1). This U-exonuclease acts on 3′-OH ends, generating UMP.[11] It is not stimulated by PPi or inhibited by pyrophosphatase.[17] These results make unlikely the proposal that the 3′-U exo activity might be the terminal-U-transferase of the editing complex running in the reverse direction.[5,6] This exonuclease appears to act efficiently on U residues at a U deletion site that were made available by the gRNA-directed endonuclease of the editing complex,[15,17] but the *in vitro* reaction also functions on 3′-U residues of unrelated RNAs not generated by the editing endonuclease[11] (see Fig. 2E,F).

The 3′-U-exonuclease is assayed using 5′ labeled RNA.[11,15] When using labeled trypanosome pre-mRNA and unlabeled U deletional gRNA, the 3′-U-exonuclease can be scored in the context of the U deletion process,[15] either with or without PP_i addition to inhibit subsequent religation.[12] The upstream mRNA initially extends to the endonuclease cleavage site but becomes sequentially shortened by removal of U residues at its 3′ end; this is visualized as a ladder of bands extending down from that representing the initial cleavage position to an intense band representing complete removal of the 3′ U residues.[15,17] The RNA is not appreciably shortened by removal of subsequent non-U residues, demonstrating U specificity of the exonuclease.

With pre-mRNA : gRNA combinations where the U residues to be deleted and a few adjoining upstream mRNA residues are in appropriate single stranded configuration, the 3′-U-exonuclease proceeds to virtual completion, removing all the 3′-U residues at the editing site; however, with molecules where these residues are in base-paired structures, abundant partial U deletion products accumulate.[12,18] This indicates that the exonuclease is selective for single strand configuration, not only of the Us being removed but also of a few abutting residues. The extent of the *in vitro* 3′-U- exonuclease reaction should therefore be diagnostic of the structure of the substrate RNA. Furthermore, full *in vitro* editing cycles generate similar amounts of partially U deleted mRNA because the ligase of U deletion is not selective for mRNAs where the 3′-U-exonuclease has gone to completion.[9,18,25] *In vivo,* partial U deletions are not commonly observed, possibly because any incompletely edited site would direct re-editing rather than progression to the next editing site.[17]

An alternative assay for the 3′-U-exonuclease that is independent of the other editing activities uses only a 5′ end-labeled RNA that ends in a run of U residues.[11] The exonuclease acts to sequentially remove the 3′-terminal U residues, yielding a ladder of partial 3′-U-exonuclease products extending down to the position corresponding to the RNA body lacking the entire U tail (Fig. 2E,F). A convenient substrate is A6 gRNA which ends in a 3′ oligo (U) (see above), 5′ end labeled using T4 polynucleotide kinase (Life Technologies, Inc.) and [γ-^{32}P]ATP. 3′-U-exonuclease reactions are for 30 min at 22° in MRB buffer (above) using ~30 fmol of substrate RNA.

Additional Relevant Assays

RNA ligase activity: This can be assessed by RNA joining or by enzyme auto-adenylylation.[11,31]

(i) *RNA joining assays:* 20 μl reactions in MRB buffer containing 60 m*M* KCl, 2–4 units RNasin, 1 m*M* ATP, 10–1000 fmol 5′ end-labeled RNA (10^4–10^5 cpm), and ~1–2 μl enzyme are incubated 30 min at 22°. We traditionally use polylinker RNA made from pBluescript or pIBI,[11,23] but it appears as if most RNAs (50–1000 nts in length) can be similarly used. Note that the ligases of the editing complex are considerably more active in joining RNA polynucleotides (yielding circularized or dimerized products, depending on the amount of input RNA)[11] than in adding dinucleotides, so it is best to avoid the [^{32}P]pCp ligation assay commonly used for other ligases.[23] Reactions are stopped by adding 10 μg tRNA, sodium acetate to 0.3 *M*, and phenol/chloroform extracting. Samples are ethanol precipitated prior to electrophoresis on 8% polyacrylamide (19 : 1 acrylamide : bisacrylamide)/8.5 *M* urea gels in 1 × TBE.

[31] L. N. Rusché, K. J. Piller, and B. Sollner-Webb, *Mol. Cell. Biol.* **15,** 2933 (1995).

(ii) *Adenylylation assays:* These are performed in 10 μl MRB buffer containing 60 mM KCl, 1 μCi [α^{32}P]ATP (3000 Ci/mmol), and 1–2 μl of the enzymatic preparation. After a 5 min incubation on ice, reactions are stopped with 5 μl 3× loading buffer (30% glycerol, 15% 2-mercaptoethanol, 0.2 M Tris-HCl, pH 6.8, 6% SDS, 0.3% bromphenol blue), heated for 3 min at 95°, and resolved on a 10% polyacrylamide (29 : 1, acrylamide : bisacrylamide) SDS–PAGE gel in Tris–glycine–SDS buffer. For initial deadenylylation with pyrophosphate (see Fig. 2, bottom right), samples are prepared as above but with ~8 mM pyrophosphate added; after 2 min on ice, 1 unit pyrophosphatase (Sigma) is added in 1 μl and the reaction incubated another 5 min on ice prior to the adenylylation reaction.

Optimized full round editing reactions[22]: These are much like the gRNA-directed cleavage reactions (above), in MRB buffer containing 10 mM KCl and 25–100 ng/μl additional protein (hexokinase, preferentially, or BSA), but omitting the PPi. They use the appropriate adenosine nucleotide concentration: U insertion reactions are optimal at 0.1–30 μM ATP (to fully charge the appropriate ligase but not inhibit the cleavage), whereas U deletion reactions are optimal at >0.5 μM ATP (to fully charge the appropriate ligase) plus 3 mM ADP or AMP-CP (to fully activate cleavage). U insertion reactions also use 150 μM UTP (to stimulate the TUTase) and <3 ng/μl torula RNA (higher concentrations inhibit insertion), whereas optimal U deletion reactions use 30 ng/μl torula RNA and no added UTP (which competes with complete U removal). These conditions provide substantially greater activity than common conditions used previously.[15,16,20,32] To analyze full round editing we generally use 1-meter-long 9% polyacrylamide, 8.5 M urea gels, run at 1600 V for 14 hr, to maximally resolve one nucleotide size differences.[18]

Conclusion

In summary, a simple protein complex from *T. brucei* catalyzes both U deletion and U insertion and accordingly contains all the activities that comprise these processes. These include endonuclease activities with distinct biochemical properties that act at U deletion and U insertion sites and an exonuclease that is specific for U residues at a 3′-OH terminus. The endonuclease activities are directed by RNA structure and cleave at (5′) single-strand–(3′) double-strand junctions, distinguished by whether the final single strand residue is a U or a different residue. These nucleases also function on heterologous RNA, allowing one to selectively cleave a desired position in an RNA by simply providing a short downstream complementary oligonucleotide.

[32] M. Burgess, S. Heidmann, and K. Stuart, *RNA* **5,** 883 (1999).

[11] Ribonuclease YI*, RNA Structure Studies, and Variable Single-Strand Specificities of RNases

By VINCENT J. CANNISTRARO and DAVID KENNELL

Introduction

The function of many RNA molecules in biological reactions is defined by their structure as well as their sequence. The most common probes of RNA structure have been specific chemicals or RNases.[1,2] Most of the endoRNases currently available for cleavage of specific phosphodiester bonds do not have the same sequence specificity at all activities, e.g., pancreatic RNase A,[3,4] RNase U2 and RNase T2,[5,6] and RNase Cl3.[7,8] RNase T1 is an exception with its stringent specificity for cleavage after unpaired G. This variable expression of bond preferences can be useful when performed over a wide range of enzyme activities, or alternatively, can lead to ambiguous results.

A second class of RNases have no obvious preference for specific phosphodiester bonds but are specific for single or double strands. RNase VI preferentially cleaves RNA double strands,[9,10] but it can also degrade single strands in stacked helical structures.[11] Many RNases and nucleases are listed as single strand specific, but for each enzyme it is for a degree of single strandedness. The specificity can range from a slight preference for single strands to an apparent requirement for a long single-stranded region.[12]

This chapter describes a yeast enzyme whose recognition specificity does not fall within either of these classes. It has a stringent preference for single strands such that it can recognize a single base-pair mismatch. We called the enzyme RNase YI*,[12] because its substrate specificity is similar to that of RNase I* of *Escherichia coli*. The *E. coli* enzyme is coded by the *rna* gene for the periplasmic RNase I,[13,14]

[1] C. Ehresmann, F. Baudin, M. Mougel, P. Romby, J.-P. Ebel, and B. Ehresmann, *Nucleic Acids Res.* **15,** 919 (1987).
[2] G. Knapp, *Methods Enzymol.* **180,** 192 (1989).
[3] B. D. McLennan and B. G. Lane, *Can. J. Biochem.* **46,** 93 (1967).
[4] H. Witzel and E. A. Barnard, *Biochem. Biophys. Res. Commun.* **7,** 295 (1962).
[5] T. Uchida, T. Arima, and F. Egami, *J. Biochem.* **67,** 91 (1970).
[6] T. Uchida and F. Egami, *Methods Enzymol.* **12,** 239 (1968).
[7] C. C. Levy and T. P. Karpetsky, *J. Biol. Chem.* **255,** 2153 (1980).
[8] M. S. Boguski, P. Hieter, and C. C. Levy, *J. Biol. Chem.* **255,** 2160 (1980).
[9] S. K. Vassilenko and V. C. Rythe, *Biokhimiya* **40,** 578 (1975).
[10] L. G. Boldyreva, A. S. Boutorine, and S. K. Vassilenko, *Eur. J. Biochem.* **121,** 587 (1978).
[11] H. B. Lowman and D. E. Draper, *J. Biol. Chem.* **261,** 5396 (1986).
[12] V. J. Cannistraro and D. Kennell, *Nucleic Acids Res.* **25,** 1405 (1997).
[13] J. Meador III and D. Kennell, *Gene* **95,** 1 (1990).

0076-6879/01 $35.00

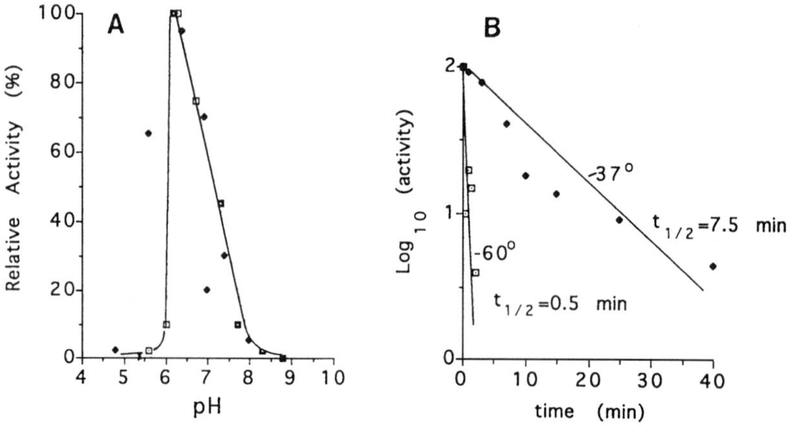

FIG. 1. The pH dependence of RNase YI* and heat inactivation of RNase YI*. (A) The maximum activity is at pH ~6.2 and falls off rapidly at lower pH. (□), MES; (◆), phosphate; (■), (Tris-Cl) buffers. The reactions contained 20 μg poly(C) and 10 units of enzyme in 100 μl at 23°. (B). Ten units of RNase YI* in 100 μl of 10 m*M* MES, pH 6.4, was incubated for the time and at the temperatures shown. Samples were brought to 0° before adding 20 μg poly(C) and incubating for an additional 10 min at 37°. From *Nucleic Acids Res.,* Vol. 25, Cannistraro and Kennell, "RNase YI* and RNA Structure Studies," p. 1405, 1997, by permission of Oxford University Press.

and it appears to be a cytoplasmic precursor of RNase I that differs not only in cell location but in enzyme specificity and physical characteristics.[15] Although both *E. coli* enzymes have no obvious bond preferences, RNase I activity causes significant nicks in double-stranded RNA that lead ultimately to the complete degradation of the RNA.[15] In contrast, RNase I* and RNase YI* are highly specific for small unstructured RNA oligonucleotides. Unlike many of the enzymes that have been used for probing RNA structure,[1,2,16] RNase I* and RNase YI* do not require divalent metal ions for activity, are weakly inhibited by NaCl, have pH optima close to neutrality, and generate 5′-OH groups on cleavage for easy end-labeling. The high selectivity of RNase YI* for unpaired nucleotides makes it a useful probe of RNA structure.[12]

Materials and Methods

Enzyme Assays

(See Ref. 12.) RNase YI* has maximum activity at pH ~ 6.2 (Fig. 1). It is assayed in 10 m*M* 2-[*N*-morpholino]ethanesulfonic acid (MES), pH 6.4. RNase I

[14] S. K. Srivastava, V. J. Cannistraro, and D. Kennell, *J. Bacteriol.* **174,** 56 (1992).

[15] V. J. Cannistraro and D. Kennell, *J. Bacteriol.* **173,** 4653 (1991).

[16] T. Ando, *Biochim. Biophys. Acta* **114,** 158 (1966).

is assayed in 20 mM Tris-HCl, pH 7.4. One unit of RNase I or RNase YI* degrades 1 μg of single-stranded homopolymer RNA ($\sim$300 nt) to <6-mers per min at 37° in a 20 μl reaction. (This definition is equivalent to $\sim$2 units based on a definition of conversion to products soluble in 5% trichloroacetic acid). Reactions with S1 nuclease or mung bean nuclease (Boehringer/Mannheim) are in 30 mM sodium acetate buffer, pH 5.0, with 1 mM $ZnSO_4$ with units defined by the vendor.

RNase YI Purification*

All steps are between 0° and 2°, since RNase YI* is heat-labile (Fig. 1). RNase YI* is a large monomer ($\sim$70 kDa) compared to many other nucleases and binds to both cation and anion exchangers. The purification steps can be modified from those outlined in the original publication.[12] The initial steps include lysis and removal of most nucleic acids by protamine sulfate precipitation followed by a 45% ammonium sulfate precipitation with the supernatant precipitated in 90% ammonium sulfate. The latter precipitate can be stored at −70°. We use a Pharmacia/LKB HPLC instrument with Pharmacia (Piscataway, NJ) column resins jacketed in ice water. These include S-Sepharose with 0 to 2 M gradient of NaCl in 20 mM MES, pH 5.7 (RNase YI* eluted in $\sim$0.7 M NaCl), HiLoad 16/90 Superdex 75 sizing column with elution in 20 mM Tris-HCl, pH 7.4 + 200 mM NaCl, and Mono Q anion exchanger with a gradient of 0 to 2 M NaCl in 20 mM Tris-HCl, pH 7.4. The final Mono Q fractionation can be omitted with sufficient purity to avoid any contaminating DNase, RNase, or phosphatase activities. The enzyme is dialyzed against 30 mM Tris, pH 7.4, 40 mM NaCl, 50% glycerol ($\sim$4-fold reduction in volume) and stored at −20° after addition of bovine serum albumin (BSA) to 100 μg/ml.

Substrates to Assay for Single-Strand Specificity

Many natural or synthetic substrates can be used for assessing the degree of single-strand specificity. An example of each class is given here. The 120 nt 5S rRNA of *E. coli* has been a model for numerous structure studies since its sequence was first determined[17,18] with a generally accepted structure, based on estimated thermodynamic stability and evolutionary conservation of specific segments.[19] However, the commercial preparation contains a fraction of intact molecules with internal "nicks." These molecules are probably present in the cell[20] and must be removed. *E. coli* 5S rRNA (Boehringer/Mannheim) is treated with heat-labile alkaline phosphatase before phenol extraction and ethanol precipitation and then

[17] G. G. Brownlee, "Determination of Sequences in RNA." North-Holland, Amsterdam, and Elsevier, New York, 1972.

[18] H. F. Noller, *Annu. Rev. Biochem.* **53,** 119 (1984).

[19] G. E. Fox and C. R. Woese, *Nature* **256,** 505 (1975).

[20] V. J. Cannistraro and D. Kennell, unpublished data (1997).

5′-end-labeled with [γ-^{32}P]ATP. The RNA is brought to 100° for 2 min to dissociate any molecules with "nicks" before loading on a 20% PAG. After electrophoresis, the gel band containing the full-length molecule is excised and crushed, and the RNA is eluted by overnight incubation in a small volume of water. The acrylamide particles are spun out and the supernatant precipitated with ethanol and the RNA resuspended in water.

The 5SrRNA is 3′ end-labeled in the RNA ligase reaction.[12,21]

RNase YI* reactions are in 20 μl of 10 mM MES, pH 6.4, at 23° for 30 min with 1 to 2 μg of [^{32}P]RNA and stopped by addition of 1 M citric acid to give 30 mM, pH 3.5. It is important that all enzyme activity be irreversibly inactivated. After addition of 10 μg poly(A), samples are brought to 100° for 1 min and then chilled to 0° before adding 8 μl of 10 M urea plus 0.2 μl of 0.1 M $ZnSO_4$ (an inactivator of RNase YI*). The tracking dyes, bromphenol blue, and xylene cyanol and glycerol (to 10%) are added and the samples loaded and run by 20% PAGE at 23°.

Mismatches in Duplex Nucleic Acids

(See Ref. 12.) Plasmid pSP73 (Promega, Madison, WI) is linearized by *Bam*HI digestion and used as a template to synthesize a runoff 58 nt RNA from the T7 promoter in the T7 RNA polymerase (Boehringer/Mannheim) reaction in 50 μl containing 40 mM Tris-HCl, pH 8.0, 6 mM $MgCl_2$, 10 mM dithiothreitol, 2 mM spermidine, 2 mM of each of four nucleoside triphosphates, 1.25 μg of linearized plasmid, and 50 units of enzyme for 60 min at 37°. A separate reaction in 20 μl contains the same concentrations of reactants plus 20 μCi [α-^{32}P]GTP to label the RNA as a marker. The labeled and unlabeled samples are fractionated by 20% PAGE. The 58 nt RNA is identified in the lanes with ^{32}P and the unlabeled bands at the same position in adjacent lanes are eluted, as described in the preceding section. The T7 RNA was then treated at pH 7.5 with heat sensitive calf alkaline phosphatase (Boehringer/Mannheim) for 5 min at 37°. The alkaline phosphatase is inactivated by heating at 100° for 2 min. The RNA is 5′ labeled with 50 μCi [γ-^{32}P]ATP in the polynucleotide kinase reaction. In order to eliminate any possible contaminating ^{32}P fragments, the ^{32}P-RNA is purified by another 20% PAGE and prepared by the preceding steps, except that 10 μg of carrier poly(A) is added before ethanol precipitating, and the final sample ($\sim$1 μg with $\sim 10^6$ cpm) is used to form an RNA–DNA hybrid with a specific 39-mer DNA oligonucleotide that contains one or more base mismatches to the first 39 nt at the 5′ end of the RNA. The RNA–DNA hybrids are purified by electrophoresis through a 20% native PAG. Reactions are at 23° for 60 min (37° for 12 min gives the same results). RNase I reactions are stopped with SDS to 0.1% and RNase YI* reactions with final concentrations of 3 M urea, 1 mM $ZnCl_2$, and 30 mM citric acid, pH 3.5. In

[21] V. J. Cannistraro, P. Hwang, and D. E. Kennell, *J. Biochem. Biophys. Meth.* **14,** 211 (1987).

both cases, the stop mixes include glycerol and dyes plus a tenfold excess of DNA oligonucleotide that has the same sequence as the 39 nt of [^{32}P]RNA. Subsequently, each mix is brought to 100°C for 90 sec to dissociate the RNA–DNA hybrid and then to 37° for 5 min to convert all the released DNA to a DNA–DNA duplex. The [^{32}P]RNA runs as single strands in the subsequent PAGE containing 7 *M* urea plus 2.5 m*M* EDTA.

Results

*Determining 5S rRNA Structure with RNase YI**

Both 5′-^{32}P-and 3′-^{32}P-labeled 5S rRNA were used as substrates with a 200-fold range of RNase YI* activities.[12] It is essential to perform such a titration with any chemical or enzymatic probe. The lowest activities show the most labile bonds to the agent. As the activity increases, bands corresponding to those bonds may disappear to be replaced by smaller bands representing secondary cleavages. Some bands may not disappear at the highest activities indicating that there are no sites for the agent more proximal to the labeled end. All three conditions can be seen for the reactions of RNase YI* with the 5S rRNA (Fig. 2). A molecular weight ladder can be constructed by digesting the labeled RNA with a range of alkali concentrations and run on the same gel to show the position of each nucleotide. Identification of the nucleotide bonds broken at each enzyme activity gave the results shown in Fig. 3. Although the probing with RNase YI* gave results consistent with the accepted structure, some additional conclusions could be made. First, the most labile part of the molecule appears to be the base of stem 1 [G9 and G10 (5′-labeled) and G-112 to G-107 from 3′-labeling]. Thus, this enzyme cleaves off stem 1 even before breaking bonds in the regions that map as single-stranded in the 2D plot. Note that the reaction products were brought to 100° and then chilled quickly to 0° and urea added to give 3 *M* before loading on the PAG to prevent cleaved molecules from remaining associated.[12] The single-stranded regions are cleaved at higher activities but susceptibility of the large loop (A-34 to U-48) is only demonstrated over a very narrow range of activities, since as expected, those bands would be eliminated by any break more proximal to the labeled end. The results from 3′-labeling differed from those from 5′-labeling in that there were fewer targets observed on the 3′ half of the molecule. Only the A-34 to U-48 loop and the cleavages to release stem 1 were obvious. This difference could result from tertiary interactions between that side of the molecule (U-48 to C-97) that make it less accessible to RNase YI* activity.

Probing 5S rRNA Structure with S1 and Mung Bean Nucleases

The results with S1 and mung bean nucleases were in agreement with each other but were markedly different from the results with RNase YI* (Fig. 4).[12] Briefly, they showed only one major cleavage region at all activities used: the large

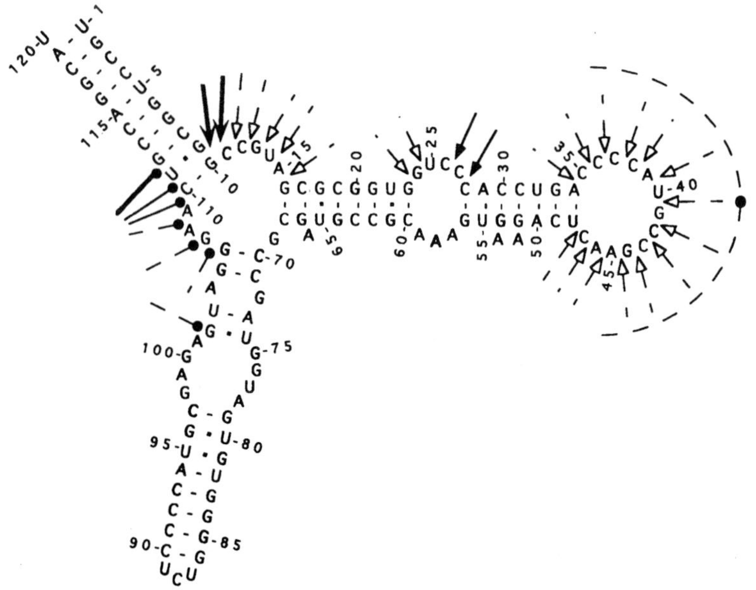

FIG. 2. Structure of 5S rRNA based on energy considerations and the conservation of secondary structures in a broad range of species.[17–19] The nucleotides are numbered starting at the 5′ end. G to U bonds are designated by a (■). The cleavage sites are determined from Fig. 3. Those from the 5′-^{32}P end are designated by pointed arrows; closed arrows show ends that persisted at the highest reaction conditions, and open arrows for bands that declined with increasing enzyme activities. Cleavage sites from 3′-^{32}P ends are shown by the ball arrows with dashed tails showing bands that declined with increasing activities. The arc in the large loop region designates ^{32}P bands that have not been resolved but must include cuts at all bonds. The darker arrows after G-9, G-10, and U-111 designate the most prominent bands, which were also the most persistent with increasing activities. The calculated total net free energy is −35.5 kcal/mol (kindly computed by Michael Zuker). From *Nucleic Acids Res.*, Vol. 25, Cannistraro and Kennell, "RNase YI* and RNA Structure Studies," p. 1405, 1997, by permission of Oxford University Press.

loop at A-34 to U-48. There was a secondary 3′-^{32}P-band mapping at G-96 that would result from cleavage between two consecutive G–U bonds. Both primary and secondary bands declined with increasing enzyme activities, but there were no major sites even at G-9 or G-112. Apparently, these enzymes are specific for extended single-stranded regions but are not sensitive to short disruptions of a duplex with the exception of the cleavage at G-96.

Recognition Specificity of RNase YI and RNase I*

The results with 5S rRNA substrate showed that there were marked differences in the recognition specificities of the single-strand-specific RNase YI* and

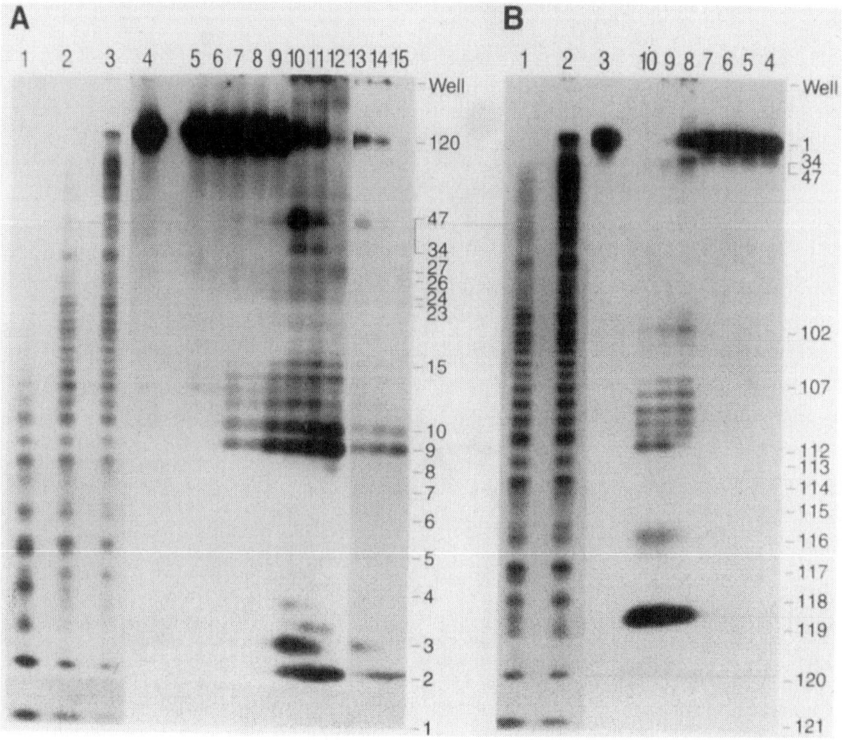

FIG. 3. ^{32}P-labeled molecules resulting from reactions of 5S-rRNA over a range of RNase YI* activities. (A) 5′-^{32}P-label, (B) 3′-^{32}P. From alkali digestions [(A) lanes 1–3 and (B) 1 and 2] the $\log(M_d)$ was plotted vs R_f to estimate nucleotide positions of bands. Except for lanes 13–15, the films were highly exposed to detect minor bands. The numbers in the right columns give the nt position from the 5′ end. The 3′ end was labeled in the RNA ligase reaction, which added a C residue (121) to that end. (A) Lane 4, no enzyme; lanes 5–12, RNase YI*, 0.07, 0.14, 0.36, 0.72, 1.4, 3.6, 7.2, 14 units, respectively. Lanes 13–15 are the same as 10–12 with film exposed a much shorter time. (B) Lane 3, no enzyme; lanes 4–10 have the same units of RNase YI* as lanes 6–12 in (A). From *Nucleic Acids Res.*, Vol. 25, Cannistraro and Kennell, "RNase YI* and RNA Structure Studies," p. 1405, 1997, by permission of Oxford University Press.

single-strand-specific nucleases, S1 and mung bean. What was recognized by RNase YI* was compared to the specificity of another single-strand-specific RNase, RNase I, by using synthetic nucleic acid duplexes with a single, or multiple, base-pair mismatch.[12]

A defined sequence of RNA was synthesized *in vitro* from the T7 promotor of plasmid pSP73 and hybrid RNA-DNA complexes formed with DNA oligonucleotides containing specific base mismatches (see Materials and Methods). The DNA–RNA hybrids with ^{32}P at the 5′ end of the RNA were reacted with a range

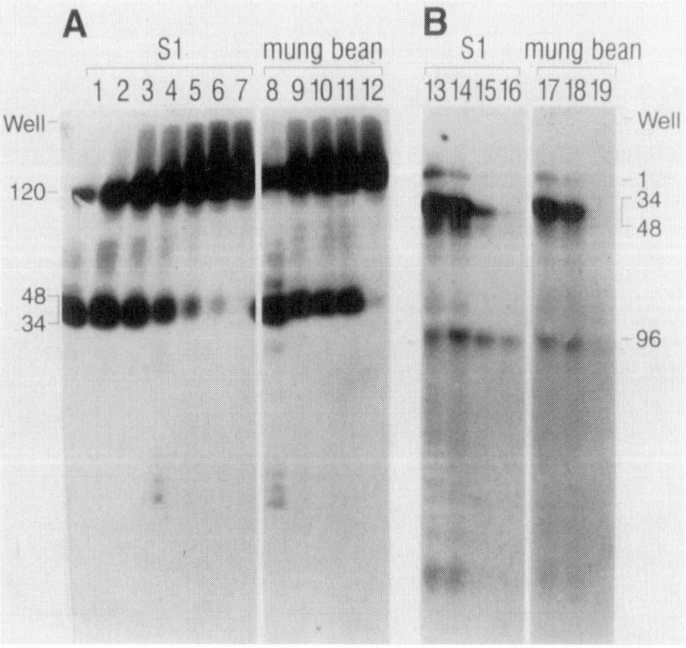

FIG. 4. The ^{32}P-labeled molecules resulting from reactions of 5S-rRNA over a range of S1 or mung bean nuclease activities. (A) [5′-^{32}P]RNA, (B) [3′-^{32}P]RNA. Lane 7, no enzyme; lanes 1–6, S1 nuclease at 20, 8, 4, 2, 0.8, 0.4 units, respectively. Lanes 8–12, mung bean nuclease at 25, 10, 5, 2.5, 1.0 units, respectively. Lanes 13–16, S1 nuclease with 8, 20, 40, 80 units, respectively, and lanes 17–19, mung bean nuclease with 5, 10, 25 units, respectively. From *Nucleic Acids Res.*, Vol. 25, Cannistraro and Kennell, "RNase YI* and RNA Structure Studies," p. 1405, 1997, by permission of Oxford University Press.

of enzyme activities. RNase I was tested in parallel with RNase YI*, for identifying either four contiguous mismatched bases or a single mismatched base pair. RNase I degraded a four nucleotide "bubble" only slightly faster than it broke phosphodiester bonds of nucleotides in perfect base pairs. As a result, the specific band was only observed in a very narrow range of enzyme activities (Fig. 5A). In contrast, RNase YI* easily identified the four base pair mismatch over a wide range of activities. The single base pair mismatch was also identified by RNase YI* but was completely missed by RNase I with the activities tested (Fig. 5B).

Discussion

Endonucleases have marked variability for single vs double strands. An ideal enzyme, or combination of enzymes, for elucidating RNA structure should (1) show

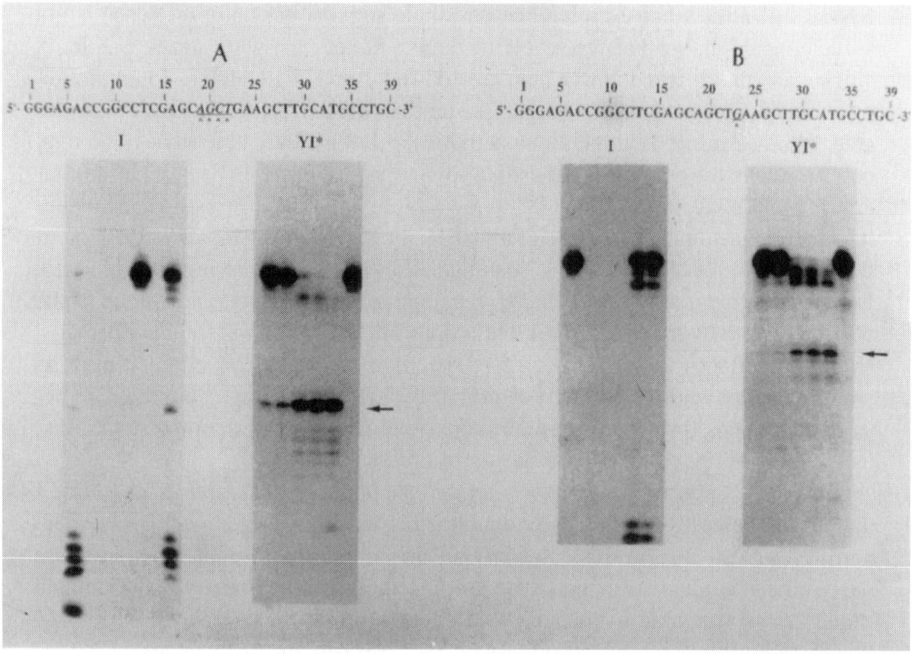

FIG. 5. Recognition of base pair mismatches in an RNA–DNA duplex. The 58 nt RNA product was:

5′-GGGAG ACCGG CCTCG AGCAG CTGAA GCTTG CATGC CTGCA GGTCG ACTCT AGAGG ATC-3′

RNA–DNA hybrids were formed with (A) a DNA oligonucleotide complementary to the first 39 nt of the RNA except for four consecutive mismatches at positions 19–22 of the RNA:

5′-GCAGG CATGC AAGCT TC**TCG A**GCTC GAGGC CGGTC TCCC-3′

or (B) one base mismatch at position 23:

5′-GCAGG CATGC AAGCT TCAGC TGCTC GAGGC CGGTC TCCC-3′

Columns with RNase YI*: no enzyme in the far right, and the other lanes from right to left: 14, 7, 3.5, 1.4 and 0.7 units, respectively. RNase I (A) left to right: 25, 50, 100, 200, zero, 12.5 units, respectively; RNase I (B): no enzyme, far left, and 200, 100, 50, 25, and 12.5 units in the next lanes from left to right, respectively. From *Nucleic Acids Res.*, Vol. 25, Cannistraro and Kennell, "RNase YI* and RNA Structure Studies," p. 1405, 1997, by permission of Oxford University Press.

high preference for any unpaired nucleotides, (2) recognize the same phosphodiester bonds at all enzyme activities, and (3) be active in reaction conditions that do not disrupt the normal RNA structure. As noted, most of the endoRNases used for probing structure have bond preferences that can be a function of the enzyme activity. Of the enzymes that have no obvious bond preferences, the periplasmic

RNase I of *E. coli* has a preference for single strands, but that preference is not sufficiently high, so that the levels of double strand nicks are unacceptable. A single base-pair mismatch was not detected and even a mismatch of four contiguous bases was degraded only slightly faster than the perfect RNA–DNA duplexes (Fig. 5). As a result, relatively low activities of RNase I are able to degrade total cell RNA, and the enzyme demonstrates this capacity *in vivo* in bacteria in stationary or death phases.[22,23] However, this weak preference for single strands makes the enzyme unsuitable for structure analysis or base-pair mismatch detection. At the other extreme, S1 and mung bean nucleases have a very stringent preference for single strands; only bonds in the presumptive long single-stranded loop of the 5S rRNA were hydrolyzed. Short disruptions of double strands were not detected (Fig. 4). Enzymes with such stringent specificity are ideal for eliminating long single-stranded regions without disturbing imperfect duplexes,[24] but they are not good candidates for detecting small loops, "bubbles," or single mismatches, and as such, are not good probes for these unpaired nucleotides.

RNase YI* specificity appears intermediate between the extremes of RNase I and these nucleases for single- *vs* double-stranded RNA preference. *Escherichia coli* RNase I* is also in this category, but its value as a probe of RNA structure has not been studied sufficiently. RNase YI* showed little activity against stable duplexes but was able to recognize and cleave short single-stranded regions of the 5S rRNA and recognized quite well the base pair mismatches tested, i.e., it degraded any phosphodiester bonds of nucleotides that were not perfectly bonded in a duplex. This sensitive recognition of imperfect duplexes was also shown by the stability of the released perfect duplexes even after multiple cleavages of the 5S rRNA. When a substrate is in great excess, all cleavages should be initial ones. At sufficiently high enzyme activities, some molecules are cleaved more than once, and larger ^{32}P-molecules decline as a result of second cleavages closer to the ^{32}P end. When RNase YI* was used to degrade 5S rRNA, the progression of band appearance and disappearance appeared to be ordered, i.e., only specific bands appeared over the entire range of activities. If one or more hits had disrupted the stable parts of the molecule sufficiently, alternative patterns of sizes from the newly generated fragments should have been observed with increasing activities. However, new bands did not appear at the highest activities, e.g., no 5′-^{32}P bands were ever seen between 3 to 9 nt or 15 to 23 nt. This suggests that the same sites that were vulnerable in the full-length substrate were the vulnerable ones in molecules that may have had two or more cleavages with the released duplex structures retaining sufficient stability. A case is stem I, which remained intact at the highest reaction activities, i.e., some cleavages were occurring at G-9 or G-10 in

[22] J. Mandelstam and H. Halvorson, *Biochim. Biophys. Acta* **40,** 43 (1960).

[23] J. Meador III, B. Cannon, V. J. Cannistraro, and D. Kennell, *Eur. J. Biochem.* **187,** 54 (1990).

[24] A. J. Berk and P. A. Sharp, *Cell* **12,** 721 (1977).

molecules that may have already had several more distal cleavages. This conclusion is apparent because RNase YI* is sufficiently selective; it does not hydrolyze bonds in perfect duplexes even at activities much greater than are needed to cleave bonds in imperfect base pairs.

This specificity makes it an ideal enzyme for probing RNA structure as well as a candidate for use in detecting base pair mismatches. It has the added advantages of optimal activity near neutral pH, no requirement for cations that might affect RNA structures, activity in the presence of salt, and the generation upon cleavage of 5′-OH ends for easy labeling.

Possible Role of RNase YI in Yeast*

The breakdown of mRNA in microorganisms represents a major metabolic activity.[25,26] In *E. coli* its rate is equivalent to half the rate of total RNA synthesis,[25] and it is probably at least that in yeast. Since the composition of mRNA reflects the heterogenity of the cell, the enzymes involved in its breakdown must have broad specificities as well as sufficient activities. The degradative enzymes for this function in *E. coli* have broad specificities as opposed to the processing RNases, which are involved in biosynthesis of specific RNA molecules and have stringent specificities. For reasons that are unclear, the endoRNases in the former class generate 5′-OH groups on cleavage, while the processing endoRNases produce 3′-P groups.[26] The acid-soluble pools of *E. coli* contain 2′,3′-cyclic mononucleotides, as well as the 5′-OH-mononucleotides derived from them,[25] which is consistent with RNase I* activity, since the 3′-exonucleases that also participate in degradation of mRNA fragments generate 5′-P-mononucleotides. RNase II is the most active of those degradative exonucleases, but its processive reaction becomes nonprocessive and much slower when the oligonucleotide size is $<\sim 12$ nucleotides.[27–29] It is these small oligonucleotide fragments that are the primary substrates for RNase I*, as was shown by an analysis of the 5′-end nucleotides on oligonucleotides as a function of their size in growing cells.[26] Thus, it was proposed that in *E. coli* the mRNA is inactivated and cleaved into fragments primarily by an endonuclease, RNase M,[26,30,31] as well as to a lesser extent by RNase I*. The resulting fragments are rapidly degraded to mononucleotides by 3′-exoribonucleases and RNase I*.[25] It remains to be seen if RNase YI* performs a similar function in yeast, but its high activity and similar specificity to RNase I* of *E. coli* make it a strong candidate for such a function.

[25] V. J. Cannistraro and D. Kennell, *J. Bacteriol.* **173,** 4653 (1991).
[26] V. J. Cannistraro and D. Kennell, *Eur. J. Biochem.* **213,** 285 (1993).
[27] V. J. Cannistraro and D. Kennell, *J. Mol. Biol.* **243,** 930 (1994).
[28] V. J. Cannistraro and D. Kennell, *Biochim. Biophys. Acta* **1433,** 170 (1999).
[29] V. J. Cannistraro and D. Kennell, *Methods Enzymol.* **342,** 26 (2001).
[30] V. J. Cannistraro, M. N. Subbarao, and D. Kennell, *J. Mol. Biol.* **192,** 257 (1986).
[31] V. J. Cannistraro and D. Kennell, *Eur. J. Biochem.* **181,** 363 (1989).

Section III

Secreted Ribonucleases

[12] Bovine Pancreatic Ribonuclease A: Oxidative and Conformational Folding Studies

By HAROLD A. SCHERAGA, WILLIAM J. WEDEMEYER, and ERVIN WELKER

Introduction

For many years, bovine pancreatic ribonuclease A (RNase A, EC 3.1.27.5) has been a model protein for studies of protein structure, energetics, disulfide-bond reactions, and conformational folding.[1–7] RNase A is a relatively small (124 amino acids) but stable protein that exhibits several remarkable structural and biochemical features (Fig. 1). RNase A possesses four disulfide bonds in the native state (26–84, 40–95, 58–110, and 65–72), an unusually large number for a protein of its size, making its oxidative folding from the fully reduced state one of the most complex ever characterized.[6,7] RNase A also possesses four X-Pro peptide bonds, of which two are cis (Tyr^{92}-Pro^{93} and Asn^{113}-Pro^{114}) and two are trans (Lys^{41}-Pro^{42} and Val^{116}-Pro^{117}) in the native state.[8] This is also unusual, since the cis isomer is present in only 6% of X-Pro peptide bonds in the native structures of proteins.[9,10] However, these unusual features may be exploited to gain insight into protein folding, as reviewed in this article.

In the first section of this article, we discuss the methods used to prepare wild-type and mutant ribonucleases, as well as variants that correspond to intermediates of oxidative folding. In the second section, we summarize the structural features of native RNase A. In the third section, we discuss the equili-brium unfolding transitions of RNase A and their structural characterization. In the fourth section, we describe the reductive unfolding and oxidative folding of RNase A, and the coupling of disulfide-bond reactions and conformational folding. In the fifth section, we discuss the effects of nonnative proline isomers on the conformational folding

[1] F. M. Richards and H. W. Wyckoff, *in* "The Enzymes" (P. D. Boyer, ed.), 3rd. Ed., Vol. 4, pp. 647–806. Academic Press, New York, 1971.

[2] P. Blackburn and S. Moore, *in* "The Enzymes" (P. D. Boyer, ed.), 3rd. Ed., Vol. 15, pp. 317–433. Academic Press, New York, 1982.

[3] J. L. Neira and M. Rico, *Folding Des.* **2,** R1 (1997).

[4] C. M. Cuchillo, M. Vilanova, and M. V. Nogues, *in* "Ribonucleases: Structures and Functions" (G. D'Alessio and J. F. Riordan, eds.), pp. 271–304. Academic Press, New York, 1997.

[5] R. T. Raines, *Chem. Rev.* **98,** 1045 (1998).

[6] W. J. Wedemeyer, E. Welker, M. Narayan, and H. A. Scheraga, *Biochemistry* **39,** 4207 (2000).

[7] M. Narayan, E. Welker, W. J. Wedemeyer, and H. A. Scheraga, *Acc. Chem. Res.* **33,** 805 (2000).

[8] A. Wlodawer, L. A. Svensson, L. Sjölin, and G. L. Gilliland, *Biochemistry* **27,** 2705 (1988).

[9] D. E. Stewart, A. Sarkar, and J. E. Wampler, *J. Mol. Biol.* **214,** 253 (1990).

[10] M. W. MacArthur and J. M. Thornton, *J. Mol. Biol.* **218,** 397 (1991).

METHODS IN ENZYMOLOGY, VOL. 341
0076-6879/01 $35.00

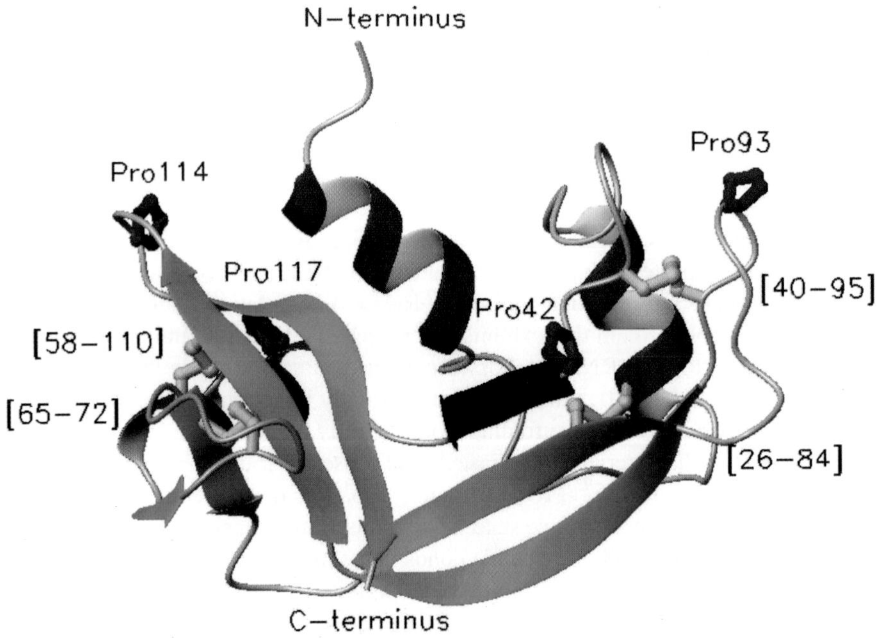

FIG. 1. The structure of wild-type RNase A, showing the four proline residues and four disulfide bonds. The 26–84 and 58–110 disulfide bonds are fully buried in the native protein, whereas the 40–95 and 65–72 disulfide bonds may be exposed by local unfolding. Both Pro-93 and Pro-114 are cis in the native structure and are located in exposed, flexible loop regions; however, the isomer of Pro-114 has little effect on conformational folding, whereas the nonnative isomer of Pro-93 retards folding roughly 500-fold. Pro-42 and Pro-117 are both trans in the native protein, although the isomeric state of Pro-42 does not appear to affect conformational folding significantly.

of RNase A. In the sixth section, we review the conformational folding of pure isomeric states of RNase A, in which the complication of proline isomerization is absent. In the final section, we present some specialized techniques for reductive unfolding and oxidative folding experiments.

Preparative Methods

Numerous preparative methods have been developed for RNase A. The earliest method involves direct purification from bovine pancreas[5]; indeed, the preparation of a large amount of RNase A by the Armour Company led directly to the adoption of RNase A as a model protein.[11] Several purification protocols have

[11] F. M. Richards, *Protein Sci.* **1,** 1721 (1992).

been developed for RNase A[12]; however, ion-exchange chromatography seems to provide high resolution and yields.[13] It is important to distinguish wild-type RNase A from sundry chemical variants that may exhibit markedly different enzymatic activities and folding behaviors.[14,15]

The RNase A gene was transferred into different expression systems to allow the preparation of site-directed mutants; however, only low yields could be obtained at first, because of the intrinsic cytotoxicity of RNase A. This technical hurdle was overcome by expressing RNase A in inclusion bodies in *Escherichia coli*.[16,17] Several protocols have been developed to reduce these inclusion-body proteins and then to regenerate the disulfide bonds of the native protein.[13]

RNase A has been synthesized chemically using several methods,[18–20] which enable unnatural amino acids to be incorporated into the protein. In principle, such unnatural substitutions can provide a powerful method for determining the general principles underlying conformational folding. For example, the effects of particular isomers of X-Pro peptide bonds on the conformational folding can be studied by substituting specific proline residues with cis- or trans-locked analogs such as the cis-locked analog L-5,5-dimethylproline[21] or the trans-locked analog L-2,4-methanoproline.[22] However, most structural studies require tens of milligrams of purified protein, which is technically challenging for these chemical synthetic methods. Recently, intein-based methods have permitted long protein fragments to be expressed in *E. coli* as activated ester intermediates; the larger size of these fragments reduces the required number of coupling steps, greatly improving the final yield of protein[23–25] and making structural studies of unnaturally substituted proteins feasible.

Several methods have been developed to isolate specific disulfide ensembles and species populated during the oxidative folding of proteins with many cysteines

[12] G. Taborsky, *J. Biol. Chem.* **234,** 2652 (1959).
[13] D. M. Rothwarf and H. A. Scheraga, *Biochemistry* **32,** 2671 (1993).
[14] T. W. Thannhauser and H. A. Scheraga, *Biochemistry* **24,** 7681 (1985).
[15] Y.-J. Li, D. M. Rothwarf, and H. A. Scheraga, *J. Am. Chem. Soc.* **120,** 2668 (1998).
[16] J. H. Laity, S. Shimotakahara, and H. A. Scheraga, *Proc. Natl. Acad. Sci. U.S.A.* **90,** 615 (1993).
[17] S. B. delCardayré, M. Ribó, E. M. Yokel, J. J. Quirk, W. J. Rutter, and R. T. Raines, *Protein Eng.* **8,** 261 (1995).
[18] B. Gutte and R. B. Merrifield, *J. Am. Chem. Soc.* **91,** 501 (1969).
[19] H. Yajima and N. Fujii, *in* "Chemical Synthesis and Sequencing of Peptides and Proteins" (T.-Y. Liu, A. N. Schechter, R. L. Heinrikson, and P. G. Condliffe, eds.), pp. 21–39. Elsevier/North-Holland, New York, 1981.
[20] D. Y. Jackson, J. Burnier, C. Quan, M. Stanley, J. Tom, and J. A. Wells, *Science* **266,** 243 (1994).
[21] S. S. A. An, C. C. Lester, J.-L. Peng, Y.-J. Li, D. M. Rothwarf, E. Welker, T. W. Thannhauser, L. S. Zhang, J. P. Tam, and H. A. Scheraga, *J. Am. Chem. Soc.* **121,** 11558 (1999).
[22] G. T. Montelione, P. Hughes, J. Clardy, and H. A. Scheraga, *J. Am. Chem. Soc.* **108,** 6765 (1986).
[23] T. W. Muir, D. Sondhi, and P. A. Cole, *Proc. Natl. Acad. Sci. U.S.A.* **95,** 6705 (1998).
[24] T. C. Evans, Jr., J. Benner, and M.-Q. Xu, *Protein Sci.* **7,** 2256 (1998).
[25] E. Welker and H. A. Scheraga, *Biochem. Biophys. Res. Commun.* **254,** 147 (1999).

such as RNase A. For example, the unblocked, structured disulfide species that have accumulated during oxidative folding may be isolated from their unstructured counterparts using a reduction pulse.[26] In this method, a short pulse (e.g., 1–2 min) of a low concentration of reducing agent (5–10 m*M* DTT^{red}) is applied to the mixture. The unstructured species are reduced rapidly under such conditions, whereas the structured species are largely unaffected, since the presence of stable tertiary structure generally protects the disulfide bonds from such a reduction pulse. Once isolated in unblocked form, the structured intermediates may be used to investigate the final steps of oxidative folding.[26]

The isolated, unblocked structured intermediates are also useful for preparing the *unstructured* disulfide ensemble having the same number of disulfide bonds. The isolation of this unstructured ensemble is important because the formation of the structured species from their unstructured precursors of this ensemble is generally the critical and rate-determining step in oxidative folding.[7] The unstructured ensemble may be prepared by transferring the structured intermediate to solution conditions that favor unfolding and allow intramolecular disulfide reshuffling reactions to occur (e.g., pH 8 at high temperatures or denaturant concentrations). On equilibration, the original disulfide species should comprise only a negligible fraction of the total ensemble, since entropy favors the population of all intermediates within the disulfide ensemble. For some proteins, it may be practical to include a preunfolding step at low pH so that, on jumping to the final conditions, the reshuffling is completed before significant air oxidation has occurred.

Native Structure of RNase A

RNase A is composed of three α helices and seven β strands arranged in two "lobes" (Fig. 2). The first lobe consists of the second α helix ($\alpha 2$, residues 24–34), the central β strand ($\beta 1$, residues 41–45) and the major β hairpin ($\beta 4$–$\beta 5$, residues 79–104), whereas the second lobe includes the third α helix ($\alpha 3$, residues 50–60), the 65–72 hairpin ($\beta 2$–$\beta 3$, residues 65–72), and the C-terminal hairpin ($\beta 6$–$\beta 7$, residues 105–124). The two lobes meet in a positively charged groove that binds the RNA substrate (Fig. 3). The N-terminal helix ($\alpha 1$, residues 3–13) rests between the two lobes and contributes the catalytically essential His-12 residue (Fig. 2). This helix is covalently connected to the rest of the protein by a poorly ordered segment (residues 16–23) that may be cleaved with no loss of activity to form RNase S[1]. Under folding conditions, the helix is bound to the rest of the protein with high affinity ($K_d \approx 30$p*M*) by noncovalent interactions.[5]

The major hydrophobic core of RNase A is associated with the 58–110 disulfide bond in the second lobe of RNase A (Fig. 1). This cluster of hydrophobic

[26] D. M. Rothwarf, Y.-J. Li, and H. A. Scheraga, *Biochemistry* **37,** 3767 (1998).

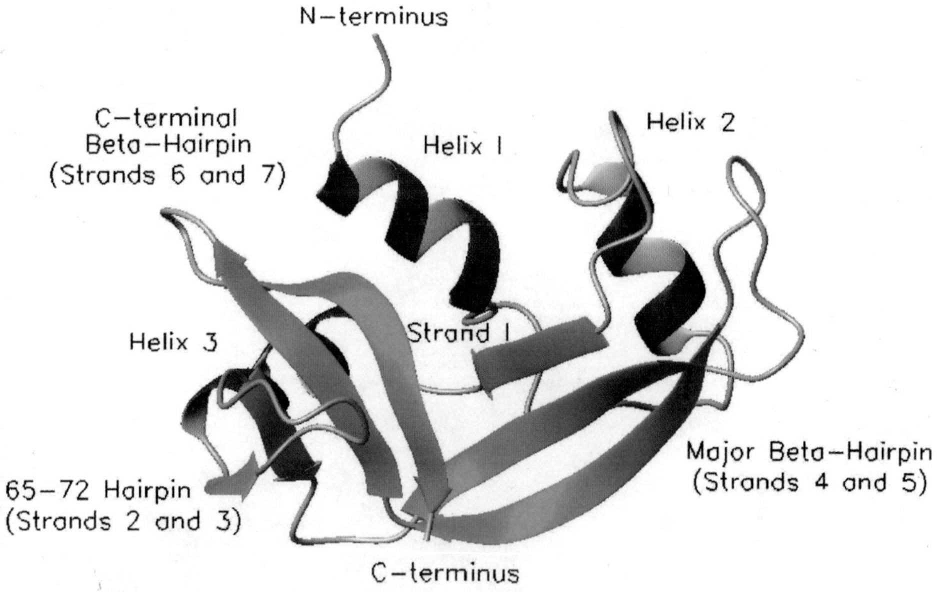

FIG. 2. The secondary structure elements of wild-type RNase A. The two lobes are clearly visible; the first lobe consists of helix 2, strand 1, and the major β hairpin, whereas the second lobe consists of helix 3 and the 65–72 and C-terminal hairpins. The N-terminal helix is poised between the two lobes. The front and back halves of the protein are also visible. The back half consists of the first β strand and the three α helices, whereas the front half consists of the three β hairpins (65–72, major, and C-terminal).

residues is composed primarily of residues from the C-terminal β-hairpin (Ile-106, Val-108, Cys-110, Pro-114–Val-118, Phe-120), as well as the third α-helix (Leu-51, Val-54, Val-57, Cys-58), and the first α helix (Ala-4, Ala-5, Phe-8, Met-13) (Fig. 2). The highly hydrophobic C-terminal hairpin has been hypothesized to be the primary chain-folding initiation site (CFIS) for RNase A.[27,28] A minor hydrophobic core of RNase A is associated with the 26–84 disulfide bond in the first lobe, composed of residues from the second α helix (Tyr-25, Cys-26, Met-29, Met-30), the major β hairpin (Cys-84, Tyr-97), and adjoining residues (Leu-35, Phe-46) (Fig. 2). The structuring of this core is presumably responsible for most of the observed change in tyrosine absorbance during folding.[29] In addition, a small third hydrophobic core is associated with the 65–72 disulfide bond in the

[27] R. R. Matheson, Jr. and H. A. Scheraga, *Macromolecules* **11,** 819 (1978).

[28] G. T. Montelione and H. A. Scheraga, *Acc. Chem. Res.* **22,** 70 (1989).

[29] D. Juminaga, W. J. Wedemeyer, R. Garduño-Júarez, M. A. McDonald, and H. A. Scheraga, *Biochemistry* **36,** 10131 (1997).

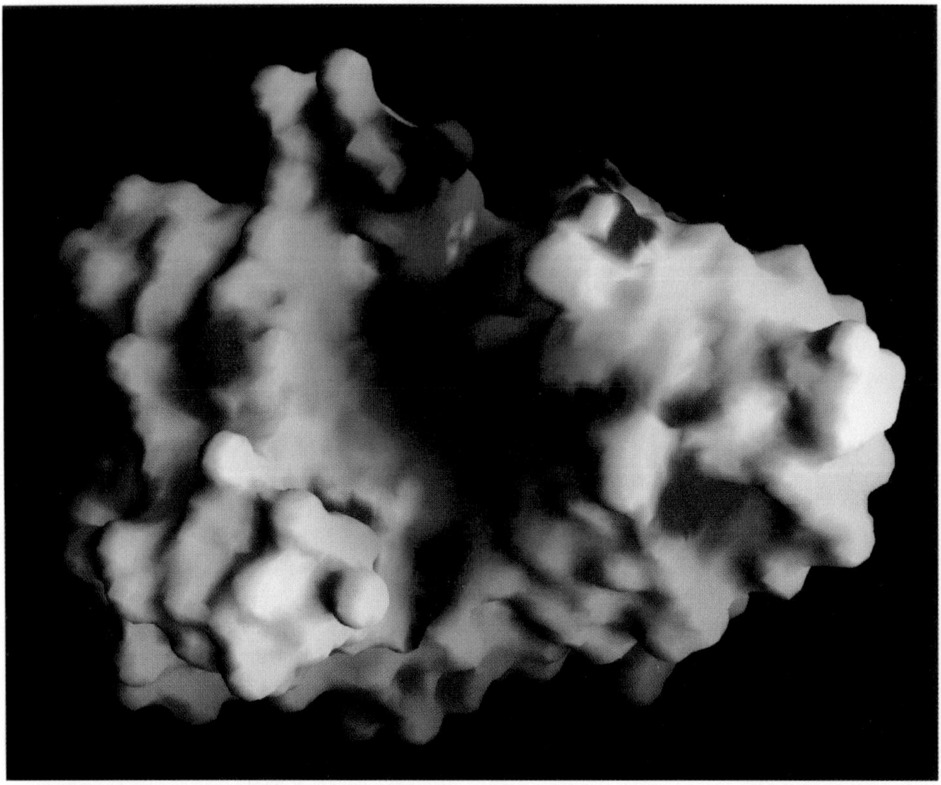

FIG. 3. GRASP diagram of wild-type RNase A, showing the two lobes of the protein separated by the positively charged "stripe" in which the RNA substrate is bound.

second lobe, and is composed primarily of residues from the 65–72 hairpin (Val-63, Cys-65, Cys-72) and from the C-terminal hairpin (Ile-107, Ala-122, Val-124) (Fig. 2).

RNase A may also be thought of as divided into two halves: an N-terminal, predominantly α-helical half (residues 1–60) and a C-terminal, predominantly β-sheet half (residues 65–124) (Fig. 2). Each lobe of RNase A is created from the disulfide bonding of an α-helical element of the first half with β-sheet elements of the second half (Fig. 2). Specifically, the 26–84 disulfide bond pins the second α helix to the major β hairpin (forming the minor hydrophobic core), whereas the 58–110 disulfide bond joins the third α helix to the C-terminal β hairpin (forming the major hydrophobic core). Interestingly, the C-terminal half appears to possess some flickering β-sheet structure even when the N-terminal half is fully unstructured, judging from fluorescence resonant energy transfer (FRET) studies of the fully reduced protein,[30] proteolytic studies of thermal denaturation,[31] and H/D exchange studies of conformational folding intermediates.[32,33]

High-resolution structures of RNase A have been determined both by X-ray crystallography and NMR techniques.[5] These studies have detected subtle changes in the internal hydrogen-bonding pattern due to several factors, e.g., changing temperatures,[34] ligand binding,[35] and the concentrations of various ions in solution.[36] In particular, the structural effects of proline isomers and disulfide bonds (and other covalent cross-links) have been investigated in X-ray and nuclear magnetic resonance (NMR) studies of chemically modified and mutant versions of RNase A.[1,37–40]

Equilibrium Unfolding Transitions of RNase A

Although the native conformation of RNase A can be characterized to atomic resolution, it is much more challenging to characterize *ensembles* of conformations to similar resolution, i.e., to determine the equilibrium distribution of conformations under a particular set of conditions. An even more demanding task is the characterization of nonequilibrium distributions of conformations, such as the distributions of conformations at various times during conformational folding. This section is concerned with equilibrium states and transitions between them, whereas the remaining sections are devoted to nonequilibrium folding studies.

Thermal Unfolding Transition

The reversible thermal unfolding transition of RNase A was first investigated by Harrington and Schellman.[41] It is an apparently two-state transition that can

[30] A. Navon, V. Ittah, P. Landsman, H. A. Scheraga, and E. Haas, *Biochemistry* **40,** 105 (2001).

[31] A. W. Burgess and H. A. Scheraga, *J. Theor. Biol.* **53,** 403 (1975).

[32] J. B. Udgaonkar and R. L. Baldwin, *Proc. Natl. Acad. Sci. USA* **87,** 8197 (1990).

[33] J. B. Udgaonkar and R. L. Baldwin, *Biochemistry* **34,** 4088 (1995).

[34] W. A. Gilbert, R. C. Lord, G. A. Petsko, and T. J. Thamann, *J. Raman Spectr.* **12,** 173 (1982).

[35] G. L. Gilliland, *in* "Ribonucleases: Structures and Functions" (G. D'Alessio and J. F. Riordan, eds.), pp. 305–341. Academic Press, New York, 1997.

[36] A. A. Fedorov, D. Joseph-McCarthy, E. Fedorov, D. Sirakova, I. Graf, and S. C. Almo, *Biochemistry* **35,** 15962 (1996).

[37] P. C. Weber, F. R. Salemme, S. H. Lin, Y. Konishi, and H. A. Scheraga, *J. Mol. Biol.* **181,** 453 (1985).

[38] M. A. Pearson, P. A. Karplus, R. W. Dodge, J. H. Laity, and H. A. Scheraga, *Protein Sci.* **7,** 1255 (1998).

[39] L. W. Schultz, S. R. Hargraves, T. A. Klink, and R. T. Raines, *Protein Sci.* **7,** 1620 (1998).

[40] Y. Xiong, D. Juminaga, G. V. T. Swapna, W. J. Wedemeyer, H. A. Scheraga, and G. T. Montelione, *Protein Sci.* **9,** 421 (2000).

[41] W. F. Harrington and J. A. Schellman, *Compt.-Rend. Trav. Lab. Carlsberg, Ser. Chim.* **30,** 21 (1956).

be divided into three stages: a pretransition stage, a cooperative-unfolding stage, and a residual-structure stage. A brief discussion of these three stages is given here. More detailed structural discussions of thermal denaturation have been presented by Burgess and Scheraga[31] (modified by Matheson and Scheraga[42]) and in a fluorescence study that employed an unusually large set of tryptophan fluorophores.[43]

In the pretransition stage, the folded state remains thermodynamically more stable than the unfolded state, but the structural fluctuations in the folded state become more pronounced as the transition temperature is approached from below. These enhanced structural fluctuations are evident from several types of experiments: proteolysis,[31,44–48] antigenicity,[49] electron paramagnetic spin resonance studies,[50] photolabeling,[42] enzymatic activity,[51] X-ray crystallography,[34] Raman,[34] and NMR data.[52,53] These structural studies also indicate that the redistribution of conformations does not occur uniformly throughout the protein, but that specific regions become labile as the temperature is increased. In particular, the data suggest that the second helix becomes destabilized at moderate temperatures well below the cooperative transition. This destabilization of the second helix in turn loosens the packing of the first helix in the major hydrophobic core (presumably because of the tight interaction between the first and second helices through the Tyr^{25}-Asp^{14} hydrogen bond[54]), leading to observable changes in Met-13,[42,53] His-48,[53] and the binding of the S-peptide. However, the first helix remains attached to the protein until the cooperative thermal transition.

The cooperative unfolding transition appears to occur near 60°C at neutral pH, corresponding to the point at which the unfolded state becomes thermodynamically stable relative to the folded state. This cooperative transition is evident in the appearance of new resonances in NMR (not continuous shifts of old resonances)[53] and in the superimposibility of the transition whether measured

[42] R. R. Matheson, Jr. and H. A. Scheraga, *Biochemistry* **18,** 2437 (1979).

[43] A. Navon, V. Ittah, J. H. Laity, H. A. Scheraga, E. Haas, and E. E. Gussakovsky, *Biochemistry* **40,** 93 (2001).

[44] J. A. Rupley and H. A. Scheraga, *Biochemistry* **2,** 421 (1963).

[45] T. Ooi, J. A. Rupley, and H. A. Scheraga, *Biochemistry* **2,** 432 (1963).

[46] W. A. Klee, *Biochemistry* **6,** 3736 (1967).

[47] A. W. Burgess, L. I. Weinstein, D. Gabel, and H. A. Scheraga, *Biochemistry* **14,** 197 (1975).

[48] U. Arnold, K. P. Rücknagel, A. Schierhorn, and R. Ulbrich-Hofmann, *Eur. J. Biochem.* **237,** 862 (1996)

[49] L. G. Chavez, Jr. and H. A. Scheraga, *Biochemistry* **16,** 1849 (1977).

[50] R. R. Matheson, Jr., H. Dugas, and H. A. Scheraga, *Biochem. Biophys. Res. Commun.* **74,** 869 (1977).

[51] R. R. Matheson, Jr. and H. A. Scheraga, *Biochemistry* **18,** 2446 (1979).

[52] C. R. Matthews and D. G. Westmoreland, *Biochemistry* **14,** 4532 (1975).

[53] F. W. Benz and G. C. K. Roberts, *J. Mol. Biol.* **91,** 345 (1975).

[54] H. A. Scheraga, *Fed. Proc.* **26,** 1380 (1967).

by optical rotation, tyrosine absorbance, viscosity, partial specific volume,[41,55–58] NMR resonances,[52,53] small-angle X-ray scattering,[59] or rapid "burst-phase" kinetics of folding and substrate binding.[60] (This "superimposability" criterion for a two-state transition was introduced by Anson.[61]) In the cooperative transition, the packing of side chains in the major hydrophobic core is disrupted, presumably allowing the solvent to penetrate and destabilize the secondary structure.

Some residual structure remains even above the cooperative thermal transition, as observed by several methods, e.g., optical rotation,[41] viscosity,[56] histidine pK_a,[52] ^{1}H NMR spectroscopy,[53] Raman spectroscopy,[34,62] near- and far-UV CD,[63] small-angle X-ray scattering,[59] Fourier transform IR spectroscopy,[64] and nuclear magnetic relaxation dispersion.[65] The thermally denatured state has long been known to have a significantly smaller radius of gyration compared to the reduced and reduced-denatured states.[41] The residual structure of the thermally denatured state appears to be localized in the C-terminal, predominantly β-sheet half of RNase A, particularly in the 65–72 and major β hairpins.[31,43] This residual structure also appears to be stable even when the temperature is increased further.[31,43] However, this residual structure can be decreased by the addition of chemical denaturants such as 6 *M* guanidinium hydrochloride (GdnHCl)[41,57,66] or completely eliminated by reduction of the disulfide bonds and 6 *M* GdnHCl.[41]

pH Dependence of Thermal Unfolding Transition

Interestingly, the thermal transition temperature of RNase A changes dramatically ($\Delta T_m \approx 30°$) as the pH is dropped from pH 5 to pH 2, apparently corresponding to the titration of one or more carboxylate groups.[55] Over the same pH range, the unfolding rate of native RNase A increases almost 200-fold.[67] The reason for this has not been established definitively, but it may result from the breaking of a strong hydrogen bond between Asp-14 of the N-terminal α helix and Tyr-25 of the second α helix.[29,54,68] Consistent with this hypothesis, the thermal stability of

[55] J. Hermans, Jr. and H. A. Scheraga, *J. Am. Chem. Soc.,* **83,** 3283 (1961).
[56] D. N. Holcomb and K. E. van Holde, *J. Phys. Chem.* **66,** 1999 (1962).
[57] C. C. Bigelow, *J. Mol. Biol.* **8,** 696 (1964).
[58] A. Ginsburg and W. R. Carroll, *Biochemistry* **4,** 2159 (1965).
[59] T. R. Sosnick and J. Trewhella, *Biochemistry* **31,** 8329 (1992).
[60] B. T. Nall and R. L. Baldwin, *Biochemistry* **16,** 3572 (1977).
[61] M. L. Anson, *Adv. Protein Chem.* **2,** 361 (1945).
[62] M. C. Chen and R. C. Lord, *Biochemistry* **15,** 1889 (1976).
[63] A. M. Labhardt, *J. Mol. Biol.* **157,** 331 (1982).
[64] S. Seshadri, K. A. Oberg, and A. L. Fink, *Biochemistry* **33,** 1351 (1994).
[65] V. P. Denisov and B. Halle, *Biochemistry* **37,** 9595 (1998).
[66] K. C. Aune, A. Salahuddin, M. H. Zarlengo, and C. Tanford, *J. Biol. Chem.* **242,** 4486 (1967).
[67] R. A. Sendak, D. M. Rothwarf, W. J. Wedemeyer, W. A. Houry, and H. A. Scheraga, *Biochemistry* **35,** 12978 (1996).
[68] H. A. Scheraga, *Carlsberg Res. Commun.* **49,** 1 (1984).

wild-type RNase A can be enhanced by a similar amount ($\Delta T_m \approx 20°$) at pH 5 by adding a covalent cross-link between Lys-7 and Lys-41 that stabilizes RNase A primarily by slowing its unfolding rate.[37,69,70] However, it has also been proposed that the pH sensitivity of the thermal transition is associated with the loss of a hydrogen bond between the C-terminal backbone carboxylate group of Val-124 and the backbone of His-105.[31] This may correlate with the roughly 20-fold decrease in the *refolding* rate in RNase A molecules with a nonnative Pro-114 isomer (U_f) as the pH is lowered from 5.0 to 3.0.[29,71,72] Finally, the pH sensitivity may result from simple electrostatic interactions within the protein,[55,73] consistent with the well-known anomalous pK_a values of certain side-chain carboxylate groups such as Asp-83 (p$K_a \approx 1.8$).[73–77]

Thermal-Folding Hypothesis

Although the unfolding transitions of RNase A are generally observed to be two-state transitions, the distribution of conformations within the folded and unfolded states varies with the solution conditions.[45] In particular, the fluctuations within the folded state increase as the unfolding transition is approached. Such fluctuations can be monitored by proteolysis,[31] disulfide-bond reduction and other chemical modifications,[26,78,79] or H/D exchange experiments,[80] which may provide estimates of the relative stabilities of specific segments of the protein against local unfolding. Proteolytic studies of the gradual destabilization of RNase A during equilibrium thermal denaturation led to the *thermal-folding hypothesis*. This hypothesis conjectures that the order in which local structures become destabilized during *equilibrium* thermal denaturation resembles (in reverse) the order in which specific structures become stabilized on the nonequilibrium folding pathway.[31] In other words, the conformational distributions populated during equilibrium thermal denaturation may be analogous to the nonequilibrium distributions populated during conformational folding. Similar hypotheses were subsequently advanced on the basis of H/D exchange experiments.[80,81] If true, the thermal-folding hypothesis

[69] S. H. Lin, Y. Konishi, M. E. Denton, and H. A. Scheraga, *Biochemistry* **23,** 5504 (1984).
[70] S. H. Lin, Y. Konishi, B. T. Nall, and H. A. Scheraga, *Biochemistry* **24,** 2680 (1985).
[71] R. W. Dodge and H. A. Scheraga, *Biochemistry* **35,** 1548 (1996).
[72] W. A. Houry and H. A. Scheraga, *Biochemistry* **35,** 11719 (1996).
[73] J. Hermans, Jr. and H. A. Scheraga, *J. Am. Chem. Soc.* **83,** 3293 (1961).
[74] H. A. Scheraga, *Biochim. Biophys. Acta* **23,** 196 (1957).
[75] C. A. Broomfield, J. P. Riehm, and H. A. Scheraga, *Biochemistry* **4,** 751 (1965).
[76] J. P. Riehm, C. A. Broomfield, and H. A. Scheraga, *Biochemistry* **4,** 760 (1965).
[77] W. R. Baker and A. Kintanar, *Arch. Biochem. Biophys.* **327,** 189 (1996).
[78] D. M. Rothwarf and H. A. Scheraga, *Biochemistry* **32,** 2698 (1993).
[79] Y.-J. Li, D. M. Rothwarf, and H. A. Scheraga, *Nature Struct. Biol.* **2,** 489 (1995).
[80] S. W. Englander and N. R. Kallenbach, *Quart. Rev. Biophys.* **16,** 521 (1984).
[81] R. Li and C. Woodward, *Protein Sci.* **8,** 1571 (1999).

would allow relatively simple equilibrium structural studies to replace their technically demanding, nonequilibrium counterparts. However, the general validity of this hypothesis has not been established.

Denaturant-Induced Unfolding of RNase A

The addition of chemical denaturants such as 6 *M* guanidine hydrochloride to disulfide-intact RNase A produces an unfolded state that exhibits less conformational order than the thermally denatured state.[82] However, the radius of gyration is significantly lower than that expected for a statistical coil, presumably because of the four disulfide bonds cross-linking the protein. It appears that few short-range or long-range interactions persist in the chemically denatured state, but the nature of its conformational distribution remains to be determined more precisely.

Unfolding of RNase A by Reducing Its Disulfide Bonds

The reduction of the four disulfide bonds of RNase A *under physiological conditions* produces an unfolded state that exhibits significant conformational order.[6] As has long been known, the reduced state of RNase A has a significantly reduced radius of gyration compared to an ideal statistical coil.[41,83,84] The relative equilibrium populations of the 28 possible one-disulfide pairings are also not well described by a simple loop–entropy formula, indicating significant conformational biases.[85,86] Moreover, FRET experiments indicate that a significant fraction of the native tertiary topology is still present, particularly in the C-terminal, predominantly β-sheet half of the protein.[30] Recent experiments suggest that the local conformational order present in the reduced state is similar to that produced in the so-called "burst phase" of conformational folding, in which a chemically denatured but disulfide-intact protein is rapidly introduced to physiological folding conditions.[6,87,88] This is another example of an equilibrium state whose conformational distribution may resemble (at least locally) that of a nonequilibrium state populated during conformational folding.

By contrast, if fully reduced RNase A is also treated with a strong denaturant such as 6 *M* GdnHCl, the *reduced-denatured state* is produced. This state is unique among the unfolded ensembles described in this section, because it appears to have little or no residual short-range or long-range structure and, hence, closely resembles a simple polymer with excluded-volume interactions (the so-called *statistical*

[82] A. Salahuddin and C. Tanford, *Biochemistry* **9,** 1342 (1970).
[83] A. Nöppert, K. Gast, M. Müller-Frohne, D. Zirwer, and G. Damaschun, *FEBS Lett.* **380,** 179 (1996).
[84] J.-M. Zhou, Y.-X. Fan, H. Kihara, K. Kimura, and Y. Amemiya, *FEBS Lett.* **430,** 275 (1998).
[85] X. Xu, D. M. Rothwarf, and H. A. Scheraga, *Biochemistry* **35,** 6406 (1996).
[86] M. J. Volles, X. Xu, and H. A. Scheraga, *Biochemistry* **38,** 7284 (1999).
[87] W. A. Houry, D. M. Rothwarf, and H. A. Scheraga, *Biochemistry* **35,** 10125 (1996).
[88] P. X. Qi, T. R. Sosnick, and S. W. Englander, *Nature Struct. Biol.* **5,** 882 (1998).

coil state). All other denaturation methods retain some short-range and/or long-range ordering, such as a low radius of gyration and circular dichroism indicating nonrandom local structure.

Other Equilibrium Unfolding Transitions of RNase A

Disulfide-intact RNase A has been unfolded by numerous other methods such as acid denaturation,[29,66] cold denaturation,[89,90] and pressure denaturation.[89,91] These methods do produce unfolded protein in a two-state transition; however, the resulting unfolded state in these cases appears to be significantly ordered.

Other Equilibrium States of RNase A

Some experiments have suggested that multiple unfolded equilibrium states may exist under unusual solution conditions.[57,92] The conformational properties of these anomalous unfolded states have not been well characterized. In some cases, e.g., at high concentrations of lithium perchlorate,[93,94] the amount of β structure appears to be decreased, perhaps by reducing the electrostatic stiffening observed in nucleic acids and other electrostatically-interacting polymers.[95] The reverse transition to β-favoring structures under low-pH, low-salt unfolding conditions has been observed in the α/β protein phosphoglycerate kinase.[96] By contrast, α-helical structures are favored in the unfolded state when trifluoroethanol and similar fluorine-substituted alcohols are added,[96,97] possibly by reducing the well-known solvent competition for hydrogen bonds to the amide groups of the protein backbone.[98]

Reductive Unfolding and Oxidative Folding of RNase A

The term *oxidative folding* describes the composite process by which a reduced, unfolded protein recovers both its native disulfide bonds (disulfide-bond regeneration) and its native structure (conformational folding). The regeneration of

[89] J. Zhang, X. Peng, A. Jonas, and J. Jonas, *Biochemistry* **34,** 8631 (1995).

[90] D. Nash, B.-S. Lee, and J. Jonas, *Biochim. Biophys. Acta* **1297,** 40 (1996).

[91] J. Torrent, J. P. Connelly, M. G. Coll, M. Ribó, R. Lange, and M. Vilanova, *Biochemistry* **38,** 15952 (1999).

[92] P. S. Sarfare and C. C. Bigelow, *Canad. J. Biochem.* **45,** 651 (1967).

[93] J. B. Denton, Y. Konishi, and H. A. Scheraga, *Biochemistry* **21,** 5155 (1982).

[94] R. M. Lynn, Y. Konishi, and H. A. Scheraga, *Biochemistry* **23,** 2470 (1984).

[95] A. Yu. Grosberg and A. R. Khokhlov, "Statistical Physics of Macromolecules." American Institute of Physics Press, New York, 1994.

[96] G. Damaschun, H. Damaschun, K. Gast, and D. Zirwer, *J. Mol. Biol.* **291,** 715 (1999).

[97] F. Chiti, N. Taddei, P. Webster, D. Hamada, T. Fiaschi, G. Ramponi, and C. M. Dobson, *Nature Struct. Biol.* **6,** 380 (1999).

[98] G. Némethy, I. Z. Steinberg, and H. A. Scheraga, *Biopolymers* **1,** 43 (1963).

the native disulfide bonds is a critical step in the maturation of many extracellular proteins. Oxidative folding is also an important stumbling block in the preparation of proteins for pharmaceutical and biotechnological applications, particularly multidomain proteins with several disulfide bonds.[99] Controlled *in vitro* experiments on relatively simple proteins such as RNase A may help to determine the factors affecting the regeneration of the native disulfide bonds both *in vivo* and for complex proteins.

Stable tertiary structure is one such factor.[6] Oxidation, reduction, and reshuffling are based upon a single chemical reaction, thiol/disulfide exchange,[100,101] in which a thiolate anion approaches a disulfide bond closely enough to delocalize its negative charge between the three sulfur atoms. Stable tertiary structure can inhibit such contacts between thiolates and disulfide bonds by controlling the accessibility, reactivity, and proximity of these reactive groups. For example, the burial of the disulfide bonds in stable tertiary structure inhibits their reduction and reshuffling, whereas the sequestering of thiol groups can inhibit their oxidation, i.e., the formation of disulfide bonds. Other structural factors include local electrostatic interactions (which may affect the local concentration, pK_a, and reactivity of the negatively charged thiolate groups) and structural propensities (which may alter the effective concentrations of the reactive groups for each other).

Definitions of Technical Terms

We adopt the following terminology to describe the covalent structure of the disulfide intermediates. A *disulfide species* refers to a protein with a particular pairing of cysteines in disulfide bonds, while a *disulfide ensemble* is any collection of disulfide species. A *des species* is a disulfide species with all but one of the native disulfide bonds; for example, the des[65–72] species of RNase A has three native disulfide bonds but lacks the 65–72 disulfide bond. The term *des* signifies only a covalent property (the absence of one native disulfide bond) and does not specify whether the species is folded. A subscript is added to characterize the conformational state; thus, des_U and des_N represent the unfolded and folded states of a des species. *n*S represents an ensemble of (typically unstructured) disulfide species with *n* disulfide bonds that are in a rapid reshuffling equilibrium among themselves; for example, 2S represents an ensemble of rapidly interconverting disulfide species with two intraprotein disulfide bonds. The rapid interconversion within such ensembles allows them to be treated as single kinetic species, although they are a heterogeneous mixture of disulfide

[99] D. R. Thatcher and A. Hitchcock, *in* "Mechanisms of Protein Folding" (R. H. Pain, ed.), pp. 229–261. IRL Press, New York, 1994.

[100] C. Huggins, D. F. Tapley, and E. V. Jensen, *Nature* **167,** 592 (1951).

[101] H. F. Gilbert, *Adv. Enzymology* **63,** 69 (1990).

species.[102] The ensemble of disulfide species with no protein disulfide bonds is denoted as R, the fully reduced ensemble (the multiplicity of R is due to the possibility of mixed disulfide bonds). Fully oxidized, nonnative species are denoted as *scrambled* disulfide species, e.g., those of the 4S ensemble in RNase A. A more general nomenclature of disulfide species (including mixed disulfide species) has been published.[103]

We introduce here the following terminology to describe the protection of disulfide bonds and thiol groups in structured disulfide species. In *disulfide-protected* species, the disulfide bonds are protected from reduction and reshuffling. Such species often accumulate since the inhibition of reshuffling in these species reduces their rate of disappearance significantly. A subset of disulfide-protected species are the *metastable* disulfide species, in which both the disulfide bonds and thiol groups are protected, inhibiting all disulfide-bond reactions (oxidation, reduction, and reshuffling). Thus, a local or global unfolding step must precede disulfide-bond reactions in such metastable species. It suffices to bury all but one thiol group in such species, since the remaining thiol group would have no partner with which to react. Two other terms (*disulfide-secure/insecure*) are introduced below to describe the relative protection of the disulfide bonds and thiol groups.

Reductive Unfolding of RNase A

In reductive unfolding experiments, a native protein is placed under strongly reducing conditions and the loss of its disulfide bonds and conformational structure is monitored. Since disulfide bonds are generally buried in native proteins,[104] the reduction of such a disulfide bond must be preceded by a local or global unfolding step that exposes the bond to the redox reagent. The $\Delta G°$ for these unfolding steps can be estimated from the rate at which the corresponding disulfide bonds are reduced[79] and measures the stability of the protective tertiary structure. The reduction of only one native disulfide bond often produces a des species with native-like structure. However, when enough disulfide bonds (usually two) become reduced or reshuffled, most disulfide-bonded proteins unfold globally because the native fold is no longer thermodynamically stable relative to the unfolded state. After global unfolding, the protein is rapidly reduced to its fully reduced state R, because all disulfide bonds are accessible to the redox reagent.

This general scenario is consistent with the reductive unfolding of RNase A. Under typical conditions (pH 8, 15–25°, with DTT^{red} as the reducing agent), the

[102] H. A. Scheraga, Y. Konishi, D. M. Rothwarf, and P. W. Mui, *Proc. Natl. Acad. Sci. U.S.A.* **84,** 5740 (1987).
[103] Y. Konishi, T. Ooi, and H. A. Scheraga, *Biochemistry* **20,** 3945 (1981).
[104] J. M. Thornton, *J. Mol. Biol.* **151,** 261 (1981).

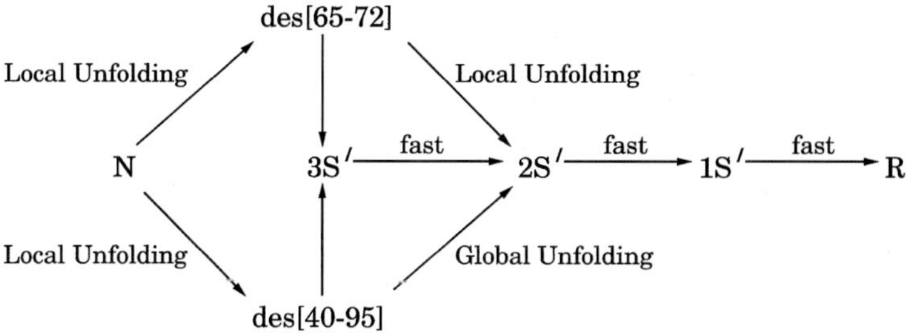

FIG. 4. The reductive unfolding of RNase A.

reductive unfolding of RNase A appears to proceed entirely through two parallel pathways, corresponding to two independent, local unfolding events exposing the 65–72 and 40–95 disulfide bonds, respectively (Fig. 4).[79] Reduction of these disulfide bonds results in two des species, des[65–72] and des[40–95], both with native-like structure.[38,105–107]

The two remaining des species, des[26–84] and des[58–110], have not been observed in the reductive unfolding of native RNase A, presumably because the activation free energies for the unfolding steps exposing these disulfide bonds is much higher than those for the unfolding steps exposing the 40–95 and 65–72 disulfide bonds. This explanation is structurally plausible, because the disulfide bonds 26–84 and 58–110 are buried in the core of the native conformation[8] and thus may require global unfolding to expose them; by contrast, the 40–95 and 65–72 disulfide bonds are only partially buried and may be exposed with local unfolding. Des[26–84] and des[58–110] also have a low conformational stability[16,108–110] and, hence, are reduced much more rapidly than des[40–95] and des[65–72]. In principle, this could also account for their apparent absence in reductive unfolding; kinetic data[79] indicate that little of the protein is reduced through such pathways, however.

The disulfide bonds of native-like species, des[40–95] and des[65–72], are also buried in protective tertiary structure and are likewise reduced through unfolding events (Fig. 4).[79] The more stable des[65–72] seems to be reduced through the same unfolding event that precedes the reduction of the 40–95 disulfide bond in the native

[105] S. Talluri, D. M. Rothwarf, and H. A. Scheraga, *Biochemistry* **33,** 10437 (1994).

[106] S. Shimotakahara, C. B. Rios, J. H. Laity, D. E. Zimmerman, H. A. Scheraga, and G. T. Montelione, *Biochemistry* **36,** 6915 (1997).

[107] J. H. Laity, C. C. Lester, S. Shimotakahara, D. E. Zimmerman, G. T. Montelione, and H. A. Scheraga, *Biochemistry* **36,** 12683 (1997).

[108] E. Welker, M. Narayan, M. J. Volles, and H. A. Scheraga, *FEBS Lett.* **460,** 477 (1999).

[109] L. K. Low, H.-C. Shin, M. Narayan, W. J. Wedemeyer, and H. A. Scheraga, *FEBS Lett.* **472,** 67 (2000).

[110] T. A. Klink, K. J. Woycechowsky, K. M. Taylor, and R. T. Raines, *Eur. J. Biochem.* **267,** 566 (2000).

protein,[79] i.e., the 40–95 disulfide bond is preferentially reduced in the des[65–72] species while the 26–84 and 58–110 disulfide bonds are largely untouched. As in the native protein, this preference probably results from a relatively low free energy to expose the 40–95 disulfide bond compared to the two other disulfide bonds. By contrast, the des[40–95] species seems to be reduced through a global unfolding step[111] that exposes all three of its disulfide bonds roughly equally.[112] The reductive unfolding of isolated des[26–84] and des[58–110] has not been studied, but these species are even less stable than des[40–95] and are likely reduced through global unfolding as well (unpublished results). Following either the reduction or reshuffling of a disulfide bond in any of these des species, the protein is unfolded conformationally and is reduced rapidly to the fully reduced state R, because its disulfide bonds are accessible to the redox reagent (Fig. 4).[79]

The nature of these local unfolding steps is currently being investigated. It has been suggested that cis/trans isomerization of the Tyr^{92}-Pro^{93} peptide bond may be involved in the local unfolding that precedes the reduction of 40–95.[111] Similarly, the 65–72 disulfide bond may be exposed by a simple unfolding of the C terminus, since Ala-122 and Val-124 contribute to the small hydrophobic core protecting this disulfide bond. It has long been known that the unstructuring of these C-terminal residues does not lead to global unfolding.[1] Experiments that test these hypotheses for the two local unfolding steps of reductive unfolding are underway.

Under strongly reducing conditions, the reshuffling pathways of the des species are relatively minor, presumably because the same unfolding step exposes the disulfide bond both to the redox reagent and to the protein thiolate groups and, under such conditions, the redox reagent competes effectively with the protein thiolate groups to reduce (not reshuffle) the protein disulfide bond.

Oxidative Folding of RNase A

The oxidative folding of RNase A is significantly more complex than its reductive unfolding, since it involves a large number of intermediates, consistent with the large entropic barrier needed to be overcome to find the native disulfide pairing. Under typical conditions (pH 8, 25°C, with DTT^{ox} as the redox reagent), the oxidative folding of RNase A is fit by the kinetic model of Fig. 5a.[26,113–115] In the early stages of oxidative folding, the disulfide ensembles 1S–4S are populated by successive oxidations, until a preequilibrium (quasi-steady-state condition) is established among these ensembles. No stable structure has been detected in these ensembles, although they do exhibit some conformational order.[6] The distribution

[111] M. Iwaoka, W. J. Wedemeyer, and H. A. Scheraga, *Biochemistry* **38,** 2805 (1999).

[112] M. Iwaoka and H. A. Scheraga, *J. Am. Chem. Soc.* **120,** 5806 (1998).

[113] D. M. Rothwarf, Y.-J. Li, and H. A. Scheraga, *Biochemistry* **37,** 3760 (1998).

[114] X. Xu and H. A. Scheraga, *Biochemistry* **37,** 7561 (1998).

[115] M. Iwaoka, D. Juminaga, and H. A. Scheraga, *Biochemistry* **37,** 4490 (1998).

(a)

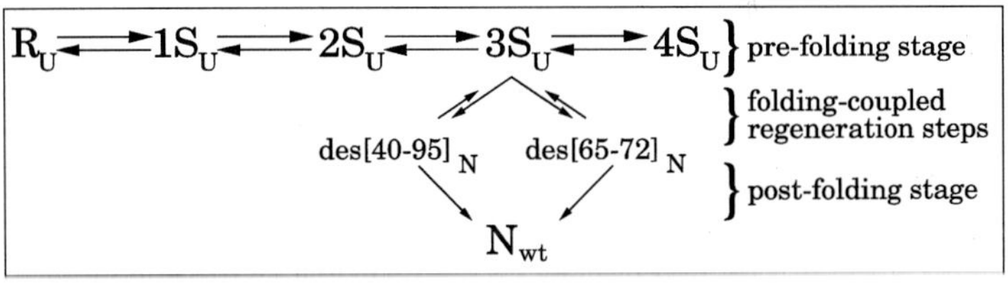

(b)

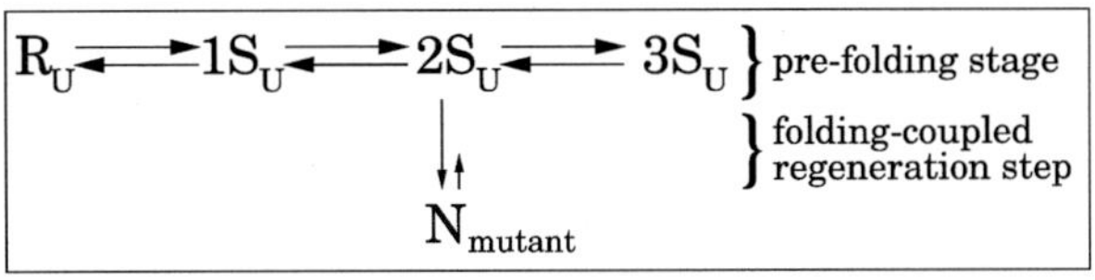

FIG. 5. (a) The oxidative folding of wild-type RNase A. (b) The oxidative folding of the three-disulfide mutants of RNase A.

of disulfide species within the 1S and 2S ensembles is enthalpically biased toward native disulfide bonds, relative to the populations predicted from loop–entropy considerations alone.[85,86,116] This preferential population of native disulfide bonds accelerates the oxidative folding relative to that of an unbiased ensemble.

The rate-determining step in the regeneration of native RNase A is the formation of two des species, des[40–95] and des[65–72], with native-like structure.[38,105–107] These species are formed largely by reshuffling from the 3S ensemble,[26,113] although a small fraction (no more than 5%) may be formed by oxidation from the 2S ensemble,[114,115] detected only by studies of oxidation of mutant analogs of the des species (Fig. 5b). On formation and folding of either des[40–95] or des[65–72], their three native disulfide bonds become protected from reduction and reshuffling ("locked in"), causing these des species to accumulate to much higher levels. However, their thiol groups remain accessible to the solvent and, hence, these species oxidize relatively rapidly to the native protein. These pathways appear to account for nearly all of the native RNase A regenerated under typical oxidative folding conditions.[26]

Stable conformational structure appears to be essential for this acceleration of oxidative folding. The regeneration becomes drastically slower when carried out under conditions that destabilize the conformational structure of the folded des species.[78,106,107] Moreover, the original rate of regeneration can be restored by adding other agents that restabilize the conformational structure (e.g., phosphate salts).[109] Thus, the protective structure of these des species is the critical

[116] D. C. Poland and H. A. Scheraga, *Biopolymers* **3,** 379 (1965).

factor in promoting oxidative folding. This type of pathway in which disulfide intermediates adopt native-like structure and protect their native disulfide bonds from rearrangement has been denoted as a "des_N-type" pathway and appears to be essential for the efficient regeneration of proteins with several disulfide bonds.[7]

The oxidative folding of mutant analogs of des[40–95] and des[65–72] (the three-disulfide mutants [C40A,C95A] and [C65S,C72S]) has also been characterized.[114,115] Consistent with the absence of structured two-disulfide intermediates in wild-type RNase A, neither mutant exhibits any folded disulfide intermediates prior to forming all three native disulfide bonds. In both mutants, the rate-determining step is the formation of the native protein by oxidation from the unstructured 2S ensemble and conformational folding (Fig. 5b). This type of pathway in which no structured disulfide intermediates are populated in oxidative folding has been denoted as a "des_U-type" pathway and is observed primarily in proteins with relatively few (≤ 3) disulfide bonds.[7]

The other two des species, des[58–110] and des[26–84], have stable conformational structure only under strongly stabilizing conditions, e.g., in the presence of stabilizing salts[109] or at low temperatures ($\leq 15°$).[108] Under such conditions, these des species are metastable dead-end species that reshuffle preferentially to the 3S ensemble rather than oxidizing to the native protein. (Accordingly, such species are said to lie on "metastable, dead-end pathways."[7]) The thiol groups and disulfide bonds of des[58–110] and des[26–84] appear to be buried; thus, oxidation as well as reduction and reshuffling are inhibited.[116a] This burial of thiol groups is structurally plausible, since these des species probably have native-like structures [judging from enzymatic activity measurements[110] and their high-performance liquid chromatography (HPLC) elution properties], and the 26–84 and 58–110 disulfide bonds are fully buried in hydrophobic cores of the native protein.

Even mixed disulfide bonds can become buried in stable conformational structure. This is observed in the oxidative folding of RNase A through the covalent adduct N′, in which a DTT molecule crosslinks cysteines-65 and -72.[15] Kinetic data indicate that N′ is formed through the burial of a mixed disulfide bond between a DTT molecule and one cysteine, probably Cys-72.[105] The remaining thiolate of the covalently bound DTT molecule forms a disulfide bond with a second DTT molecule, which is attacked by the remaining protein thiolate to form the adduct. Such protected mixed disulfide bonds seem more likely to form with linear redox reagents such as glutathione[117] than with rapidly recyclizing reagents such as DTT.

General Scenario for Oxidative Folding

A general scenario for oxidative folding can be developed from the results for RNase A. The key steps in oxidative folding are those in which structured disulfide

[116a] E. Welker, M. Narayan, W. J. Wedemeyer, and H. A. Scheraga, *Proc. Natl. Acad. Sci. U.S.A.* **98,** 2312 (2001).

[117] J. S. Weissman and P. S. Kim, *Science* **256,** 112 (1992).

species are generated from unstructured precursors. Such folding-coupled regeneration steps divide the oxidative folding into three stages: (1) a prefolding stage in which the disulfide species are largely unstructured and the disulfide-bond reactions occur at rates determined largely by loop–entropy and conformational biases; (2) a postfolding stage in which the disulfide species are largely structured and the disulfide-bond reactions occur at rates determined largely by the accessibility and proximity of the reactive groups; and (3) the folding-coupled regeneration steps themselves, in which conformational folding and unfolding events compete with disulfide bond reactions.

Insights into Conformational Folding from Oxidative Folding Experiments

Experiments carried out in each stage of oxidative folding yield complementary insights into the process of conformational folding. First, in the prefolding stage, the relative populations of disulfide species can be used to assess the conformational biases present in the unstructured ensembles relative to an idealized ensemble in which only loop entropy would determine the disulfide-bond reaction rates.[85,86] These measured biases help to characterize the conformational order present in the earliest stages of conformational folding of the protein.[6] In fact, FRET experiments[30] demonstrate that the reduced protein, under folding conditions, has a considerable fraction of the native backbone topology, but a large separation between the N- and C-terminal halves. Thus, the molecule is already poised for the conformational folding that occurs when three of the four disulfide bonds are formed. Second, in the postfolding stage, the relative accessibility and proximity of different cysteines in the native protein can be assessed from their relative reaction rates, similar to the situation described above for reductive unfolding. Third, the competition between conformational folding and disulfide-bond reactions in folding-coupled regeneration steps can be exploited to characterize the structures populated in conformational folding. In principle, chemical analysis of the disulfide species produced in such steps can report on the transient distributions of conformations present in folding through the relative reactivities of different disulfide bonds and thiol groups. We now describe these three types of experiments in more detail.

In *folding-coupled regeneration steps,* conformational folding can be investigated by setting up a competition between conformational folding and disulfide reshuffling or between conformational folding and disulfide reduction. In the former *reshuffling–competition* protocol, the initially unfolded protein (with at least one free thiol group) is jumped to conditions that favor both conformational folding and disulfide-bond reshuffling (e.g., from low to high pH), followed by a quench of the reshuffling reactions. In the analogous *reduction–competition* protocol, the initially unfolded protein (with no free thiol groups, i.e., disulfide-intact protein) is jumped to highly reducing conditions (e.g., 100 m*M* DTT^{red}) that otherwise strongly favor conformational folding, followed by a rapid quenching of the reduction reactions. These methods may reveal differences in the rates at which different parts of the protein become protected from reduction during slow conformational folding. Moreover, the relative rates of reduction/reshuffling may also help to characterize the distributions of conformations populated during such folding.

In the *prefolding stage* of oxidative folding, the distributions of disulfide bonds in the 1S and 2S disulfide ensembles indicate the presence of significant conformational order. The relative concentrations of disulfide bonds in these ensembles show significant deviations from random probabilities,[85,86] indicating significant enthalpic contributions to the free energy in addition to the loop entropy. In particular, the 65–72 disulfide bond is strongly favored, comprising fully 40% of the 1S ensemble,[85] and exists in an even larger fraction of the 2S ensemble.[86] However, these enthalpic contributions appear to be dominated by local interactions, i.e., interactions between residues near the cysteines of the disulfide bonds, since nearly identical enthalpic contributions are observed in homologous oligopeptide model systems.[118–120] Moreover, the enthalpic contributions observed in the 1S ensemble appear to account for those observed in the 2S ensemble (unpublished results). An NMR study of two-disulfide mutant analogs of RNase A[121] also indicated that a disulfide species with two native disulfide bonds may exhibit no more structure than those with two nonnative disulfide bonds. These results suggest that the conformational order does not increase strongly with the number of disulfide bonds, prior to the appearance of native-like structure in the folding-coupled regeneration steps.

The rapid reshuffling of disulfide species in the prefolding ensembles casts doubt on recent "rugged funnel" models of protein folding. These models propose that stable, misfolded states populate the energy landscape of proteins densely, and that a protein will collapse globally into a slowly rearranging, "glassy" state whenever it is prevented from entering the "native folding funnel." Given that the prefolding ensembles generally sample most of the possible disulfide species,[6] such rugged-funnel models are inconsistent with the absence of long-lived disulfide species with nonnative disulfide bonds and nonnative stable tertiary structure.

To study the *postfolding stage,* we isolate the structured disulfide species as described above and place them under oxidizing conditions that maintain their folded structure (e.g., 100 m*M* DTT^{ox}, pH 8.0, 15°). A small amount of reducing agent is added (e.g., 0.25 m*M* DTT^{red}) as well; this concentration is chosen so that the structured species will not be reduced directly at a significant rate, but if the structured species should reshuffle to an unstructured species, the latter species will

[118] P. J. Milburn and H. A. Scheraga, *J. Protein Chem.* **7,** 377 (1988).

[119] K.-H. Altmann and H. A. Scheraga, *J. Am. Chem. Soc.* **112,** 4926 (1990).

[120] S. Talluri, C. M. Falcomer, and H. A. Scheraga, *J. Am. Chem. Soc.* **115,** 3041 (1993).

[121] C. C. Lester, X. Xu, J. H. Laity, S. Shimotakahara, and H. A. Scheraga, *Biochemistry* **36,** 13068 (1997).

be reduced more rapidly than it will interconvert to another structured species. This precaution allows us to distinguish the oxidations of the structured species from their interconversions by disulfide reshuffling.[116a]

The results indicate that the two more stable des species, des[40–95] and des[65–72], oxidize directly to the native protein, whereas the other two des species, des[26–84] and des[58–110], preferentially reshuffle to their unstructured ensembles. This is structurally plausible, since the former two disulfide bonds are relatively exposed in the native structure (and, hence, their thiols may be exposed by local unfolding events), whereas the two latter disulfide bonds are fully buried (and, hence, the exposure of their thiols may require global unfolding). These two types of structured disulfide species are denoted as *disulfide-secure* and *disulfide-insecure,* respectively, since the disulfide bonds remain buried while the thiol groups are exposed in the former case, but not in the latter. The directly oxidizing species des[40–95] exhibits a significantly increased oxidation rate relative to the unstructured intermediates, consistent with an enhanced proximity of these cysteines in the des species. However, no significant enhancement is observable in the des[65–72] species, possibly because the mixed disulfide bond may not be fully exposed, consistent with NMR studies[105] and the N′ results[15] cited above.[116a]

These oxidative folding results (in which the exposure of thiol groups in the structured des species is the critical conformational event) are consistent with those of reductive unfolding (in which the exposure of disulfide bonds in the native protein is the critical conformational event). In particular, this agreement in the results between oxidative folding and reductive unfolding in the des[26–84] and des[58–110] species supports our hypothesis[6] that these species are not observed in reductive unfolding because of the high free-energy fluctuation (global unfolding) required to expose these disulfide bonds in the native protein, rather than because of their rapid subsequent reduction.

Effect of Proline cis/trans-Isomers in Disulfide-Intact Folding

In its native state, RNase A has two *cis*-X-Pro peptide bonds (Tyr^{92}-Pro^{93} and Asn^{113}-Pro^{114}) and two *trans*-X-Pro peptide bonds (Lys^{41}-Pro^{42} and Val^{116}-Pro^{117}). On unfolding, these four X-Pro peptide bonds undergo cis/trans-isomerization to form a heterogeneous mixture of 16 (2^4) isomeric states. (As defined here, the isomeric state of a protein specifies the cis/trans-isomers of its peptide bonds, especially X-Pro peptide bonds.) Under typical conditions, the isomerization between these states is much slower than the equilibration of conformations within any such state (in which only the ϕ, ψ, and χ dihedral angles are varied). Therefore, different isomeric states can be treated as though they were transient site-directed mutants. In particular, it is possible in principle to study the effects on conformational folding that result from the localized disruptions of the backbone conformation and side-chain packing caused by nonnative proline isomers. In contrast to normal site-directed mutants, such "isomeric mutants" do not change the chemical bonding of the protein in any way and introduce no new atoms or interactions. However, although isomeric mutants are highly conservative by these criteria, they can lead to profound changes in the conformational folding, as demonstrated in RNase A. The sensitivity of conformational folding to the isomeric state is all the more remarkable, since prolines are generally not buried and do not belong to regular secondary structure elements. Hence, their isomerization should generally not disrupt the packing of any hydrophobic core(s).

The effects of nonnative proline isomers on the conformational folding is often observable as a kinetic heterogeneity; in particular, slow-refolding phases are often attributed to the parallel conformational folding of different isomeric states. However, several key points should be noted. First, such kinetic heterogeneity can also arise from intermediates of purely conformational folding. Therefore, it is critical to establish that different isomeric states are indeed responsible for the observed kinetic heterogeneity and, more specifically, to determine which kinetic phases are due to proline isomerization. (It should be noted that a given kinetic phase may be composed of two or more phases of similar time constants; the component exponential phases can often be discerned by varying the conditions and/or by making specific mutations.[122]) Second, it must be established whether these refolding phases are rate-limited by proline isomerization or whether the nonnative isomeric state merely retards the conformational folding. Third, it must be determined which isomeric state(s) is (are) responsible, i.e., to correlate the observed kinetics with specific structural features, namely, the nonnative proline isomers. Only in this way is it possible to clearly establish a link between specific isomeric states and specific conformational refolding behaviors, the goal of such studies.

For clarity, the investigation of the conformational folding of the isomeric states of RNase A can be divided into five steps. In the *first step,* it is necessary to ascertain how many distinct refolding phases are present, i.e., how many subpopulations with distinct refolding behaviors exist in the equilibrium unfolded state. Hence, the conformational folding of RNase A was investigated under a wide variety of refolding conditions. Five exponential phases were discerned in these single-jump refolding experiments: U_{vf}, U_f, U_m, U_S^{II}, and U_S^{I}, corresponding to the very fast, fast, medium, slow, and very slow refolding phases. This kinetic heterogeneity was conjectured to result from nonnative proline isomers,[123] based on the following observations. First, each of these phases produces fully native protein, not a kinetic folding intermediate. Second, the equilibrium ratios of the amplitudes are independent of the refolding conditions (indicating that they correspond to a heterogeneity of the unfolded state) and agree in magnitude with those expected from the proline-isomerization hypothesis. Third, the interconversion rates under unfolding conditions between these phases are similar to that of

[122] D. Juminaga, W. J. Wedemeyer, and H. A. Scheraga, *Biochemistry* **37,** 11614 (1998).
[123] J. F. Brandts, H. R. Halvorson, and M. Brennan, *Biochemistry* **14,** 4953 (1975).

proline isomerization and exhibit several hallmarks of proline isomerization, e.g., an activation energy of 20 kcal/mol, catalysis by strong acids, and an indifference to pH and denaturant concentrations.

In the *second step,* it is necessary to determine which prolines are responsible ("essential") for the kinetic heterogeneity observed in the first step, and which proline isomers are irrelevant to conformational folding. Hence, the single-jump refolding experiments were repeated on all possible single proline-to-alanine mutants of RNase A, i.e., P42A, P93A, P114A, and P117A. It was shown that prolines-93, -114, and -117 are essential, i.e., significantly affect conformational folding, whereas Pro-42 appears to be nonessential.[71,124] However, it is also possible that the relative cis fraction of the Lys^{41}-Pro^{42} peptide bond is too small under unfolding conditions to allow its effect on conformational folding to be resolved. Lys-Pro peptide bonds are well-known to favor the trans isomer heavily.[125–127]

In the *third step,* it is necessary to characterize how the refolding heterogeneity varies with the unfolding time, both to establish that proline isomerization is indeed responsible and to estimate the cis/trans ratios and isomerization rates of the essential prolines under unfolding conditions. Hence, double-jump refolding experiments were carried out on the wild-type protein, in which the folded native protein was unfolded for various times and then reintroduced to folding conditions. The amplitudes of the five refolding phases were monitored as a function of the unfolding time. These data were fit exhaustively to all possible kinetic models of three independently isomerizing prolines (so-called "box" models). The resulting best-fit box model is shown in Fig. 6, along with the equilibrium percentages of each isomeric state and the time constants for the cis/trans isomerization for each proline under unfolding conditions.

In the *fourth step,* it is necessary to connect the prolines of the kinetic box model with the essential prolines identified in the second step. To this end, independent fluorescence assays of the isomerizing prolines were developed.[122] These assays depend on the change in fluorescence of a tyrosine residue due to the isomerization of an adjacent proline, possibly due to ring–ring interactions[21,128,129]; the nearby 40–95 disulfide bond does not contribute to the quenching.[111] The time constants measured directly by these assays[122] agreed with those determined in the kinetic fitting of the third step,[29] providing independent evidence supporting the box model.

[124] R. W. Dodge, J. H. Laity, D. M. Rothwarf, S. Shimotakahara, and H. A. Scheraga, *J. Protein Chem.* **13,** 409 (1994).

[125] H. J. Dyson, M. Rance, R. A. Houghten, R. A. Lerner, and P. E. Wright, *J. Mol. Biol.* **201,** 161 (1988).

[126] D. P. Raleigh, P. A. Evans, M. Pitkeathly, and C. M. Dobson, *J. Mol. Biol.* **228,** 338 (1992).

[127] U. Reimer, G. Scherer, M. Drewello, S. Kruber, M. Schutkowski, and G. Fischer, *J. Mol. Biol.* **279,** 449 (1998).

[128] J. Yao, H. J. Dyson, and P. E. Wright, *J. Mol. Biol.* **243,** 754 (1994).

[129] W.-J. Wu and D. P. Raleigh, *Biopolymers* **45,** 381 (1998).

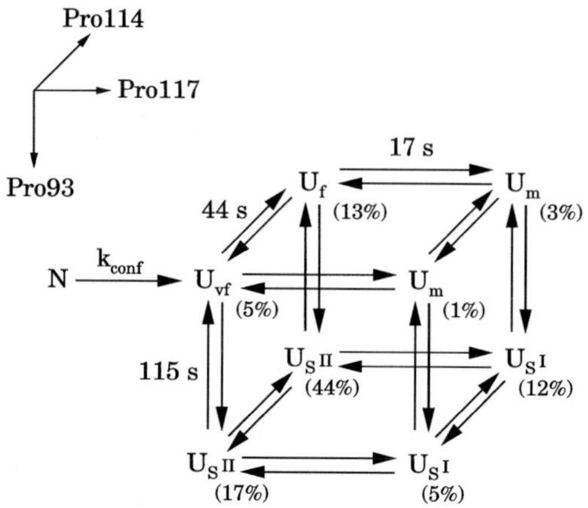

FIG. 6. The best-fitting "box" model for the isomerization of Pro-93, Pro-114, and Pro-117 under unfolding conditions.

In the *fifth step,* the box model (with identified prolines) should be cross-validated. Hence, the double-jump experiments were repeated for the proline-to-alanine mutants.[71] It should be noted that most, but not all, of these double-jump experiments agreed with the proposed box model. In particular, contrary to the box model of Fig. 6, the P93A and P117A mutants both exhibited a U_m refolding phase. However, in other respects, the mutant data agreed well; for example, the P114A mutant exhibited no U_{vf} refolding phase, and the P117A mutant exhibited no U_S^I refolding phase. The time constants also agreed roughly with those predicted from the box model.

The conclusion of these studies was that nonnative isomers of Pro-114, Pro-117, and Pro-93 cause increasing disruption of conformational folding. However, most isomeric states do fold conformationally (i.e., their folded state is still thermodynamically more stable than their unfolded state), except for those states in which both Pro-93 and Pro-117 have adopted nonnative isomers. Specifically, the U_{vf}, U_f, U_m, and U_s^{II} and U_s^I refolding phases have been assigned as follows: U_{vf} corresponds to the single native isomeric state. U_f corresponds to the single isomeric state in which only Pro-114 is nonnative trans. U_m corresponds to the two isomeric states in which Pro-93 is native cis, Pro-117 is nonnative cis and Pro-114 may be either cis or trans. U_s^{II} corresponds to the two isomer states in which Pro-93 is nonnative trans, Pro-117 is native trans, and Pro-114 may be either cis or trans. Finally, U_s^I corresponds to the two isomeric states in which Pro-93 is nonnative trans, Pro-117 is nonnative cis, and Pro-114 may be cis or trans.

It is remarkable that a nonnative isomer of Pro-93 slows folding roughly 500-fold (U_s^{II}), although it is located in a flexible loop region, whereas a nonnative isomer of Pro-117 retards folding roughly 25-fold (U_m), although it is fully buried in the major hydrophobic core.[29,71] This indicates that the major β hairpin is critical for conformational folding, consistent with other mutagenesis studies.[29] It seems likely that the disruption of conformational folding is due primarily to the anomalous backbone structure of the trans X-Pro bond, since the local region is not tightly packed in the native strtucture and since cleavage of the nearby 40–95 disulfide bond does not affect conformational folding significantly (unpublished results). Equally remarkably, the native peptide bond at Tyr^{92}-Pro^{93} remains cis even when Pro-93 is mutated to alanine.[38,40,71]

The extraordinary sensitivity of the conformational folding to the isomeric state of Pro-93 casts doubt on the recently proposed smooth funnel models of protein folding, which propose that there are no well-defined pathways for folding. In particular, these models predict that small structural changes are incapable of slowing folding significantly since, if one pathway is blocked, many other pathways remain open. This is evidently contradicted by our results for the nonnative isomer of the Tyr^{92}-Pro^{93} peptide bond. To quantify the effects of nonnative isomeric states on the conformational folding, a Φ_{isomer} value could be defined, in analogy with Φ_F[130] and the Φ_{link} value discussed below,[6] where Φ_{isomer} is defined as $\Delta\Delta G^\circ_{TS\text{-}U}/\Delta\Delta G^\circ_{N\text{-}U}$ and these $\Delta\Delta G^\circ$'s represent the changes in free energy differences associated with the change in the isomeric state of the X-Pro peptide bond. Values of Φ_{isomer} close to one (zero) indicate that the isomeric states is critical (not critical) for conformational folding.

Admittedly, the five-step protocol described above is a rather complicated procedure to assess the effects of nonnative isomeric states on the conformational folding, although it requires only relatively simple kinetic and mutagenesis steps. Moreover, the anomalous U_m observations of the P93A and P117A mutants point out that its results are not absolutely certain. A more direct but technically more challenging method is to use the semisynthetic methods described above to introduce cis- and trans-locked analogs of prolines into the protein. In principle, no proline isomerization would occur in such isomerically locked molecules; hence, the conformational folding of pure isomeric states could be determined directly. Such experiments are underway.

Conformational Folding of RNase A

Refolding of U_{vf}

The conformational folding of a pure isomeric state can be studied by the following double-jump procedure.[67,87,131–133] Starting in the native state (which,

[130] A. Fersht, "Structure and Mechanism in Protein Science." W. H. Freeman, New York, 1999.
[131] W. A. Houry, D. M. Rothwarf, and H. A. Scheraga, *Biochemistry* **33,** 2516 (1994).
[132] W. A. Houry, D. M. Rothwarf, and H. A. Scheraga, *Nature Struct. Biol.* **2,** 495 (1995).
[133] W. A. Houry and H. A. Scheraga, *Biochemistry* **35,** 11734 (1996).

by assumption, is isomerically pure), the protein is jumped to strongly unfolding conditions (e.g., 4.2 *M* GdnHCl at pH 2, 15° for RNase A), in which the conformational unfolding proceeds much more rapidly than the proline isomerization. The protein is held under these conditions long enough to allow complete unfolding of the protein but not long enough to populate any nonnative isomeric state significantly (roughly 0.8 s for RNase A at 15°). The protein is then returned to folding conditions and its refolding is monitored by sundry probes such as absorbance, fluorescence, circular dichroism, and H/D exchange. By repeating such experiments under different conditions and with different structural probes, the conformational folding of the pure native isomeric state can be investigated.

The double-jump protocol has been combined with several structural probes to study the U_{vf} refolding phase of RNase A. The pH and GdnHCl dependence of the refolding kinetics indicated that a conformational intermediate (denoted I_{ϕ}) is populated in refolding.[132] The intermediate appears to be favored under stabilizing conditions, e.g., at low denaturant conditions or neutral pH.[132] Subsequent circular dichroism (CD) studies revealed a burst phase[87] that is not apparent in the wild-type protein by absorbance or fluorescence under similar conditions. Analogous experiments on the fully reduced protein R suggest that similar amounts of local secondary structure are induced in U_{vf} and R by the jump to physiological folding conditions.[6,88] H/D exchange experiments have suggested that the amide protons of the minor hydrophobic core (near the 26–84 disulfide bond) become strongly protected, whereas those of the major hydrophobic core (near the 58–110 disulfide bond) receive medium protection and others receive little or no protection.[133] The burst phase may even be observed by tyrosine absorbance in the wild-type protein if the refolding is carried out in the presence of sulfate (unpublished results).

Site-directed mutants also indicate the importance of the minor hydrophobic core in the refolding of U_{vf}. The Y25F and Y97F mutants slow the folding drastically,[29] suggesting that these residues are required for this folding. The Y92W mutant [67] exhibits a burst phase by fluorescence, a relatively small burst phase in the refolding of U_{vf}, and a larger burst phase in single-jump refolding experiments from the equilibrium unfolded state. Subsequent experiments (unpublished results) have indicated that this fluorescence burst phase is composed of two well-resolved exponentials on the submillisecond time scale, plus a still-unresolved burst phase that likely corresponds to ring–ring interactions between the tryptophan and the adjacent proline.

Conformational Folding of U_s^{II}

The double-jump protocol may be simplified by examining the single-jump refolding from the equilibrium unfolded state of a single phase that is well separated kinetically from all other phases, e.g., the U_S^{II} phase of RNase A. This simplified protocol is much more economical than the double-jump protocol in terms of the amount of protein required. However, this single-jump protocol is also less certain, since there is no guarantee that the phase being examined is indeed a single phase corresponding to a structurally homogeneous refolding. For example, multiple isomeric states may refold along structurally different pathways, but at rates so similar that their separate phases cannot be separated. Such a confounding of structurally heterogeneous refolding phases may explain the apparently anomalous results observed using this protocol in the U_S^{II} phase of RNase A,[33] consistent with other experiments indicating that the U_S^{II} phase comprises the refolding of two distinct isomeric states.[29,122]

The conformational folding of U_S^{II} has been studied by pulsed H/D exchange methods and correlated with optical probes.[32,33] The results suggest that the C-terminal, predominantly β-sheet half of RNase A is almost fully formed in an intermediate of conformational folding, and that the rate-determining step consists of the ordering of the N-terminal, predominantly α-helical half onto the β sheet.

The conformational refolding of U_S^{II} has also been studied using a novel cross-linking protocol.[69] Lysines-7 and -41 were cross-linked through their N^{ε} atoms using 1,5-difluoro-2,4-dinitrobenzene. The cross-linked protein exhibits the same native structure as wild-type RNase A,[37] and similar refolding phases.[70] The cross-linked protein is thermodynamically much more stable than wild-type RNase A,[69] and this change in stability appears to result largely from entropic not enthalpic factors, consistent with the high conformational lability observed in the loop region (residues 16–23) and the close similarity between RNase A and S. Therefore, the cross-linked protein may be used to assess the association of lysines 7 and 41 in the folding transition state of the wild-type protein.[6,70] The results suggested that the N-terminal α helix is not associated with the β sheet in the transition state for conformational folding[6,70] of U_S^{II}. To characterize the changes in relative free energies upon the addition of the crosslink more quantitatively, one may define $\Phi_{link} \equiv \Delta\Delta G^{\circ}_{TS\text{-}U}/\Delta\Delta G^{\circ}_{N\text{-}U}$ in analogy with the quantity Φ_F defined for site-directed mutations.[130] Values of Φ_{link} close to one (zero) indicate that the two residues are (not) associated in the folding transition state of the wild-type protein.

Conformational Folding of U_f, U_m, and U_s^I

The conformational folding of U_f is generally similar to that of U_{vf} for pH values above 5. This is consistent with experimental and theoretical studies indicating that the cis/trans isomerization of the Asn[113]-Pro[114] peptide bond makes relatively minor, localized changes that do not affect the structure of the adjacent 58–110 hydrophobic core.[71,134–140] However, at low pH, the conformational folding of U_f differs significantly from that of U_{vf}, as described above.

It is worth noting that the nonnative cis isomer of Pro-117 (which is fully buried in the major hydrophobic core) causes less disruption of conformational folding than the nonnative trans isomer of Pro-93, which occurs in a relatively exposed, flexible segment of the protein. Similarly, the P117A mutant causes a large increase in the conformational unfolding rate but relatively little change in the refolding rates,[71] indicating a low Φ_F value.[130] However, the local amide protons in the C-terminal hairpin are well protected against H/D exchange.[32,33] Taken together, these data suggest that the C-terminal β hairpin has formed in the folding transition state but that the major (58–110) hydrophobic core has not yet become packed.

The refolding of U_S^I appears to be rate-limited by proline isomerization.[72] Hence, the conformational folding of the corresponding isomeric states cannot be studied, although the results indicate that nonnative isomers of Pro-93 and Pro-117 suffice to make the native state thermodynamically unstable relative to the unfolded state.

Conformational Folding Experiments on Homologous Proteins

Experiments similar to those described above may be carried out on homologous proteins. RNase A belongs to a large homologous superfamily for which more than 40 members have been sequenced,[141] including angiogenin, frog onconase, and bovine seminal ribonuclease. These ribonucleases are also similar to the family of microbial ribonucleases, which includes ribonuclease U_2, ribonuclease T1, ribonuclease Sa, and barnase. The number of disulfide bonds in these proteins ranges from zero (e.g., barnase) up to four (e.g., RNase A). Comparative studies of the conformational and oxidative folding of these ribonucleases will be helpful in determining which features of RNase A folding are specific to that protein and which are representative of a broader class of proteins. This in turn should help to uncover the general principles governing protein folding. In particular, such experiments will help determine how the polypeptide chain may possibly fold progressively along multiple pathways, beginning with local folding in hydrophobic regions known as CFISs,[27,28,142] which promote the formation of

[134] E. R. Stimson, G. T. Montelione, Y. C. Meinwald, R. K. E. Rudolph, and H. A. Scheraga, *Biochemistry* **21,** 5252 (1982).

[135] M. R. Pincus, F. Gerewitz, H. Wako, and H. A. Scheraga, *J. Protein Chem.* **2,** 131 (1983).

[136] G. T. Montelione, E. Arnold, Y. C. Meinwald, E. R. Stimson, J. B. Denton, S.-G. Huang, J. Clardy, and H. A. Scheraga, *J. Am. Chem. Soc.* **106,** 7946 (1984).

[137] M. Oka, G. T. Montelione, and H. A. Scheraga, *J. Am. Chem. Soc.* **106,** 7959 (1984).

[138] S. Ihara and T. Ooi, *Biochim. Biophys. Acta* **830,** 109 (1985).

[139] D. A. Schultz, F. X. Schmid, and R. L. Baldwin, *Protein Sci.* **1,** 917 (1992).

[140] D. A. Schultz, A. Friedman, and R. O. Fox, *Protein Sci.* **2** (Suppl. 1), 67 (1993).

[141] J. J. Beintema, C. Schüller, M. Irie, and A. Carsana, *Prog. Biophys. Mol. Biol.* **51,** 165 (1988).

[142] P. E. Wright, H. J. Dyson, and R. A. Lerner, *Biochemistry* **27,** 7167 (1988).

medium- and long-range interresidue interactions in subsequent conformational stages. This consecutive formation of ordered structures (and consequently the interactions that lead to them) determines the pathways along which the polypeptide folds to the native protein.[143,144] By observing the development of ordered structure along these pathways, the interresidue interactions that determine this progression and, thus, the final folded native structure can be characterized.

Such comparisons of conformational folding have been carried out for several homologs of RNase A.[145] These studies indicated that homologous proteins with even relatively low sequence identity nevertheless exhibit similar folding kinetics and folding phases, suggesting that they are conformationally analogous. These homologous protein studies also demonstrated that the introduction of new proline residues may induce new slow-refolding phases that are rate-limited by proline isomerization, as in porcine ribonuclease A.[146] Such a "proline-scanning" method may reveal critical regions that must be structured in the transition state of conformational folding. In particular, proline substitution could disrupt regular secondary structure regardless of its isomer, while the isomeric dependence could identify critical turn structures.

Methodology of Reductive Unfolding and Oxidative Folding Studies

Reductive unfolding and oxidative folding experiments require special methods that are not generally known. In this section, we discuss the factors affecting the choice of redox reagent and the experimental conditions (especially pH), as well as the methods for quenching the disulfide-bond reactions and subsequent chemical analysis. It should be noted that these methods work well for RNase A, but may not always be directly applicable to other proteins.

Redox Reagent

The two types of redox reagents most commonly used are linear disulfide reagents such as glutathione and cyclic disulfide reagents such as dithiothreitol (DTT). Air oxidation is not generally used, because it is complicated by metal-ion catalysis and its rate is difficult to control precisely; indeed, oxygen is generally eliminated from the reactions (e.g., by sparging with inert gases such as argon) for the same reasons. Novel redox reagents are an ongoing area of research.[147,148]

[143] S. Tanaka and H. A. Scheraga, *Macromolecules* **10,** 291 (1977).
[144] G. Némethy and H. A. Scheraga, *Proc. Natl. Acad. Sci. U.S.A.* **76,** 6050 (1979).
[145] K. Lang, A. Wrba, H. Krebs, F. X. Schmid, and J. J. Beintema, *FEBS Lett.* **204,** 135 (1986).
[146] R. Grafl, K. Lang, A. Wrba, and F. X. Schmid, *J. Mol. Biol.* **191,** 281 (1986).
[147] G. V. Lamoureux and G. M. Whitesides, *J. Org. Chem.* **58,** 633 (1993).
[148] K. J. Woycechowsky, K. D. Wittrup, and R. T. Raines, *Chem. Biol.* **6,** 871 (1999).

We presently favor the cyclic redox reagent DTT because such rapidly recyclizing reagents ensure that mixed disulfide species are populated to negligible levels. This simplifies the interpretation of the experimental results, since one need not consider the structure and folding behavior of the covalently modified proteins with mixed disulfide bonds. By contrast, linear redox reagents are potentially problematic since such mixed disulfide species can be extremely numerous for proteins with many cysteines[149–151] and, hence, such species may predominate. Moreover, at high concentrations of the linear redox reagent, fully blocked species may be populated, i.e., those in which all free thiol groups are involved in mixed disulfide bonds. Despite these difficulties, linear redox reagents have several advantages. For example, glutathione is a much stronger oxidizing agent than DTT and is likely the chief redox reagent for oxidative folding *in vivo*.

The protein itself may act as a redox reagent[152] through intermolecular disproportionation reactions such as $3S + 3S \rightleftharpoons 2S + 4S$. However, the effects of such disproportionation reactions should be generally negligible in the presence of a redox reagent or at most physiological concentrations of the protein.

Experimental Conditions, especially pH

The experimental conditions are often dictated by the scientific question being addressed in the redox experiments. However, it should be borne in mind that changes in the experimental conditions can profoundly affect the regeneration of the native protein, both by influencing the conformational stability of disulfide intermediates and by changing the relative rate of reduction/oxidation and disulfide reshuffling. Since many disulfide intermediates are unstructured (and, thus, have relatively low solubility), the conditions should be chosen to obtain a high rate of producing structured species such as the native protein, in order to minimize aggregation.

High pH (pH $\geq$ typical cysteine $pK_a \approx 8.7$) may be helpful in accelerating oxidative folding by increasing the concentration of the (reactive) thiolate anion relative to the (unreactive) thiol group. High pH may also have the advantage of destabilizing metastable (kinetically trapped) species, since the burial of a charged thiolate group is strongly disfavored.[153] However, higher pH may also have unwanted side effects such as deamidation,[14] conformational destabilization of structured species, and aggregation/polymerization, especially for proteins with high

[149] W. Kauzmann, *in* "Symposium on Sulfur in Proteins" (R. Benesch, R. E. Benesch, P. D. Boyer, I. M. Klotz, W. R. Middlebrook, A. G. Szent-Györgyi, and D. R. Schwarz, eds.), p. 93. Academic Press, New York, 1959.

[150] M. Sela and S. Lifson, *Biochim. Biophys. Acta* **36,** 471 (1959).

[151] Y. Konishi, T. Ooi, and H. A. Scheraga, *Biochemistry* **20,** 3945 (1981).

[152] M.-C. Song and H. A. Scheraga, *FEBS Lett.* **471,** 177 (2000).

[153] K. A. Dill, *Biochemistry* **29,** 7133 (1990).

isoelectric points such as bovine pancreatic trypsin inhibitor and RNase A. Our experiments are generally carried out at pH 8.0.

Quenching of Disulfide-Bond Reactions

Disulfide-bond regeneration can be arrested at different times, either by acid quenching or by a rapid thiol-blocking agent such as aminoethylmethane thiosulfonate (AEMTS)[154] and similar thiosulfonates. Acid quenching relies on the slowness of thiol/disulfide exchange at low pH (only the thiolate anion is reactive[101]) and has the advantage that the protein itself is not modified chemically. However, we prefer AEMTS blocking, which is exceptionally rapid compared to other thiol-blocking reagents (10^5 times faster than iodoacetate[155]). AEMTS blocking facilitates the subsequent fractionation (by adding a positively charged cysteamine group for every blocked thiol) and does not require the maintenance of low pH during the subsequent analysis. Unfortunately, neither acid quenching nor AEMTS blocking is ideal, since both methods may perturb the conformational structure of the intermediate disulfide species (relative to its structure under oxidative folding conditions near neutral pH).

Common thiol-blocking reagents such as iodoacetate and iodoacetamide should be used with caution in oxidative folding experiments, since they are often too slow to prevent disulfide reshuffling during the blocking reaction. These reagents can produce results that disagree with the majority of other methods, even for relatively simple measurements such as the K_{eq} between DTT and glutathione.[156,157] These blocking reagents have also produced anomalous results for the oxidative folding of bovine pancreatic trypsin inhibitor (BPTI), when compared to experiments carried out under identical conditions using acid quenching.[117,158] At higher concentrations, these reagents may also chemically modify other residues of the protein.[159]

When using reversible chemical blocking agents such as AEMTS, sufficiently high concentrations of the blocking agent must be used to ensure that all the thiols are rapidly blocked. If the blocking of thiols is too slow, the blocking agent may act as an oxidizing agent; the unblocked thiols may attack the disulfide bond of the blocked thiols and form an intraprotein disulfide bond. Therefore, it is crucial to establish experimentally that a sufficiently high concentration of the blocking agent has been used.

[154] T. W. Bruice and G. L. Kenyon, *J. Protein Chem.* **1,** 47 (1982).

[155] D. M. Rothwarf and H. A. Scheraga, *J. Am. Chem. Soc.* **113,** 6293 (1991).

[156] D. M. Rothwarf and H. A. Scheraga, *Proc. Natl. Acad. Sci. U.S.A.* **89,** 7944 (1992).

[157] W. J. Lees and G. M. Whitesides, *J. Org. Chem.* **58,** 642 (1993).

[158] T. E. Creighton, *Science* **256,** 111 (1992).

[159] G. E. Means and R. E. Feeney, "Chemical Modification of Proteins." Holden-Day, San Francisco, 1971.

Analysis of Intermediates

The use of a chemical blocking group such as the cysteamine of AEMTS facilitates the subsequent analysis of the quenched disulfide ensembles, since it introduces a positively charged group to every free thiol present prior to blocking. For example, ion-exchange chromatography can be used to separate the intermediates by charge; ion-exchange columns show very little loss of protein and, hence, the measured concentrations of the eluted fractions can be regarded as quantitatively accurate. The individual fractions can also be collected and analyzed to determine the number of thiol groups and disulfide bonds,[160] using a novel methodology that exploits the AEMTS blocking.[161] The results clearly indicate that the cation-exchange columns separate the blocked disulfide species according to the number of disulfide bonds. Ion-exchange columns can also be used at pH 5.0 to isolate unblocked disulfide species. However, a wide spectrum of analytic separation methods can be applied, e.g., 2D gel electrophoresis,[162] capillary electrophoresis,[163] gel filtration,[164] and reversed-phase columns.[165] The structured species can be separated from the unstructured ensembles using the reduction-pulse protocol[26,113] described above in the Preparative Methods section. Peptide mapping methods can be used to determine the disulfide bonding present in a disulfide species or ensemble.[166]

Typical Protocol for Oxidative Folding

A typical oxidative folding study might be carried out as follows. The fully reduced protein may be prepared by incubating purified protein with a strong reducing reagent (e.g., 100 m*M* DTT^{red}) and a strong denaturant (e.g., 6 *M* GdnHCl) that unfolds the protein, ensuring that its disulfide bonds are exposed to the reducing reagent. The use of a denaturant is critical, since the disulfide bonds of many proteins are well protected and are reduced extremely slowly under folding conditions. Unfortunately, this precaution is not always taken, leading to some erroneous conclusions.[59,83,84]

Oxidative folding may be initiated by introducing the fully reduced protein to conditions that favor conformational folding and the formation of disulfide bonds (e.g., 100 m*M* DTT^{ox} at pH 8 and 15–37°). Aliquots are withdrawn at various times after initiating oxidative folding, and thiol/disulfide exchange is arrested by acid quenching or AEMTS blocking. The mixture of disulfide species is then fractionated by reversed-phase or ion-exchange HPLC. Structured species with protected

[160] T. W. Thannhauser, Y. Konishi, and H. A. Scheraga, *Methods Enzymol.* **143,** 115 (1987).
[161] T. W. Thannhauser, R. W. Sherwood, and H. A. Scheraga, *J. Protein Chem.* **17,** 37 (1998).
[162] T. E. Creighton, *Methods Enzymol.* **131,** 83 (1986).
[163] T. W. Thannhauser, D. M. Rothwarf, and H. A. Scheraga, *Biochemistry* **36,** 2154 (1997).
[164] A. Galat, T. E. Creighton, R. C. Lord, and E. R. Blout, *Biochemistry* **20,** 594 (1981).
[165] J. S. Weissman and P. S. Kim, *Science* **253,** 1386 (1991).
[166] T. W. Thannhauser, C. A. McWherter, and H. A. Scheraga, *Anal. Biochem.* **149,** 322 (1985).

disulfide bonds (which play a critical role in oxidative folding) may be detected and isolated by exposing the mixture to a short reduction pulse[26,113] that leaves the disulfide-protected species largely untouched but reduces disulfide-unprotected species. The isolated species may then be identified by peptide mapping[166] and studied further by equilibrium transitions,[105–107,167] folding and unfolding behavior,[111,167] and high-resolution structural methods such as NMR[105–107,167] and X-ray crystallography.[38] Kinetic fitting is used to model the variations in the concentrations of the disulfide-protected species and unstructured *n*S ensembles with time[26,168,169] Site-directed mutagenesis[16,17] can be exploited to make analogs of these intermediates or to study the role(s) of various cysteine residues in the regeneration.

[167] J. H. Laity, G. T. Montelione, and H. A. Scheraga, *Biochemistry* **38,** 16432 (1999).
[168] H. A. Scheraga, Y. Konishi, and T. Ooi, *Adv. Biophys.* **18,** 21 (1984).
[169] D. M. Rothwarf and H. A. Scheraga, *Biochemistry* **32,** 2680 (1993).

[13] Purification of Engineered Human Pancreatic Ribonuclease

By MARC RIBÓ, ANTONI BENITO, ALBERT CANALS, M. VICTÒRIA NOGUÉS, CLAUDI M. CUCHILLO, and MARIA VILANOVA

Introduction

Human pancreatic ribonuclease (HP-RNase, RNase 1, EC 3.1.27.5) is a secreted ribonuclease that belongs to the ribonuclease superfamily. This enzyme is considered the human counterpart of the well-known bovine pancreatic ribonuclease A (RNase A) with which it shares 70% sequence identity and the ability to cleave RNA specifically on the 3′ side of pyrimidine bases. Both enzymes show an increased activity with poly(C) over that with poly(U) and also an optimal activity near pH 8.0. However, HP-RNase presents a higher activity both on double-stranded RNA and on destabilizing DNA.

Ribonucleases are considered potential alternative toxins because of their ability to degrade RNA molecules. As toxins most of these enzymes catalytically abolish protein synthesis when they translocate into the cytosol.[1] Onconase, a ribonuclease isolated from oocytes of *Rana pipiens,* is of outstanding interest because of its preferential cytotoxicity to cancer cells. At present it is in phase II/III

[1] S. M. Rybak and D. L. Newton, *Exp. Cell. Res.* **253,** 825 (1999).

0076-6879/01 $35.00

clinical trials. The interest in developing engineered or conjugated HP-RNases with cytoxic properties is based on the expectation that clinical applications might be facilitated by a reduced antigenicity. Other members of the ribonuclease superfamily are also being investigated for this purpose.[2] The purification of HP-RNase from its natural source has been previously described.[3,4] However, because of the minute amounts of enzyme present in the natural source (10 μg/g tissue) and because of the heterogeneity due to different extents of glycosylation, purification of HP-RNase from this source cannot yield the amounts of protein needed for carrying out structure–function studies.

Thus, larger amounts of protein are required for either performing function–structure studies or engineering HP-RNase to take advantage of its potential cytotoxicity as a monomer or when conjugated with other molecules. For these reasons, the production of recombinant HP-RNase in heterologous systems has been used by several research groups as an alternative solution.

Heterologous Production of HP-RNase

When attempting to produce any recombinant protein with the aim of performing either biological or structural studies, one has to face two main challenges: (1) obtaining an adequate amount of the polypeptide, and (2) recovering as much pure, homogeneous protein in its three-dimensional native conformation as possible. The synthesis of foreign proteins is largely empirical to the extent that a particular protein is almost as likely to be the exception to, as it is to follow, the rule. This is also true for properly folding any polypeptide chain or an engineered variant of it.

Heterologous production of HP-RNase has been achieved using either eukaryotic and prokaryotic expression systems. Expression of this enzyme in eukaryotic hosts such as chinese hamster ovary (CHO) cell[5] or *Saccharomyces cerevisiae*[6] resulted in low and heterogeneous production of the enzyme, and the procedures were more time-consuming and expensive. Most research groups since then have used *Escherichia coli* as a suitable host to produce a recombinant form of HP-RNase.[6–12]

[2] G. D'Alessio, A. Di Donata, R. Piccoli, and N. Russo, *Methods Enzymol.* **341,** [15] 2001 (this volume).

[3] J. L. Weickman, M. Elson, and D. G. Glitz, *Biochemistry* **20,** 1272 (1981).

[4] M. Ribó, J. J. Beintema, M. Osset, E. Fernández, J. Bravo, R. de Llorens, and C. M. Cuchillo, *Biol. Chem. Hoppe-Seyler* **375,** 357 (1994).

[5] N. Russo, M. de Nigris, A. Ciardiello, A. Di Donato, and G. D'Alessio, *FEBS Lett.* **333,** 233 (1993).

[6] M. Ribó, S. B. delCardayré, R. T. Raines, R. de Llorens, and C. M. Cuchillo, *Prot. Express. Purif.* **7,** 253 (1996).

[7] Y. Wu, S. K. Saxena, W. Ardelt, M. Gadina, S. M. Mikulski, V. De Lorenzo, G. D'Alessio, and R. J. Youle, *J. Biol. Chem.* **270,** 17476 (1995).

[8] J. Futami, M. Seno, M. Kosaka, H. Tada, S. Seno, and H. Yamada, *Biochem. Biophys. Res. Commun.* **216,** 406 (1995).

The source of the coding sequence for HP-RNase has mainly been a synthetic gene, although cloning of cDNA has also been used successfully.[8] The former strategy is advantageous since the specific codon usage for a heterologous host can be used when designing the gene. The preferred option to express HP-RNase in *E. coli* has been the T7 system vectors harboring the T7 RNA polymerase promoter developed by Studier *et al.*,[13] and in all cases but one where HP-RNase gene was fused to the *pelB* leader sequence and results in a final low yield,[6] the recombinant protein has been expressed directly into the cytosol in an insoluble form. This approximation has the advantage that isolation of inclusion bodies provides a purification step of the protein in an inactive form that is not cytotoxic to the producer cell. Protein insolubility, however, necessitates the use of a purification protocol that must include a solubilization/denaturation step and a subsequent refolding/oxidation of the denatured protein. This protocol will be described in the next sections.

Several steps can be distinguished in the process of producing a reasonable yield of pure fully active HP-RNase: (1) expression, (2) solubilization, (3) refolding and reoxidation, and (4) purification. In this section, the basic protocols used in our laboratory to produce and characterize the integrity of HP-RNase will be described in detail. For clarity, pM and PM will be used to denote plasmid constructs and the encoded protein variant, respectively. We will refer to other described protocols only for comparison. Also, after each of the main steps, we will summarize approaches by which the engineering of HP-RNase has helped in obtaining either a higher yield or to improve the homogeneity of the final product.

Obtaining High Levels of HP-RNase

HP-RNase Expression Procedure

Small cultures are set up by inoculating 10–15 ml of Luria–Bertani (LB) medium supplemented with 400 μg/ml ampicillin, with a single colony of BL21 (DE3) cells carrying the desired expression vector. The cultures are allowed to grow overnight with an agitation of 250 rpm at 37°. It is highly recommended

[9] E. Boix, Y. Wu, V. M. Vasandani, S. K. Saxena, W. Ardelt, J. Ladner, and R. J. Youle, *J. Mol. Biol.* **257,** 992 (1996).

[10] H. P. Bal and J. K. Batra, *Eur. J. Biochem.* **245,** 465 (1997).

[11] R. Piccoli, S. Di Gaetano, C. De Lorenzo, M. Grouso, C. Monaco, D. Spallazetti-Cernia, P. Laccetti, J. Cintál, J. Matousek, and G. D'Alessio, *Proc. Natl. Acad. Sci. USA* **96,** 7768 (1999).

[12] A. Canals, M. Ribó, A. Benito, M. Bosch, E. Mombelli, and M. Vilanova, *Prot. Express. Purif.* **17,** 169 (1999).

[13] F. W. Studier, A. H. Rosenberg, J. J. Dunn, and J. W. Dubendorff, *Meth. Enzymology* **185,** 60.

that freshly transformed or plate streaked clones be used. Dilutions of the single colony inoculate (1 : 10^2 and 1 : 10^4) may yield a log-phase culture with enriched plasmid-harboring cell population. Alternatively, small overnight cultures can be centrifuged for 5 min at 3000 *g* at room temperature and the cells resuspended in fresh LB medium to minimize β-lactamase activity.

The culture is used to inoculate, in a 1 : 100 dilution, 1 liter of LB supplemented with 100 μg/ml of ampicillin and incubated at 37° with stirring at 250 rpm until the culture reaches an OD_{550} around 1.0. Terrific broth (TB) medium has also been used when growing cells for expression. However, we have not observed any difference in expression levels when compared to LB medium. Expression is then induced by the addition of sterile isopropylthiogalactoside (IPTG) to a final concentration of 1 m*M*. After 4 h of growth, cells are harvested by centrifugation at 10,000*g* at 4° for 10 min in a Sorvall GSA rotor, and cell pellets are stored at −20° until protein processing. In our hands, 1 liter of culture grown for 4 h after induction yields cell pellets of around 3–4 g. Overexpression can be estimated from densitometric analysis of Coomassie blue–stained SDS–PAGE of a whole extract of induced cell pellet. Cell fractionation experiments with all the variants assayed by our group have shown that the protein is present in the insoluble fraction.

Engineered HP-RNases: Strategies to Increase Expression Level

Originally, expression of HP-RNase in our laboratory was carried out from a synthetic gene cloned into pET22b(+) vector.[6] Analysis of the produced HP-RNase showed that the *pelB* signal sequence fused at the N terminus of the protein was correctly processed. However, the protein was always found in the insoluble fraction. A densitometric analysis of whole-cell induced extracts subjected to electrophoresis showed that only 1% of the total protein corresponded to HP-RNase. This yield contrasted with results published for other ribonucleases. For instance, expression of bovine seminal ribonuclease (BS-RNase) in the pET17b vector (which directs expression to the cytosol) yielded 15–20% of the total protein.[14] High levels of transcript RNA molecules are ensured provided T7 promoter is used,[13] so other factors were considered to be responsible for the low expression of the construction bearing the sequence coding for the human enzyme. A secondary structure prediction analysis of the 5′ end of the HP-RNase mRNA (nucleotides −71 to +171) was performed using the program MFOLD.[15] When the generated optimal and suboptimal mRNA secondary structures were drawn using the programs LOOPVIEWER and LOOPDLOOP,[16] it was shown that the AUG start codon and the Shine–Dalgarno sequence were not accessible as they

[14] J.-S. Kim and R. T. Raines, *J. Biol. Chem.* **268,** 17392 (1993).

[15] M. Zuker, *Science* **244,** 48 (1989).

[16] D. G. Gilbert, "*loopDloop,*" a Macintosh program for visualizing RNA secondary structure (1992). Published electronically on the Internet; available via anonymous ftp to *ftp.bio.indiana.edu*.

showed a high content of Watson–Crick base pairing, in contrast to what is observed for BS-RNase mRNA.[14] A first HP-RNase mutant was constructed (named pM5) by substituting the sequence coding for the first 20 amino acids with the corresponding sequence for the BS-RNase gene.[12] Although a DNA sequence coding for a total of 20 residues was substituted, only five of them were different from the original HP-RNase amino acid sequence (R4A, K6A, Q9E, D16G, and S17N). Besides the inaccessibility of the Shine–Dalgarno and start codon sequences, other possible reasons that could account for a decrease in the yield of recombinant protein, pM5 construction overcome: (i) expression of pM5 is targeted to the cytoplasm, and different research groups have successfully used this strategy when producing HP-RNase[6–12]; (ii) in pM5, a Met residue is encoded at position −1 just before the original N-terminal Lys. Protein half-lives of only 2 min have been reported when one of the residues Lys, Arg, Phe, Leu, Trp, or Tyr was present at the amino terminus. All other amino acids confer half-lives longer than 10 h.[17] Growth of BL21 (DE3) cells harboring pM5 for 4 h after induction as described in the previous section allowed for the expression of this engineered HP-RNase variant up to a level comparable to that observed for BS-RNase. Protein estimated by densitometry of SDS–PAGE of total cell extract accounted for about 15% of the total protein.

Redesigning Wild-Type HP-RNase Coding Sequence

Overexpression of the pM5 construct prompted us to redesign the DNA sequence coding for residues 1 to 20, in order to make a new construct coding for wild-type HP-RNase, which was named pM9. The new sequence was designed by substituting the minimum number of nucleotides in the gene coding for PM5, with nucleotides that not only would provide the appropiate residue change, but also would preserve the accessibility of the start codon and of the Shine–Dalgarno determinant in the mRNA sequence. Predicted secondary structure of the mRNA that encodes PM9 showed great similarity with the predicted mRNA secondary structures for BS-RNase mRNA and PM5 mRNA but was remarkably different from that of the formerly used pET22b(+)-based construct.[12,14] *Escherichia coli* BL21(DE3) cells harboring the pM9 construct showed protein expression levels similar to those observed for constructions coding for BS-RNase and the hybrid pM5.

Native Conformation: From Solubilization to Purification

Solubilization of HP-RNase

Frozen induced cell pellets are thawed in warm water, resuspended in 30 ml of lysis buffer (50 m*M* Tris-HCl, pH 8, 10 m*M* EDTA) at 4°, and disrupted by three

[17] J. W. Tobias, T. E. Shrader, G. Rocap, and A. Varchavsky, *Science* **254,** 1374 (1991).

passages through a French pressure cell (SLM-Aminco, Spectronic Instruments, Rochester, NY) at 1000 psi. Alternatively, when handling small volumes, sonication is the method of choice for cell lysis. Inclusion bodies are obtained from crude extract by centrifugation at 12,000*g* in a Sorvall SS34 rotor at 4° for 10 min. Solubilization of the aggregates is accomplished by resuspending the insoluble fraction in 100 m*M* Tris–acetate, pH 8.5, buffer containing 6 *M* guanidinium chloride (Gdm-Cl) (Fluka, Ronkonkoma, NY), and 2 m*M* EDTA. The solubilized sample is reduced by the addition of solid reduced glutathione (GSH) (Roche) to a final concentration of 100 m*M*, and the pH is immediately adjusted to pH 8.5 by the addition of solid Tris. The reduced and denatured sample is incubated at room temperature for 2 h under N_2 atmosphere to allow for the complete unfolding and to facilitate reduction of the disulfide bonds. Insoluble material is removed by centrifugation (it might be necesary to repeat this step twice) at 12,000*g* for 30 min, at room temperature in a Sorvall SS34 rotor. The supernatant contains the reduced denatured HP-RNase, which undergoes refolding and reoxidation as described below.

Refolding and Reoxidation

On the way to reaching its three-dimensional native conformation, a polypeptide chain may undergo side reactions that lead to the formation of aggregates or stable intermediates. Thus, refolding conditions have to favor the pathway to the native state of the protein or, at least, minimize the formation of nonproductive reactions. As mentioned above, this step in the process of producing any recombinant protein is as critical as the step of expression itself. Unproductive aggregation processes may originate from nonspecific (hydrophobic) interactions of predominantly unfolded polypeptide chains as well as from incorrect interactions of partially structured folding intermediates. These aggregation reactions are second- (or higher) order processes, whereas correct folding is generally determined by first-order reactions. As a consequence, an increase of the rate of unproductive aggregation is observed by increasing the concentration of unfolded polypeptide chains. For many proteins, the yield of renaturation decreases above a limiting concentration of about 10 μg/ml, thus requiring excessive reaction volumes.

In order to avoid this, the refolding reaction is initiated by a 1 : 100 drop-by-drop dilution of the unfolded protein into a 100 m*M* Tris–acetate buffer at pH 8.5 containing 0.5 *M* L-Arg (Sigma, St. Louis, MO), 8 m*M* oxidized glutathione (GSSG) (Serva), and 2 m*M* EDTA at 10°, and is completed after incubation without stirring for 24 h at this temperature. The protein concentration in the refolding buffer ranges from 50 to 100 μg/ml. Alternatively, unfolded HP-RNase can be purified before refolding by using a cation-exchange chromatography column under denaturing conditions.[8] This strategy is not recommended when expressing the recombinant protein in an insoluble form because of the need for using high concentrations of denaturing agent. All the refolding protocols described in the literature for HP-RNase renaturation are very similar.[6–12] Refolding is accomplished by diluting the protein in basic pH-buffered solutions (7.5–8.5) to facilitate the formation of the four disulfide bonds that are necessary to reach the native conformation. Diluting the protein is critical since it has been shown that low concentrations of HP-RNase minimize protein aggregation and precipitation during refolding. To avoid excessive reaction volumes, refolding is carried out by the slow addition of the unfolded protein. Stirring of the sample during refolding leads to the formation of more precipitate, probably because of undesirable intermolecular reactions that cause aggregation. Most of the buffers used to refold HP-RNase include L-arginine and EDTA. Arginine has been proposed to act as a low molecular mass enhancer of refolding by acting as a chaotropic agent. Although it contains a guanidium group, destabilization of the native folded structure is not as effective as when prompted by other chaotropes such as Gdm-Cl. Yet, the exact mechanism by which L-arginine acts as an enhancer during protein folding remains unknown.[18,19] In the absence of EDTA we have observed about an 80% decrease in the final yield, which is probably due to the presence of traces of proteolytic enzymes. Considerable variability is found in the ratio of the GSH : GSSG redox pair used to enhance the reoxidation of cysteines. The ratios range from 1 : 8 to 10 : 1, in contrast to the general agreement on the time needed for refolding (12–24 h) or the temperature at which refolding has to be carried out (10°).

Refolded HP-RNase Purification Procedure

After incubation, the pH of the refolded sample is adjusted to pH 5.0 with glacial acetic acid to stop further oxidation. Insoluble material is removed by centrifugation in a Sorvall GSA rotor at 12,000 *g* for 10 min at 4°. The sample is concentrated to 30–40 ml by ultrafiltration in a Minitan II (Millipore, Bedford, MA) apparatus with a fixed transmembrane average pressure of 15 psi and a filter area of 240 cm^2. The concentrated solution is dialyzed at 4° against the column equilibration buffer (buffer A: 50 m*M* sodium acetate pH 5.0) with several changes. At this stage when arginine is removed, a white precipitate appears after 12 h of dialysis and is removed by centrifugation at 12,000*g* at 4° for 30 min. Native HP-RNase is purified by cation-exchange chromatography on a Mono S HR 5/5 FPLC column (Pharmacia, Piscataway, NJ) with a linear gradient from 0 to 600 m*M* NaCl in buffer A. Native HP-RNase elutes at about 500 m*M* NaCl. This chromatography allows for the separation of native HP-RNase from other

[18] J. Buchner, I. Pastan, and U. Brinkmann, *Anal. Biochem.* **205,** 263 (1992).
[19] R. Rudolph and H. Lilie, *FASEB J.* **10,** 49 (1996).

nonnative isoforms lacking some of the disulfide bonds, as they elute at a lower NaCl concentration (Fig. 1A). These latter forms contain mixed disulfide bonds with glutathione. Each additional glutathione brings an extra negative charge to the molecule. Fractions containing the pure and homogeneous ribonucleases (peak 3, Fig. 1A) are dialyzed against ultrapure water and lyophilized before being stored at −20° for further analysis. The yield of pure and homogeneous enzyme from 1 liter of induced culture is around 5 mg of protein as measured by the Bradford[20] assay. This represents 5–10% of the total HP-RNase present in the induced culture.

Role of Extra Pro Residues in Recovery of Native-Form Protein

Overexpressed HP-RNase (PM9) is purified to homogeneity with a final yield of 5 mg per liter of culture. However, this represents a low percentage of the total protein present in the crude extract. On the other hand, when its homolog from bovine pancreas (RNase A) is produced following the same protocol, a higher yield is obtained. However, it has been reported that the presence of two *cis*-prolyl peptide bonds (Pro-93 and Pro-114) in the RNase A molecule substantially complicates the folding process by creating slow-folding conformers.[21,22]

In comparing the primary structures of HP-RNase and RNase A, it is seen that three extra prolines at positions 19, 50, and 101 are present in the former protein in addition to the four proline residues present in the bovine enzyme. Two single variants named PM7 (P50S) and PM8 (P101Q) were made by substitution of the proline residues at positions 50 and 101 by the corresponding residues present at the same positions in the RNase A molecule. These variants were made in order to evaluate whether these proline residues play a significant role in the refolding process and consequently affect the final recovery yield. Pro-19 in HP-RNase was not considered because this residue is located in a disordered region that is unlikely to play a critical role in the protein folding process. In addition, Pro-19 is a conserved residue in other ribonucleases such as BS-RNase, that can be successfully produced by heterologous expression. Induction of BL21(DE3) cells carrying plasmids pM7 or pM8 lead to a recombinant protein expression level similar to that obtained for pM5.[12] When purification of PM7 protein was performed, a yield of about 10 mg per liter of culture was obtained, whereas the yield of purified PM8 was very similar to that obtained for PM5 and PM9.

These results suggest that Pro-50 could be an important residue for the refolding process of HP-RNase since the recovery yield of PM7 protein (P50S) is twice that

[20] M. M. Bradford, *Anal. Biochem.* **72,** 248 (1976).

[21] R. W. Dodge and H. A. Scheraga, *Biochemistry* **35,** 1548 (1996).

[22] C. M. Cuchillo, M. Vilanova, and M. V. Nogués, *in* "Ribonucleases: Structures and Functions" (G. D'Alessio and J. F. Riordan, eds.), p. 271. Academic Press, New York, 1997.

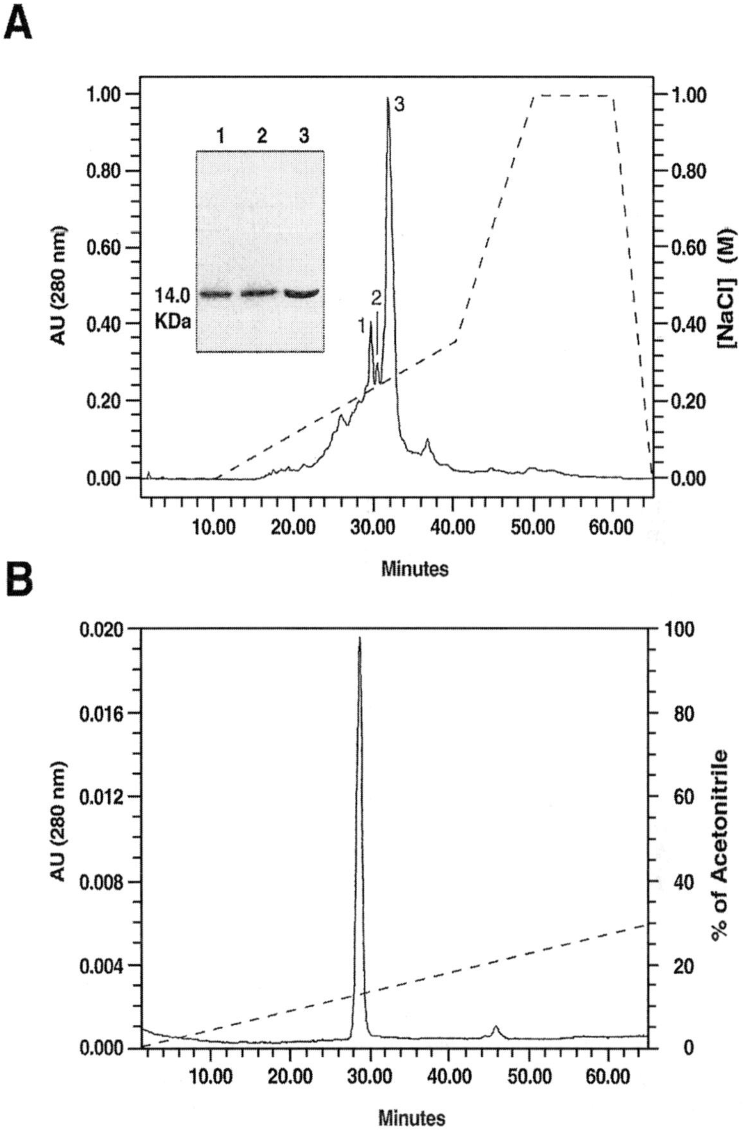

FIG. 1. Purification of PM5 protein. (**A**) Cation-exchange chromatography on a Mono S column of the concentrated sample obtained from the refolding step following the protocol described in the text. Peak 3 corresponds to native PM5 ribonuclease and peaks 1 and 2 correspond to different HP-RNase isoforms. *Inset:* SDS–PAGE analysis of the same peaks of the chromatogram. Lanes 1, 2, and 3 correspond to peaks 1, 2, and 3, respectively. (**B**) Reversed-phase chromatography on a Vydac C_4 column of peak 3 eluted from the cation-exchange chromatography.

obtained for the PM5 protein. The recoveries of PM8 (P101Q) and PM5 proteins are similar, suggesting that Pro-101 does not significantly contribute to the refolding process. From the three-dimensional structure of the PM7 variant,[23] it is known that the X-Ser-50 and X-Pro-101 peptide bonds are both in the *trans* conformation. It is possible that Pro-50 could play a comparable role to that proposed for Pro-117 in RNase A, which is also in the *trans* conformation, which complicates the refolding process.[24] This could explain, at least in part, the origin of the differences in the final recovery yield between the PM7 HP-RNase variant and the other recombinant pancreatic ribonucleases so far produced in our laboratory.

The procedure just described might be applied with specific modifications to the production of HP-RNase conjugates. However, the main goal of this procedure is that its use has allowed us to produce pure, homogeneous wild-type HP-RNase, and for any of the above-mentioned engineered variants, to obtain crystals and solve the three-dimensional structure of the PM7 variant.[23]

Characterization: Analysis of Protein Homogeneity and Structural Integrity

Each engineered HP-RNase variant was obtained with the purpose of studying either its structure or any functional aspect derived from its activity. Thus, it was essential to assess the purity, homogeneity, and structural and functional integrity of the final product.

Protein Homogeneity

Cation exchange-purified ribonuclease variants are shown to be pure by SDS–PAGE analysis (insert, Fig. 1A). In order to assess the homogeneity of the sample, about 50–100 μg of the protein as determined by Bradford assay[20] is subjected to reversed-phase chromatography in a 214TP Vydac C_4 HPLC column. A linear gradient is performed from 0.1% trifluoroacetic acid (TFA) to 100% acetonitrile, 0.1% TFA, in 100 min at a flow rate of 0.5 mL/min. HP-RNase, as well as all the engineered variants, elutes as a single peak between 20 and 35% acetonitrile, indicating that the final purified product is chromatographically homogeneous (Fig. 1B).

In addition, the molecular masses of the recombinant protein, as well as the absence of isoforms with mixed disulfides with gluthathione, are confirmed by MALDI-TOF (matrix-assisted laser desorption/ionization time-of-flight) mass spectrometry according to Wu and Watson,[25] using a Bruker-Biflex instrument

[23] J. Pous, A. Canals, S. S. Terzyan, A. Guasch, A. Benito, M. Ribó, M. Vilanova, and M. Coll, *J. Mol. Biol.* **303,** 49 (2000).

[24] D. Juminaga, W. J. Wedemeyer, R. Garduño-Juárez, M. A. McDonald, and H. A. Scheraga, *Biochemistry* **36,** 10131 (1997).

[25] J. Wu and J. T. Watson, *Protein Sci.* **6,** 391 (1997).

TABLE I
KINETIC AND THERMOSTABILITY PROPERTIES OF HP-RNase VARIANTS AND RNase A[a]

	C>p substrate		Poly(C) substrate		
Enzyme	K_m (mM)	k_{cat} (min^{-1})	K_m (mg/ml)	$(V_{max}/[E_0])^b_{rel}$ (%)	$T_{1/2}$ (°C)
PM5	1.80	46.6	0.442	100	56.3 ± 1
PM7	1.93	55.5	0.350	82.4	56.1 ± 1
PM8	1.57	48.0	0.399	104.4	55.4 ± 1
PM9	1.08	61.6	0.225	64.3	51.4 ± 1
RNase A	0.81	87.1	0.430	84.7	58.0 ± 1

[a] The rate of cleavage of the high molecular mass substrate poly(C) and the hydrolysis of the low molecular mass substrate C > p were assayed at 25° in 0.2 *M* sodium acetate, pH 5.5.
[b] $[E_0]$: Total concentration of enzyme in the assay. PM5 is taken as reference.

from Bruker (Billerica, MA). The purified proteins have molecular masses identical to those expected for wild-type HP-RNase or the corresponding variant.

Determination of Steady-State Kinetic Parameters

Ribonuclease cytotoxicity is dependent on its catalytic activity. Thus structure–function studies require appropiate activity measurements. To this end, spectrophotometric assays according to the methodology described by Boix *et al.*[26] can be used to determine the kinetic parameters for the cleavage of poly(C) and the hydrolysis of cytidine 2′,3′-cyclic monophosphate (C > p) by wild-type and mutant HP-RNases. For C > p, the concentration of enzyme is 0.1 μM, the initial concentration of C > p ranges from 0.1 to 5.5 mM, and the activity is measured by recording the increase in absorbance at 296 nm. For assays with poly(C), the concentration of enzyme is 5 nM, the initial concentration of poly(C) ranges from 0.1 to 2.5 mg/ml, and the decrease in absorbance at 294 nm is monitored. Both assays are carried out at 25° in 0.2 *M* sodium acetate buffer, pH 5.5, using 1 cm path length quartz cells for C > p and 0.2 cm path length quartz cells for poly(C). Steady-state kinetic parameters are obtained by nonlinear regression analysis using the program ENZFITTER.[27] The values in Table I are the average of three determinations, with a standard error of less than 10%. Whithin the error of the assay

[26] E. Boix, M. V. Nogués, C. H. Schein, S. A. Benner, and C. M. Cuchillo, *J. Biol. Chem.* **269,** 2529 (1994).
[27] R. J. Leatherbarrow (ed.), "ENZFITTER: A Non-Linear Regression Data Analysis Program for the IBM-PC," Elsevier Biosoft, Cambridge, 1987.

method, no significative differences were found between the HP-RNase variants and the wild-type enzyme (Table I).

Circular Dichroism

In order to make sure that both wild-type and HP-RNase variants showed no significant structural alterations, CD spectra are recorded at 37° using a Jasco-J715 spectropolarimeter equipped with a thermostatted cell holder (Jasco, Tokyo, Japan). The instrument is calibrated with an aqueous solution of d_{10}-(+)-camphorsulfonic acid.[28] All CD spectra are obtained at a scan speed of 20 nm/min, with a 2 nm band width, and a response time of 1 sec. Spectra are signal-averaged over four scans and the solvent contribution is subtracted using the Jasco software. For all the variants mentioned in this section no significant differences were observed when comparing the individual spectra, thus indicating comparable contents of secondary-structure elements and a high degree of similarity in the three-dimensional conformation.

Thermal Denaturation Experiments

Among the requirements for any ribonuclease to be cytotoxic, thermal stability is a parameter that has to be evaluated since, as potential toxins, these proteins must be active at around 37°. In order to determine the temperature denaturation midpoint ($T_{1/2}$), enzymes are dissolved at a concentration of 1 mg/ml in 50 m*M* sodium acetate buffer pH 5.0. Absorption spectra are recorded using the modified absorption Cary spectrometer described in Lange *et al.*[29] between 250 and 310 nm as a function of temperature from 20° to 80° in 4° steps. Following each temperature increment, the system is equilibrated for 2 min before measurement. Each spectrum is corrected for the temperature-dependent change in volume and enzyme concentration before the 4th derivative is calculated.[30] It is worth mentioning (see Table I) that HP-RNase is about 5° less stable than the PM5 variant, which differs in the N terminus, and about 7° less stable than the bovine enzyme whose sequence is 70% identical.

In summary, the production of the different HP-RNase variants using the methodology described in this section is highly reproducible and yields an electrophoretically and chromatographically homogeneous product. The characterization by means of steady-state kinetic parameter determination and by far- and near-UV CD spectra suggests that all the human ribonucleases described in this section exhibit closely similar three-dimensional structures that are also very similar to that of their bovine counterpart.

[28] J. T. Yang, C.-S. C. Wu, and H. M. Martinez, *Methods Enzymol.*, **130,** 208 (1986).

[29] R. Lange, J. Frank, J. L. Saldana, and C. Balny, *Eur. Biophys. J.* **24,** 277 (1996).

[30] J. Torrent, J. P. Connelly, M. G. Coll, M. Ribó, R. Lange, and M. Vilanova, *Biochemistry* **38,** 15952 (1999).

Future Prospects and Applications of HP-RNase Engineering

One of the aims of structural biology is to understand the relationships between the primary sequence of proteins and their three-dimensional structure and functional properties. Protein engineering offers a potent tool for investigating the interplay between sequence and three-dimensional determinants, and their eventual effect on the biological function of the molecule.

Despite the high degree of primary structure and three-dimensional structural identity shared between HP-RNase and bovine RNase A, significant differences are found in comparisons of either their stabilities or their behavior when refolding under the same conditions.[12] The availability of two polypeptide chains, 70% identical in sequence, but differing in their structural properties, together with the vast amount of knowledge on the bovine enzyme,[31,32] makes the human counterpart a good model for obtaining a deeper insight into the relationship between protein primary structure and native conformation.

On the other hand, several secreted ribonucleases are endowed with special biological activities,[1,2,33] but the nature and the role of the corresponding primary structure determinants responsible for these features are only partly known. The engineering of RNase A, based on the analysis of the interactions of this enzyme with the cytosolic ribonuclease inhibitor,[12,34,35] has conferred nonnatural toxicity on the G88R RNase A variant.[36] When the same substitution is introduced in the human enzyme or even when the loop that contains position 88 is reproduced by means of site-directed mutagenesis, the cytotoxicity of HP-RNase is not increased at all.[23]

Resolution of the three-dimensional structure of HP-RNase PM7 variant has allowed the application of molecular dynamics techniques to generate a molecular model by superimposing the atomic coordinates of PM7 onto those of RNase A in the porcine ribonuclease inhibitor–RNase A complex.[34,37] It can be inferred from the model, in spite of the amino acid sequence similarity shared by the human and the bovine pancreatic ribonucleases, that to increase the K_i (2.0×10^{-13} M)[9] for HP-RNase sufficiently to avoid the RI action, it might be necessary to disrupt additional determinants to those that have to be modified to allow the bovine enzyme to avoid RI action. To the best of our knowledge, toxicity has been

[31] G. D'Alessio and J. F. Riordan (eds.), "Ribonucleases: Structures and Functions," Academic Press, New York, 1997.

[32] R. T. Raines, *Chem. Rev.* **98,** 1045 (1998).

[33] R. J. Youle and G. D'Alessio, *in* "Ribonucleases: Structures and Functions" (G. D'Alessio and J. F. Riordan, eds.), p. 491, Academic Press, New York, 1997.

[34] R. Shapiro, *Methods Enzymol.* **341,** [39] 2001 (this volume).

[35] J. Hofsteenge, *in* "Ribonucleases: Structures and Functions" (G. D'Alessio and J. F. Riordan, eds.), p. 621. Academic Press, New York, 1997.

[36] P. A. Leland, L. W. Schultz, B.-M. Kim, and R. T. Raines, *Proc. Natl. Acad. Sci. U.S.A.* **95,** 10407 (1998).

[37] B. Kobe and J. Deisenhofer, *Nature* **366,** 751 (1993).

conferred on HP-RNase only when it is engineered to produce a dimeric protein analogous to BS-RNase,[11] or when conjugated to other molecules that facilitate the internalization of the ribonuclease activity.[38,39]

Engineered variants of monomeric HP-RNase designed with the purpose of being used as new chemotherapeutic agents have to show a balanced equilibrium between ribonuclease activity, lowered affinity for ribonuclease inhibitor, and thermostability. The efficiency of the HP-RNase variants as cytotoxic agents will depend on a rational design of these parameters. Much work has to be done in the design of these potential chemotherapeutic agents for directing them to a chosen target. The target specificity of these proteins will depend on the characterization of the determinants reponsible for cellular attachment and internalization.

Acknowledgments

This work has been supported by Grants PB96-1172-CO2-01/02 from DGES, Ministerio de Educación y Cultura, Spain, and SGR 1998-65 and SGR 2000-64 from CIRIT, Generalitat de Catalunya, Spain. A.B. acknowledges a postdoctoral (RED) from CIRIT, Generalitat de Catalunya, Spain. We are also indebted to Fundació "M. F. de Roviralta," Barcelona, Spain, for equipment purchasing grants.

[38] K. Psarras, M. Ueda, T. Yamamuda, S. Ozawa, M. Kitajima, S. Aiso, S. Komatsu, and M. Seno, *Protein Eng.* **11,** 1285 (1998).

[39] M. Suzuki, S. K. Saxena, E. Boix, R. J. Prill, V. M. Vasandani, J. E. Ladner, C. Sung, and R. J. Youle, *Nature Biotechnol.* **17,** 265 (1999).

[14] Degradation of Double-Stranded RNA by Mammalian Pancreatic-Type Ribonucleases

By MASSIMO LIBONATI and SALVATORE SORRENTINO

Introduction

The term double-stranded RNA (dsRNA) refers to high molecular weight RNA in the "A" form, having the following properties: (1) a base composition expected for an RNA duplex composed of two complementary, antiparallel strands stabilized by hydrogen bonds and hydrophobic interactions; (2) a molar absorbance (per phosphodiester group) lower than that of single-stranded RNA (ssRNA); (3) an absolute hypochromicity remarkably larger than that of ssRNA; (4) temperature transition profiles of a cooperative type, with ionic strength-dependent T_m values; and (5) sedimentation coefficients ($s_{20,w}$), above 8–9 S.[1]

[1] M. A. Billeter and C. Weissmann, *in* "Procedures in Nucleic Acid Research" (G. L. Cantoni and D. R. Davies, eds.), Vol. 1, p. 498. Harper and Row, New York, 1966.

METHODS IN ENZYMOLOGY, VOL. 341
0076-6879/01 $35.00

These properties belong (a) to the genomes of several types of animal, plant, fungal, and bacterial viruses, as well as to the double-stranded RNA once called the "replicative form" and later considered as a side product of the replication of phage RNA, devoid of physiological significance[2]; or (b) to synthetic high molecular weight double-stranded polyribonucleotides, such as the poly(A) · poly(U) or poly(I) · poly(C) complexes.

To these canonical double-stranded RNA species the well-known acidic forms of polyadenylate and polycytidylate can be added. These polyribonucleotides at acidic pH assume a well-characterized [and for poly(A) particularly stable] double-stranded structure, due to the protonation of the bases at pH values below the p*K* values of adenine and cytosine.

All other types of RNA, such as rRNA, mRNA, tRNA, single-stranded viral RNA, or viroid RNA, also bear more or less abundant self-complementary sequences able to form double-helical secondary structures, which however are incomplete and/or irregular. These RNA species will not be considered here.

Properties of Double-Stranded RNA

Although all ssRNAs are extremely sensitive to the action of even minute amounts (ng) of bovine pancreatic RNase A, double-stranded RNA shows a remarkable resistance to RNase A under defined experimental conditions: 50 μg of RNase A per ml, 30 min incubation at 25° in 0.15 *M* NaCl, 0.015 *M* sodium citrate (SSC), pH 7, or different buffers with similar, physiological ionic strength.[1]

The well-known "in line" mechanism of ssRNA degradation by mammalian RNases requires that the 2′ oxygen of the ribose hydroxyl group acting as an internal nucleophile and the 5′ oxygen leaving group occupy simultaneously the two apical positions within the trigonal bipyramidal phosphorane intermediate.[3] This steric requirement is impossible with an RNA double helix (Fig. 1) and explains why dsRNA is greatly resistant to the action of bovine pancreatic RNase A, provided the stability of the duplex secondary structure is ensured by proper salt concentrations. The resistance of dsRNA to RNase A is indeed highly dependent on ionic strength and may disappear when salt concentrations become too low. However, resistance of dsRNA is never absolute, and a recovery of about 95% of this RNA species is typical when digestion with RNase A is carried out under physiological ionic strength conditions, i.e., under the assay conditions[1] indicated above. Therefore, bovine RNase A cannot be considered to be an enzyme "specific" for ssRNA, but only a "single-strand-preferring" ribonuclease.

[2] C. Weissmann, *FEBS Lett.* **40,** S10 (1974).

[3] A. W. Nicholson, *Progr. Nucleic Acids Res. Mol. Biol.* **52,** 1 (1996).

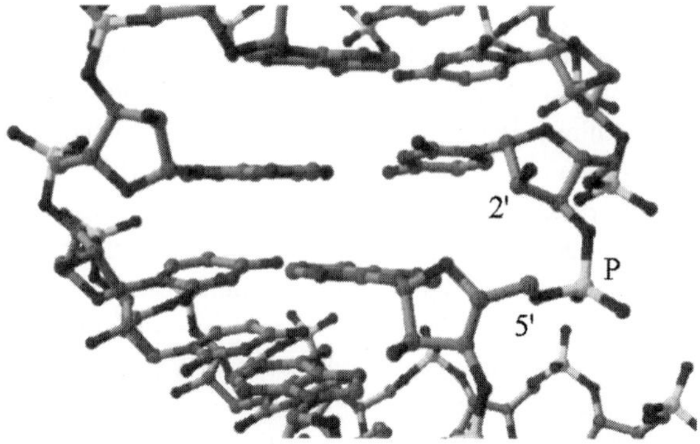

FIG. 1. Structural constraint of double-stranded RNA. The 2′ oxygen and 5′ oxygen cannot simultaneously occupy the apical positions within a trigonal bipyramidal phosphorane intermediate required by the transesterification mechanism of RNA degradation catalyzed by pancreatic-type ribonucleases.

Mechanism of Double-Stranded RNA Degradation

Double-stranded RNA, although it shows a remarkable resistance to bovine RNase A under proper salt conditions, is promptly and specifically degraded by *Escherichia coli* RNase III and related enzymes.[3] These ribonucleaes will not be considered here. We will instead refer on the cleavage of dsRNA by various mammalian pancreatic-type (pt) RNases such as human, horse, pig, hamster, pronghorn, and whale pancreas ribonucleases, as well as bovine seminal RNase. These enzymes preferentially degrade ssRNA, but are also able to depolymerize dsRNA more or less efficiently under experimental conditions where the action of bovine RNase A is essentially undetectable.

How can these ribonucleases, a number of which are highly similar to bovine RNase A (70–80% of homology[4]), degrade dsRNA?

It must be mentioned that, although monomeric RNase A is "unable" to cleave dsRNA, 30 years ago it was found that dimers of bovine RNase A, prepared according to Crestfield *et al.*[5] by lyophilizing solutions of the enzyme in 30–50% acetic acid, acquire the ability to degrade dsRNA under conditions where the native monomer is inactive.[6] At the same time, bovine seminal RNase was shown to be the first naturally dimeric[7] mammalian ribonuclease able to efficiently cleave dsRNA.[6] Therefore, it was originally proposed that degradation of dsRNA could reflect the simultaneous action on the RNA duplex of the two active sites present

[4] J. J. Beintema, W. M. Fitch, and A. Carsana, *Mol. Biol. Evol.* **3,** 262 (1986).

[5] A. M. Crestfield, W. H. Stein, and S. Moore, *Arch. Biochem. Biophys.*, Suppl. **1,** 217 (1962).

[6] M. Libonati and A. Floridi, *Eur. J. Biochem.* **8,** 81 (1969).

[7] G. D'Alessio, A. Parente, C. Guida, and E. Leone, *FEBS Lett.* **27,** 285 (1972).

in a naturally or artificially dimeric ribonuclease.[6,8] This idea held until it was found that a monomeric derivative of bovine seminal ribonuclease, obtained by reduction and iodoacetate alkylation of the two disulfide bonds joining the two subunits, maintained about 70% of the activity of the native dimer on dsRNA.[9]

The observation that all mammalian, single-strand-preferring ribonucleases active on RNA duplexes have in common a higher basicity than bovine RNase A led to the hypothesis that the ability of a ribonuclease to cleave dsRNA could be related to the number of positive charges present on the RNase molecule. The original idea was that the mechanism of dsRNA degradation could consist of an RNase-induced "destabilization" of the RNA duplex, followed by degradation of the single-stranded RNA sequences formed.[10] More recently,[11–13] the destabilization event was interpreted according to the model proposed by Jensen and von Hippel[14] to explain the destabilization of DNA by RNase A, as originally reported by Felsenfeld *et al.*[15] In this model, the RNase binds to segments of the double-stranded nucleic acid transiently made single-stranded by the continuous thermal fluctuation of the DNA secondary structure. This model has been extended to the interaction of dsRNA with bovine pancreatic RNase A and other pancreatic-type ribonucleases[11–13] since spontaneous base-opening reactions also occur in double-helical RNA, as shown by Leroy *et al.*[16] As a first approximation we can therefore assume that the binding efficiency of any ptRNase to short RNA sequences transiently made single-stranded by thermal fluctuation of the RNA duplex, or even to single nucleotides "wound off" the double helix,[17] would depend on the number of basic charges present on the RNase molecule.[11] The first event would be a local "destabilization" of the RNA duplex as the consequence of the higher, charge-dependent affinity of the ribonuclease to single-stranded compared to double-stranded RNA. The second event would be the cleavage of the bound single-stranded RNA sequences.[11–13,17] This also can explain the low-level activity of native RNase A toward dsRNA, detectable under physiological salt conditions.

The helix-unwinding ability of various mammalian pancreatic-type ribonucleases active on dsRNA was demonstrated by analyzing the thermal transition profiles of DNA or double-stranded polydeoxyribonucleotides, such as

[8] G. D'Alessio, J. Doskocil, and M. Libonati, *Biochem. J.* **141,** 317 (1974).

[9] M. Libonati, M. C. Malorni, A. Parente, and G. D'Alessio, *Biochim. Biophys. Acta* **402,** 83 (1975).

[10] M. Libonati, A. Furia, and J. J. Beintema, *Eur. J. Biochem.* **69,** 445 (1976).

[11] M. Libonati and S. Sorrentino, *Mol. Cell. Biochem.* **117,** 139 (1992).

[12] S. Sorrentino and M. Libonati, *Arch. Biochem. Biophys.* **312,** 340 (1994).

[13] S. Sorrentino and M. Libonati, *FEBS Lett.* **404,** 1 (1997).

[14] D. E. Jensen and P. H. von Hippel, *J. Biol. Chem.* **251,** 7198 (1976).

[15] G. Felsenfeld, G. Sandeen, and P. H. von Hippel, *Proc. Natl. Acad. Sci. U.S.A.* **50,** 644 (1963).

[16] J. L. Leroy, D. Broseta, and M. Gueron, *J. Mol. Biol.* **184,** 165 (1985).

[17] G. I. Yakovlev, G. P. Moiseyev, S. Sorrentino, R. De Prisco, and M. Libonati, *J. Biomol. Struct. Dyn.* **15,** 243 (1997).

poly(dA-dT) · poly(dA-dT), in the presence or absence of the RNases. The event was found to be indeed a function of the basicity of the enzyme protein.[12,18] It has to be pointed out that for a ribonuclease to be able to unwind the nucleic acid double helix, it is not merely the simple existence of basic charges on the surface of the enzyme protein that matters, but their specific location, as demonstrated by Jensen and von Hippel[14] and Karpel *et al.*[19] In bovine RNase A—the prototype of ptRNases—nine basic amino acid residues (essentially lysines and arginines) can constitute an extended, multisite, cationic region, as suggested by McPherson *et al.*[20] In their model these basic amino acids (at positions 7, 41, 66, 85, 39, 91, 98, 33, 31) are arranged in a linear array that is complementary to the phosphate groups of the polyribonucleotide chain. Some of them (Lys-7 and Lys-41) have a role in the endonucleolytic mechanism of degradation; the others may only contribute, as noncatalytic subsites, to the binding of the substrate. The nine residues mentioned above, together with other basic amino acids (Arg-10, Lys-37, Lys-104), are generally well conserved in all ptRNases. In conclusion, the helix-unwinding efficiency of the various ptRNases toward double-helical polydeoxy- or polyribonucleotides[16] as well as their consequent (in the latter case) cleaving ability appear to be correlated with the multiplicity and cooperativity of electrostatic interactions between "specifically" located positively charged residues of the RNase molecule and the phosphates of the substrate. In particular, the remarkable activity of some ptRNases can be ascribed to the presence of additional basic amino acids that could improve the binding of the enzymes to the short, transiently exposed, single-stranded sequences of the substrate.[11–13,21] Data supporting a possible correlation between specifically located basic residues in ptRNases and enzyme activity toward dsRNA are presented in Table I. For example, human ptRNase 1, whose activity on dsRNA is 200–300 times higher than that of ribonuclease A, has four basic residues (Arg-4, Lys-6, Arg-32, Lys-102) at positions where neutral residues are instead present in bovine RNase A. These positively charged amino acids are close to some of the nine basic residues constituting the multisite cationic region[20] mentioned above, and therefore strengthen the local electrostatic potential of that region. Moreover, Arg-4 and Lys-102 are located at the active site and can be directly involved in the RNase–dsRNA interaction as additional, noncatalytic phosphate binding subsites. It must be noticed that Arg-39 is one of the nine basic amino acids important for the helix-unwinding ability of a ribonuclease. Therefore, the presence of a neutral amino acid at position 38 in all RNases listed in Table I whose relative activity on dsRNA is definitely above

[18] A. Carsana, A. Furia, and M. Libonati, *Mol. Cell. Biochem.* **56,** 89 (1983).

[19] R. L. Karpel, D. J. Merkler, B. K. Flowers, and M. D. Delahunty, *Biochim. Biophys. Acta* **654,** 42 (1981).

[20] A. McPherson, G. Brayer, D. Cascio, and R. Williams, *Science* **232,** 765 (1986).

[21] S. Sorrentino, *Cell. Mol. Life Sci.* **54,** 785 (1998).

TABLE I
CORRELATION BETWEEN LOCATION OF BASIC RESIDUES IN ptRNases AND ACTIVITY ON dsRNA

	Positions							Relative activity on	
Ribonucleases	4	6	32	34	38	39	102	dsRNA	Yeast RNA
Ox pancreas	Ala	Ala	Ser	Asn	Asp	**Arg**	Ala	1	1
Giraffe pancreas	Ala	Ala	Ala	Asn	Asp	**Arg**	Ala	1	1
Human pancreas	**Arg**	**Lys**	**Arg**	Asn	Gly	**Arg**	**Lys**	300	0.3
Whale pancreas	Pro	Met	**Arg**	**Lys**	Gly	**Arg**	**Lys**	32	0.3
Pig pancreas	Pro	**Lys**	**Arg**	Asn	Gly	**Arg**	Glu	24	0.5
Ox brain	Ala	Ala	**Arg**	**Arg**	Gly	**Arg**	**Lys**	22	0.8
r-Giraffe brain	Ala	Ala	**Arg**	**Arg**	Gly	**Arg**	**Lys**	21	0.7
Horse pancreas	Pro	Met	**Arg**	Asn	Gly	Trp	**Lys**	22	0.6
Ox semen	Ala	Ala	Cys	**Lys**	Gly	**Lys**	Val	14	0.5
Hamster pancreas	Ser	Met	**Arg**	Asn	Gly	Tyr	Leu	12	0.6
Pronghorn pancreas	Ala	Ala	Ser	Asn	Gly	**Arg**	Ala	7	1
r-Human RNase 4	Met	Gln	**Arg**	**Lys**	Tyr	His	Ser	1.3	0.2

Basic amino acid residues are indicated in boldface type.

1 can be responsible for a stronger interaction between those enzymes and the polyanionic substrate, and therefore for their more efficient "destabilizing" and cleaving action toward dsRNA. This idea[21] is in agreement with the observation that an RNase A mutant having a glycine at position 38 shows an increased activity toward dsRNA.[22]

The importance of the specific location of basic amino acids on the ribonuclease surface for efficient "destabilization" and cleavage of dsRNA is indicated by the observation (Fig. 2) that while ptRNases show a remarkable helix-unwinding ability, non-pancreatic-type (npt) ribonucleases—such as human RNases 2 and 3—are totally inactive as "destabilizers" as well as degrading enzymes, notwithstanding their high basicity.[12,13,21] In these ribonucleases only two of the nine basic residues constituting the multisite cationic region suggested by McPherson *et al.*[20] are in fact conserved.[13,21] In Fig. 3 the distribution of basic amino acid residues present on the surface of human ptRNase 1 is compared with that of human nptRNase 2.

On the basis of these facts, npt RNases could rightly deserve to be considered as "single-strand-specific" ribonucleases.

Some results obtained with bovine brain and recombinant giraffe brain ribonucleases (see Table I) emphasize the importance of number and location of basic charges in the ribonuclease molecule for an efficient activity toward dsRNA. These

[22] J. G. Opitz, M. I. Ciglic, M. Haugg, K. Trautwein-Fritz, S. A. Raillard, T. M. Jermach, and S. A. Benner, *Biochemistry* **37,** 4023 (1998).

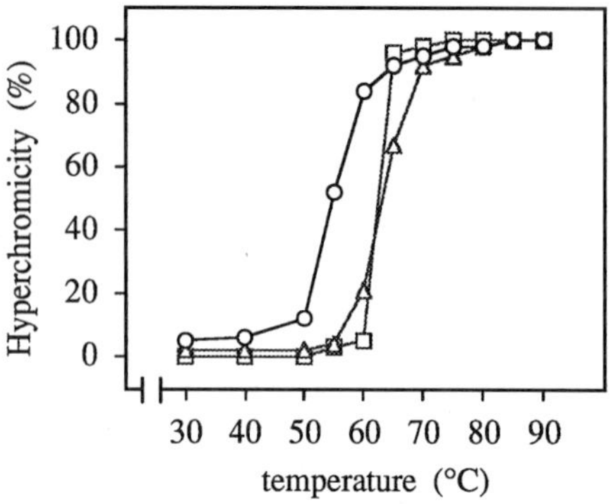

FIG. 2. Thermal transition profiles of double-stranded poly(dA-dT) · poly(dA-dT) in the presence or absence of human pancreatic-type or nonpancreatic-type RNases. 11 μg of poly(dA-dT) · poly(dA-dT) were mixed with 12 μg of ptRNase 1 (○) or nptRNase 2 (Δ) in 1 ml of 10 m*M* MOPS, pH 7.5, containing 0.1 *M* NaCl. (□), poly(dA-dT) · poly(dA-dT) in the absence of protein. Modified from S. Sorrentino and M. Libonati, *Arch. Biochem. Biophys.* **312,** 340 (1994).

ribonucleases, having two arginine residues at positions 32 and 34,[23] where two neutral residues are present in bovine ribonuclease A, are 20 times more efficient than RNase A in cleaving double-stranded poly(A) · poly(U).[24]

In contrast, recombinant human RNase 4, although having an arginine and a lysine at positions 32 and 34 and a neutral amino acid at position 38, shows on poly(A) · poly(U) an activity no different from that of bovine RNase A. This could be due to the lack of Lys-31 (not shown in Table I) and Arg-39, two of the nine basic residues characterizing pancreatic-type ribonucleases.

Although the idea[6,8] that the dimeric structure of bovine seminal RNase could be determinant for the action of the enzyme toward dsRNA has been resumed by Opitz *et al.*,[22] it is worth pointing out that all other ribonucleases active on dsRNA (Table I) are monomeric proteins. Therefore, at least in the case of these monomeric RNases, the model advanced by Opitz *et al.*[22] cannot be applied, and no correlation between degrading activity on dsRNA and dimeric structure of a ribonuclease is of course possible. But also for bovine seminal RNase this correlation becomes doubtful if one takes into account that, as already mentioned, a reduced and alkylated monomeric derivative of bovine seminal RNase maintained about 70% of the activity shown by the native dimeric enzyme.[9] Moreover, the cleaving activity of the monomeric derivative could be restored to at least 92% that of the native dimer when alkylation was performed with iodoacetamide instead of iodoacetate (i.e., without introducing two negative charges into the enzyme

[23] J. J. Beintema, H. J. Breukelman, A. Carsana, and A. Furia, *in* "Ribonucleases: Structures and Functions" (G. D'Alessio and J. F. Riordan, eds.), p. 245. Academic Press, San Diego, CA, 1997.

[24] S. Sorrentino, "Fifth Intern. Meet. on Ribonucleases," Abstr. Book, p. 9, 1999.

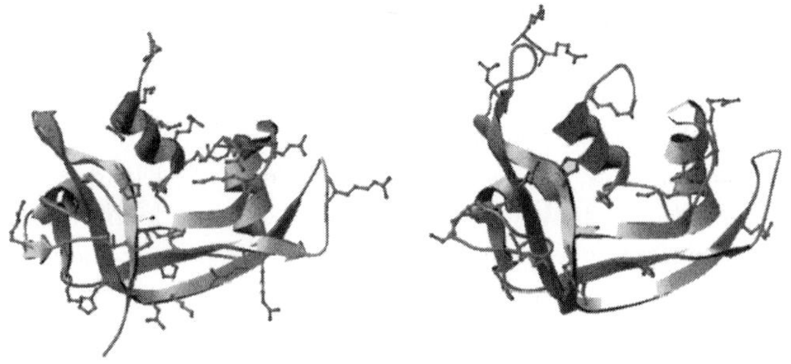

FIG. 3. Distribution of basic amino acid residues on the surface of human ptRNase 1 or nptRNase 2. *Left:* Model of human ptRNase 1 sequence superimposed on bovine RNase A structure. *Right:* X-ray structure of r-EDN (nptRNase 2). Modified from S. C. Mosimann, D. L. Newton, R. J. Youle, and M. N. G. James, *J. Mol. Biol.* **260,** 540 (1996).

protein).[25] This result also supports the view that a relation exists between basic charges of an RNase and its activity toward double-stranded polyribonucleotides. Recently, the monomeric derivative of bovine seminal RNase mentioned above has been shown to be even more active than the native dimeric enzyme toward the poly(A) · poly(U) complex in 0.1 *M* MOPS, pH 7.5/0.1 *M* NaCl.[26] Moreover, the hypothesis that the dimeric nature of a ribonuclease might be necessary for enzymic degradation of dsRNAs cannot be easily reconciled with the two following experimental data. (1) Dimeric RNase A, obtained by the procedure outlined by Crestfield *et al.*,[5] consists of two conformational isomers,[27,28] a minor form and a major form (in the ratio of about 1 : 3), that can be isolated by gel filtration and/or ion-exchange chromatography. The crystal structure of one of them, very probably the minor species,[29] has been determined by Liu *et al.*[30] as a 3D domain-swapped dimer, the two subunits of which exchange their N-terminal helices. This structure, which is in agreement with the model proposed by Crestfield *et al.*,[5] shows that the two subunits of the dimer are related by a rotation of 160°, so that the two active sites are opposite to each other, whereas in bovine seminal RNase they are oriented in the same direction, which is required for the interaction between the enzyme and double-stranded RNA in the model proposed by Opitz *et al.*[22] (2) RNase A oligomers higher than dimers, i.e., trimers, tetramers, and pentamers, also form when bovine RNase A is lyophilized from 40% acetic acid

[25] A. Parente, M. Palmieri, R. De Prisco, and M. Libonati, *Boll. Soc. It. Biol. Sper.* **53,** 465 (1977).

[26] S. Sorrentino and G. D'Alessio, unpublished results, 1999.

[27] R. G. Fruchter and A. M. Crestfield, *J. Biol. Chem.* **240,** 3868 (1965).

[28] G. Gotte and M. Libonati, *Biochim. Biophys. Acta* **1386,** 106 (1998).

[29] S. Sorrentino, R. Barone, E. Bucci, G. Gotte, N. Russo, M. Libonati, and G. D'Alessio, *FEBS Lett.* **466,** 35 (2000).

[30] Y. Liu, P. J. Hart, M. P. Schlunegger, and D. Eisenberg, *Proc. Natl. Acad. Sci. U.S.A.* **95,** 3437 (1998).

solutions.[31] They are definitely more active than monomeric or dimeric RNase A toward poly(A) · poly(U), showing an activity that actually increases as a function of the mass of the oligomer.[31] Moreover, as occurs for the RNase A dimers,[27,28] each oligomeric species is composed of two different conformers, isolated by ion-exchange chromatography.[32] The two conformers of each aggregated species exhibit differential cleavage of poly(A) · poly(U) or viral double-stranded RNA, the more basic conformer being more active than its less basic counterpart.[32]

These last findings clearly show that a correlation exists between number (or rather exposure) and availability of basic charges present in the oligomer(s) and enzymatic activity toward dsRNA.

In conclusion, although it appears to be improbable that the dimeric nature of a ribonuclease could generally be the sole or the main factor determining an efficient enzyme attack on dsRNA, it could have a role in the case of bovine seminal RNase or other possible dimeric ribonucleases having a structure similar to that of the seminal enzyme.

Influence of Ionic Strength on dsRNA Degradation by Mammalian ptRNases

Among other possible variables[11,33] that can play a role in the mechanism of enzymatic degradation of dsRNA, the concentration of counterions has a remarkable influence on the process since it obviously has a great influence on the structure of dsRNA. The secondary structure of dsRNA is tightened by high salt concentrations (0.15–0.2 *M*), while it may become unstable under lower ionic strength conditions. However, it must be noted that the thermal transition profile of double-stranded RNA isolated from *E. coli* K-38 cells infected with f2 *sus11* bacteriophage retains its cooperative nature and a high T_m value (89°) in a salt solution (0.015 *M* NaCl, 0.0015 *M* sodium citrate, pH 7) 10 times less concentrated than the standard one (SSC).[34] Similarly, the T_m values reported for double-stranded MS2 RNA are 103° in SSC, 87° in 10 times diluted SSC, and about 80° in 100 times diluted SSC.[1] This indicates that the stability of the secondary structure of dsRNA does not change very much within that range of salt concentrations; this could be particularly true at the temperature of a standard ribonuclease assay (25°).

A series of experiments performed with dimeric bovine seminal RNase, monomeric RNase A and its cross-linked (with dimethyl suberimidate[35]) dimers, and

[31] M. Libonati, M. Bertoldi, and S. Sorrentino, *Biochem. J.* **318,** 287 (1996).

[32] G. Gotte, M. Bertoldi, and M. Libonati, *Eur. J. Biochem.* **265,** 680 (1999).

[33] A. Carsana, A. Furia, A. Gallo, J. J. Beintema, and M. Libonati, *Biochim. Biophys. Acta* **654,** 77 (1981).

[34] M. Libonati and M. Palmieri, *Biochim. Biophys. Acta* **518,** 277 (1978).

[35] D. Wang, G. Wilson, and S. Moore, *Biochemistry* **15,** 660 (1976).

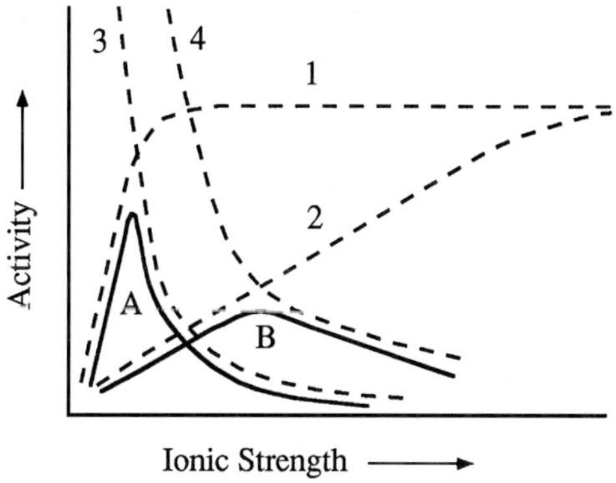

FIG. 4. Model for the ionic control of ribonuclease activity on single- or double-stranded RNA. Curve A, activity pattern of monomeric and/or less basic ribonucleases. Curve B, activity pattern of oligomeric and/or more basic ribonucleases. Both curves reflect the result of two effects: (i) the influence of "specific" enzyme–substrate interactions (depending on enzyme "rigidity") on the activity of (curve 1) monomeric or less basic RNases, or (curve 2) oligomeric or more basic RNases; (ii) the influence on enzyme activity of the solvation-dependent strength of "nonspecific" interactions between polyribonucleotides and (curve 3) monomeric or less basic RNases, or (curve 4) oligomeric or more basic RNases. Modified from S. Sorrentino, G. I. Yakovlev, and M. Libonati, *Eur. J. Biochem.* **124,** 183 (1982).

viral dsRNA, poly(A) · poly(U) or the acidic, double-stranded forms of poly(A) and poly(C) as substrates have shown that the influence of ionic strength on ribonuclease activity toward dsRNAs is indeed rather complex.[36] Enzyme activities determined as a function of ionic strength exhibit bell-shaped curves with activity maxima that do not occur at salt concentrations where the stability of the secondary structure of the double-helical substrates should be minimal. Moreover, the curves and their maxima are progressively shifted to higher ionic strength values as a function of the enzyme protein basicity.[36] The same is true for the bell-shaped curves expressing the helix-unwinding activities of several mammalian ribonucleases differing in their basic net charge.[18] A model is shown in Fig. 4. An interpretation of these facts (see Fig. 4) is that salt influences both "specific" (involving base, ribose, and phosphate binding subsites of the RNase[37,38]) and "nonspecific" (electrostatic) enzyme–substrate interactions. Thus, the influence of salt not only on the nucleic acid(s) but also on the enzyme protein(s) must be taken into account. In other words, whereas at low salt concentrations the enzyme

[36] S. Sorrentino, A. Carsana, A. Furia, J. Doskocil, and M. Libonati, *Biochim. Biophys. Acta* **609,** 40 (1980).

[37] R. de Llorens, C. Arus, X. Parés, and C. M. Cucillo, *Protein Eng.* **2,** 417 (1989).

[38] X. Parés, M.V. Noués, R. de Llorens, and C. M. Cucillo, *in* "Essays in Biochemistry" (K. F. Tipton, ed.), p. 89. Portland Press, London, 1991.

molecule has a greater "rigidity," with the consequence that its "specific" interactions with the substrate can be less productive, at higher concentrations of salt the "rigidity" of the enzyme protein becomes lower and the catalytic activity due to "specific" enzyme–substrate interactions can increase. On the other hand, increasing salt concentrations progressively screen the charged groups of both enzyme and substrate, with the consequent weakening and/or reduction of "nonspecific" enzyme–substrate interactions, and of the activity depending on them.[11,36] In conclusion (see Fig. 4), the ascending arm of the bell-shaped curves should express the influence of salt on enzyme activity "controlled" by "specific" interactions, while the descending arm of the curves should be the expression of the influence of salt on the enzyme activity "controlled" by "nonspecific" enzyme–substrate interactions.[11,36] In this regard, it might be worth pointing out that the ionic control of enzyme reactions described above for the RNase(s)–dsRNA system could have a more general significance. In other cases in which the substrates were polyelectrolytes we have in fact observed phenomena quite similar to those summarized above: for example, in studying the activities of DNase I and its cross linked dimer, native, dimeric DNase II, egg-white lysozyme and its cross-linked dimer, papain and its cross-linked dimer on their polyelectrolyte substrates as a function of ionic strength.[39]

Various assay procedures could be available and used to determine the activity of mammalian pancreatic-type ribonucleases on dsRNA. The spectrophotometric method we will describe here is based on the hyperchromic effect due to the splitting by the RNase of the phosphodiester bonds of double-stranded polyribonucleotides used as substrates. Our choice is justified by the accuracy, reliability, and rapidity of the procedure proposed, and successfully used by us and others for many years.

Since mammalian pancreatic-type ribonucleases are able to "destabilize" the double-helical structure of synthetic or natural double-stranded polydeoxyribonucleotides, we will also describe the procedure routinely used by us to determine the helix-unwinding activity of the RNases considered, which could also be used to study the destabilizing activity of any other ribonuclease of interest.

Experimental Procedure

Materials

Synthetic, double-stranded polyribonucleotides used as substrates can be purchased from several biochemical supply companies. Poly(A) · poly(U) and poly(I) · poly(C) should be high molecular weight (sodium salt) Miles (Kaukakee, IL) or Sigma (St. Louis, MO) products. A double-stranded, alternating copolymer

[39] S. Sorrentino, G. I. Yakovlev, and M. Libonati, *Eur. J. Biochem.* **124,** 183 (1982).

such as the sodium salt of poly(dA-dT) · poly(dA-dT), or calf thymus DNA, both suitable substrates to determine the helix-unwinding activity of the various ribonucleases, should also be Boehringer (Mannheim, Germany), Miles, or Sigma products. Double-stranded viral RNA, ^{3}H-labeled or unlabeled, can be prepared according to a procedure already described.[1,8]

Bovine pancreatic ribonuclease A (Type XII-A) should be purchased from Sigma. All other ribonucleases can be purified according to published procedures.

Reagents

All chemicals should be of analytical grade. Water to be used for the preparation of buffer(s) must be ultrapure (e.g., Millipore Milli-RO Plus reverse osmosis) or at least twice distilled. The assay buffer solution (0.1 *M* MOPS, pH 7.5, containing 0.1 *M* NaCl) has to be freshly prepared, and always filtered before use.

Assay of RNase Activity

Although ribonuclease activity toward double-stranded poly(A) · poly(U) could also be measured by a precipitation assay based upon the procedure of Anfinsen *et al.*,[40] or according to Billeter and Weissmann[1] if the substrate is tritiated viral double-stranded RNA, our aim is to describe here a very rapid, accurate, and reproducible spectrophotometric procedure, which we[6,12,13,17,28,29,31,32] or others[22,35] have successfully used for many years. It is based on a method originally described by Libonati and Floridi[6] that has been only slightly modified. The method can be used with the poly(A) · poly(U) or poly(I) · poly(C) complexes as well as with unlabeled viral double-stranded RNA.[32] We will describe the procedure routinely followed in the assay on poly(A) · poly(U). Various substrate concentrations (in the range of 20–50 μg/ml, corresponding to 0.06–0.16 m*M* in phosphate) have been used in the past by us or others.[22,35] The same is true for buffers, or for specific activity values (units/mg protein) that often were calculated[6,12,31,32] by using different arbitrary enzyme units (for instance, the increase in absorbance at 260 nm per minute/the total measurable change). In the following assay we convert the arbitrary units used in the past into enzyme units as defined and recommended by the IUBMB.

Procedure

The substrate is gently dissolved, under sterile/RNase-free conditions, in a suitable volume (20–40 ml) of cold buffer (0.1 *M* MOPS, pH 7.5, containing 0.1 *M* NaCl) at a concentration of about 40 μg/ml. This operation could require

[40] C. B. Anfinsen, R. R. Redfield, W. L. Choate, J. Page, and W. R. Carrol, *J. Biol. Chem.* **207,** 201 (1954).

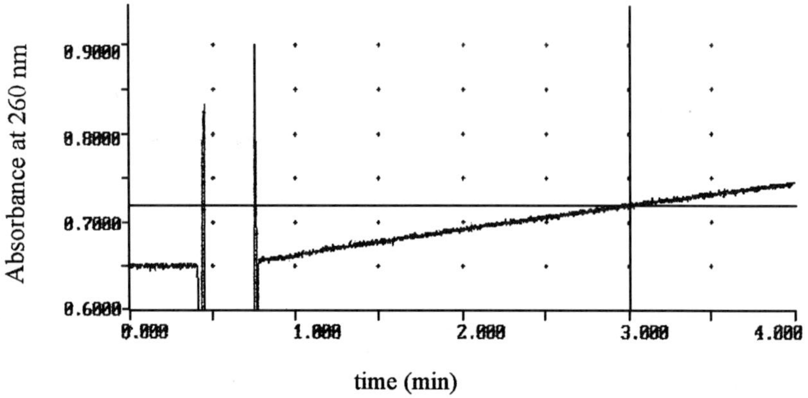

FIG. 5. A typical spectrophotometric assay to determine the enzymatic cleavage of double-stranded polyribonucleotides. Reaction mixture: poly(A) · poly(U), 0.1 m*M*, and human ptRNase, 0.28 μg, in 1 ml of 0.1 *M* Mops, pH 7.5, containing 0.1 *M* NaCl. The increase in absorbance due to substrate degradation is monitored as a function of time at 25° using a Varian Cary 1E spectrophotometer.

1–2 hr. The substrate solution should be always kept at 0–4° during the preparation of the experiments. The concentration of the substrate in phosphate (i.e., in phosphodiester groups) is determined spectrophotometrically using the following extinction coefficients calculated at pH 7.5: $\varepsilon_{260} = 6.5$ mM^{-1} cm^{-1} for poly(A) · poly(U); $\varepsilon_{260} = 4.4$ mM^{-1} cm^{-1} for poly(I) · poly(C). The substrate solution is then diluted to 0.1 m*M* with the same buffer. A quartz cuvette (1 ml, path length 1 cm) containing the substrate solution (1 ml) is preequilibrated at 25° for 3–5 min. When its absorbance at 260 nm becomes stable the reaction is initiated by the addition of an aliquot of enzyme solution (2–5 μl, containing 10–20 μg of bovine RNase A or 0.1–0.2 μg of human ptRNase 1); as rapidly as possible the cuvette, temporarily sealed with Parafilm, is inverted three times to mix the contents, and the increase in absorbance at 260 nm is continuously monitored for 3–5 min at 25° (Fig. 5). The change in absorbance per minute (deduced from the slope of the linear part of the recording) is converted to initial reaction velocity by using the following differences in absorption coefficients, calculated at pH 7.5 taking into account the millimolar absorptivity (in phosphate) of the substrate and that of the degradation products measured after about 8–12 hr at 25°, i.e., at the end of the RNase-catalyzed transphosphorylation reaction: $\Delta\varepsilon_{260} = 3.4$ mM^{-1} cm^{-1} for poly(A) · poly(U) or $\Delta\varepsilon_{260} = 1.8$ mM^{-1} cm^{-1} for poly(I) · poly(C). One enzyme unit is defined as the amount of RNase that catalyzes the cleavage of 1 μmol of phosphodiester linkages of the double-helical substrate per minute at 25°. Specific activity is expressed as units per mg protein. As an example, under the experimental conditions described in the legend to Fig. 5, the specific activity of the human RNase sample assayed was 30 units/mg (about 200 times higher than that of bovine RNase A). This value was calculated from an increase in absorbance per minute of 0.0029, corresponding to 8.5×10^{-3} enzyme units. Protein concentration of human ptRNase 1 was determined spectrophotometrically using an absorbance coefficient of 6.5 at 280 nm for a 1% solution.[41,42]

The activity of a ribonuclease on both single- and double-stranded substrates is highly dependent on the molecular size and homogeneity of chain length of the polymeric substrate. For this reason, in a comparative kinetic investigation the same substrate preparation should be used in all determinations of enzyme activity.

Assay of DNA Helix-Unwinding Activity of ptRNases

The effect of a ribonuclease on the thermal transition profile of a synthetic double-stranded polydeoxyribonucleotide, such as poly(dA-dT) · poly(dA-dT) or native calf thymus DNA,[43] can be analyzed by following the hyperchromicity at 260 nm as the temperature is increased from 25° to 90°. In a typical experiment to be carried out under the conditions described in the legend to Fig. 2, the thermal transition profile of the nucleic acid–RNase complex is determined in comparison with that of the double-stranded polydeoxyribonucleotide in the absence of ribonuclease. The buffer to be used can be 0.01 *M* MOPS, pH 7.5, containing 0.1 *M* NaCl. However, the buffer molarity as well as its pH are not critical: any salt concentration or pH values (of course within reasonable limits) can be used depending on circumstances. For example, to analyze the thermal transition of calf thymus DNA complexed with ox pancreas or ox semen RNases,[43] 0.0015 *M* sodium phosphate buffer, pH 7, containing 0.002 *M* NaCl (T_m 75°) was used. The thermal transition of the mixture of poly(dA-dT) · poly(dA-dT) (11–12 μg/ml) with ptRNase 1 (11–12 μg/ml) must be analyzed in stoppered cuvettes with a spectrophotometer equipped with a thermostatically controlled water bath. A suitable initial absorbance of the nucleic acid at 260 nm can be 0.270 (12 μg/ml of the double-helical polynucleotide). Before starting the experiment, the absorbance spectrum of the solution containing the nucleic acid/RNase complex should be determined, in comparison with that of the polymer in the absence of protein, within the range from 245 to 320 nm, in order to ascertain the absence of any turbidity or spectral changes in the range 290–320 nm. The contribution of the protein (usually added at a final concentration of about 11–12 μg/ml; see Fig. 2) to the absorbance of the mixture at 260 nm is nil. Protein concentration of ptRNase 1 is calculated on the basis of the extinction coefficient[41,42] indicated above. The concentration of

[41] T. Morita, Y. Niwata, K. Ohgi, M. Ogawa, and M. Irie, *J. Biochem.* **99,** 17 (1986).

[42] M. Iwama, M. Kunihiro, K. Ohgi, and M. Irie, *J. Biochem.* **89,** 1005 (1981).

[43] M. Libonati and J. J. Beintema, *Biochem. Soc. Trans.* **5,** 470 (1977).

nptRNase 2 (see Fig. 2) can be calculated by using an absorbance coefficient of 15.5 for a 1% solution.[44] At the end of the analysis, the increase in absorbance (as percentage of maximum increase) at 260 nm is plotted as a function of temperature.

Acknowledgment

This work was supported by the Italian MURST-PRIN 1999. We are grateful to Drs. E. Confalone and A. Furia for the kind gift of recombinant giraffe brain RNase, and to Drs. V. Cafaro and A. Di Donato for having kindly supplied us with recombinant human RNase 4.

[44] G. J. Gleich, D. A. Loegering, M. P. Bell, J. L. Checkell, S. J. Ackerman, and D. J. McKean, *Proc. Natl. Acad. Sci. U.S.A.* **83,** 3146 (1986).

[15] Seminal Ribonuclease: Preparation of Natural and Recombinant Enzyme, Quaternary Isoforms, Isoenzymes, Monomeric Forms; Assay for Selective Cytotoxicity of the Enzyme

By GIUSEPPE D'ALESSIO, ALBERTO DI DONATO, RENATA PICCOLI, and NELLO RUSSO

Introduction

1. The Protein

Among the more than 100 recognized members of the tetrapod RNase superfamily,[1] with bovine pancreatic RNase A as prototype, the only enzyme endowed with a quaternary structure is bovine seminal RNase (BS-RNase).[2] In the BS-RNase dimer, the subunits are linked by two disulfide bridges as well as by noncovalent forces.

The dimeric structure is not the only unusual feature of BS-RNase, as the dimer is in fact a mixture of two quaternary forms, which differ in the position of the N-terminal α-helical segments of the two subunits.[3] The three-dimensional structure of one isoform (henceforth denoted as M$\times$M), resolved by X-ray diffraction

[1] J. J. Beintema, H. J. Breukelman, A. Carsana, and A. Furia, *in* "Ribonucleases: Structures and Function" (G. D'Alessio and J. F. Riordan, eds.), p. 245. Academic Press, San Diego, 1997.
[2] G. D'Alessio, A. Di Donato, L. Mazzarella, and R. Piccoli, *in* "Ribonucleases: Structures and Function" (G. D'Alessio and J. F. Riordan, eds.), p. 383. Academic Press, San Diego, 1997.
[3] R. Piccoli, M. Tamburrini, G. Piccialli, A. Di Donato, A. Parente, and G. D'Alessio, *Proc. Natl. Acad. Sci. U.S.A.* **89,** 1870 (1992).

0076-6879/01 $35.00

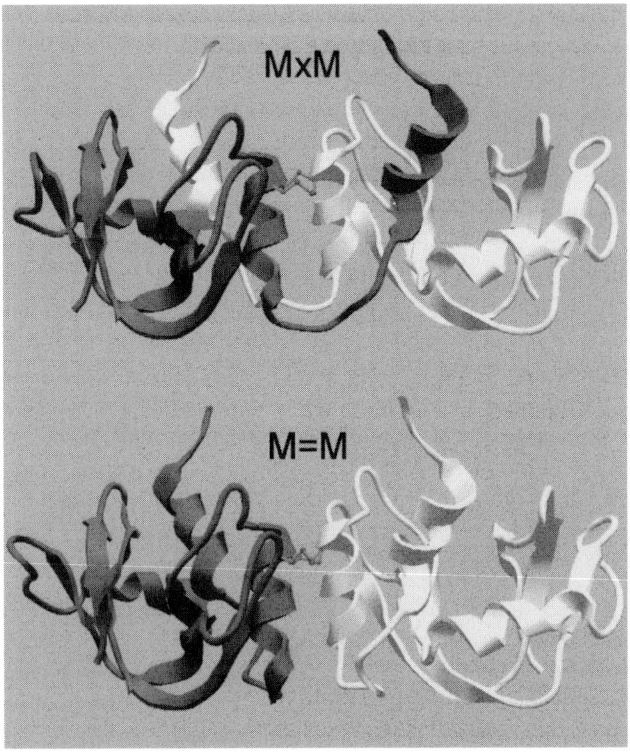

FIG. 1. The two quaternary structures of BS-RNase.

analyses,[4] is that of a swapped dimer, in which the two subunits exchange their N-terminal arms (see Fig. 1). This generates two composite active sites, made up of residues that belong to different subunits (see below). In the other quaternary isoform (henceforth denoted as M = M) there is no exchange of parts between subunits, and each subunit is spatially organized as an independent entity.[3] The isolated forms interconvert until the equilibrium ratio M×M/M = M of 2 is reached.

Another singular feature of seminal RNase is that the protein spontaneously generates isoenzymic subforms.[5] The native, secreted isoenzyme (termed α_2) has a full complement of amide groups, whereas in the other isoenzymes, $\alpha\beta$ and β_2, one or both subunits, respectively, are selectively deamidated at Asn-67. This deamidation can be reproduced *in vitro*.[5]

[4] L. Mazzarella, S. Capasso, D. Demasi, G. Di Lorenzo, C. A. Mattia, and A. Zagari, *Acta Crystallogr.* **D49,** 389 (1993).

[5] A. Di Donato and G. D'Alessio, *Biochemistry* **20,** 7232 (1981).

The atypical structural features of BS-RNase are reflected in its unusual (for a ribonuclease) enzymatic and biological properties.

2. The Enzyme

BS-RNase cleaves RNA at pyrimidine sites with a transesterification–hydrolytic two-step mechanism, as described for RNase A[2]; however, it also can effectively degrade double-stranded polyribonucleotides.[2] More unusual, in fact unique among known RNases of its superfamily, are the allosteric properties of BS-RNase, which is modulated by substrate in the second rate-limiting step of the reaction, as well as by the reaction product.[6,7] Nonhyperbolic saturation curves and binding studies have revealed a mixed-type cooperativity between the active sites, with negative cooperativity at low, and positive cooperativity at higher substrate concentrations.[6,7] When the kinetics of the isolated M×M and M=M quaternary forms were investigated, it was clear that the allosteric behavior was the unique property of the M×M isoform, which is the only isoform with composite active sites.[3]

3. BS-RNase Special Bioactions

BS-RNase was first discovered as an RNA degrading enzyme, then independently rediscovered as an antispermatogenic factor.[8] It has since been tested for its ability to affect several biological systems and found to exhibit several peculiar activities (for a review see references 2 and 9 and references therein): (1) a strong and selective cytotoxic action toward tumor cells, both *in vitro* and *in vivo;* (2) an immunosuppressive activity, as BS-RNase inhibits the proliferation of activated *T* cells, and prolongs the survival of skin grafts transplanted into allogenic mice; (3) an antispermatogenic activity, resulting in reversible infertility in male mice and rabbits; (4) an embryotoxic activity. The immunosuppressive action of BS-RNase may have a physiological relevance in bull semen, as BS-RNase may act as a protecting factor of sperm cells against the immune system of the female organism.[9] All these activities are strictly dependent on the ribonucleolytic activity of the protein.

Experimental Procedures

1. Preparation of Seminal RNase from Natural Sources

When bull seminal plasma is available, moderate amounts of BS-RNase can be obtained by a single-step purification procedure that yields about 1.2 mg of

[6] A. Di Donato, R. Piccoli, and G. D'Alessio, *Biochem. J.* **241,** 435 (1987).
[7] R. Piccoli, A. Di Donato, and G. D'Alessio, *Biochem. J.* **253,** 329 (1988).
[8] J. Matousek, *J. Reprod. Fert.* **20,** 189 (1969).
[9] G. D'Alessio, *Trends Cell Biol.* **3,** 106 (1993).

protein per ml of plasma.[10] Larger amounts of BS-RNase can more conveniently be prepared from bull seminal vesicles, as detailed in the following protocol, which is a modification of the procedure of Tamburrini *et al.*[10] designed to yield over 200 mg of BS-RNase from 500 g of seminal vesicles in 3 days.

Seminal vesicles should be collected from bulls within few hours of their death, freed of the surrounding connective tissue and fat, and either processed immediately or stored at −20° for up to 1 year. Unless otherwise specified, the following steps are performed at 4°. Seminal vesicles (500 g of tissue from about 10 couples of vesicles) are minced in a meat grinder, mixed with 500 ml of 0.75 *N* H_2SO_4, then homogenized with a Waring blendor at full speed for 3 min. The homogenate is stirred occasionally over a 1 h period, then centrifuged at 14,000*g* for 20 min. The supernatant is filtered through cheesecloth and set apart. The pellet is suspended in 500 ml of 0.75 *N* H_2SO_4, rehomogenized, centrifuged, and filtered as described above. The two filtered samples are then combined, supplemented with solid ammonium sulfate to 50% saturation (291 g/liter), and stirred for 1 h. The precipitate is removed by centrifugation as above, and the supernatant is taken to 100% saturation with ammonium sulfate (348 g/liter). After 2–16 h of stirring, the sample is centrifuged as above and both the pellet, containing about two thirds of the total amount of BS-RNase, and the supernatant are saved. The latter is taken to pH 9 by dropwise addition of 10 *M* NaOH and stirred for 3 h. This precipitates the residual fraction of BS-RNase, which is collected by centrifugation as above. The pellets from the last two centrifugations are dissolved in a total volume of 80 ml of water and dialyzed extensively against water, using tubes with a cutoff <12,000 Da.

The dialyzed preparation is clarified by centrifugation at 11,000*g* for 40 min, supplemented with Tris-HCl pH 8 and NaCl to a final concentration of 0.1 *M* each, and loaded onto a 2.5 × 6 cm column of carboxymethylcellulose (Whatman Clifton, NJ CM-32 or -52) previously equilibrated in the same buffer. Chromatography is performed at room temperature at a flow rate of 1 ml/min, with collection of 5-ml fractions. The resin is then washed with the equilibration buffer until the absorbance at 280 nm of the eluate is stably below 0.1. Under these washing conditions, all proteins but BS-RNase are eluted from the column. BS-RNase is then eluted from the resin with 0.1 *M* Tris-HCl pH 8 containing 0.3 *M* NaCl. After the salt is removed by dialysis against water, the protein is concentrated by ultrafiltration and quantitated spectrophotometrically by using a value of $E^{0.1\%}$ at 278 nm of 0.465. The final preparation can be stored at −20° for at least a year.

Purity of BS-RNase samples can be assessed by SDS–PAGE[11] with 12% polyacrylamide gels. Enzymatic activity on yeast RNA as a substrate, tested

[10] M. Tamburrini, R. Piccoli, R. De Prisco, A. Di Donato, and G. D'Alessio, *Ital. J. Biochem.* **35,** 22 (1986).

[11] U. Laemmli, *Nature* **227,** 680 (1970).

by the spectrophotometric assay of Kunitz,[12] should yield a specific activity of 30–40 U/mg.

2. Preparation of Recombinant Seminal RNase

2.1. Preparation of cDNA Coding for Seminal RNase. A cDNA coding for the subunit chain of seminal RNase has been cloned by screening a cDNA library from bovine seminal vesicles,[13] or constructed by total synthesis[14] or by semisynthesis.[15] The last procedure will be given in detail as it can be conveniently used also for the preparation of hybrid recombinants made up of segments from an available cDNA (coding for seminal RNase) fused with *ad hoc* synthesized DNA segments.

In the procedure reported here the starting material was a double-stranded fragment coding for the amino acid sequence 49–124 of the seminal RNase subunit chain. This was fused with a DNA segment constructed by synthesis of three couples of complementary chemically phosphorylated oligonucleotides. Their design[16] was based on the retro-translation of the amino acid sequence 1–49 of seminal RNase. In the design of the oligonucleotides care was taken to use as a first choice the *Escherichia coli* codon usage, and to accommodate compatible restriction sequences at the 5′ end of the first and 3′ end of the third couple.

Each couple of oligonucleotides (10–15 nmol) was annealed in 200 m*M* Tris-HCl pH 7.4, 20 m*M* $MgCl_2$, 0.5 *M* NaCl by boiling for 3 min in sealed capillaries, followed by slow cooling. The annealed mixtures were then added to the dephosphorylated DNA coding for the amino acid sequence 49–124. Ligation was carried out in the presence of T4 DNA ligase (3 units) for 12 h at room temperature. Ligation products were directly cloned into pUC18 vector between *Eco*RI and *Bam*HI sites by addition to the ligation mixture of pUC18 cleaved with the appropriate restriction enzymes. After 12 h at 15° in the presence of T4 DNA ligase, the ligation mixture was used to transform JM101 competent cells. Selection of recombinant clones containing the coding sequence of BS-RNase subunit was made by DNA sequencing. The cDNA coding for BS-RNase was excised from pUC18 with *Eco*RI and *Hin*dIII, and subjected to polymerase chain reaction (PCR) site-directed mutagenesis to convert its *Eco*RI 5′ end into an *Nde*I site for proper frame positioning in the pET22b(+) vector.[17] The DNA fragment resulting from PCR amplification (402 bp) was isolated, digested with *Nde*I and *Sal*I, ligated with the vector previously cut with the same enzymes, and used to transform JM101 competent cells.

[12] M. Kunitz, *J. Biol. Chem.* **164,** 563 (1946).

[13] K. D. Preuss, S. Wagner, J. Freudenstein, and K. H. Scheit, *Nucleic Acids Res.* **18,** 1057 (1990).

[14] J. S. Kim and R. T. Raines, *J. Biol. Chem.* **268,** 17392 (1993).

[15] M. De Nigris, N. Russo, R. Piccoli, G. D'Alessio, and A. Di Donato, *Biochem. Biophys. Res. Commun.* **193,** 155 (1993).

[16] G. Libertini and A. Di Donato, *Protein Eng.* **5,** 821 (1992).

[17] F. W. Studier, A. H. Rosemberg, J. J. Dunn, and J. W. Dubendorff, *Methods Enzymol.* **185,** 60 (1990).

Selection of a recombinant clone containing the BS-RNase coding sequence was achieved by direct sequencing of recombinant clones.

In the total synthesis of a cDNA coding for BS-RNase[14] the starting material was 12 overlapping oligonucleotides, coding for both strands. The construct was designed to maximize the production of the protein in prokaryotes, and to incorporate unique recognition sites for restriction enzymes. The chemically synthesized oligonucleotides, purified by denaturing polyacrylamide gels, were annealed, ligated in the presence of T4 DNA ligase, and inserted into pBluescript II SK(−) vector between *Eco*RI and *Bam*HI sites.

2.2. Expression in E. coli and Purification of Recombinant BS-RNase (rBS-RNase). Given the intracellular toxicity of the enzyme, and its relatively high number of disulfides (four intrachain per each subunit, and two intersubunit), the organism harboring the heterologous gene survives by sequestering the protein in inclusion bodies as a denatured protein, with randomly reoxidized disulfides. Thus a native protein can be only recovered after denaturation and full reduction of the disulfides, followed by their reoxidation under controlled conditions. This can be accomplished (*Method 1,* below) by air reoxidation,[14] or (*Method 2*) by the use of a glutathione buffer,[15] which has been shown to selectively react with and protect the cysteine residues (Cys-31 and Cys-32) involved in the two intersubunit disulfides. This will generate an intermediate in the refolding of recombinant dimeric BS-RNase, i.e., a monomeric species with Cys-31 and Cys-32 protected through mixed disulfides with glutathione moieties. The latter procedure has thus the advantage to be of use also for the construction of hybrid dimers of seminal RNase, in which the two subunits have been separately expressed from distinct cDNAs and contain different structural features.

2.2.1. Refolding with glutathione/air disulfide oxidation. The expression plasmid containing the sequence coding for BS-RNase, prepared as described above, is used to transform BL21(DE3) pLysS cells.[17] Fresh recombinant colonies are always used to inoculate 1 liter of Terrific broth medium,[18] containing 100 μg/ml ampicillin. Cells are grown at 37° until OD equals 4.0 at 600 nm; then isopropylthiogalactoside (IPTG) is added to 0.4 m*M* final concentration and bacterial growth is continued for 3 h.

Cells are collected by centrifugation, suspended in 80 ml of 50 m*M* Tris-HCl pH 8.0, and disrupted by sonication. The insoluble fraction is recovered by centrifugation, suspended in 30 ml of 50 m*M* Tris-HCl pH 8.0, containing 4% Triton X-100 and 2 *M* urea, and extracted by sonication. This procedure is repeated one more time in the same buffer and three times in H_2O to obtain a clean preparation of inclusion bodies, freed of any membranes and cellular debris.

[18] J. Sambrook, E. F. Fritsch, and T. Maniatis, eds., "Molecular Cloning, A Laboratory Manual." Cold Spring Harbor Laboratory Press, Cold Spring Harbor, NY, 1989.

Inclusion bodies are dissolved in 40 ml of 0.1 *M* Tris-HCl, pH 8.4 containing 6 *M* guanidine hydrochloride, 2 m*M* sodium EDTA, and 25 m*M* dithiothreitol (DTT). After 3 h at 37°, the protein mixture is dialyzed against 20 m*M* acetic acid at 4°, and any insoluble material (which does not contain BS-RNase) is removed by centrifugation.

For renaturation, the protein solution is first diluted to 0.3 mg/ml in 0.1 *M* Tris-HCl pH 8.4 containing 0.5 *M* L-arginine, and then purged with N_2. Then oxidized and reduced glutathione are added to final concentrations of 0.6 and 3 m*M*, respectively. Maximal renaturation is achieved after 17–18 h at room temperature, as judged by the recovery of activity in RNase assays.[12] After any insoluble material is removed by centrifugation, the preparation is gel filtered on an FPLC apparatus (Amersham Pharmacia Biotech) equipped with a HiLoad Superdex 75 column (Amersham Pharmacia Biotech). More than 85% of the protein elutes as a monomer.

Dimerization is obtained through removal of the protecting glutathione moieties[15] from the half-cystine residues at positions 31 and 32, followed by air oxidation of the freed sulfhydryls. The first step is accomplished by selective reduction with DTT added at a 10-fold molar excess over the monomeric protein[19]; the second by dialysis for 24 h at room temperature against 0.1 *M* Tris-HCl, pH 8.4. Finally, unreacted monomers are removed by gel filtration as described above. The final yield of dimeric BS-RNase is 15–20 mg/liter of original cell culture. The procedure described above is summarized in Fig. 2.

2.2.2. Refolding with air disulfide oxidation. Escherichia coli strain BL21 (DE3) pLys S cells are transformed with plasmid pBluescript II SK(−) harboring the synthetic BS-RNase gene prepared according to Kim and Raines,[14] grown at 37° in Terrific broth[18] containing 100 μg/ml ampicillin until the absorbance at 600 nm reaches 4 units. IPTG is then added at 0.4 m*M* final concentration, and the cells grown overnight. Cells are collected by centrifugation and suspended in 20 m*M* sodium phosphate buffer at pH 7.0, containing 10 m*M* EDTA, and lysed by two passages through a French press. Inclusion bodies are collected by centrifugation at 12,000 *g* for 30 min, and washed with the same buffer. The pellet is then solubilized in 3 ml of 10 m*M* Tris-HCl pH 8.0 containing 6 *M* guanidine hydrochloride, 10 m*M* EDTA, and 100 m*M* DTT. After 3 h of incubation at room temperature, the solution is dialyzed against 20 m*M* acetic acid and centrifuged to remove any insoluble material.

Renaturation is obtained by dilution of the dialyzed solution in 0.1 *M* Tris–acetate buffer pH 8 at a protein concentration lower than 0.7 mg/ml. The renaturing solution is incubated in an open container at room temperature in the presence of 10 m*M* iodoacetamide. After 24 h the solution is concentrated to 10 ml using a Centriprep 10 concentrator and loaded onto a Sephadex G-75 superfine column (100 × 2.5 cm) equilibrated with 20 m*M* sodium phosphate buffer pH 7.

[19] G. D'Alessio, M. C. Malorni, and A. Parente, *Biochemistry* **14,** 1116 (1975).

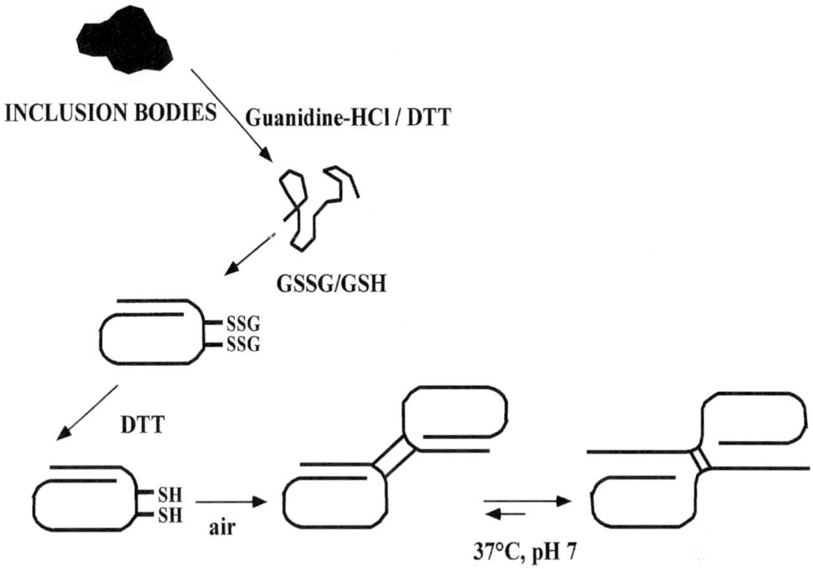

FIG. 2. Scheme for the preparation of recombinant dimeric RNases. Denatured, fully reduced RNase chains from *E. coli* inclusion bodies are refolded in the presence of a GSH/GSSG redox mixture to yield monomeric intermediates, in which cysteine residues 31 and 32 (involved in intersubunit disulfides) are blocked by glutathione moieties. The removal of glutathione by a mild reduction with DTT, followed by air reoxidation, yields a dimeric RNase. On incubation in the conditions as described in the text the equilibrium mixture of swapped and nonswapped RNase isoforms is obtained.

Dimeric rBS-RNase is collected in the peak eluting at the volume predicted for the elution of a dimer, but it may be contaminated by monomeric and aggregated species.

2.2.3. Characterization of recombinant protein. Electrospray mass measurements and amino acid sequence analyses indicate that the recombinant protein (Met-1)rBS-RNase as obtained retains the N-terminal methionine. However, the incorrect processing of the sequence does not affect the catalytic activity of (Met-1) rBS-RNase. When tested on RNA as a substrate, its specific activity is about 40 units/mg of protein by the Kunitz assay,[12] identical to that of the enzyme isolated from seminal plasma or seminal vesicles (see above). The removal of the N-terminal methionine is instead mandatory to obtain a biologically fully active protein, in particular endowed with a cytotoxic activity as high as that of the protein isolated from natural sources.[20]

The removal of the Met(−1) residue from recombinant BS-RNase is carried out by treating the protein at 10 μM final concentration with *Aeromonas*

[20] B. S. Adinolfi, V. Cafaro, G. D'Alessio, and A. Di Donato, *Biochem. Biophys. Res. Commun.* **213,** 525 (1995).

aminopeptidase (6 nM) in 100 mM Tris–acetate pH 7 containing 10 mM zinc sulfate. After 48 h at 37°, the protein is separated from the proteolytic enzyme and zinc ions by affinity chromatography on a uridine 2′,5′- and 3′,5′-diphosphate-agarose (Sigma, St. Louis, MO) column (2 ml). Electrospray mass measurements of the molecular weight and amino acid sequence analyses indicate that more than 96% of N-terminal methionine is removed with this procedure.

Recombinant BS-RNase obtained with this procedure is made up mainly of M = M dimers. An equilibrium BS-RNase (see above), containing two-thirds exchanging (M×M) and one-third nonexchanging (M = M) dimers, is obtained on incubation of the recombinant protein at 37° in 0.1 M Tris-HCl at pH 7.3 for 96 h. Alternatively, an identical result can be obtained by subjecting the protein solution (0.6 mg/ml in 0.1 M Tris–acetate, pH 7.0) to a much faster three-step thermal treatment. In the first step the protein is heated to 60° for 50 min; after 5 min at 60°, the temperature is lowered to 25° for 50 min.

After the thermal treatment, an analysis of rBS-RNase by ion exchange chromatography, as described below, reveals the characteristic $\alpha_2 : \alpha\beta : \beta_2$ isoenzymic pattern.

2.3. Expression of Recombinant BS-RNase in Eukaryotic Cells. Recombinant dimeric RNase, with all the structural and biological properties of natural BS-RNase as isolated from bovine seminal plasma or seminal vesicles, can be readily produced in eukaryotic expression systems. Yields, however, are low.

The eukaryotic system[21] described here is based on Chinese hamster ovary cells (CHO K1, from American Type Culture Collection, Manassas, VA) and on the pLEN expression vector[22] equipped with the inducible metallothionein promoter.

The recipient cells are cotransfected with the expression vector, containing the cDNA encoding the subunit chain of BS-RNase fused to its own secretion signal sequence, and with the pSV2NEO vector (Amersham Pharmacia Biotech) which induces resistance to the neomycin analog G418. After selection for resistance to G418, stable transformants are isolated, subcultured, and induced with 40 μM zinc sulfate in the presence of 0.5% fetal calf serum. The clone expressing the highest level of BS-RNase-like immunoreactivity (~0.4 mg/liter on the basis of an ELISA assay with anti-BS-RNase polyclonal antibodies) is then amplified in 150 mm plates and subjected to induction, performed under the same conditions reported above with replacement of the medium every 2–3 days.

Medium from induced cultures, concentrated 10-fold by ultrafiltration on an Amicon (Denvers, MA) concentrator (equipped with YM3 membranes), is dialyzed against 40 mM Tris-HCl pH 7.4, freed of any insoluble material by centrifugation, and loaded on a heparin-agarose column (1.4 × 4 cm) equilibrated

[21] N. Russo, M. de Nigris, A. Di Donato, and G. D'Alessio, *FEBS Lett.* **318,** 242 (1993).

[22] M. R. W. Ehlers, Y. P. Chen, and J. F. Riordan, *Protein Exp. Purif.* **2,** 1 (1991).

in the same buffer. The column is washed with the equilibrating buffer containing 50 m*M* NaCl and then eluted with 0.1 *M* Tris-HCl, pH 8, containing 0.4 *M* NaCl. The eluate is then diluted 1 : 2 with the same buffer (without NaCl) and loaded on a Mono S column equilibrated with the same buffer containing 0.2 *M* NaCl. After a 30 min wash with this buffer, the NaCl concentration is raised to 0.24 *M*, and after 35 min to 0.4 *M*. This procedure elutes a single peak of homogeneous rBS-RNase as tested by SDS–gel electrophoresis. The final yield of the protein, desalted and concentrated by centrifugation on a Centricon concentrator, is about 0.1 mg per liter of conditioned culture medium.

The secreted recombinant protein is undistinguishable from natural BS-RNase when analyzed by the following criteria: molecular mass; N-terminal amino acid sequence; isoenzymic $\alpha_2 : \alpha\beta : \beta_2$ pattern; specific activity on RNA as a substrate, and cytotoxic activity selective for malignant cells.[22]

3. *Preparation of Quaternary Isoforms of BS-RNase*

The isolation of homogeneous preparations of BS-RNase M = M and M×M quaternary forms (see Introduction) is easily obtained by the following protocol, illustrated in Fig. 3.

Step 1: Selective Reduction of Intersubunit Disulfides. BS-RNase (20 mg in 2 ml of 0.1 *M* Tris-HCl buffer, pH 8.4) is incubated for 20 min at room temperature under a nitrogen barrier in the presence of DTT in a 10-fold molar excess over the dimeric protein. To verify the completeness of disulfide cleavage, aliquots of the reacted protein: (i) can be tested with dithionitrobenzoate (DTNB)[23]; the test should result in four sulfhydryls per dimeric protein molecule; or (ii) can be treated for 20 min at room temperature in the dark with iodoacetamide, added at a twofold molar excess over the concentration of DTT. The selectively carboxymethylated protein, analyzed by SDS–gel electrophoresis in the absence of reducing agents, should migrate as a single, monomeric species. The expected ratio of 2 carboxymethylcysteine residues per monomer, verified by amino acid analyses, can confirm that the reductive cleavage has not affected intrachain disulfides.

Step 2: Separation of BS-RNase Reduction Products. Given the coexistence in BS-RNase at equilibrium of both M×M and M = M isoforms, the reductive cleavage of the intersubunit disulfides generates a monomeric species from the M = M form, whereas a noncovalent dimeric (NCD) species is produced from M×M, with the two subunits still held together through the exchange of their N-terminal segments. Monomers and dimers can be separated by gel-filtration on a Sephadex G-75 column (1.9 × 80 cm), equilibrated in 50 m*M* ammonium acetate buffer at pH 5, run at a flow rate of 6 ml/h. The fractions containing the

[23] G. L. Ellman, *Arch. Biochem. Biophys.* **82,** 70 (1959).

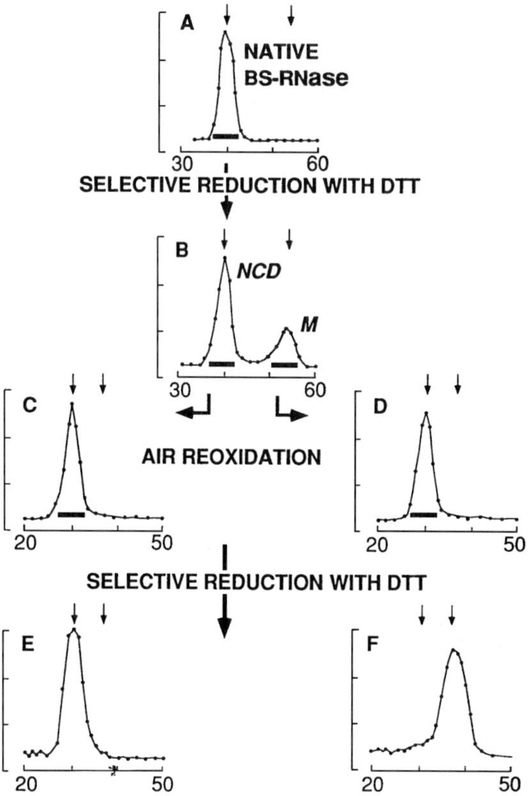

FIG. 3. Scheme for the preparation of the isolated M × M and M = M quaternary forms of BS-RNase. Gel-filtration pattern: (A) of BS-RNase in the dimeric, native state; (B) after selective reduction of the intersubunit disulfides with DTT (*NCD,* noncovalent dimers; *M,* monomers); (C and D) of *NCD* and *M,* respectively, after air reoxidation of the intersubunit disulfides; (E and F) of the protein species produced by cleavage of the intersubunit disulfides in the dimers produced in C and D, respectively. The arrows indicate the elution volumes of dimeric and of monomerized BS-RNase. For the separations illustrated in A and B a Sephadex G-75 column (220 ml) was used; for the other separations a Sephadex G-75 column (37 ml) was used. Ordinates: absorbancy at 280 nm (A and B) or at 230 nm (C–F); abscissas: fraction number. Reproduced from R. Piccoli, M. Tamburrini, G. Piccialli, A. Di Donato, A. Parente, and G. D'Alessio, *Proc. Natl. Acad. Sci. U.S.A.* **89,** 1870 (1992).

NCD noncovalent dimers, corresponding to about two-thirds of the total protein, and the monomeric (M) species, are separately pooled. The noncovalent nature of the dimer is confirmed by gel electrophoresis under denaturing conditions in SDS, in which NCD migrates as a monomer. Titration experiments with DTNB will give four sulfhydryls per dimeric molecule in NCD and 2 per monomer in M. The separated M and NCD fractions must be immediately processed.

Step 3: Regeneration of Covalent Dimers. Tris base (1 *M*, 1/10, vol/vol) is added to the NCD and M preparations to bring the pH to 8.4, and each preparation is transferred into a 10-cm cell culture dish, to increase the surface at contact with air, and incubated at 4° for 24–48 h with gentle shaking. During the reoxidation, aliquots are withdrawn and analyzed with DTNB and by SDS–gel electrophoresis (in the absence of reducing agents) to test the disappearance of free sulfhydryls and the production of covalently linked dimeric species.

The reoxidized products are analyzed as follows. Aliquots of each product are treated with a 10-fold molar excess of DTT as in step 1 and analyzed by gel filtration on a 37-ml Sephadex G-75 column or on the FPLC (fast protein liquid chromatography) apparatus equipped with a Superdex Hi-load 10/30 column (Amersham Pharmacia Biotech). The columns are equilibrated in 50 m*M* ammonium acetate buffer at pH 5 containing 0.3 *M* NaCl. Reoxidized NCD will consist of homogeneous M×M, and upon reduction with DTT will yield a single dimeric species of noncovalent dimers; reoxidized M, i.e., homogeneous M=M, will instead yield on reduction a single monomeric species.

Homogeneous preparations of the two dimeric forms M×M and M=M are stable at 4° for up to 3 months. They cannot be frozen. Each isoform can be converted into the other form up to the equilibrium ratio of the native protein (M×M/M=M = 2) when incubated at 37° in 0.1 *M* Tris–acetate buffer at pH 7. The kinetics of interconversion can be followed by assaying, at appropriate times, aliquots (30 μg) of the incubation mixtures on analytical FPLC column Superdex Hi-load 10/30 as described above. Interconversion is estimated from the percent of M and NCD species in the selectively reduced samples (see above). At each time interval aliquots of 5 μg of protein are analyzed by SDS–gel electrophoresis to verify the completeness of intersubunit disulfide reduction.

4. Preparation of Seminal RNase Isoenzymes

Bovine seminal ribonuclease is isolated as a mixture of isoenzymic subforms in which one or both subunits are deamidated at a single Asn residue, located at position 67 (see above). These isoenzymes can be separated by ion-exchange chromatography on carboxymethylcellulose,[5] or Mono S[10] cation exchangers.

A carboxymethyl-cellulose column (0.9 × 17 cm) is equilibrated in 0.1 *M* Tris-HCl pH 8.4, containing 0.1 *M* NaCl. The protein is eluted at a flow rate of 20 ml/h, at room temperature, with a linear gradient (100 + 100 ml) made up with 0.1 *M* Tris-HCl containing 0.1 *M* NaCl in the mixing chamber, and the same buffer containing 0.3 *M* NaCl in the reservoir. Up to 100 mg of BS-RNase can be resolved by this procedure in its three isoenzymic subforms.

For analytical purposes, or for separations of up to 5 mg of BS-RNase, a faster procedure can be used, at high pressure, using an FPLC apparatus[10] equipped with the Mono S (Amersham Pharmacia Biotech) cation exchanger. The column is equilibrated with 0.1 *M* Tris-HCl pH 8, containing 0.1 *M* NaCl. The same buffer is in elution chamber A; chamber B is filled with the same buffer containing 0.3 *M* NaCl (buffer B). Elution is started with a 1 : 1 mixture of the buffers A and B, at a flow rate of 1 ml/min. After 5 min, a two-step gradient elution is carried out, from 50 to 80% of buffer B in 4 min, followed by a raise of buffer B concentration from 80 to 100% in 6 min. The gradient elutes the three isoenzymic forms, with the same resolution observed in the low-pressure system.

The fully amidated β_2 isoenzyme can be converted *in vitro* into the hybrid $\alpha\beta$ isoenzyme in which only one subunit is deamidated at Asn-67, and eventually into the α_2 isoenzyme (with Asn-67 deamidated in both subunits). This occurs when the protein (10 mg/ml) is incubated in 1% ammonium bicarbonate at 37°. The progress of deamidation is estimated by analyzing aliquots of the protein by ion exchange on the Mono S column as described above. A complete transformation into homogeneous α_2 isoenzyme is obtained in about 72 h.

5. *Preparation of Monomeric Forms of Seminal RNase*

The high differential sensitivity of the intersubunit disulfides of dimeric BS-RNase to thiol reagents is the basis for the production of stable, catalytically active monomers of the protein.[19] The general procedure for obtaining these derivatives is performed in two steps: (1) selective reduction of the intersubunit disulfides; and (2) stabilization of the exposed sulfhydryls with electrophilic reagents. The first step is performed by treating BS-RNase (10 mg/ml) for 20 min at room temperature under a nitrogen barrier with a 10-fold molar excess of dithiothreitol in 0.1 *M* Tris-HCl, pH 8.4. The exposed sulfhydryls can be immediately reacted with an alkylating agent to produce stable monomers. Alternatively, they can be reacted with a thiosulfonate reagent; this will produce monomers linked through mixed disulfides to small moieties (see Fig. 4).

Carboxymethylation[19] or carboxamidomethylation[24] of the exposed sulfhydryls is carried out by adding a 20-fold molar excess (with respect to total -SH concentration) of freshly recrystallized iodoacetic acid or iodoacetamide, respectively. The alkylating reagent, dissolved in H_2O, is previously neutralized to pH 7.0 with concentrated NaOH. After 60 min at room temperature in the dark the protein is freed of excess reagents and by-products by gel-filtration through a Bio-Gel P2 or a Sephadex G-25 column (1.5 × 20 cm) equilibrated in 0.1 *M* ammonium acetate pH 5.

By this procedure however only about one-third of the protein is obtained as a monomer, since the prevalent M×M quaternary isoform in which subunits are still held together through the interchange of their N-terminal ends is recovered,

[24] A. Parente, D. Albanesi, A. M. Garzillo, and G. D'Alessio, *Ital. J. Biochem.* **26,** 451 (1977).

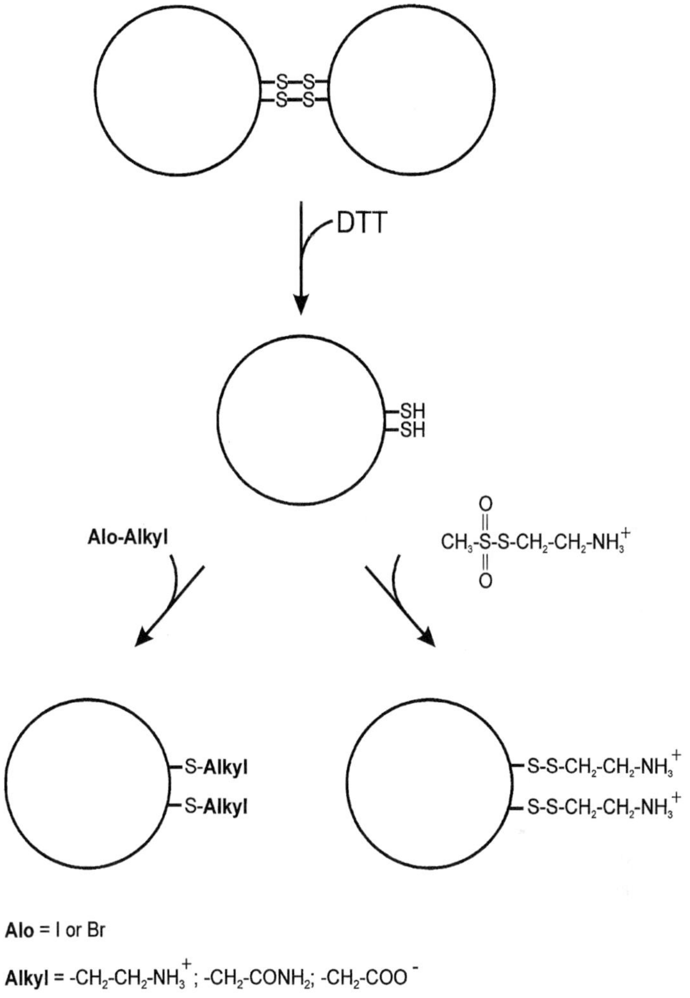

FIG. 4. Scheme for the preparation of monomeric derivatives of BS-RNase. On selective reduction with DTT of BS-RNase intersubunit disulfides, the exposed sulfhydryls were reacted with the indicated reagents.

after the cleavage of intersubunit disulfides, as a noncovalent dimer (see above). To obtain the whole protein in a monomeric form the gel-filtration step is performed with 5 *M* urea in the column buffer. This will dissociate the noncovalent dimers. On gel filtration on a Sephadex G-75 column (2.5 $\times$ 80 cm), equilibrated and eluted with 0.1 *M* ammonium acetate pH 5, more than 90% of the protein is recovered as a monomeric form of seminal RNase.

An aminoethyl derivative of BS-RNase is prepared by reaction with bromoethylamine of the protein selectively reduced as described above, after the pH is adjusted to 8.6 by addition of Tris base. The reaction is carried out with a 40-fold molar excess of bromoethylamine over free sulfhydryls for 16 h at 37°. After acidification with acetic acid, the protein is freed of excess reagents and by-products by gel-filtration on a Sephadex G-25 column as described above. Also the purification of the S-aminoethylated monomeric derivative of BS-RNase is carried out as described above for the S-alkylated derivatives.

For the preparation of a monomeric form of BS-RNase in which the Cys residues involved in the intersubunit disulfides are linked with disulfides with a small molecule, the dimeric protein is selectively reduced as described above with a 10-fold molar excess of DTT. Unreacted dithiothreitol is then removed by gel-filtration on a Sephadex G-25 column equilibrated in 50 m*M* ammonium acetate. The protein (diluted to 1–1.5 mg/ml in the same buffer) is cooled to 0° into an ice bath and reacted with a 2.5 molar excess over free sulfhydryls of *S*-(4-methoxyphenyl)-4-methoxybenzene thiosulfonate dissolved in 32 m*M* Tris base and 44% dioxane (final apparent pH 7.2). After 15 min at 0°, the protein is freed of excess and unreacted reagents by gel filtration on a Sephadex G-25 column equilibrated and eluted in 50 m*M* ammonium acetate pH 5. The protein is then lyophilized and subjected to a second gel filtration on the same column, but equilibrated in 50 m*M* ammonium acetate pH 5 containing 5 *M* urea. This will dissociate any noncovalent dimers present in the mixture (see above). The combined protein fractions eluted from the Sephadex G-25 column are then directly loaded onto a Sephadex G-75 column equilibrated in 100 m*M* ammonium acetate at pH 5. The final yield of the monomeric mixed disulfide derivative, $M(SSA)_2$, is about 40% of the original dimeric BS-RNase.

6. Tests for Selective Cytotoxicity of Seminal RNase

The cytotoxic activity of BS-RNase and its selectivity for malignant cells can be tested *in vitro* by the use of two well-characterized murine cell lines from the American Type Culture Collection. These are BALB/c 3T3 fibroblasts and a transformed cell line (SVT2 fibroblasts) derived from the 3T3 cells by infection with the SV40 virus. The 3T3 cells are not affected by BS-RNase, whereas the growth of SVT2 fibroblasts is strongly inhibited by BS-RNase.

SVT2 cells are cultured in Dulbecco's modified Eagle's medium (DMEM) supplemented with 10% fetal calf serum, 50 units/ml penicillin, and 50 μg/ml streptomycin, at 37°, under a 10% CO_2 in air atmosphere. At the beginning of a cytotoxicity assay, the fibroblasts are trypsinized, collected by centrifugation at 200*g* for 5 min, and resuspended in fresh medium at cell densities 1–2 $\times$ 10^5 cells/ml. Aliquots of the cell suspension (100 μl) are then seeded in a 96-well plate to give four groups of three wells each. Three groups are supplemented with sterile solutions of BS-RNase in phosphate-buffered saline (PBS), to reach protein concentrations of 10, 20, and 40 μg/ml, respectively. The fourth group of wells is teated with PBS. After 48 h of growth at 37°, cells are trypsinized and treated with Trypan blue, and the unstained cells are counted. A dose–response curve is then obtained by plotting the average number of viable cells—expressed as percent of the average cell number in the PBS-treated control—versus the concentration of BS-RNase. A typical dose–response curve should yield a value of IC_{50} (the protein concentration for a 50% effect on survival) of 20–30 μg/ml.

An identical, parallel experiment run with 3T3 cells will reveal that BS-RNase has no effect on these cells, because of its selective toxicity for malignant cells. In fact the finding of a growth inhibitory effect may well be considered as a sign of partial transformation of the 3T3 cell line, an event that can spontaneously occur on repeated passages of the cells in culture.

[16] Angiogenin

By JAMES F. RIORDAN

Introduction

Angiogenin (Ang) is one of the members of the pancreatic ribonuclease (RNase) superfamily that have unusual biological properties (for reviews see Refs. 1–3). It was discovered by searching for a so-called "tumor angiogenesis factor"—a product secreted by tumor cells that would elicit the growth of new blood vessels that, in turn, would allow the cells to proliferate and become a solid tumor. Angiogenin was first isolated from medium conditioned by human colon cancer cells (HT-29) grown in culture, and was purified based solely on its capacity to induce blood vessel formation in the rabbit cornea and on the chorioallantoic membrane (CAM) of the chicken embryo.[4] The assumption that it was a tumor-derived angiogenesis factor seemed to be confirmed by the fact that it is, indeed, secreted by a large variety of tumor cells. However, it was soon discovered that it is also present in normal human blood plasma[5] and expressed in a wide variety of human cells.[6]

[1] J. F. Riordan, *in* "Ribonucleases: Structures and Functions" (G. D'Alessio and J. F. Riordan, eds.), p. 45. Academic Press, New York, 1997.

[2] B. L. Vallee and J. F. Riordan, *Cell. Mol. Life Sci.* **53,** 803 (1997).

[3] S. A. Adams and V. Subramanian, *Angiogenesis* **3,** 189 (1999).

[4] J. W. Fett, D. J. Strydom, R. R. Lobb, E. M. Alderman, J. L. Bethune, J. F. Riordan, and B. L. Vallee, *Biochemistry* **24,** 5480 (1985).

[5] R. Shapiro, D. J. Strydom, K. A. Olson, and B. L. Vallee, *Biochemistry* **26,** 5141 (1987).

[6] M. Moenner, M. Gusse, E. Hatzi, and J. Badet, *Eur. J. Biochem.* **226,** 483 (1994).

0076-6879/01 $35.00

This led to the suggestion that Ang has other biological functions. Indeed, one such function may be to inhibit the degranulation of polymorphonuclear leukocytes.[7] Angiogenins have also been found to be present in other species including cow, pig, rabbit, and mouse.[8,9]

The amino acid sequence of angiogenin revealed that it was homologous to RNase A.[10] It shares 33% sequence identity with the bovine enzyme, including counterparts to the two histidines and one lysine that constitute the catalytic active site. It was therefore surprising when initial studies using conventional RNase A assays showed it to be devoid of detectable ribonucleolytic activity. Later, however, it was found to degrade ribosomal RNA, albeit weakly, and to abolish cell-free protein synthesis.[11,12] Both its angiogenic and ribonucleolytic activities are blocked by the 50 kDa RNase inhibitor that has broad specificity for the pancreatic RNase superfamily and occurs in most mammalian tissues.[13] A more quantitative enzymatic assay was developed based on the hydrolysis of tRNA,[14] but since this activity was extremely low compared to that of RNase A, elaborate purification procedures were required to ensure the absence of contamination. Creation of active site variants of Ang (H13A, K40N, or H114A) eventually established that its weak enzymatic activity was necessary for its angiogenic activity.[15]

The most significant sequence difference between Ang and all other known mammalian RNases is the lack of a short disulfide loop that in RNase A encompasses residues 65–72.[16] In RNase, this region constitutes part of the (B_2) substrate binding site, whereas in Ang it is thought to be involved in binding to cell surface receptors critical to the angiogenic response. Accordingly, an Ang variant in which residues 59–73 of RNase A were substituted for residues 58–70 of Ang had 300- to 600-fold increased enzymatic activity (a better substrate binding site?) but essentially no angiogenic activity (no cell surface binding?).[17]

The three-dimensional structure of Ang, determined both by X-ray crystallography and NMR spectroscopy,[18,19] confirmed a close similarity to RNase A,

[7] H. Tschesche, C. Kopp, W. H. Horl, and U. Hempleman, *J. Biol. Chem.* **269,** 30274 (1994).

[8] M. D. Bond and B. L. Vallee, *Biochemistry* **27,** 6282 (1998).

[9] M. D. Bond, D. J. Strydom, and B. L. Vallee, *Biochim. Biophys. Acta* **1162,** 177 (1993).

[10] D. J. Strydom, J. W. Fett, R. R. Lobb, E. M. Alderman, J. L. Bethune, J. F. Riordan, and B. L. Vallee, *Biochemistry* **24,** 5486 (1985).

[11] R. Shapiro, J. F. Riordan, and B. L. Vallee, *Biochemistry* **27,** 3527 (1986).

[12] D. K. St. Clair, S. M. Rybak, J. F. Riordan, and B. L. Vallee, *Biochemistry* **27,** 7263 (1988).

[13] R. Shapiro and B. L. Vallee, *Proc. Natl. Acad. Sci. U.S.A.* **84,** 2238 (1987).

[14] R. Shapiro, S. Weremowicz, J. F. Riordan, and B. L. Vallee, *Proc. Natl. Acad. Sci. U.S.A.* **84,** 8783 (1987).

[15] R. Shapiro and B. L. Vallee, *Biochemistry* **28,** 7401 (1989).

[16] J. J. Beintema, C. Schuller, M. Irie, and A. Carsana, *Prog. Biophys. Mol. Biol.* **51,** 165 (1988).

[17] J. W. Harper and B. L. Vallee, *Biochemistry* **28,** 1875 (1989).

[18] K. R. Acharya, R. Shapiro, J. F. Riordan, and B. L. Vallee, *Proc. Natl. Acad. Sci. U.S.A.* **91,** 2915 (1994).

[19] O. Lequin, H. Thuring, M. Robin, and J.-Y. Lallemand, *Eur. J. Biochem.* **250,** 712 (1997).

including the positions of the active site residues, and revealed that the region of the missing disulfide loop has undergone reorganization. The most startling difference was in the pyrimidine-binding (B_1) site. In RNase A this is an open pocket, but in Ang the "pocket" does not exist: it is occupied by the side chain of Gln-117. These changes in both the B_1 and the B_2 sites likely contribute to the markedly attenuated ribonucleolytic activity of Ang. Nevertheless, it does have finite activity. It cleaves preferentially on the 3′ side of pyrimidines to generate a cyclic phosphate intermediate and, thus, has much the same substrate specificity as RNase A. Hence, substrates must still bind to its B_1 site. Clearly, this would require Ang to undergo a change in conformation, but how this occurs is not yet known.

Methods of Assay

Angiogenic Activity

The CAM assay is the preferred means to assess the angiogenic activity of Ang, and several different alternatives have been described. Some laboratories studying angiogenesis employ a method in which the developing chicken embryo is removed from the shell and transferred to a petri dish. The author's laboratory uses intact eggs. Access to the CAM is achieved by cutting a window into the shell. A sample can be applied to the CAM, the membrane that surrounds the embryo, either directly as a solid (for crude samples), implanted in a slow-release polymer pellet, or dried onto a plastic coverslip. The time of application, postfertilization, is rather critical, as is the time at which blood vessel growth is recorded. Some laboratories attempt to score the angiogenic response on some arbitrary scale, e.g., 0–4, but ours reads the results as positive or negative. Our protocol, modified from that of Knighton *et al.*,[20] is as follows:

1. On day 1, fertilized chicken eggs (Spafas, Inc., Norwich, CT) maintained at 4° or 18° by the supplier, are transferred on arrival to a humidified egg incubator (Leahy Manufacturing Co., Higginsville, MO) at 37° and kept for 3 days. The eggs can be kept at 18° for 2–3 days, if necessary, prior to starting the assay.
2. On day 4, a hole is drilled in the narrow end of the egg and 1–2 ml albumen (egg white) is aspirated with a syringe.
3. On day 5, a 1–2 cm "window" is cut through the shell and covered with clear tape. The eggs are incubated window-side up for another 6 days at 37° before samples are applied.
4. On day 11, 5 μl samples containing 1–10 ng of Ang in water are applied to sterile Thermanox 15-mm disks (Flow Laboratories, Inc., Rockville, MD) and

[20] D. Knighton, D. Ausprunk, D. Tapper, and J. Folkman, *Br. J. Cancer* **35,** 347 (1977).

dried under laminar flow conditions. The loaded disks are placed on the CAM of the windowed eggs, sample side down.

5. On day 14, 68 ± 4 hr after applying the sample, the growth of blood vessels is observed microscopically through the window and recorded as either positive or negative. A positive response has a characteristic "spokewheel" appearance, readily recognized by an experienced analyst. At least 3 sets of multiple eggs (10–25 per set) are required for calculation of statistical significance (χ^2), based on comparison with water controls tested simultaneously. A $p < 0.05$ indicates a positive sample.

A more complicated, but also more dramatic, angiogenesis assay involves implantation of the sample, incorporated into a slow-release pellet, into a pocket created in the normally avascular cornea of the rabbit eye at a fixed distance from the peripheral blood vessels.[21] Capillary growth extending through the cornea toward the sample can be seen after several days. Despite the stark evidence of angiogenesis, this method has serious disadvantages. It is difficult to quantify, requires an investigator skilled in microsurgery with an ability to insert samples reproducibly at precisely the right distance from the corneoscleral junction, is not amenable to screening large numbers of samples, is highly variable despite technical expertise, and is very expensive.

A more popular method, often referred to as *in vitro* angiogenesis, examines the formation of capillary-like tubes in a matrigel, collagen, or laminin matrix on addition of an angiogenic factor to endothelial cells in culture.[22] This assay is not a strict measure of angiogenesis, however, as it is not based on the induction of new capillaries from preexisting blood vessels. Moreover, the full complement of cell types involved in angiogenesis is not part of the assay, and results depend on the type of endothelial cell employed. The assay does provide some information and is certainly more convenient than the CAM assay.

Angiogenesis has come under intense scrutiny in recent years, yet it is remarkable that none of the many assays in use has been accepted as the standard of reference. We have found that repetitive CAM assays, over a range of concentrations of Ang or other angiogenic substance, are critical for interpretation of significance and seem to give reliable results. However, it is not uncommon to find conclusions in the literature based on results with just a few eggs. The fact that all angiogenesis assays are extremely sensitive to artifacts may account for the lack of general agreement. It should be noted, nevertheless, that there has been remarkable progress in the field, often based on proxy assays such as mitogenesis or cell proliferation.

[21] R. Langer and J. Folkman, *Nature (London)* **263,** 797 (1976).

[22] S. Jimi, K. Ito, K. Kohno, M. Ono, M. Kuwano, Y. Itagaki, and H. Ishikawa, *Biochem. Biophys. Res. Commun.* **211,** 476 (1995).

Ribonucleolytic Activity

The most convenient assay for the routine measurement of the ribonucleolytic activity of Ang is based on the hydrolysis of tRNA.[14] All solutions should be prepared in disposable, sterile plastic tubes (Ang adheres tightly to glass) or in plasticware treated with 0.1 *N* NaOH to remove traces of RNase. Reagents should be prepared in Sep-Pak filtered water. The HEPES buffer can also be filtered through a Sep-Pak C_{18} cartridge (Waters, Milford, MA).

1. To each of a series of 1.5-ml Eppendorf tubes, add in this order: 110 − *x* μl Sep-Pak filtered water, 100 μl assay buffer (0.1 *M* HEPES–NaOH, 0.1 *M* NaCl, pH 7.0) (do not mix), 30 μl 1 mg/ml BSA in water (RNase-free, Worthington, Freehold, NJ) (mix by repeated pipetting) and *x* μl of sample to be assayed (mix several times).
2. At time 0, add 60 μl of tRNA solution (10 mg/ml in water, Type X, Sigma, St. Louis, MO) to tube 1, mix extensively by pipetting, and place in a 37° water bath.
3. At convenient (20 or 30 sec) intervals add 60 μl of tRNA solution to the remaining tubes, mixing extensively after each addition, and place tubes in 37° water bath.
4. Terminate the reactions after an appropriate incubation time by adding 700 μl of ice-cold 3.4% perchloric acid at the same intervals used for the tRNA additions. Typical Ang standards are 5, 10, 20, 30, 40, and 50 μl in water and the incubation time is 2 hr. It is necessary to use a range of concentrations because the standard curve is not linear.
5. Vortex each tube thoroughly and place in ice or ice–water for 10 min.
6. Centrifuge each tube at 4° for 10 min and put tubes in ice.
7. Determine the absorbance of each supernatant at 260 nm (zero the instrument with water). The blank is the reaction obtained with no sample (in duplicate).
8. To obtain activities of test samples, first plot a standard curve of absorbancies of Ang standards (minus the blank) vs Ang concentrations. Calculate from the Ang concentration required to produce the same absorbance as the sample.

A sensitive HPLC method has been developed to measure kinetic parameters with dinucleoside phosphates as well as to demonstrate that Ang catalyzes the hydrolysis of both cytidine and uridine cyclic phosphates. Quantitative studies demonstrated that Ang favors C over U by about 10-fold in the N site of NpN′ substrates and A over G by about 3-fold in the N′ site.[23]

[23] T. P. Curran, R. Shapiro, and J. F. Riordan, *Biochemistry* **32,** 2307 (1993).

Isolation of Angiogenin

Angiogenin was originally isolated from serum-free culture medium conditioned by HT-29 human colon carcinoma cells maintained in cell factories.[4] After acidification, clarification, and buffer exchange of the medium, it was applied to a CM-cellulose column and eluted with high salt. Subsequent purification by HPLC resulted in a yield of only 0.5 μg of pure protein per liter of medium. Although this was not a very efficient source for purification, it provided enough material for initial characterization as almost all of it was consumed in the many CAM assays needed to establish angiogenic activity.

Recombinant human Ang is now routinely prepared with an expression system that adds the *Escherichia coli phoA* signal sequence to the N terminus of Ang. This generates a soluble protein in the native pyroGlu form in yields of several milligrams per liter. A detailed protocol for this procedure has been reported[24] and will not be described here.

A very convenient source of Ang turns out to be conventional cow's milk obtained, e.g., from a supermarket. Bovine Ang shares many of the same chemical and biological properties as the human protein, and its isolation is more or less applicable to Ang's from any source. It was isolated originally by Denefle *et al.*[25] and more recently, as a heparin-binding growth factor purified from bovine colostral fat-globule membranes.[26] The following procedure is based on one developed independently in this laboratory.[8] It should be noted that the product may contain traces of RNase A.

Materials

Cow's milk (~15 liters) containing 1% fat or less. 5× CPD: 131.5 g sodium citrate dihydrate, 16.35 g citric acid monohydrate, 11.1 g monobasic sodium phosphate monohydrate, 116.2 g anhydrous dextrose in 1 liter water. This is the standard citrate–phosphate–dextrose anticoagulant used to make blood plasma. CM 52 loading buffer: 50 m*M* dibasic sodium phosphate heptahydrate titrated with HCl to pH 6.6. CM 52 elution buffer: loading buffer plus 1 *M* NaCl. CM 52 washing buffer: loading buffer plus 3 *M* NaCl.

Methods

All of the following steps prior to Mono S chromatography are performed at 4°.

[24] J. F. Riordan and R. Shapiro, *in* "Methods in Molecular Biology: Nuclease Methods and Protocols" (C. H. Schein, ed.) Vol. 160, p. 375. Humana Press, Totowa, NJ, 2001.

[25] P. Denefle, S. Kovarik, J.-D. Guitton, T. Cartwright, and J.-F. Mayaux, *Gene* **56,** 61 (1987).

[26] T. Hironaka, H. Ohishi, T. Araki, and T. Masaki, *Milchwissenschaft* **52,** 508 (1997).

1. Before starting the purification, prepare 600 ml of CM 52 cellulose. Suspend the resin in 2 liters loading buffer, allow to settle, and pour off fines. Resuspend in 600 ml loading buffer and transfer to a 2-liter sintered glass funnel attached to a 4-liter filter flask. Wash under house- or water-vacuum with 2 liters washing buffer and equilibrate with loading buffer.

2. To ~15 liters milk add 28 ml/liter of 5× CPD. Check the pH to be sure that it is no higher than that of the loading buffer.

3. Apply the milk mixture semibatchwise to the resin at a flow rate of ~3 liter/hr.

4. After all the milk has been loaded, wash the resin with 5–6 liters loading buffer, with occasional resuspension, and then transfer it to a 5 × 50 cm chromatography column. Allow the column to pack under flow for about 1 hr and then let it sit overnight.

5. The next morning, replace the top buffer (which is cloudy) with fresh buffer and start the flow again (~5 ml/min). When the A_{280} of the effluent is <0.05 (1–2 column volumes), replace the buffer with eluting buffer. After 200 ml has passed through the column, begin collecting 8-ml fractions in glass test tubes. Pool fractions with an A_{280} of at least 0.08.

6. Concentrate the pooled fractions by ultrafiltration in an Amicon 400-ml device. When the volume has been reduced to <10 ml, add 5 volumes of water, reconcentrate to <10 ml, add another 5 volumes of water, and concentrate to ~10 ml. Rinse the concentrator twice with 1–2 ml water, and add the washings to the concentrate. Centrifuge in Eppendorf tubes for 15 min at 4° immediately before loading onto a Mono S column.

7. Equilibrate the Mono S column (0.5 × 5 cm) with 15% buffer B (Buffer A: 10 m*M* Tris-HCl, pH 8.0; Buffer B: 10 m*M* Tris-HCl, 1 *M* NaCl, pH 8.0). Inject ~2 μg of an Ang standard in water (50–100 μl) and run a 60 min linear gradient from 10 to 45% Buffer B at 0.8 ml/min. Monitor absorbance at 214 nm, 0.05 AU full-scale. Angiogenin should elute at ~36 min.

8. Run a pilot sample of the preparation (~0.5% of the total) and monitor with 1.0 AU full-scale. Estimate the yield from the pilot run with that of the standard by comparing peak areas. Make preparative runs with no more than ~4 mg per run.

9. Equilibrate the C_{18} column (4.6 × 250 mm) with 30% buffer B [Buffer A: 0.1% trifluoroacetic acid (TFA) in water; Buffer B: 0.08% TFA in 2-propanol : acetonitrile : water, 3 : 2 : 2].

10. Run an Ang standard as for the Mono S column, but with a 25-min gradient from 30 to 50% B followed by 5 min at 50% B. Angiogenin elutes at ~ 24 min.

11. Run an aliquot of the preparation as above with the same gradient as for the standard. Run up to 4 mg of the preparation at a time. Collect the Ang fractions, dialyze vs Milli-Q water, lyophilize and reconstitute with Sep-Pak water. Aliquot and store at −70° (stable for >5 years).

Biological Function

Wound Healing

Angiogenin is a potent angiogenic molecule: as little as 0.5 ng can induce blood vessel formation on the CAM. Paradoxically, it circulates in normal human plasma at a concentration of 250–500 ng/ml[27–29] and yet fails to cause unrestrained capillary proliferation. Angiogenin isolated from plasma is fully active on the CAM, and there is no indication that it exists as an inactive complex in blood. A reasonable explanation for its apparent lack of activity is that the endothelial cells that constitute the blood vessels are not sensitive to Ang. These cells are confluent and for the most part have an extremely long half-life. In culture, confluent endothelial cells are unresponsive to exogenous Ang.[30] They neither bind it nor are stimulated by it. Actually, Ang does bind to an endothelial cell-surface protein that has been identified as a type of actin, but the Ang/actin complex dissociates from the plasma membrane.[31] In contrast, sparse cultures of endothelial cells specifically bind Ang, which induces DNA synthesis and cell proliferation. This suggests that the putative Ang receptor is only expressed on nonconfluent cells. Hence, one role of plasma Ang might be to promote wound healing when it becomes extravascular, e.g., as a result of injury.[32,33] Any loss of vascular integrity would disrupt endothelial cell confluence, and the resulting local high concentration of Ang could facilitate rapid blood vessel growth and tissue repair. In this regard, Ang has been shown to be a class 2 acute phase protein whose synthesis is induced by interleukin 6 (IL-6) in response to inflammation.[29]

The neovascularization of the healing process cannot be allowed to proceed indefinitely. Hence, a second level of regulation of Ang-induced angiogenesis would seem to be required. Hu has proposed that the fibrin clot that accompanies wound repair might serve as the basis for this control.[32] The clot must be degraded before angiogenesis can occur. Lysis is initiated by plasmin in conjunction with the neutrophil elastase that is released from the recruited polymorphonuclear leukocytes that adhere to polymerized fibrin. Plasmin is generated from plasminogen by tissue plasminogen activator, a reaction that is markedly stimulated by the Ang/actin complex.[34] An active fragment of plasmin, called miniplasmin, is also generated

[27] S. Shimoyama, F. Gansauge, S. Gansauge, G. Negri, T. Oohara, and H. G. Berger, *Cancer Res.* **56,** 2703 (1996).

[28] J. Blaser, S. Triebl, C. Kopp, and H. Tschesche, *Eur. J. Clin. Chem. Clin. Biochem.* **31,** 513 (1993).

[29] S. J. Verselis, K. A. Olson, and J. W. Fett, *Biochem. Biophys. Res. Commun.* **259,** 178 (1999).

[30] G.-F. Hu, J. F. Riordan, and B. L. Vallee, *Proc. Natl. Acad. Sci. U.S.A.* **94,** 2204 (1997).

[31] G.-F. Hu, D. J. Strydom, J. W. Fett, J. F. Riordan, and B. L. Vallee, *Proc. Natl. Acad. Sci. U.S.A.* **90,** 1217 (1993).

[32] G.-F. Hu, *J. Protein Chem.* **16,** 669 (1997).

[33] E. Hatzi and J. Badet, *Eur. J. Biochem.* **260,** 825 (1999).

[34] G.-F. Hu and J. F. Riordan, *Biochem. Biophys. Res. Commun.* **197,** 682 (1993).

from plasminogen by neutrophil elastase, which itself can degrade fibrin. Once these enzymes degrade the clot, endothelial cells can be stimulated to migrate, proliferate, and form tubules that become blood vessels.

Among the many levels of control that function during this process, one involves the plasminogen-stimulated, proteolytic inactivation of Ang by elastase.[32] Cleavage of a single bond between Ile-29 and Met-30 abolishes angiogenic activity but not ribonucleolytic activity. Another control could be the formation of angiostatin, a peptide fragment of plasminogen generated by elastase that is a potent inhibitor of endothelial cell growth.[35] One countermeasure that may also operate is the inhibition of the spontaneous and/or stimulated degranulation of polymorphonuclear leukocytes by Ang, which would limit the supply of elastase.[7] Many other factors are involved in the restoration of normalcy and, clearly, the fine-tuning of the system remains to be established. Nevertheless, the functional and regulatory properties of Ang indicate that it could be a critical component of the overall wound healing mechanism.

Mechanism of Action

Angiogenin is a ribonuclease, and its enzymatic activity has been shown to be essential for angiogenesis. It is also an extracellular protein and since it induces blood vessel formation, its principal target has been assumed to be endothelial cells. Cross-linking experiments have shown that Ang binds to a molecule on the surface of these cells, but to date the identity of this receptor has not been established. On the other hand, limited proteolysis experiments have localized a presumptive cell-binding region of Ang to residues 60–68.[36] Once Ang binds to the plasma membrane it undergoes receptor-mediated endocytosis and is rapidly translocated to the nucleus, where it accumulates in the nucleolus.[37,38] A nuclear localization signal has been found that encompasses the basic triad Arg-31/Arg-32/Arg-33.[39] Very little is known about the details of the process by which Ang is endocytosed, transported to the nuclear membrane, released into the nucleus, localized to the nucleolus, and bound to its ultimate target molecule(s). Translation is independent of lysosomes and microtubules[40] and may therefore follow a unique pathway. Binding to the cell surface also stimulates the activities of phospholipases C and A_2 and initiates a cascade of second messengers.[41,42]

[35] M. S. O'Reilly, L. Holmgren, Y. Shing, C. Chen, R. A. Rosenthal, M. Moses, W. S. Lane, Y. Cao, E. H. Sage, and J. Folkman, *Cell* **79,** 315 (1994).

[36] T. W. Hallahan, R. Shapiro, and B. L. Vallee, *Proc. Natl. Acad. Sci. U.S.A.* **88,** 2222 (1991).

[37] J. Moroianu and J. F. Riordan, *Proc. Natl. Acad. Sci. U.S.A.* **91,** 1677 (1994).

[38] J. Moroianu and J. F. Riordan, *Biochemistry* **33,** 12535 (1994).

[39] J. Moroianu and J. F. Riordan, *Biochem. Biophys. Res. Commun.* **203,** 1765 (1994).

[40] R. Li, J. F. Riordan, and G.-F. Hu, *Biochem. Biophys. Res. Commun.* **238,** 305 (1997).

[41] R. Bicknell and B. L. Vallee, *Proc. Natl. Acad. Sci. U.S.A.* **85,** 5961 (1988).

[42] R. Bicknell and B. L Vallee, *Proc. Natl. Acad. Sci. U.S.A.* **86,** 1573 (1989).

Again, the molecular details of the cascade are not known, including how the actions of the end products converge with those of nuclear Ang. In any event, the final result is to induce cell proliferation.

It seems apparent that Ang exerts its ribonucleolytic activity once it is inside the nucleus. A 45-mer oligodeoxynucleotide obtained by selective exponential enrichment,[43] abolished both the angiogenic and ribonucleolytic activities of Ang but did not interfere with its nuclear translocation. Importantly, the inhibitor cotranslocates to the nucleus with Ang in $\sim 1:1$ stoichiometry, indicating that the intranuclear ribonucleolytic activity of Ang is essential for angiogenesis.

Nuclear Ang is firmly bound to its target. No appreciable amount is released from isolated nuclei on incubation in PBS for several hours in the absence or presence of RNase A.[44] Significant release only occurs in the presence of DNase I, an indication that Ang binds to DNA. Whether or not it acts as a regulator of gene transcription has yet to be established. The fact that Ang localizes to the nucleolus, the site of ribosomal RNA transcription, would suggest that it plays a role in ribosome biosynthesis, either as a transcription activator or in rRNA processing, or both.

Ang also interacts with smooth muscle cells, another cell type associated with blood vessels. It binds to a cell surface receptor and induces rapid phosphorylation of SAPK/JNK.[45] It undergoes nuclear translocation and stimulates cell proliferation. There are some differences in the response of smooth muscle versus endothelial cells that likely reflect the sequence of events leading to the formation of a mature vessel. Exploring these differences will be the object of future investigations.

In addition to its intranuclear functions, Ang exhibits several other properties associated with angiogenesis. It activates cell-associated proteases,[34] stimulates cell migration and invasion of the extracellular matrix,[46] supports cell adhesion,[47] and promotes tube formation by cultured endothelial cells.[22] All of these individual cellular events are necessary for the formation of new blood vessels.

Inhibition and Suppression of Tumor Growth

Olson *et al.*[48] have shown that the growth of human colon adenocarcinoma (HT-29) cells in athymic mice is dependent on their production of Ang. A specific neutralizing monoclonal antibody to human Ang, in microgram doses, significantly delays or prevents the appearance of HT-29 tumors. Up to 65% of treated animals

[43] V. Nobile, N. Russo, G.-F. Hu, and J. F. Riordan, *Biochemistry* **37,** 6857 (1998).
[44] G.-F. Hu, C.-J. Xu, and J. F. Riordan, *J. Cell. Biochem.* **76,** 452 (2000).
[45] Z.-P. Xu, D. Monti, and G.-F. Hu, submitted.
[46] G.-F. Hu, J. F. Riordan, and B. L. Vallee, *Proc. Natl. Acad. Sci. U.S.A.* **91,** 12096 (1994).
[47] F. Soncin, D. J. Strydom, and R. Shapiro, *J. Biol. Chem.* **272,** 9818 (1997).
[48] K. A. Olson, T. C. French, B. L. Vallee, and J. W. Fett, *Cancer Res.* **54,** 4576 (1994).

remain tumor-free for at least 30 days. Similar results were obtained using actin, the Ang binding protein, in place of the monoclonal antibody.[49] Neither of the therapeutic agents had any cytotoxic effect on the tumor cells. Hence, it was concluded that the suppression of tumor growth was due to inhibition of Ang-induced neovascularization. Subsequent studies extended these findings to several other tumor cell types,[50] and combining the therapeutic agent with conventional cytotoxic therapy (cisplatin and suramin) was more effective than either treatment alone.[49]

An antisense phosphorothioate oligodeoxynucleotide directed at the AUG translational start site of the Ang mRNA completely inhibited the growth of human prostate tumor cells implanted subcutaneously in athymic mice. It also had a significant effect on the development of regional metastases in an orthotopic mouse model.[51] All of these studies confirm the importance of Ang in the growth and maintenance of primary and metastatic tumors.

In summary, while great strides have been made in elucidating the structure, function, and full range of biological potential of Ang in the 15 years since its discovery, many mechanistic details remain unknown. Recently emerging data suggest that it plays a key role in wound healing as well as in the growth of solid tumors. But it may well serve an even more fundamental role in promoting ribosome biosynthesis as a prelude to cell proliferation. The coming years should witness a greater appreciation of why this ribonuclease was coopted into being so much more than a tool to digest RNA.

[49] K. A. Olson, J. W. Fett, T. C. French, M. E. Key, and B. L. Vallee, *Proc. Natl. Acad. Sci. U.S.A.* **92,** 442 (1995).

[50] K. A. Olson and J. W. Fett, *Proc. Amer. Assoc. Cancer Res.* **37,** 57 (1996).

[51] K. A. Olson and J. W. Fett, *Proc. Amer. Assoc. Cancer Res.* **39,** 98 (1998).

[17] Eosinophil-Derived Neurotoxin

By HELENE F. ROSENBERG and JOSEPH B. DOMACHOWSKE

1. Background

Eosinophil-derived neurotoxin (EDN, RNase 2) is a major component of human eosinophilic leukocytes that is also found in human liver, lung, and spleen. The name EDN is a historical one, given by Durack and colleagues[1] to the 18.4 kDa

[1] D. T. Durack, S. J. Ackerman, D. A. Loegering, and G. J. Gleich, *Proc. Natl. Acad. Sci. U.S.A.* **78,** 1443 (1981).

0076-6879/01 $35.00

protein purified from eosinophils that appeared to promote the eosinophil-mediated syndrome of cerebellar dysfunction known as the "Gordon phenomenon."[2] Gleich and colleagues[3] were first to report the amino-terminal sequence of EDN and to note the similarity of this sequence to that of human pancreatic RNase. Molecular cloning confirmed EDN as a member of the emerging RNase A gene superfamily,[4,5] a relationship that has been emphasized by the recently reported crystallographic structure.[6] Native, eosinophil-derived EDN was shown to be an enzymatically active ribonuclease,[7] capable of degrading single-stranded RNA substrates at high efficiency.[8] Human liver ribonuclease[9] and RNase Us[10] are ribonucleases that are identical to EDN. EDN's ribonuclease activity has been exploited for the construction of several specific fusion immunotoxins,[11–14] a subject that is covered in more detail in other sections of this volume.

Among the most interesting features of EDN is its unusual pattern of evolution. Primate EDNs are among the most rapidly evolving functional coding sequences known with nonsynonymous (nonsilent) mutations accumulated at a rate of 1.9×10^{-3} per site per million years.[15] Rodent EDNs have evolved into species-specific clusters in a pattern reminiscent of that seen in major histocompatibility complex (MHC), immunoglobulin (Ig) and T cell receptor gene families.[16] Taken together, these results suggested to us that the physiologic function of EDN must take into account both ribonuclease activity and the propensity for generating diversity. With this in mind, we have initiated a series of studies designed to link eosinophils, EDN, and host defense against single-stranded RNA viral pathogens, specifically those infecting the respiratory tract.[17] We have characterized the eosinophilic

[2] M. H. Gordon, *Br. Med. J.* **1,** 641 (1933).

[3] G. J. Gleich, D. A. Loegering, M. P. Bell, J. L. Checkel, S. J. Ackerman, and D. J. McKean, *Proc. Natl. Acad. Sci. U.S.A.* **83,** 3146 (1986).

[4] H. F. Rosenberg, D. G. Tenen, and S. J. Ackerman, *Proc. Natl. Acad. Sci. U.S.A.* **86,** 4460 (1989).

[5] K. J. Hamann, R. L. Barker, D. A. Loegering, L. R. Pease, and G. J. Gleich, *Gene* **83,** 161 (1989).

[6] S. C. Mosimann, D. L. Newton, R. J. Youle, and M. N. G. James, *J. Mol. Biol.* **260,** 540 (1996).

[7] N. R. Slifman, D. A. Loegering, D. J. McKean, and G. J. Gleich, *J. Immunol.* **137,** 2913 (1986).

[8] S. Sorrentino and M. Libonati, *Arch. Biochem. Biophys.* **312,** 340 (1994).

[9] S. Sorrentino, D. G. Glitz, K. J. Hamann, D. A. Loegering, J. L. Checkel, and G. J. Gleich, *J. Biol. Chem.* **267,** 14859 (1992).

[10] J. J. Beintema, J. Hofsteenge, M. Iwama, T. Morita, K. Ohgi, M. Irie, R. H. Sugiyama, G. L. Schieven, C. A. Dekker, and D. G. Glitz, *Biochemistry* **27,** 4530 (1988).

[11] D. L. Newton, P. J. Nicholls, S. M. Rybak, and R. J. Youle, *J. Biol. Chem.* **269,** 26739 (1994).

[12] M. Zewe, S. M. Rybak, S. Dubel, J. F. Coy, M. Welschof, D. L. Newton, and M. Little, *Immunotechnology* **3,** 127 (1997).

[13] M. Gadina, D. L. Newton, S. M. Rybak, Y. N. Wu, and R. J. Youle, *Ther. Immunol.* **1,** 59 (1994).

[14] M. Suzuki, S. K. Saxena, E. Boix, R. J. Prill, V. M. Vasandani, J. E. Ladner, C. Sung, and R. J. Youle, *Nature Biotechnol.* **17,** 265 (1999).

[15] H. F. Rosenberg, K. D. Dyer, H. L. Tiffany, and M. Gonzalez, *Nature Genet.* **10,** 219 (1995).

[16] J. Zhang, K. D. Dyer, and H. F. Rosenberg, *Proc. Natl. Acad. Sci. U.S.A.* **97,** 4701 (2000).

[17] H. F. Rosenberg and J. B. Domachowske, *Immunol. Res.* **20,** 261 (1999).

response to infection with a species-specific RNA virus pathogen in mice[18,19] and have demonstrated that recombinant EDN can reduce the infectivity of RNA virus pathogens in experiments performed *in vitro*.[20–22] These intriguing but preliminary results suggest that EDN might be an unrecognized component of innate and specific antiviral host defense.

2. GenBank Accession Numbers for EDN

There are several GenBank listings for human EDN and quite a few for nonhuman primate and rodent variants as well. These accession numbers and appropriate commentary have been assembled as a reference resource in Table I.

3. Recombinant EDN from Bacterial Expression Systems

There are two methods currently in use for the production and isolation of recombinant EDN (rEDN) from bacteria. The first to be reported was that described by Newton and colleagues,[11,23] which has since been used by others to produce and purify milligram quantities of rEDN for experimental use.[6,14,24] This method involves the isolation of insoluble rEDN from bacterial inclusion bodies followed by purification and refolding; this method has been described in detail elsewhere in this volume by Boix.[25]

The second method, used primarily by our laboratories, uses the commercial expression vector pFLAG-CTS (Sigma, St. Louis, MO) with protein expression driven by the inducible *tac* promoter. The recombinant vector contains a PCR-amplified cDNA encoding "mature" EDN (without signal peptide) inserted into *Hin*dIII and *Eco*RI sites in-frame with an amino-terminal bacterial secretion sequence (ompA) and a C-terminal "FLAG" octapeptide. When placed at the C terminus of rEDN, the FLAG octapeptide has no apparent effect on folding or enzymatic activity. The secretion sequence directs the recombinant protein through

[18] J. B. Domachowske, C. A. Bonville, K. D. Dyer, A. J. Easton, and H. F. Rosenberg, *Cell. Immunol.* **200,** 98 (2000).

[19] J. B. Domachowske, C. A. Bonville, J.-L. Gao, P. M. Murphy, A. J. Easton, and H. F. Rosenberg, *J. Immunol.* **165,** 2677 (2000).

[20] J. B. Domachowske, K. D. Dyer, C. A. Bonville, and H. F. Rosenberg, *J. Infect. Dis.* **177,** 1458 (1998).

[21] J. B. Domachowske, C. A. Bonville, K. D. Dyer, and H. F. Rosenberg, *Nucl. Acids Res.* **26,** 5327 (1998).

[22] S. Lee-Huang, P. L. Huang, Y. Sun, P. L. Huang, H.-F. Kung, D. L. Blithe, and H.-C. Chen, *Proc. Natl. Acad. Sci. U.S.A.* **96,** 2678 (1999).

[23] D. L. Newton, S. Walbridge, S. M. Mikulski, W. Ardelt, K. Shogen, S. J. Ackerman, S. M. Rybak, and R. J. Youle, *J. Neurosci.* **14,** 538 (1994).

[24] D. L. Newton and S. M. Rybak, *J. Natl. Cancer Inst.* **90,** 1787 (1998).

[25] E. Boix, *Methods Enzymol.* **341,** [19] 2001 (this volume).

TABLE I
GENBANK ACCESSION NUMBERS AND SOURCE OF EDN SEQUENCES

GenBank no.	Organism	Base pairs	Source (reference)
M24157	Human, mRNA	709	[a]
M30510	Human, mRNA	694	[b]
X16546	Human, gene	1489	[c]
X55988	Human, mRNA	735	Simonsen *et al.*, unpublished
X55987	Human, gene	1501	Simonsen *et al.*, unpublished
X55989	Human, pseudogene	1347	Simonsen *et al.*, unpublished
U24102	*P. troglodytes,* chimpanzee	483	[d]
U24100	*G. gorilla,* gorilla	483	[d]
U24104	*P. pygmaeus,* orangutan	483	[d]
U24096	*M. fascicularis,* macaque	480	[d]
U24099	*S. oedipus,* tamarin	474	[d]
U88827	*A. trivirgatus,* owl monkey	490	[e]
AF078127	*C. aethiops,* African green monkey	399	Deming *et al.*, unpublished
	M. musculus, house mouse:		
U72032	mEAR-1	468	[f]
U72031	mEAR-2	471	[f]
AF017258	mEAR-3	471	[g]
AF017259	mEAR-4	468	[g]
AF017260	mEAR-5	468	[g]
AF017261	mEAR-6p[j]	387	[g]
AF171649	mEAR-7	465	[h]
AF171650	mEAR-8	465	[h]
AF171651	mEAR-9	468	[h]
	M. saxicola, spiny mouse:		
AF238395	spmEAR-2	465	[i]
AF238396	spmEAR-4	462	[i]
AF238397	spmEAR-6	468	[i]
AF238398	spmEAR-10	459	[i]
AF238399	spmEAR-11	468	[i]
AF238400	spmEAR-21	468	[i]
AF238401	spmEAR-23	468	[i]
AF238423	spmEAR-1p	454	[i]
AF238424	spmEAR-7p	451	[i]
AF238425	spmEAR-8p	448	[i]
AF238426	spmEAR-9p	450	[i]
AF238427	spmEAR-14p	v449	[i]
AF238428	spmEAR-16p	449	[i]
AF238429	spmEAR-26p	466	[i]
	M. pahari, shrew mouse:		
AF238402	shmEAR-2	465	[i]
AF238403	shmEAR-7	468	[i]
AF238404	shmEAR-8	468	[i]
AF238405	shmEAR-12	468	[i]

TABLE I
(*continued*)

GenBank no.	Organism	Base pairs	Source (reference)
AF238430	shmEAR-4p	452	*i*
AF238431	shmEAR-5p	460	*i*
AF238432	shmEAR-6p	463	*i*
AF238433	shmEAR-10p	461	*i*
AF238434	shmEAR-11p	466	*i*
AF238435	shmEAR-15p	467	*i*
AF238436	shmEAR-16p	467	*i*
AF238437	shmEAR-17p	468	*i*
AF238438	shmEAR-18p	469	*i*
AF238439	shmEAR-19p	468	*i*
AF238440	shmEAR-23p	451	*i*
AF238441	shmEAR-24p	467	*i*
AF238442	shmEAR-25p	452	*i*
	M. caroli, ricefield mouse:		
AF238406	rfm-EAR-1	471	*i*
AF238407	rfm-EAR-2	468	*i*
AF238408	rfm-EAR-5	465	*i*
AF238409	rfm-EAR-7	471	*i*
AF238410	rfm-EAR-10	471	*i*
AF238411	rfm-EAR-12	468	*i*
AF238412	rfm-EAR-13	471	*i*
AF238413	rfm-EAR-14	471	*i*
AF238414	rfm-EAR-16	471	*i*
AF238415	rfm-EAR-18	471	*i*
AF238416	rfm-EAR-21	468	*i*
AF238417	rfm-EAR-22	471	*i*
AF238418	rfm-EAR-24	471	*i*
AF238419	rfm-EAR-28	468	*i*
AF238420	rfm-EAR-29	468	*i*
AF238421	rfm-EAR-32	468	*i*
AF238422	rfm-EAR-33	468	*i*
AF238443	rfm-EAR-25p	456	*i*
AF238444	rfm-EAR-27p	464	*i*
	R. norvegicus:		
AF171641	rR-1	465	*h*
AF171642	rR-2	462	*h*
AF171643	rR-4	465	*h*
AF171644	rR-5	465	*h*
AF171645	rR-7	465	*h*
AF171646	rR-8	462	*h*
AF171647	rR-12	462	*h*
AF171648	rR-14	465	*h*

(*continued*)

TABLE I
(*continued*)

GenBank no.	Organism	Base pairs	Source (reference)
	C. griseus, hamster:		
AF238385	EAR-7	474	[i]
AF238386	EAR-10	477	[i]
AF238387	EAR-11	477	[i]
AF238388	EAR-15	477	[i]
AF238389	EAR-24	474	[i]
AF238445	EAR-30p	448	[i]
	M. unguiculatus, gerbil:		
AF238390	EAR-25	474	[i]
AF238391	EAR-34	474	[i]
AF238392	EAR-36	474	[i]
AF238393	EAR-11	474	[i]
AF238394	EAR-44	474	[i]

[a] H. F. Rosenberg, D. G. Tenen, and S. J. Ackerman, *Proc. Natl. Acad. Sci. U.S.A.* **86,** 4460 (1989).
[b] K. J. Hamann, R. L. Barker, D. A. Loegering, L. R. Pease, and G. J. Gleich, *Gene* **83,** 161 (1989).
[c] K. J. Hamann, R. M. Ten, D. A. Loegering, R. B. Jenkins, M. T. Heise, C. R. Schad, L. R. Pease, G. J. Gleich, and R. L. Barker, *Genomics* **7,** 535 (1990).
[d] H. F. Rosenberg, K. D. Dyer, H. L. Tiffany, and M. Gonzalez, *Nature Genet.* **10,** 219 (1995).
[e] H. F. Rosenberg and K. D. Dyer, *Nucl. Acids Res.* **25,** 3532 (1997).
[f] K. A. Larson, E. V. Olson, B. J. Madden, G. J. Gleich, N. A. Lee, and J. J. Lee, *Proc. Natl. Acad. Sci. U.S.A.* **93,** 12370 (1996).
[g] D. Batten, K. D. Dyer, J. B. Domachowske, and H. F. Rosenberg, *Nucl. Acids Res.* **25,** 4235 (1997).
[h] N. A. Singhania, K. D. Dyer, J. Zhang, M. S. Deming, C. A. Bonville, J. B. Domachowske, and H. F. Rosenberg, *J. Mol. Evol.* **49,** 721 (1999).
[i] J. Zhang, K. D. Dyer, and H. F. Rosenberg, *Proc. Natl. Acad. Sci. U.S.A.* **97,** 4701 (2000).
[j] All genes marked "p" are pseudogenes.

the cell membrane into the bacterial periplasm, thus facilitating the harvest of correctly folded, ribonucleolytically active rEDN directly from the bacterial cell pellets. The commercially available M2 monoclonal antibody (M2 MAb, Sigma) can be used to detect recombinant protein and for immunoaffinity purification (M2 agarose, Sigma). Although we initially purified the protein by gentle osmotic lysis to release the periplasm, we have since found that our yields improved by reducing the inducing stimulus (isopropyl-β-D-thiogalactopyranoside, IPTG; Sigma) and harvesting by quantitative bacterial cell lysis.[26] We have found this system to be quite versatile, and have expressed human EDN as well as several primate EDN orthologs by this method.[21,27,28]

Production and Isolation of rEDN Expressed in pFCTS Vector in Escherichia coli DH5α

1. A 1 liter culture of bacterial transformants is grown overnight in a shaker–incubator in Luria–Bertani (LB) broth with 100 μg/ml ampicillin (utilizing the vector-encoded ampicillin-resistance gene). The stationary phase culture is then diluted 1 : 5 (final volume 5 liters) in LB-ampicillin, and regrown to late exponential phase, $OD_{600} = 0.6$–0.7.

2. Protein expression is induced with IPTG added to a final concentration of 10 to 50 μM. This reduced concentration permits some additional bacterial growth, and reduces (if not eliminates) accumulation of intracellular unprocessed forms of rEDN. Induction continues for 6 to 8 hr, after which bacteria are harvested by centrifugation, washed once with phosphate-buffered saline (PBS), and frozen (-80°) prior to purification.

3. To isolate rEDN, the frozen pellets from a 5 liter bacterial culture are resuspended while on ice in total ~20 ml cold PBS + 2 m*M* phenylmethylsulfonyl fluoride (PMSF, protease inhibitor) and then subjected to sonication. The bacterial debris is removed by high-speed centrifugation (38,000*g*, or 18,000 rpm in the SS34 rotor, Sorvall RC5C centrifuge.)

4. For batch immunoaffinity purification, M2 agarose (0.3 to 0.4 ml) is added to the clarified supernatant, along with sodium azide (final concentration 0.1%) to prevent bacterial overgrowth, and equilibrated end-over-end overnight at 4°. The resin is then washed 3 times with a total of 150 to 200 ml cold PBS, and rEDN is eluted from the M2 agarose by brief incubation with 100 m*M* glycine, pH 3.0 (1 ml). The eluate is neutralized immediately with 50 to 100 μl 1 *M* Tris, pH 8.0, and frozen in small aliquots. Protein concentration is determined by probing a serial-dilution Western blot probed with the M2 MAb (1 : 200 dilution in antibody buffer, followed by a 1 : 1000 dilution of alkaline phosphatase conjugated goat anti-mouse IgG and standard developing reagents[27]), comparing the unknown quantity of rEDN with a known-concentration standard (recombinant carboxy-FLAG-bacterial alkaline phosphatase, 1.0 to 3.0 μg/μl, Sigma) with correction for relative molecular mass. The apparent molecular mass of the secreted, FLAG-tagged bacterial-derived rEDN on a 14% acrylamide Tris–glycine gel (Invitrogen, San Diego, CA) is ~18,000–20,000 Da.[27]

[26] H. F. Rosenberg, *Biotechniques* **24,** 188 (1998).

[27] H. F. Rosenberg and K. D. Dyer, *J. Biol. Chem.* **270,** 21539 (1995).

[28] H. F. Rosenberg and K. D. Dyer, *Nucl. Acids Res.* **25,** 3532 (1997).

Overall, the advantages of this procedure are its speed (3 days from bacterial culture to protein) and reproducible specific activity (as each molecule isolated has been folded *in vivo*). However, the relatively small yields—at best, microgram quantitites from 5 liter of bacteria—can be a drawback for many applications. We have used this procedure to prepare rEDN for ribonuclease assays and for our *in vitro* antiviral assays, described in Sections 7 and 8 below. We have also purified bacterial rEDN via a double-column fast protein liquid chromatography (FPLC) method, which will be outlined in the section on baculovirus-derived proteins (Section 4).

4. Recombinant EDN from Baculovirus

We first explored the baculovirus system as a means for preparing recombinant EDN, ECP, and RNase k6 for use in our antiviral assays, and we have been successful with fairly standard methodology.[21,29] Briefly, the rEDN vector was constructed by cloning the full-length coding sequence of EDN (including the native signal peptide) with the C-terminal FLAG (see Section 3) into the *Bam*HI/*Xba*I sites of the commercial transfer vector pVL1393 (Invitrogen, San Diego, CA). Three micrograms of the recombinant vector and 0.5 μg of linearized wild-type baculovirus AcNPV (Pharmingen, San Diego, CA) were used to cotransfect *Spodoptera frugiperda* (Sf9) cells by the lipofectin method (Life Technologies, Gaithersburg, MD). Polyhedrin-deficient plaques were identified by microscopic inspection, and rEDN was detected by immunoblotting with the M2 anti-FLAG MAb (see Section 3). To obtain recombinant protein, 100 ml cultures of log-phase Sf9 cells are infected with recombinant virions, and supernatant containing rEDN is harvested when >50% of the cells stain positively with trypan blue, usually at 72 to 96 hours postinfection. The supernatants can be frozen (-80°) for long-term storage or maintained at 4° with sodium azide (0.1%) until purification.

Isolation of rEDN from Baculovirus-Infected Cell Supernatants

Although batch immunoaffinity purification (see Section 3) can be used here, it is somewhat clumsy given the larger volumes of supernatant involved. To this end, we have developed a double-column fast protein liquid chromatography (FPLC) isolation method. This method is also applicable for the purification of bacterial rEDN described in Section 3.

[29] J. B. Domachowske, K. D. Dyer, A. G. Adams, T. L. Leto, and H. F. Rosenberg, *Nucl. Acids Res.* **26,** 3358 (1998).

1. A heparin-Sepharose column (HiTrap Heparin, 5 ml total volume, Sigma) is washed with 25 ml (5 column volumes) of distilled water followed by 25 ml of Buffer A (50 m*M* Tris, pH 8.0 + 1 m*M* NaCl) at a flow rate of 5.0 ml/min. All buffers used in FPLC are degassed by wall-vacuum suction immediately before use.

2. Four hundred fifty milliliters of baculovirus culture supernatant are clarified via high-speed centrifugation (38,000*g*, or 18,000 rpm in SS34 rotor in Sorvall RC5C centrifuge) and loaded into an FPLC superloop, 150 ml volume (Amersham Pharmacia Biotech, Piscataway, NJ). The clarified supernatant is passed over the heparin-Sepharose column (5.0 ml/min) and the superloop reloaded two times to complete the 450 ml volume.

3. The column is then washed with 50 ml of Buffer A. Bound proteins are eluted by a linear gradient of Buffer A to Buffer B (50 m*M* Tris pH 8.0 + 1 *M* NaCl) 50 ml total volume, eluting at 2.0 ml/min, collecting 2 ml fractions for assay. Recombinant EDN from baculovirus typically elutes in fractions 7 to 10 (~300–400 m*M* NaCl) and is detected by ribonuclease assay (see Section 7). The ribonucleolytically active fractions are combined (typically 4 fractions) and the 8 ml volume is reduced to ~2 ml by centrifugation in a Centricon 10 concentrator column (Millipore Corporation, Bedford, MA) at 4500*g* (6300 rpm in the SS34 rotor of an RC5C Sorvall centrifuge) at 4°.

4. The second column is a preparative grade Superdex 75 size-selection column (Amersham Pharmacia Biotech, Piscataway, NJ) equilibrated with buffer C (50 m*M* Tris pH 8.0 + 150 m*M* NaCl). The rEDN-containing concentrate is loaded via a 2 ml loop onto the column with the buffer C mobile phase running at 1.5 ml/min. The first 60 ml of eluate is discarded, and 1.5 ml fractions are collected thereafter, stopping at 90 ml. Ribonucleolytically active rEDN generally elutes in a broad band at 67 through 83 ml. The highest activity fractions are combined, and concentration determined by serial dilution Western blot as described in Section 3.

Unlike the bacterially derived recombinant protein, the baculovirus-derived protein is heavily glycosylated, and migrates on a 14% acrylamide Tris–glycine gel as a heterogeneous cluster of molecular mass ~25,000 to 35,000 Da. The yield of rEDN from 450 ml of Sf9 culture supernatant is typically in the low microgram range. This form of rEDN has also been used in our antiviral studies (Section 8). In addition to rEDN, we have used this method to produce recombinant EDNdK38 (mutant), wild-type ECP, RNase k6,[21] and the mouse ortholog of EDN, mouse eosinophil-associated ribonuclease-2.[30]

[30] A. L. McDevitt, M. S. Deming, H. F. Rosenberg, and K. D. Dyer, *Gene* **267,** 23 (2001).

5. Recombinant EDN from Eukaryotic Cell Culture

As part of our ongoing work on the mechanism of EDN's antiviral activity, we have recently engineered a human epithelial cell line (HEp-2, American Type Culture Collection, Manassas, VA) that produces and secretes rEDN. In collaboration with Dr. Jan Hofsteenge of the Friedrich Miescher Institut, we co-transfected 40 μg of the linearized expression vector pSCMCi-EDN [full-length EDN under control of a cytomegalovirus (CMV) promoter][31] with 4 μg of pSV-Neo by electroporation into log-phase HEp-2 cells, and began selection with 1.5 mg/ml G418 (Geneticin, Life Technologies, Gaithersburg, MD) after 48 hr in culture. Single colonies were picked and expanded, and positive colonies identified by ribonucleolytic activity in the supernatant. The resulting cell line, HEp-2/103, is G418 resistant and has incorporated two additional copies of the gene encoding EDN. This cell line is remarkably resistant to viral infection and secretes a very highly glycosylated form of rhEDN (apparent molecular mass ~35,000–40,000 Da[32]). We have had similar success in engineering cell lines that secrete the C-terminal FLAG-tagged form of EDN as well as FLAG-tagged, ribonuclease-minus forms containing mutations of crucial residues H^{15} and H^{129}.

6. Natural Forms of EDN

As noted in Section 1, EDN has been identified as a major component of human eosinophilic leukocytes, but proteins identical to EDN have been identified in lung, liver, and spleen. Interestingly, EDN (most probably from liver) is a major contaminant of a commercial preparation of β-chorionic gonadotropin (β-HCG),[33] a hormone preparation derived from the urine of pregnant women. As part of an investigation into the anti-Kaposi's sarcoma activity attributed to this preparation, Newton and Rybak[24] reported the isolation of "-4EDN," a variant of EDN in which the N terminus includes 4 amino acids initially considered as part of the signal peptide.

The purification of EDN from eosinophils was described in great detail in the original paper by Gleich and colleagues.[3] Eosinophils are generally in short supply, given that the normal human leukocyte population contains at most 1 to 5% of these cells. β-HCG, on the other hand, is a good commercial source for EDN, and purification of EDN from this source is presented.

[31] J. Krieg, W. Glasner, A. Vincentini, M. A. Doucey, A. Loffler, D. Hess, and J. Hofsteenge, *J. Biol. Chem.* **272,** 26687 (1997).

[32] J. B. Domachowske, C. A. Bonville, K. D. Dyer, J. Hofsteenge, and H. F. Rosenberg, manuscript in preparation.

[33] A. Kardana, K. D. Bagshawe, B. Coles, D. Read, and M. Taylor, *Br. J. Cancer* **67,** 686 (1993).

In a recent publication, Kaufmann and colleagues[34] determined the EDN concentrations of various sources of β-HCG. We have used the material supplied by Sigma for the FPLC purification scheme described here.

Isolation of EDN from Commercial Preparations of β-HCG

1. Treating the human-source material as a biohazard, each 10,000 I.U. vial is hydrated with 10 ml distilled water. Material from 10 to 15 vials can be combined to a total volume of 100 to 150 ml, which is then dialyzed overnight against 4 liters of FPLC Buffer A (see Section 5).

2. Equilibrate a 1 ml heparin-Sepharose column (Sigma) with 5 ml distilled water followed by 5 ml Buffer A at a flow rate of 1.0 ml/min (see Section 5). Load the dialyzed material into a 150 ml Superloop and run over column at a flow rate of 1.0 ml/min. Wash column with 10 ml Buffer A.

3. Elute bound proteins with a 25 ml linear gradient of Buffer A and Buffer D (50 m*M* Tris, pH 8.0 + 2 *M* NaCl) at a flow rate of 1.0 ml/min, collecting 1 ml fractions.

4. Ribonucleolytically active fractions (see assay, Section 7) are concentrated (Centricon 10, Amicon, Danvers, MA) and subjected to size selection via a preparative Superdex-75 column as described in Section 5. The EDN from this preparation typically elutes at ~80–83 ml. It migrates on a 14% acrylamide Tris–glycine gel as a doublet of apparent molecular mass ~18,000 Da.

Recently, the EDN contaminant of β-HCG has received significant attention as a potential source of the anti-Kaposi's sarcoma activity noted for this preparation. Although the results of these studies have been mixed and controversial,[35,36] this potential therapeutic modality remains intriguing.

7. Ribonuclease Assay

The literature abounds with ribonuclease assays that can be adapted to detect enymatic activity in both solid (gel, membrane) and solution phase. We have reviewed several of these assays in a recent publication.[37] For general purposes, such as assaying column fractions and comparing relative enzymatic activity, we

[34] H. F. Kaufmann, H. Hovenga, H. W. de Bruijn, and J. J. Beintema, *Eur. J. Obstet. Gynecol. Reprod. Biol.* **82,** 111 (1999).

[35] S. J. Griffiths, D. J. Adams, and S. J. Talbot, *Nature* **390,** 568 (1997).

[36] Y. Lunardi-Iskandar, J. L. Bryant, W. A. Blattner, C. L. Hung, L. Flamand, P. Gill, P. Hermans, S. Birken, and R. C. Gallo, *Nature Med.* **4,** 428 (1998).

[37] H. F. Rosenberg and J. B. Domachowske, *in* "Methods in Molecular Biology" (C. H. Schein, ed.). Vol. 160, p. 355. Humana Press Inc., Totowa, NJ, 2001.

use a solution-phase assay based on that originally described by Anfinsen and colleagues[38] as modified by Slifman and colleagues.[7]

Materials

Diethyl pyrocarbonate-treated, ribonuclease-free water for all solutions
100 m*M* Sodium phosphate, pH 7.4
40 m*M* Lanthanum nitrate (Sigma)
6% Perchloric acid (stock 60% solution, Sigma)
Yeast tRNA (Sigma), diluted to 4 mg/ml and stored at −80°
Ribonuclease containing solution to be assayed
UV spectrophotometer, quartz cuvettes, multichannel timer, autoclaved tips and microcentrifuge tubes

Method

1. Prepare stop solution (1:1 vol/vol mixture of the lanthanum nitrate and perchloric acid listed above, made fresh and kept on ice during the assay.)
2. Prepare assay solutions in microcentrifuge tubes—0.3 ml sodium phosphate buffer, 0.5 ml water—in triplicate for each time point to be measured (generally 3 to 4 per assay) and one for a $t = 0$ control.
3. Add solution to be assayed in a volume of 50 μl or less to each tube. Add 0.5 ml of stop solution to the $t = 0$ control tube only.
4. Defrost tRNA immediately prior to use. Add 10 μl tRNA to the $t = 0$ tube, cover, invert, and store on ice during assay.
5. Add 10 μl tRNA to each assay tube, three per time point. Start one channel timing as tRNA is added to the first tube at a given time point, and invert tubes individually immediately after tRNA is added.
6. As specific times have elapsed (3, 5, 10 min, etc.) stop each set of triplicates with 0.5 ml cold stop buffer, invert, and store on ice. After the final set, keep all tubes on ice for an additional 10 minutes.
7. Precipitate the acid-insoluble, undigested tRNA by centrifugation for 5 min at room temperature in a microcentrifuge at highest setting.
8. Read the optical density (OD) of the supernatant containing the digested, solubilized tRNA at $A_{260\text{ nm}}$. OD units can be converted into nmol ribonucleotides using the average molecular weight of tRNA (28,100) and the conversion of 1 OD unit = 40 μg RNA.[27]

The plot of nmol ribonucleotides per unit time should be linear (initial rates). If the plot appears to plateau, repeat using smaller time intervals and/or less

[38] C. B. Anfinsen, R. R. Redfield, W. L. Choate, J. Page, and W. R. Carroll, *J. Biol. Chem.* **207,** 201 (1954).

ribonuclease. Note that this assay as described can be used to determine comparative rates and ultimately to construct a double-reciprocal plot with varying substrate concentrations (see Ref. 28). Singlets (with a $t = 0$ control for each fraction) at one time point are sufficient for assaying column fractions.

8. Biological Assays: Antiviral Activity

The physiologic function of EDN remains somewhat mysterious. EDN-mediated neurotoxicity, a syndrome of cerebellar dysfunction relating to the loss of Purkinje cells, is the result of the artificial introduction of microgram quantities of protein directly into the cerebrospinal fluid of experimental rabbits or guinea pigs.[1,23] Although this is probably telling us something important about EDN that we do not yet completely understand, the physiologic relevance of this phenomenon is unclear. This is particularly noteworthy given the evolutionary diversity of this gene lineage and the fact that neither rabbits nor guinea pigs have genes that encode anything with more than ~60% amino acid sequence homology to the human EDN used in these experiments.[15,16]

Based on our studies of eosinophil ribonucleases and evolutionary diversity, we are exploring a role for EDN and its host cell, the eosinophilic leukocyte, in innate antiviral host defense. To facilitate these studies, we have developed an assay for viral infectivity known as the Quantitative Shell Vial Assay.[39] This assay has permitted us evaluate the infectivity of the human pathogen respiratory syncytial virus (RSV) and to determine the degree to which rEDN does or does not interfere with this primary infection.

Quantitative Shell Vial Assay: Materials

Dram shell vials (Fisher Scientific, Pittsburgh, PA)
Glass coverslips (Fisher Scientific)
HEp-2 epithelial cell line (ATCC)
Cell culture: Iscove's Modified Dulbecco's medium (Life Technologies) with 10% heat-inactivated fetal calf serum, 2 m*M* glutamine, and penicillin/streptomycin; trypsin–EDTA solution; sterile PBS
Recombinant EDN and controls
Respiratory syncytial virus (RSV-B, ATCC) grown in HEp-2 cells and stored in aliquots at $>10^6$ plaque-forming units/ml)
FITC-tagged anti-RSV blend (Chemicon International, Temecula, CA, cat. #5022)
Fluorescent microscope, thumb-click counter, glass slides, mounting fluid

[39] J. B. Domachowske and C. A. Bonville, *Biotechniques* **25,** 644 (1998).

Method

1. Prepare shell vials for cells by rinsing with distilled water and drying inverted on paper toweling. Add one glass coverslip to each, cover with foil and autoclave.
2. Pass log phase HEp-2 cells into shell vials, 10^5 cells/ml, 1 ml per vial. Cover with plastic cap and incubate overnight at 37° in a CO_2 incubator.
3. When cells reach ~80% confluence (usually 24 to 48 hr postplating), vials are ready for assay. Prepare aliquots of complete culture medium with varying concentrations of rEDN or diluent control for assay (no more than 20% volume of the medium). Prepare enough for 0.3 to 0.5 ml volume per vial, with triplicate vials assayed at each rEDN concentration.
4. Remove culture medium by suction with sterile glass pipet; add medium containing rEDN or control to each vial. Cover and let sit at room temperature for 5 to 10 min.
5. Add 5 to 10 μl of a 1 : 100 to 1 : 200 dilution of RSV-B (dilution adjusted so as to obtain ~300–500 foci per coverslip.) Cover and spin amplify (700*g*, or 1800 rpm in standard table-top centrifuge) for 1 hr at room temperature. Place in 37° CO_2 incubator overnight.
6. Next morning (optimally, at $t = 16$ to 18 hr), remove virus-containing culture medium, wash 2–3 times with PBS, and fix monolayer with cold acetone for 5 min. Remove acetone by suction, let monolayer air-dry, and then stain with 4–5 drops of anti-RSV blend (contains anti-RSV antibodies and methylene blue counterstain). Let stain for 1 hr at 37°, remove, wash twice with PBS, and mount the coverslip in inverted fashion in mounting solution on a glass slide (protect from direct light).
7. Count all fluorescently labeled single cells (= fluorescent foci) covering the entire coverslip (containing ~4 to 5 × 10^5 cells.) By comparing various concentrations of rEDN with volume-equivalent diluent controls, the reduction in infectivity (#foci/monolayer) mediated by EDN and variants can be determined.[20,21]

Although this technique is somewhat more expensive than a traditional viral plaque assay (because of the cost of the anti-RSV antibody), it makes up for this disadvantage in speed (overnight vs 1 week) and use of reduced quantities of recombinant protein. As the actual counting process is a bit tedious, we are currently attempting to adapt this into a FACS (fluorescent-activated cell sorting) assay.

[18] Eosinophil Cationic Protein

By ESTER BOIX

The eosinophil cationic protein (ECP) is one of the basic proteins contained in the matrix of secondary eosinophil granules. ECP is synthesized as a preprotein with a signal peptide of 27 residues, and a mature protein of 133 amino acids. The native protein is a single polypeptide of 15.5 kDa, with a p*I* of 10.8, and three glycosylated forms, ranging from 18 to 21 kDa. ECP mRNA has been detected in eosinophil-enriched peripheral blood cells and in HL-60 cells induced to eosinophilic differentiation. ECP, also known as RNase-3, belongs to the bovine pancreatic RNase A superfamily, and has a 67% amino acid identity to another eosinophil RNase, the eosinophil derived neurotoxin (EDN, or RNase-2), and a 31% primary sequence identity to RNase A (Fig. 1). The gene for ECP (RNS3) is located on the q24-q31 region of the human chromosome 14 in the same region as the EDN gene and is composed of a noncoding exon and a coding exon, separated by a single intron. ECP was generated from an ECP/EDN precursor gene about 30 million years ago, at an evolutionary rate among the highest of all studied primate genes. The sequence has incorporated 12 additional arginines, creating a mature protein with a much higher p*I* and new anti-pathogen properties. ECP exhibits a modest ribonucleolytic activity, and a marked cytotoxicity against parasites, bacteria, single-stranded RNA viruses, and mammalian cells. Noncytotoxic properties have also been attributed *in vitro,* such as histamine and tryptase release, upregulation of various epithelial cell receptors, regulation of the complement pathway, and proteoglycan production. These activities may play an immunoregulatory role *in vivo.* Although its mechanism of action is unknown, a role as a host defense protein is suggested. For a recent review, see Venge *et al.*[1]

In this article, we describe the methodology for ECP purification and characterization, including specific protocols, from the purification of the native or recombinant protein, to most of the available functional assays to analyze ECP ribonucleolytic and biological properties. A review of its clinical application as a marker of eosinophil activation in the diagnosis of inflammatory diseases is also included. Finally, an analysis of the three-dimensional structure is provided to indicate the putative key residues at the catalytic and substrate binding sites.

[1] P. Venge, J. Bystrom, M. Carlson, L. Hakansson, M. Karawacjzyk, C. Peterson, L. Seveus, and A. Trulson, *Clin. Exp. Allergy* **29,** 1172 (1999).

0076-6879/01 $35.00

```
               1      10         20            30         40       50        60
ECP    : --MRPPQFTRAQWFAIQHISL-N-PP----RCTIAMR-AINNYRWRCKNQNTFLRTTFANVVNVCGNQSIRC
EDN    : --MKPPQFTWAQWFETQHIN--MTSQ----QCTNAMQ-VINNYQRRCKNQNTFLLTTFANVVNVCGNPNMTC
RNaseA : KE------TAAAKFERQHMDSSTSAASSSNYCNQMMKSRNLTK-DRCKPVNTFVHESLADVQAVCSQKNVAC
         1               10       20        30         40        50        60

                 70        80        90        100       110        120       130
ECP    : PHNRTLNNCHRSRFRVPLLHCDLINPGAQNISNCRYADRPGRRFYVVACDNRDP-RDSPRYPVVPVHLDTTI
EDN    : PSNKTRKNCHHSGSQVPLIHCNLTTPSPQNISNCRYAQTPANMFYIVACDNRDQRRDPPQYPVVPVHLDRII
RNaseA : -KNG-QTNCYQSYSTMSITDCRETG--SSKYPNCAYKTTQANKHIIVACEGN---------PYVPVHFDASV
               70        80         90        100       110                120
```

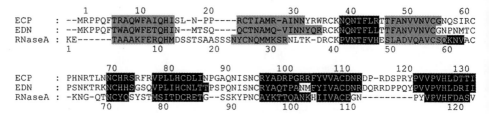

FIG. 1. Structure- based sequence alignment of ECP (1QMT.pdb) EDN and RNase A (7RSA.pdb); [E. Boix, D. D. Leonidas, Z. Nikolovski, M. V. Nogués, C. M. Cuchillo, and K. R. Acharya, *Biochemistry* **38,** 16794 (1999)]. Secondary structural elements, α helices and β strands, are in gray and black boxes, respectively. Every tenth residue in the ECP and RNase A sequences is numbered.

Purification of ECP

Purification of Native ECP from Eosinophil Granule Fraction

Assay 1. Peripheral blood eosinophils arc isolated from blood samples of normal volunteers.[2–4] Heparinized blood is sedimented with 6% dextran and the leukocyte-enriched buffy coats are harvested and subjected to Ficoll-Paque density centrifugation (Pharmacia, Piscataway, NJ). After centrifugation at 400*g* for 20 min the red blood cells comigrating with the granulocyte pellet are removed by hypotonic lysis. Granulocytes are washed twice with phosphate-buffered saline (PBS) and supplemented with 0.5% bovine serum albumin (BSA) and 1 m*M* ethylenediaminetetraacetic acid (EDTA), and neutrophils are removed by incubation with CD16-conjugated magnetic beads and isolation by magnetic activated cell sorting using a magnetic cell separation system, MACS (Miltenyi Biotec, Auburn, CA). Eosinophils can be isolated to 97–99% purity as determined by microscopic examination with Quik-Diff staining (Fisher Scientific, Hampton, NH) and more than 95% are viable cells, as checked by trypan blue exclusion.

The granule proteins are then extracted from the eosinophil fraction.[5] Eosinophils are washed twice in 0.15 *M* NaCl and once in 0.34 *M* sucrose, 5 m*M* EDTA, and 12 m*M* $NaHCO_3$. The pellet is resuspended in 4 volumes of ice-cold PBS containing 5 m*M* EDTA and 12 m*M* $NaHCO_3$ and pressurized with nitrogen for 30 min at 500 psi at 4° under constant stirring in a cell disruption bomb. The lysate is collected into 1 volume of the same buffer containing 0.7 *M* sucrose and 0.3 *M* NaCl and centrifuged at 800 *g* to eliminate cell nuclei and disrupted cells. Granules are isolated from the supernatant by centrifugation at 10,000*g* for 15 min at 4°. After three cycles of freezing and thawing, the pellet is extracted with 5 volumes of

[2] J. B. Domachowske and H. F. Rosenberg, *J. Leukocyte Biol.* **62,** 363 (1997).
[3] B. Olszewska-Pazdrak, K. Pazdrak, P. L. Ogra, and R. P. Garofalo, *J. Immunol.* **160,** 4889 (1998).
[4] H. F. Rosenberg and J. B. Domachowske, *Methods Enzymol.* **341,** [18] 2001 (this volume).
[5] C. M. Reimert, P. Venge, A. Kharazmi, and K. Bendtzen, *J. Immunol. Methods* **138,** 285 (1991).

50 mM acetic acid under constant stirring for 1 h at 4°. An equal volume of 0.4 M ammonium acetate is then added and the extraction continued for 3 h. The extract is centrifuged at 20,000g for 30 min at 4°C and the supernatant containing granular proteins is concentrated 30-fold by diafiltration using YM10 filters (Amicon, Danvers, MA).

ECP is purified from the granular protein extract by a Sephadex G-75 superfine column (Pharmacia) equilibrated with 0.2 M ammonium acetate, pH 4.[5] Fractions containing 16–22 kDa bands on a sodium dodecyl sulfate–polyacrylamide gel electrophoresis (SDS–PAGE) are pooled, concentrated and equilibrated by diafiltration on YM10 filter (Amicon) with 0.2 M ammonium acetate, 0.5 M NaCl, 10% (v/v) glycerol, pH 8, and applied to a Zn-chelating Sepharose 6B column (Pharmacia). ECP elutes as a single peak following application of a pH gradient up to 0.2 M ammonium acetate, 0.5 M NaCl, 10% glycerol, pH 4.5, with pure ECP eluting at about pH 5.[6] The protein concentration of the purified ECP is determined using an extinction coefficient ($E^{1\%}{}_{1\ \mathrm{cm}}$) of 15.45 at 280 nm, using the data from the protein amino acid analysis. The purified sample as analyzed by SDS–PAGE (16% gel) contains three forms of 18.5, 20, and 22 kDa, with different degrees of glycosylation. The overall recovery of the protein is about 55% from the initial content of ECP in the granule protein fraction.[6]

The isolated fractions from the Zn-chelating Sepharose, corresponding to the different isoforms of ECP, can be further purified by a Mono S high-resolution cation-exchange column (Pharmacia).[7] The column is equilibrated with 0.5 M sodium acetate at pH 5 at a flow rate of 1 ml/min, and fractions are eluted with a gradient to 1.5 M sodium acetate, pH 5, with the following steps. After 5 min of elution with the equilibration buffer (0.5 M sodium acetate, pH 5), a gradient up to 0.9 M sodium acetate, pH 5, is achieved in 5 min, followed by an isocratic elution at 0.9 M for 15 min, and a final gradient up to 1.5 M sodium acetate, pH 5, in 10 min. To achieve homogeneous samples for each of the ECP native isoforms, peak fractions should be rechromatographed twice.

Assay 2. Alternatively, eosinophils are lysed with sucrose and heparin, and the granules are collected in the pellet fraction after centrifugation.[8] The granules are then lysed by exposure to 0.01 M HCl, pH 2, and sonication. A detergent [0.5% (v/v) Nonidet P-40 (NP-40), 1–2% (v/v) Triton X-100 or 1% (v/v) cetyltrimethyl ammonium bromide (CTAB) in either 5.5 mM NaOH or 10 mM HCl] can also be used in the lysis buffer. Considerable differences in the final yield are observed for the other eosinophil granule proteins, but not for ECP, although extraction with 0.5% NP-40 and water instead of HCl is slightly more efficient. Slightly better

[6] C. G. B. Peterson, H. Jornvall, and P. Venge, *Eur. J. Haematol.* **40,** 415 (1988).

[7] I. Olsson, A.-M. Persson, and I. Winqvist, *Blood* **67,** 498 (1986).

[8] R. I. Abughazaleh, S. L. Dunnette, D. A. Loegering, J. L. Checkel, H. Kita, L. L. Thomas, and G. J. Gleich, *J. Leukocyte Biol.* **52,** 611 (1992).

yields for ECP recovery from eosinophil lysates with 1% Triton X-100 have been reported.[9] The effect of the pH has also been analyzed, with the maximum recovery for ECP at a pH of 5.6, and a dramatic reduction at pH 10. After removal of the insoluble material by centrifugation, the supernatant lysate is applied to a Sephadex G-50 column (Pharmacia), equilibrated in 25 m*M* sodium acetate, 0.15 *M* NaCl, pH 4.2. The second elution peak containing EDN and ECP is dialyzed against PBS for 4 h and applied to heparin-Sepharose column (Pharmacia). The protein is eluted using a linear gradient of 0.075 to 1.5 *M* NaCl in 5 m*M* phosphate buffer, pH 7, with ECP eluting at 0.35–0.65 *M* NaCl.

Purification of Recombinant ECP Expressed in Escherichia coli from Inclusion Bodies

A synthetic gene for human ECP is synthesized using *E. coli* codon bias and the sequence is subcloned in the pET11c vector (Novagen, Madison, WI).[10] *E. coli* Novablue competent cells are used for the subcloning procedures and *E. coli* BL21(DE3) cells for protein expression. Transformed cells are grown at 37° in 1 liter of Terrific broth (TB)[11] containing 400 μg/ml of carbenicillin. Expression in minimal medium M9[11] gives a yield of about 25% in comparison with the TB media values. Cells are induced with 1 m*M* isopropyl-β-D-thiogalactopyranoside (IPTG) at an $A_{600} = 0.6$–0.8. After 4 h following IPTG induction, about 100 mg of rECP is aggregated in inclusion bodies, representing 70% of the total cell protein. Longer incubations lead to accumulation of other proteins and a lower yield in the refolding step. After centrifugation at 1500*g* for 15 min at 4°, the cell pellet is resuspended in 10 m*M* Tris-HCl, 2 m*M* EDTA, pH 8, and sonicated thoroughly. After centrifugation at 15,000*g* for 30 min at 4°, the pellet fraction is washed with 50 m*M* Tris-HCl, 2 m*M* EDTA, 0.3 *M* NaCl, pH 8 and centrifuged at 20,000*g* for 30 min at 4°. The pellet fraction, containing the inclusion bodies, is dissolved in 12 ml of 6 *M* guanidine hydrochloride, 0.1 *M* Tris–acetate, 2 m*M* EDTA, pH 8.5, containing 80 m*M* GSH. After 2 h incubation at room temperature, the protein is refolded by a 1/50–1/200 dilution into cold 0.1 *M* Tris–acetate, pH 8.5, containing 0.5 *M* L-arginine and oxidized glutathione (GSSG) to obtain a reduced oxidized glutathione (GSH/GSSG) ratio of 4. Dilution in the refolding buffer is adjusted to obtain a final protein concentration of 30–150 μg/ml. Protein can be added in serial additions, each at about 50 μg/ml, allowing 1-h intervals, to optimize the refolding process. The protein is incubated in the refolding buffer for 48–72 h

[9] S. Sur, D. G. Glitz, H. Kita, S. M. Kujawa, E. A. Peterson, D. A. Weiler, G. M. Kephart, J. M. Wagner, T. J. George, G. J. Gleich, and K. M. Leiferman, *J. Leukocyte Biol.* **63,** 715 (1998).

[10] E. Boix, Z. Nikolovski, G. P. Moiseyev, H. F. Rosenberg, C. M. Cuchillo, and M. V. Nogués, *J. Biol. Chem.* **274,** 15605 (1999).

[11] J. Sambrook, E. F. Fritsch, and T. Maniatis, "Molecular Cloning: A Laboratory Manuual." Cold Spring Harbor Laboratory Press, Cold Spring Harbor, NY, 1989.

at 4° and then concentrated and buffer exchanged against 0.15 *M* sodium acetate, pH 5, using a Pellicon tangential filtration system (Millipore, Bedford, MA). The sample is loaded onto a 6 ml cation exchange Resource S column (Pharmacia) equilibrated with 0.15 *M* sodium acetate, pH 5, and the protein is eluted at 2 ml/min with a linear gradient from the initial buffer A (0.15 *M* sodium acetate, pH 5) to the final buffer B (0.15 *M* sodium acetate, pH 5, 2 *M* NaCl) with the following steps: 15 min with 100% buffer A, 7 min to 20% of buffer B, 40 min to 35% buffer B, and 12 min to 100% buffer B. Peak fractions can be identified by RNase activity or by SDS–PAGE; rECP elutes at about 35% of buffer B. For further purification, the peak fraction is pooled, equilibrated again in 0.15 *M* sodium acetate, pH 5, and applied to a cation exchange 1 ml Mono S column (Pharmacia). Protein is eluted at 1 ml/min with a NaCl gradient from 0 to 2 *M*, with the following steps: 10 min with buffer A (0.15 *M* sodium acetate, pH 5), 5 min to 25% buffer B (0.15 *M* sodium acetate, pH 5, 2 *M* NaCl), 20 min to 45% buffer B, and 5 min to 100% buffer B. The protein concentration is determined spectrophotometrically using the calculated extinction coefficient 17,460 M^{1} cm^{-1} or using the microBCA assay (Pierce, Rockford, IL). Homogeneity of the protein is checked by SDS–15% PAGE, and by N-terminal sequencing, and the molecular weight is confirmed by mass spectrometry. The described expression system for rECP has the highest yield for rECP reported so far in the literature, with a final yield of 5–10 mg of purified protein per liter of culture. The recombinant protein has a formylmethionine at the N terminus. However, the N-terminal Met does not interact with any other residues of the protein, as indicated by the three-dimensional structure.[12]

Purification of rECP Using Other Expression Systems

Recombinant ECP can also be expressed in baculovirus.[4,13] The recombinant vector is constructed by inserting the ECP coding sequence into the pVL1393 transfer vector and *Spodoptera frugiperda* (Sf9) cells are cotransfected with the recombinant vector and the wild-type baculovirus AcNPV. The purified rECP migrates as a single band of 22 kDa by SDS–PAGE, with a molecular weight similar to that of the heavily glycosylated form of native ECP, and deglycosylates upon treatment with peptide *N*-glycosidase F (PNGase F). The final yield of purified protein is about 1–2 μg/ml of infected Sf9 cell supernatant.

Alternatively, ECP can be expressed in the pFCTS bacterial expression system as a soluble protein.[4,14,15] Briefly, ECP cDNA is inserted behind the

[12] E. Boix, D. D. Leonidas, Z. Nikolovski, M. V. Nogués, C. M. Cuchillo, and K. R. Acharya, *Biochemistry* **38,** 16794 (1999).

[13] J. B. Domachowske, K. D. Dyer, A. G. Adams, T. L. Leto, and H. F. Rosenberg, *Nucleic Acids Res.* **26,** 3358 (1998).

[14] H. F. Rosenberg and K. D. Dyer, *J. Biol. Chem.* **270,** 21539 (1995).

[15] H. F. Rosenberg and K. D. Dyer, *Nucleic Acids Res.* **25,** 3532 (1997).

N-terminal bacterial secretion peptide, so that the FLAG octapeptide is attached at the C terminus. Recombinant protein can be purified by a M2 anti-FLAG mAb–agarose affinity chromatography, with a final yield of few micrograms per liter of bacteria culture.

Protein Deglycosylation

The biosynthesis of the different glycosylated forms of ECP in marrow cells has been analyzed *in vitro*.[7] Hyperglycosylation of ECP has been observed in a promyelocytic leukemia cell line and in differentiated peripheral blood progenitor cells.[16] ECP isoforms may have differentiated biological properties.[1] A higher cytotoxicity and RNase activity for the 22 kDa glycosylated rECP, expressed in the baculovirus system, when compared to the recombinant nonglycosylated protein has been observed.[13] Interestingly the main glycosylated form of bovine pancreatic RNase (RNase B), with (Man)5–9$(GlcNAc)_2$ oligosaccharide linked to Asn-34, displays the same RNase activity as the nonglycosylated RNase A.[17] However, the glycosylated pancreatic RNase B is more stable and resistant to proteolysis.[18] The increased cytotoxicity of the glycosylated form of ECP might be a consequence of an increased intracellular stability.

Deglycosylated ECP migrates as a single band of about 16 kDa. PNGase F removes all types of complex and hybrid asparagine-linked oligosaccharides, and eliminates all the heterogeneity of ECP forms, while endoglycosidase H, which cleaves high-mannose N-linked oligosaccharides, does not degrade ECP isoforms. The protein sample is heat/detergent denatured prior to deglycosylation and treated with endoglycosidase H (endo H) and PNGase F (New England Biolabs, Beverly, MA).[19] An aliquot of 0.5 μg of ECP is heated to 95° in 0.5% SDS, and cooled on ice. NP-40 is added to a final concentration of 1%, followed by reaction buffer to a final concentration of 50 m*M* sodium citrate, pH 7.5. Digestion is performed with 2000 units of PNGase F for 2 h at 37°. The samples are analyzed by SDS–PAGE.

RNase Activity

ECP has a low catalytic efficiency for all the common assayed RNase substrates. The RNase activity of ECP is required for its neurotoxic and antiviral properties but not for its antibacterial and antihelminthic activities. The literature reported values are summarized in Table I; catalytic efficiencies are only comparative within each protocol. Comparative substrate preference shows that ECP

[16] H. L. Tiffany and H. F. Rosenberg, *J. Leukocyte Biol.* **58,** 49 (1995).

[17] M. E. Eftink and R. L. Biltonen, *in* "Hydrolytic Enzymes," (A. Neuberger and K. Brocklehurst, eds.), p. 333. Elsevier Science Publishers, Amsterdam, New York and Oxford, 1987.

[18] U. Arnold, A. Schierhorn and R. Ulbrich-Hofmann, *Eur. J. Biochem.* **259,** 470 (1999).

[19] H. F. Rosenberg and H. L. Tiffany, *J. Leukocyte Biol.* **56,** 502 (1994).

TABLE I
RIBONUCLEASE ACTIVITY OF ECP

Substrate	K_{cat} (s^{-1})	k_m (mM)	K_{cat}/k_m (M^{-1} s^{-1})	U/mg	Reference and comments
C > p	1.4×10^{-2}	1.5	11		[a]
U > p	4.3×10^{-3}	1.0	4		[a]
CpA	4.2	2.4	1750		[a]
UpA	6.2	5.4	1150		[a]
UpG			Not detected		[a]
UpU > p			38		[a]
$(Up)_2$ U > p	0.56	1.4	400		[a]
$(Up)_3$ U > p	1.2	0.7	1714		[a]
$(Up)_4$ U > p	1.4	0.17	8235		[a]
Yeast tRNA	0.0024	0.0041	590		[b]
Yeast tRNA	0.1	0.0019	49,000		[c]
Yeast RNA				1.3 (2.4[e])	[d]
Phage f2 RNA				125	[f]
Poly(U)				12.2	[f]
Poly(C)				3.1	[f]
Poly(A)				Not detected	[f]
Poly(A,U)				232	[f]
Poly(A,C)				205	[f]
Poly(A) : Poly(U)				Not detected	[f]
Poly(I) : Poly(C)				Not detected	[f]

[a] E. Boix, Z. Nikolovski, G. P. Moiseyev, H. F. Rosenberg, C. M. Cuchillo and M. V. Nogués, *J. Biol. Chem.* **274,** 15605 (1999). Values obtained by the spectrophotometric method using rECP. Reaction conditions are 50 mM MES–NaOH, 2 mM EDTA, pH 6.2, at 25°. $(Up)_n$U > p oligonucleotides are obtained by HPLC as described in the procedures.

[b] H. F. Rosenberg and K. D. Dyer, *Nucleic Acids Res.* **25,** 3532 (1997). Values obtained using rECP by the measurement of the generation of acid soluble nucleotides as described in the procedures. Assays are performed in 40 mM sodium phosphate, pH 7 at 25°. It is assumed that the average molecular weight of tRNA is about 28,100 and that the $A_{260} = 1$ corresponds to 40 μg of RNA.

[c] J. B. Domachowske, K. D. Dyer, A. G. Adams, T. L. Leto and H. F. Rosenberg, *Nucleic Acids Res.* **26,** 3358 (1998). Values obtained using N-glycosylated rECP. Assays are performed in 40 mM sodium phosphate, pH 7.5 at 25°, measuring the generation of acid soluble nucleotides as previously described.

[d] S. Sorrentino and D. G. Glitz, *FEBS Lett.* **288,** 23 (1991). Values obtained using eosinophil purified ECP by the spectrophotometric method as described in the procedures. Assay in 50 mM Mops, pH 5, 0.1 M NaCl at 25°.

[e] Assay in 50 mM Mops, pH 7.5, 0.1 M NaCl. Relative units defined as in S. Sorrentino and D. G. Glitz, *FEBS Lett.* **288,** 23 (1991).

[f] S. Sorrentino and D. G. Glitz, *FEBS Lett.* **288,** 23 (1991). Values obtained using eosinophil purified ECP by the spectrophotometric method as described in the procedures. Assay in 0.1 M MES, pH 7, 0.1 M NaCl at 25°. Units defined as (change in A_{260}/min)/total measurable change in A_{260}.

has a clear preference for single-stranded RNA, followed by poly(U) and poly(C), and a lower activity for yeast RNA. Whereas human pancreatic RNase is active on poly(A) and ds RNA, ECP is totally inactive. The inability of ECP to attack double-stranded substrates is correlated with its lack of helix-destabilizing activity.[20] The CpA/UpA ratio is comparable to the one reported for RNase A and EDN. No activity could be detected with UpG, using up to 2 m*M* substrate concentration and up to 20 μM final enzyme concentration. The catalytic efficiency for C > p and U > p hydrolysis is very low, explaining why this activity was not detected in previous studies.

A different cleavage pattern for polynucleotide digestion is observed when the distribution of the reaction products of ECP[10] is compared with that of RNase A.[21] The HPLC elution profile for ECP digestion products suggests a predominant exonucleolytic mechanism, whereas RNase A displays a characteristic endonucleolytic mechanism.[22] The favored exonuclease activity and low efficiency endonuclease activity is more pronounced for poly(C) than for poly(U). The relative percentage of U > p to 3′UMP during the digestion of poly(U), as determined by the HPLC methodology, indicates a very low efficiency for the hydrolysis step in comparison to the transphosphorylation step.[10] This ratio is even lower than the value reported for RNase A.[22]

Characterization of ECP Catalytic and Substrate Binding Sites

Kinetic characterization of rECP[10] and the analysis of the recently solved three-dimensional structure[12] have enabled the identification of some of the potential key residues that participate in the ECP catalytic and substrate binding sites (Fig. 2). The catalytic and substrate binding sites have been defined according to the nomenclature originally designated for RNase A.[23] RNase A binding sites for the phosphate, base, and ribose of the RNA substrate are designated as P_n, B_n, and R_n, respectively. The main sites are p_1, at which phosphodiester bond scission occurs; B_1, which interacts with the base, with the ribose attached to the 3′-oxygen for the p_1 phosphate; and B_2, which binds the base whose ribose provides the 5′ oxygen to the p_1 phosphate. B_1 has nearly absolute specificity for pyrimidines, whereas B_2 strongly prefers purines.

The catalytic triad that forms the RNase catalytic site in ECP, His-15, His-128, and Lys-38, occupies equivalent positions as His-12, His-119, and Lys-41 in RNase A. Site-directed mutagenesis of His-128 and Lys-38 confirmed that these two

[20] S. Sorrentino and D. G. Glitz, *FEBS Lett.* **288,** 23 (1991).

[21] M. Moussaoui, M. V. Nogués, A. Guasch, T. Barman, F. Travers, and C. M. Cuchillo, *J. Biol. Chem.* **273,** 25565 (1998).

[22] M. Moussaoui, A. Guasch, E. Boix, C. M. Cuchillo, and M. V. Nogués, *J. Biol. Chem.* **271,** 4687 (1996).

[23] F. M. Richards and H. W. Wyckoff, *in* "The Enzymes" **4,** (P. D. Boyer, ed.), p. 647. Academic Press, New York, 1971.

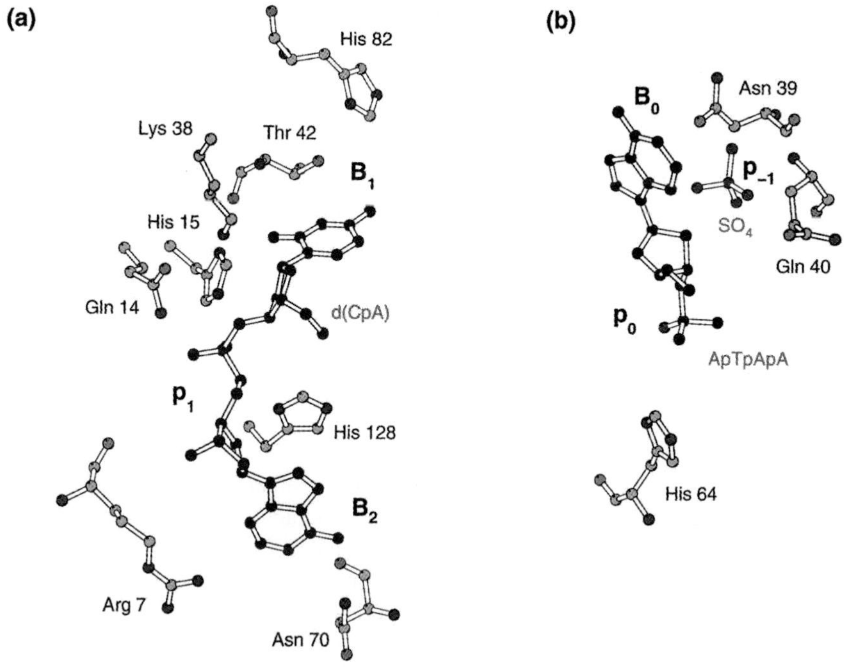

FIG. 2. Putative residues involved in the catalytic and substrate binding sites of ECP according to ECP structure [E. Boix, D. D. Leonidas, Z. Nikolovski, M. V. Nogués, C. M. Cuchillo, and K. R. Acharya, *Biochemistry* **38,** 16794 (1999)] and the superimposed structures of RNase A–substrate analog complexes: (a) RNase A-d (CpA) [1RPG.pdb; I. Zegers, D. Maes, M.-H. Dao-Thi, F. Poortmans, R. Palmer, and L. Wyns. *Protein Sci.* **31,** 2322 (1994)]; (b) RNase A-d (ApTpApA) [1RCN.pdb; J. C. Fontecilla-Camps, R. de Llorens, M. H. le Du, and C. M. Cuchillo, *J. Biol. Chem.* **269,** 21526 (1994)] and EDN-SO_4 [S. C. Mosimann, D. L. Newton, R. J. Youle, and M. N. G. James, *J. Mol. Biol.* **260,** 540 (1996)]. Portions of substrate analogs are drawn in black.

residues are necessary for the RNase activity.[24] ECP exhibits a slight preference for cytosine over uracil in the B_1 position similar to RNase A (Table I). The position of Thr-42 is equivalent to Thr-45 in RNase A. His-82 might be involved in the flexibility of B_1 to interact either with cytosine or uracil, as Asp-83 in RNase A. The B_2 purine binding site shows a strong preference for purines. The position of Asn-70 in ECP is moved slightly away from the purine pocket, compared to its counterpart in RNase A, Asn-71, and cannot interact with the base unless it shifts considerably upon ligand binding. However, the stacking interactions of the catalytic His-128 with the adenine ring observed in RNase A–substrate analog complexes should be

[24] H. F. Rosenberg, *J. Biol. Chem.* **270,** 7876 (1995).

feasible in ECP. The side chain of Asp-112 in ECP cannot mimic the interactions of Glu-111 in RNase A with the adenine. The position of Arg-7 side chain might hamper the binding of a guanine at B_2 and would explain the undetectable activity toward UpG. There are no equivalent residues in ECP to the key p_2 site residues in RNase A, Lys-7, and Arg-10. Superposition of the ECP structure with the RNase A–d(ApTpApA) complex[25] suggests that His-64 can contribute to substrate binding at the p_0 position. The His-64 side chain might provide a stronger interaction than the corresponding p_0 residue in RNase A, Lys-66.[26,27] Analysis of the EDN-sulfate bound structure revealed a conserved secondary sulfate ascribed to a p-1 site, interacting with residues Arg-36, Asn-39, and Gln-40.[28] Arg-34, Asn-39, and Gln-40 residues in ECP occupy equivalent positions, whereas in RNase A Pro-42 and Val-43 cannot perform similar interactions.

Spectrophotometric Method

Assay 1. Kinetic parameters are determined by a spectrophotometric method.[10] Assays are carried out in 50 m*M* 2-(*N*-morpholino)ethanesulfonic acid (MES)–NaOH, 2 m*M* EDTA, pH 6.2, at 25°. Polyuridylic acid, poly(U); polycytidylic acid, poly(C), cytidine 2′,3′-cyclic monophosphate, C > p; uridine 2′,3′-cyclic monophosphate, U > p; cytidylyl(3′,5′)adenosine, CpA; uridylyl(3′,5′)adenosine, UpA; uridylyl(3′,5′)guanosine, UpG; and oligouridylic acids, $(Up)_nU > p$, $n = 1$–4, are used as substrates. Cuvettes of 0.2 cm path length are used, except for $(Up)_4U > p$ substrate, where 1 cm cuvettes are used. Substrate concentration is determined spectrophotometrically using the following extinction coefficients: $\varepsilon_{268} = 6200 M^{-1}cm^{-1}$ for poly(C) nucleotide residue; $\varepsilon_{260} = 9430\ M^{-1}cm^{-1}$ for poly(U) nucleotide residue; $\varepsilon_{260} = 10{,}000\ M^{-1}cm^{-1}$ for U > p; $\varepsilon_{268} = 8400\ M^{-1}cm^{-1}$ for C > p; $\varepsilon_{265} = 21{,}000\ M^{-1}cm^{-1}$ for CpA; $\varepsilon_{261} = 23{,}500\ M^{-1}cm^{-1}$ for UpA; and $\varepsilon_{261} = 20{,}600\ M^{-1}cm^{-1}$ for UpG. For oligouridylic acids ($(Up)_n$ U > p) the ε_{260} values in each case are determined according to the following formula: $\varepsilon_{260} = 10{,}000(n+1)/n$, where n is the number of noncyclic phosphodiester bonds. Oligouridylic acids are obtained by separation of the reaction products of poly(U) digestion by HPLC, as described below. The activity is measured by following the initial reaction velocities using the difference molar absorbance coefficients, in relation to the cleaved phosphodiester bonds in the transphosphorylation reaction: $\Delta\varepsilon_{250} = 2380\ M^{-1}cm^{-1}$ for poly(C) and $\Delta\varepsilon_{282} = 829\ M^{-1}cm^{-1}$ for poly(U); $\Delta\varepsilon_{286} = 1450\ M^{-1}cm^{-1}$ for CpA; $\Delta\varepsilon_{286} = 570\ M^{-1}cm^{-1}$ for UpA;

[25] J. C. Fontecilla-Camps, R. de Llorens, M. H. le Du, and C. M. Cuchillo, *J. Biol. Chem.* **269,** 21526 (1994).

[26] M. V. Nogués, M. Moussaoui, E. Boix, M. Vilanova, M. Ribó, and C. M. Cuchillo, *Cell. Mol. Life Sci.* **54,** 766 (1998).

[27] B. M. Fisher, J. H. Ha, and R. T. Raines, *Biochemistry* **37,** 12121 (1998).

[28] S. C. Mosimann, D. L. Newton, R. J. Youle, and M. N. G. James, *J. Mol. Biol.* **260,** 540 (1996).

$\Delta\varepsilon_{280} = 480\ M^{-1}\text{cm}^{-1}$ for UpG; and the following coefficients for the hydrolysis reaction: $\Delta\varepsilon_{286} = 1450\ M^{-1}\text{cm}^{-1}$ for C > p and $\Delta\varepsilon_{280} = 1000\ M^{-1}\text{cm}^{-1}$ for U > p. For the transphosphorylation reaction of oligouridylic acids $(\text{Up})_n$ U > p, $\Delta\varepsilon_{280} = 700\ M^{-1}\text{cm}^{-1}$ is used, considering that the $\Delta\varepsilon_{280}/n$ is equivalent for all the assayed oligonucleotides where n = 1–4. Substrate concentration ranges are 0.1 to 1 m*M* for cyclic mononucleotides and dinucleotides, and from 0.01 to 0.1 m*M* for oligouridylic acids. Final rECP concentration is in the range from 0.1 to 10 μM depending on the substrate assayed. EDTA is included in the incubation buffer, as it was found that Zn^{2+} reduces ECP activity. Zn^{2+}/ECP molar ratios from 1 to 80 were tested and a reduction of the enzymatic activity up to 50% was observed, using poly(U) as a substrate.

No accurate k_{cat} and K_m parameters can be calculated for polynucleotides, because of the sigmoidal behavior of the cleavage progress curve, as followed by spectrophotometry. The progressive steps of the enzymatic reaction of polynucleotide degradation is analyzed by HPLC of the digestion products, as described below.

Assay 2. The activity using yeast RNA substrate is determined spectrophotometrically by following the hypochromicity at 300 nm in 50 m*M* 3-(*N*-morpholino) propanesulfonic (MOPS), pH 5 or 7.5, 0.1 *M* NaCl at 25°.[20] The degradation of viral ssRNA and dsRNA, as well as single-stranded polynucleotides [poly(U), poly(C), poly(A), poly(A,U) and poly(A,C)] and double-stranded homopolymers [poly(A) : poly(U) and poly(I) : poly(C)], is measured by following the absorbance increase at 260 nm.[20] Substrate and enzyme are mixed in 1 ml 0.1 *M* MES, pH 7, 0.1 *M* NaCl, at 25°. Single-stranded viral RNA is obtained from bacteriophage f2.[29] The concentration of each substrate calculated at pH 7 is 0.1 m*M* in phosphodiester groups (calculated on the basis of A_{260}).

Analysis by HPLC of Polynucleotide Digestion Products

Polynucleotide digestion is analyzed by HPLC.[10] Poly(C) or poly(U) substrate, 25 μl of 5 mg/ml in 10 m*M* Tris-HCl, pH 7.5, is incubated with 5 μl of 7 μM of rECP for poly(U) and 28 μM for poly(C) digestion. Aliquots are taken at different digestion times between 0 and 90 min. An aliquot (15 μl) of the reaction mixture is injected on a reverse phase HPLC column, Novapack C_{18} (3.9 × 150 mm) (Waters, Milford, MA), equilibrated with 10% ammonium acetate (w/v) and 1% acetonitrile in water. After washing the column with the equilibrating buffer for 10 min, a 50 min linear gradient is applied from 0 to 90% of elution buffer (10% ammonium acetate (w/v) and 11% acetonitrile (v/v) in water). Product elution is detected by following the absorbance at 260 nm. Peak identification is performed by mass spectrometric determination of the molecular masses. The mononucleotides

[29] D. G. Glitz, *Biochemistry* **7,** 927 (1968).

3′UMP and U > p elute sequentially but very close in the first peak. A better separation of both mononucleotides can be performed by injecting the first eluted peak in an anion-exchange Nucleosil 10SB column (Macherey-Nagel, Easton, PA).[22] The relative amount of each nucleotide is determined by area peak integration at 260 nm, taking into consideration the number of nucleotide residues. The elution profile cannot distinguish between consecutive sizes after a length of 9–10 nucleotides. For large preparation of $(Np)_nN > p$ oligonucleotides, 500 μl of 10 mg/ml of substrate and 100 μl of enzyme dilution are used. Oligonucleotide fractions are pooled and freeze-dried. Molecular mass of the oligonucleotides is determined by MALDI-TOF mass spectrometry on a Brucker Biflex mass spectrometer (Bremen, Germany). Freeze-dried samples are resuspended in deionized water. An aliquot of 1 μl of NH_4^+ cation-exchange polymer beads Dowex 50W-X8 (Bio-Rad, Hercules, CA) is loaded onto the target inert metal surface, and after removal of excess solvent, 1 μl of 300 m*M* 3-hydroxypicolinic acid in 50% acetonitrile is added, along with 0.7 μl of 1–10 m*M* of the oligonucleotide. All spectra are taken in the reflectron positive ion mode.

Generation of Acid-Soluble Nucleotides

Baker's is yeast RNA (type XI, Sigma) is dissolved (1.7 mg/ml final concentration) in 50 m*M* MES–NaOH, pH 6.5, 3 m*M* EDTA, and 0.01% (w/v) bovine serum albumin (BSA).[30] A volume of 0.3 ml of diluted substrate is mixed with 0.2 ml of sample containing ECP. The mixture is incubated for 30 min at 37°. The reaction is stopped with addition of quenching buffer (0.25 ml); the quenching buffer is freshly prepared by mixing equal volumes of perchloric acid (25%, v/v) containing 0.75% (w/v) phosphotungstic acid and a BSA solution in deionized water (0.6%, w/v) and kept on ice before use. After addition of the quenching buffer, the reaction mixture is maintained on ice for 10 min and centrifuged at 5000*g* for 20 min at 4°. The supernatant is diluted 25 times with distilled water and the concentration of acid-soluble nucleotides is measured at 260 nm. This assay has been used for the analysis of chromatographic fractions during the purification protocol of rECP.

Yeast tRNA degradation is assayed by the measurement of solubilized nucleotides after the precipitation of the undigested tRNA with lanthanum nitrate and 3% perchloric acid.[4,24]

Activity Staining Gels

Samples are dissolved in nonreducing buffer (60 m*M* Tris-HCl, pH 6.8, 10% (v/v) glycerol, 3%(w/v) SDS, 0.015% bromphenol blue) and loaded on an

[30] E. Boix, M. V. Nogués, C. H. Schein, S. A. Benner, and C. M. Cuchillo, *J. Biol. Chem.* **269,** 2529 (1994).

SDS–15% PAGE containing 0.6 mg/ml of either poly(U) or poly(C) as substrate.[31] The polynucleotide should be diluted in deionized water and heated to 55° for 5 min before mixing with the polyacrylamide solution, preheated for 1–2 min at 55°. The stacking gel contains 3.5% acrylamide. Samples are heated for 5 min at 37° before loading. After electrophoresis, SDS is eliminated by washing the gels with 10 m*M* Tris-HCl, pH 7.5, containing 20% cold 2-propanol. Gels are incubated for 15–90 min in 100 m*M* Tris-HCl, pH 7.5. After a wash with 10 m*M* Tris-HCl, pH 7.5, the gels are stained with 0.2% toluidine blue (w/v) in 10 m*M* Tris-HCl, pH 8 and washed with deionized water to eliminate the colorant excess. Using poly(U) as a substrate, up to 15 ng of rECP can be detected.

ECP Biological Properties

Antibacterial Activity

Eosinophils are activated during bacterial infection and a specific release of ECP is observed.[1] ECP is active against both gram-positive and gram-negative species.[32] The antibacterial activity is not shared either by EDN or by the "ancestral" EDN/ECP precursor, the mEDN from *Saguinus oedipus*.[14]

Overnight cultures of *Escherichia coli* strain HB101 and *Staphylococcus aureus* strain 502 A (ATCC) are washed twice and resuspended at 1 : 100 or 1 : 1000 in 10 m*M* sodium phosphate, pH 7.5; 20 μl of bacterial suspension is then added to several solutions containing up to 2.5 μM of ECP in the same phosphate buffer.[24,32] After 5 h incubation at 37°, the remaining colony-forming unit (cfu)/ml are determined by serial plating of 10-fold dilutions of the treated bacterial suspensions on Luria–Bertani (LB) agar plates incubated overnight at 37°. Growth of both *E. coli* (strain HB101) and *S. aureus* (strain SA502A) are inhibited by micromolar concentration of native ECP, purified from eosinophil granules; about 0.1% and 10% of cfu remain after exposure of 2.3 μM of ECP to *S. aureus* and *E. coli* strains, respectively. Nonglycosylated rECP obtained from the pFCTS prokaryote expression system (see above) shows slightly lower antibacterial activity than the eosinophil purified ECP; the K38R + H128D mutant, which lacks RNase activity, shows similar antibacterial capacity as the wild-type ECP.[24] The influence of external factors in the antibacterial activity of ECP has been analyzed.[32] Antibacterial activity of ECP is highly dependent on the growth conditions. The activity is enhanced if mid-logarithmic phase bacteria are replaced by stationary-phase organisms. Nutrient-free conditions or the presence of proton ionophores, DNP, and carbonyl cyanide *m*-chlorophenyl hydrazone (CCCP) inhibit ECP antibacterial activity. Equivalent activity is observed from pH 6 to pH 7.5, but no activity

[31] J. Bravo, E. Fernandez, M. Ribó, R. de Llorens, and C. M. Cuchillo, *Anal. Biochem.* **219,** 82 (1994).

[32] R. I. Lehrer, D. Szklarek, A. Barton, T. Ganz, K. J. Hamann, and G. J. Gleich, *J. Immunol.* **142,** 4428 (1989).

is detected at a pH lower than 5. Addition of 140 m*M* NaCl, 1 m*M* Ca^{2+} or 1 m*M* Mg^{2+} also reduces the antibacterial activity. The activity is related to the observed capacity of ECP to permeabilize the inner and outer membranes of *E. coli* ML-35 strain.[32]

Neurotoxicity

When eosinophils are injected intrathecally to experimental animals a neurotoxic reaction is triggered. The syndrome referred as the *Gordon phenomenon* consists of muscular rigidity, ataxia, and paralysis. Histopathology of the affected animals shows a loss of cerebellar Purkinje cells, and a spongiform change in the white matter of the cerebellum, pons, and spinal cord, whereas the gray matter remains normal. The *Gordon phenomenon* is also induced by ECP and EDN. Neurotoxicity of the eosinophil granule proteins has been related to the neurologic abnormalities detected in hypereosinophilia.

A dose of 60 ng of ECP injected into guinea pigs is sufficient to destroy the Purkinje cells of the cerebellum, ECP being 100 times more potent than EDN.[33] Similar effects for either protein are observed when injected into rabbits, EDN being effective down to 0.15 μg and ECP to 2 μg.[34] By inactivation of RNase activity of both EDN and ECP by carboxymethylation of the catalytic histidines, it is concluded that the RNase activity is necessary to induce the neurotoxic effect.[35]

The neurotoxic activity has been assayed by injecting a dose of 0.15–10 μg in a volume of 0.2 ml intrathecally to New Zealand White rabbits.[34] Rabbits are observed daily for signs of neurotoxic reaction.

Antiviral Activity

Eosinophil recruitment and degranulation is associated with the pathogenesis of respiratory syncytial virus (RSV) bronchiolitis.[36] The RSV is a nonsegmented single stranded RNA virus of the Paramyxoviridae family and is one of the major cause of severe bronchial epithelium infection in infants.[3]

The antiviral activity assay is described in detail elsewhere.[4,13] Briefly, rECP, from 60 to 500 n*M*, is added to viral stocks of RSV and after 2 h of incubation the treated viral stocks are used to infect target HEp-2 cells (human pulmonary epithelial/laryngeal carcinoma cells). Primary RSV-infected cells are detected by

[33] K. Fredens, R. Dahl, and P. Venge, *J. Allergy Clin. Immunol.* **70,** 361 (1982).

[34] G. J. Gleich, D. A. Loegering, M. P. Bell, J. L. Checkel, S. J. Ackerman, and D. J. McKean, *Proc. Natl. Acad. Sci. U.S.A.* **83,** 3146 (1986).

[35] S. Sorrentino, D. G. Glitz, K. J. Hamann, D. A. Loegering, J. A. Checkel, and G. J. Gleich, *J. Biol. Chem.* **267,** 14859 (1992).

[36] A. M. Harrison, C. A. Bonville, H. F. Rosenberg, and J. B. Domachowske, *Am. J. Respir. Crit. Care Med.* **159,** 1918 (1999).

immunofluorescence staining. No toxicity to the HEp-2 monolayer is observed at any of the enzyme concentrations used. To quantify the intact RNA virus genome left after the treatment with ECP, reverse transcriptase–polymerase chain reaction (RT-PCR) is performed.[2] An aliquot of 500 n*M* of glycosylated rECP obtained in the baculovirus expression system reduces the virus infectivity 6-fold, whereas the same quantity of rEDN reduces the infectivity by more than 50-fold.[13]

Helminthotoxicity

Release of eosinophil granule proteins has been detected with several species of trematodes, nematodes, and cestodes; deposition of the granular content onto microfilariae also has been observed.[37,38] Filariasis, like many helminth infections, is marked by blood and tissue eosinophilia. ECP is a potent helminthotoxin, with a much higher toxicity than its close homolog EDN on *Schistosoma mansoni*. ECP kills the larva *in vitro* and the microfilariae of *Brugia pahangi* and *Brugia malayi* in a dose-related manner. Concentration range from 10^{-6} to 10^{-4} *M* have been reported to increase the mortality of several helminth species to 50% or even higher values. Heparin inhibits the toxicity, but not placental RNase inhibitor.[37]

Helminth cultures are prepared with microfilaria (mf) obtained from peritoneal washings of infected mongolian jirds *(Meriones unguiculatus)*, resupended to a density of 200–300 mf/50 μl. After various periods of incubation with the granule proteins, up to 72 h, the percentage of dead mf is examined microscopically.[37]

Cytotoxicity on Mammalian Cells

Eosinophils are regarded as the primary leukocytes responsible for tissue damage in bronchial asthma. Bronchial biopsies of patients observed by electron microscopy show widening of the intracellular spaces and opening of tight junctions where eosinophils are present. The toxicity of ECP has been analyzed on guinea pig tracheal epithelium *in vitro*. Deposits of granule proteins are found on myoblasts incubated with eosinophils infected with *Trypanosoma cruzi*. Myoblast injury appears to be mediated by eosinophil secretion products. Monolayers of rat myoblast cells are prepared, and myoblast lysis or detachment and the deposit of granule proteins is analyzed.[39] The formation of transmembrane pores both in cell membranes and artificial lipid vesicles is observed on incubation with 0.1 to 1 μg of ECP.[40]

[37] K. J. Hamann, G. J. Gleich, J. L. Checkel, D. A. Loegering, J. W. McCall, and R. L. Barker, *J. Immunol.* **144,** 3166 (1990).

[38] S. Motojima, *in* "Eosinophils. Biological and Clinical Aspects" (S. Makino and T. Fukuda, eds.), p. 75. CRC Press, Boca Raton, FL, 1993.

[39] H. A. Molina and F. Kierzenbaum, *Immunology* **66,** 289 (1989).

[40] J. D. Young, C. G. B. Peterson, P. Venge, and Z. A. Cohn, *Nature* **321,** 613 (1986).

Quantification of ECP Levels in Biological Fluids

ECP is currently used as a marker in clinical trials for the diagnosis and monitoring of eosinophilia and inflammatory disorders. However, ECP measurements have to be considered as a complement in the diagnosis and monitoring of inflammation diseases, and cannot replace other clinical methods for the assessment of the severity of inflammation. Table II summarizes some of the literature reported reference values for ECP content of control (healthy) donors.

TABLE II
ECP CONTENT IN HEALTHY DONORS

Sample	ECP concentration	Ref.
Eosinophils	3.4 μg/ml/10^6 cells	a
	5 μg/10^6 cells	b
Neutrophils	0.16 μg/10^6 cells	a
	0.05 μg/10^6 cells	b
Basophils	0.08 μg/10^6 cells	b
Serum	7–20 μg/liter	c
Sputum	50/100 μg/liter	d
Lower airway secretions	5–35 μg/liter/mg protein	e
Nasal secretions	3–50 ng/ml	f
Urine	0.65 μg/g creatinine	g

[a] S. Sur, D. G. Glitz, H. Kita, S. M. Kujawa, E. A. Peterson, D. A. Weiler, G. M. Kephart, J. M. Wagner, T. J. George, G. J. Gleich, and K. M. Leiferman, *J. Leukocyte Biol.* **63,** 715 (1998).

[b] R. I. Abughazaleh, S. L. Dunnette, D. A. Loegering, J. L. Checkel, H. Kita, L. L. Thomas, and G. J. Gleich, *J. Leukocyte Biol.* **52,** 611 (1992).

[c] Average values from 51 studies from healthy donors in several countries. P. Venge, J. Bystrom, M. Carlson, L. Hakansson, M. Karawacjzyk, C. Peterson, L. Seveus, and A. Trulson, *Clin. Exp. Allergy* **29,** 1172 (1999).

[d] Supernatant/cell lysate ratio in sputum. P. G. Gibson, K. L. Woolley, K. Carty, K. Murree-Allen, and N. Saltos, *Clin. Exp. Allergy* **28,** 1081 (1998).

[e] A. M. Harrison, C. A. Bonville, H. F. Rosenberg, and J. B. Domachowske, *Am. J. Respir. Crit. Care Med.* **159,** 1918 (1999).

[f] L. Klimek and G. Rasp, *Clin. Exp. Allergy* **29,** 367 (1999).

[g] F. W. Tischendorf, N. W. Bratting, G. D. Burchard, T. Kubica, G. Kreuzpaintner, and M. Lintzel, *Acta Tropica* **72,** 157 (1999).

ECP immunodetection has been performed in many clinical studies using EG1 and EG2 monoclonal antibodies (Pharmacia).[41] EG1 and EG2 antibodies are reported to distinguish between storage and secreted eosinophil proteins. EG1 recognizes both the storage and secreted forms of ECP; EG2 binds to only the secreted form of ECP and also recognizes EDN. EG1 can recognize the three glycosylated forms of ECP, but can only be used in nonreducing conditions, whereas EG2 detects only one of the three glycosylated forms of ECP.[19] However, cell staining using EG2 antibody has been reported to be mainly dependent on the immunohistochemical technique.[42] An increasing number of clinical studies currently use ECP as a marker, and Pharmacia & Upjohn has developed an improved detection fluorescence immunoenzymatic assay (FEIA) and radioimmunoassay (RIA), Pharmacia CAP System, that can detect up to 2 ng/ml concentration of ECP.

Eosinophils

Eosinophils are tissue-dwelling cells, and their peripheral circulation is regarded as "passing through." Eosinophil half-life in the blood is about 6–18 h, whereas its lifespan in the tissues can range from 2 to 5 days.[43] Table II lists the ECP content of eosinophils and neutrophils. ECP content detected in neutrophils has been suggested to be due to active uptake by the neutrophils of the protein secreted by the eosinophils.[1] When blood specimens from patients with eosinophilia are compared to those of normal individuals, two populations of eosinophils are revealed, "hypodense" and "normodense" cells. The number of light density eosinophils correlates with the degree of blood eosinophilia. "Hypodense" eosinophils have small granules and contain less granule proteins and are regarded as "activated eosinophils."[43] A reduction in intracellular expression of ECP in peripheral eosinophils has been detected in asthmatic children.[44]

Blood

ECP levels in EDTA–plasma are considered to indicate more accurately the circulation levels, whereas serum levels would reflect the secretory activity of the eosinophils and their propensity to release the granule proteins. ECP serum levels can be a result of spontaneous release during blood processing, depending on the

[41] P. C. Tai, C. J. F. Spry, C. Peterson, P. Venge, and I. Olsson, *Nature* **309,** 182 (1984).

[42] H. Nakajima, D. A. Loegering, H. Kita, G. M. Kephart, and G. J. Gleich, *J. Leukocyte Biol.* **66,** 447 (1999).

[43] G. J. Gleich, H. Kita and C. R. Adolphson, *in* "Immunologic Diseases" (M. M. Frank, K. F. Austen, H. N. Claman, and E. R. Unanue, eds.), p. 205. Little, Brown & Co., Boston, 1995.

[44] N. Krug, U. Napp, I. Enander, E. Eklund, C. H. L. Rieger, and U. Schauer, *Clin. Exp. Allergy* **29,** 1507 (1999).

presence of divalent cations, time, and temperature of incubation.[45] An increase relative to serum control values (Table II) to 200–500 μg/liter has been reported in inflammatory diseases.[46]

Blood samples, with or without EDTA, are collected and after incubation at different temperatures, or at various time intervals, plasma and serum are recovered by centrifugation at 3000g for 10 min at 4° and frozen at −20°.[45]

Sputum

Sputum is a noninvasive marker of airway passage inflammation. The following assay allows the measurement in sputum samples of total ECP levels and of ECP released by eosinophil degranulation.[47] Sputum is induced by hypertonic saline inhalation and sputum plugs are isolated from the sample, diluted 1/10 in 0.1% dithiothreitol (DTT), and mixed at room temperature for 30 min to disperse the cells. The samples are then filtered through a 50 μm nylon mesh filter and centrifuged at 700g for 10 min at 4°, and the supernatant decanted for detection of ECP. Stored sputum supernatant samples are stable at −20° for at least 6–9 months. The sputum cell pellet is further processed by dilution in PBS to 1×10^{-6} ml^{-1} for eosinophil counts. The sputum cell pellet (1×10^{-6} cells) is lysed with 1 ml of 0.1% Triton X-100 in 0.5 *M* Tris-HCl, pH 8, and kept at room temperature for 30 min. Samples are then centrifuged 10 min at 700g at 4°, and the cell-free supernatant is stored at −20° for measurement of the ECP levels in the sputum cell lysate. The ECP in the lysate fraction provides an estimate of the eosinophil number, and the ECP supernatant/cell lysate ECP ratio provides a measure of eosinophil degranulation. Sputum ECP levels can be affected by high levels of saliva contamination.[48] It is recommended to measure the salivary contamination by the count of squamous cell percentage and discard the samples with a 40% of squamous cells. Sputum supernatant/cell lysate ratio ECP values increases from 50/100 μg/liter in normal donors (Table II) to 600/700 μg/liter in stable asthma, and to 4600/1600 μg/liter in asthma exacerbation.

Other Biological Fluids

Bronchoalveolar lavage (BAL) has been used to study inflammation in the lung airways. A positive correlation is observed between ECP levels and the severity

[45] C. M. Reimert, L. K. Poulsen, C. Bindslev-Jensen, A. Kharazmi, and K. Bendtzen, *J. Immunol. Methods* **166,** 183 (1993).

[46] B. Zimmerman, A. Lanner, I. Enander, R. S. Zimmerman, C. G. B. Peterson, and S. Ahlstedt, *Clin. Exp. Allergy* **23,** 564 (1993).

[47] P. G. Gibson, K. L. Woolley, K. Carty, K. Murree-Allen, and N. Saltos, *Clin. Exp. Allergy* **28,** 1081 (1998).

[48] E. Bacci, S. Cianchetti, L. Ruocco, M. L. Bartoli, S. Carnevali, F. L. Dente, A. Di Franco, D. Giannini, P. Macchioni, B. Vagaggini, M. C. Morelli, and P. L. Paggiaro, *Clin. Exp. Allergy* **28,** 1237 (1998).

of asthma. The ECP content in BAL has been also assayed in other inflammatory diseases and acute rejection in allografts.[1] An increase in the lower airway secretions is detected from control values (Table II) to about 1000 μg/liter/mg protein in RSV infection.[36]

Nasal lavage (NL) has mostly been used for the study of allergic rhinitis. Increased levels of ECP are observed during allergen exposure of patients with rhinitis.[1,49]

ECP levels in urine can only be used as marker of eosinophiluria, whereas EDN urine levels are a good marker for evaluation of inflammatory diseases.[50]

Acknowledgments

I thank Victòria Nogués and Claudi Cuchillo for general discussions and for their contribution in the kinetic characterization methodology; Zoran Nikolovski for his help in obtaining recombinant protein; and Gennady Moyseyev for his work on the determination of the kinetic parameters and for constructive criticisms of the manuscript. I also thank Ravi Acharya and Demetres Leonidas for their contribution in the ECP structure determination. I am also very grateful to Helene Rosenberg and Richard Youle for their involvement when I started my work on ECP at the NIH.

[49] L. Klimek and G. Rasp, *Clin. Exp. Allergy* **29,** 367 (1999).
[50] F. W. Tischendorf, N. W. Brattig, G. D. Burchard, T. Kubica, G. Kreuzpaintner, and M. Lintzel, *Acta Tropica* **72,** 157 (1999).

[19] Deciphering the Mechanism of RNase T1

By STEFAN LOVERIX and JAN STEYAERT

Introduction

Ribonucleases have been the subject of four Nobel Prize lectures in chemistry for landmark work on the structure, folding, stability, and molecular evolution of proteins.[1–3] Bovine pancreatic ribonuclease A (RNase A, EC 3.1.27.5), a member of a homologous superfamily of vertebrate ribonucleases, was the first enzyme to be sequenced. Ribonuclease T1 (RNase T1, EC 3.1.27.3) of the slime mold

[1] S. Moore and W. H. Stein, *Science* **180,** 458 (1973).
[2] C. B. Anfinsen, *Science* **181,** 223 (1973).
[3] B. Merrifield, *Biosci. Rep.* **5,** 353 (1985).

0076-6879/01 $35.00

Aspergillus oryzae[4–7] is the best known representative of another family of homologous microbial ribonucleases with members in the prokaryotic and the eukaryotic world.[8–11] These microbial ribonucleases span the greatest evolutionary divide of all known protein families.[9]

RNase T1, a guanosine specific endoribonuclease, was first isolated by Sato and Egami[4] from an enzyme extract (Taka-diastase) of the culture medium of *A. oryzae* used in the malting process of sake brewing. The enzyme consists of a single polypeptide chain of 104 residues of known sequence[12,13] and contains two disulfide bridges.[14] Since its first structural analysis,[15] a large number of crystal structures of RNase T1 in different liganded states became available.[16–28] The general fold consists of a 4.5-turn α helix and two antiparallel β sheets, connected through a series of wide loops[15,16] (Fig. 1). The residues implicated in catalysis

[4] K. Sato and F. Egami, *J. Biochem.* **44,** 753 (1957).
[5] K. Takahashi, T. Ushida, and F. Egami, *Adv. Biophys.* **1,** 53 (1970).
[6] T. Ushida and F. Egami, *Enzymes* **4,** 205 (1971).
[7] K. Takahashi and S. Moore, *Enzymes* **15,** 435 (1982).
[8] R. W. Hartley, *J. Mol. Evol.* **15,** 355 (1980).
[9] C. Hill, G. Dodson, U. Heinemann, W. Saenger, Y. Mitsui, K. Nakamura, S. Borisov, G. Tischenko, K. Polyakov, and S. Pavlovsky, *Trends. Biochem. Sci.* **8,** 364 (1983).
[10] J. Sevcik, R. G. Sanishvilli, A. G. Pavlovsky, and K. M. Polyakov, *Trends. Biochem. Sci.* **15,** 158 (1990).
[11] R. Loris, U. Langhorst, S. De Vos, K. Decanniere, J. Bouckaert, D. Maes, T. R. Transue, and J. Steyaert, *Proteins* **36,** 117 (1999).
[12] K. Takahashi, *J. Biol. Chem.* **240,** 4117 (1965).
[13] K. Takahashi, *J. Biochem.* **98,** 815 (1985).
[14] C. N. Pace, U. Heinemann, U. Hahn, and W. Saenger, *Angew. Chem., Int. Ed. Engl.* **30,** 343 (1991).
[15] U. Heinemann and W. Saenger, *Nature* **299,** 27 (1982).
[16] R. Arni, U. Heinemann, R. Tokuoka, and W. Saenger, *J. Biol. Chem.* **263,** 15358 (1988).
[17] S. Sugio, A. Takashi, H. Ohishi, and K.-I. Tomita, *J. Biochem.* **103,** 354 (1988).
[18] J. Martinez-Oyanedel, H. W. Choe, U. Heinemann, and W. Saenger, *J. Mol. Biol.* **222,** 335 (1991).
[19] T. Hakoshima, T. Itoh, K.-I. Tomita, K. Goda, S. Nishikawa, H. Morioka, S.-I. Uesugi, E. Ohtsuka, and M. Ikehara, *J. Mol. Biol.* **223,** 1013 (1992).
[20] G. Koellner, H. W. Choe, U. Heinemann, H. P. Grunert, A. Zouni, U. Hahn, and W. Saenger, *J. Mol. Biol.* **224,** 701 (1992).
[21] I. Zegers, P. Verhelst, H. W. Choe, J. Steyaert, U. Heinemann, W. Saenger, and L. Wyns, *Biochemistry* **31,** 11317 (1992).
[22] A. Heydenreich, G. Koellner, H. W. Choe, F. Cordes, C. Kisker, H. Schindelin, R. Adamiak, U. Hahn, and W. Saenger, *Eur. J. Biochem.* **218,** 1005 (1993).
[23] A. Lenz, H. W. Choe, J. Granzin, U. Heinemann, and W. Saenger, *Eur. J. Biochem.* **211,** 311 (1993).
[24] K. Gohda, K.-I. Oka, K.-I. Tomita, and T. Hakoshima, *J. Biol. Chem.* **269,** 17531 (1994).
[25] J. Pletinckx, J. Steyaert, I. Zegers, H. W. Choe, U. Heinemann, and L. Wyns, *Biochemistry* **33,** 1654 (1994).
[26] J. Doumen, M. Gonciarz, I. Zegers, R. Loris, L. Wyns, and J. Steyaert, *Protein Sci.* **5,** 1523 (1996).
[27] S. De Vos, J. Doumen, U. Langhorst, and J. Steyaert, *J. Mol. Biol.* **275,** 651 (1998).
[28] U. Langhorst, R. Loris, V. P. Denisov, B. Halle, P. Roose, and J. Steyaert, *Protein Sci.* **8,** 722 (1999).

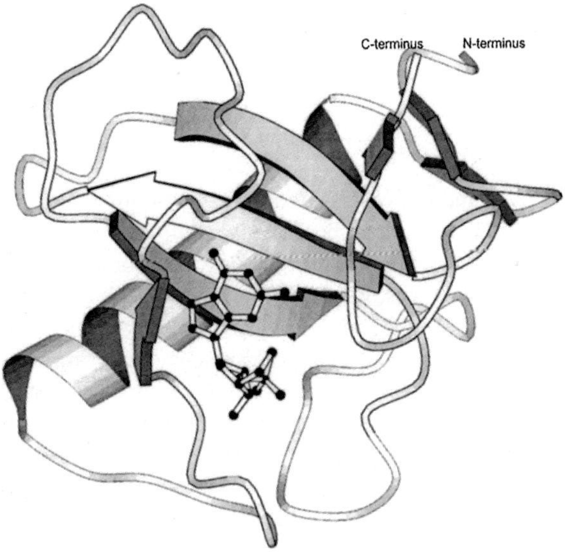

FIG. 1. Ribbon drawing of the three-dimensional structure of RNase T1 in complex with the competitive inhibitor 2′-GMP.[15]

are anchored in the β sheets, while residues in loop regions establish guanosine recognition.

RNase T1 enhances the rate of GpC cleavage about 10^{11}-fold by binding to the transition state of the reaction with a dissociation constant of $3 \times 10^{-15} M$.[29] This paper focuses on the nature of this transition state and reviews the structural and chemical principles underlying the remarkable affinity of RNase T1 for this highest energy intermediate. The ultimate goal is to generate a comparative picture of *all* intermolecular interactions of the RNase–RNA complex in the ground state *versus* the transition state. The introduction of delicate changes in the RNA substrate (chemical synthesis)[30–34] or at the active site of RNase T1 (site-directed mutagenesis)[26,27,35,36] and the subsequent biochemical and structural analysis of

[29] J. E. Thompson, T. G. Kutateladze, M. C. Schuster, F. D. Venegas, J. M. Messmore, and R. T. Raines, *Bioorg. Chem.* **23,** 471 (1995).

[30] F. Eckstein, H. H. Schulz, H. Rüterjans, W. Haar, and W. Maurer, *Biochemistry* **11,** 3507 (1972).

[31] J. Steyaert, A. F. Haikal, L. Wyns, and P. Stanssens, *Biochemistry* **30,** 8666 (1991).

[32] J. Steyaert, C. Opsomer, L. Wyns, and P. Stanssens, *Biochemistry* **30,** 494 (1991).

[33] S. Loverix, A. Winqvist, R. Strömberg, and J. Steyaert, *Nat. Struct. Biol.* **5,** 365 (1998).

[34] S. Loverix, A. Winqvist, R. Strömberg, and J. Steyaert, *Chem. Biol.* **7,** 651 (2000).

[35] S. Nishikawa, H. Morioka, H. J. Kim, K. Fuchimura, T. Tanaka, S. Uesugi, T. Hakoshima, K. Tomita, E. Ohtsuka, and M. Ikehara, *Biochemistry* **26,** 8620 (1987).

[36] J. Steyaert, K. Hallenga, L. Wyns, and P. Stanssens, *Biochemistry* **29,** 9064 (1990).

the complex allowed construction of an increasingly detailed picture of the assembly of protein side chains used to control the recognition between the enzyme and the substrate undergoing transphosphorylation.[37,38]

Reactivity of Scissile Bond

Phosphate esters and anhydrides dominate the living world. Biological information is stored in DNA and can be expressed into proteins. RNA serves as the conduit: DNA→RNA→protein. By catalyzing the synthesis and degradation of RNA, two classes of enzymes, RNA polymerases and ribonucleases, control this flow. Ribonucleases accelerate the hydrolysis of the P–O5′ phosphodiester bonds in single-stranded RNA by nucleophilic displacements of oxygens at the phosphorus involving reaction intermediates (or transition states) in which phosphor is transiently engaged in five bonds to oxygen atoms. This chemistry is hard to accomplish. Indeed phosphoric acid is specially adapted for its role in nucleic acids because it can link two nucleotides and still ionize.[39] The negative charge serves to stabilize the diester against nucleophilic attack. The P–O phosphodiester bonds in DNA have a half-life of at least 22 million years at 25° in the absence of catalysts.[40] The uncatalyzed conversion of the dinucleoside phosphate UpA to 2′,3′-cyclic UMP and A was found to proceed at a rate of 5×10^{-9} s^{-1} indicating that the phosphodiester bond in RNA has a half-life of about 4 years.[29]

The classical reaction scheme involves a reversible transphosphorylation yielding a 2′,3′-cyclophosphate (Fig. 2a). In a second, separate step, this cyclic product is hydrolyzed to yield a free 3′-phosphate.[41–43] So far, most authors have considered 3′- nucleotides as the quasi irreversible product of the ribonuclease-catalyzed reactions because the high water concentration (55 *M* under physiological conditions) pushes the equilibrium toward the fully hydrolyzed reactant. In reality, ribonucleases cycle continuously between the 3′-nucleotide and the 2′,3′-cyclophosphate. Indeed, older literature indicates that 2′,3′-cyclic nucleotides can be synthesized enzymatically from 3′-nucleotides by RNase A.[44,45] Isotope exchange and intermediate trapping experiments also show that the hydrolysis step is readily reversible in

[37] J. Steyaert, *Eur. J. Biochem.* **247,** 1 (1997).

[38] M. Irie, *in* "Ribonucleases: Structures and Functions" (G. D'Alessio and J. F. Riordan, eds.), p. 101. Academic Press, New York, 1997.

[39] F. H. Westheimer, *Science* **235,** 1173 (1987).

[40] R. Wolfenden, C. Ridgway, and G. Young, *J. Am. Chem. Soc.* **120,** 833 (1998).

[41] C. M. Cuchillo, X. Pares, A. Guasch, T. Barman, F. Travers, and M. V. Nogues, *FEBS Lett.* **333,** 207 (1993).

[42] J. E. Thompson, F. D. Venegas, and R. T. Raines, *Biochemistry* **33,** 7408 (1994).

[43] S. Loverix, G. Laus, J. C. Martins, L. Wyns, and J. Steyaert, *Eur. J. Biochem.* **257,** 286 (1998).

[44] J. B. Bahr, R. E. Cathou, and G. G. Hammes, *J. Biol. Chem.* **240,** 3372 (1965).

[45] E. J. del Rosario and G. G. Hammes, *Biochemistry* **8,** 1884 (1969).

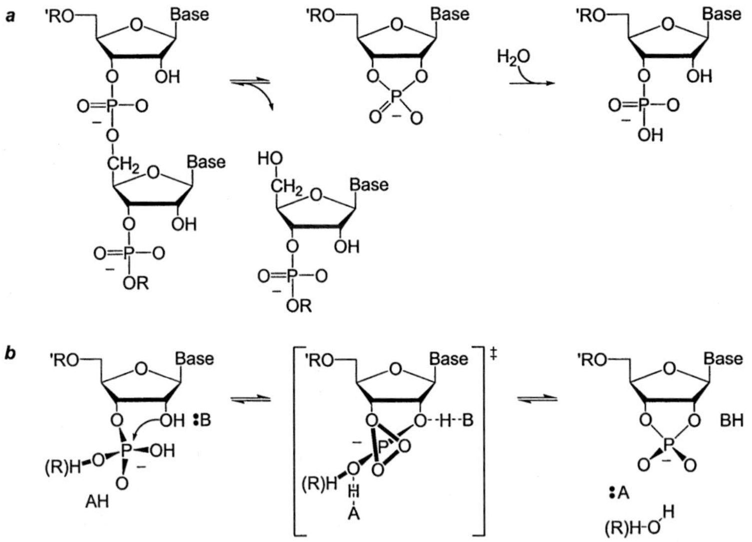

FIG. 2. The chemistry of ribonuclease-catalyzed reactions. (a) Classic mechanism involving a reversible transphosphorylation followed by an irreversible hydrolysis step. (b) Single transferase equilibrium between a phosphodiester and a 2′,3′-cyclic phosphate.

the presence of RNase A or RNase T1.[43] A single equilibrium between a phosphodiester and a 2′,3′-cyclophosphate accounts for all catalyzed reactions (Fig. 2b), even if the leaving/attacking group is a water molecule. The equilibrium constant for the catalyzed interconversion is close to 1 *M*. It follows that the activation energies of the cyclization and the decyclization reaction are very similar.

The forward and reverse nucleophilic substitution reactions (Fig. 2b) in RNA hydrolysis oppose two different types of molecular forces.[43] Considerable ring strain is released on decyclization of the 2′,3′-cyclophosphate. The rate of hydrolysis of cyclic phosphodiesters is 10^7 times greater[46] than that for dimethyl phosphate. The cyclization step is favored by intramolecular nucleophilic attack by the 2′-OH group of RNA. Loverix *et al.* estimated the effective concentration of the 2′-OH nucleophile at 10^7 *M* by comparing the second-order rate of the intermolecular hydrolysis of dimethyl phosphate and the first-order cyclization rate of the RNA phosphodiester bond, using the hydrolysis rate of 2′,3′ cyclic nucleosides as an internal standard.[43] The effective concentration of the 2′-OH nucleophile can also be calculated $\approx 3 \times 10^8$ *M* from the ratio of the uncatalyzed bimolecular rate of hydrolysis of dimethyl phosphate (corrected for the concentration of water)[40] and the uncatalyzed intramolecular cyclization rate of UpA.[29] In total, it appears

[46] J. A. Gerlt, F. H. Westheimer, and J. M. Sturtevant, *J. Biol. Chem.* **250,** 5059 (1975).

that the high effective concentration of the vicinal nucleophile counterbalances the energetic cost of forming a strained pentacyclic phosphodiester to reach an equimolar equilibrium. The reversibility of the cyclization reaction also implies that ribonucleases bind 3′-nucleotides to form a genuine minimal sized ES complex that is relevant to the reaction occurring during RNA depolymerization. Many crystal structures of RNases have been solved in complex with 3′- nucleotides.[22,47] These structures should be considered as genuine enzyme–substrate complexes.

Chemical Nature of Transition State

The ribonuclease-catalyzed reaction consists of a nucleophilic in-line inversion displacement of the 5′-leaving group at the phosphorus atom by the entering 2′-oxygen.[48,49] The classical acid–base mechanism[50,51] is shown in Fig. 3a. For many years, all ribonuclease catalyzed reactions (including RNase A and RNase T1) have been thought to be concerted, with a monoanionic trigonal bipyramidal phosphorane structure (TBP) along the reaction pathway, implying a base and an acid located on either side of the scissile bond[7] (Fig. 3a). In this reaction scheme, the nonbridging oxygens of the TBP structure are not protonated and the catalytic acid and base interact exclusively with the apical oxygens in a concerted fashion. However, many aspects regarding the microscopic details of the catalytic mechanism, including catalytic proton transfer steps, remain unproven.

For RNase A, the actual character of the TBP phosphorane structure, including its protonation state and whether it is a transition state or a true intermediate, has become an issue of considerable debate.[51,52] New triester-like mechanisms have been proposed that are characterized by early proton transfer to a nonbridging oxygen rendering the phosphate neutral (Fig. 3b). Protonation of a nonbridging oxygen facilitates nucleophilic attack by the 2′-alkoxide. In the Breslow mechanism[52] this proton comes from the catalytic acid (Fig. 3b, right route). In an internal proton transfer mechanism,[53,54] the proton comes from the 2′-OH group (Fig. 3b, left route). Either way, the result is a monoanionic phosphorane. Experimental results[34] suggest that the microbial enzyme RNase T1 follows a triester-like rather than the classical concerted mechanism. These data are summarized below. Following, we will survey how RNase T1 stabilizes the transition state of this reaction.

[47] I. Zegers, A. F. Haikal, R. Palmer, and L. Wyns, *J. Biol. Chem.* **269,** 127 (1994).
[48] D. A. Usher, E. S. Erenrich, and F. Eckstein, *Proc. Natl. Acad. Sci. U.S.A.* **69,** 115 (1972).
[49] F. Eckstein, W. Saenger, and D. Suck, *Biochem. Biophys. Res. Commun.* **46,** 964 (1972).
[50] J. E. Thompson and R. T. Raines, *J. Am. Chem. Soc.* **116,** 5467 (1994).
[51] D. Herschlag, *J. Am. Chem. Soc.* **116,** 11631 (1994).
[52] R. Breslow and W. H. Chapman, Jr., *Proc. Natl. Acad. Sci. U.S.A.* **93,** 10018 (1996).
[53] T. M. Glennon and A. Warshel, *J. Am. Chem. Soc.* **120,** 10234 (1998).
[54] B. D. Wladkowski, L. A. Svensson, L. Sjolin, J. E. Ladner, and G. L. Gilliland, *J. Am. Chem. Soc.* **120,** 5488 (1998).

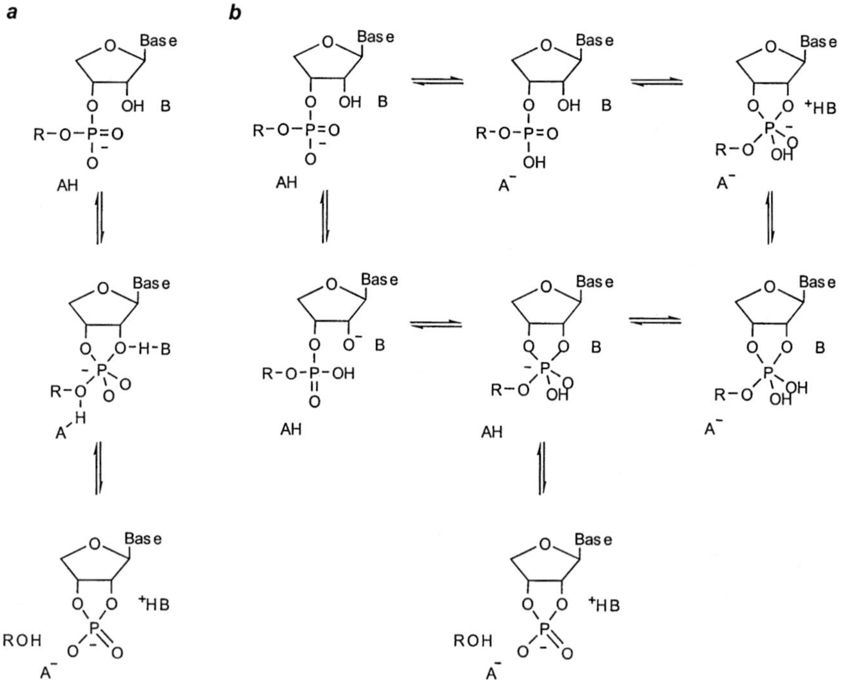

FIG. 3. Proposals for RNase catalyzed reactions. **(a)** Classic acid–base mechanism. **(b)** Triester-like mechanism.

Experimental Basis for In-Line Acid–Base Catalysis

Elegant experiments with the chiral substrate R_p $-2',3'$-cyclic guanosine phosphorothioate[30] (see further) and analysis of the pH dependence of the catalytic parameters[36,55] show that the RNase T1 catalyzed transphosphorylation reaction follows an in-line mechanism implying a base and an acid at either side of the scissile bond. RNase A accelerates the cyclization reaction by general base/general acid catalysis involving two histidines identified as His-12 and His-119.[50,56] Kinetic and protein engineering studies[7,935,36,57] have shown that Glu-58 and His-92 fulfill these functions in RNase T1. In the crystal complexes of RNase T1 and RNase A with their minimal substrates 3′-GMP and 3′-CMP (Fig. 4), the catalytic bases (Glu-58, His-12) and the catalytic acids (His-92, His-119) are positioned at opposite ends of the catalytic site. The bases are in hydrogen bonding distance to the 2′-hydroxyl group of the substrate, whereas the acids bind the nonbridging

[55] H. L. Osterman and F. G. Walz, *Biochemistry* **17,** 4124 (1978).
[56] D. Findlay, D. G. Herries, A. P. Mathias, B. R. Rabin, and C. A. Ross, *Nature* **190,** 781 (1961).
[57] H. L. Osterman and F. G. Walz, *Biochemistry* **18,** 1984 (1979).

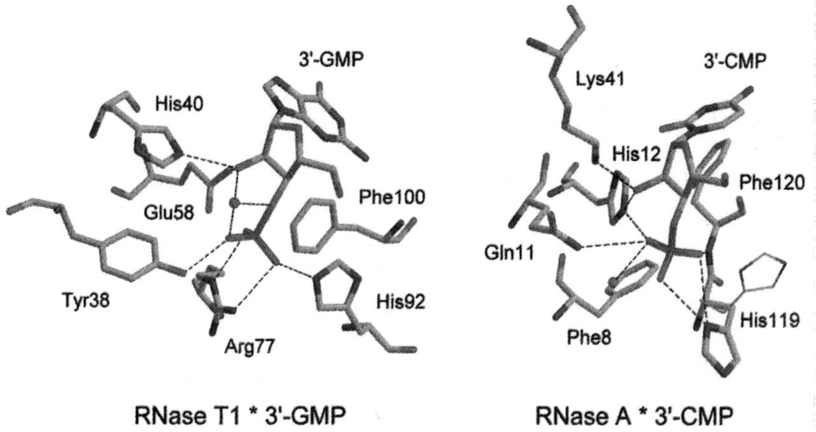

FIG. 4. Active sites of RNase T1 and RNase A in complex with their minimal substrates 3′-GMP and 3′-CMP. Carbon, nitrogen, oxygen, and phosphorus are depicted in gray, blue, red, and magenta, respectively. The general acid His-119 has 2 conformations; most probably His-119 is active in the A conformation.[79]

phosphate oxygen that is furthest away from the attacking nucleophile. This recurrent structural arrangement is entirely consistent with the acid–base catalyzed in-line nucleophilic displacement mechanism at phosphorus, as proposed for both evolutionary unrelated enzymes.[30,49] A time-resolved X-ray crystallographic study[58] of the hydrolysis of the slow substrate *exo*-guanosine 2′,3′-cyclophosphorothioate (a structural analog of 2′,3′-cGMP) by RNase T1 demonstrates that the acid (His-92) and the base (Glu-58) are properly positioned to reciprocally act as a base and an acid in the decyclization reaction as well.

Thio Substitutions to Map Catalytic Interactions with Nonbridging Oxygens

The classical mechanism (Fig. 3a) and the triester-like mechanism (Fig. 3b) differ essentially in the protonation state of the nonbridging oxygens during transition state formation. Thio substitutions on phosphorus have been used to extensively analyze the properties of these oxygens in the RNA–RNase T1 complex.

Internucleotidic phosphorothioate diastereomers (R_p, S_p) resulting from substitutions of one nonbridging phosphoryl oxygen with sulfur are common tools

[58] I. Zegers, R. Loris, G. Dehollander, H. A. Fattah, F. Poortmans, J. Steyaert, and L. Wyns, *Nat. Struct. Biol.* **5,** 280 (1998).

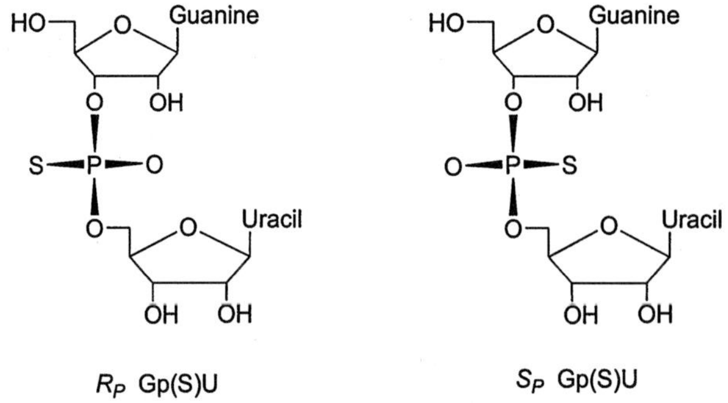

FIG. 5. Thio-substituted diastereomers of the dinucleoside phosphate substrate GpU for the RNase T1 catalyzed transphosphorylation reaction.

to investigate reaction stereochemistry.[59,60] Indeed, a prochiral phosphate diester can be made chiral by replacing an oxygen atom with a sulfur atom. The stereochemical course of RNase T1 action has been established using the R_P diastereomer of 2′,3′-cyclic guanosine phosphorothioate.[30] Alternatively, phosphorothioate (oligo)nucleotides can be applied to identify protein–nucleic acid interactions.[33,61,62] Thio substitutions have also proven to be particularly useful in the mechanistic investigation of ribozymes where they have been applied predominantly to probe the role of metal ions in phosphoryl transfer.[63–65] Nonprotein catalysts such as ribozymes have in common with enzymes an architecture that anchors functional groups in positions suitable for exerting a rate-enhancing effect. Thio substitutions constitute a standard technique in ribozyme research that is comparable to site-directed mutagenesis in enzymology.

Thio-substituted substrates as depicted in Fig. 5 have been used to extensively probe catalytic interactions of RNase T1 with the nonbridging oxygens of the phosphodiester linkage in RNA.[33,34,58] These nonbridging oxygens take up equatorial positions in the TBP transition state of the catalyzed reaction. Quantum chemical studies on the conformational flexibility of phosphate and phosphorothioate dimethyl esters in aqueous solution have been performed.[66] The calculated similarities in the conformational behavior of these esters and the equivalent electronic

[59] F. Eckstein, *Annu. Rev. Biochem.* **54,** 367 (1985).
[60] P. A. Frey, *Adv. Enzymol. Relat. Areas Mol. Biol.* **62,** 119 (1989).
[61] J. F. Milligan and O. C. Uhlenbeck, *Biochemistry* **28,** 2849 (1989).
[62] D. Dertinger, L. S. Behlen, and O. C. Uhlenbeck, *Biochemistry* **39,** 55 (2000).
[63] S. C. Dahm and O. C. Uhlenbeck, *Biochemistry* **30,** 9464 (1991).
[64] D. Herschlag, J. A. Piccirilli, and T. R. Cech, *Biochemistry* **30,** 4844 (1991).
[65] J. M. Warnecke, R. Held, S. Busch, and R. K. Hartmann, *J. Mol. Biol.* **290,** 433 (1999).
[66] J. Florian, M. Strajbl, and A. Warshel, *J. Am. Chem. Soc.* **120,** 7959 (1998).

configuration of oxygen and sulfur justify the use of thio-substitution experiments to study the role of intermolecular interactions involving phosphate oxygens of phosphodiesters. The effects of thio substitutions on the kinetic parameters (the so-called thio effect) are caused by differences between sulfur and oxygen including size, electronegativity, and hydrophobicity. The van der Waals radius of sulfur, 1.85 Å, exceeds that of oxygen by 0.45 Å and the P–S bond, 1.8 Å, is about 0.3 Å longer than P–O.[67] Both features add, and the sulfur projects outward about 0.8 Å more than oxygen. As the electronegativities of oxygen (3.5) and sulfur (2.5) differ appreciably, thio substitutions also affect the distribution of the electrons on phosphate and its substituents. In addition, sulfur is more easily polarized and more hydrophobic than oxygen.[68] Thiophosphate has lower pK_a values than phosphate: 1.7, 5.4, and 10.1 for thiophosphate as compared to 2.1, 7.2, and 12.3 for phosphate.[69] Yet, these differences in pK_a values underestimate the poorer proton affinity of sulfur versus oxygen because protonation occurs preferentially on oxygen for thiophosphate.[70] We have used the R_p and S_p diastereomers of guanylyl-3′,5′-uridine phosphorothioate [Gp(S)U], two optically pure phosphorothioate analogs of the dinucleoside phosphate GpU in which the *pro*-R_p oxygen and the *pro*-S_p oxygen are replaced by sulfur, respectively (Fig. 5). GpU was chosen as the parent dinucleoside phosphate substrate because its enzymatic turnover is limited by the rate of bond making and breaking.[71] This is important because no mechanistic conclusion can be drawn from kinetic thio effects if substrate association of product dissociation are rate-limiting. Because individual thio substitutions of the nonbridging oxygens have little effect on the nonenzymatic cleavage rate of dinucleoside phosphates,[72–74] any thio effect observed on k_{cat}/K_m reflects changes in the enzyme–substrate complementarity of the transition state. To confine this effect to a particular part of the enzyme active site, we analyzed thio effects for wild-type enzyme and a number of active-site mutants. The results are summarized in Table I. The energetic coupling between a stereospecific thio effect and a particular active site mutation (quantified as the apparent coupling energy $\Delta\Delta G$ in Table I) is indicative of a direct or indirect catalytic contact between the nonbridging oxygen of the RNA substrate and the enzymatic side chain under investigation. Loverix *et al.*[33] found that there is a significant interdependence of the R_p thiosubstitution and the Tyr38Phe mutation. It was concluded from these

[67] W. Saenger and F. Eckstein, *J. Am. Chem. Soc.* **92,** 4712 (1970).

[68] W. Saenger, "Principles of Nucleic Acid Structure." Springer-Verlag New York, 1984.

[69] W. P. Jencks and J. Regenstein, *in* "Handbook of Biochemistry and Molecular Biology" (G. D. Fasman, ed.), p. 305. CRC Press, Cleveland, OH, 1976.

[70] P. A. Frey and R. D. Sammons, *Science* **228,** 541 (1985).

[71] J. Steyaert, L. Wyns, and P. Stanssens, *Biochemistry* **30,** 8661 (1991).

[72] P. M. J. Burgers and F. Eckstein, *Biochemistry* **18,** 592 (1979).

[73] H. Almer and R. Strömberg, *Tetrahedron Lett.* **32,** 3723 (1991).

[74] M. Oivanen, M. Ora, H. Almer, R. Strömberg, and H. Lönnberg, *J. Org. Chem.* **60,** 5620 (1995).

TABLE I
S_p AND R_p THIO EFFECTS ON CATALYSIS VERSUS ACTIVE SITE MUTATIONS

	GpU		S_p Gp(S)U			
Mutation	k_{cat}/K_m ($mM^{-1}s^{-1}$)	ΔG^b (kcal/mol)	k_{cat}/K_m ($M^{-1}s^{-1}$)	ΔG^b (kcal/mol)	Thio effect[a]	Coupling[c] $\Delta\Delta G$ (kcal/mol)
Wild type	1000	/	11.3	/	88500	/
Tyr38Phe	15.1	2.56	0.458	1.96	33000	0.6 ± 0.2
His40Ala	0.153	5.38	ND[d]	ND	ND	ND
Glu58Ala	27.40	2.20	7520	−3.98	3.6	6.2 ± 0.2
His92Gln	0.117	5.54	ND	ND	ND	ND
Phe100Ala	1.38	4.03	8.65	0.17	160	3.9 ± 0.3
			R_p Gp(S)U			
			k_{cat}/K_m ($mM^{-1}s^{-1}$)	ΔG^b (kcal/mol)		
Wild type	1000	/	665	/	1.50	/
Tyr38Phe	15.1	2.56	210	0.70	0.072	1.86 ± 0.13
His40Ala	0.153	5.38	0.029	6.15	5.33	−0.77 ± 0.07
Glu58Ala	27.40	2.20	5.35	2.95	5.12	−0.75 ± 0.07
His92Gln	0.117	5.54	0.028	6.17	4.21	−0.63 ± 0.09
Phe100Ala	1.38	4.03	0.508	4.39	4.29	−0.36 ± 0.07

[a] The thio effect is defined as $(k_{cat}/K_m)^{phosphate}/(k_{cat}/K_m)^{phosphorothioate}$.

[b] Apparent interaction free energy between the deleted side chain and the substrate in $ES^{\ddagger}$, as calculated from the ratio of the k_{cat}/K_m values of wild-type enzyme and the mutant under investigation.[95]

[c] Energetic coupling between the mutation under investigation and the substrate's thiosubstitution. Values have been calculated from the difference in apparent interaction energy of a particular side chain with GpU and Gp(S)U, respectively.

[d] ND, Not determined due to extremely slow kinetics.

results that RNase T1 interacts with the *pro-R_p* oxygen via donation of a hydrogen bond by Tyr-38 OηH. Surprisingly, a large coupling was found between the *pro-S_p* thiosubstitution and deletion of the carboxylate of the general base Glu-58.[34] A coupling between the *pro-S_p* thiosubstitution and the removal of the Phe-100 side chain has also been observed. These experimental observations provide new mechanistic insights in RNase T1 catalyzed phosphoryl transfer.

Triester-Like Mechanism Involving Internal Proton Transfer

Historically, the classical acid/base mechanism (Fig. 3a) has been proposed to explain the pH dependence of the reactions catalyzed by RNase A[56] and RNase

T1.[36,55] This simple general acid–general base mechanism has also been convenient to rationalize protein engineering studies on RNase A[50] and RNase T1.[36] The nonadditive nature of the kinetic effects of the single Glu58Ala and His92Gln mutants in the corresponding double mutant[75] reflects the interdependent action of the base and the acid in RNase T1. This concerted behavior has been used as an argument in favor of the classical concerted acid-base mechanism involving a monoanionic phosphorane (Fig. 3a).

However, recapitulation of older kinetic data on two guanosine 2′,3′-cyclophosphorothioate diastereomers[30] and a recent kinetic analysis using the phosphorothioate substrate S_p Gp(S)U[34] support a more complex triester-like mechanism for RNase T1 (Fig. 3b). Both studies indicate that substitution of the *pro*-S_p oxygen by sulfur has a remarkable effect on the kinetics of the cyclization and the decyclization reactions. S_p-Guanosine 2′,3′-cyclophosphorothioate is a very poor substrate of RNase T1.[30,58] S_p thio substitution similarly reduces the cyclization of the model substrate GpU 10^5-fold (Table I). Such large thio effects are not observed when the *pro*-R_p oxygen is replaced by sulfur.[30,33] Because sulfur has a much lower affinity for protons than oxygen, a large thio effect would be predicted for at least one of the thio isomers if the enzyme follows a triester-like mechanism.[51,76] The fact that RNase T1 exhibits an exclusive S_p thio effect reflects the protonation of the *pro*-S_p oxygen during the rate-limiting step. Accordingly, the *pro*-R_p oxygen appears to remain unprotonated in the transition state.

A second piece of evidence in favor of a triester-like mechanism comes from the remarkable observation that the general base Glu-58 interacts with the *pro*-S_p oxygen in the rate-limiting step (Table I). Indeed, the S_p thio effect is largely dependent on the presence of the Glu-58 side chain.[34] This rate limiting interaction between the catalytic base and the nonbridging *pro*-S_p oxygen is inconsistent with the classical acid–base mechanism[50,51] as in Fig. 3a. In the latter mechanism, the catalytic base interacts exclusively with the apical 2′-oxygen. The magnitude of the S_p thio effect and the observation that this thio effect is relieved when Glu-58 is replaced by alanine can satisfactorily be explained by a repulsive interaction between the Glu-58 carboxylate and the S_p sulfur atom. This interaction is productive in the case of unsubstituted phosphodiesters, indicating that there is a close contact between Glu-58 and the S_p oxygen in the reference transition state complex. Such a contact can only occur when a hydrogen atom is positioned between both charged oxygens.

Various triester-like mechanisms found in literature differ essentially in the origin of the (catalytic) proton that is initially transferred to a nonbridging oxygen. In a Breslow-like mechanism[52] (right route in Fig. 3b) the proton originates

[75] J. Steyaert and L. Wyns, *J. Mol. Biol.* **229,** 770 (1993).

[76] R. B. Silverman, ed., "The Organic Chemistry of Enzyme Catalyzed Reactions," p. 76. Academic Press, San Diego, 2000.

from the catalytic acid, whereas the proton comes from the 2′-OH group in an internal proton transfer mechanism[54,54,77] (left route in Fig. 3b). Structural limitations of the RNase T1 active site render the substrate's 2′-OH groups as the evident source of the catalytic proton that has to be transferred to the *pro*-S_p nonbridging oxygen. Our finding that Glu58, being the catalytic base of RNase T1,[36] interacts with the *pro*-S_p oxygen in the transition state suggests that Glu-58 catalyzes the proton transfer from the 2′-OH to this nonbridging oxygen. Phe-100 may contribute to this protonation step by tailoring the local dielectric environmen.[26] All these considerations point to a triester-like internal proton transfer mechanism as depicted schematically in Fig. 3b (left route). The individual proton transfer steps are represented in more detail in Fig. 6a. In this stepwise mechanism, the catalytic base Glu-58 combines two functions. First, Glu-58 transfers the catalytic proton from the nucleophile to the equatorial *pro*-S_p oxygen of the TBP phosphorane. Then, Glu-58 becomes a general base and abstracts the proton from the equatorial *pro*-S_p oxygen, leading to the expulsion of the leaving group.

A Three-Centered Hydrogen Bond for Concerted Phosphoryl Transfer

In stepwise triester-like mechanisms, nucleophile activation, nucleophilic attack, and leaving group expulsion are more or less separate events resulting in more or less stable intermediates on the reaction coordinate (Fig. 6a). The presence of such intermediates has never been established, neither in RNase T1,[75] nor in RNase A or any other ribonuclease. Building on this consideration, a concerted alternative (Fig. 6b) that relies on a three-centered hydrogen bond in the transition state has been presented.[34] In this three-centered hydrogen bond, the nucleophilic 2′-oxygen, the *pro*- S_p oxygen, and the catalytic base cluster around the proton that has to be abstracted from the 2′-OH group. We postulate that the proton becomes symmetrically bound in this three-centered configuration during the formation of the transition state. If the hydrogen bonds are short and the proton is symmetrically located in the plane of the three donor/acceptor atoms, the barrier to proton transfer may be relatively low.[78] Simultaneous proton transfer from 2′-OH toward the *pro*-S_p oxygen and the catalytic base could allow nucleophilic attack on an activated phosphate and general base catalysis to occur in the same transition state (Fig. 6b). In this concerted triester-like mechanism, formation of the three-centered hydrogen bond accompanies the expulsion/protonation of the 5′-leaving group.

[77] J. N. Ladner, B. D. Wladkowski, L. A. Svensson, L. Sjolin, and G. L. Gilliland, *Acta Cryst.* **D53,** 290 (1997).

[78] G. A. Jeffrey and W. Saenger, "Hydrogen Bonding in Biological Structures." Springer-Verlag, Berlin, 1991.

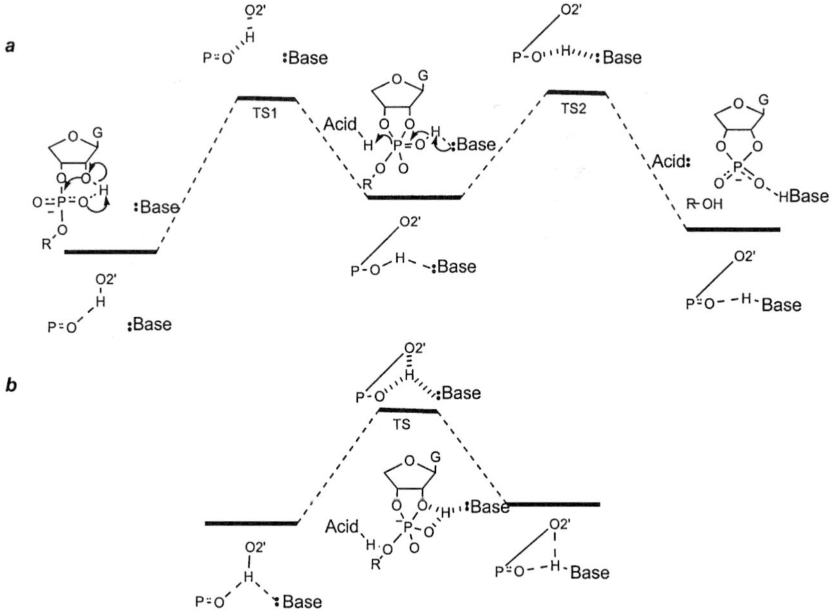

FIG. 6. Triester-like mechanism featuring internal proton transfer. (**a**) Sequential phosphoryl transfer. (**b**) Three-centered hydrogen bond enables concerted phosphoryl transfer.

Nucleophile Activation by Cooperative Hydrogen Bonding

In the complex of RNase T1 with its minimal substrate 3′-GMP (Fig. 4), the protonated His-40 imidazole[36] engages the 2′-OH group in a cooperative His40-Nε2H$^+$•••O2′-H•••Oε1-Glu58 hydrogen bond with the catalytic base Glu-58.[22,47] Whereas Glu-58 abstracts a proton from the 2′-oxygen, thus polarizing the 2′-hydroxyl, and catalyzes the transfer of the catalytic proton to the *pro-S*$_p$ nonbridging oxygen, His-40 most probably orients the nucleophile for attack on the phosphorus. The functional relevance of this structural observation is supported by the synergy among the kinetic effects of the His40Ala and Glu58Ala mutants.[75] Together, His-40 and Glu-58 raise the second-order rate constant for turnover (k_{cat}/K_m) of dinucleoside phosphate substrates by 40,000-fold. In the complex of RNase A with 3′-CMP, the general base His-12 and Lys-41 are involved in a similar hydrogen bonding system (His12-Nε2•••H-02′•••$^+$HNζ-Lys41) with the 2′-hydroxyl. When the geometry of the active sites of RNase T1 and RNase A are compared,[79] it is obvious that the general acid/general base pairs

[79] I. Zegers, D. Maes, M. H. Dao-Thi, F. Poortmans, R. Palmer, and L. Wyns, *Protein Sci.* **3,** 2322 (1994).

(Glu-58/His-92; His-12/His-109) and the electrostatic catalysts (His-40, Lys-41) superimpose structurally. Other residues feature more variation. It thus appears that nucleophile activation via proton abstraction by a base (Glu, His) in concert with an electrostatic catalyst ($HisH^+$, $LysH^+$) to stabilize negative charge buildup on the nucleophile is a recurring catalytic strategy in the ribonuclease world.

Considering the critical role of His-40 in RNase T1 catalysis, it is quite remarkable that this residue is not conserved in the bacterial RNases of which barnase is the best known representative. Superposition of the RNase T1 and barnase crystal structures indicates that the Lys-27 side chain of barnase and the His-40 imidazole of RNase T1 occupy a similar position in the respective active sites.[36] Deletion of the side chains of Lys-27 in barnase and His-40 in RNase T1 lowers k_{cat} for dinucleotide transphosphorylation by a factor of about 4000.[80] In total, these observations suggest that His-40 of RNase T1 and Lys-27 of barnase are functional analogs, having a positive charge in common that contributes to electrostatic catalysis. It appears that His-40, although strictly conserved only in the fungal RNases, has a functional counterpart in the bacterial enzymes.

Positioning of Proper Imidazole Tautomer Contributes to Leaving Group Activation

In the active site of RNase A, the catalytic acid His-119 forms a short 2.6 Å hydrogen bond with Asp-121 (His199- $N\varepsilon_2H^+\bullet\bullet\bullet O\delta_2$-Asp121) generally referred to as a catalytic dyad.[81] It has been shown that the role of Asp-121 is merely to position the proper tautomer of His-119. Considering the clustering of positive charges (His-40, Arg-77, His-92) in the active site of RNase T1,[15] the pK_a of the catalytic acid His-92 is surprisingly high.[36,82] A lowered pK_a would render the enzyme inactive at physiological pH. The high affinity of the His-92 imidazole for the exchangeable proton predominantly results from the His92-$N\delta_1H^+\bullet\bullet\bullet O$-Asn-99 hydrogen bond that preferentially forms if the histidine becomes protonated.[27] In the case where the imidazole is protonated, the $N\delta_1$ nitrogen is protonated and partially charged, making it an excellent hydrogen bonding partner for the Asn-99 backbone carbonyl. This interaction indirectly binds the catalytic proton required for cyclization and may similarly contribute to the positioning of the proper tautomer of the catalytic acid His-92. Thus, the active site of RNase T1 contains a catalytic dyad composed of the catalytic acid His-92 and the backbone oxygen of Asn-99.

[80] D. E. Mossakowska, K. Nyberg, and A. R. Fersht, *Biochemistry* **28,** 3843 (1989).

[81] L. W. Schultz, D. J. Quirk, and R. T. Raines, *Biochemistry* **37,** 8886 (1998).

[82] M. McNutt, L. S. Mullins, F. M. Raushel, and C. N. Pace, *Biochemistry* **29,** 7572 (1990).

Transition State Stabilization by Specific Solvation/Desolvation

Solvation is expected to have a critical effect on the activation energy of (ribo)nuclease catalyzed reactions because a charged transition state is formed from two single charged species.[83] Enzymes solvate the highest energy species better than water does by providing an active site environment prearranged with a polarity that is complementary to the charge distribution of the transition state.[84] In RNase T1, the side chains of Tyr-38, Arg-77, and Phe-100 may contribute to the optimal solvation/desolvation of the TBP phosphorane.

The kinetic effect of the Tyr38Phe mutation depends significantly on thio-substitution of the *pro-*R_p oxygen (see Table I and Ref. 33). This observation reflects a catalytic hydrogen bond between the Tyr-38 OηH and the equatorial *pro-*R_p oxygen, that contributes 70-fold to turnover.[31] Consistent with this, Oη of Tyr38 forms a short hydrogen bond with O2P in the RNase T1/3′-GMP complex (Fig. 4). If O3P corresponds to the leaving oxygen, O2P takes an equatorial position in the transition state (Fig. 2b). Because the Tyr38Phe mutation exerts only minor effects on substrate binding (K_s) and because the aromatic ring of Tyr-38 is buried in the interior of the hydrophobic core of RNase T1, we suggest that the interaction of O2P with the spatially fixed Tyr38 Oη improves when the former moves to its equatorial position in the transition state.

The Arg77-NεH•••OIP and/or Arg77-Nη2H••• O3P hydrogen bonds may render the phosphorus more electropositive, facilitating nucleophilic attack[22] (Fig. 4). Unfortunately, the interactions of Arg-77 with the transition state remain obscure because the function of this residue cannot be probed by site-directed mutagenesis. The guanidinium group of this residue has a unique structural role that can not be fulfilled by any other natural amino acid. This may explain why nobody has been able to purify recombinant variants of RNase T1 containing a mutation at position 77.

The side chain of Phe-100 is an apolar catalytic element at the bottom of the active site of RNase T1 that controls the dielectric environment and stabilizes the unique charge separations that occur in the transition state.[26] From the coupling between the Phe100Ala mutation and the S_p thio substitution, it appears that Phe-100 exerts its function at the level of the *pro-*S_p nonbridging oxygen. Because of its hydrophobic nature, Phe-100 may contribute to the desolvation of a particular part of the transition state. We speculate that Tyr-38, Arg-77, and Phe-100 form a prearranged structural and dielectric microenvironment that is complementary in shape, charge, and hydrogen bonding capacity to the equatorial oxygens of the transition state.

[83] A. Dejaegere and M. Karplus, *J. Am. Chem. Soc.* **115,** 5316 (1993).

[84] A. Warshel, J. Aqvist, and S. Creighton, *Proc. Natl. Acad. Sci. U.S.A.* **86,** 5820 (1989).

Subsite-Binding Effects on Turnover

The second-order rate constants for the ribonuclease-catalyzed transphosphorylation vary largely on the nature of the leaving group of the substrate. For RNase T1, the rate of transphosphorylation drops almost 5 orders of magnitude if the leaving nucleoside of GpN substrates is replaced by a hydrogen (Fig. 2b), indicating that RNase T1 contains a subsite that contributes considerably (≈6.5 kcal/mol) to substrate turnover.[43,71] Similar observations have been made for RNase A[85] and barnase.[86] Because the equilibrium constant of the reaction in Fig. 2b is close to 1 *M*[43] and enzymes accelerate the forward and the reverse reactions of equilibria to the same extent, it follows that the differences between the depolymerization rate and the hydrolysis rate are exclusively due to these subsite interactions (and small differences in the pK_a of the leaving group).

The stereochemical features of the minimal enzyme–substrate complexes of RNase A and RNase T1 indicate how interactions with the leaving group may contribute to turnover. The catalyzed in-line nucleophilic attack requires that the attacking 2′-oxygen, the leaving oxygen, and the phosphorus are linearly arranged in the transition state.[87] These conditions are not met in the complex of RNase T1 with the minimal substrate 3′-GMP (Fig. 4). In addition, theoretical studies on the phosphodiester electronic structure[88,89] suggest that the strength of the scissile bond depends considerably on the torsion angles α and ζ (Fig. 7). Binding interactions with larger leaving groups most probably contribute to fulfill all these stereochemical constraints, thus contributing to chemical turnover.

Part of the subsite contribution may also arise from the exact location of catalytic groups relative to the substrate. This results in a loss of translational and rotational entropy, which constitutes a large fraction of the energy barrier for the chemical reaction.[90] For example, in unliganded RNase A and in its complex with the minimal substrate 3′-CMP, the catalytic acid His-119 takes multiple conformations (Fig. 4).

Perfect Match between RNase T1 and Transition State

In 1948, Pauling postulated that enzymatic catalysis is established by molecular recognition of the transition state of the catalyzed reaction by the enzyme.[91] By

[85] A. Tarragona-Fiol, H. J. Eggelte, S. Harbron, E. Sanchez, C. J. Taylorson, J. M. Ward, and B. R. Rabin, *Protein Eng.* **6,** 901 (1993).

[86] G. A. Day, D. Parsonage, S. Ebel, T. Brown, and A. R. Fersht, *Biochemistry* **31,** 6390 (1992).

[87] F. H. Westheimer, *Acc. Chem. Res.* **1,** 70 (1968).

[88] D. G. Gorenstein, B. A. Luxon, and J. B. Findlay, *J. Am. Chem. Soc.* **101,** 5869 (1979).

[89] D. G. Gorenstein, B. A. Luxon, and E. M. Goldfield, *J. Am. Chem. Soc.* **102,** 1757 (1980).

[90] W. P. Jencks, *Adv. Enzymol.* **43,** 219 (1975).

[91] L. Pauling, *Am. Sci.* **36,** 51 (1948).

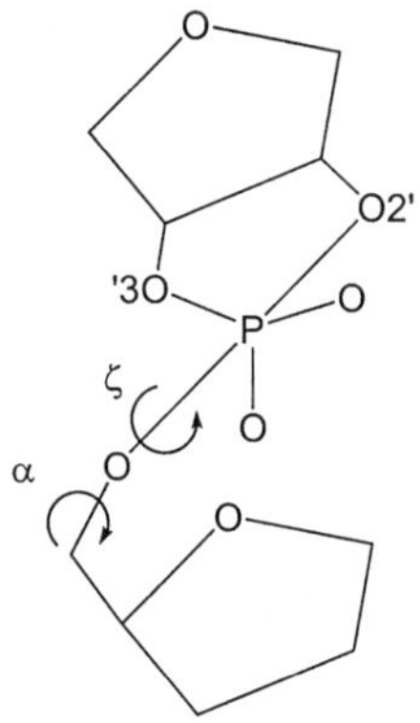

FIG. 7. In RNA phosphodiester linkages, the strength of the PO5′-scissile bond depends considerably on the torsion angles α and ζ.

comparing the uncatalyzed rate[29] versus the RNase T1 catalyzed second-order rate constant of dinucleoside phosphates,[71] the dissociation constant of the enzyme for its transition state can be estimated as 5×10^{-15} *M*. This value is mirrored by an interaction free energy of about 20.2 kcal/mol and illustrates that there is a perfect match between the transition state and the enzyme. The sum of energetic contributions of the active site residues and the subsite (Table II) exceeds this value by 5.7 kcal/mol. This sum is still an underestimate because the energetic contribution of Arg-77 has not been taken into account. Anyhow, this significant discrepancy illustrates that a number of active site residues function in a mutually

TABLE II
CATALYTIC CONTRIBUTIONS IN RNASE T1 FOR TRANSPHOSPHORYLATION OF DINUCLEOSIDE SUBSTRATE GpU[a]

Side-chain mutation	ΔG_{app} (kcal/mol)	Reference
Tyr38Phe	2.6	71
His40Ala	5.4	71
Glu58Ala	2.4	33
His92Gln	5.5	71
Phe100Ala	3.8	33
Subsite	6.3	See text
Total	25.9	

[a] Apparent free energies were calculated as $\Delta G_{app} = -RT \ln(k_{cat}/K_m{}^{mutant}/k_{cat}/K_m{}^{wild\ type})$.

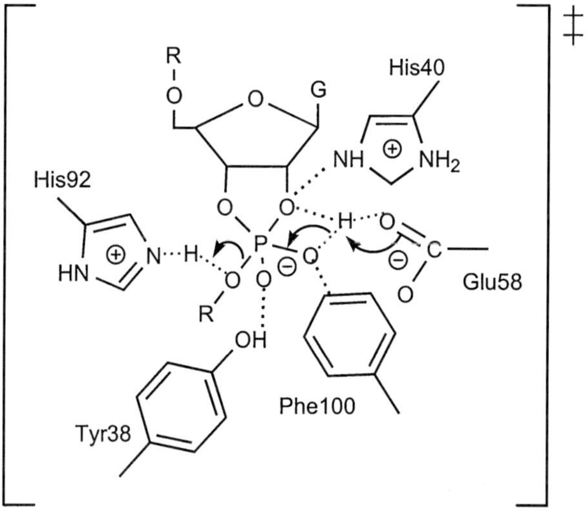

FIG. 8. Schematic view of the transition state interactions in RNase T1 catalyzed phosphoryl transfer.

dependent manner. Consistent with this view, it has been shown that the energetic contributions of His-40, Glu-58, and His-92 are highly cooperative.[75]

Half a century of research on RNase T1 has led to an almost complete picture of the intermolecular interactions that constitute the perfect match between the enzyme and the transition state of the reaction (Fig. 8). In contrast to other systems such as orotidine 5′-phosphate decarboxylase and staphylococcal nuclease,[92] the origin of catalysis by RNase T1 can be traced to regular intermolecular interactions in the transition state (e.g., hydrogen bonds, van der Waals contacts). There is no need to invoke the existence of low barrier hydrogen bonds[93] or important medium effects.[94] The increasingly detailed picture of the transition state structure of the RNA/RNase T1 complex may provide a starting point for quantum chemical calculations that can tell us more about the structural and electronic changes occurring during the reaction.

[92] A. Radzicka and R. Wolfenden, *Science* **267,** 90 (1995).

[93] W. W. Cleland, P. A. Frey, and J. A. Gerlt, *J. Biol. Chem.* **273,** 25529 (1998).

[94] W. R. Cannon and S. J. Benkovic, *J. Biol. Chem.* **273,** 26257 (1998).

[95] A. R. Fersht, *Biochemistry* **27,** 1577 (1988).

[20] Mitogillin and Related Fungal Ribotoxins

By RICHARD KAO, ANTONIO MARTÍNEZ-RUIZ, ALVARO MARTÍNEZ DEL POZO, RETO CRAMERI, and JULIAN DAVIES

Introduction

The name ribotoxin was first proposed to describe a group of fungal ribosome-inactivating proteins (RIPs), α-sarcin, restrictocin, and mitogillin, which are among the most potent inhibitors of translation known.[1] These toxins, produced by the aspergilli, are basic proteins of ~17 kDa. Restrictocin differs from mitogillin by only one amino acid while α-sarcin and restrictocin share 86% amino acid sequence identity. The ribotoxins were thought to be produced exclusively in *Aspergillus* sp. (*A. giganteus, A. restrictus,* and *A. fumigatus*) but recent reports suggest that a family of α-sarcin-like ribotoxins are produced by various fungal species including some members of the genus *Penicillium*.[2]

Ribotoxins block protein synthesis by inhibiting both the elongation factor 1 or Tu (EF-1 or EF-Tu)-dependent binding of aminoacyl-tRNA and the GTP-dependent binding of elongation factor 2 or G (EF-2 or EF-G) to ribosomes. Eukaryotic ribosomes are extremely sensitive to the fungal ribotoxins: α-sarcin can inhibit amino acid incorporation in wheat germ extracts at 0.1 n*M* concentration.[1] The ability of ribotoxins to inhibit protein synthesis always correlates with the production of the α-fragment, a 3′-terminal cleavage product from the rRNA of the large ribosomal subunit.[3] All large ribosomal subunits tested *in vitro,* including those from ribotoxin-producing species such as *A. giganteus* and *A. restrictus,* are inactivated by incubation with the ribotoxins.[4,5]

The specific cleavage site of α-sarcin in 28S rRNA was determined to be on the 3′ side of G4325[6] located in the universally conserved α-sarcin/ricin loop (SRL) of the large subunit rRNAs. The SRL is a distinctive region of large subunit rRNA and has been identified on the surface of the prokaryotic ribosome.[7] A 35-mer oligoribonucleotide, which mimics the SRL of 28S rRNA, has been used to study the substrate specificity of toxins such as α-sarcin and ricin.[8] Extensive

[1] B. Lamy, J. Davies, and D. Schindler, *in* "Genetically Engineered Toxins" (A. E. Frankel, ed.), p. 237. Marcel Dekker, New York, 1992.

[2] A. Lin, C. Ciou-Jau, and S. S. Tzean, EMBL/GenBank/DDBJ databases, accession numbers AF012812-AF012817 (1997).

[3] D. G. Schindler and J. E. Davies, *Nucl. Acids Res.* **4,** 1097 (1977).

[4] A. N. Hobden, Ph.D. Dissertation, University of Leicester, Leicester, U.K. (1978).

[5] S. P. Miller and J. W. Bodley, *FEBS Lett.* **229,** 388 (1988).

[6] Y. L. Chan, Y. Endo, and I. G. Wool, *J. Biol. Chem.* **258,** 12768 (1983).

[7] N. Ban, P. Nissen, J. Hansen, M. Capel, P. Moore, and T. A. Steitz, *Nature* **400,** 841 (1999).

[8] Y. Endo, Y. L. Chan, A. Lin, K. Tsurugi, and I. G. Wool, *J. Biol. Chem.* **263,** 7917 (1988).

0076-6879/01 $35.00

mutational studies of the 35-mer synthetic oligoribonucleotide have shed light on the rRNA identity elements for the toxins, and the results suggest that although ribotoxins and ricin act on adjacent nucleotides, their rRNA identity elements are different.[9]

The complete amino acid sequences for α-sarcin, mitogillin, and restrictocin have been determined and it has been demonstrated that motifs common to the ribotoxins and other microbial ribonucleases are required for their catalytic activity.[10,11] Additional sequences/motifs found in the ribotoxins represent domains that determine substrate (ribosome) specificity.[10,12,13]

The incidence of fungal infections, especially aspergillosis, has increased drastically over the past decade as the number of immunocompromised patients has risen as a result of the use of immunosuppressive agents. Invasive aspergillosis has become an important cause of hospital mortality following major surgery, especially during organ and tissue replacement procedures.[14] The major human pathogenic species of *Aspergillus* is *Aspergillus fumigatus,* and studies have shown that restrictocin is the major antigen detected in the urine of patients with disseminated aspergillosis.[15–17] Antibody studies have shown the accumulation of restrictocin in the vicinity of nodes of fungal infection.[17] Because of the clinical significance of the fungal ribotoxins, a section of this chapter will be devoted to the investigation of the immunological properties of the toxins.

Assay Methods

Specific Cleavage of Rabbit Ribosomes (in Vitro α-Fragment Release)

a. Using Purified Proteins

1. Incubate 30 μl reaction mixtures [20.0 μl of untreated rabbit reticulocyte lysate (Promega, Madison, WI), 600 n*M* purified mitogillin, 15 m*M* Tris-HCl (pH 7.6), 15 m*M* NaCl, 50 m*M* KCl, 2.5 m*M* EDTA] at 37° for 15 min.
2. Stop the reaction by adding 3.0 μl 10% sodium dodecyl sulfate (SDS). After the addition of 20.0 μl phenol/chloroform, vortex the mixture for 30 sec and add 20.0 μl TE (pH 8.0) to enhance the extraction of RNA; vortex again for 30 sec.

[9] I. G. Wool, *in* "Ribonucleases: Structures and Functions" (G. D'Alessio and J. F. Riordan, eds.), p. 131. Academic Press, New York, 1992.

[10] R. Kao and J. Davies, *Biochem. Cell Biol.* **73,** 1151 (1995).

[11] R. Kao, J. E. Shea, J. Davies, and D. W. Holden, *Mol. Microbiol.* **29,** 1019 (1998).

[12] R. Kao and J. Davies, *J. Biol. Chem.* **274,** 12576 (1999).

[13] R. Kao and J. Davies, *FEBS Lett.* **466,** 87 (2000).

[14] P. L. Hibberd and R. H. Rubin, *Clin. Infect. Dis.* **19,** S33 (1994).

[15] L. K. Arruda, T. A. Platts-Mills, J. W. Fox, and M. D. Chapman, *J. Experi. Med.* **172,** 1529 (1990).

[16] L. K. Arruda, B. J. Mann, and M. D. Chapman, *J. Immunol.* **149,** 3354 (1992).

[17] B. Lamy, M. Moutaouakil, J. P. Latge, and J. Davies, *Mol. Microbiol.* **5,** 1811 (1991).

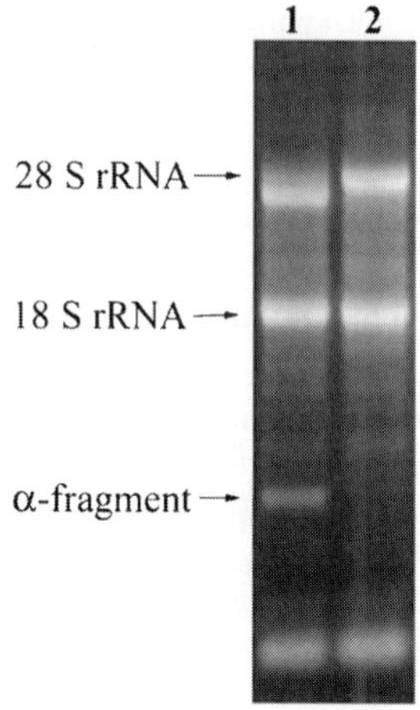

FIG. 1. Specific cleavage of rabbit ribosomes (*in vitro* α-fragment release) by mitogillin. Lane 1, rabbit reticulocyte treated with 600 n*M* mitogillin; lane 2, no mitogillin added. 28S rRNA, 28S ribosomal RNA; 18S rRNA, 18S ribosomal RNA; α-fragment, the 393 nucleotides cleavage product of 28S rRNA by mitogillin. Electrophoresis was carried out on a 2% agarose gel at 200 V in TAE buffer (40 m*M* Tris–acetate, 1 m*M* EDTA, pH 8.0) for 20 min.

3. The reaction mixture is centrifuged at 14,000 rpm in a bench-top centrifuge (5415C Eppendorf) for 15 min at room temperature; remove 20.0 μl of aqueous layer and mix with 1.0 μl electrophoresis loading buffer [0.025% bromphenol blue (Bio-Rad, Hercules, CA), 0.025% xylene cyanol FF (Bio-Rad), 3% glycerol in water].

4. Heat the sample at 95° for 5 min before loading onto a 2.0% agarose gel. Electrophoresis is carried out at 200 V in TAE buffer (40 m*M* Tris–acetate, 1 m*M* EDTA, pH 8.0) for 20 min.

5. Stain the gel with 0.5 μg/ml ethidium bromide for 20 min. RNA species can be detected on a UV transluminator (312 n*m* Variable Intensity Transluminator, Fisher Biotech) and photographed using a Photo-documentation Camera (Fisher Biotect). The mitogillin-treated sample should produce a 393-nucleotide RNA species (the α-fragment) that can be detected on the gel (Fig. 1)

b. Using Induced Bacterial Culture Supernatants. For the direct detection of the ribosome-cleavage activity of mitogillin from induced bacterial culture supernatants (see below), a similar protocol as described in (a) can be adopted.

1. Mix 3.0 μl rabbit reticulocyte lysate and 3.0 μl induced culture supernatant in a 500 μl Eppendorf tube and incubate for 10 min.

2. Stop the reaction by adding 0.5 μl 10% sodium dodecyl sulfate (SDS); extract the RNAs with 7.0 μl phenol/chloroform and 20.0 μl TE (pH 8.0) employing steps 2–5 described in (a); remove 15.0 μl of the aqueous layer for electrophoresis.

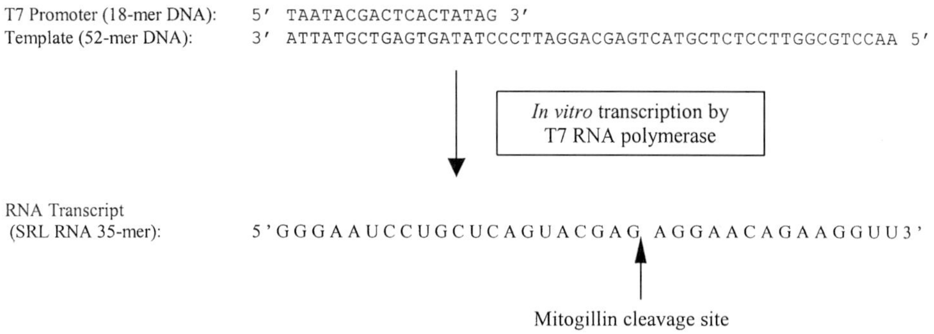

FIG. 2. Synthesis of the SRL using T7 RNA polymerase and synthetic DNA templates. The specific cleavage site by fungal ribotoxins is indicated.

Specific Cleavage of Synthetic SRL RNA

a. SRL (35-mer RNA) Synthesis Using T7 RNA Polymerase and Synthetic DNA Templates. The protocol described by Endo *et al.*[8] can be adopted, with some modifications, to carry out *in vitro* transcription of a SRL-encoding synthetic DNA fragment using T7 RNA polymerase (Fig. 2). Deoxyribonucleotides are available commercially and any commercial DNA synthesizer can be employed.

1. Purify the synthetic SRL DNA templates by preparative denaturing 15% polyacrylamide gel electrophoresis (PAGE) according to Sambrook *et al.*[18] The DNA bands can be detected by illuminating the gel with short wavelength UV (254 nm) on a fluorescent background (silica gel 60 F_{254} TLC plates). The gel bands are excised, and the DNA eluted by soaking the gel slices (homogenized with a glass rod) in deionized water for 4 hr. Eluted DNA can be concentrated using Centricon-3 concentrators (Amicon, Danvers, MA).

2. Prepare the DNA templates for transcription by heating the bottom strand (a 52-mer) which encodes the T7 promoter region and the SRL sequences, with the top strand (an 18-mer) which contains only the promoter region, at 90° for 2 min and then cooling the DNA on ice to allow annealing to occur.

3. *In vitro* transcription is carried out in 40 m*M* Tris-HCl (pH 8.0), 8.0 m*M* $MgCl_2$, 2 m*M* spermidine trihydrochloride, 25 m*M* NaCl, 5 m*M* dithiothreitol (DTT), 0.01% (v/v) Triton X-100, and 80 mg/ml polyethylene glycol 8000. Following the addition of 2.5 m*M* NTPs (a mixture of GTP, ATP, UTP, and CTP) (Pharmacia, Piscataway, NJ), 100 nM DNA templates, and 5U/μl T7 RNA polymerase (Pharmacia), incubate the reaction mixtures (100 μl per tube) at 37° for 3 hr.

[18] J. Sambrook, E. F. Fitsch, and T. Maniatis, *"Molecular Cloning: A Laboratory Manual."* Cold Spring Harbor Laboratory Press, Cold Spring Harbor, NY, 1989.

4. Stop the reactions by the addition of 100 μl deionized formamide (Bio-Rad) and 0.1% (SDS). Purification of the RNA can be achieved by employing the same protocol used to purify the DNA templates,[18] except that the percentage of the gel is increased to 19% and 10 m*M* Tris- HCl (pH 7.4) is used to elute the oligoribonucleotides.

b. Ribotoxin Cleavage of Synthetic SRL RNA

1. Combine purified synthetic SRL 35-mer (1.0 μM) with mitogillin (600 n*M*) in a total volume of 6.0 μl in 10 m*M* Tris-HCl (pH 7.4), and incubate at 37° for 15 min.
2. Stop the reaction by the addition of 4.0 μl deionized formamide followed by heating at 95° for 3 minutes. Cleavage products can be separated by denaturing polyacrylamide gel electrophoresis.[18]
3. Stain the gel with SYBR-GOLD nucleic acid stain (Molecular Probes, Eugene, OR). RNA species can be detected with a UV transilluminator (312 nm, Variable Intensity Transilluminator, Fisher Biotec). The mitogillin treated sample will produce a 21-mer and a 14-mer (Fig. 3).

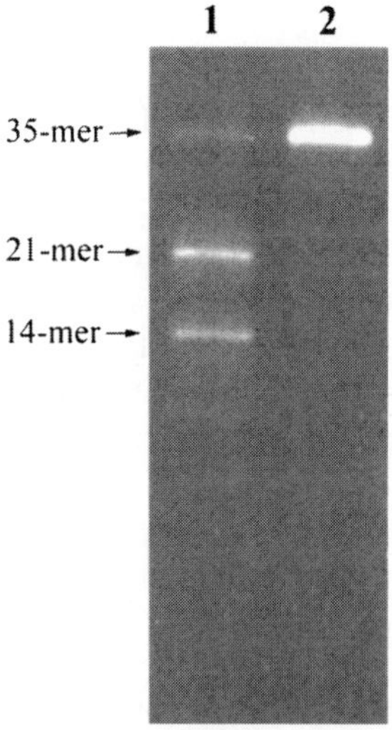

FIG. 3. Specific cleavage of the synthetic SRL RNA by mitogillin. Lane 1, synthetic SRL RNA (35-mer) treated with 600 n*M* mitogillin; lane 2, no mitogillin added. 21-mer and 14-mer are the cleavage products of SRL 35-mer by mitogillin. Electrophoresis was carried out on a denaturing polyacrylamide gel (19%) at 20 V/cm in TBE buffer (89 m*M* Tris-borate, 89 m*M* boric acid, 2 m*M* EDTA, pH 8.0) until the bromphenol blue had migrated two-thirds of the length of the gel.

Nonspecific Ribonucleolytic Activity against Synthetic Homopolynucleotides

Mitogillin and related fungal ribotoxins possess very weak general ribonuclease activity. Ribonucleolytic activity against synthetic homopolynucleotides can be assayed by a method employing Microcon-30 concentrators (Amicon). Because poly (I) is the most effective substrate for mitogillin of homopolynucleotides tested,[10] a protocol employing poly(I) as the substrate will be described.

1. Incubate poly(I) (Sigma, St. Louis, MO) (200μM) with mitogillin (3.0 μM) in 30 μl 10 mM Tris-HCl (pH 7.4) at 37° for 30 min.
2. Stop the reaction by the addition of 30 μl 20 mM Tris-HCl (pH 11.5); place the samples on ice.
3. Load the samples on prewashed Microcon-30 concentrators (molecular cutoff of 60 nucleotides for single-stranded RNA) and centrifuge at 4° at 13,800g for 15 min. It is important to note that the concentrators have to be spun to dryness before applying samples to obtain consistent results.
4. Collect the filtrates (containing degradation products smaller than 60 nucleotides) and determine absorbance at 248.5 nm. The percentages of digestion can be calculated from the molar absorption coefficient of poly(I).[18]

Zymogram Electrophoresis and Activity Staining

The possibility of contaminating ribonucleases in ribotoxin preparations can be assessed by using an activity staining assay described by Blank *et al.*[19]

1. Standard SDS–PAGE is performed using 1.0 μg of mitogillin under nonreducing condition on a 0.1% SDS/15% polyacrylamide gel containing 0.3 mg/ml poly(I).
2. Wash the gel extensively according to the protocol described by Blank *et al.*[19] and incubate the gel in 100 mM Tris-HCl, pH 7.4, at 37° for 1 hr.
3. Stain the gel with 0.2% toluidine blue for 10 min at room temperature, and wash extensively with deionized water for 1 hr. RNase activity is indicated by the appearance of a clear zone on a dark blue background.

Overexpression and Purification of Mitogillin

The mitogillin gene used was chemically synthesized by Better *et al.*[20] and cloned into *Escherichia coli* W3110 employing the expression vector pING3522, an efficient inducible *E. coli* secretion vector. Transcription from the *Salmonella*

[19] A. Blank, R. H. Sugiyama, and C. A. Dekker, *Anal. Biochem.* **120,** 267 (1982).

[20] M. Better, S. L. Bernhard, S. P. Lei, D. M. Fishwild, and S. F. Carroll, *J. Biol. Chem.* **267,** 16712 (1992).

typhimurium araB promoter is very tightly regulated; the tight regulation, coupled with an efficient leader peptide sequence, is required to avoid the lethal effect of the toxin expression in *E. coli.* On induction by arabinose, the produced mitogillin is directed through the cytoplasmic membrane of the host and secreted into the culture medium by the *Erwinia carotovora pelB* leader sequence. The secreted mitogillin is biologically active. Such controlled heterogeneous expression of ribotoxins has proved to be extremely valuable in producing pure material for molecular studies.

1. Inoculate *E. coli* host containing the mitogillin gene in 10 ml of tryptone broth (containing 10 μg tetracycline/ml); culture at 37° overnight with shaking.
2. Transfer the culture into 1 liter of fresh medium (tryptone broth containing 10 μg tetracycline/ml and incubate at 37° with shaking to $A_{600} = 0.4$ (approximately 5 hr).
3. Induce the culture by adding 0.1% L-arabinose (Sigma) and grow for another 18 hr.
4. Dilute the culture supernatant with an equal volume of deionized distilled water (adjust the final pH to 7.0 with 10 *M* NaOH); add 15 ml of SP-Sepharose Fast Flow (Pharmacia Biotech) to the diluted supernatant and agitate the mixture at 4° for 90 min.
5. Elute the bound proteins with 2 *M* NaCl. The eluate can be concentrated and desalted using a Centriprep-10 column (Amicon). Dilute the resulting solution into 5 ml of 50 m*M* sodium phosphate buffer (pH 7.2) and load onto a Mono S HR 5/5 column (Pharmacia Biotech) preequilibrated with 50 m*M* sodium phosphate buffer (pH 7.2). Elute mitogillin with a 0–500 m*M* NaCl linear gradient (mitogillin will be eluted at 300–350 m*M* NaCl). Fractions containing mitogillin can be concentrated and exchanged into 50 m*M* sodium phosphate buffer (150 m*M* NaCl, pH 7.4) using a Centricon-10 column (Amicon).
6. Load the concentrated protein solution onto a Superose 12 HR 10/30 gel filtration column (Pharmacia Biotech) and collect fractions corresponding to mitogillin (monitored by zymogram electrophoresis and activity staining described in previous section). The described protocol should yield 1–2 mg purified mitogillin per liter of culture.

Conformational Analysis of Mitogillin

Secondary and Tertiary Structures

The far-UV circular dichroism spectrum of native recombinant mitogillin is coincident with the one reported for the natural fungal mitogillin and restrictocin.[21] It shows a minimum ellipticity value of about 6000 degree $\times$ cm^2 $\times$ dmol^{-1} around

[21] J. G. Gavilanes, *J. Protein Chem.* **2,** 251 (1983).

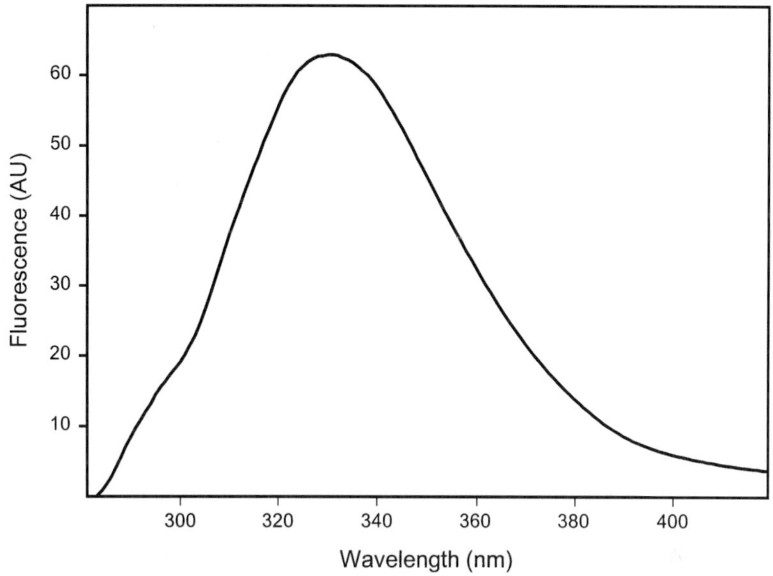

FIG. 4. Fluorescence emission spectrum for excitation at 275 nm of mitogillin. Fluorescence intensity is expressed in arbitrary units.

220 nm. Mitogillin is an α/β protein, the β-structure content being about 3-fold that of the α helix, which is in agreement with the three-dimensional structure of restrictocin.[22]

The near-UV CD spectrum of mitogillin shows two ellipticity maxima centered at 287 and 292 nm and two broad extreme values at 271 and 302 nm, corresponding to contributions of the tryptophan residues and disulfide bridges of the protein.

The fluorescence emission spectrum of mitogillin shows a maximum at 330 nm due to tryptophan contribution and a small shoulder around 300 nm related to the tyrosine emission, on excitation at 275 nm (Fig. 4).

The thermal stability of mitogillin has been studied by measuring the temperature dependence of the ellipticity at 220 nm. Protein samples were prepared in 50 m*M* phosphate buffer, containing 0.1 *M* NaCl, pH 7.0, to avoid buffering variations due to the temperature changes. The thermal denaturation of the native protein shows a single transition, the midpoint (T_m) being centered at 58–59°.

Comments

The procedures and conditions employed to study the conformational properties of mitogillin are included for those interested in the structural analysis of mitogillin and related fungal ribotoxins.

[22] X. Yang and K. Moffat, *Structure* **4,** 837 (1996).

Circular dichroism (CD) spectra were obtained using a Jasco J-715 spectropolarimeter. The instrument can be calibrated by using (+)-10-camphorsulfonic acid, $\Delta\varepsilon = +2.37\, M^{-1}\, \text{cm}^{-1}$ at 290.5 nm, and $\Delta\varepsilon = -4.95\, M^{-1}\, \text{cm}^{-1}$ at 192.5 nm. Measurements were performed at 25° with thermostatted cylindrical cells of 0.10 cm path length in the peptide bond region (far-UV, 195–250 nm), and 1 cm optical path, in the aromatic residues region (near-UV, 250–350 nm). Data were collected every 0.2 nm at 50 nm/min; CD results are expressed in units of degree × $\text{cm}^2 \times \text{dmol}^{-1}$ of amino acid residue. The value of the mean residue mass is 114 based on the primary structure of native mitogillin. Deconvolution of the far-UV CD curves to obtain a quantitative estimation of the secondary structure was performed according to the convex constraint analysis.[23] Thermal denaturation of the protein samples was studied by analyzing the ellipticity variation at 220 nm (2 nm band width) upon increase of the temperature at a rate of 30°/hr (0.2° resolution) by using a water-circulating RT-111 bath (Neslab). Fluorescence emission spectra, for excitation at 275 and 295 nm, was recorded in the ratio mode, at 1 nm/spc scanning rate on a SLM Aminco 8000 spectrofluorimeter. The excitation optical path of the cells was 0.2 cm. The slit widths for both excitation and emission beams were 4 nm. Fluorescence emission was expressed in arbitrary units in terms of the same protein concentration. Baseline substraction and correction for instrumental response was done using the standard software provided by the manufacturer.

Immunological Properties

Aspergillus species are considered as the most important opportunistic pathogens. In particular, *Aspergillus fumigatus* is the etiological agent isolated in about 80% of the *Aspergillus* infections in humans and is associated with an impressive number of pulmonary complications.[24] Among these, IgE-mediated allergic asthma and allergic bronchopulmonary aspergillosis (ABPA) are severe allergic pulmonary complications caused by *Aspergillus* species.[25] Extracts of *A. fumigatus* used to diagnose allergic reactions are complex mixtures containing up to 200 different proteins and glycoproteins and low molecular weight compounds[26] and are, as a consequence of their complexity, almost impossible to standardize.[24] The first major *A. fumigatus* allergen identified by biochemical methods and formally termed Asp f 1 is a member of the mitogillin family of cytotoxins.[15] Asp f 1 was the first recombinant allergen tested *in vivo*[27] and has subsequently been

[23] G. Tusnady and G. D. Fasman, *Protein Eng.* **4,** 669 (1991).

[24] R. Crameri, *Int. Arch. Allergy Immunol.* **115,** 99 (1988).

[25] V. P. Kurup and A. Kumar, *Clin. Microbiol. Rev.* **4,** 439 (1991).

[26] J. E. Piechura, C. J. Huang, S. H. Cohen, J. M. Kidd, V. P. Kurup, and J. J. Calvanico, *Immunology* **49,** 657 (1983).

[27] M. Moser, R. Crameri, G. Menz, T. Schneider, T. Dudler, C. Virchow, M. Gmachl, K. Blaser, and M. Suter, *J. Immunol.* **149,** 454 (1992).

investigated in clinical studies enrolling large numbers of allergic asthmatics with or without ABPA,[28] patients with cystic fibrosis and *A. fumigatus* sensitization,[29] and individuals suffering from atopic dermatitis concomitant with *A. fumigatus* sensitization.[30] Moreover, rAsp f 1 has been immobilized to the Pharmacia CAP System, allowing a fully automated analysis of rAsp f 1-specific IgE in patients' sera.[31,32] The experience accumulated in skin tests involving more than 400 individuals suffering from various *A. fumigatus*-related diseases and 40 healthy control persons demonstrated complete concordance between *in vivo* tests and serology with recombinant Asp f 1.[24,28,32] These findings and the excellent correlation between serologic and skin test data suggest the possibility of relying on serologic analyses with recombinant allergens for a fully automated diagnosis of allergic diseases, thus avoiding allergen challenges.[28,32] Such an improvement in the reliability of the *in vitro* diagnosis of allergy can only be obtained with the use of highly pure, completely standardized recombinant allergen preparations.[24]

The IgE-mediated reactions to ribotoxins have immediate clinical and practical implications. The rAsp f 1 ribotoxin produced as a major allergen by *A. fumigatus*[27] shares a high degree of sequence identity, ranging from 86 to 100% to the ribotoxins restrictocin and mitogillin produced by *A. restrictus*, α-sarcin produced by *A. giganteus* and clavin produced by *A. clavatus*.[32] Therefore, as described for other allergens showing sequence similarity to related proteins of different origin,[33–35] cross-reactivity at the T and B cell level between these ribotoxins should be expected. In fact, rAsp f 1, mitogillin, clavin, and α-sarcin fully cross-react in IgE-binding studies using sera of patients sensitized to rAsp f 1 (unpublished results). Interestingly, *A. restrictus, A. giganteus,* and *A. clavatus* have not so far been identified as opportunistic human pathogens. The IgE cross-reactivity detected among homologous ribotoxins produced by different species suggests that the allergenicity of these three *Aspergilli* should be clinically reevaluated in the light of the ribotoxin-mediated ability to provoke strong allergic reactions.

[28] M. Moser, R. Crameri, E. Brust, M. Suter, and G. Menz, *J. Allergy Clin. Immunol.* **93,** 1 (1994).

[29] W. H. Nikolaizik, R. Crameri, K. Blaser, and M. H. Schöni, *Int. Arch. Allergy Immunol.* **111,** 403 (1996).

[30] R. Disch, G. Menz, K. Blaser, and R. Crameri, *Int. Arch. Allergy Immunol.* **108,** 89 (1995).

[31] R. Crameri, J. Lidholm, H. Gronlund, D. Stuber, K. Blaser, and G. Menz, *Clin. Exp. Allergy* **26,** 1411 (1996).

[32] S. Hemmann, G. Menz, C. Ismail, K. Blaser, and R. Crameri, *J. Allergy Clin. Immunol.* **104,** 601 (1999).

[33] A. Martinez-Ruiz, R. Kao, J. Davies, and A. Martinez del Pozo, *Toxicon.* **37,** 1549 (1999).

[34] R. Crameri, A. Faith, S. Hemmann, R. Jaussi, C. Ismail, G. Menz, and K. Blaser, *J. Exp. Med.* **184,** 265 (1996).

[35] C. Mayer, U. Appenzeller, H. Seelbach, G. Achatz, H. Oberkofler, M. Breitenbach, K. Blaser, and R. Crameri, *J. Exp. Med.* **189,** 1507 (1999).

Procedures for Detection of IgE Antibodies against Asp f 1

Immunoglobulin E (IgE) antibodies raised against Asp f 1 in serum of individuals sensitized to *A. fumigatus* can be detected by enzyme-linked immunosorbent assay (ELISA).[28]

1. Coat polystyrene Maxisorp microtiter plates (Nunc, Roskilde, Denmark) with 100 μl/well of Asp f 1 solution (10 μg/ml) in phosphate-buffered saline solution (PBS) pH 8.0 for 2 hr at 37° or overnight at 4°.
2. Block the remaining free sites with PBS, pH 7.4, containing 2% (w/v) nonfat dry milk powder (blocking buffer) at 37° for 1 hr. The wells are then washed three times with PBS, pH 7.4, containing 0.05% Tween 20 (washing buffer).
3. Sera are prediluted 1 : 5 in blocking buffer containing 5% Tween 20 (dilution buffer) and further serially diluted twofold in the coated plate, and incubated for 2 hr at 37°.
4. After washing three times, add 100 μl/well of TN-142 mouse anti-human IgE monoclonal antibody (MAb) at a concentration of 400 ng/ml in dilution buffer and incubate the plates for 2 hr at 37°.
5. Add 100 μl/well alkaline phosphatase (AP)-conjugated goat anti-mouse IgG, H + L chains (Pierce, Rockford, IL) (500 ng/ml in dilution buffer); incubate for 1 hr at 37°. Thereafter, ELISA plates are washed three times and developed for 45 min with 100 μl/well of 1.5 mg/ml 4-nitrophenyl phosphate disodium salt hexahydrate (Merk, Darmstadt, Germany) dissolved in 1 *M* diethanolamine buffer (pH 9.8) containing 0.5 m*M* magnesium chloride.
6. Stop the enzyme reaction by adding 50 μl 2 *M* NaOH. Absorbance readings are measured at 405 nm with a Molecular Devices Reader (Menlo Park, CA). As reference standard, a serum pool from positive patients is used, with absorbance and dilution values converted into ELISA units (EU) and arbitrarily set as 100 EU/ml. From the serum dilution series for each patient, an average EU value is calculated from the values within the titratable region and values below 5 EU/ml are considered as negative according to the skin test results.[28]

Concluding Remarks and Future Directions

The ribotoxin family of proteins are thought to have evolved from more general ribonucleases,[10] and their highly specific cleavage reactions have been attributed to the presence of specific targeting domains in the ribotoxin structure that permit ribosome binding and cleavage of the SRL.[12,13] Ribotoxins have been employed in RNA footprinting studies and they may have general applications for this purpose.[36] In addition the identification of the targeting sequences suggests a

[36] P. W. Huber and I. G. Wool, *Methods Enzymol.* **164,** 468 (1988).

method by which nucleases and other proteins may be conveniently engineered to target specific macromolecules.

Acknowledgments

This work was supported by the Natural Sciences and Engineering Research Council of Canada (R.K. and J.D.), and by Grant PB96/0601 from the Dirección General de Enseñanza Superior (Spain). A.M.-R. was recipient of a fellowship from the Ministerio de Educación y Cultura (Spain). Work at the Swiss Institute of Allergy and Asthma Research was supported by the Swiss National Science Foundation (Grant 31-50515.97).

[21] RNase U2 and α-Sarcin: A Study of Relationships

By ANTONIO MARTÍNEZ-RUIZ, LUCÍA GARCÍA-ORTEGA, RICHARD KAO, JAVIER LACADENA, MERCEDES OÑADERRA, JOSÉ M. MANCHEÑO, JULIAN DAVIES, ÁLVARO MARTÍNEZ DEL POZO, and JOSÉ G. GAVILANES

Introduction

Elsewhere in this volume,[1] it was mentioned that it would be interesting to insert the targeting motifs of ribotoxins (presumably responsible for the ability to gain access to the cytoplasm of eukaryotic cells, and the exquisite specificity toward ribosomes) into other enzyme sequences, in order to generate specific targeted toxins. In this regard, α-sarcin, one of the most representative ribotoxins,[1] and RNase U2 seem to be the best candidates for exchange of selected sequence motifs. RNase U2 is a nontoxic extracellular enzyme of the RNase T1 family[2] which shows the closest phylogenetic relationship with ribotoxins.[3,4] Although α-sarcin and RNase U2 display different biological actions, they share structural (amino acid sequence and global fold) and enzymatic similarities. This suggests that both proteins may have been descended from a common ancestor. It is therefore tempting to speculate on the features that distinguish α-sarcin as a cytotoxic ribonuclease.

[1] R. Kao, A. Martínez-Ruiz, A. Martínez del Pozo, R. Crameri, and J. Davies, *Methods Enzymol.* **341,** [21] 2001 (this volume).

[2] M. Irie and K. Ohgi, *Methods Enzymol.* **341,** [3] 2001 (this volume).

[3] G. Sacco, K. Drickamer, and I. G. Wool, *J. Biol. Chem.* **258,** 5811 (1983).

[4] J. M. Mancheño, M. Gasset, J. Lacadena, A. Martínez del Pozo, M. Oñaderra, and J. G. Gavilanes, *J. Theor. Biol.* **172,** 259 (1995).

0076-6879/01 $35.00

RNase U2

The smut fungus *Ustilago sphaerogena* secretes at least four different RNases, U1, U2, U3, and U4.[5,6] RNase U1 is a guanylic acid-specific enzyme of the RNase T1 type,[5–9] RNase U4 is a nonspecific RNA exonuclease,[5,6,10] and RNases U2 and U3 are two similar enzymes that display specificity for purines [5,6,11].

RNase U2 [EC 3.1.27.4] is a single polypeptide chain of 114 amino acids (12,490 molecular mass).[12] It shows a strong preference for 3′ linked purine nucleotide phosphodiester linkages ($A > G \gg C > U$),[13,14] which is rather unusual within the group of microbial extracellular RNases[11,15] (RNase T1 from *Aspergillus oryzae,* for example, shows strict specificity for the guanylyl group).[2] RNase U2 also differs in its optimum pH (4.5) from RNase T1,[6] but both are cyclizing enzymes, cleaving RNA in two separate stages, transphosphorylation and hydrolysis.[16]

The amino acid sequence of RNase U2[12,17,18] shows that the enzyme lacks Lys residues and displays a high content of acidic residues (p*I* of 2.8–3.3). It contains one Trp residue, although its fluorescence emission is strongly quenched under native conditions.[19] A summary of its physicochemical and kinetic properties is reported elsewhere.[19–21] RNase U2 exists as a mixture of two isoforms, U2A (the one originally isolated, which is easily converted to the other species after alkali treatment)[22] and U2B, which are different in terms of specific activity and secondary structure content due to the presence of an isoaspartate bond in RNase U2B.[17,22] A third isoform, RNase U2C, containing another isoaspartate

[5] T. Arima, T. Uchida, and F. Egami, *Biochem. J.* **106,** 601 (1968).
[6] T. Arima, T. Uchida, and F. Egami, *Biochem. J.* **106,** 609 (1968).
[7] K. Sato and F. Egami, *J. Biochem.* **44,** 753 (1957).
[8] D. G. Glitz and C. A. Dekker, *Biochemistry* **2,** 1185 (1963).
[9] W. C. Kenney and C. A. Dekker, *Biochemistry* **10,** 4962 (1971).
[10] A. Blank and C. A. Dekker, *Biochemistry* **11,** 3956 (1972).
[11] J. M. Adams, P. G. N. Jeppeson, F. Sanger, and B. G. Barrell, *Nature* **223,** 1009 (1969).
[12] S. Sato and T. Uchida, *Biochem. J.* **145,** 353 (1975).
[13] G. W. Rushizky, J. H. Mozejko, D. J. Rogerson, Jr., and H. Sober, *Biochemistry* **9,** 4966 (1970).
[14] T. Uchida, T. Arima, and F. Egami, *J. Biochem.* **67,** 91 (1970).
[15] T. Uchida, L. Bonen, H. W. Schaup, B. J. Lewis, L. Zablen, and C. Woese, *J. Mol. Evol.* **3,** 63 (1974).
[16] T. Yasuda and Y. Inoue, *Biochemistry* **21,** 364 (1982).
[17] S. Kanaya and T. Uchida, *Biochem. J.* **240,** 163 (1986).
[18] S. Kanaya and T. Uchida, *J. Biochem.* **118,** 681 (1995).
[19] S. Minato and A. Hirai, *J. Biochem.* **85,** 327 (1979).
[20] T. Uchida and F. Egami, *in* "The Enzymes" (P. D. Boyer, ed.), Vol. 4, pp. 205–250. Academic Press, New York, 1971.
[21] F. Egami, T. Oshima, and T. Uchida, *in* "Molecular Biology, Biochemistry, and Biophysics" (F. Chapeville and A.-L. Haenni, eds.), Vol. 32, p. 250. Springer-Verlag, Heidelberg, 1980.
[22] T. Uchida and Y. Shibata, *J. Biochem.* **90,** 463 (1981).

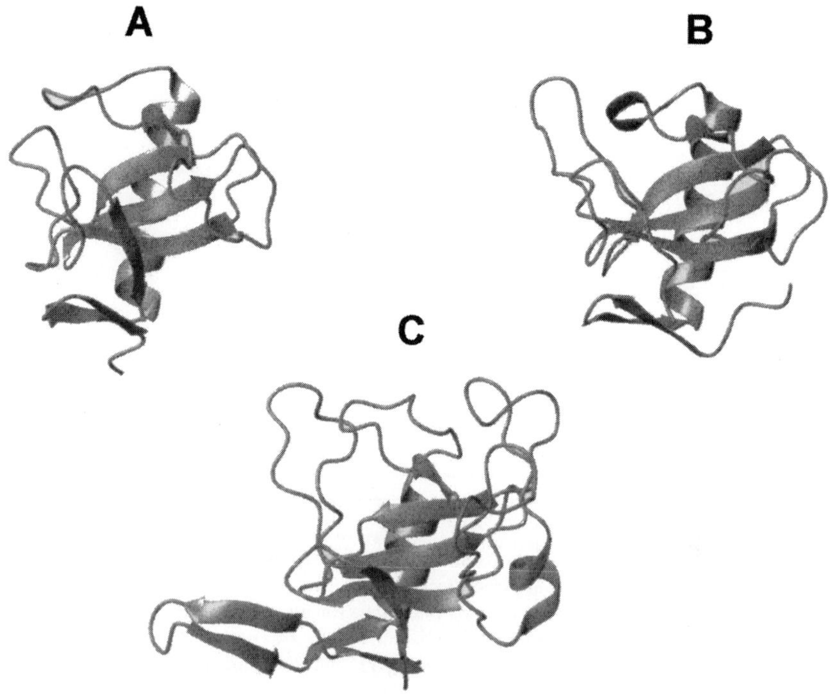

FIG. 1. Schematic representation of the three-dimensional structures of (A) RNase T1,[24] (B) RNase U2,[23] and (C) α-sarcin.[25]

modification, was found during the resolution of the three-dimensional structure of the protein (at 1.8 Å resolution).[23]

The structure of RNase U2 is very similar to that of RNase T1[24] and α-sarcin[25] (Fig. 1). The three proteins display a common structural core constituted by a central β sheet packed against a small α helix. A long, protruding loop, containing residues Tyr-67 to Gly-82, and three disulfide bridges (located at positions 1–53, 9–112, and 54–95)[12,26] are distinctive features of the RNase U2 structure (Fig. 2).

α-Sarcin

α-Sarcin is a microbial ribonuclease secreted by the mold *Aspergillus giganteus.*[27,28] It is extremely specific since it only cleaves a single phosphodiester

[23] S. Noguchi, Y. Satow, T. Uchida, C. Sasaki, and T. Matsuzaki, *Biochemistry* **34,** 15583 (1995).
[24] C. N. Pace, U. Heinemann, U. Hahn, and W. Saenger, *Angew. Chem. Int. Ed. Engl.* **30,** 343 (1991).
[25] J. M. Pérez-Cañadillas, J. Santoro, R. Campos-Olivas, J. Lacadena, A. Martínez del Pozo, J. G. Gavilanes, M. Rico, and M. Bruix, *J. Mol. Biol.* **299,** 1061 (2000).
[26] S. Sato and T. Uchida, *J. Biochem.* **77,** 795 (1975).

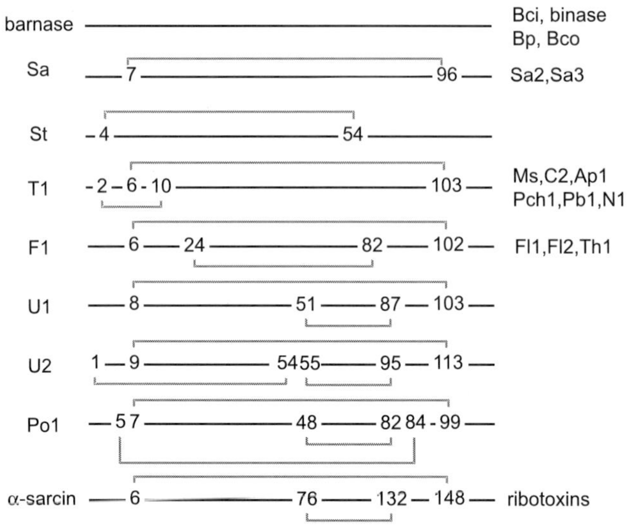

FIG. 2. Disulfide bridge patterns of different microbial RNases.[4,103] Ap1, RNase Ap1 from *Aspergillus palidus;* Barnase, RNase from *Bacillus amyloquefaciens;* Bci, RNase Bci from *Bacillus circulans;* Bco, RNase Bco from *Bacillus coagulans;* Binase, RNase from *Bacillus intermedius;* Bp, RNase Bp from *Bacillus pumilus;* C2, RNase C2 from *Aspergillus clavatus;* F1, RNase F1 from *Fusarium moniliforme;* Fl1 and Fl2, RNases Fl1 and Fl2 from *Fusarium lateritium;* Ms, RNase Ms from *Aspergillus saitoi;* N1, RNase N1 from *Neurospora crassa;* Pb1, RNase Pb1 from *Penicillium brevicompactum;* Pch1, RNase Pch1 from *Penicillium chrysogenum;* Po1, RNase Po1 from *Pleurotus ostreatus;* Sa, Sa2, and Sa3, RNases Sa, Sa2, and Sa3 from *Streptomyces aureofaciens;* α-sarcin, ribotoxin from *Aspergillus giganteus;* T1, RNase T1 from *Aspergillus oryzae;* Th1, RNase Th1 from *Trichoderma harzianum:* U1 and U2, RNases U1 and U2 from *Ustilago sphaerogena.*

bond of the large-subunit rRNA.[1,29] This inhibits protein biosynthesis and leads to cellular death.[1,30–32] Similar to RNase U2, α-sarcin is an acid cyclizing RNase that can hydrolyze A/GpA dinucleotides, although with low specific activity.[33–35]

---

[27] B. H. Olson and G. L. Goerner, *Applied Microbiol.* **13,** 314 (1965).

[28] B. H. Olson, J. C. Jennings, V. Roga, A. J. Junek, and D. M. Schuurmans, *Applied Microbiol.* **13,** 322 (1965).

[29] D. G. Schindler and J. E. Davies, *Nucleic Acid Res.* **4,** 1097 (1977).

[30] Y. Endo, P. W. Hubert, and I. G. Wool, *J. Biol. Chem.* **258,** 2662 (1983).

[31] I. G. Wool, *Trends Biochem. Sci.* **9,** 14 (1984).

[32] I. G. Wool, A. Glück, and Y. Endo, *Trends Biochem. Sci.* **17,** 266 (1992).

[33] J. Lacadena, A. Martínez del Pozo, V. Lacadena, A. Martínez-Ruiz, J. M. Mancheño, M. Oñaderra, and J. G. Gavilanes, *FEBS Lett.* **424,** 46 (1998).

[34] J. M. Pérez-Cañadillas, R. Campos-Olivas, J. Lacadena, A. Martínez del Pozo, J. G. Gavilanes, J. Santoro, M. Rico, and M. Bruix, *Biochemistry* **37,** 15865 (1998).

[35] J. Lacadena, A. Martínez del Pozo, A. Martínez-Ruiz, J. M. Pérez-Cañadillas, M. Bruix, J. M. Mancheño, M. Oñaderra, and J. G. Gavilanes, *Proteins* **37,** 474 (1999).

The target rRNA sequence for α-sarcin, and its related ribotoxins, is conserved in all ribosomes so far studied.[1,36,37] However, α-sarcin is quite selective against intact cells. For example, it does not kill bacteria or fungi,[28] and it is especially active against virus-infected[38,39] and transformed eukaryotic cells.[28,40,41] This selectivity is probably related to the ability of α-sarcin to destabilize bilayers containing negatively charged phospholipids since no membrane protein receptors have been found. Such an interaction with membranes promotes their aggregation and fusion[42–46] resulting in protein translocation across the bilayers of model vesicles.[47–49]

Although α-sarcin is a basic protein ($pI > 8.0$)[3] and longer (150 amino acids) than RNase U2, both proteins display a high degree of sequence similarity,[3,4] the two disulfide bridges of α-sarcin being conserved in RNase U2 (Fig. 2). An even greater degree of similarity can be seen in comparing the three-dimensional structures. As mentioned above, α-sarcin[25,50] exhibits the same global fold as RNase U2 (Fig. 1). Significant differences are present in the long unstructured loops of α-sarcin, which contain a number of basic residues. This evidence suggests that the specific action of the ribotoxin on ribosomes and membranes is related to these loops.[1,4,51–55]

The three-dimensional structure similarity extends to the residues considered as catalytic.[25,34,35,50] His-137 and Glu-96 of α-sarcin act as general acid and base

[36] S. P. Miller and J. W. Bodley, *FEBS Lett.* **229,** 388 (1988).

[37] A. A. Szewczak, P. B. Moore, Y.-L. Chan, and I. G. Wool, *Proc. Natl. Acad. U.S.A.* **90,** 9581 (1993).

[38] L. Carrasco and M. Esteban, *Virology* **117,** 62 (1982).

[39] C. Fernández-Puentes and L. Carrasco, *Cell* **20,** 769 (1980).

[40] N. Olmo, J. Turnay, M. A. Lizarbe, and J. G. Gavilanes, *STP Pharma Sci.* **3,** 93 (1993).

[41] J. Turnay, N. Olmo, J. Jiménez, M. A. Lizarbe, and J. G. Gavilanes, *Mol. Cell. Biochem.* **122,** 39 (1993).

[42] M. Gasset, A. Martínez del Pozo, M. Oñaderra, and J. G. Gavilanes, *Biochem. J.* **258,** 569 (1989).

[43] M. Gasset, M. Oñaderra, P. G. Thomas, and J. G. Gavilanes, *Biochem. J.* **265,** 815 (1990).

[44] J. M. Mancheño, M. Gasset, J. Lacadena, F. Ramón, A. Martínez del Pozo, M. Oñaderra, and J. G. Gavilanes, *Biophys. J.* **67,** 1117 (1994).

[45] J. M. Mancheño, M. Gasset, J. P. Albar, J. Lacadena, A. Martínez del Pozo, M. Oñaderra, and J. G. Gavilanes, *Biophys. J.* **68,** 2387 (1995).

[46] J. M. Mancheño, A. Martínez del Pozo, J. P. Albar, M. Oñaderra, and J. G. Gavilanes, *J. Peptide Res.* **51,** 142 (1998).

[47] M. Oñaderra, J. M. Mancheño, M. Gasset, J. Lacadena, G. Schiavo, A. Martínez del Pozo, and J. G. Gavilanes, *Biochem. J.* **295,** 221 (1993).

[48] M. Gasset, J. M. Mancheño, J. Lacadena, J. Turnay, N. Olmo, M. A. Lizarbe, A. Martínez del Pozo, M. Oñaderra, and J. G. Gavilanes, *Curr. Topi. Peptide Protein Res.* **1,** 99 (1994).

[49] M. Oñaderra, J. M. Mancheño, J. Lacadena, V. De los Ríos, A. Martínez del Pozo, and J. G. Gavilanes, *Mol. Membr. Biol.* **15,** 141 (1998).

[50] R. Campos-Olivas, M. Bruix, J. Santoro, A. Martínez del Pozo, J. Lacadena, J. G. Gavilanes, and M. Rico, *FEBS Lett.* **399,** 163 (1996).

[51] A. Martínez del Pozo, M. Gasset, M. Oñaderra, and J. G. Gavilanes, *Biochim. Biophys. Acta* **953,** 280 (1988).

catalytic residues of the transphosphorylation step, respectively, while His-50 stabilizes the transition state.[34,35] The equivalent conserved amino acids in RNase U2 would be His-101, Glu-62, and His-41.[4,56] In fact, crystal structures of RNase U2 complexed with different substrate (di- and trinucleotides) and analogs,[57] and chemical modification with ethoxyformic anhydride[26] indicate that either His-41 or Glu-62 must act as a general base while His-101 behaves as a general acid in the first step of the RNA cleavage reaction.

The spectroscopic characteristics of α-sarcin are consistent with the presence of 8 Tyr and 2 Trp residues.[51,58] Recent site-directed mutagenesis experiments have revealed that the near-UV circular dichroism spectrum of the protein is dominated by Trp-51 while most of the fluorescence emission is due to Trp-4.[59]

Methods

The methods outlined below focus on the production and characterization of natural and recombinant forms of RNase U2 and α-sarcin in order to produce and study these chimeric proteins with exchanged loops.

Cloning and Expression

Synthesis of cDNA. RNase U2-producing *Ustilago sphaerogena* strain can be obtained either from the American Type Culture Collection (Manassas, VA, ATCC 12421) or the Centraalbureau voor Schimmelcultures (CBS 534.71). The standard growth medium (SGM) consists of 2.0% (w/v) glucose, 0.2% (w/v) glycine, 0.05% (w/v) KH_2PO_4, 0.05% (w/v) K_2HPO_4, 0.01% (w/v) $MgSO_4$, 0.01% (w/v) KCl, and 0.01% (w/v) $CaCl_2$[5]. The most widely used α-sarcin producing *A. giganteus* strain, originally isolated in Michigan[27] (MDH 18894), is not available from any of the well-known microorganism collections. The presence of the protein has been detected in other isolates,[60,61] but they have not been used for its production and

[52] B. Lamy, J. Davies, and D. Schindler, *in* "Genetically Engineered Toxins" (A. E. Frankel, ed.), pp. 237–258, Marcel Dekker, New York, 1992.
[53] R. Kao and J. Davies, *J. Biochem. Cell Biol.* **73,** 1151 (1995).
[54] R. Kao and J. Davies, *J. Biol. Chem.* **274,** 12576 (1999).
[55] R. Kao and J. Davies, *FEBS Lett.* **466,** 87 (2000).
[56] A. Martínez-Ruiz, A. Martínez del Pozo, J. Lacadena, M. Oñaderra, and J. G. Gavilanes, *J. Invertebr. Pathol.* **74,** 96 (1999).
[57] N. Noda, S. Noguchi, and Y. Satow, *Nucl. Acids Symp. Series* **37,** 285 (1997).
[58] J. G. Gavilanes, D. Vázquez, F. Soriano, and E. Méndez, *J. Protein. Chem.* **2,** 251 (1983).
[59] C. De Antonio, A. Martínez del Pozo, J. M. Mancheño, M. Oñaderra, J. Lacadena, A. Martínez-Ruiz, J. M. Pérez-Cañadillas, M. Bruix, and J. G. Gavilanes, *Proteins* **41,** 350 (2000).
[60] S. Wnendt, H. Felske-Zech, P.-P. C. Henze, N. Ulbrich, and U. Stahl, *Gene* **124,** 239 (1993).
[61] A. Martínez-Ruiz, R. Kao, J. Davies, and A. Martínez del Pozo, *Toxicon* **37,** 1549 (1999).

purification. The culture medium employed[28,62] contains 1.5% (w/v) beef extract, 2% (w/v) peptone, 2% (w/v) corn starch, and 0.5% (w/v) sodium chloride.

RNA is isolated from the mycelial culture at the time of maximum protein production.[62,63] Therefore, the method employs strong denaturing conditions to inactivate all present RNases.[62–64] Mycelium (300 mg wet weight) of *U. sphaerogena* or *A. giganteus* is suspended in 2 ml of 4 *M* guanidinium isothiocyanate, 25 m*M* sodium citrate, 0.5% (w/v) lauryl sarcosinate, pH 7.5, and 10 μl of 2-mercaptoethanol and 2 μl of 289-Sigma antifoam is added. This mixture is homogenized by three cycles of 20 sec with a Polytron (Brinkmann Instruments, Westbury, NY), previously sterilized with chloroform, avoiding foam formation. The soluble fraction is collected by centrifugation (3000*g*, 10 min at room temperature) and loaded onto a CsCl cushion. The remaining steps are described in detail elsewhere.[65] The procedure yields a RNA preparation which is used (5 μg) to synthesize the cDNA by a conventional reverse transcriptase-polymerase chain reaction (RT-PCR) amplification. Appropriate primers, including the stop codon, are 5′TGCGACATCCCTCAGTCCA3′ and 5′TTACGAGCACTGGGTGAAGC3′, and 5′GCGGTGACCTGGACCTGCT3′ and 5′CTAATGAGAGCAGAGCTTA3′ for cDNA of RNase U2[63] and α-sarcin,[62] respectively. Extensions containing convenient restriction sites for each construct planned should be also included.

Construction of Expression Plasmids. Native recombinant α-sarcin has been produced in *Escherichia coli*.[62,66] RNase U2 has been expressed as a recombinant protein in the methylotrophic yeast *Pichia pastoris*.[63] Attempts to produce large amounts of native protein in *E. coli* have not been so far successful. A system to produce α-sarcin in *P. pastoris* has been also reported[67] but the yield is not as good as that obtained in bacteria.

The approach for RNase U2 production is based on the plasmid pHILS1 (Invitrogen), which allows fusion of the gene to the signal peptide of the *P. pastoris* alkaline phosphatase. This construct, named pHSU2, produces extracellular RNase U2 that possesses three extra amino acids (Arg-Glu-Leu) at its NH_2-terminal end.[63] For this purpose, the RNase U2 cDNA is amplified with the oligonucleotides described above but containing convenient extensions, to

[62] J. Lacadena, A. Martínez del Pozo, J. L. Barbero, J. M. Mancheño, M. Gasset, M. Oñaderra, C. López-Otín, S. Ortega, J. L. García, and J. G. Gavilanes, *Gene* **142,** 147 (1994).

[63] A. Martínez- Ruiz, L. García-Ortega, R. Kao, M. Oñaderra, J. M. Mancheño, J. Davies, A. Martínez del Pozo, and José G. Gavilanes, *FEMS Microbiol. Lett.* **189,** 165 (2000).

[64] P. Chomczynski and N. Sacchi, *Anal. Biochem.* **162,** 156 (1987).

[65] J. Sambrook, E. F. Fritsch, and T. Maniatis, "Molecular Cloning: A Laboratory Manual." Cold Spring Harbor Laboratory Press, Cold Spring Harbor, New York, 1989.

[66] L. García-Ortega, J. Lacadena, V. Lacadena, M. Masip, C. De Antonio, A. Martínez-Ruiz, and A. Martínez del Pozo, *Lett. Applied Microbiol.* **30,** 298 (2000).

[67] A. Martínez-Ruiz, A. Martínez del Pozo, J. Lacadena, J. M. Mancheño, M. Oñaderra, C. López-Otín, and J. G. Gavilanes, *Protein Expr. Purif.* **12,** 315 (1998).

harbor *Mfe*I and *Bgl*II restriction sites, and cloned into the corresponding compatible *Eco*RI and *Bam*HI sites of pHILS1. *Pichia pastoris* GS115 cells are transformed by standard procedures.[67–69] Screening for gene replacement of the RNase U2 constructs yielding a (His^+ Mut^s) phenotype is performed by patching the His^+ colonies on two different types of His-deficient media, one containing glucose (MD) and the other with methanol (MM) as the only carbon source[63,67–69]

The best RNase-producing His^+ Mut^s strains are further selected according to their ability to cleave RNA embedded in agar plates.[63,70] Agar minimal medium containing methanol (MM), 2 g/liter *Torula* RNA (Type VI Sigma), and 50 mg/ml toluidine blue is used to grow the colonies for this purpose. Those secreting RNase U2 produce a red halo, its diameter being proportional to the ribonucleolytic activity present.

The procedure employed to clone α-sarcin is not so straightforward as that used for RNase U2 and has been described in detail before.[62] Essentially, the α-sarcin sequence is fused to a modified version of the OmpA signal peptide. This construct, pINPGαS, produces α-sarcin that is exported to the periplasm, thus avoiding the toxicity of the recombinant protein. Other systems to produce α-sarcin in *E. coli* have been described,[71–75] some of them yielding higher amounts of protein.[74,75] However, a slightly modified protein was produced in some of these examples[72,74] and the recombinant protein obtained in all these cases was not characterized in such detail as that obtained through the method herein described.[25,50,62]

Production and Purification

Fungal RNase U2. A sample (100 ml) of 40 h PDB *U. sphaerogena* culture is used to inoculate 3 liters of SGM distributed in Erlenmeyer flasks (4 : 1 flask volume to medium volume ratio). This culture is incubated at 30° with strong shaking, and aliquots are taken every 12 h for measuring ribonucleolytic activity at pH 4.5 as an indicator of RNase U2 production. The fungus must grow with a "yeastlike" morphology since an aggregated mycelium corresponds with low RNase levels.

[68] M. D. Rose, F. Winston, and P. Hieter, "Methods in Yeast Genetics: A Laboratory Manual." Cold Spring Harbor Laboratory Press, Cold Spring Harbor, New York, 1990.

[69] S. Huecas, M. Villalba, E. González, A. Martínez-Ruiz, and R. Rodríguez, *Eur. J. Biochem.* **261,** 539 (1999).

[70] R. Quaas, O. Landt, H.-P. Grunert, M. Beineke, and U. Hahn, *Nucl. Acids Res.* **17,** 3318 (1989).

[71] P. P. C. Henze, U. Hahn, V. A. Erdmann, and N. Ulbrich, *Eur. J. Biochem.* **192,** 127 (1990).

[72] T. Oka, Y. Aoyana, Y. Natori, T. Katano, and Y. Endo, *Biochim. Biophys. Acta* **1130,** 182 (1992).

[73] D. Rathore, S. K. Nayak, and J. K. Batra, *Gene* **190,** 31 (1997).

[74] I. D. Sylvester, L. M. Roberts, and J. M. Lord, *Biochim. Biophys. Acta* **1358,** 53 (1997).

[75] D. Parente, G. Raucci, B. Celano, A. Pacilli, I. Zanoni, S. Canevari, E. Adorate, M. I. Colnaghe, F. Dosio, S. Arpicco, L. Catell, A. Mele, and R. de Santis, *Eur. J. Biochem.* **239,** 272 (1998).

Once the production is maximum (about 100 h), the extracelluar medium is collected by filtration through two layers of Whatman (Clifton, NJ) 3MM paper and dialyzed against 20 m*M* piperazine, pH 6.0, and mixed overnight, under continuous and gentle stirring, at 4° with 10 ml of DEAE-cellulose (DE32, Whatman) equilibrated in the same buffer. The resin is packed in a column, washed with 20 m*M* piperazine, pH 6.0, and eluted with a 0–0.5 *M* NaCl gradient in the same buffer. In the resulting fractions, ribonuclease activity is measured at pH 4.5 and 7.5, to distinguish among the different RNases produced by the fungus. Those containing the acidic activity are concentrated using Centriprep 3 (Amicon, Danvers, MA) devices, and loaded on a Sephadex G-75 column (1.8 × 82 cm) equilibrated in 50 m*M* ammonium acetate, pH 4.5. Fractions containing acidic RNase activity are collected and concentrated, yielding about 0.3 mg of homogenous RNase U2.

A fermentation procedure involving 10 liter of medium cultured for 4 days at 30° in a New Brunswick fermentor has been also described.[13] A large-scale fermentation (3,000 liter) of *U. sphaerogena* grown in 0.8% ammonium sulfate, 0.1% (w/v) urea, and 5.0% (w/v) corn meal produced more than 10 g of purified RNase U2.[19] Many of the studies with this enzyme were obtained from this batch of protein.

Recombinant RNase U2. Production of RNase U2 from *P. pastoris* is regulated by the *AOX1* promoter, which is fully induced when methanol is the sole carbon source, but biomass production is first achieved by growing the yeast on a different medium. A volume of 2 liter of buffered minimal glycerol medium (BMG; 10 g/liter glycerol, 13.4 g/liter yeast nitrogen base without amino acids, 400 μg/l biotin, 0.1 *M* potassium phosphate buffer, pH 6.0; this medium is advised so as to avoid anionic nonprotein contaminants that were difficult to eliminate during the later purification steps) is inoculated with 50 ml preinoculum in the same medium. After 2 days of incubation at 30°, the cells are harvested by centrifugation, resuspended in 400 ml of buffered minimal methanol medium (BMM; as BMG using 5 g/liter methanol instead of glycerol), and further incubated for 2 days in two baffled 1-liter flasks covered with two layers of Miracloth. Methanol is replenished up to 0.5% twice each day. Finally, the extracellular medium is collected by centrifugation. Recombinant RNase U2 is purified as described for the fungal protein but replacing the last gel filtration step by affinity chromatography[22,63] to remove a contamination responsible for an apparent Trp fluorescence. Thus, the pooled fractions from the DEAE cellulose column step are dialyzed against 50 m*M* sodium acetate, pH 4.5, 50 m*M* NaCl, and loaded on a 2′,5′-ADP-Sepharose (Pharmacia, Uppsala, Sweden) column. The elution is performed with a linear gradient of 50 m*M* sodium acetate, pH 4.5/50 m*M* Tris-HCl, pH 7.0, both containing 50 m*M* NaCl. RNase activity-containing fractions are concentrated and dialyzed against 50 m*M* ammonium acetate, pH 4.5. The recombinant RNase U2 obtained (1.5 mg/liter of crude extract) is homogeneous according to its SDS–PAGE behavior, amino acid composition, kinetic parameters, and spectroscopic features, including the fluorescence spectrum.[63]

Finally, it is noteworthy that only unreduced RNase U2 behaves as a monomer, whereas reduction results in a single SDS–PAGE band with mobility corresponding to the protein dimer (unpublished results). This anomalous behavior is observed for both fungal and recombinant RNase U2.

Fungal α-Sarcin. Production and purification of the protein are performed essentially as described,[27] with some modifications.[62] Using the original culture medium described[28] (see above) is crucial since small changes in its composition result in a dramatic drop in the final yield.

Aspergillus giganteus spores are seeded and precultured for 24 h at 30° in 250 ml of medium. A sample of this culture is employed for inoculation of a second identical culture that is further used to inoculate the final production medium (usually 5 liter distributed in 10 1-liter Erlenmeyer flasks). This is incubated at 30°, with rapid shaking, for about 80 h. The production of α-sarcin is maximum at the time when the pH of the culture reaches pH 8.0–8.5. The culture medium is filtered through two layers of Whatman 3M paper, dialyzed against 50 m*M* sodium phosphate, pH 7.0, and mixed overnight with 80 ml of Amberlite IRC50 equilibrated in the same buffer. The resin is packed on a column and washed with phosphate buffer containing 0.2 *M* NaCl, and the protein is eluted with the same buffer but containing 0.6 *M* NaCl. α-Sarcin-containing fractions are pooled, dialyzed against water and lyophilized. The resulting sample is dissolved in 10 ml of 50 m*M* Tris-HCl, pH 7.0, containing 0.1 *M* NaCl and loaded on a Bio-Gel P10 column (2 × 150 cm, Bio-Rad) equilibrated in the same buffer. The α-sarcin-containing fractions are treated as above, and finally dissolved in 4 ml of 50 m*M* acetic acid. This sample is loaded on a Bio-Gel P2 column (1 × 20 cm) to completely remove any salts, and the protein-containing fractions are lyophilized. All the lyophilization steps can be substituted by concentrating in Centriprep-10 devices (Amicon). This procedure yields 5.0–20.0 mg of homogeneous α-sarcin per liter of original culture.

Recombinant α-Sarcin. Production of recombinant α-sarcin[62,66] is achieved using *E. coli* BL21 (DE3) harboring two plasmids, pINPGαS for α-sarcin and pT-Trx for thioredoxin,[76] in order to improve its solubility. One colony is used to inoculate 250 ml of LB medium containing ampicillin (100 μg/ml) and chloramphenicol (34 μg/ml), which is incubated overnight at 37°. This preinoculum is used to seed 10 1-liter Erlenmeyer flasks (25 ml of preinoculum per flask) each containing 0.5 liter of M9 minimal medium,[65] supplemented with both antibiotics and 0.2% (w/v) glucose. The culture is incubated at 37° until an OD_{600} of 0.7 is reached and then induced with IPTG (1 m*M* final concentration) and cultured with shaking for 16–18 h at 37°.

[76] T. Yasukawa, C. Kanei-Ishii, T. Mackaura, J. Fujimoto, T. Yamamoto, and S. Ishii, *J. Biol. Chem.* **270,** 25328 (1995).

The extracellular medium is collected by centrifugation (5000*g*, 15 min, 25°). Its pH is adjusted to 7.0 with 0.1 *M* NaOH and maintained at 4°. The cellular pellet is *carefully* suspended in 150 ml of 33 m*M* Tris-HCl, pH 7.0, containing 33 m*M* EDTA, mixed with 150 ml of 40% (w/v) sucrose, and *gently* shaken for 15 min at room temperature to release the periplasmic fraction by osmotic shock. The solution is centrifuged (10,000*g*, 15 min, 25°) and the sucrose-containing supernatant is also maintained at 4°. The cellular pellet is *very carefully* suspended in 300 ml of ice-cold water and shaken for 15 min at 4°C. The supernatant obtained after centrifugation (12,000*g*, 30 min, 4°) is enriched in the periplasmic fraction. The pellet is once more suspended in 50 m*M* sodium phosphate, pH 7.0, and sonicated (seven pulses of 20 KHz for 1 min in an ice bath). The intracellular soluble fraction is obtained by centrifugation at 12,000*g* for 30 min at 4°. The three soluble cellular fractions (sucrose-enriched, periplasmic, and intracellular) are mixed and the pH adjusted to 7.0.

Amberlite IRC-50, equilibrated in 50 m*M* sodium phosphate, pH 7.0, is added to the extracellular medium and to the mixed soluble cellular fractions (15 and 60 ml of resin, respectively). The following steps are as described previously for separation of the fungal protein. This procedure yields 25–55 mg of homogeneous native α-sarcin.[66]

Activity Assays

Ribonucleolytic Activity. As explained above, RNase U2 can be considered a nonspecific ribonuclease when compared with α-sarcin. RNase U2 at catalytic concentration (nanomolar) extensively degrades RNA with a strong preference for purines, while α-sarcin cleaves a single phosphodiester bond of the larger rRNA. However, at higher concentrations (micromolar), α-sarcin completely digests RNA[30,77] and slowly cleaves G/ApA substrates[33] while RNase U2 can also cleave phosphodiester bonds flanking pyrimidine nucleotides.[13,14,16] Both enzymes display optimum pH around 5.0.[6,34,35] The assays for the characteristic activity of α-sarcin have been described elsewhere in this volume[1] and several chapters are also dedicated to methods to analyze RNase activities.[78,79] Consequently, only three types of activity assays are described here. Both apply to RNase U2 and α-sarcin, but much larger enzyme concentrations must be used when the ribotoxin is assayed.

The first assay, employed in RNase purification protocols, uses *Torula* RNA (Type VI Sigma) as substrate[80] (it can also be used for α-sarcin[77] but it is not very

[77] A. Martínez del Pozo, M. Gasset, M. Oñaderra, and J. G. Gavilanes, *Int. J. Peptide Res.* **33,** 406 (1989).

[78] T. Yasuda, T. Takeshita, and K. Kishi, *Methods Enzymol.* **341,** [7] 2001 (this volume).

[79] K. Korn, T. Dreiner-Stöffele, and U. Hahn, *Methods Enzymol.* **341,** [10] 2001 (this volume).

[80] J. M. García-Segura, M. M. Orozco, J. M. Fominaya, and J. G. Gavilanes, *Eur. J. Biochem.* **158,** 367 (1986).

useful because the culture medium contains more active nonspecific ribonucleases). The RNA substrate has to be fractionated to remove contaminants and small oligonucleotides. For this purpose, 100*g* of *Torula* RNA is slowly dissolved in 500 ml of ice-cold distilled water maintaining the pH around 7.0. Sodium acetate and acetic acid are added to obtain pH 5.5 and 3 *M* final salt concentration and the solution is maintained overnight at −20°. The brownish pellet, containing high molecular weight insoluble RNA, is recovered by centrifugation (13,000*g*, 30 min, at −20°) and then resuspended. This procedure is repeated twice. The resulting RNA pellet is dissolved in 500 ml of water, and 1 liter of ethanol (previously cooled to −20°) is added, maintaining the suspension overnight at −20°. A white RNA pellet is recovered by centrifugation under the above conditions, which is dissolved in 25 m*M* EDTA and dialyzed against 20 volumes of first 25 m*M* EDTA, then 0.15 *M* NaCl, and finally water. This solution is freeze-dried and maintained at −20°. About 3 g of purified RNA is obtained.

The assay is performed in tubes containing 50 μl of 0.5 *M* KCl, 0.25 *M* sodium acetate, 0.25 *M* Tris-HCl, 25 m*M* EDTA, pH 4.5, 25 μl of distilled water, 25 μl of 0.1% (w/v) BSA, and 25 μl of the protein sample. The mixture (125 μl) is incubated for 15 min at 37° and the reaction is started by the addition of 125 μl of the purified RNA (4 mg/ml) and vigorous shaking. In a standard assay this mixture is incubated for 15 min at 37°, and the reaction is stopped by addition of 250 μl of ice-cold 10% (v/v) perchloric acid, containing 0.25% (w/v) uranyl acetate. The resulting suspension is maintained for 30 min in an ice bath and then centrifuged at 14,000*g* for 15 min at 4°. The supernatant is diluted 10-fold and absorbance at 260 nm is measured. This value is dependent (the A_{260} value is linear up to 0.6) on the concentration of soluble oligonucleotides produced by the ribonuclease activity. One activity unit is defined as the enzyme amount required to produce an absorbance increment of 0.4 in the conditions of a standard assay.

Ribonuclease activity against homopolynucleotides can be assayed by an electrophoretic procedure. This method, designated as *zymogram,* has been described in this volume.[1] In this particular case, the only differences are the homopolynucleotide used, poly(A), and the buffer employed, 0.1 *M* sodium acetate, pH 4.5.

The third type of assay involves measurement of the cleavage of dinucleotides. This is carried out by spectrophotometry for RNase U2,[16] and HPLC for α-sarcin.[33] In the former case, 780 μl of dinucleotide substrate (10–60 μM) in 50 m*M* sodium acetate, pH 4.5, containing 50 m*M* NaCl, is placed in a 1-cm optical-path cell (or smaller if the absorbance value becomes higher than 1.0) in a thermostated spectrophotometer. The reaction is started by adding enzyme (in 20 μl of the same buffer; approximately 200 n*M* final concentration) and the absorbance change at 262 nm is recorded for 5–10 min. The difference in molar extinction coefficients at 262 nm between the substrate and the products[16] is used to evaluate the initial rates, which allows calculation of K_m and K_{cat} by hyperbolic fitting to the Michaelis–Menten equation.

To assay α-sarcin against dinucleotides,[33–35] the enzyme (about 2 μM) and the substrate (5–100 μM) are placed in microcentrifuge tubes (150 μl final volume) and the mixture is incubated at room temperature for 14–15 h. The reaction is stopped by freezing the tubes in liquid nitrogen (keep the tubes frozen at −20°). Substrate and products are fractionated by HPLC,[33,81] injecting the thawed, centrifuged samples onto a reversed phase C_{18} column (5 μm, 0.46 × 15 cm). Elution (1.0 ml/min flow rate) is performed with a 35 min linear gradient from 100 m*M* potassium phosphate buffer, pH 7.0, to 90 m*M* potassium phosphate buffer, pH 7.0, in 32.5% (v/v) methanol, at room temperature. The eluting products are detected by continuous reading of absorbance at 254 nm. The amount of product is calculated from the area under the corresponding peak using calibration plots constructed with convenient standards. Such amounts are used to estimate the initial rates.

Interaction with Phospholipid Vesicles

The interaction of α-sarcin and its mutant forms with lipid membranes has been analyzed by measuring the effects on large unilamellar vesicles (LUVs). In particular we have monitored (i) vesicle aggregation; (ii) lipid mixing from different bilayers; (iii) leakage of vesicular aqueous contents; and (iv) changes in the thermotropic behavior of the lipid. This set of assays provides a general picture about the potential ability of a protein to interact with membranes.

Preparation of Unilamellar Vesicles. A phospholipid (1–5 mg) solution in 2 : 1 (v/v) chloroform:methanol is dried under flow of nitrogen for at least 60 min. The resulting lipid film is stored in the darkness at −20° until use. A lipid dispersion is prepared by adding the required aqueous buffer over the lipid film (at 1–5 mg/ml concentration), briefly vortexing the sample, and incubating for 60 min above the melting temperature (T_c) of the corresponding phospholipid. The heterogeneous (in size and lamellae per vesicle) suspension of multilamellar vesicles (MLVs) is then introduced into an Extruder (Lipex Biomembranes Inc., Vancouver, B.C., Canada), containing two opposed polycarbonate filters (Nuclepore Costar, Cambridge, MA) with a defined pore size, usually between 100 and 400 nm. Five to ten cycles of extrusion through the polycarbonate filters are sufficient to obtain a fairly homogeneous population of LUVs with the average diameter of the filter pores.[82–84] The phospholipid concentration is determined from its phosphorus content.[85] The absence of lysophospholipids is verified by chromatographic methods.[86]

[81] R. Shapiro, J. W. Fett, D. J. Strydom, and B. L. Vallee, *Biochemistry* **25,** 7255 (1986).
[82] L. D. Mayer, M. J. Hope, P. R. Cullis, and A. S. Janoff, *Biochim. Biophys. Acta* **817,** 193 (1985).
[83] L. D. Mayer, M. J. Hope, and P. R. Cullis, *Biochim. Biophys. Acta* **858,** 161 (1986).
[84] M. J. Hope, M. B. Bally, G. Webb, and P. R. Cullis, *Biochim. Biophys. Acta* **812,** 55 (1985).
[85] G. R. Bartlett, *J. Biol. Chem.* **234,** 466 (1959).
[86] J. G. Gavilanes, M. A. Lizarbe, A. M. Municio, and M. Oñaderra, *Biochemistry* **20,** 5689 (1981).

Protein Binding to Vesicles. Analysis of protein binding to vesicles is performed by ultracentrifugation on an Airfuge (Beckman, Palo Alto, CA) under conditions in which vesicle-bound protein is sedimented. Different protein/vesicle samples (in the 0–200 lipid/protein molar ratio range) are prepared by mixing a protein solution (0.2 mg/ml, final constant protein concentration) and the required amount of the LUVs suspension in a total volume of 250 μl. Prior to addition of protein, the vesicle suspension is maintained at a temperature above the T_c of the phospholipid, and after mixing, incubation is continued for 60 min in a thermostatted water bath under gentle agitation. Control samples with no vesicles are also included.

Aliquots (175 μl if polyallomer tubes and a Beckman A-110 rotor are used) from the above samples are centrifuged for 60 min at 4° at a pressure of 25 psi (164,000 relative centrifugal field). Under these conditions, lipid vesicles sediment, as previously verified by using traces of radioactive phospholipids incorporated in the vesicles. The absence of nonspecific protein adsorption to the centrifugation tubes is also verified with the control samples.

The concentration of free unbound protein is determined from Coomassie blue stained SDS–PAGE gels of the resulting supernatants. The volumogram (density or quantity of a spot calculated from its volume made of the sum of all pixel intensities composing the spot) of each protein band is obtained on a Photodocumentation System (UVItec, Cambridge, UK). A linear relationship between protein amount loaded on the gel and value of the corresponding volumogram is previously verified with control samples.

Aggregation of Phospholipid Vesicles. Aggregation of phospholipid vesicles is analyzed by measuring the increase of light scattering of a vesicle sample on addition of the protein.[42–45] The aggregation of individual vesicles induced by α-sarcin occurs in a millisecond to second time scale, as revealed by stopped-flow light-scattering measurement,[44] but further aggregation of large protein–lipid complexes occurs in a minute time scale. Monitoring this last process by measuring the increase in apparent absorbance at 360 nm reveals the ability of the protein to aggregate model membranes.

LUVs (1 mg/ml lipid concentration) are prepared as described above. An aliquot of the protein solution (20 μl) is added to a vesicle suspension at a temperature above the T_c of the corresponding lipid in 1-cm optical path cuvettes (1 ml final volume; 0.02 mg/ml final lipid concentration). The time-dependent change in A_{360} is recorded until a constant value is attained. Control analyses in the absence of protein (no absorbance variation was observed) are also carried out. This kinetic study must be performed under continuous stirring of the reaction mixture to avoid sedimentation of large protein–lipid aggregates in the cuvette, which would result in an aberrant decrease of the recorded A_{360} value.

Lipid Mixing of Vesicles. Perturbation of bilayers leading to the merging of lipids from different vesicles[42–46,87] can be readily revealed by a fluorescence assay.[88] This is based on the measurement of fluorescence resonance energy transfer (RET) between a donor and acceptor pair, in this case, two derivatives of phosphatidylethanolamine (PE), labeled in their polar region with *N*-(7-nitro-2-1,3-benzoxadiazol-4-yl) (NBD-PE), and *N*-(Lissamine) rhodamine B sulfonyl (Rh-PE), respectively.[42–46,87,89–91] These probes are nonexchangeable between vesicles even when they are aggregated.[91] In a typical assay, both probes are incorporated in the same bilayer at proportions displaying high RET efficiency, i.e., low fluorescence emission of NBD-PE (donor). Fusion of the labeled bilayer with unlabeled membranes results in a decrease of the RET efficiency due to dilution of the probes into the newly formed membrane, and hence in an increase of the donor fluorescence.

Vesicles containing both NBD-PE and Rh-PE are prepared essentially as described above and maintained in the darkness. Both probes are incorporated into the organic solution containing the phospholipids (usually 1% NBD-PE and 0.6% Rh-PE mol%, although a linear response between donor fluorescence and probe dilution must be checked in order to accurately measure lipid mixing[92]). In addition, vesicles containing only the donor probe are prepared for calibration of the 100% donor fluorescence.

Fluorescence-labeled and unlabeled vesicles are mixed (1 : 9 ratio; 0.1 mg/ml total lipid concentration), and the fluorescence emission at 530 nm for excitation at 450 nm is recorded by using polarizers to eliminate the potential contribution of the sample turbidity to the fluorescence signal. The protein sample to be tested (100 μl) is added to the thermostated and magnetically stirred vesicle sample (2 ml total volume). The RET efficiency (%*E*) is calculated once the fluorescence intensity becomes constant. (%*E*) is defined as $100 \times (1\text{-}F/F_0)$, where F and F_0 are the fluorescence intensities at 530 nm in presence and absence of Rh-PE (acceptor), respectively. The (%*E*) value of the initial vesicle mixture must be 75–80%.

Leakage of Intravesicular Aqueous Contents. The breakdown of the permeability barrier of lipid bilayers can be analyzed using an assay employing the fluorescence probe 8-aminonaphthalene-1,3,5-trisulfonic acid (ANTS) and its collisional quencher *p*-xylene bis(pyridinium) bromide (DPX).[87,93–95] When both are encapsulated into lipid vesicles, the release of the intravesicular contents to the

[87] J. M. Mancheño, M. Oñaderra, A. Martínez del Pozo, P. Díaz-Achirica, D. Andreu, L. Rivas, and J. G. Gavilanes, *Biochemistry* **35,** 9892 (1996).

[88] D. K. Struck, D. Hoekstra, and R. E. Pagano, *Biochemistry* **20,** 4093 (1981).

[89] A. Walter and D. P. Siegel, *Biochemistry* **32,** 3271 (1993).

[90] J. Wilschut, S. Nir, J. Scholma, and D. Hoekstra, *Biochemistry* **24,** 4630 (1985).

[91] J. Wilschut, *in* "Membrane Fusion" (J. Wilschut and D. Hoekstra, eds.), pp. 89–125. Marcel Dekker, New York, 1991.

[92] A. J. M. Driessen, D. Hoekstra, G. Scherphof, R. D. Kalicharan, and J. Wilschut, *J. Biol. Chem.* **260,** 10880 (1985).

[93] H. Ellens, J. Bentz, and F. C. Szoka, *Biochemistry* **24,** 3099 (1985).

[94] H. Ellens, J. Bentz, and F. C. Szoka, *Biochemistry* **25,** 4141 (1986).

[95] V. De los Ríos, J. M. Mancheño, M. E. Lanio, M. Oñaderra, and J. G. Gavilanes, *Eur. J. Biochem.* **252,** 284 (1998).

external medium results in a dilution of both probe and quencher and a concomitant increase of the ANTS fluorescence. This assay allows the study of both the extent and mechanism (*all or none,* or graded) of leakage.[96,97]

Vesicles encapsulating ANTS/DPX are prepared as described above by hydrating the lipid films in aqueous buffer containing both fluorophore and quencher (15 m*M* Tris, pH 7.5, 20 m*M* NaCl, 12.5 m*M* ANTS, and 45 m*M* DPX). To increase the encapsulation efficiency, MLVs are subjected to 5 to 10 cycles of freezing and thawing with liquid nitrogen prior to 5 cycles of extrusion (100 nm average filter pore diameter). Unencapsulated material is separated from the vesicles by gel filtration on a Sephadex G-75 column equilibrated in 15 m*M* Tris, pH 7.0, containing 0.1 *M* NaCl and 1 m*M* EDTA, the buffer for the leakage assays, which is isosmotic with the hydration buffer.

The fluorescence intensity, measured through a 3-68 Corning cutoff filter (>530 nm) on excitation at 386 nm of the vesicle sample in the absence of protein is considered as 0% leakage (background signal). The fluorescence measured after addition of 50 μl of 10% Triton X-100 (total vesicle lysis) corresponds to 100% leakage. Polarizers are used to eliminate potential contribution of sample turbidity to the fluorescence. The extent of leakage (%*L*) is defined as $(\%L) = 100\ (F_P - F_0)/(F_T - F_0)$, where F_p and F_0 are the fluorescence intensity values after and prior to the addition of protein, respectively, and F_T is the value after detergent addition. The stability of the vesicles must be verified prior to any leakage assay since unstable vesicles may generate a continuous increase in the fluorescence intensity.

Thermotropic Behavior of Lipids. Protein-induced structural and dynamic membrane changes can be studied by fluorescence techniques after selective probe-labeling of the bilayers. The assumption is made that the order and dynamics of the fluorophore moiety reflect the behavior of the surrounding lipid molecules.[98,99] The most popular fluorophores are 1,6-diphenyl-1,3,5-hexatriene (DPH; probes the center of the bilayer[100]) and its derivative 1-(4-trimethylammonium-phenyl)-6-phenylhexatriene (TMA-DPH; probes the polar headgroup region[101,102]).

DPH-labeled vesicles are used to characterize the effect of α-sarcin on the thermotropic gel-to-liquid crystal phase transition of phospholipid vesicles.[42–46] Vesicles containing these probes are prepared essentially as described above and

[96] W. C. Wimley, M. E. Selsted, and S. H. White, *Protein Sci.* **3,** 1362 (1994).
[97] A. S. Ladokhin, W. C. Wimley, and S. H. White, *Biophys. J.* **69,** 1964 (1995).
[98] L. W. Engel and F. G. Prendergast, *Biochemistry* **20,** 7338 (1981).
[99] P. K. Wolber and B. S. Hudson, *Biochemistry* **20,** 2800 (1981).
[100] L. Davenport, R. E. Dale, R. H. Bisby, and R. B. Cundall, *Biochemistry* **24,** 4097 (1985).
[101] B. R. Lentz, *Chem. Phys. Lipids* **50,** 171 (1989).
[102] F. G. Prendergast, R. P. Haugland, and P. J. Callahan, *Biochemistry* **20,** 7333 (1981).
[103] M. Irie, *in* "Ribonucleases: Structures and Functions" (G. D'Alessio and J. F. Riordan, eds.), pp. 101–130. Academic Press, New York, 1997.

maintained in the dark until use. DPH is dissolved in tetrahydrofuran and added to the initial lipid solution (1000 : 1 lipid/probe weigh ratio).

The thermotropic behavior of the phospholipid is shown by the thermal variation around the T_c of the fluorescence polarization degree of the probe at 425 nm for excitation at 365 nm. This is analyzed by 1° stepwise increase (10 min equilibration at each temperature) from temperatures below T_c. Any change in order and mobility of the phospholipids produced by the protein results in a modification of the thermotropic plot.

Acknowledgments

This work was supported by grant from the MEC (Spain) and NSERC (Canada; R.K., J.D.). A.M.-R. and L.G.-O. are recipients of fellowships from the M.E.C. (Spain).

[22] Secretory Acid Ribonucleases from Tomato, *Lycopersicon esculentum* Mill.

By STEFFEN ABEL and MARGRET KÖCK

Introduction

The first secretory ribonuclease from tomato was identified in isolated vacuoles of cultured tomato cells by Abel and Glund.[1] It was shown later that cultured tomato cells starved for inorganic phosphate (P_i) induce a limited set of ribonucleases, which comprise extracellular RNase LE, three vacuolar enzymes, RNases LV-1, LV-2, and LV-3, and microsomal RNase LX. Subsequently, these ribonucleases were purified, their primary structures directly determined by protein sequencing, and their enzymatic properties characterized. Isolation of cDNAs coding for RNase LE and RNase LX confirmed the chemically determined primary structures. The deduced secretory targeting signals are in agreement with previous cellular localization studies.

Plant ribonuclease proteins structurally related to RNases LE and LX were purified and characterized from seeds of bitter gourd (*Momordica charantia*), RNase MC1, and from leaves of *Nicotiana glutinosa,* RNase NW.[2,3] A number of similar plant ribonucleases were identified by gene cloning such as RNS1, RNS2, and

[1] S. Abel and K. Glund, *Physiol. Plant.* **66,** 79 (1986).

[2] H. Ide, M. Kimura, M. Arai, and G. Funatsu, *FEBS Lett.* **284,** 161 (1991).

[3] T. Kariu, K. Sano, H. Shimokawa, R. Itoh, N. Yamasaki, and M. Kimura, *Biosci. Biotechnol. Biochem.* **62,** 1144 (1998).

0076-6879/01 $35.00

RNS3 from *Arabidopsis thaliana* or ZRNase I and ZRNase II from *Zinnia elegans*. This group of related plant enzymes was designated S-like ribonucleases.[4,5] S-like RNases share conserved sequence motifs that are also found in S-gene products, which play a role in self-incompatibility cell-to-cell interactions,[6] in fungal RNases typified by RNase T2 as well as in animal and bacterial RNases with acidic pH optima.[7] Amino acid residues important to the catalytic activity of S-like RNases, foremost conserved histidine residues, are present in these motifs. All the above-mentioned enzymes are members of the superfamily of acid T2-type ribonucleases,[7] which is also known as the superfamily of T2/S ribonucleases.[4]

It became clear that S-like RNases likely have functions in a number of physiological processes other than in self-incompatibility interactions, which is indicated by differential gene expression in response to a variety of exogenous factors and during plant development. Such physiologic processes include leaf and flower senescence and responses to plant hormones, inorganic phosphate (P_i) starvation, or wounding.[4,5]

This article focuses on the analysis of secretory acid ribonucleases from tomato, which are synthesized in remarkably high amounts in cultivated tomato cells. We describe enzyme purification from cultured tomato cells and biochemical analysis of several protein parameters of the highly similar RNases. We further discuss structural relationships between the tomato enzymes with regard to their subcellular locations.

Source of Enzymes

All studies are conducted with purified ribonucleases prepared from cell suspension cultures of tomato (*Lycopersicon esculentum* Mill. cv. Lukullus). Cell suspension cultures provide an advantage in (i) producing and purifying the enzymes of interest in large amounts, and (ii) conveniently carrying out biochemical and immunological analyses in cellular localization studies.

The cell suspension culture is propagated in a modified Murashige–Skoog medium.[8] Suspensions contain mostly small aggregates of 20–50 cells and plasma-rich, single cells to a lesser extent. Cultures are inoculated with 2×10^7 cells from 3-day-old exponentially growing cell cultures and are cultivated for 2–6 days in complete medium depending on purpose. The P_i concentration of the

[4] P. J. Green, *Annu. Rev. Plant Physiol. Mol. Biol.* **45,** 421 (1994).

[5] P. A. Bariola and P. J. Green, *in* "Ribonucleases: Structures and Functions" (G. D'Alessio and J. F. Riordan, eds.), p. 163. Academic Press, New York, 1997.

[6] S. K. Parry, Y.-H. Liu, A. E. Clarke, and E. Newbigin, *in* "Ribonucleases: Structures and Functions" (G. D'Alessio and J. F. Riordan, eds.), p. 192. Academic Press, New York, 1997.

[7] M. Irie, *Pharmacol. Ther.* **81,** 77 (1999).

[8] A. Tewes, K. Glund, R. Walther, and H. Reinbothe, *Z. Pflanzenphysiol.* **113,** 141 (1984).

liquid medium drastically decreases until day 3 of cultivation. Subsequently, cells enter the stationary growth phase during which secretory acid RNases and other P_i-regulated enzymes are induced.[1,9,10] Alternatively, to induce more efficiently secretory ribonucleases, 3-day-old cells of the logarithmic growth phase can be transferred into P_i-free medium and subcultured for 3–4 days.[11]

Assay Methods

Spectrophotometric Assay of Ribonucleolytic Activity

Principle. Ribonuclease activity is assayed essentially according to the method of Ambellan and Hollander[12] by following the absorbance at 260 nm of the time-dependent release of ethanol-soluble hydrolysis products from high molecular weight RNA substrates.

Reagents

RNA substrate: High molecular weight fraction of yeast RNA (from *Torula* yeast, Type II-S; Sigma), which is prepared by gel filtration on Sephadex G-25, ethanol precipitation and dissolving of the RNA precipitate to a concentration of 5 mg/ml in water

Assay buffer: 150 m*M* Sodium acetate (pH 5.6)

Precipitation reagent: 50% (v/v) ethanol containing 50 m*M* sodium acetate (pH 5.5), 10 m*M* magnesium acetate, and 0.8 m*M* $La(NO_3)_3$

Procedure. Mix 50 μl of assay buffer with 50 μl of appropriate amounts of enzyme in 1.5 ml Eppendorf tubes and preincubate for 5 min at 37°. Start the assay by adding 50 μl of RNA substrate (250 μg) to the reaction mixture. Continue the incubation for 0–60 min at 37° and stop the reaction with 1.35 ml of precipitation reagent. Briefly mix and chill for at least 45 min at −20°. Sediment the precipitating RNA substrate by centrifugation at room temperature (12,000*g*, 20 min). Carefully remove the supernatant and measure its absorbance at 260 nm against a blank taken at zero time. The ribonuclease assay is linear with respect to time and enzyme concentration for up to an absorbance of A_{260} of 1.0. One ribonuclease unit is defined as the amount of enzyme causing an increase in ΔA_{260} of 1.0 min^{-1} cm^{-1} ml^{-1} according to Wilson.[13]

[9] T. Nürnberger, S. Abel, W. Jost, and K. Glund, *Plant Physiol.* **92,** 970 (1990).

[10] S. Abel, T. Nürnberger, V. Ahnert, G.-J. Krauß, and K. Glund, *Plant Physiol.* **122,** 543 (2000).

[11] A. Löffler, S. Abel, W. Jost, J. J. Beintema, and K. Glund, *Plant Physiol.* **98,** 1472 (1992).

[12] V. Ambellan and V. P. Hollander, *Anal. Biochem.* **17,** 474 (1966).

[13] M. Wilson, *in* "Isoenzymes: Current Topics in Biological and Medical Research" (M. C. Ratazzi, J. G. Scandalios, and G. S. Whitt, eds.), Vol. 6, p. 33. Alan R. Liss Inc., New York, 1982.

Determination of Relative Base Specificity by HPLC Analysis of RNA Hydrolysis Products

Principle. The enzyme- and time-dependent release of mononucleotides from yeast RNA is followed by isocratic reversed-phase high-performance liquid chromatography (HPLC).[14]

Reagents

RNA substrate: High molecular weight fraction of yeast RNA (from *Torula* yeast, Type II-S; Sigma), which is prepared as described above

Assay buffer: 50 m*M* Sodium acetate (pH 5.6)

HPLC system: Merck-Hitachi LiChroGraph (Darmstadt, Germany), prepacked Octadecyl-Si 100 Polyol column (4.6 × 250 mm, 5 μm, Serva Feinbiochemica, Heidelberg, Germany), 20 m*M* $(NH_4)H_2PO_4$ (pH 6.2) as mobile phase buffer

Procedure. Mix 10 mg of RNA and 0.5 units of purified ribonuclease in a total volume of 2 ml of assay buffer and incubate at 37° under moderate shaking. After appropriate intervals (0–60 min), remove 100 μl of the reaction mixture and add to 900 μl of cold ethanol. Briefly mix and chill for at least 2 h at −20°. Centrifuge at room temperature for 20 min at 12,000 *g* to sediment the precipitating RNA substrate. Carefully remove and evaporate supernatant to dryness. Dissolve residue in 100 μl of mobile phase buffer. Inject sample volumes of 5 μl to 30 μl onto the HPLC column, develop column in the isocratic mode at a flow rate of 1.5 ml min^{-1}, and monitor eluates at 254 nm. Identify 2′,3′-cyclic NMP and 3′(2′)-NMP products by comparing their retention times with those of authentic standards. For quantification of the time-dependent release of mononucleotide products, i.e., 2′,3′-cyclic NMP and 3′(2′)-NMP for each nucleobase, generate calibration curves for each authentic compound. Calibration curves should be linear in the range of 0.1 nmol to at least 4.0 nmol.

Determination of Relative Base Specificity by HPLC Analysis of Diribonucleoside Monophosphate Hydrolysis Products

Principle. The enzyme- and time-dependent hydrolysis of the 16 diribonucleoside monophosphate substrates (NpN) is followed by isocratic boronate affinity HPLC.

Reagents

Substrates: Diribonucleoside monophosphates (Sigma, St. Louis, MO)

Assay buffer: 50 m*M* Sodium acetate (pH 5.6)

[14] S. Abel, G.-J. Krauß, and K. Glund, *J. Chromatogr.* **446,** 187 (1988).

HPLC system: Merck-Hitachi LiChroGraph, prepacked Dihydroxyboronyl-Si 100 Polyol column (4.6 × 250 mm, 5 μm, Serva Feinbiochemica), 10 mM KH_2PO_4 (pH 6.0) as mobile phase buffer

Procedure. Reactions are carried out in a total volume of 100 μl containing assay buffer, 1 mM of the respective NpN substrate, and appropriate amounts of enzyme activity (0.005–0.5 units). Preincubate the enzyme in assay buffer for 5 min at 37°. Start the assay by adding the substrate to the reaction mixture and continue the incubation for 0–60 min at 37°. Use heat-inactivated enzyme for each substrate in control incubations. Terminate the reactions by injecting a 10-μl aliquot directly onto the HPLC column. Develop the column in the isocratic mode at a flow rate of 1.5 ml min^{-1} and monitor eluates at 254 nm. Identify NpN substrates and 2′,3′-cyclic NMP, 3′(2′)-NMP and nucleoside hydrolysis products by comparing their retention times with those of authentic standards. Calculate reaction velocities of enzymatic NpN hydrolysis by quantitating 2′,3′-cNMP generation. Use calibration curves as described above.

Assay of Intrinsic Cyclic Nucleotide Phosphodiesterase Activity by HPLC

Principle. The enzyme- and time-dependent hydrolysis of 2′,3′-cyclic NMP substrates is followed by isocratic reversed-phase HPLC.

Reagents

Substrates: 2′,3′-cyclic NMP (Sigma)
Assay buffer: 50 mM Sodium acetate (pH 5.6)
HPLC system: Merck-Hitachi LiChroGraph, prepacked Octyl-Si 100 Polyol column (4.6 × 250 mm, 5 μm, Serva Feinbiochemica), 20 mM $(NH_4)H_2PO_4$ (pH 6.2) as mobile phase buffer

Procedure. Reactions are carried out in a total volume of 100 μl containing assay buffer, 1 mM of the respective 2′,3′-cyclic NMP substrate, and appropriate amounts of enzyme activity (0.005–0.5 units) essentially as described for the hydrolysis of NpN substrates (see above). Terminate the reactions by injecting a 10-μl aliquot directly onto the HPLC column. Develop the column in the isocratic mode at a flow rate of 1.5 ml min^{-1} and monitor eluates at 254 nm. Identify 2′,3′-cyclic NMP substrates and 3′(2′)-NMP hydrolysis products by comparing their retention times with those of authentic standards. Calculate reaction velocities of enzymatic 2′,3′-cyclic NMP hydrolysis by quantitating 3′(2′)-NMP generation based on calibration curves (see above).

Test for 3′- and 5′-Nucleotidase Activity by HPLC

Principle. The enzyme- and time-dependent hydrolysis of 3′-NMP and 5′-NMP substrates is followed by isocratic reversed-phase HPLC.

Reagents

Substrates: 3′-NMP, 5′-NMP (Sigma)
Assay buffer: 50 m*M* Sodium acetate (pH 5.6); 50 m*M* Tris-HCl (pH 8.8)
HPLC system: Merck-Hitachi LiChroGraph, prepacked Octyl-Si 100 Polyol column (4.6 × 250 mm, 5 μm, Serva Feinbiochemica), 20 m*M* $(NH_4)H_2PO_4$ (pH 6.2) as mobile phase buffer

Procedure. Reaction mixtures (total volume of 100 μl) contain assay buffer, 5 m*M* of the respective 3′-NMP or 5′-NMP substrate, and 4 units of purified enzyme. Incubate for 24 h at 37°. Terminate the reactions by injecting a 10-μl aliquot directly onto the HPLC column. Develop the column in the isocratic mode at a flow rate of 1.5 ml min^{-1} and monitor eluates at 254 nm. Identify 3′-NMP and 5′-NMP substrates and nucleoside hydrolysis products by comparing their retention times with those of authentic standards. Purified secretory ribonucleases from tomato should not contain any detectable nucleotidase activity.

Determination of Ribonuclease Activity by in-Gel Assays

Disc gel electrophoresis on native polyacrylamide slab gels (12% or 15% acrylamide) is performed without SDS using the discontinuous buffer system according to Laemmli.[15] Nondenatured proteins in sample buffer lacking SDS and 2-mercaptoethanol are loaded. After electrophoresis, RNases are detected by negative activity staining. The gels are (i) equilibrated with 150 m*M* sodium acetate buffer (pH 5.6) for 2 × 10 min, (ii) incubated in substrate solution (150 m*M* sodium acetate buffer [pH 5.6], 2.5 m*M* EDTA, 0.4% yeast RNA) for 30 min at 37°, (iii) briefly rinsed in equilibration buffer, (iv) stained in 0.2% (w/v) toluidine blue, 0.5% (v/v) acetic acid for 5 min, and (v) destained in 0.5% (v/v) acetic acid until transparent activity bands appear in the otherwise intensely stained gel. The addition of 2.5 m*M* EDTA to the substrate solution improves the staining pattern by inhibiting nuclease activities and unspecific reactions. The quality of the RNA used is not critical. We successfully used crude technical preparations of yeast RNA. The substrate and staining solutions can be used several times. Store the substrate solution at −20°.

Alternatively, SDS–polyacrylamide gel electrophoresis followed by activity staining was described for separating and detecting ribonucleases according to their molecular weight.[16] This assay incorporates the RNA substrate into the gel prior to electrophoresis, which reduces the time required for activity staining. However, since the molecular weights of secretory RNases from tomato are very similar, native gel electrophoresis is the preferred method for in-gel activity assays.

[15] U. K. Laemmli, *Nature (London)* **227,** 680 (1970).
[16] Y. Yen and P. J. Green, *Plant Physiol.* **97,** 1487 (1991).

Purification Procedures

Purification of Intracellular RNases (RNase LX, RNases LV-1, LV-2, LV-3)

All steps of the purification protocol are assessed by activity staining in native gels for the presence and enrichment of the individual RNases. The spectrophotometric assay is used to measure total RNase activity.

Cell Extraction. Cells for preparation of protein extracts are grown for 3 to 4 days in P_i-free medium[11] and washed extensively with water to remove extracellular RNase LE. The washed cells are rinsed with acetone and dried at room temperature. All the following steps are performed at 4°. To prepare cellular extracts, cells (1 kg wet weight) are resuspended in 4 liters of extraction buffer (100 m*M* citric acid–Na_2HPO_4 buffer [pH 7.0] 1 m*M* phenylmethylsulfonyl fluoride (PMSF), 10 m*M* EDTA, 0.1 % [w/v] Triton X-100) and are disrupted by sonication (200-ml aliquots, 3 × 1 min, 400 W). The homogenate is centrifuged to remove cell debris (4000*g*, 20 min). The protein fraction of the supernatant precipitating between 60% and 75% acetone is collected (10,000*g*, 20 min). The proteins are dissolved in water, dialyzed against water, and concentrated using a rotary evaporator under mild conditions.

Gel Filtration. A 1.5 ml aliquot of the concentrated protein extract is applied to a Sephadex G-50 column (1.5 × 80 cm) and eluted at a flow rate of 30 ml h^{-1} with 75 m*M* sodium acetate (pH 5.6) containing 0.5 *M* NaCl. This step removes all acid phosphatase and nuclease activities. Active fractions eluting in the molecular mass range of 18 to 25 kDa are pooled and concentrated by precipitation with 75% acetone. After this step, two approaches can be used to purify cellular ribonucleases to homogeneity. In a first approach, method (A), purification by affinity chromatography is followed by preparative native electrophoresis, which yields highly pure RNase preparations although in low amounts.[11] In a second approach, method (B), a series of conventional chromatographic purification procedures are used (Table I).[17]

Method (A)

Affinity Chromatography. Acetone-precipitated proteins of the gel filtration step are dissolved in 5 m*M* sodium acetate (pH 5.6), and 10-ml aliquots are applied to a UMP-agarose column (0.6 × 5 cm) equilibrated with the same buffer. Elution of the column is performed with 5 m*M* sodium acetate (pH 5.6) containing 1 *M* NaCl at a flow rate of 30 ml h^{-1}. The enzyme preparation is free of contaminating phosphodiesterase and phosphomonoesterase activities and is subjected to preparative native polyacrylamide gel electrophoresis on 15% slab gels to separate the different intracellular ribonucleases.

[17] A. Löffler, Doctoral Dissertation, University of Halle (1993).

TABLE I
PURIFICATION OF INTRACELLULAR RIBONUCLEASES FROM TOMATO CELL CULTURE[a]

Fraction	Total protein (mg)	Total activity $(U \times 10^{-3})^d$	Specific activity $(U\ mg^{-1})^d$	Purification factor (-fold)	Yield (%)
Cell homogenate (pH 7)[b]	350,000	5920	16.9	1	100
4000*g* supernatant	298,000	5660	19.0	1.1	96
60–75% Acetone	52,000	5356	103.0	6.1	90
Sephadex G-50	16,300	4194	257.0	15.2	71
DEAE-Toyopearl 650M[c]					
RNase LX	68.1	873.5	12,827	759[e]	14.8[e]
RNase LV-1	17.1	109.1	6380	377[e]	1.8[e]
RNase LV-2	31.6	364.9	11,550	683[e]	6.2[e]
RNase LV-3	66.2	1000.2	15,080	892[e]	16.9[e]
UMP-Agarose					
RNase LX	10.8	603.6	55,890	3307[e]	10.2[e]
RNase LV-1	2.2	68.1	30,950	1832[e]	1.1[e]
RNase LV-2	3.6	202.7	56,300	3331[e]	3.4[e]
RNase LV-3	8.1	603.2	74,470	4407[e]	10.2[e]

[a] From Ref. 17.
[b] The homogenate is prepared from 3 kg cells (fresh weight).
[c] Calculations take into account both chromatographic steps on DEAE-Toyopearl 650M.
[d] One ribonuclease unit is defined as the amount of enzyme causing an increase in ΔA_{260} of 1.0 min^{-1} cm^{-1} ml^{-1} at 37° according to Wilson.[13]
[e] Data refer to the total cell homogenate and do not consider the percentage of each enzyme in the homogenate.

Method (B)

Ion-Exchange Chromatography (I). Acetone-precipitated proteins of the gel filtration step (approximately 400 mg) are dissolved in 20 ml of 20 m*M* sodium phosphate buffer (pH 6.7) and are applied to a DEAE-Toyopearl 650M column (1.5 × 70 cm) equilibrated with the same buffer. After washing with 100 ml buffer, the column is developed with a linear gradient of 0–0.1 *M* NaCl (1200 ml) at a flow rate of 60 ml h^{-1}. This step substantially increases the specific activities of RNases LV-1 and LV-2, which elute at 20 m*M* and 35 m*M* NaCl, respectively. Both RNase LV-3 and RNase LX elute between 45 m*M* and 80 m*M* NaCl.

Ion-Exchange Chromatography (II). Rechromatography of the active fractions from the above step on DEAE-Toyopearl 650M under the same conditions and at the same flow rate of 60 ml h^{-1}, although at a different pH, further improves purification of the RNase isoenzymes. *RNase LV-1:* Active fractions are pooled and rechromatographed on the same column using a linear gradient of 0–0.1 *M* NaCl (1200 ml) in 20 m*M* sodium phosphate (pH 7.6). The enzyme elutes at 55 m*M* NaCl. Active fractions are concentrated using a rotary evaporator and desalted by gel filtration on a Sephadex G-25 column (1.5 × 50 cm) equilibrated with 50 m*M* triethylamine/acetate buffer, pH 8.0. *RNase LV-2:* Active fractions are pooled and rechromatographed on the same column using a linear gradient of 0–0.1 *M* NaCl (1200 ml) in 20 m*M* sodium phosphate (pH 6.7). The enzyme elutes at 35 m*M* NaCl. Active fractions are concentrated and desalted as described for RNase LV-1. *RNase LV-3 and RNase LX:* Active fractions are pooled and rechromatographed on the same column using a linear gradient of 0–0.1 *M* NaCl (1200 ml) in 20 m*M* sodium phosphate (pH 5.6). RNase LV-3 elutes at 65 m*M* NaCl whereas RNase LX elutes at 40 m*M* NaCl. Active fractions of each enzyme are concentrated and desalted as described above.

Affinity Chromatography. After dialysis against 10 m*M* sodium acetate (pH 5.6), the active RNase fractions of the above step are applied to a UMP-agarose column (1 × 8 cm) equilibrated with the same buffer. After washing with 20 ml of buffer, the column is developed with a salt gradient (60 ml, 0–2.5 *M* NaCl, 30 ml h^{-1}). RNase isoenzymes elute between 1.0 *M* and 1.5 *M* NaCl. After desalting on Sephadex G-25, the purified proteins are lyophilized and stored at −20°.

Purification of RNase LV-3 (Alternative Protocol)

An alternative protocol for purifying RNase LV-3 is given below, which is based on preferential extraction of the enzyme at low pH.[18]

Cell Extraction. Cells of a 7-day-old stationary phase culture or of a -P_i culture are used as the source of enzyme. Cells (150 g wet weight) are extensively washed with distilled water to remove RNase LE, resuspended in 1.5 liter extraction buffer (150 m*M* citric acid/sodium phosphate [pH 3], 0.1 m*M* PMSF), and disrupted by sonication. After centrifugation of the cell homogenate, the supernatant is adjusted to pH 5 with 2 *N* NaOH, and the proteins precipitating between 50% and 70% ammonium sulfate saturation are collected. The pellet is dissolved in 60 ml of 100 m*M* citric acid/sodium phosphate (pH 5.0). The protein solution is incubated for 20 min at 50°, cooled to 2–4°, and cleared by centrifugation.

Gel Filtration. The supernatant is applied to a Sephadex G-75 column (5.5 × 80 cm) equilibrated with 100 m*M* sodium acetate (pH 5.0). Proteins are eluted at a flow rate of 70 ml h^{-1}. Active fractions are pooled and adjusted to 10% glycerol.

Ion-Exchange Chromatography. Pooled fractions are loaded onto a DEAE-Sephadex A-25 column (1.5 × 10 cm) equilibrated with 50 m*M* sodium acetate (pH 5.0), 10% glycerol. Proteins are eluted at a flow rate of 30 ml h^{-1} with a descending discontinuous pH gradient of the same buffer in the following order: 50 ml each of pH 4.6 and pH 4.3, and 100 ml of pH 3.7. RNase LV-3 elutes between pH 4.3 and pH 3.7.

Hydroxyapatite Chromatography. The pooled fractions are adjusted to 5 m*M* sodium acetate (pH 5.5), 10% glycerol and are loaded on a hydroxyapatite column

[18] S. Abel and K. Glund, *Planta* **172,** 71 (1987).

(1 × 8 cm) equilibrated with the same buffer. After washing, the column is developed with a 200-ml linear gradient (0–200 m*M*) of sodium phosphate (pH 5.5) in equilibration buffer (flow rate 80 ml h^{-1}). RNase activity elutes in a sharp peak at 30 m*M* sodium phosphate.

Affinity Chromatography. Active fractions from the previous step are pooled, dialyzed against 20 m*M* sodium acetate (pH 5.0), 10% glycerol, and loaded on a UMP-agarose column (0.6 × 5 cm) equilibrated with the same buffer. After washing, the proteins are eluted at a flow rate of 30 ml h^{-1} with a 60-ml linear gradient (0–3.0 *M*) of NaCl in the same buffer. RNase LV-3 elutes between 1.0 *M* and 1.5 *M* NaCl. Active fractions are pooled and dialyzed against 5 m*M* sodium acetate (pH 5.0), 10% glycerol and are stored at −20°.

Purification of Extracellular RNase LE

Initial Concentration Step. Spent culture medium of early stationary cell cultures (4–5 d) is filtered and and cleared by centrifugation (10,000*g*, 15 min, 4°). Proteins precipitating in the range of 50% to 70% ammonium sulfate saturation are collected (20,000*g*, 30 min). The pellet is dissolved in 5 m*M* sodium acetate (pH 5.5) containing 1 m*M* EDTA, 1 m*M* PMSF, 1 m*M* iodoacetic acid, and the protein solution is dialyzed against this buffer.

Ion-Exchange Chromatography. The dialyzed sample is applied to a DE 52-cellulose column (1.5 × 15 cm) equilibrated with 50 m*M* sodium acetate (pH 5.5). The column is eluted at a flow rate of 30 ml h^{-1} with a discontinuous pH gradient of the same buffer in the following order: 25 ml each of pH 4.5, pH 4.2, and pH 3.8. RNase activity elutes between pH 4.2 and pH 3.8. Active fractions are pooled, concentrated by ultrafiltration, and adjusted to pH 5.5 with 1 *N* NaOH.

Gel Filtration. The protein sample is applied to a Sephadex G-75 column (5.5 × 80 cm) equilibrated with 20 m*M* sodium acetate (pH 5.5), and the proteins are eluted at a flow rate of 70 ml h^{-1}. RNase LE elutes as a monodisperse activity peak.

Affinity Chromatography. Pooled fractions are loaded on a UMP-agarose column (0.5 × 7.5 cm) equilibrated with 20 m*M* sodium acetate (pH 5.5). The proteins are eluted at a flow rate of 30 ml h^{-1} with a discontinuous ascending gradient of NaCl (0.5–2.5 *M*, increments of 0.5 *M*, 10 ml each) in sodium acetate (pH 5.5). RNase LE elutes between 1 and 2 *M* NaCl. Active fractions are pooled and desalted. This procedure follows the method described by Nürnberger *et al.*[9]

Notes on Protein Purification

The best source of RNase enzymes are 3- to 4-day old -P_i cell cultures. Stationary phase cells (7 d) of normal cultures are not recommended because of the high content of copurifying secondary metabolites that interfere with biochemical analyses. To reduce loss of RNase activity, all purification steps should be carried out at 4°, unless indicated otherwise.

The cell extraction at pH 3 provides a powerful enrichment step that removes most of the cellular protein (approximately 99%) but recovers most of the vacuolar RNase LV-3 activity. However, this procedure is not advised for purification of RNases LV-1, LV-2, or LX because of severe isoenzyme activity loss. The most convenient analytical method to reliably distinguish between the different secretory tomato RNases is native polyacrylamide gel electrophoresis followed by in-gel activity assays. Therefore, all steps of the purification protocols are assessed by activity staining in native gels for the presence and enrichment of the individual RNases. When purifying intracellular RNases, care has to be taken not to contaminate the preparations with extracellular RNase LE that accounts for about 80% of the total RNase activity of cultivated cells. Excessive washing of the cell cake with at least 20 volumes of distilled water is recommended. Intracellular RNases are more stable in citric acid/Na_2HPO_4 (pH 7.0) or in sodium acetate (pH 5.5) than in other buffers. All final enzyme preparations are essentially homogeneous as judged by SDS–PAGE and protein silver staining.

Subcellular Localization of Intracellular Ribonucleases

Step 1: Protoplast Isolation

Reagents

Cell protoplast washing solution (CPW)[19] (10 m*M* $CaCl_2$, 1 m*M* KI, 1 m*M* KNO_3, 1 m*M* $MgSO_4$, 0.2 m*M* KH_2PO_4, 0.1 m*M* $CuSO_4$)

Solution A (CPW, 0.5 *M* sorbitol); Solution B (CPW, 0.4 *M* sorbitol); Solution C (CPW, 0.4 *M* sucrose, 5% [w/v] Ficoll 400).

Procedure. Incubate 4.8×10^7 cells in 12 ml solution A containing 1% (w/v) Driselase (Sigma) and 0.01% (w/v) HUPc-cellulase on a gyratory shaker at 75 rpm for 10 to 12 h at 23° in the dark.[8] Filter the digested material through a nylon net (45 μm mesh size). Centrifuge the filtered crude protoplast suspension at 100 *g* for 10 min. Resuspend the pellet in solution B and wash the protoplasts three times in this solution by repeated centrifugation. Further purify protoplasts by an additional flotation step. After the final sedimentation step, resuspend the pellet (containing about 10^7 protoplasts) in 5 ml solution C and overlay with 2 ml of a 4 : 1 (v/v) mixture of solution C and B, followed by 2 ml of a 4 : 1.5 (v/v) mixture of solution C and B and by 1 ml of solution B only. After centrifugation at 100*g* for 10 min, collect purified protoplasts from the upper interphase. Dilute

[19] E. M. Frearson, J. B. Power, and E. C. Cocking, *Dev. Biol.* **33,** 1 (1973).

collected protoplasts into solution B and sediment protoplasts (100*g*, 10 min) to remove Ficoll 400.

Step 2: Isolation of Vacuoles

Reagents

Buffer I (5 m*M* MES–Tris [pH 6.0], 0.4 *M* sorbitol); Buffer II (15 m*M* KH_2PO_4/K_2HPO_4 [pH 8.0], 0.4 *M* sorbitol); Buffer III (15 m*M* KH_2PO_4/K_2HPO_4 [pH 8.0], 0.4 *M* sucrose, 5% [w/v] Ficoll 400)

Procedure. Resuspend the sedimented protoplasts in buffer II supplemented with 5 m*M* EDTA to accelerate protoplast lysis. Vigorously shake protoplast suspension at 200 rpm for about 10 min at 30°. Centrifuge the suspension for 30 s at 500*g* to remove protoplast debris. Transfer supernatant to a new tube, resuspend the lysed protoplast pellet in the above buffer, and repeat the centrifugation step. Combine the supernatants, which contain intact vacuoles but are largely free of protoplasts. Mix 3 volumes of the crude vacuole suspension with 2 volumes buffer III. Overlay the diluted vacuole suspension sequentially with 2 ml each of the following mixtures of buffer II and buffer III: first, 3 volumes buffer II/1.5 volumes buffer III and second, 3 volumes buffer II/0.5 volume buffer III. Finally, overlay the step gradient with 0.5 ml of buffer II. After centrifugation at 2500*g* for 10 min, collect the purified vacuoles from the upper interphase and directly analyze for RNase activity and vacuolar and extravacuolar marker enzymes.[20]

Step 3: Isolation of Microsomal Fractions

Reagents

Lysis buffer (10 m*M* HEPES [pH 7.5], 1 m*M* 2-mercaptoethanol)
Gradient buffer (50 m*M* HEPES [pH 7.5], 10 m*M* $MgCl_2$, 2.5 m*M* EDTA, 5 m*M* 2-mercaptoethanol, 10% [v/v] glycerol)

Procedure. Incubate cultured tomato cells from -P_i cell cultures with cell-wall degrading enzymes supplemented with 1% (w/v) cellulase Onozuka R-10 in solution A for 2 h as described in *Step 1* (see above). Wash protoplasts three times in solution B and resuspend in 0.2 ml of 50% sucrose. Carefully mix protoplast suspension with 1.6 ml of lysis buffer on ice. Protoplast lysis can be assisted mechanically by repeatedly pressing the dense suspension through a thin needle. After centrifugation at 1500*g* for 5 min at 4°, load 3 ml of the pooled supernatants on

[20] K. Glund, A. Tewes, S. Abel, V. Leinhos, R. Walther, and H. Reinbothe, *Z. Pflanzenphysiol.* **113,** 151 (1984).

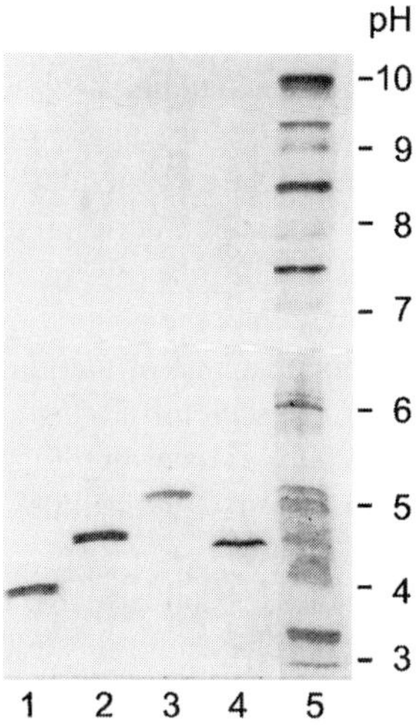

FIG. 1. Isoelectric focusing of purified intracellular ribonucleases. Lanes (1) RNase LV-3, (2) RNase LV-2, (3) RNase LV-1, (4) RNase LX, and (5) IEF marker proteins. Aliquots of purified enzymes (10 μg each) and marker proteins (100 μg Protein Test Mixture, Serva) were separated on SERVALYT PRECOTES 3–10 (Serva) using the buffer system recommended by the manufacturer (anode buffer: 25 m*M* L-aspartic acid, 25 m*M* L-glutamic acid; cathode buffer: 25 m*M* L-arginine, 2.5 m*M* L-lysine, 1.5 *M* ethylenediamine) for 4 h (starting voltage at 200 V, 1000 V max.). Proteins were stained with Coomassie blue (B. J. Davis, *Ann. N. Y. Acad. Sci.* **121,** 407, 1964).

a discontinuous sucrose gradient consisting of 7 ml 13% and 1.5 ml 50% sucrose in gradient buffer. After centrifugation of the sucrose gradient (100,000*g*, 2 h at 4°) in a SW 41 Ti rotor (Beckman), collect the turbid fraction at the interphase (total microsomal fraction) according to the method of Sticher *et al.*[21] Directly analyze fractions for total RNase activity and microsomal marker enzymes [antimycin A-insensitive NADH cytochrome-*c* reductase (ER), inosine diphosphatase (Golgi apparatus)][22] in addition to above-mentioned marker enzymes of cell organelles.

Properties of Tomato Ribonucleases LE, LX, LV-1, LV-2, and LV-3

The secretory RNases from tomato share very similar enzymic properties, but differ slightly in protein chemical parameters such as molecular mass and isoelectric point (Fig. 1). The different cellular locations are the most conspicuous differences between the tomato enzymes.[9,11,17,18]

[21] L. Sticher, U. Hinz, A. D. Meyer, and F. Meins Jr., *Planta* **188,** 559 (1992).
[22] G. Shore and G. A. MacLachlan, *J. Cell Biol.* **64,** 557 (1975).

Enzymatic Properties

The secretory tomato ribonucleases studied, RNase LE, LX, and LV-1 to LV-3, are endoribonucleases that are specific for single-stranded RNA substrates. During RNA hydrolysis, the tomato RNases generate by a phosphotransferase reaction 2′,3′-cyclic NMP as obligate monomeric products. Further hydrolysis of 2′,3′-cyclic NMP to 3′-NMP is a side reaction of the tomato enzymes, which proceeds significantly more slowly than the generation of 2′,3′-cyclic NMP from synthetic diribonucleoside monophosphate or natural RNA substrates.[9,23] Based on these mechanistic properties, the secretory tomato ribonucleases are classified as of the RNase I-type (EC 3.1.27.1). According to Wilson,[13] RNase I enzymes are endoribonucleases that hydrolyze RNA substrates via 2′,3′-cyclic NMP intermediates.

All secretory tomato ribonucleases studied have an acidic pH optimum (between pH 5 and pH 6) with yeast RNA as the substrate and preferentially release purine, in particular guanine, mononucleotides from diribonucleoside monophosphate or RNA substrates (relative nucleobase specificity: G≫A>U>C).[9,23,24] As determined for RNase LE, RNase LX, and RNase LV-3, hydrolysis of RNA does not require divalent metal ions (Mg^{2+}, Mn^{2+}, Ca^{2+}, Co^{2+}) and is not affected by the presence of 2.5 m*M* EDTA. However, RNase activities are slightly stimulated by the presence of 25 m*M* sodium citrate (170%) or 25 m*M* sodium phosphate (153%), when compared to 25 m*M* sodium acetate (100%). Secretory tomato RNase activities are completely inhibited by incubation in 10 m*M* Ag^{+}, Zn^{2+}, Hg^{2+}, Cu^{2+}, or Al^{3+} ions, and 1 m*M* *p*-chloromercuribenzoic acid causes 80% inhibition of RNase LE activity. Possible end products of RNA hydrolysis and derivatives at concentrations of 1 m*M* do not affect the enzyme activities.[1,9,17,18]

Stability

Secretory tomato RNases can be stored without measurable loss of activity for at least 2 months at −20°. The enzymes are less thermostable in comparison to other RNases. Total inactivation is observed for all secretory tomato RNases within 5 min by incubating the enzymes at 100° in buffers of different pH and ionic strength.[9,17] Five disulfide bridges are present in RNase LE, one of which is unique to the subfamily of acid ribonucleases (Cys^{25}– Cys^{81}). This disulfide bridge appears to stabilize the structure of RNase LE and to protect the enzyme against proteolytic cleavage.[25]

[23] S. Abel, G.-J. Krauß, and K. Glund, *Biochim. Biophys. Acta* **998,** 145 (1989).

[24] K. Ohgi, Y. Shiratori, A. Nakajima, M. Iwama, H. Kobayashi, N. Inokuchi, T. Koyama, M. Köck, A. Löffler, K. Glund, and M. Irie, *Biosci. Biotech. Biochem.* **61,** 432 (1997).

[25] N. Tanaka, J. Arai, N. Inokuchi, T. Koyama, K. Ohgi, M. Irie, and K. T. Nakamura, *J. Mol. Biol.* **298,** 859 (2000).

Primary and Secondary Structures

The primary structures of RNase LE and RNase LX were directly determined by sequencing of the proteins, which allowed determination of the signal peptidase cleavage site. Both enzymes consist of a single polypeptide chain. Mature RNase LE contains 205 amino acid residue, whereas RNase LX is composed of 213 amino acids.[26,27] Although other members of the superfamily of T2/S ribonucleases are glycoproteins,[6] conserved glycosylation sites do not occur in the primary structures of both enzymes. Likewise, experimental tests for glycosylation such as adsorption of RNases LE and RNase LX to concanavalin A or digestion with endoglycosidase H suggested that the tomato enzymes are not modified by glycosylation.[17]

The primary structures of RNase LE and RNase LX share 62% amino acid sequence identity. Amino acid residues are particularly conserved in the active site cassettes important for catalysis. Amino acid sequence identity between tomato RNases and other members of the T2/S superfamily ranges from 25% (fungal RNases) to 30% (S-RNases). RNase LE shares the highest amino acid sequence identity with RNase NE and RNase NW from *Nicotiana* species (about 85%), with ZRNase II from *Zinnia elegans,* and with RNS1 from *Arabidopsis thaliana* (about 70%). RNase LX is closely related to ZRNase I from *Zinnia elegans* and to RNase NGR3 (*Nicotiana glutinosa*) (EMBL AB032257), and the sequence similarity extends to the C terminus.

Partial amino acid sequences, including N-and C-terminal sequences, were determined for RNases LV-1, LV-2, and LV-3.[11,28] Enzyme purification and characterization studies by Abel and Glund[18] and Nürnberger *et al.*[9] as well as direct sequencing of the proteins[11,26] strongly suggest that RNase LE and RNase LV-3 are identical enzymes. The two proteins (i) comigrate in native and denaturing polyacrylamide gels, (ii) have identical protein chemical (CD spectra, IEP)[17] and enzymic properties, (iii) give similar peptide fingerprints after tryptic digestion, and (iv) share identical amino acid sequences in N-terminal and internal tryptic peptides. About 50% of the RNase LV-2 and 20% of the RNase LV-1 protein were sequenced, including the N and C terminus of RNase LV-2. Except for the C-terminal peptides, tryptic maps of both RNases were identical. Comparison of the primary structure of RNase LX with the vaculoar isoenzymes reveals differences only at the truncated C termini of RNases LV-2 and LV-1.[28]

Tertiary Structures

The crystal structure of RNase LE has been determined at 1.65 Å resolution.[25] The enzyme is composed of seven α helices and seven β strands. Five disulfide

[26] W. Jost, H. Bak, K. Glund, P. Terpstra, and J. J. Beintema, *Eur. J. Biochem.* **198,** 1 (1991).

[27] A. Löffler, K. Glund, and M. Irie, *Eur. J. Biochem.* **214,** 627 (1993).

[28] M. Köck, A. Löffler, S. Abel, and K. Glund, *Plant Mol. Biol.* **27,** 477 (1995).

bridges have been identified but only two are common to all acid T2-type RNases. Another two disulfide bridges are common to the animal/plant subfamily (including S-RNases),[29] but their position differs in the fungal RNase Rh.[30] One disulfide bridge ($Cys^{25}-Cys^{81}$) and an adjacent *cis*-peptide bond (Pro-82) are unique to RNase LE.[25] Since the related tomato enzyme, RNase LX, and other members of the plant S-like RNase subfamily have conserved amino acid residues at identical positions, it is very likely that they share similar secondary structures, which may explain similar biochemical properties of these enzymes. It is expected that the three-dimensional structure of RNase LE will provide a basic framework for the animal/plant subfamily of RNase T2 enzymes, including S-RNases. Based on the sequence alignment with RNase Rh, the catalytically important residues of RNase LE were predicted to be His-39, Trp-42, His-92, Glu-93, Lys-96, and His-97, which has been confirmed by crystal structure analysis of RNase LE.[25]

Gene Structures

By screening a cDNA library prepared from P_i-starved cells, we isolated cDNA clones for RNase LE and RNase LX. The deduced amino acid sequences for RNase LE and RNase LX contain N-terminal signal sequences of 25 and 24 amino acids, respectively. The hydrophobic core and other properties of the predicted signal peptides are consistent with a function in cotranslational transport into the endoplasmic reticulum.[28] Southern blotting analysis revealed that both RNases are encoded by single-copy genes in tomato.[28,31] cDNA library screening and Southern analysis under low stringency conditions did not indicate the presence of additional, related genes in the tomato genome. Therefore, it is likely that the extracellular and vacuolar RNases are the products of posttranscriptional processing and/or posttranslational modification reactions.

Localization

When cultured tomato cells and the spent liquid medium were analyzed for ribonuclease activity, four intracellular RNases and one extracellular RNase were identified by native gel electrophoresis. The most mobile intracellular enzyme, RNase LV-3, and the enzyme secreted into the cell culture medium, RNase LE, share identical electrophoretic properties and likely share the same primary structure (see above). Nonetheless, both proteins have different destinations in the secretory pathway. The absence of a vacuolar targeting signal[32] in RNase LE and its high levels in the cell culture medium, which severalfold exceed the amount

[29] D. Oxley and A. Bacic, *Eur. J. Biochem.* **242,** 75 (1996).

[30] H. Kurihara, Y. Mitsui, K. Ohgi, M. Irie, H. Mizuno, and K. T. Nakamura, *FEBS Lett.* **306,** 189 (1992).

[31] A. Lers, A. Khalchitski, E. Lomaniec, S. Burd, and P. J. Green, *Plant Mol. Biol.* **36,** 439 (1998).

[32] F. Marty, *Plant Cell* **11,** 587 (1999).

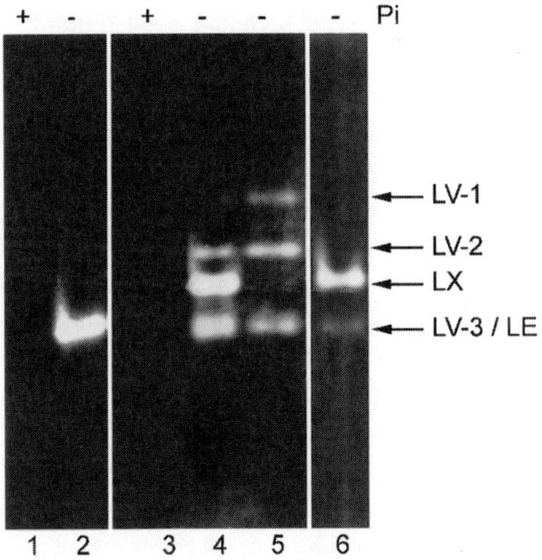

FIG. 2. Electrophoretic separation of ribonucleases from different cellular fractions using native PAGE. RNases are detected by negative activity staining with RNA as substrate. Positions of the different RNases in the gel are marked on the right. Medium corresponding to 5×10^5 cells grown for 48 h under $+P_i$ (lane 1) or $-P_i$ (lane 2), and cell extracts of 5×10^5 cells grown for 48 h under $+P_i$ (lane 3) or $-P_i$ (lane 4), isolated vacuoles corresponding to 5×10^5 cells (lane 5), and the isolated microsomal fraction (lane 6). [+/− P_i] indicates whether cells were grown with or without phosphate in the cultivation medium. Gradient fractions of isolation procedures for vacuoles and microsomes were characterized by calculation of contaminating cellular enzymes and by cofractionation with marker enzymes according to Glund *et al.*,[20] Shore and MacLachlan,[22] and Kaletta *et al.*[34]

of RNase LV-3 in the vacuoles, strongly suggest that the extracellular space is the location of RNase LE function. This is supported by studies in transgenic tomato plants that constitutively overexpress RNase LE. As predicted, RNase LE is almost exclusively detectable in the extracellular fluid collected from the intercellular spaces of tomato leaves.[33]

Using isolated vacuoles prepared from protoplasts of cultured tomato cells, it was shown by in-gel activity assays under native and denaturing conditions that three of the four intracellular RNase activities copurify with the vacuolar organelles. These enzymes were designated RNase LV-1, LV-2, and LV-3. The fourth RNase activity is of extravacuolar origin and was named RNase LX (see Fig. 2). Further localization studies demonstrated that RNase LX is retained in the endoplasmic reticulum (ER) of tomato cells by virtue of the C-terminal amino acids, HDEF. The C terminus of RNase LX was shown to function as a novel ER retention signal.[34] Interestingly, as revealed by C-terminal sequencing

[33] M. Köck, K. Kaletta, and I. Stenzel, *in* "Abstracts of 5th Int. Meet. on Ribonucleases", p. 36. Warrenton, USA, 1999.

[34] K. Kaletta, I. Kunze, G. Kunze, and M. Köck, *FEBS Lett.* **434,** 377 (1998).

with carboxypeptidase Y, RNase LV-2 lacks the C-terminal ER retention signal.[28] Although the C-terminal amino acids of RNase LV-1 could not be identified, it is likely that the enzyme lacks more amino acid residues than the HDEF motif at the C terminus. This assumption is based on electrophoretic properties of RNase LV-1, which has a lower molecular weight but migrates more slowly native gels than RNase LV-2. The electrophoretic behavior of RNase LV-1 is possibly caused by the absence of two aspartic acid residues, which are present at the very C terminus of RNase LV-2. We suspect that a minor fraction of the RNase LX population that can escape the endoplasmic reticulum is transported into vacuoles. In this lytic environment, RNase LX is proteolytically cleaved, which results in loss of the C-terminal ER retention signal peptide, and further degradation may occur.

Acknowledgment

We are grateful to Dr. A. Löffler for making available Fig. 1. Parts of this work was supported by grants of the Deutsche Forschungsgemeinschaft, SFB 363, to M.K.

[23] Leczyme

By KAZUO NITTA

Introduction

We have isolated two sialic acid-binding lectins (SBLs) from frog eggs (SBL-C, from *Rana catesbeiana*; SBL-J, from *R. japonica*) that not only agglutinate a large variety of tumor cells[1–5] but also hydrolyze RNA,[6,7] and have named them leczyme.[8] Leczyme from *R. catesbeiana* (SBL-C) was found to inhibit

[1] H. Kawauchi, F. Sakakibara, and K. Watanabe, *Experientia* **31,** 364 (1975).

[2] K. Nitta, G. Takayanagi, H. Kawauchi, and S. Hakomori, *Cancer Res.* **47,** 4877 (1987).

[3] K. Titani, K. Takio, M. Kuwada, K. Nitta, F. Sakakibara, H. Kawauchi, G. Takayanagi, and S. Hakomori, *Biochemistry* **26,** 2189 (1987).

[4] F. Sakakibara, H. Kawauchi, G. Takayanagi, and H. Ise, *Cancer Res.* **39,** 1347 (1979).

[5] Y. Kamiya, F. Oyama, R. Oyama, F. Sakakibara, K. Nitta, H. Kawauchi, Y. Takayanagi, and K. Titani, *J. Biochem.* **108,** 139 (1990).

[6] Y. Okabe, N. Katayama, M. Iwama, H. Watanabe, K. Ohgi, M. Irie, K. Nitta, H. Kawauchi, Y. Takayanagi, F. Oyama, K. Titani, Y. Abe, T. Okazaki, N. Inokuchi, and T. Koyama, *J. Biochem.* **109,** 786 (1991).

[7] K. Nitta, F. Oyama, R. Oyama, K. Sekiguchi, H. Kawauchi, Y. Takayanagi, S. Hakomori, and K. Titani, *Glycobiology* **3,** 37 (1993).

[8] K. Nitta, K. Ozaki, Y. Tsukamoto, M. Hosono, Y. Ogawa-Konno, H. Kawauchi, Y. Takayanagi, S. Tsuiki, and S. Hakomori, *Int. J. Oncol.* **9,** 19 (1996).

0076-6879/01 $35.00

murine leukemia P388 cell proliferation.[9,10] We have demonstrated that introduction of exogenous SBL-C as RNase into the tumor cells causes inhibition of the tumor cell proliferation.[8] The antiproliferative effect of SBL-C is based on intracellular RNA degradation following receptor binding and internalization. SBL-C induces apoptotic morphological changes, such as nuclear condensation and disappearance of microvilli, to P388 cells.[11] We can therefore utilize information regarding SBLs and their receptors for examining the relationship between SBL receptors and apoptosis related antigens and for developing methods to specifically introduce nucleases or anticancer agents into target tumor cells.[8,12–17] This article describes the correlation of tumor cell agglutination with inhibition of cell growth, and the molecular mechanism of SBL-C-induced apoptosis.

Materials and Cells

SBL-C and SBL-J are purified by sequential chromatographies on Sephadex G-75, DEAE-cellulose, heparin-Sepharose, and SP-Sepharose columns as described previously.[2,4,9]

Murine leukemia P388 cells (wild-type) are maintained in RPMI 1640 supplemented with 10% fetal calf serum. P388 cells are cultured in the presence of SBL-C at a concentration gradually increasing up to 150 μM SBL during ~24 passages. The resulting SBL-C-resistant mutant is termed RC150.[10] RC150 cells are found to be internalization-defective.[10] These cells are cultured synchronously with aphidicolin, a tetracyclic diterpenoid obtained from *Cephalospolium aphidicola* Petch.[18]

[9] K. Nitta, K. Ozaki, M. Ishikawa, S. Furusawa, M. Hosono, H. Kawauchi, K. Sasaki, Y. Takayanagi, S. Tsuiki, and S. Hakomori, *Cancer Res.* **54,** 920 (1994).

[10] K. Nitta, K. Ozaki, Y. Tsukamoto, S. Furusawa, Y. Ohkubo, H. Takimoto, R. Murata, M. Hosono, N. Hikichi, K. Sasaki, H. Kawauchi, Y. Takayanagi, S. Tsuiki, and S. Hakomori, *Cancer Res.* **54,** 928 (1994).

[11] K. Nitta, Y. Ogawa, M. Hosono, Y. Takayanagi, and S. Hakomori, *in* "Proceeding of the Vth International Meeting on Ribonucleases," p. 74, Warrenton, USA, 1999.

[12] M. Irie, K. Nitta, and T. Nonaka, *Cell. Mol. Life Sci.* **54,** 775 (1998).

[13] R. J. Youle and G. D'Alessio, *in* "Ribonucleases: Structures and Functions" (G. D'Alessio, and J. F. Riordan, eds.), p. 491. Academic Press, New York, 1997.

[14] Y. Wu, S. K. Saxena, W. Ardelt, M. Gadina, S. M. Mikulski, C. DeLorenzo, G. D'Alessio, and R. J. Youle, *J. Biol. Chem.* **270,** 17476 (1995).

[15] D. L. Newton, P. J. Nicholls, S. M. Rybak, and R. J. Youle, *J. Biol. Chem.* **269,** 26739 (1994).

[16] D. L. Newton, D. Pollock, P. DiTullio, Y. Echelard, M. Harvey, B. Wilburn, J. Williams, H. R. Hoogenboom, J. C. Raus, H. M. Meade, and S. M. Rybak, *J. Immunol. Methods* **231,** 159 (1999).

[17] M. Suzuki, S. K. Saxena, E. Boix, R. J. Prill, V. M. Vasandani, J. E. Ladner, C. Sung, and R. J. Youle, *Nat. Biotechnol.* **17,** 265 (1999).

[18] S. Ikegami, T. Taguchi, M. Ohashi, M. Oguro, H. Nagano, and Y. Mano, *Nature* **275,** 458 (1978).

Tumor Cell Agglutination and Growth Inhibition

Lectin Activity and Its Inhibition

Tumor cell agglutination is determined using P388 and RC150 cells. A solution of SBL in 0.15 *M* NaCl is successively double-diluted from a concentration of 500 μg/100 μl to 0.1 μg/100 μl. A tumor cell suspension ($1 \times 10^6/100$ μl) is added to each tube and incubated for 10 min under gentle rotation at room temperature, then allowed to stand for 30 min.[2,9,19] SBL-C can agglutinate P388 and RC150 cells at concentrations of 0.5–1.0 μg/ml and 4.0–31 μg/ml, respectively. Cells at the logarithmic phase rather than confluent cells are agglutinated by SBL-C.[10] SBL-C and -J can also agglutinate mouse Ehrlich carcinoma, rat hepatoma AH109A, and human gastric cancer MKN45 in concentration ranges of 0.5–5.0 μg/ml.[2,4]

Inhibition of SBL-induced tumor cell agglutination is determined based on effects of serial dilutions of each potential inhibitor. Candidate inhibitors, such as mono- and oligosaccharides, glycoproteins, glycosaminoglycans, and nucleotides, are serially diluted with 0.15 *M* NaCl, and each tube is incubated with 8-fold concentration of the minimum amount of SBL-C or -J that causes obvious agglutination (total volume, 100 μl) at room temperature for 30 min, followed by addition of the cell suspension ($1 \times 10^6/100$ μl) as described above. Sialomucin from bovine submaxillary gland (BSM), heparin, 2′,(3′)-CMP and 2′,(3′)-UMP can inhibit SBL-induced agglutination at low concentrations.[6,9] However, tumor cell agglutination is not inhibited by asialo- BSM.[9] In addition, a number of mono- and oligosaccharides do not inhibit this agglutination.[9]

Identification of Glycoprotein Receptor(s)

To determine whether SBL-C binds to sialyl oligosaccharide chains of cell surface glycoconjugates and triggers inhibitory signaling of cell growth, cells are treated with sialidase from *Arthrobacter ureafaciens*.[2,9] Both tumor cell agglutination and inhibition of cell proliferation induced by SBL-C are greatly reduced by sialidase pretreatment. Pretreatment with 2,3-dehydro-2-deoxy-*N*-acetylneuraminic acid, a sialidase inhibitor, blocks this reduction.[2,9]

To confirm carrier protein(s) of sialyl oligosaccharide chains on P388 cell membrane, cells are suspended in ice-cold extraction buffer consisting of 20 m*M* Tris-HCl (pH 7.5), 5 m*M* EDTA, 5 m*M* EGTA, 2 m*M* phenylmethylsulfonyl fluoride, and 2-mercaptoethanol, disrupted by sonication at 4°, and centrifuged at 100,000*g* for 1 h. The pellet is referred to as "membrane fraction." This fraction is resolved by sodium dodecyl sulfate–polyacrylamide gel electrophoresis (SDS–PAGE) in an 8% gel under reducing conditions according to the method of Laemmli.[20] The

[19] H. Kawauchi, M. Hosono, Y. Takayanagi, and K. Nitta, *Experientia* **49,** 358 (1993).
[20] U. K. Laemmli, *Nature* **227,** 680 (1970).

gels are equilibrated in Towbin's transfer buffer (25 mM Tris: 192 mM glycine) for 10 min. Electrophoretic transfer of (glyco)proteins onto nitrocellulose membrane is performed in this transfer buffer at 60 V for 2 h.[21] After transfer, the membrane is sequentially incubated with SBL-C for 4 h, with anti-SBL-C antibody for 4 h, and with horseradish peroxidase (HRP)-labeled goat anti-rabbit IgG for 2 h at room temperature. SBL-C-reactive glycoproteins are visualized using Konica HRP-1000. Separately, the membrane is incubated with HRP-tagged *Limax flavus* agglutinin (LFA), which binds to sialic acid in cell surface carbohydrate chains,[22] for 4 h. LFA receptors are visualized by the same method described above. SBL-C can bind to 105 kDa and 80 kDa bands derived from P388 but not from RC150 cell membrane. The LFA receptor is the 80 kDa sialoglycoprotein.

Cytotoxicity to Tumor Cells and RNA Degradation

To evaluate the cytotoxic effect of SBL-C, cells ($2 \times 10^4/100$ μl) are exposed to varying concentrations of SBL-C at 37° for 24 h. SBL-C-treated cells in 96-well plates are incubated with Alamar Blue for 4 h.[23,24] Optical density (OD) is determined using a microplate reader MPR-A4i (Tosoh, Tokyo, Japan) at 570 nm/600 nm. The 50% inhibitory concentration for P388 cell growth is approximately 3.1–6.25 μM, whereas RC150 cells are only 30% inhibited even by 150 μM.

To determine how SBL-C affects the growth of P388 cells, cells (5×10^5/ml) are incubated with dansylcadaverine (DC)-labeled SBL-C (final concentration, 12.5 μM) in culture either at 0° for 30 min or at 37° for 3 h. The cells are washed with phosphate-buffered saline (PBS) containing NaN_3 (0.1%) and fetal calf serum (2%), and resuspended in this buffer. Bound and internalized fluorescence is detected by a digital imaging microscope (Argus 100/CA) equipped with a computer-assisted image processor (Hamamatsu Photonics, Hamamatsu, Japan).[10] DC-labeled SBL-C can bind to the cell surface of not only P388 but also RC150 cells, whereas it can invade only P388 cells.

To confirm whether the internalized SBL-C acts as an RNase in P388 cells, cells are incubated with SBL-C (12.5 μM) at 37° for a specified period. Total RNA is extracted from SBL-C-treated cells by the use of RNeasy Total RNA Kit (Qiagen, Chatsworth, CA) according to the manufacturer's protocol. The extracted RNA (OD at 260 nm corresponded to 3–5 μg of RNA) is analyzed by electrophoresis on 1.5% agarose gels in 0.04 M Tris–acetate (pH 8.0) containing 2 mM EDTA, and visualized with ethidium bromide.[8] Degradation of intracellular RNA can be

[21] H. Towbin, T. Staehelin, and J. Gordon, *Proc. Natl. Acad. Sci. U.S.A.* **76,** 4350 (1979).

[22] R. L. Miller, *Methods Enzymol.* **138,** 527 (1987).

[23] B. Page, M. Page, and N. Noël, *Int. J. Oncol.* **3,** 473 (1993).

[24] G. R. Nakayama, M. C. Caton, M. P. Nova, and Z. Parandoosh, *J. Immunol. Methods* **204,** 205 (1997).

observed after 1 h treatment with SBL-C in P388 cells. Destruction of RNA into very small fragments is completed by 24 h. In the case of RC150 cells, RNA fragmentation does not occur even after exposure to SBL-C for 24 h.

Inhibition of Leczyme-Induced Cell Death by Sialomucin

We can check whether binding of SBL-C to cell surface sialyl oligosaccharide chains fulfills the requirements for causing P388 cell death. After SBL-C (3.1 μM) is mixed with BSM or asialo-BSM (25 μg/ml) at 37° for 30 min, P388 cells (2×10^5) are treated with the mixture for 24 h. Cell viability is assessed by the Alamar Blue assay method described above. Pretreatment of SBL-C with BSM inhibits SBL-C-induced cell death. No inhibition is observed when SBL-C is pretreated with asialo-BSM. The prerequisite condition for SBL-C-induced cell death is in turn binding and incorporation in the cells.

Leczyme-Induced Apoptosis and Expression of Apoptosis-Related Antigens

DNA Fragmentation

To test whether DNA is fragmented in SBL-C-treated apoptotic cells, P388 and RC150 cells are incubated with SBL-C (3.1 μM). After washing with PBS, the cells (1×10^6) are lysed in Tris-HCl buffer (50 m*M*, pH 7.8) containing 10 m*M* EDTA and 0.5% w/v sodium-*N*-lauroylsarcosinate (20 μl). RNase A (1 μl, 10 mg/ml) is added and the mixture is then incubated at 50° for 30 min. Then, proteinase K (1 μl, 10 mg/ml) is added and the mixture is treated at 50° for 60 min. DNA samples are run on 1.6% agarose gel and visualized under UV light after staining with ethidium bromide (5 μg/ml).[25] A preparation of *Hae*III-digested ϕX 174 DNA is run as a marker. DNA fragmentation is observed after 7 h treatment in P388 cells but not in RC150 cells. In the case of P388 cells, a DNA ladder pattern becomes very clear after 24 h treatment.

Annexin V and Propidium Iodide Staining

After SBL-C treatment (3.1 μM), cells are washed with ice-cold PBS. The cells are incubated with fluorescein isothiocyanate (FITC)–annexin V and propidium iodide (PI) at room temperature for 30 min according to the user's manual of the Early Apoptosis Detection Kit (Kamiya Biochemical Co., Seattle, WA).[26,27] Stained cells are analyzed using an FACScan flow cytometer (Becton Dickinson,

[25] C. A. Smith, G. T. Williams, R. Kingston, E. J. Jenkinson, and J. J. Owen, *Nature* **337,** 181 (1989).

[26] S. J. Martin, C. P. M. Reutelingsperger, A. J. McGohan, J. A. Rader, R. C. A. A. van Schie, D. M. LaFace, and D. R. Green, *J. Exp. Med.* **182,** 1545 (1995).

[27] I. Vermesa, C. Haanena, H. Steffens-Nakkena, and C. Reutelingsperger, *J. Immunol. Methods* **184,** 39 (1995).

San Jose, CA) at a laser setting of 36 mW and an excitation wavelength of 488 nm. The number of P 388 cells binding annexin V and incorporating PI increases after 5 and 7 h treatments with SBL-C, respectively.

Caspase Activation

To solve the signal transduction mechanism of SBL-C-induced apoptosis, cells are cultured in the presence of SBL-C (3.1 μM) for 24 h. The hydrolysis of caspase substrate is monitored colorimetrically at 405 nm according to the Caspase Colorimetric Protease Assay Kit protocol (MBL Co., Nagoya, Japan).[28] SBL-C-treated cells are incubated with cell lysis buffer on ice for 10 min. The cytosolic extract is then mixed with 2× reaction buffer containing 10 mM dithiothreitol (total volume, 100 μl), and reacted with the substrate–*p*-nitroanilide (final concentration, 200 μM) at 37° for 3 h. SBL-C can activate caspase-8 and caspase-3, and then induce cell death. Maximal activation of caspase-8 is detected at 3 h after treatment with SBL-C.

To confirm whether SBL-C-induced cell death needs caspase activation, cells are preincubated for 2 h with acetyl-tyrosyl-valinyl-alanyl-aspartyl-aldehyde (Ac-YVAD-CHO, inhibitor of caspase-1),[29] acetyl-aspartyl-glutamyl-valinyl-aspartyl-aldehyde (Ac-DEVD-CHO, inhibitor of caspase-3),[28] and carbobenzoxy-isoleucyl-glutamyl-threonyl-aspartyl-fluoromethylketone (Z-IETD-fmk, inhibitor of caspase-8)[28,30] at varying concentrations, and exposed to SBL-C (2.5 μM) for 24 h. Cell viability is assessed by the Alamar Blue assay method described above. Addition of Ac-DEVD-CHO inhibits the antiproliferative effect and caspase-3-activation ability of SBL-C.

Expression of Fas Antigen or Tumor Necrosis Factor Receptor

Given that many cell surface antigens are changed by induction of apoptosis,[31,32] we can test whether apoptosis-related antigens, such as Fas antigen and tumor necrosis factor receptor (TNFR), are expressed on the cell surface of SBL-C-treated cells. Cells (2×10^5/ml) are treated with SBL-C (final concentration, 3.1 μM) for 24 h, in a six-well plate. Then, the cells are incubated with anti-Fas antibody or anti-TNFR p55 antibody at 4° for 30 min, washed with PBS three times, and reacted with FITC-labeled second antibodies at 4° for 30 min. The

[28] N. A. Thornberry, T. A. Rano, E. P. Peterson, D. M. Rasper, T. Timkey, M. Garcia-Calvo, V. M. Houtzager, P. A. Nordstrom, S. Roy, J. P. Vaillancourt, K. T. Chapman, and D. W. Nicholson, *J. Biol. Chem.* **272,** 17907 (1997).

[29] S. M. Molineaux, F. J. Casano, A. M. Rolando, E. P. Peterson, G. Limjuco, J. Chin, P. R. Griffin, J. R. Calaycay, G. J.-F. Ding, T.-T. Yamin, O. C. Palyha, S. Luell, D. Fletcher, D. K. Miller, A. D. Howard, N. A. Thornberry, and M. J. Kostura, *Proc. Natl. Acad. Sci. U.S.A.* **90,** 1809 (1993).

[30] Z. Han, E. A. Hendrickson, T. A. Bremner, and J. H. Wyche, *J. Biol. Chem.* **272,** 13432 (1997).

[31] J. Savill, V. Fadok, P. Henson, and C. Haslett, *Immunol. Today* **14,** 131 (1993).

[32] K. Hiraishi, K. Suzuki, S. Hakomori, and M. Adachi, *Glycobiology* **3,** 381 (1993).

TABLE I
DIFFERENCES OF LECZYME ACTION AGAINST P388 CELLS AND RC150 CELLS

	Leczyme only		Leczyme pretreated with sialomucin
Action of leczyme	P388	RC150	P388
Inhibition of cell growth	Yes	No	No
Agglutination of cells	Yes	Yes	No
Binding to cell surface	Yes	Yes	No
Internalization into cells	Yes	No	NT[a]
Degradation of intercellular RNA	Yes	No	NT
Activation of caspase-8 and -3	Yes	No	NT
Enhancement of annexin V binding	Yes	No	NT
Induction of DNA fragmentation	Yes	No	No
Expression of Fas antigen and TNFR	Yes	No	NT

[a]NT, Not tested.

expression of Fas antigen and TNFR on P388 or RC150 cells is analyzed using by flow cytometry. The data from 10^4 cells/sample are collected, stored, and analyzed using CellQuest software. Both P388 and RC150 cells are Fas- and TNFR-negative. Treatment with SBL-C increases the expression of Fas antigen and TNFR only on P388 cell membrane (see Table I).

Conclusion

Binding and internalization of leczyme to leczyme-sensitive P388 leukemia cells underlie the correlation of increase rate of not only RNA but also DNA degradation with inhibition of cell proliferation. Treatment of P388 cells with leczyme causes activation of caspase-8 and caspase-3, binding of FITC–annexin V, DNA fragmentation, and then apoptotic morphological changes. It is possible that leczyme-induced apoptosis occurs through the up-regulation of Fas antigen and TNFR, leading to Fas ligand/Fas and TNF/TNFR interactions and cell death (see Table I). The application of methods described here can provide the way to create a new multifunctional protein with anticancer activity and to solve relationship between RNA degradation and DNA fragmentation in the process of leczyme-induced apoptosis.

Acknowledgments

I thank Drs. M. Hosono and Y. Ogawa for support throughout the development of this work and helpful discussions. A part of this work was supported by Grant·09240228 from the Ministry of Education, Science and Culture of Japan and by the Science Research Promotion Fund from the Promotion and Mutual Aid Corporation for Private Schools of Japan.

Section IV

Ribonucleases H

[24] Prokaryotic Type 2 RNases H

By SHIGENORI KANAYA

Introduction

Ribonuclease H (RNase H) specifically hydrolyzes the RNA strand of RNA/DNA hybrids.[1] The enzyme is universally present in various organisms. The *rnh* (*rnhA*) gene encoding RNase H was first cloned from *Escherichia coli*.[2] Because it was later shown that the *E. coli* genome contains another *rnh* (*rnhB*) gene, the first and the second RNases H have been designated as RNases HI and HII, respectively.[3] These two proteins show a poor sequence similarity. Therefore, the *rnhA* and *rnhB* genes are probably not paralogous. Database analyses indicated that the *rnh* genes so far identified are orthologous to either the *rnhA* or the *rnhB* gene. Thus, RNases H that show sequence similarities to *E. coli* RNase HI (*Ec*-RNase HI) have been classified as the Type 1 enzymes, and the others have been classified as the Type 2 enzymes.[4] Interestingly, certain bacterial genomes contain two paralogous genes encoding Type 2 RNases H. The enzyme that shows higher sequence similarity to *E. coli* RNase HII (*Ec*-RNase HII) has been designated as RNase HII and the other has been designated as RNase HIII.[5]

Type 1 RNases H have been structurally and functionally well studied (see Refs. 6–9 for reviews). The crystal structures have been determined for *Ec*-RNase HI,[10,11] RNase H from *Thermus thermophilus* HB8,[12] and the RNase H domain of

[1] R. J. Crouch and M.-L. Dirksen, *in* "Nuclease" (S. M. Linn and R. J. Roberts, eds.), p. 211. Cold Spring Harbor Laboratory Press, Cold Spring Harbor, NY, 1982.

[2] S. Kanaya and R. J. Crouch, *J. Biol. Chem.* **258,** 1276 (1983).

[3] M. Itaya, *Proc. Natl. Acad. Sci. USA* **87,** 8587 (1990).

[4] N. Ohtani, M. Haruki, M. Morikawa, and S. Kanaya, *J. Biosci. Bioeng.* **88,** 12 (1999).

[5] N. Ohtani, M. Haruki, M. Morikawa, R. J. Crouch, M. Itaya, and S. Kanaya, *Biochemistry* **38,** 605 (1999).

[6] R. J. Crouch, *New Biologist* **2,** 771 (1990).

[7] Z. Hostomsky, Z. Hostomska, and D. A. Matthews, *in* "Nucleases" 2nd Ed. (S. M. Linn and R. J. Robert, eds.), p. 341. Cold Spring Harbor Laboratory Press, Cold Spring Harbor, NY, 1993.

[8] S. Kanaya and M. Ikehara, *in* "Subcellular Biochemistry" (B. B. Biswas and S. Roy, eds.), Vol. 24, p. 377. Plenum Press, New York, 1995.

[9] S. Kanaya, *in* "Ribonucleases H" (R. J. Crouch and J. J. Toulme, eds.), p. 1. INSERM, Paris, 1998.

[10] K. Katayanagi, M. Miyagawa, M. Matsushima, M. Ishikawa, S. Kanaya, M. Ikehara, T. Matsuzaki, and K. Morikawa, *Nature* **347,** 306 (1990).

[11] W. Yang, W. A. Hendrickson, R. J. Crouch, and Y. Satow, *Science* **249,** 1398 (1990).

[12] K. Ishikawa, M. Okumara, K. Katayanagi, S. Kimura, S. Kanaya, H. Nakamura, and K. Morikawa, *J. Mol. Biol.* **230,** 529 (1993).

METHODS IN ENZYMOLOGY, VOL. 341
0076-6879/01 $35.00

human immunodeficiency virus type 1 (HIV-1) reverse transcriptase.[13,14] Mutational studies of *Ec*-RNase HI revealed that Asp-10, Glu-48, Asp-70, His-124, and Asp-134 are involved in catalytic function[15–17] and a positive-charge cluster in a basic protrusion is important for substrate binding.[18,19] Substrate titration observed by heteronuclear two-dimensional NMR spectra also facilitated the identification of the amino acid residues involved in substrate binding.[20] Based on this information, a model for the complex between *Ec*-RNase HI and RNA/DNA hybrid,[20,21] as well as a catalytic mechanism of *Ec*-RNase HI,[22–24] has been proposed. The physiological role of *Ec*-RNase HI has also been studied extensively using *E. coli rnhA* mutants, and the involvement of the enzyme in DNA replication and repair has been proposed.[25] In contrast, much less is known of the structures and functions of Type 2 RNases H.

Type 2 RNases H are universally present in various organisms, including bacteria, archaea, and eukaryotes. They are classified into four subfamilies, bacterial RNase HII, archaeal RNase HII, eukaryotic RNase H2, and bacterial RNase HIII (Table I). A phylogenetic tree based on a multiple sequence alignment created by the ClustalW program of DNA Data Bank Japan (DDBJ) showed that the members of each subfamily form an independent cluster (Fig. 1).[4] In this chapter I focus on the biochemical properties of *Ec*-RNase HII, *Bacillus subtilis* RNases HII and HIII (*Bs*-RNases HII and HIII), and RNase HII from *Thermococcus kodakaraensis* KOD1 (*Tk*-RNase HII) in comparison with that of *Ec*-RNase HI. *Ec*-RNase HII

[13] L. A. Kohlstaedt, J. Wang, J. M. Friedman, P. A. Rice, and T. A. Steitz, *Science* **256,** 1783 (1992).

[14] A. Jacobo-Molina, J. Ding, R. G. Nanni, A. D. Clark, Jr., X. Lu, C. Tantillo, R. L. Williams, G. Kamer, A. L. Ferris, P. Clark, A. Hizi, S. H. Hughes, and E. Arnold, *Proc. Natl. Acad. Sci. USA* **90,** 6320 (1993).

[15] S. Kanaya, A. Kohara, Y. Miura, A. Sekiguchi, S. Iwai, H. Inoue, E. Ohtsuka, and M. Ikehara, *J. Biol. Chem.* **265,** 4615 (1990).

[16] Y. Oda, M. Yoshida, and S. Kanaya, *J. Biol. Chem.* **268,** 88 (1993).

[17] M. Haruki, E. Noguchi, C. Nakai, Y-Y. Liu, M. Oobatake, M. Itaya, and S. Kanaya, *Eur. J. Biochem.* **220,** 623 (1994).

[18] S. Kanaya, C. Katsuda-Nakai, and M. Ikehara, *J. Biol. Chem.* **266,** 11621 (1991).

[19] J. L. Keck and S. Marqusee, *J. Biol. Chem.* **271,** 19883 (1996).

[20] H. Nakamura, Y. Oda, S. Iwai, H. Inoue, E. Ohtsuka, S. Kanaya, S. Kimura, C. Katsuda, K. Katayanagi, K. Morikawa, H. Miyashiro, and M. Ikehara, *Proc. Natl. Acad. Sci. USA.* **88,** 11535 (1991).

[21] K. Katayanagi, M. Miyagawa, M. Matsushima, M. Ishikawa, S. Kanaya, H. Nakamura, M. Ikehara, T. Matsuzaki, and K. Morikawa, *J. Mol. Biol.* **223,** 1029 (1992).

[22] S. Kanaya, M. Oobatake, and Y-Y. Liu, *J. Biol. Chem.* **271,** 32729 (1996).

[23] T. Kashiwagi, D. Jeanteur, M. Haruki, M. Katayanagi, S. Kanaya, and K. Morikawa, *Protein Eng.* **9,** 857 (1996).

[24] Y. Uchiyama, Y. Miura, H. Inoue, E. Ohtsuka, Y. Ueno, M. Ikehara, and S. Iwai, *J. Mol. Biol.* **243,** 782 (1994).

[25] T. Kogoma and P. L. Foster, *in* "Ribonucleases H" (R. J. Crouch and J. J. Toulme, eds.), p. 39. INSERM, Paris, 1998.

TABLE I
SUBFAMILIES OF TYPE 2 RNASES H

Sources	Abbreviation	Accession no.	Identity (%)[a]	
			Family	Subfamily
Bacterial RNase HII				
Escherichia coli	*Eco*	P10442	100	
Haemophilus influenzae	*Hin*	P43808	69	
Bacillus subtilis	*Bsu*	Z99112	47	
Chlorobium tepidum	*Cte*	Site[b]	47	
Magnetospirillum sp.	*Msp*	D32253	46	
Streptococcus pneumoniae	*Spn*	Site[b]	46	
Thermotoga maritima	*Tma*	Site[b]	46	
Enterococcus faecalis	*Efa*	Site[b]	46	
Mycobacterium tuberculosis	*Mtu*	Q10793	44	
Rickettsia prowazekii	*Rpr*	3860766	44	
Aquifex aeolicus	*Aae*	AE000765	44	
Mycobacterium leprae	*Mle*	Z97369	43	
Brucella melitensis	*Bme*	AF054610	43	
Chlamydia trachomatis	*Ctr*	AE001277	42	
Synechocystis sp. PCC6803	*Ssp*	D90899	42	
Deinococcus radiodurans	*Dra*	Site[b]	38	
Borrelia burgdorferi	*Bbu*	AE001118	38	
Streptomyces coelicolor	*Sco*	AL022374	37	
Helicobacter pylori	*Hpy*	P56121	33	
Archaeal RNase HII				
Thermococcus kodakaraensis[c]	*Tko*	AB012613	29	100
Pyrococcus furiosus	*Pfu*	Site[b]	29	56
Pyrococcus horikoshii	*Pho*	AP000006	28	55
Archaeoglobus fulgidus	*Afu*	AE001062	27	42
Methanobacterium thermoautotrophicum	*Mth*	AE000875	27	40
Methanococcus jannaschii	*Mja*	U67470	26	40
Eukaryotic RNase H2				
Saccharomyces cerevisiae	*Sce*	Z71348	24	100
Schizosaccharomyces pombe	*Spo*	Q10236	25	39
Caenorhabditis elegans	*Cel*	Z66524	23	33
Arabidopsis thaliana	*Ath*	AL031369	26	32
Homo sapiens	*Hsa*	Z97029	24	32
Bacterial RNase HIII				
Bacillus subtilis	*Bsu*	Z75208	22	100
Streptococcus pneumoniae	*Spn*	U93576	20	39
Mycoplasma genitalium	*Mge*	P47441	22	30
Mycoplasma pneumoniae	*Mpn*	P75446	22	28
Chlamydia trachomatis	*Ctr*	AE001275	18	26
Aquifex aeolicus	*Aae*	AE000755	18	24

[a] The amino acid sequence identities were calculated based on a multiple sequence alignment created by the ClustalW program provided by the DNA Data Bank Japan (DDBJ). The sequence of *E. coli* RNase HII (family) or the first member of each subfamily was arbitrarily set at 100%.

[b] http://www.ncbi.nlm.nih.gov/BLAST/unfinishedgenome.html

[c] Previously reported as *Pyrococcus kodakaraensis* [M. Haruki, K. Hayashi, T. Kochi, A. Muroya, Y. Koga, M. Morikawa, T. Imanaka, and S. Kanaya, *J. Bacteriol.* **180,** 6207 (1998)].

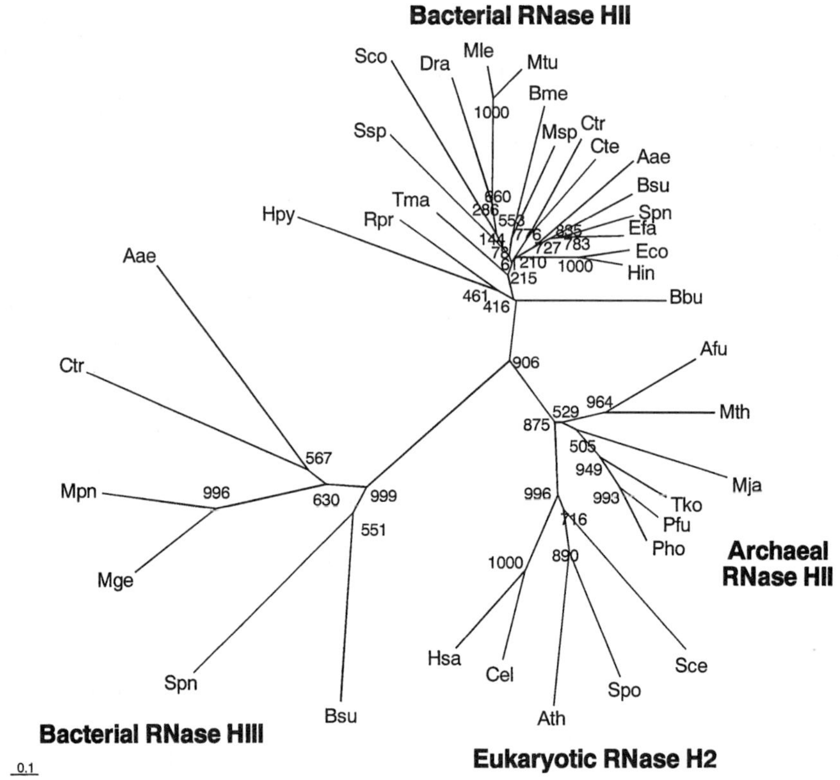

FIG. 1. Phylogenetic trees derived from Type 2 RNase H sequences. Phylogenetic trees were constructed by using the TreeView program [R. D. M. Page, *Comp. Appl. Biosci.* **12,** 357 (1996)] based on a multiple sequence alignment created by the ClustalW program [J. D. Thompson, D. G. Higgins, and T. J. Gibson, *Nucl. Acids Res.* **22,** 4673 (1994)] and a distance matrix produced by the PHYLIP program [J. Felsenstein, Phylogeny Inference Package, version 3.5c, distributed by the author. Department of Genetics, University of Washington, Seattle, WA (1993)]. The sequences used for these analyses are listed in Table I. The numbers show how often the target sequence is located at the same position of the tree when the analyses were repetitively performed 1000 times. The distance of the branch from a diverging point is proportional to the frequency on average of the amino acid substitution at each position. The scale bar corresponds to this frequency of 0.1. [Adapted with permission from N. Ohtani, M. Haruki, M. Morikawa, and S. Kanaya, *J. Biosci. Bioeng.* **88,** 12 (1999).]

and *Bs*-RNase HII represent bacterial RNase HII, *Tk*-RNase HII represents archaeal RNase HII, and *Bs*-RNase HIII represents bacterial RNase HIII. *Ec*-RNase HII shows 47% amino acid sequence identity to *Bs*-RNases HII, 29% to *Tk*-RNase HII, and 22% to *Bs*-RNases HIII. *Bs*-RNases HII and *Tk*-RNase HII show 19% and 12% amino acid sequence identities to *Bs*-RNases HIII, respectively. The sequence identity between *Bs*-RNases HII and *Tk*-RNase HII is 29%.

Overproduction and Purification

Ec-RNase HII, which is encoded by the *rnhB* gene, is composed of 198 amino acid residues (molecular weight of 21,525 and p*I* of 6.9). This enzyme was first reported to be composed of 213 amino acid residues,[3] based on an incorrect nucleotide sequence of the *rnhB* (ORF_{23}) gene previously determined,[26] in which a thymine base is inserted within the codon for Leu-194. *Ec*-RNase HII is overproduced mostly in an insoluble form when *E. coli* BL21 (DE3), harboring a plasmid in which the expression of the *rnhB* gene is under the control of the T7 promoter, is used as the overproducing strain.[3] However, when *E. coli* HMS174(DE3) or MIC2067(DE3) harboring the same plasmid is used as an overproducing strain, the yield of the protein recovered in a soluble form is dramatically increased.[27] MIC2067(DE3) is an *rnhA* and *rnhB* double mutant[28] of *E. coli* lysogenized with λDE3. All purification procedures are carried out at 4°. To simplify the purification procedures, the enzyme is overproduced as a fusion protein with the self-cleavable intein and the chitin-binding domain in *E. coli* MIC2067(DE3) using IMPACT system of New England Biolabs (Beverly, MA).[27] Cells are harvested, suspended in 20 m*M* Tris-HCl (pH 8.0) containing 0.5 *M* NaCl, 0.1 m*M* EDTA, and 0.1% Triton X-100 (buffer A), disrupted by sonication, and centrifuged at 15,000*g* for 30 min. The supernatant is applied to a chitin column (New England Biolabs) equilibrated with buffer A. The fusion protein bound to chitin is self-cleaved between the C terminus of *Ec*-RNase HII and the N terminus of intein, by incubating in 20 m*M* Tris-HCl (pH 8.0) containing 50 m*M* NaCl, 0.1 m*M* EDTA, and 30 m*M* dithiothreitol (DTT) at 4° for 20 h. *Ec*-RNase HII is then eluted from the chitin column with 20 m*M* Tris-HCl (pH 8.0) containing 50 m*M* NaCl and 0.1 m*M* EDTA. The fraction containing the pure enzyme is dialyzed against 20 m*M* sodium acetate (pH 5.5) containing 150 m*M* NaCl and used for further biochemical characterizations.

Bs-RNase HII is composed of 255 amino acid residues (molecular weight of 28,322 and p*I* of 5.6). The *rnhB* gene encoding this enzyme was originally identified as the *rnh* gene, when the complete genome sequence of *B. subtilis* was determined. *Bs*-RNases HII is overproduced in a soluble form when *E. coli* MIC3009 harboring a plasmid, in which the expression of the *rnhB* gene is under the control of the bacteriophage λ promoters P_R and P_L, is used as an overproducing strain.[5] MIC3009 is an *rnhA* mutant of *E. coli.*[29] Cells are harvested, suspended in 10 m*M* Tris-HCl (pH7.5) containing 1 m*M* EDTA (TE-buffer), disrupted by sonication, and centrifuged at 15,000*g* for 30 min. The supernatant (crude lysate)

[26] H. G. Tomasiewicz and C. S. McHenry, *J. Bacteriol.* **169,** 5735 (1987).

[27] N. Ohtani, M. Haruki, A. Muroya, M. Morikawa, and S. Kanaya, *J. Biochem.* **127,** 895 (2000).

[28] M. Itaya, A. Omori, S. Kanaya, R. J. Crouch, T. Tanaka, and K. Kondo, *J. Bacteriol.* **181,** 2118 (1999).

[29] M. Itaya and R. J. Crouch, *Mol. Gen. Genet.* **227,** 424 (1991).

is pooled and the enzyme is precipitated at 70% saturation with ammonium sulfate to remove nucleic acids. The precipitates are collected by centrifugation at 10,000*g* for 20 min, dissolved in 10 m*M* Tris-HCl (pH 8.0), and applied to a column (1 ml) of DE-52 (Whatman, Clifton, NJ) equilibrated with the same buffer. The enzyme is eluted from the column by a linear gradient of NaCl (0 to 0.5*M*). Fractions containing the enzyme are combined and applied to a gel filtration column (1.6 cm × 60 cm) of HiLoad 16/60 Superdex 200 pg (Pharmacia LKB Biotechnology, Piscataway, NJ) equilibrated with 10 m*M* Tris-HCl (pH 7.5), 150 m*M* NaCl. Fractions containing the pure enzyme are combined and used for further biochemical characterization.

Tk-RNase HII, which is encoded by the *rnhB* gene, is composed of 228 amino acid residues (molecular weight of 25,799 and p*I* of 5.4). *Tk*-RNase HII is overproduced in a soluble form by using an overproduction system,[30] which is similar to that for *Bs*-RNase HII. The crude lysate, which is prepared as mentioned above for *Bs*-RNase HII, is incubated at 90° for 15 min to precipitate most of the proteins derived from host cells and centrifuged at 30,000*g* for 30 min. The supernatant is pooled and the enzyme is precipitated at 70% saturation with ammonium sulfate to remove nucleic acids. The precipitates are collected by centrifugation at 10,000*g* for 20 min, dissolved in TE-buffer, and applied to a column of DE-52 equilibrated with the same buffer. The enzyme is eluted from the column by a linear gradient of NaCl (0 to 0.5 *M*). Fractions containing the enzyme are combined and applied to a column of HiLoad 16/60 Superdex 200 pg equilibrated with TE buffer containing 0.1 M NaCl. Fractions containing the pure enzyme are combined and used for further biochemical characterization.

Bs-RNase HIII is composed of 313 amino acid residues (Molecular weight of 34,037 and p*I* of 10.5). The *rnhC* gene encoding this enzyme was originally identified as the *ysgB* gene, when the complete genome sequence of *B. subtilis* was determined. *Bs*-RNase HIII is overproduced in a soluble form in HMS174(DE3)pLysS cells that harbor a plasmid in which expression of the *rnhC* gene is under control of a T7 promoter.[5] The crude lysate, which is prepared as mentioned above for *Bs*-RNase HII, is applied to a column (1 ml) of DE-52 equilibrated with 10 m*M* Tris-HCl (pH 7.5). The flow-through fraction containing the enzyme is applied to a column (1 ml) of P-11 (Whatman) equilibrated with 10 m*M* Tris-HCl (pH 7.5). The flow-through fraction containing the enzyme is applied to a column of HiLoad 16/60 Superdex 200 pg equilibrated with 10 m*M* Tris-HCl (pH 7.5), 150 m*M* NaCl. Fractions containing the pure enzyme were combined and the enzyme precipitated at 70% saturation with ammonium sulfate to remove nucleic acids. The precipitates are collected by centrifugation at 10,000*g* for

[30] M. Haruki, K. Hayashi, T. Kochi, A. Muroya, Y. Koga, M. Morikawa, T. Imanaka, and S. Kanaya, *J. Bacteriol.* **180,** 6207 (1998).

20 min, dissolved in 10 m*M* Tris-HCl (pH 7.5), and dialyzed against the same buffer. This solution containing the pure enzyme is used for further biochemical characterization.

The production level of the protein is roughly 60 mg/liter culture for *Ec*-RNase HII (as a fusion protein), 25 mg/liter culture for *Bs*-RNases HII and *Tk*-RNase HII, and 40 mg/liter culture for *Bs*-RNases HIII. The amounts of the proteins purified from one liter of culture are equally 5–6 mg.

Molecular Properties

The molecular weights of four prokaryotic Type 2 RNases H were estimated from gel filtration column chromatography as 22,000 for *Ec*-RNase HII,[27] 36,000 for *Bs*-RNase HII,[5] 33,000 for *Tk*-RNase HII,[30] and 40,000 for *Bs*-RNase HIII.[5] Because these values are slightly larger than but comparable to the calculated ones, these enzymes are monomeric like bacterial Type 1 RNases H. It is noted that recombinant *Tk*-RNase HII is composed of 226, instead of 228, amino acid residues (from Met-1 to Arg-226) with a calculated molecular weight of 25,577. The C-terminal two residues of this protein were eliminated probably by a protease from *E. coli*.

The far-UV CD spectra of *Ec*-RNase HII and *Tk*-RNase HII are shown in Fig. 2 in comparison with that of *Ec*-RNase HI. The spectra of *Ec*-RNase HII and *Tk*-RNase HII, which are similar to each other but distinct from that of *Ec*-RNase HI, exhibit a broad trough with a minimum [θ] value of $\sim-11{,}000$ at 209 nm, accompanied by a shoulder or another minimum at ~220 nm. *Bs*-RNase HII and *Bs*-RNase HIII also exhibited circular dichroism (CD) spectra similar to that of *Ec*-RNase HII (data not shown). These results strongly suggest that the three-dimensional structures of prokaryotic Type 2 RNases H are basically identical with one another.

T. kodakaraensis KOD1 is a hyperthermophilic archaeon, which can grow at temperatures ranging from 65° to 100°.[31] The optimal temperature for growth of this strain is 95°. To compare the stabilities of Type 2 RNases H from this hyperthermophile and from a mesophile, the stabilities of *Tk*-RNase HII and *Ec*-RNase HII toward irreversible heat inactivation were analyzed by incubating the enzyme solution (0.1–1 μg/ml) in 10 m*M* Tris-HCl (pH 7.5), 0.1 *M* NaCl, 1 m*M* EDTA, 10% glycerol, and 0.1 mg/ml of bovine serum albumin (BSA) at 90° and 70°, respectively. An aliquot of the enzyme solution was withdrawn at appropriate intervals, chilled on ice, and determined for remaining activity using M13 RNA/DNA hybrid as a substrate as described below. As expected, *Tk*-RNase

[31] M. Morikawa, Y. Izawa, N. Rashid, T. Hoaki, and T. Imanaka, *Appl. Environ. Microbiol.* **60,** 4559 (1994).

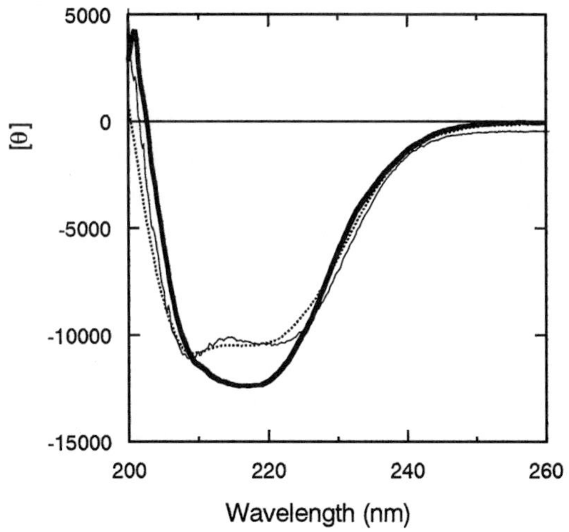

FIG. 2. CD spectra of RNases H. The CD spectra of *Ec*-RNase HI (thick line), *Ec*-RNase HII (dotted line), and *Tk*-RNase HII (thin line) were measured on a J-720W spectropolarimeter (Japan Spectroscopic Co., Ltd.). The far UV (200–260 nm) CD spectrum was obtained at 25° by using solution containing enzyme at ~0.1 mg/ml in 20 m*M* sodium acetate (pH 5.5) (*Ec*-RNase HI and *Ec*-RNase HII) or 10 m*M* Tris-HCl (pH 7.5) (*Tk*-RNase HII) containing 150 m*M* NaCl in a cell with an optical path of 2 mm. The mean residual ellipticity [θ], which has the units of deg cm^2 $dmol^{-1}$, was calculated by using an average amino acid molecular weight of 110.

HII was fairly stable toward irreversible heat inactivation at 90°, whereas *Ec*-RNase HII rapidly lost activity even at 70° with half-life of 9 min (data not shown).

Enzymatic Activities

RNase H activity is determined by measuring the radioactivity of the acid-soluble digestion product of ^{3}H-labeled M13 RNA/DNA hybrid.[32] This substrate is prepared by transcription of M13mp18 single- stranded DNA with *E. coli* RNA polymerase. For this reaction, 34 μg of M13mp 18 single-stranded DNA (Toyobo Co., Ltd.) and 20 μg of *E. coli* RNA polymerase (Sigma) are incubated in 500 μl of 44 m*M* Tris-HCl (pH 8.0), 14 m*M* $MgCl_2$, 14 m*M* 2-mercaptoethanol, 20 m*M* NaCl, 2% glycerol, and 35 μg/ml acetylated bovine serum albumin (Ac-BSA) (Bethesda Research Laboratories) in the presence of 37 MBq of [2,5′,8-^{3}H]ATP (Amersham Pharmacia Biotech), 1350 nmol of ATP, 500 nmol of GTP, 800 nmol of UTP, and 300 nmol of CTP at 30° for 30 min. The reaction is terminated by the addition of

[32] S. Kanaya, C. Katsuda, S. Kimura, T. Nakai, E. Kitakuni, H. Nakamura, K. Katayanagi, K. Morikawa, and M. Ikehara, *J. Biol. Chem.* **266,** 6038 (1991).

TABLE II
ASSAY BUFFERS

Enzyme	Assay buffer[a]
Ec-RNase HI	10 m*M* Tris-HCl (pH 8.0), 10 m*M* $MgCl_2$, 50 m*M* NaCl
Ec-RNase HII	10 m*M* Tris-HCl (pH 8.5), 10 m*M* $MnCl_2$, 50 m*M* NaCl
Tk-RNase HII	10 m*M* Tris-HCl (pH 8.0), 10 m*M* $CoCl_2$, 50 m*M* NaCl
Bs-RNase HII	10 m*M* Tris-HCl (pH 8.5), 10 m*M* $MnCl_2$, 50 m*M* KCl
Bs-RNase HIII	10 m*M* Tris-HCl (pH 8.5), 50 m*M* $MgCl_2$, 100 m*M* NaCl

[a] All buffers contain 1 m*M* 2-mercaptoethanol and 50 μg/ml acetylated BSA in addition to the chemicals listed.

5 μl of 10% SDS. After extraction with an equal volume of chloroform, the aqueous solution is applied to a column of Sephadex G-50 (Pharmacia LKB Biotechnology) equilibrated with 10 m*M* Tris-HCl (pH 8.0), 50 mM NaCl. The flow-through fractions containing ^{3}H-labeled M13 RNA/DNA hybrid are combined and used as the substrate. The substrate concentration is calculated as the RNA nucleotide phosphate concentration, assuming that the content of adenine in the RNA strand of the substrate is identical with that in the RNA strand, which is complementary to the entire single- stranded DNA of M13mp18.

For the assay, 1–2 μM substrate and an appropriate amount of the enzyme are incubated in 20 μl of the assay buffer listed in Table II at 30° for 15 min. The reaction is terminated by the addition of 20 μl of 1 mg/ml Ac-BSA and 200 μl of 20% trichloroacetic acid. After standing on ice for 5 min, the acid-soluble digestion product is recovered by centrifugation (15,000 rpm $\times$ 10 min) and measured for radioactivity. One unit is defined as the amount of enzyme producing 1 μmol of acid-soluble material/min at 30°. The specific activity is defined as the enzymatic activity/mg of protein. The protein concentrations are determined from the UV absorption with $A_{280}^{0.1\%}$ values of 2.01 for *Ec*-RNase HI, 0.56 for *Ec*-RNase HII, 0.63 for *Tk*-RNase HII, 0.93 for *Bs*-RNases HII, and 0.62 for *Bs*-RNases HIII. These values, except that of *Ec*-RNase HI which is determined by amino acid analysis,[33] are calculated by using $\varepsilon = 1,576 M^{-1}$ cm^{-1} for Tyr and $5,225\ M^{-1}$cm^{-1} for Trp at 280 nm.[34]

Like Type 1 RNases H, Type 2 RNases H require divalent metal ions for activity. However, the effect of divalent metal ions on the enzymatic activity varied for different enzymes. The specific activities of the Type 2 enzymes, as well as that of *Ec*-RNase HI, determined in the presence of various concentrations of $MgCl_2$, $MnCl_2$, $CoCl_2$, and $NiCl_2$ are summarized in Table III. These data clearly show that *Ec*-RNase HI and *Bs*- RNase HIII exhibit Mg^{2+}-dependent RNase H activity,

[33] S. Kanaya, S. Kimura, C. Katsuda, and M. Ikehara, *Biochem. J.* **271,** 59 (1990).
[34] T. W. Goodwin and R. A. Morton, *Biochem. J.* **40,** 628 (1946).

TABLE III
SPECIFIC ACTIVITIES OF ENZYMES IN PRESENCE OF VARIOUS METAL IONS

Metal ion and concentration	Specific activity (units/mg)[a]				
	Ec-RNase HI	*Ec*-RNase HII	*Tk*-RNase HII	*Bs*-RNase HII	*Bs*-RNase HIII
$MgCl_2$					
1 m*M*	7.0	0.01	0.1	<0.01	1.2
10 m*M*	**9.5**	0.03	1.1	0.02	4.0
50 m*M*	7.5	0.01	1.5	<0.01	**10**
$MnCl_2$					
1 m*M*	0.05	0.25	0.4	0.05	0.4
10 m*M*	<0.01	**0.5**	0.6	**0.5**	0.5
50 m*M*	<0.01	0.01	0.5	0.05	0.4
$CoCl_2$					
1 m*M*	<0.01	<0.01	0.2	0.05	0.2
10 m*M*	<0.01	<0.01	**2.1**	0.08	0.4
50 m*M*	<0.01	<0.01	2.0	0.03	0.5
$NiCl_2$					
1 m*M*	<0.01	<0.01	0.1	<0.01	0.05
10 m*M*	<0.01	<0.01	0.6	<0.01	0.07
50 m*M*	<0.01	<0.01	0.7	<0.01	0.07

[a] The hydrolysis of the M13 RNA/DNA hybrid with the enzyme was carried out at 30° for 15 min in an appropriate buffer, which contains 1 m*M* 2-mercaptoethanol and 50 μg/ml of acetylated BSA, in the presence of $MgCl_2$, $MnCl_2$, or $CoCl_2$ at the concentration indicated. The assay buffers were 10 m*M* Tris-HCl (pH 8.0) containing 50 m*M* NaCl for *Ec*-RNase HI and *Tk*-RNase HII, 10 m*M* Tris-HCl (pH 8.5) containing 50 m*M* NaCl for *Ec*-RNase HII, 10 m*M* Tris-HCl (pH 8.5) containing 50 m*M* KCl for *Bs*-RNase HII, and 10 m*M* Tris-HCl (pH 8.5) containing 100 m*M* NaCl for *Bs*-RNase HIII. Errors are within 20% of the values reported. The highest value for each enzymes is set in boldface type.

Ec-RNase HII and *Bs*-RNase HII exhibit Mn^{2+}-dependent RNase H activity, and *Tk*-RNase HII exhibits rather broad metal ion specificities. The levels of the Type 2 RNase H activities determined in the presence of the most preferable metal ion (Co^{2+} for *Tk*-RNase HII) relative to that of *Ec*-RNase HI were 5.3% for *Ec*-RNase HII and *Bs*-RNase HII, 22% for *Tk*-RNase HII, and 105% for *Bs*-RNase HIII. It has been reported that *Ec*-RNase HI exhibits Mn^{2+}-dependent RNase H activity at low metal ion concentrations (2–20 μM).[19,35] The specific activity of this enzyme determined in the presence of 5 μM $MnCl_2$ was ~20% of that determined in the presence of 5 m*M* $MgCl_2$.[19] However, none of the Type 2 enzymes exhibits such an RNase H activity at low metal ion concentrations.

The dependence of the enzymatic activity on the concentration of NaCl or KCl also varied for different enzymes. These salts are always inhibitory for activity

[35] J. L. Keck, E. R. Goedken, and S. Marqusee, *J. Biol. Chem.* **273,** 34128 (1998).

at concentrations higher than 100 m*M* (200 m*M* for the *Bs*-RNase HIII activity) but enhance all the enzymatic activities, except for the *Tk*-RNase HII activity, at concentrations lower than 100 m*M* (200 m*M* for the *Bs*-RNase HIII activity). The *Tk*-RNase HII activity was not seriously changed by the addition of these salts, unless their concentrations exceed 100 m*M*. In addition, *Bs*-RNase HII prefers KCl to NaCl for activity, whereas the activities of other enzymes respond equally to NaCl and KCl. In summary, *Ec*- RNase HI and *Ec*-RNase HII exhibited the highest activities in the presence of 50 m*M* NaCl or KCl. *Tk*-RNase HII, *Bs*-RNase HII, and *Bs*-RNase HIII exhibited the highest activities in the presence of 0–100 m*M* NaCl or KCl, 50 m*M* KCl, and 100 m*M* NaCl or KCl, respectively.

Type 2 RNases H exhibited enzymatic activities at alkaline pH, similar to *Ec*-RNase HI. However, unlike the enzymatic activity of *Ec*-RNase HI, which reaches a maximum at pH 8–9,[36] the activities of Type 2 enzymes continue to increase as the pH increases.[5,27] *Tk*-RNase HII[30] and *Bs*-RNase HII[5] have been reported to have an optimum pH at pH 9.5. However, the activities of these enzymes may decrease at pH values higher than 9.5, probably because of a decrease in the solubility of the divalent cations at these pH values. Therefore, the enzymatic activities of the Type 2 enzymes were determined at relatively mild pH (pH 8.0 or 8.5).

The dependence of the enzymatic activity on the temperature has not been analyzed, because the substrate is unstable at high temperatures. Therefore, optimum temperatures for the activities of the Type 2 enzymes remain to be determined. However, preliminary studies suggest that *Ec*-RNase HI exhibits the highest activity at 50–60°. This result is consistent with the stability of *Ec*-RNase HI previously reported.[37] *Ec*-RNase HI has been shown to lose 50% of its activity after heating at 53–54° for 10 min.

Cleavage of Oligomeric Substrates

To examine whether and how Type 2 RNases H hydrolyze the substrate, a 12 bp RNA/DNA hybrid and a 29 bp DNA–RNA–DNA/DNA duplex, which are ^{32}P-labeled at their 5′-ends, are used as a substrate. The sequence is 5′-cggagaugacgg-3′ for the 12-base RNA and 5′-AATAGAGAAAAAGaaaaAAGATGGCAAAG-3′ for the 29-base DNA–RNA–DNA, in which DNA and RNA are represented by uppercase and lowercase letters, respectively. For the hydrolytic reaction, 10 pmol of substrate and an appropriate amount of the enzyme are incubated at 30° for 15 min in 10 μl of assay buffer listed in Table II. The reaction is terminated by the addition of 10 μl of gel electrophoresis loading buffer containing 50 m*M* EDTA. The products are analyzed on a 20% polyacrylamide gel containing 7 *M* urea, and quantified using the Instant Imager (Packard). The products are characterized by

[36] I. Berkower, J. Leis, and J. Hurwitz, *J. Biol. Chem.* **248,** 5914 (1973).
[37] S. Kanaya and M. Itaya, *J. Biol. Chem.* **267,** 10184 (1992).

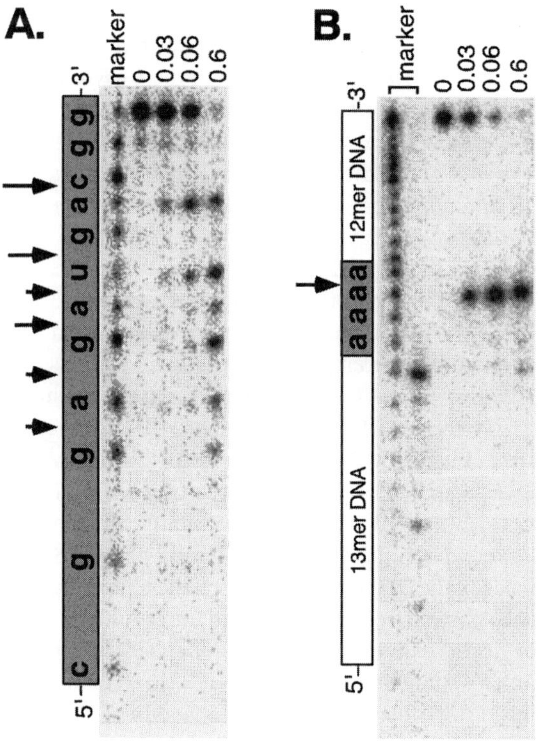

FIG. 3. Autoradiograph of cleavage reaction products. Hydrolyses of the 5′ end-labeled 12-base RNA hybridized to the 12-base DNA (A) and the 5-end-labeled 29-base DNA–RNA–DNA hybridized to the 29-base DNA (B) with *Tk*-RNase HII were carried out at 30° for 15 min and the hydrolyzates were separated on a 20% polyacrylamide gel containing 7 *M* urea as described in the text. The amount of the substrate in the reaction mixture is 10 pmol. Marker: Partial digest of the 12-base RNA with snake venom phosphodiesterase. Numbers indicate the amount (picomole) of the enzyme used for hydrolysis of the substrate. The sequence of each substrate is schematically shown alongside the autoradiogram, in which the 3′-terminal ribonucleotide residue of each oligonucleotide generated by the partial digestion with snake venom phosphodiesterase is shown by lowercase *letters*. Shaded and open boxes represent RNA and DNA, respectively. Arrows indicate the cleavage sites. Differences in the size of the arrows reflect the relative cleavage intensities at the indicated position.

comparing their migration on the gel with those of the oligonucleotides generated by partial digestion of the ^{32}P-labeled 12-base RNA or 29-base DNA–RNA–DNA with snake venom phosphodiesterase.[38]

Figure 3 shows representative cleavage reactions of *Tk*-RNase HII. When the 12 bp RNA/DNA substrate is completely cleaved with *Tk*-RNase HII, various

[38] E. Jay, R. Bambara, P. Padmanabham, and R. Wu, *Nucleic Acids Res.* **1,** 331 (1974).

5′-terminal oligonucleotides, such as 5′-cgg-3′ (r3-mer), 5′-cgga-3′ (r4-mer), 5′-cggag-3′ (r5-mer), 5′-cggaga-3′ (r6-mer), 5′-cggagau-3′ (r7-mer), and 5′-cggagauga-3′ (r9-mer) accumulate. Because the r9-mer is first produced when the substrate is hydrolyzed in a limited manner, the phosphodiester bond between the 9th adenosine and 10th cytidine (a9-c10) is most preferably hydrolyzed by the enzyme. The final products generated by the extensive hydrolysis of the substrate with the enzyme remained to be determined. However, there is no doubt that these oligonucleotides hybridized to 12-base DNA are much less susceptible to the hydrolysis with *Tk*-RNase HII as compared to the 12-base RNA hybridized to the 12-base DNA. In contrast, only a single oligonucleotide accumulated upon complete hydrolysis of the 29 bp DNA-RNA-DNA/DNA substrate with *Tk*-RNase HII. This product is generated by the cleavage of the phosphodiester bond between the third and fourth adenosine (a16-a17). Apparently, this product is not further cleaved by the enzyme. It is noted that the products produced by *Tk*-RNase HII comigrate with those produced by snake venom phosphodiesterase. These results indicate that the Type 2 enzymes cleave the substrate endonucleolytically at the P-O3′ bond as do Type 1 RNases H.

The cleavage reactions of other Type 2 enzymes and *Ec*-RNase HI, together with those of *Tk*-RNase HII, are summarized in Fig. 4. When the 12 bp RNA/DNA hybrid was used as a substrate, all of these enzymes cleaved it at multiple sites. Each enzyme generated products that reflected differences in site-specificity and relative frequencies of cleavage. All enzymes, except for *Bs*-RNase HII, cleaved this substrate preferentially at a 9-c10, but showed a different second site preference. In contrast, *Bs*-RNase HII cleaved the substrate in an almost random manner at all sites, except for c1-g2 and g2-g3, with nearly equal frequencies. Although some of the initial reaction products, such as the r9-mer hybridized to the 12-base DNA, are often recognized as a substrate and further cleaved to produce smaller products upon extensive hydrolysis, complete hydrolysis of the substrate usually resulted in an accumulation of a number of the products. The relative amount and the species of these products varied for different enzymes. Interestingly, 5′-CMP is apparently the only final product generated by the extensive hydrolysis with *Bs*-RNase HIII. However, it remained to be determined whether the substrate was completely hydrolyzed to mononucleotides, because this substrate was ^{32}P-labeled only at the 5′-end. When the 29 bp DNA-RNA-DNA/DNA substrate was used as a substrate, *Ec*-RNase HII, *Tk*-RNase HII, and *Bs*-RNase HII specifically cleaved the phosphodiester bond between the third and fourth adenosines (a16-a17). In contrast, *Ec*-RNase HI and *Bs*-RNase HIII specifically cleaved at the middle of the tetra-adenosine sequence (a15-a16). *Bs*-RNase HIII also cleaved this substrate at a14-a15 and a16-a17, but with lower efficiency. The cleavage of this substrate at these sites was confirmed by use of a substrate in which the 29-base DNA–RNA–DNA was ^{32}P-labeled at the 3′-end.[5,30]

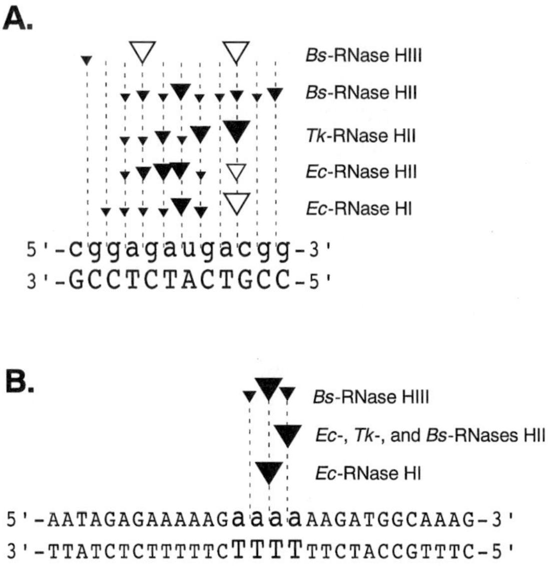

FIG. 4. Graphical representation of the sites and extent of cleavage by RNases H. Cleavage sites of the 12 bp RNA/DNA hybrid (A) and those of the 29 bp DNA–RNA–DNA/DNA substrate (B) with various RNases are shown by inverted triangles. Differences in the size of inverted triangles reflect the relative cleavage intensities at the indicated positions. The 5′-terminal oligonucleotides generated by the cleavage at the positions indicated by open and solid inverted triangles represent those that did not and did accumulate upon complete hydrolysis of the substrate, respectively. Therefore, the size of inverted triangles does not reflect the amount of the products accumulated upon complete hydrolysis of the substrate. For example, r9-mer and r4-mer were primarily produced from the 12 bp RNA/DNA hybrid by *Bs*-RNase HIII, but were further cleaved to produce 5′-CMP upon complete hydrolysis of the substrate. Deoxyribonucleotides are shown by *capital letters* and ribonucleotides are shown by lowercase *letters*.

The type 2 enzymes were unable to cleave 12-base 5′-^{32}P-RNA, 29-base 5′-^{32}P-DNA–RNA–DNA, and 12 bp 5′^{32}P-RNA/RNA duplex (data not shown). In addition, none of these enzymes cleaved the DNA strand of the 12 bp DNA-RNA-DNA/DNA substrate. These results are consistent with the previous ones, which show that *Ec*-RNase HI does not cleave single-stranded DNA, single-stranded RNA, double-stranded DNA, or double-stranded RNA.[3] Thus, Type 2 RNases H cleave the RNA strand only when it is hybridized to the DNA strand, like Type 1 enzymes.

Catalytic Mechanism

The amino acid residues involved in the catalytic function of Type 2 RNases H remained to be determined. However, multiple sequence alignment of the Type 2 enzymes indicated that five acidic residues, two basic residues, three glycine

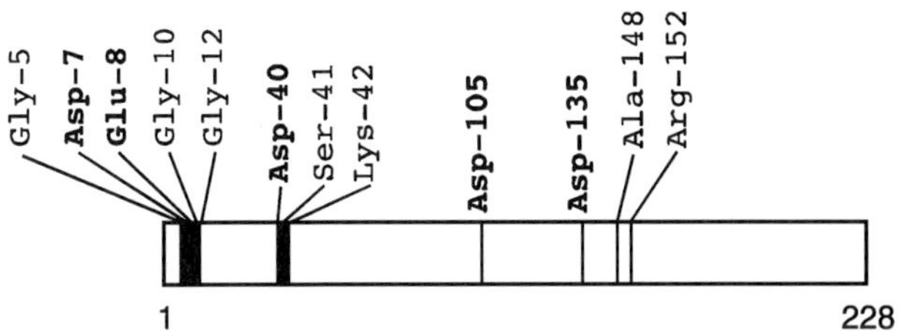

FIG. 5. Schematic representation of the amino acid sequence of *Tk*-RNase HII. The positions of the amino acid residues, which are well conserved in the Type 2 RNase H sequences, are shown. Of them, five conserved acidic residues are designated in boldface. Ser-41, Asp-135, and Arg-152 are occasionally replaced by Thr, Glu, and Lys, respectively. Numbers represent the positions of the amino acid residues, relative to the N-terminal methionine.

residues, one alanine residue, and one serine or threonine residue (12 in total) are fully conserved.[4] The localization of these residues in the *Tk*-RNase HII sequence is schematically shown in Fig. 5. A preliminary crystallographic analysis of *Tk*-RNase HII shows that a folding topology of the catalytic domain of *Tk*-RNase HII is similar to that of *Ec*-RNase HI, despite their poor sequence similarity.[39] It also shows that, of the fully conserved acidic residues shown in Fig. 5, Asp-7, Glu-8, Asp-105, and Asp-135 are clustered in a region corresponding to the active site of *Ec*-RNase HI. A preliminary mutational analysis suggests that these residues are important for activity.[40] These results, together with the similarity in the mode of action and requirement for activity, strongly suggest that Type 1 and Type 2 RNases H share a common catalytic mechanism.

Two different catalytic mechanisms have been proposed for Type 1 RNases H. One is a two-metal ion mechanism, in which the hydrolysis of the P-O3′ bond is promoted by two metal ions, such that the first metal ion activates an attacking H_2O molecule, and the second metal ion stabilizes the transient pentacovalent phosphorus intermediate and facilitates the release of the 3′-oxyanion.[11] Four fully conserved acidic residues, which correspond to Asp-10, Glu-48, Asp-70, and Asp-134 of *Ec*-RNase HI, have been suggested to provide ligands for the binding of these metal ions. The other is a general acid–base mechanism. According to this mechanism proposed for *Ec*-RNase HI,[22] His-124 activates an H_2O molecule that acts as a general base. Asp-134 coordinates the H_2O molecule. Glu-48 anchors an H_2O molecule that acts as a general acid. Asp-70 governs the conformation of Asp-10 through negative charge repulsion, and thereby facilitates the binding of the metal ion. The major difference in these two mechanisms is the number of metal ions required for activity. According to the general acid–base mechanism, the single required metal ion stabilizes a transient intermediate by interacting with the 2′-hydroxyl group of the substrate through the formation of an outer-sphere

[39] A. Muroya and K. Morikawa, unpublished results (2000).
[40] T. Kochi, M. Haruki, M. Morikawa, and S. Kanaya, unpublished results (2000).

complex with H_2O molecules.[24] Although no direct evidence has been provided, the data obtained from structural[41,42] and biochemical[35,43,44] studies support a general acid–base mechanism.

Lack of the conserved histidine residue in the Type 2 RNase H sequences does not necessarily mean that a catalytic mechanism of Type 2 RNases H is different from that of Type 1 RNases H. It has been reported that His-124 is involved in the catalytic function of *Ec*-RNase HI, but in an auxiliary manner, as mutation of this residue did not completely abolish enzymatic activity.[16] Thus, the *Ec*-RNase HI mutant, in which His-124 was replaced by Ala, retained 3% of the enzymatic activity of the wild-type protein, and its activity showed a pH-dependence similar to those of Type 2 RNases H.[23] In addition, the truncated protein of *Ec*-RNase HI, which lacks His-124 and Asp-134, still exhibited a weak enzymatic activity.[45] These results suggest that Type 2 RNases H cleave the substrate by a general acid–base mechanism, using a side chain (probably Asp or Glu) for activation of the attacking H_2O molecule different from that (His-124) proposed for *Ec*-RNase HI. Phylogenetic analyses of the RNase H sequences indicated that Type 2 RNases H are universally present in various organisms, whereas Type 1 RNases H are not.[4] Therefore, the catalytic mechanism of Type 2 RNases H may represent the basic one for the RNases H. The catalytic mechanism of Type 1 RNases H may represent an advanced one, in which a histidine residue is utilized as the general base.

Possible Physiological Functions

Escherichia coli strains MIC3001 with *rnhA* and *recB*(Ts) mutations and MIC3037 with *rnhA* and *recC*(Ts) mutations show an RNase H-dependent temperature-sensitive growth phenotype.[29] *Ec*-RNase HII,[3] *Tk*-RNase HII,[30] and *Bs*-RNases HII and HIII[28] can function in *E. coli* to complement the temperature-sensitive growth phenotypes of these strains. The requirement for RNase H activity in strains defective for RecBCD is not understood. However, these results suggest that removal of RNA–DNA hybrids is sufficient to permit growth of strains mutated in *recB* or *recC* and that the differences in the biochemical properties reported for the Type 2 RNases do not affect their ability to remove these "undesirable" RNA/DNA hybrids. Physiological functions of RNase H remained to be determined. However, *E. coli* strain MIC2067 with the *rnhA* and *rnhB* double mutations showed a temperature-sensitive growth phenotype.[28] In addition, the *rnhB* and *rnhC* double mutations make *B. subtilis* unable to grow.[28] These results suggest that RNase H is involved in biological processes that are important for cell

[41] K. Katayanagi, M. Okumura, and K. Morikawa, *Proteins: Struct. Funct. Genet.* **17,** 337 (1993).
[42] Y. Oda, H. Nakamura, S. Kanaya, and M. Ikehara, *J. Biomol. NMR* **1,** 247 (1991).
[43] H.-W. Huang and J. A. Cowan, *Eur. J. Biochem.* **219,** 253 (1994).
[44] Y. Uchiyama, S. Iwai, Y. Ueno, M. Ikehara, and E. Ohtsuka, *J. Biochem.* **116,** 1322 (1994).
[45] E. C. Goedken, T. M. Raschke, and S. Marqusee, *Biochemistry* **36,** 7256 (1997).

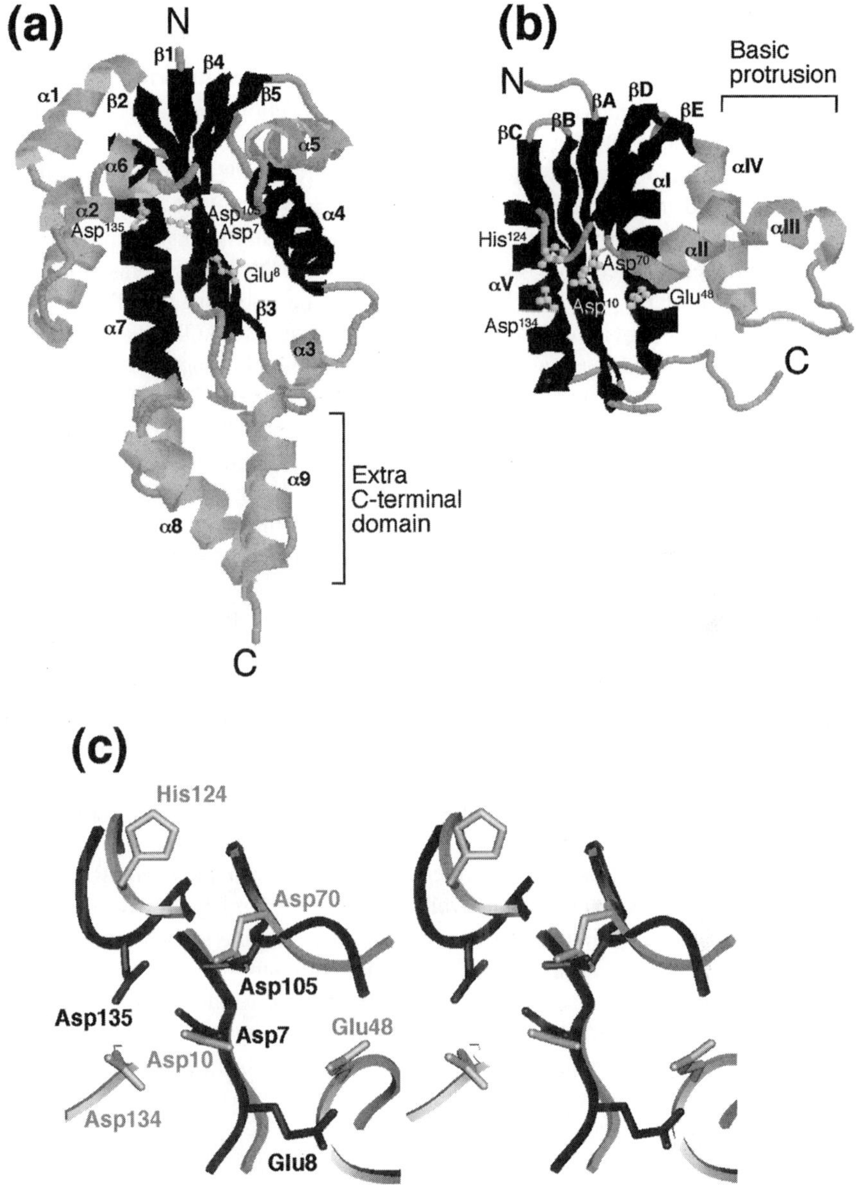

FIG. 6. Comparison of the crystal structures of Type 1 and Type 2 RNases H. The backbone structure of *Tk*-RNase HII drawn with the program RasMol (a) is shown in comparison with that of *E. coli* RNase HI (b). A main chain fold shared by these two structures, which consists of a five-stranded β sheet and two α helices, are black. The N and C termini are labeled. The active-site residues (Asp-7, Glu-8, Asp-105, and Asp-135 for *Tk*-RNase HII, and Asp-10, Glu-48, Asp-70, His-124, and Asp-134 for *E. coli* RNase HI) are shown in ball-and-stick form. (c) The active-site residues in the crystal structures of *Tk*-RNase HII (black) and *E. coli* RNase HI (grey) are superimposed. This figure was kindly prepared by A. Muroya.

growth. The physiological significance for the multiplicity of the *rnh* genes in a single bacterial or eukaryotic genome may be a protection of cells from a lethal mutation introduced into the *rnh* gene.

The most important physiological function of RNase H may be removal of RNA primers from Okazaki fragments during lagging strand DNA synthesis. In bacteria, *in vitro* experiments have indicated that Okazaki primers can be completely removed by the $5' \rightarrow 3'$-exonuclease activity of DNA polymerase I (pol I) in the absence of RNase H.[46] However, it cannot be ruled out that Type 1 and Type 2 RNases H play a role in removal of RNA primers from Okazaki fragments *in vivo* in a concerted manner with pol I. Another possible function of RNase H might be removal of ribonucleotides misincorporated into the DNA strand during DNA synthesis. However, *Ec*-RNase HI and all Type 2 RNases H are unable to cleave the 29 bp DNA–RNA–DNA/DNA substrate at either the DNA–RNA or RNA–DNA junction. Thus, neither prokaryotic Type 1 nor Type 2 RNase H may have a repair function by which single ribonucleotides misincorporated into DNA are removed, and they cannot eliminate single ribonucleotides attached to the 5′ end of the DNA strand in Okazaki fragments.

Note added in proof: The crystal structure of *Tk*-RNase HII has been determined.[47] This structure, which highly resembles that of RNase HII from *Methanococus jannaschii* recently reported,[48] is shown in Fig. 6, in comparison with that of *E. coli* RNase HI. Despite the poor amino acid sequence identity, these two proteins share a main chain fold, which consists of a five-stranded β sheet and two α-helices. The geometrical arrangements of the four acidic active-site residues are similar in these two proteins. These results are consistent with the proposal made in this paper that Type 1 and Type 2 RNases H share a common catalytic mechanism. The *Tk*-RNase HII structure seems to differ from the *E. coli* RNase HI structure in the location of a domain involved in substrate binding. The former lacks the basic protrusion, which is important for substrate binding in the *E. coli* RNase HI structure.[18] Instead, it has an extra C-terminal domain, which is composed of $\alpha 8$ and $\alpha 9$ helices. Because this domain is rich in basic residues and deletion of the $\alpha 9$ helix seriously affects the substrate binding affinity of the enzyme,[47] this extra C-terminal domain may be somehow involved in substrate binding.

Acknowledgment

We thank M. Itaya for providing *rnh* mutant of *E. coli* and ID Biomedical Corp. for providing the 29-base DNA-RNA-DNA for the RNase H assay. We also thank M. Haruki, M. Morikawa, N. Ohtani, and R. J. Crouch for helpful discussions, and R. Matsumoto for technical assistance.

[46] V. Lyamichev, M. A. D. Brow, and J. E. Dahlberg, *Science* **260,** 778 (1993).

[47] A. Muroya, D. Tsuchiya, M. Ishikawa, M. Haruki, M. Morikawa, S. Kanaya, and K. Morikawa, *Protein Sci.* **10,** 707 (2001).

[48] L. Lai, H. Yokota, L.-W. Hung, R. Kim, and S.-H. Kim, *Structure* **8,** 897 (2000).

[25] RNase H1 of *Saccharomyces cerevisiae*: Methods and Nomenclature

By ROBERT J. CROUCH, ARULVATHANI ARUDCHANDRAN, and SUSANA M. CERRITELLI

The yeast *Saccharomyces cerevisiae* is currently one of the best systems for studying cellular processes common to all eukaryotic organisms. For studies concerning ribonucleases H, the extensive genetic manipulations and knowledge of the DNA sequence of the entire genome make *S. cerevisiae* extremely useful. To date, there are three RNases H described in *S. cerevisiae* that exhibit the specificity of cleaving RNA of RNA–DNA hybrids.[1–3] Two of these enzymes have sequence similarity to RNases HI and HII of *Escherichia coli* while the third has been examined uniquely in *S. cerevisiae.*[4,5] Here, we will concentrate on the *S. cerevisiae* RNase H1 enzyme and the *RNH1* gene from which it is expressed. However, there is evidence that the two RNases H of *S. cerevisiae* related to the *E. coli* enzymes share some common substrates and these data and related methods will be mentioned as well.[6]

The presence of more than one type of ribonuclease H per cell was first reported by Büsen and Hausen[7] with considerable follow up work to demonstrate the differences between these two classes of enzymes. As is common in naming multiple activities, the designations RNase H1 and RNase H2 or sometimes class I RNase H and class II RNase H were used. Quite logically, Büsen and Hausen chose to name the enzymes based on the order of discovery. The study of RNases H in *E. coli* during this early period focused on a very active small (18 kDa) protein.[8–10]

[1] R. J. Crouch and S. M. Cerritelli, *in* "Ribonucleases H" (R. J. Crouch and J. J. Toulmé, eds.), Chapter 4, pp. 79–100. INSERM, Paris, 1998.

[2] U. Wintersberger, *Pharmac. Ther.* **48,** 259 (1990).

[3] U. Wintersberger, C. Kühne, and R. Karwan, *Biochim. Biophys. Acta* **951,** 322 (1988).

[4] P. Frank, C. Braunshofer-Reiter, A. Karwan, R. Grimm, and U. Wintersberger, *FEBS Lett.* **450,** 251 (1999).

[5] P. Frank, C. Braunshofer-Reiter, and U. Wintersberger, *FEBS Lett.* **421,** 23 (1998).

[6] A. Arudchandran, S. M. Cerritelli, S. K. Narimatsu, M. Itaya, D. Y. Shin, Y. Shimada, and R. J. Crouch, *Genes Cells* **5,** 789 (2000).

[7] W. Büsen and P. Hausen, *Eur. J. Biochem.* **52,** 179 (1975).

[8] R. J. Crouch and M. L. Dirksen, Ribonucleases H, *in* "Nucleases" (S. M. Linn and R. J. Roberts, eds.), pp. 211–241. Cold Spring Harbor Laboratory Press, Cold Spring Harbor, NY, 1982.

[9] S. Kanaya and R. J. Crouch, *J. Biol. Chem.* **258,** 1276 (1983).

[10] T. Kogoma and P. L. Foster, *in* "Ribonucleases H" (R. J. Crouch and J. J. Toulmé, eds.), Chapter 2, pp. 39–66. INSERM, Paris, 1998.

The gene was cloned, its nucleic acid sequence was determined, and some of the phenotypes of strains with defects in this RNase H were described. Discovery of synthetic lethality when mutants in the *rnh* and *recB* (or *recC*) genes were combined led to the development of a temperature-sensitive lethal phenotype by using an *rnhA::cat* insertion mutation with *recB* or *recC* ts mutants, the latter being ts for Rec functions but not ts for growth. Itaya used one of these strains to clone a second *rnh* gene in *E. coli.*[11] Now being confronted with two RNases H in *E. coli,* Itaya chose to designate the RNase H first described as RNase HI (gene name *rnhA*) and his new enzyme as RNase HII (gene name *rnhB*). Each of the *E. coli* proteins was purified and the amino acid sequence deduced from the nucleic acid sequence. Thus, we were now also faced with the problem of defining the *in vivo* function of not just one but two RNases H in *E. coli*. The clear similarity in structure and mechanism of *E. coli* RNase HI and the RNase H of retroviral reverse transcriptase[12] provided impetus for examining the most prominent RNase H of *E. coli*. However, a surprise awaited us when the DNA sequence of complete genomes started to become available. No recognizable homolog of *E. coli* RNase HI is present in a wide variety of bacteria and archaea.[13] Some of these organisms have only an enzyme related by sequence to *E. coli* RNase HII and many of them have two genes of the RNase HII type. *Bacillus subtilis* is an example of an organism with two proteins related by amino acid sequence to *E. coli* RNase HII. One of the two enzymes from *B. subtilis* resembles *E. coli* RNase HI in activity and divalent metal ion requirements while the second has properties similar to *E. coli* RNase HII,[13] and each of the *B. subtilis* RNases H is capable of suppressing the ts-phenotype of the *rnhA339::cat recB270*(Ts) strains of *E. coli*.[14] Based on the differences in enzymatic properties, the two *B. subtilis* RNases H are called RNase HII and RNase HIII, with RNase HII having enzymatic characteristics similar to those of *E. coli* RNase HII.[13,15] In spite of the lack of amino acid sequence similarity, the two types of RNases H seem to be structurally and mechanistically similar.[16]

The first eukaryotic RNase H gene cloned[17] was that of *S. cerevisiae*. Using the ts phenotype of the *rnhA339::cat recB270* (Ts), plasmids from a genomic library were selected by plating the *E. coli* strain at the nonpermissive temperature (see section below on Complementation). The DNA sequence for the gene obtained was determined, and comparison of the predicted protein with that of *E. coli* RNase

[11] M. Itaya, *Proc. Nat. Acad. Sci. U.S.A.* **87,** 8587 (1990).

[12] S. H. Hughes, E. Arnold, and Z. Hostomsky, *in* "Ribonucleases H" (R. J. Crouch and J. J. Toulmé, eds.), Chapter 10, pp. 195–224. INSERM, Paris, 1998.

[13] N. Ohtani, M. Haruki, M. Morikawa, R. J. Crouch, M. Itaya, and S. Kanaya, *Biochemistry* **38,** 605 (1999).

[14] M. Itaya, A. Omori, S. Kanaya, R. J. Crouch, T. Tanaka, and K. Kondo, *J. Bacteriol.* **181,** 2118 (1999).

[15] N. Ohtani, M. Haruki, M. Morikawa, and S. Kanaya, *J. Biosci. Bioeng.* **88,** 12 (1999).

[16] L. H. Lai, H. Yokota, L. W. Hung, R. Kim, and S. H. Kim, *Structure* **8,** 897 (2000).

[17] M. Itaya, D. McKelvin, S. K. Chatterjie, and R. J. Crouch, *Mol. Gen. Genet.,* 1 (1991).

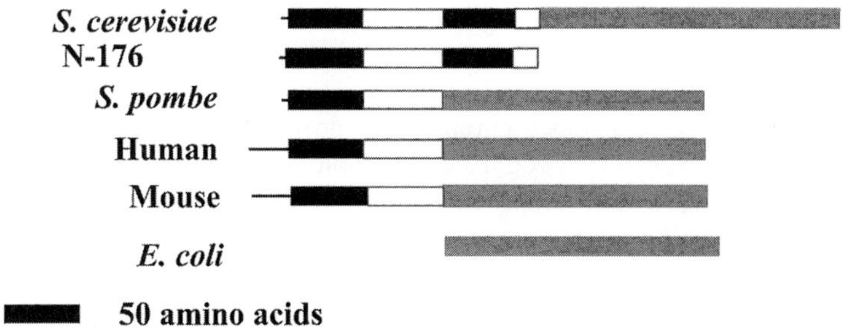

FIG. 1. RNase H1 of *S. cerevisiae* compared with *E. coli* RNase HI and other eukaryotic RNases H1. The RNases H1 of *S. cerevisiae*, *Schizosaccharomyces pombe* (*S. pombe*), human, and mouse are represented with the RNase H domain of each shown in gray, as is that of *E. coli* RNase HI. The areas of the proteins shown in black correspond to the duplex RNA-binding domains. The areas filled with white are nonconserved regions.

HI showed that the two are related. Alignment of the two amino acid sequences demonstrates the similarity in sequence between *E. coli* RNase HI and the C-terminal domain of *S. cerevisiae* RNase H1 (a diagram of which is shown in Fig. 1). Because this was the first RNase H encoding gene from *S. cerevisiae,* the name given to the gene was *RNH1* and the enzyme RNase H1. Genes encoding several other RNases H related to *S. cerevisiae* RNase H1 were cloned and were designated RNase H1; this was true even for *Drosophila melanogaster,*[18] for which gene names are notoriously flamboyant.

The relationship between bacterial RNases HI and eukaryotic RNases H1 to the two classes of mammalian RNases H remained unclear until recently when genomic sequencing projects uncovered cDNA clones derived from human[19,20] and mouse[20] encoding RNases H with strong sequence similarity to *E. coli* RNases HI (Fig. 1). Analogous searches for eukaryotic RNases H related to *E. coli* RNase HII also revealed cDNA encoding similar proteins in eukaryotic organisms, including mammalian and fungal sources. The properties of the mouse and human RNases H related by amino acid sequence to *E. coli* RNase HI are those of Class II enzymes.[21] Likewise, Class I RNases H are related to *E. coli* RNase HII. From the perspective of consistency of nomenclature, it was disappointing that the names failed to match. Comparison of DNA sequences to uncover the sequence of eukaryotic RNase H was not the only means of finding the identity of the class II enzymes. Frank *et al.*[22] purified the enzyme from calf thymus and obtained a partial amino acid

[18] V. Filippov, M. Filippova, and S. S. Gill, *Biochem. Biophys. Res. Commun.* **240,** 844 (1997).

[19] H. J. Wu, W. F. Lima, and S. T. Crooke, *Antisense Nucleic Acid Drug Devel.* **8,** 53 (1998).

[20] S. M. Cerritelli and R. J. Crouch, *Genomics* **53,** 300 (1998).

[21] W. Büsen and P. Frank, *in* "Ribonucleases H" (R. J. Crouch and J. J. Toulmé, eds.), Chapter 6, pp. 113–146. INSERM, Paris, 1998.

[22] P. Frank, C. Braunshofer-Reiter, A. Poltl, and K. Holzmann, *Biol. Chem.* **379,** 1407 (1998).

sequence, which they then used to search the databases. They were able to find cDNA sequences for the 5′ and 3′ ends of human EST sequences, from which they generated a cDNA clone.

When Frank and Wintersberger reported their work on the *S. cerevisiae* genes encoding RNases H,[4,5] they recognized the problems with nomenclature and designated the proteins as RNase H(MW) where MW means molecular mass. For example, RNase H(70) refers to the 63 kDa RNase H, and RNase H(35) refers to the 35 kDa *S. cerevisiae* equivalent of *E. coli* RNase HII.

The problems generated by inconsistency of names for the various RNases H are significant and need to be resolved. The proposal we are making here is based on some papers using terminology close to that suggested here.[6,13–15] RNases H of bacteria and archaea would be designated as I, II, or III (Roman numerals) based on similarity in sequence to *E. coli* and *B. subtilis* RNases H. Eukaryotic RNases H would be a little more complicated. All RNases H related by sequence to the typical enzymes present in a wide variety of eukaryotes and to *E. coli* RNase HI would be RNase H1 (arabic numeral). The gene names for mouse and human RNases H1 are *Rnaseh1* and *RNASEH1,* respectively (the *RNH* gene in humans encodes a ribonuclease inhibitor). After considerable effort by Frank and Büsen, it is now clear that the other type of RNase H in mammalian cells contains two subunits, one of about 35 kDa and the other around 22 kDa.[21,23] The 35 kDa subunit has strong sequence similarity to *E. coli* RNase HII while the nature of the second subunit has not yet been defined. Thus, the enzyme should be RNase H2 and the subunits RNase H2L and RNase H2S, where L is the large and S the small subunit. The genes equivalent to *RNH2L* of *S. cerevisiae* of mouse, human, and *Caenorhabditis elegans* have molecular masses of 33.4, 33.3, and 33.2 kDa, respectively, and the size of any second subunit for each of these enzymes is unknown at this time, but the small subunit of calf thymus RNase H2 migrates with a mobility of a 21 kDa protein in SDS–polyacrylamide gels.[23] Thus, "L" and "S" seem appropriate to describe the large and small subunits, accommodating variations in sizes when comparing enzymes from different organisms.

Roles for RNases H in Cells

In *E. coli,* loss of RNase HI activity permits initiation of DNA replication at sites other than oriC, normally the only replicative origin in *E. coli*. New origins for DNA replication are likely to arise because of R loops that remain stable in the absence of RNase HI with either the primer being the RNA of the R loop or the displaced DNA strand providing an assembly site for DNA replication enzyme complexes.[10] Only modest decreases in RNase HI activity are required for new

[23] P. Frank, C. Braunshofer-Reiter, U. Wintersberger, R. Grimm, and W. Büsen, *Proc. Nat. Acad. Sci. USA* **95,** 12872 (1998).

origins to become active, whereas complete loss of RNase HI activity is necessary to gain a requirement for RecBCD.[24] How does RNase HII when expressed from a multicopy plasmid suppress the lethality of *rnhA-339::cat recB270* (Ts) but fails to help when expressed at its typical level? Insufficient RNase HII activity is the most likely explanation with the limitation due to either inherently low activity of RNase HII or sequestration of RNase HII (for example, interaction of RNase HII with another protein or complex of proteins). Suppression by RNase HII of the synthetic-lethal phenotype of *rnhA-recB* mutants demonstrates RNases HI and HII can recognize the same substrates *in vivo*. This latter point is also supported by the requirement for RNase HI and HII in some strains for the cells to grow at 42°. In *S. cerevisiae*, strains deleted for both *RNH1* and *RNH2L (RNH35)* are viable. But when both genes are absent, the cells are sensitive to caffeine, a drug that affects DNA checkpoints, and alkylating agents such as ethylmethane sulfonate.[6] *RNH1*Δ strains are actually more tolerant of caffeine than either wild-type or *RNH2L*Δ strains. An insertion mutation near the promoter region of *RNH1* is hypersensitive to caffeine.[25] We have found that RNase H1 is still expressed in strains carrying the insertion mutation, indicating that hypersensitivity to caffeine is not due to the absence of RNase H1 (unpublished, A.A. and R.C., 2000). Alkylating agent sensitivity is increased in either *RNH1* or *RNH2L* strains with the double deletion strain being very sensitive. For both *E. coli* and *S. cerevisiae*, it seems that RNA–DNA hybrids can be recognized by either of the two RNases H with perhaps different routes of resolution of the hybrids. *RNH2L*Δ strains also exhibit a slight increase in frequency of mutagenesis.[26]

Solution-Based Assay

RNase H1 activity is undetectable in crude extracts of *S. cerevisiae* using a solution-based assay and measuring acid-soluble products.[6] For the same protein expressed in *E. coli*, this type of assay is quite useful. RNase H1 activity is readily detected when using the gel-renaturation assay described below.

To make extracts of *S. cerevisiae*, a 5 ml mid log phase culture of cells is centrifuged at 3000 rpm, resuspended in ice-cold sterile water, and recentrifuged. The pellet is resuspended in an equal volume (for 5 ml of culture usually 200 μl) of extraction buffer (50 m*M* Tris-HCl pH 7.9, 1 m*M* dithiothreitol (DTT), 1 m*M* phenylmethylsulfonyl fluoride (PMSF), containing leupeptin, pepstatin, and aprotenin each at 5 μg/ml). Acid-washed glass beads (Sigma, St. Louis, MO) are added

[24] M. Itaya and R. J. Crouch, *Mol. Gen. Genet.* **227,** 433 (1991).

[25] M. Lussier, A. M. White, J. Sheraton, T. diPaolo, J. Treadwell, S. B. Southard, C. I. Horenstein, J. Chen Weiner, A. F. J. Ram, J. C. Kapteyn, T. W. Roemer, D. H. Vo, D. C. Bondoc, J. Hall, W. W. Zhong, A. M. Sdicu, J. Davies, F. M. Klis, P. W. Robbins, and H. Bussey, *Genetics* **147,** 435 (1997).

[26] J. J. Z. Chen, J. Z. Qiu, B. H. Shen, and G. P. Holmquist, *Nucleic Acids Res.* **28,** 3649 (2000).

until the liquid is absorbed. Cells are broken by rapid mixing in a vortex with best breakage when performed using glass tubes. Mixing is for 1 min duration followed by 1 min on ice. The process is repeated until the cells are substantially broken as observed in a microscope. This normally requires four 1-min periods of agitation. At this point 200 μl of extraction buffer is added, and one additional period of agitation is performed. The liquid (together with a small amount of glass beads and cell debris) is removed with a Pasteur pipette and placed in a clean tube. A second addition of 200 μl of extraction buffer, vortexing, and removal of the liquid is carried out. The combined fractions are centrifuged at 4° for 5 min at 3000 rpm. For soluble-based assays, 1 μl of a 1:100 dilution of the extract is placed in a 10 μl reaction mixture as described below or for the gel-renaturation assay 20 μl of the crude extract is used. Most of the activity found in crude extracts when assessed using the soluble-based assay is derived from RNase H2 as shown by a decrease of 70% in activity when using strains deleted for *RNH2L*. No discernible difference in activity is observed when *RNH1*Δ strains are used.

Gel-Renaturation Assay

One of the more useful assays for RNases H goes by various names (*in situ* gel assay, activity gel assay, etc.). We described this assay in our laboratory as the gel-renaturation assay because it emphasizes the need for the protein to renature in order to be detected. Figure 2 A shows the results obtained from an assay of crude extracts of XBE12 and related tissue culture cells and the result of exposure of the gel to X-ray film (bottom of Fig. 2 A) and the reverse image (top of Fig. 2A). The reverse image makes weak bands more obvious. There are several reasons for showing this particular assay. First, it points out that the activity of RNase H can be detected without any purification: simply boil the cells in the SDS- sample buffer and load on the gel. Second, the subunit molecular weight of the enzyme can be approximated from its mobility in the gel. There are several important aspects of knowing the molecular mass of the protein. First, it can be distinguished from other RNases H in the cell (e.g., if *E. coli* RNase H is expressed in a human cell line, the size of the human cellular enzyme and that of *E. coli* are so different that identifying the presence of the bacterial enzyme is simple). Second, it is clear that the activity observed is contributed by a single polypeptide, either as a monomer or a multimer of a single polypeptide since the gel is a denaturing SDS gel. Of course, there is the outside chance that two different subunits of the enzyme comigrate. Third, the amount of protein applied to the gel is important. There is a general tendency to attribute some quantitative aspect to the data. However, it is difficult to do so as shown by the relatively uniform appearance of the band marked "U" while the band marked "L" changes from sample to sample in a manner somewhat related to the amount of protein applied to the gel. Fourth, the smear of degradation of the substrate, seen at the highest levels of protein applied

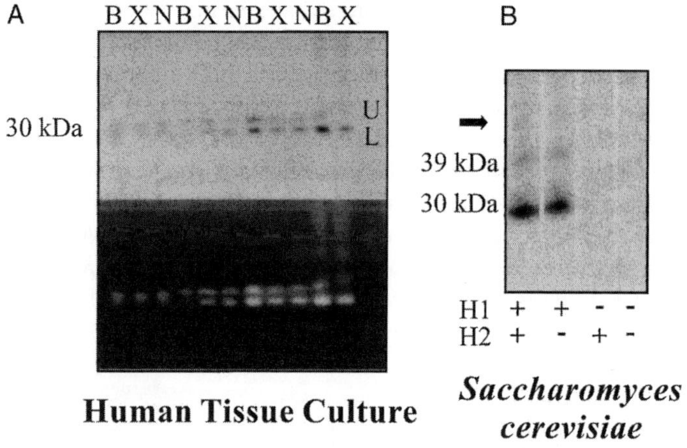

FIG. 2. Gel-renaturation assay. (A) Samples from human tissue culture cells were obtained as described in the text and applied to the gel. Each set of three from left to right is from B = XP12BE, X = XPA, and N = Normal. U and L refer to upper and lower bands as described in the text. The samples correspond to 1 (lanes 1–3), 5 (lanes 4–6), 10 (lanes 7–9), and 20 μl (lanes 10–11) of extract applied to the gel. XPA is from a xeroderma pigmentosa group A cell line, XP12BE is a Group A cell line that is hypersensitive to UV irradiation, and N is a normal cell line. (B) Extracts were prepared from *S. cerevisiae*, wild-type and various deletion strains (noted as "+" or "−"), by the procedure described in the text for soluble-based assays. The nominal molecular masses of the proteins are indicated with an arrow pointing to a weak band, most likely a multimer of RNase H1, an interpretation based on its absence in *RNH1*Δ strains.

to the gel, is common when high levels of RNase H activity are present. Often, we see complete loss of radioactivity trailing the RNase H when we are examining expression of an enzyme in *E. coli*. Another problem with attempts to quantify the results arises from the problems seen when protein purification is being followed. We have shown that the detection limits of *E. coli* RNase HI can be increased for the purified enzyme when we add a protein that migrates faster in the gel than *E. coli* RNase HI, presumably due to protection of the RNase H from chemicals present in the gel.[27] Thus, it often requires less protein to detect RNase H activity in crude extracts compared with highly purified fractions.

For most organisms when crude extracts are examined, there is one major activity detected, but in some cases such as for *S. cerevisiae,* more than one band is visible (Fig. 2B): one at 30 kDa and a slower migrating activity at 39 kDa. Both bands are absent in *rnh1*Δ strains and are therefore RNase H1. The 39 kDa protein is the primary protein product, whereas the 30 kDa protein is derived from cleavage of the 39 kDa protein, about 5 kDa upstream of the RNase H domain. The ratio of the two bands is variable depending on the method of extraction. With the use of extracts made by treatment with zymolase (as shown in Fig. 2B), most of

[27] L. Y. Han, W. P. Ma, and R. J. Crouch, *Biotechniques* **23,** 920 (1997).

the activity is found in the 30 kDa region. To have the highest proportion of activity in the 39 kDa band, cells need to be boiled directly in the gel loading buffer, and for greater extraction of protein, glass beads normally used for breaking yeast cells need to be included. The protease-sensitive region also is cleaved at or near the same site as that seen for extracts from *S. cerevisiae* when the protein is expressed in *E. coli*. The properties of the 30 kDa and 39 kdDa forms of the enzyme are quite different, particularly noteworthy is the strong binding of the 39 kDa protein to duplex RNA (see Northwestern assay). The two forms of the protein also differ in detection in the gel-renaturation assay, in which the 30 kDa protein gives a stronger signal on a weight basis than the 39 kDa enzyme, possibly due to a higher rate of refolding of the former.

Substrate Preparation

Two types of substrates are used for the gel-renaturation assay.

^{32}P-labeled poly(A) is synthesized using *E. coli* RNA polymerase with poly(dT) as template producing poly(rA)-poly(dT) as follows. First, mix

423 μl of H_2O
5 μl of 1 *M* $MgCl_2$
0.5 μl of 1 *M* 2- mercaptoethanol
10 μl of polydT (20 A_{260} units)
5 μl of 0.01 *M* ATP
50 μl of [α-^{32}P] ATP (10 mCi/ml)
2 μl of *E. coli* RNA polymerase (8 Units/μl) USB or recently Epicentre (core enzyme) (Madison, WI)

Remove a 5 μl aliquot into 455 μl of 0.01 *M* Tris-HCl pH 7.5 and count 10 μl of the dilution to determine the total radioactivity present.

Incubate the reaction for 1 hr at 37°. Add 10 μl of 10% sodium dodecyl sulfate (SDS) and 500 μl of chloroform. Vortex the mixture to inactivate any proteins present. The aqueous phase is applied to a Sephadex G-50 column (0.8 × 27 cm) equlibrated with 0.05 *M* NaCl, 0.01 *M* Tris, pH 7.9, 0.1 m*M* EDTA. We use plastic 10 ml pipettes for the column with glass wool placed at the bottom to secure the Sepharose in the column. Fractions of approximately 0.5 ml are collected with the poly(rA)-poly(dT) product emerging in fraction 8–12. A second peak of unincorporated ATP will follow. When the drop size decreases dramatically (due to elution of the SDS), chromatography is stopped. Aliquots of the fractions are counted in a liquid scintillation counter (usually 5 μl of each fraction in 5 ml of scintillation fluid). Peak fractions are combined and an aliquot is counted. The radioactive ATP added to the reaction mixture contributes a negligible quantity of ATP compared with that of the unlabeled ATP (5 μl of 0.01 *M*). Typically, we are adding 1–2 × 10^9 cpm and 5 × 10^4 pmol of ATP producing a product with

$2–4 \times 10^4$ cpm/pmol AMP residue. Yields are usually in the range of 0.5×10^8 cpm (1.7 nmol AMP).

A 10 μl aliquot is assessed for acid solubility by precipitating two 5 μl aliquots in separate 1.5 ml Eppendorf tubes. To the poly(rA)poly(dT) aliquot, add 10 μl of bovine serum albumin (BSA, 10 mg/ml), immediately mix and add 100 μl of cold (4°) 20% trichloroacetic acid (TCA). Tubes are mixed by vortexing and placed on ice for a minimum of 5 min. The samples are centrifuged in a microcentrifuge for 15 min; 100 μl of the supernatant is removed and counted in a scintillation counter. Typical levels of nonprecipitable radioactivity are around 0.1–0.3%.

Fluorescent Labeled RNA–DNA Hybrid Substrate.

Synthesis of fluorescent RNA. Digest plasmid DNA (T7 promoter followed by about 1 kbp of DNA prior to restriction enzyme digestion site) with restriction enzyme of choice (depends on plasmid DNA). We have used the *RNH1* gene of *S. cerevisiae* cloned into pET3a and digested with *Eco*RI. Depending on the method of preparation of the plasmid DNA it may contain RNase A. To eliminate the RNase activity, the DNA can be treated with proteinase K after digestion of the DNA with the restriction enzyme, followed by phenol extraction.[28] Precipitate DNA with ethanol and suspend in H_2O at a concentration of 0.5–1 mg/ml. See Ref. 27.

Reaction mixture

10 μl of H_2O
2 μl of 10× buffer (Ambion MAXIscript kit, Austin, TX)
1 μl of RNase Inhibitor (Ambion)
1 μl of 10 m*M* NTP mix (Ambion)
3 μl of 1 m*M* BODIPY-TR-14-UTP (Molecular Probes, Eugene, OR)
1 μl of T7 RNA polymerase (Ambion kit)

Mix and incubate 2 hr at 37°. The reaction is terminated by addition of 1 μl of 0.5 *M* EDTA. The reaction is adjusted to 0.05% SDS by adding 10% SDS (1 μl) followed by phenol extraction and ethanol precipitation. The RNA is suspended in H_2O to a concentration 1 μg/μl. Using 1 μg of linearized DNA, we obtain between 5 and 20 μg of RNA.

RNA–DNA hybrid synthesis

To a 1.5 ml microfuge tube add:

12 μl of H_2O
1 μl of DNA oligo nucleotide primer (6 pmol/μl)
1.5 μl of BODIPY-TR-14 RNA

[28] J. S. Crowe, H. J. Cooper, M. A. Smith, M. J. Sims, D. Parker, and D. Gewert, *Nucleic Acids Res.* **19,** 184 (1991).

Heat to 70° for 10 min and chill quickly by placing tube in ice water.
Centrifuge briefly to collect contents in bottom of tube.
Add:

4 μl of 5X Buffer (250 mM Tris-HCl (pH 8.3), 375 mM KCl, 15 mM $MgCl_2$, 50 mM DTT)
1 μl of 10 mM dNTP (adjusted to pH 7)

After mixing, place the tube in a 42° water bath for 2 min.

Add 1 μl of reverse transcriptase (200 units SuperScript II; Life Technologies, Gaithersburg, MD).

Incubate at 42° for an additional 50 min, then heat the sample to 70° for 15 min to stop the reaction and inhibit the reverse transcriptase. The RNA–DNA hybrid is then phenol extracted and ethanol precipitated. The yield is typically 3 μg, and the hybrid is stored at −20° in 0.05 M NaCl, 0.01 M Tris-HCl, pH 7.9.

RNA–DNA hybrid formation can be tested by demonstrating sensitivity to RNase HI of *E. coli,* followed by SDS–gel electrophoresis of the RNA–DNA hybrid using undigested hybrid as control. See section on solution-based assay for details of conditions.

Gel Polymerization

For fluorescently labeled RNA–DNA hybrids, we prepare 15% SDS–polyacrylamide gels according to Laemmli[29] except that BODIPY-TR-14 RNA–DNA hybrid (1.5 μg) is included in the running gel. Gels for this amount of RNA–DNA hybrid are 8 × 8 × 0.1 cm with the resolving gel being 6 ml and the stacking gel 1.5 ml. Electrophoresis is carried out at 80 V for the first 40 min and then at 150 V for an additional 90 min. For ^{32}P-labeled poly(rA)-poly(dT), the gels are 10 × 10 × 0.2 cm and contain about 0.25 nmol (AMP) of substrate—about 2×10^6 cpm. Electrophoresis of the samples in the ^{32}P-containing gels is for 18 hr at 50 V. In both instances, we include one lane with Rainbow Dye Molecular Weight Markers (Amersham, Arlington Heights, IL). The difference in electrophoresis times and voltages is for convenience.

Sample Preparation

For *E. coli* expressed proteins, cells are grown and extracts are made from cells expressing RNases H in a constitutive manner or following induction. Cells are centrifuged and 2× SDS-sample buffer [0.1 M Tris-HCl, 0.2% SDS (w/v), 0.28 M 2-mercaptoethanol, 20% glycerol (v/v)] is added. For each ml of culture, the cells are resuspended in 100 μl of 2× SDS-sample buffer and 3–20 μl is

[29] U. Laemmli, *Nature* **227,** 680 (1970).

loaded on the gel. For *S. cerevisiae* cells, samples are produced as described for the soluble-based assay or by addition of 2× SDS-sample buffer to cells together with acid-washed glass beads treated alternately by placing in a boiling water bath and then removing every 15 sec and briefly agitated using a vortex mixer for 15 sec. This process is repeated several times. The latter procedure is more rapid and a higher proportion of the activity is found in the 39 kDa region of the gel. Problems in the renaturation gel assay are sometimes noted which we attribute to contamination of dyes (bromphenol blue or xylene cyanol FF) with nucleases. Therefore, no dye is present in any of the samples.

Processing of Gels after Electrophoresis

The gels are removed from the glass plates and placed in a buffer containing 0.05 *M* NaCl, 0.01 *M* $MgCl_2$ (or 0.002 *M* $MnCl_2$),[30] 20% glycerol, 0.0001 *M* EDTA, 0.01 *M* Tris-HCl (pH 7.9), 0.001 *M* 2-mercaptoethanol in a glass baking dish. Processing can be enhanced if 2-propanol (10%) is included in the first two changes of buffer. Gels are incubated at room temperature for 2 hr and the buffer is replaced with a fresh solution. Since the fluorescent-labeled hybrid is in a thinner gel, renaturation occurs faster. Thus the process can be completed in a few hours and scanning takes only 15–20 min. For the ^{32}P-labeled hybrid, the gels are removed in the morning and processed until evening and the gel can be exposed to X-ray film overnight. In either case, the gels can be returned to the glass dish and incubation continued since neither detection method inhibits the appearance of new bands. When a result is obtained, the gel can be stained with Coomassie blue and the relative amounts of proteins in each lane can be observed.

Complementation of Temperature-Sensitive Growth Defect of *rnhA* Mutants of *Escherichia coli*

Strains of *E. coli* carrying the *rnhA-339::cat* mutation can grow at both 32° and 42°,[31] but when a temperature-sensitive *recB* mutation is placed in combination with the *rnhA-339::cat,* the strain becomes temperature sensitive for growth (Fig. 3). We do not understand the synthetic lethality resulting from mutations in genes that are not required for growth when mutated in the absence of the other. Combining the *rnhA-339::cat* mutation with mutations in *recG, topA,* or, in some genetic backgrounds, *rnhB* also results in lethality. However, only the *rnhA-339::cat, rnhB716::kan,* and *rnhA-339::cat recB270*(Ts) double-mutant strains have been used to examine function of RNases H obtained from genes cloned

[30] P. Frank, C. Cazenave, S. Albert, and J. J. Toulmé, *Biochem. Biophys. Res. Commun.* **196,** 1552 (1993).

[31] M. Itaya and R. J. Crouch, *Mol. Gen. Genet.* **227,** 424 (1991).

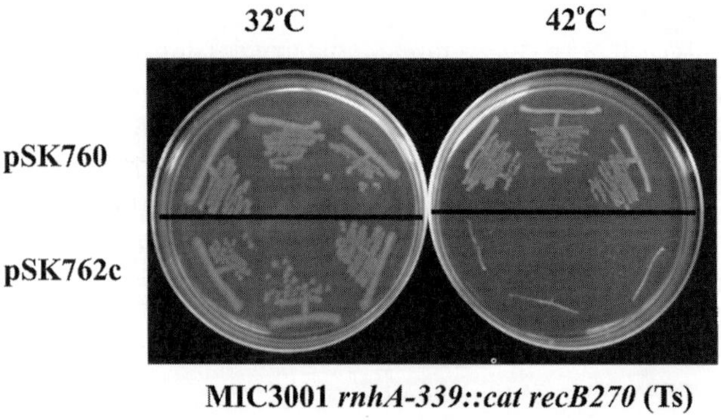

FIG. 3. Complementation assay for genes producing RNase H activity in *E. coli*. LB plates containing ampicillin were incubated at 32° or 42° following streaking of MIC3001 containing plasmids pSK760 (*rnhA*$^+$) or pSK276c (*rnhA*$^-$).

from various organisms, with the last being the one most extensively used. Here, we will describe the use of a strain such as MIC3001 for testing of genes for production of RNases H.

The results shown in Fig. 3 are from transformants of MIC3001 with plasmids containing the wt *rnhA* gene (pSK 760) or a gene in which the *Bam*HI site in the middle of the *rnhA* gene was digested and religated after filling in the site (pSK760c). In most instances, we observe growth at 42° in the first streak with no growth of individual colonies in the cross streak, unless a functional RNase H is supplied.

There are some points to remember when using MIC3001. First, the temperature sensitivity of the strain can be overcome by addition of NaCl to the LB plate; therefore, we use a recipe for LB plates in which the NaCl concentration is 5 g/liter. Second, temperature-sensitive growth is not found when large numbers of cells are spread on a plate. It appears that cells in a colony produce a product that is useful to its neighbors, or degrades a growth inhibitor. When sufficient numbers of cells are nearby, the inhibitor is decreased permitting growth of the colony. Spreading 10^4 cells on a standard 8.5 cm diameter plate gives a lawn of cells, but 100 colonies per plate will yield very few if any colonies. Another manifestation of the cell-density contribution to growth is the growth of all cells to form colonies if the plates on which the cells are placed are too cold. As the plate warms, cells form small colonies that continue to expand at 42°. Therefore, plates should be prewarmed to 42°. Third, for reasons yet to be determined, *rnhA* mutant strains are sensitive to ampicillin,[32] which may account for problems when using high copy plasmids for examination of RNase H activity by the complementation assay. Itaya prefers

[32] T. Katayama and T. Nagata, *Mol. Gen. Genet.* **223,** 353 (1990).

to use pBR322 as a vector rather than the higher copy pUC-like vectors (personal communication, 2000), but if a gene has been cloned that expresses RNase H, plamids such as pET15b (Novagen, Madison, WI) do exhibit complementation of the temperature-sensitive growth defect.

The *RNH1* gene of *S. cerevisiae* was obtained from a library of *S. cerevisiae* DNA made by cloning partial *Sau*3A DNA fragments into YEp13 at the *Bam*HI site. Almost the complete coding sequence was obtained. However, transcription was probably initiated from the Tet promoter of the vector and the first methionine codon encountered in the gene was near the beginning of the RNase H domain. In contrast, the *RNH2L* gene was not obtained from the initial screen but when later cloned and expressed in MIC1066, the protein failed to complement the ts-defect (A.A and R.C., unpublished, 2000). MIC1066 is a strain with *rnhA-339::cat recB270*(Ts) mutations and λ(DE3) integrated into the chromosome. When induced by addition of ITPG, the T7 RNA polymerase transcribes from plasmid-borne genes linked to a T7 promoter and the product is highly expressed. In the absence of induction, the basal level of transcription often produce enough RNase H to permit growth of *rnhA-339::cat recB270*(Ts) strains at 42°.

Northwestern Assay

Northwestern assays detect protein–nucleic acid interactions after separation of the proteins in a denaturing SDS–PAGE, and subsequent transfer to a membrane. The membrane-bound proteins are then renatured and incubated with labeled nucleic acid. After incubation for a few hours the unbound nucleic acid is removed, and the membrane is washed and exposed to X-ray film or scanned by instruments capable of detecting fluorescence or chemiluminescence to observe binding. This procedure was first developed to determine protein–RNA interactions, and from that came the name Northwestern. However, any nucleic acid can be used, and we have used this technique to check for binding of *S. cerevisiae* RNase H1 to dsDNA, ssDNA, RNA–DNA hybrids, ssRNA, and dsRNA. We found the strongest binding was to dsRNA and RNA–DNA hybrids, although the protein also binds weakly to dsDNA and ssDNA, and almost not at all to ssRNA.[33]

The Northwestern assay, such as the one shown in Fig. 4, is a convenient, fast, and easy procedure for detection of protein–nucleic acid interactions and can even use crude extracts of cells as the source of protein, obviating the need to purify the protein in each sample prior to the assay. As shown in Fig. 4, N-176 protein (Fig. 1) binds equally well when using purified protein or protein found in crude extracts of *E. coli* lysates. Some protein–nucleic acid complexes may persist throughout purification, making it difficult if not impossible to determine the types of nucleic acids to which the protein will bind. The denaturation step separates

[33] S. M. Cerritelli and R. J. Crouch, *RNA* **1,** 246 (1995).

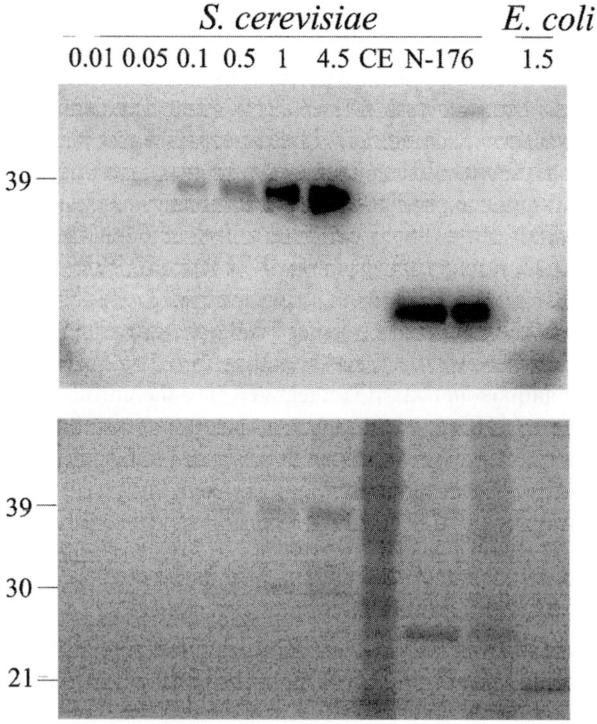

FIG. 4. Northwestern Assay for duplex RNA binding. Northwestern assays are shown for *S. cerevisiae* RNase H1, N-176, and *E. coli* RNase HI. Various quantities (μg) of purified *S. cerevisiae* RNase H1 as indicated were examined by the Northwestern assay as described in the text. The renaturation shown here was for 1 hr at room temperature. In addition, N-176 (Fig. 1), crude extract of *S. cerevisiae* (CE), and purified *E. coli* RNase HI were present on the gel. The upper part of the figure shows the binding of the proteins to ^{32}P-labeled poly(rA)-poly(dT); the bottom half shows a gel run in parallel with the same proteins stained with Coomassie blue. Nominal molecular masses (in kDa) are indicated.

the two components, thereby improving association of the labeled nucleic acid. When we started purifying *S. cerevisiae* RNase H1, we observed that the full-length protein copurified with an unknown nucleic acid component. Attempts to dissociate the complexes were unsuccessful at the time, making the Northwestern assay the method of choice for measuring the interaction of the protein with various nucleic acids. Moreover, the molecular weight of the protein binding to the nucleic acid can be determined, simplifying the identification of the protein(s) that interacts with a particular nucleic acid from a mixture of polypeptides. The 39 kDa *S. cerevisiae* RNase H1 is specifically cleaved, producing a 30 kDa fragment that, as mentioned above (see Gel Renaturation Assay), retains RNase H activity. A mixture of the two forms is observed in crude extracts as well as in highly purified fractions as a result of proteolytic cleavage of ~80 amino acids from the N terminus, removing the major motif through which the protein interacts with duplex RNA (Fig. 4). Standard gel-shift procedures would not have yielded this information with the ease of the Northwestern assay.

What are the limits of detection of *S. cerevisiae* RNase H1 by the Northwestern assay? Various amounts of RNase H1 were used for SDS–PAGE separation and then transferred to a membrane for incubation with ^{32}P-labeled poly(rA)-poly(dT). In Fig. 4, it is possible to see binding to as little as 0.05 μg of *S. cerevisiae* RNase H1, and in longer exposures even the binding of 0.01 μg of protein can be detected (data not shown). The relative quantities of the nucleic acid bound to the *S. cerevisiae* RNase H1 applied to the gel suggests that there is no great effect of damage of the protein upon electrophoresis as we have noted for *E. coli* RNase HI in the gel-renaturation assay. We have not examined the influence of adding a fast-migrating protein to the sample prior to denaturation in SDS-sample buffer, and, thus, it is possible that the sensitivity of the Northwestern assay could be increased. Nevertheless, no binding to poly(rA)-poly(dT) was observed when we used crude extracts of *S. cerevisiae* (Fig. 4, CE). The protein is not very abundant, and it is very unstable, making it difficult to detect. However, gel-renaturation assays are the best way to see RNase H1 activity from extracts of *S. cerevisiae*, indicating that gel-renaturation assays are more sensitive than Northwestern assays. Some of the difference in level of detection may be due to the extra step of transfer of the proteins to the membrane required for the Northwestern assay, a procedure that is never 100% efficient. The N-176 protein comprises the first 176 amino acids of the protein (Fig. 1), including the duplex-RNA binding domain, and it is sufficient for binding as we can see in Fig. 4, demonstrating that binding occurs via the duplex-RNA binding domain. Note that 1.5 μg of *E. coli* RNase HI bound very weakly. Although originally we were unable to detect binding of *E. coli* RNase HI to RNA–DNA hybrids in Northwestern assays, now, with overnight renaturation of the protein on the membrane, we have been able to observe weak but detectable binding of the *E. coli* protein.

There are limitations of the Northwestern assay that restrict its usefulness. It requires the protein to renature after separation in denaturing SDS–PAGE and transfer to a membrane, it requires the protein to be composed of a single polypeptide or two subunits that comigrate in the gel, and it generally requires strong and stable binding for an interaction to be detected. The first two limitations are also found for the gel-renaturation assay detecting RNase H activity, while the third is unique to the Northwestern assay and is illustrated when comparing the results of *E. coli* and *S. cerevisiae* RNases H. Both enzymes degrade RNA of RNA–DNA hybrids, yet only the full-length *S. cerevisiae* enzyme binds strongly to duplex nucleic acids in the Northwestern assay. These assays are not quantitative, because the binding detected depends not only on the affinity of the protein for a particular nucleic acid, but also on the ability of the protein to renature after removal of SDS and refold into an active form. However, it is possible to compare the binding affinity of a protein to different nucleic acids. For binding estimation, the intensity of phosphor emission from nucleic acid bound to proteins attached to the membrane can be quantified using a PhosphorImager and then related to the amount of protein present in the membrane.

Substrate Preparation

We have used various nucleic acids labeled in different manners in our Northwestern assays. The nucleic acid can be labeled and purified by any established procedure, either internally or at the 5′ end. We will briefly describe our most commonly used procedures.

To compare the binding of RNA–DNA hybrids and dsRNA of same length and sequence, both nucleic acids can be prepared by *in vitro* transcription, using the plasmid pSPORT 1 (Life Technologies, Rockville, MD), which contains two promoters in opposite orientations. One to 2 μg of plasmid DNA in 30 μl of water is linearized with *Nar*I [New England Biolabs (NEB), Beverly, MA] for transcription using the SP6 promoter, or with *Sph*I (NEB) for transcription from the T7 promoter by incubation overnight with the corresponding enzyme in the buffer recommended by the manufacturer. Completion of DNA digestion is checked by running 2 μl of the digested DNA in a 1% agarose gel. Linear plasmids are purified by bringing the volume to 100 μl and adding an equal volume of phenol–chloroform (1 : 1 mixture). The mixture is vortexed to inactivate any proteins present and centrifuged. The DNA present in the aqueous phase is precipitated by adding 2.5 volumes of chilled 99.9% ethanol and kept at −70° for 20 min. The DNA is centrifuged for 10 min at 4° and washed with 100 μl of 75% ice-cold ethanol, dried in a DNA SpeedVac (Savant, Farmingdale, NY), and resuspended in 10 μl of DEPC (diethyl pyrocarbonate)-treated water. Purified linear DNA is used for *in vitro* transcription with the corresponding polymerase according to a procedure previously described by Milligan and Uhlenbeck.[34,35] Only the RNA made from the SP6 promoter is usually labeled with [α-^{32}P]CTP.

Mixture for SP6 transcript labeled with [α-^{32}P]CTP:

10 μl DEPC water
10 μl purified DNA linearized with *Nar*I
20 μl of 5× Transcription Optimized buffer (Promega, Madison, WI)
10 μl of 100 m*M* DTT (Promega)
6 μl of 40 U/μl rRNasin (Promega)

[34] J. F. Milligan and O. C. Uhlenbeck, *Methods Enzymol.* **180,** 51 (1989).

[35] J. F. Milligan, D. R. Grobe, G. W. Witherell, and O. C. Uhlenbeck, *Nucleic Acids Res.* **15,** 8783 (1987).

13 μl of 2.5 m*M* NTPs minus CTP; the 2.5 m*M* NTP mixture is prepared by mixing equal volumes of water, 0.1 *M* NaOH, ATP, GTP, and UTP, all 25 μmol, pH 7.5 (Amersham Pharmacia Biotech, Piscataway, NJ)
2.5 μl of 0.25 m*M* CTP prepared by diluting 1:10 CTP 25 μmol, pH 7.5 (Amersham Pharmacia Biotech)
20 μl [α-^{32}P]CTP, 3000 Ci/mmol
8 μl SP6 RNA Polymerase 80 U/μl (Promega).

Mixture for unlabeled T7 transcript (same as SP6 mixture except):

33 μl DEPC water
10 μl purified DNA linearized with *Sph*I
13 μl 2.5 m*M* mixture of all NTPs
8 μl T7 RNA Polymerase 80 U/μl (Promega)

Each mixture is incubated at 37° for 2 hr. The DNA is degraded by adding 2 μl of 1 U/μl RQ1 RNase-Free DNase (Promega) and incubating at 37° for 20 min. Proteins are inactivated in both mixtures by adding 100 μl phenol–chloroform (1:1 mixture), vortexing, and centrifuging. The aqueous phase of the radiolabeled mixture is applied to a Nuc Trap Probe Purification Column (Stratagene, La Jolla, CA) equilibrated in STE buffer: 100 m*M* NaCl, 20 m*M* Tris-HCl pH 7.5, and 10 m*M* EDTA. The column is then rinsed with 70 μl of STE buffer. The liquid from the sample and the rinse are combined, 12 μl of 3 *M* sodium acetate is added, and the mixture is precipitated in 2.5 volumes of ice-cold 99.9% ethanol. The pellet is washed with 100 μl of 75% chilled ethanol, dried in a DNA SpeedVac (Savant), and resuspended in 100 μl of DEPC-treated water. Incorporated radioactivity is determined by counting a 1 μl of sample in 5 ml of scintillation fluid in a liquid scintillation counter. Acid-soluble material is measured by precipitating two 1 μl aliquots in separate 1.5 ml Eppendorf tubes. To the sample is added 10 μl of BSA (10 mg/ml) and 100 μl of cold 20% trichloroacetic acid. Tubes are mixed and placed on ice for 15 min, then centrifuged in a microcentrifuge for 15 min. A sample (100 μl) of the supernatant is removed and counted in a scintillation counter. Usually no more than 1–2% is acid soluble.

The unlabeled transcript after phenol–chloroform extraction is ethanol precipitated in the same manner as the radiolabeled sample, omitting the NucTrap purification step. The amount of cold transcript is determined by measuring A_{260} of 1 : 100 dilution. Assuming that 1 A_{260} is equal to 40 μg of RNA, we usually obtain more than 3 mg/ml.

To form RNA–DNA hybrids, the labeled SP6 transcript is used as template and the T7 "primer" oligonucleotide complementary to the transcript as primer. To form a primer/template duplex, 1 μl T7 primer at 68 pmol/μl and 11 μl SP6 radiolabeled transcript are mixed and heated at 70° for 10 min and quickly chilled on ice. Synthesis of the cDNA is the same as that for preparation of fluorescent-labeled RNA–DNA described above under Gel-Renaturation Assay.

To form dsRNA, equal amounts (typically 10 μl) of the SP6 (labeled) and T7 (unlabeled and in an excess) runoff transcripts are annealed in 50 m*M* Tris-HCl pH 8.3, 75 m*M* KCl, and 3 m*M* $MgCl_2$ by heating to 95° for 5 min and slow cooling.

RNA–RNA and RNA–DNA duplexes are treated with RNase A and T1, to remove unpaired ends, and make both substrates of equal length. Twenty μl of dsRNA or RNA–DNA hybrids, both in 50 m*M* Tris-HCl pH 8.3, 75 m*M* KCl, and 3 m*M* $MgCl_2$ buffer is incubated with 1 μl of 20 μg/ml RNase A diluted from the 10 mg/ml stock solution from 5 Prime-3 Prime (Boulder, CO) and 1 μl of 12 U/μl RNase T1 diluted from 1292 U/μl stock solution from 5 Prime-3 Prime. The 1:100 dilution of both enzymes is performed in buffer: 50% glycerol, 10 m*M* Tris-HCl pH 7.9, 40 m*M* DTT, 0.1 mg/ml BSA. After incubation for 30 min at 37°, followed by phenol extraction and ethanol precipitation, the RNA duplexes are resuspended in 50 μl of 100 m*M* NaCl, 20 m*M* Tris-HCl pH 7.5, and 10 m*M* EDTA. To check the integrity of the nucleic acid, a 1-μl sample in 1× RNA gel loading solution (Quality Biological Inc., Gaithersberg, MD, USA) is fractionated in a nondenaturing 12% polyacrylamide–0.5xTris–borate–EDTA (TBE) 10 × 10 × 0.2 cm. Electrophoresis is carried out at 150 V until the bromphenol blue in the loading buffer migrates two-thirds of the way through the gel. The gel is dried and exposed to X-ray film. For maximum purity of the nucleic acid, a preparative gel can be made with a larger loading well, and the complete sample can be fractionated. After electrophoresis as above, the wet gel is briefly exposed to X-ray film and the radioactive band of RNA–DNA hybrids or dsRNA is cut out of the gel and extracted and purified using the RNaid kit protocol (Bio 101, Vista, CA, USA):

Crush gel and add three volumes of RNA binding salt (Bio 101).
Incubate for 30 min at 37° with shaking.
Filter with Acrodisk PF (0.8 μm/0.2 μm) to remove acrylamide.
Add 10 μl RNA matrix and mix gently for 10 min at room temperature.
Spin down and wash with 400 μl RNase-free wash. This process is repeated for a total of three washes.
Dry the pellet in SpeedVac for 2 min.
Resuspend pellet in 50 μl DEPC water.
Incubate at 42° for 5 min, mixing on a platform rotating shaker.
Spin down and recover supernatant.
Extract pellet with 50 μl DEPC water as before.
Combine both supernatants and store at −20° until use.

To assess binding of RNA–DNA hybrids, we have used the polyrA-polydT substrate, which was prepared as described above (see Gel Renaturation). Short duplex nucleic acids are formed from chemically synthesized oligonucleotides (either ribo or deoxyribo) with one oligonuclotide labeled at its 5′-end with ^{32}P using T4 polynucleotide kinase and [γ-^{32}P] ATP as described by the manufacturer (NEB).

Northwestern Procedure

We use 0.5–5 μg of purified protein, 10 μg of crude extract of cells overexpressing the protein of interest (or larger amounts of extract if the protein is in low copy) boiled in 2× SDS sample buffer (0.1 *M* Tris-HCl pH, 0.2% SDS, 0.28 *M* 2-mercaptoethanol, 20% glycerol) for 5 min, then separated in a 10–20% polyacrylamide gel containing SDS. One lane is used for Rainbow Dye Molecular Weight Markers (Amersham, Arlington Heights, IL). Electrophoresis is carried out at 150 V for about 60 min. We run two gele in parallel loaded identically. When electrophoresis is concluded, we stain one gel with Coomassie blue to visualize proteins, or with SYPRO Orange (Molecular Probes), a fluorescent dye that can be quantified using the STORM machine (Molecular Dynamics, Sunnyvale, CA). The other gel is electroblotted using a semidry blotter onto Immobilon P membranes (Millipore) in Tris–glycine–SDS buffer[29] for 2 hr at 5 V. The denatured proteins bound to the membrane are renatured at room temperature by incubation in a 150 × 25 mm Falcon tissue culture dish containing 70 ml of buffer (50 m*M* Tris-HCl pH 7.9, 100 m*M* NaCl, and 10 m*M* 2-mercaptoethanol). The renaturation step can be at room temperature for 1 hr with changes of buffer every 10 min, but greater sensitivity can be obtained when renaturation is at room temperature for 2 hr with four changes of buffer, followed by an overnight incubation at 4°. The following morning the membrane is blocked at room temperature for 1 hr (two 30-min incubations with one change of buffer) in a 150 × 25 mm Falcon tissue culture dish containing 70 ml of blocking buffer: 10 m*M* HEPES, pH 8.0, 100 m*M* NaCl, 1 m*M* dithiothreitol, 0.1 m*M* EDTA, and 0.2% I-Block (Tropix, Bedford, MA). The membrane is then placed in a 100 × 15 mm Falcon petri dish with 10 ml of blocking buffer containing ^{32}P-labeled nucleic acid at 2×10^5 cpm/ml (1–10 pmol/ml), and incubated at room temperature for 2 to 4 hr. Subsequently, the membranes are washed in a 150 × 25 mm Falcon tissue culture dish containing 70 ml of blocking buffer with changes at 10-min intervals for a total of 40 min. The membrane is wrapped in plastic wrap and exposed to X-ray film for 5 to 16 hours and, in most cases, the filters are also exposed to a storage phosphor screen (Molecular Dynamics). After exposure, the screen is scanned into the PhosphorImager and quantified using the ImageQuant program of Molecular Dynamics.

Although as mentioned above, Northwestern assays are not quantitative, estimation and comparison of binding strength can be done by relating the intensity of phosphor emission from nucleic acid bound to the protein to the amount of protein determined by staining a parallel gel with SYPRO Orange (Molecular Probes), and scanning the gel in the STORM machine using the blue-excited laser.

[26] Ribonucleases H of the Budding Yeast, *Saccharomyces cerevisiae*

By ULRIKE WINTERSBERGER and PETER FRANK

Introduction

The budding yeast, *Saccharomyces cerevisiae,* is a highly useful eukaryotic model organism because of its easy accessibility to genetic experimentation, including classical as well as recombinant DNA methods, and because of the availability of the complete genome sequence. For proteins from other organisms with unclear functions but known amino acid sequences, orthologs of *S. cerevisiae* may help elucidate biological roles. On the other hand, sequences of yeast genes with known functions may serve as guides for searching for orthologs from other organisms (e.g., *Homo sapiens*) with similar properties. Both approaches were successfully used for the isolation and characterization of ribonucleases H.

In yeast cell extracts the presence of ribonuclease H (RNase H) activity, defined as the ability to specifically hydrolyze the RNA strand of a RNA/DNA hybrid, was shown as early as 1973.[1] Attempts to purify the active protein(s) were hampered by the high concentration of proteases in crude yeast cell extracts and by the extreme sensitivity of the enzymes to proteolysis. Using a yeast strain deficient in one protease (strain 20B-12, deficient in Pep4, from the American Type Culture Collection, Manassas, VA) in a classical purification procedure led to enrichment of a RNase H exhibiting a molecular weight around 70,000 in a glycerol gradient, which therefore was called RNase H(70).[2] As soon as recombinant DNA techniques became available the new method was employed for searching for a yeast RNase H gene by complementation of an *Escherichia coli* strain lacking its own gene for the then single known bacterial RNase H (which, since the finding of a second bacterial RNase H[3], has been called RNase HI). This approach led to the isolation of a yeast gene related to that of bacterial RNase HI and therefore was called *RNH1,* coding for the yeast enzyme RNase H1.[4] A further yeast gene coding for a RNase H was found by its relationship to mammalian RNase HI,[5] and, because of difficulties arising with nomenclature, was called *RNH35* according to the molecular mass of its product, RNase H(35).[6] The gene, *RNH70,* for RNase H(70) was cloned only recently.[7]

[1] F. Wyers, A. Sentenac, and P. Fromageot, *Eur. J. Biochem.* **35,** 270 (1973).

[2] R. Karwan, H. Blutsch, and U. Wintersberger, *Biochemistry* **22,** 5500 (1983).

[3] M. Itaya, *Proc. Natl. Acad. Sci. U.S.A.* **87,** 8587 (1990).

[4] M. Itaya, D. McKelvin, S. K. Chatterjie, and R. J. Crouch, *Mol. Gen. Genet.* **227,** 438 (1991).

[5] P. Frank, C. Braunshofer-Reiter, U. Wintersberger, R. Grimm, and W. Büsen, *Proc. Natl. Acad. Sci. U.S.A.* **95,** 12872 (1998).

[6] P. Frank, C. Braunshofer-Reiter, and U. Wintersberger, *FEBS Lett.* **421,** 23 (1998).

0076-6879/01 $35.00

Table I gives a summary of yeast proteins known to exhibit RNase H activity. In addition to the enzymes mentioned above, Rad27p and exonuclease 1 are included in Table I. Rad27p and the mammalian ortholog flap endonuclease, FEN1,[8] cleave specifically at branched DNA structures, in which the single stranded branch, the flap, may also be RNA. This ability is responsible for their involvement in Okazaki fragment processing during DNA replication. In addition these enzymes are also engaged in DNA recombination and repair. The evolutionarily related exonuclease 1, a double-stranded DNA $5' \rightarrow 3'$-exonuclease, was originally found to be involved in DNA mismatch repair and homologous recombination.[9] It is, however, able to also hydrolyze RNA of an Okazaki fragment-like substrate.

The question of why a cell needs more than one enzyme possessing RNase H activity remains to be answered. Remarkably, deletions of the genes *RNH1, RNH35,* and *RNH70* separately, or in pairs, resulted in viable mutants under laboratory conditions.[6,7] This is even the case for yeast cells missing all three genes. Because the RNase H activity, as determined in cell extracts, is significantly lower in cells deleted for *RNH35* than in wild-type cells it was hypothesized that RNase H(35) may be the main RNase H of *S. cerevisiae*.[6] The suggestive idea of RNase H(35) being responsible for hydrolysis of the RNA part of Okazaki fragments during DNA replication was recently tested: although RNase H(35) is able to process model substrates *in vitro,* Rad27p, most probably, is the central enzyme for RNA primer removal *in vivo*.[10] The frequency of reversion of a frame shift mutation was found slightly increased in a *rnh35* deletion mutant,[10] indicating a role of RNase H(35) in mutation avoidance. Deletion of *RNH70* does not change RNase H activity in cell extracts. It was, however, reported recently that such a deletion mutant is defective in 5S rRNA and tRNA-Arg3 maturation.[11] This result indicates that RNase H(70) possesses a second ribonuclease activity, namely that of a highly specific $3' \rightarrow 5'$-exonuclease. Therefore the *RNH70* gene and its product are now also called *REX1* and Rex1p.[11]

Assays for Determining RNase H Activity

For following RNase H activity during purification procedures a simple solution assay is needed, which uses a substrate recognized by many, if not all,

[7] P. Frank, C. Braunshofer-Reiter, A. Karwan, R. Grimm, and U. Wintersberger, *FEBS Lett.* **450,** 251 (1999).

[8] M. R. Lieber, *BioEssays* **19,** 233 (1997).

[9] P. Fiorentini, K. N. Huang, D. X. Tishkoff, R. D. Kolodner, and L. S. Symington, *Mol. Cell. Biol.* **17,** 2764 (1997).

[10] J. Qiu, Y. Qian, P. Frank, U. Wintersberger, and B. Shen, *Mol. Cell. Biol.* **19,** 8361 (1999).

[11] A. van Hoof, P. Lennertz, and R. Parker, *EMBO J.* **19,** 1357 (2000).

TABLE I
PROTEINS OF *S. CEREVISIAE* EXHIBITING RNase H ACTIVITY

Enzyme	Apparent molecular weight[a]	Molecular mass[b]	p*I*[c]-(calculated)	Known enzymatic activities	Preferred divalent cation	Purification from *S. cerevisiae* extracts	Evolutionarily related proteins from other organisms[d]
RNase H1	39 kDa	39,445	9.45	RNase H (Endo)	Mg^{2+}	–	*Hs* RNase HII, prokaryotic RNase HI, retroviral RT associated RNase H
RNase H(35)	35 kDa	34,880	9.09	RNase H (Endo)	Mn^{2+}or Mg^{2+}	–	*Hs* RNase HI, prokaryotic RNase HII
RNase H(70)	70 kDa	62,845	7.01	RNase H (Endo), Exonuclease	Mg^{2+}	+	*Pt* GOR antigen
Rad27p	43 kDa	43,288	9.16	Flap-Endonuclease RNase H (Exo)	Mg^{2+}	+	*Hs* FEN1, bacteriophage T4 RNase H, *Hs* Exo1
Exol	42 kDa	80,170	8.69	Exonuclease, RNase H	Mn^{2+}or Mg^{2+}	+	*Hs* Exo 1, *Hs* ERCC5, *Hs* FEN1

[a] As determined from SDS–gel electrophoresis.
[b] Relative molecular mass as calculated from the deduced amino acid sequence.
[c] p*I*, isoelectric point.
[d] *Hs*, *Homo sapiens; Pt*, *Pan troglodytes;* RT, reverse transcriptase.

polypeptides possessing RNase H activity. In this assay the enzyme is incubated with a RNA/DNA hybrid (^{3}H-labeled RNA synthesized with denatured calf thymus DNA as template). The decrease of trichloroacetic acid precipitable radiolabeled hybrid is then determined by liquid scintillation counting.[2] For the determination of substrate specificity, various defined substrates are used in a liquid assay, and the degradation products are analyzed on a sequencing gel.[10] For testing purity and for discrimination between enzymatically active polypeptides of different molecular mass a gel renaturation assay will be described.[12]

Preparation of Substrates

We will use the following notations: RNA–DNA for a single-stranded polynucleotide comprised of a stretch of RNA covalently joined with a DNA sequence; RNA/DNA (hybrid) for a double-stranded polynucleotide, one strand of which is RNA hybridized to a complementary DNA strand.

^{3}H-Labeled RNA/DNA Hybrid. The synthesis is performed as described[2] (with several modifications) in 200 μl 50 mM Tris-HCl (pH 8.0), 8 mM $MgCl_2$, 75 mM KCl, 1 mM $MnCl_2$, 2 mM dithiothreitol (DTT), 0.2 mM ATP, CTP, GTP, and UTP, respectively, using 100 μg heat denatured calf thymus DNA (95°, 5 min, then shock-cooled in ice) as a template, and 100 μCi [^{3}H]UTP ([5-^{3}H]uridine 5′-triphosphate, 15.6 Ci/mmol, 1 mCi/ml). Polymerization is executed by adding 15.2 units of a commercially available RNA polymerase from *Escherichia coli,* and incubated at 37° for 30 min. The yield of the reaction is calculated by determination of the ratio of acid-precipitable radioactivity to total radioactivity. The hybrid is purified using the Geneclean II Kit (Bio 101 Inc., La Jolla, CA): a sample of 100 μl is mixed with 450 μl 6 M NaI and 100 μl Glassmilk and stored for 10 min at room temperature. Samples are centrifuged and washed three times with 450 μl NEW (NaCl/ethanol/water), which is included in the kit. Elution takes place by four subsequent steps of incubation with 100 μl 10 mM Tris-HCl (pH 8.0) at 40° for 5 min. Eluates, obtained after centrifugation, are pooled and adjusted to 100 mM NaCl. The specific radioactivity is 2.4×10^6 cpm/nmol UTP (3.7×10^6 Bq/46.41 nmol UTP), or 6×10^4 cpm/100 pmol rNTPs (if one assumes that all four nucleotides occur with equal frequency).

RNA–DNA/DNA Hybrid Substrates.[10] We outline the construction of nine RNA–DNA/DNA hybrid and six DNA substrates, as used in the assays described below. These substrates mimic the Okazaki fragment in different dynamic situations (see Fig. 1).[10] They are labeled on the 5′ end using T4 polynucleotide kinase (New England Biolabs, Beverly, MA) and [γ-^{32}P]ATP (New England Nuclear, Boston, MA). Oligonucleotides used are synthesized chemically by an

[12] P. Frank, C. Cazenave, S. Albert, and J. J. Toulmé, *Biochem. Biophys. Res. Commun.* **196,** 1552 (1993).

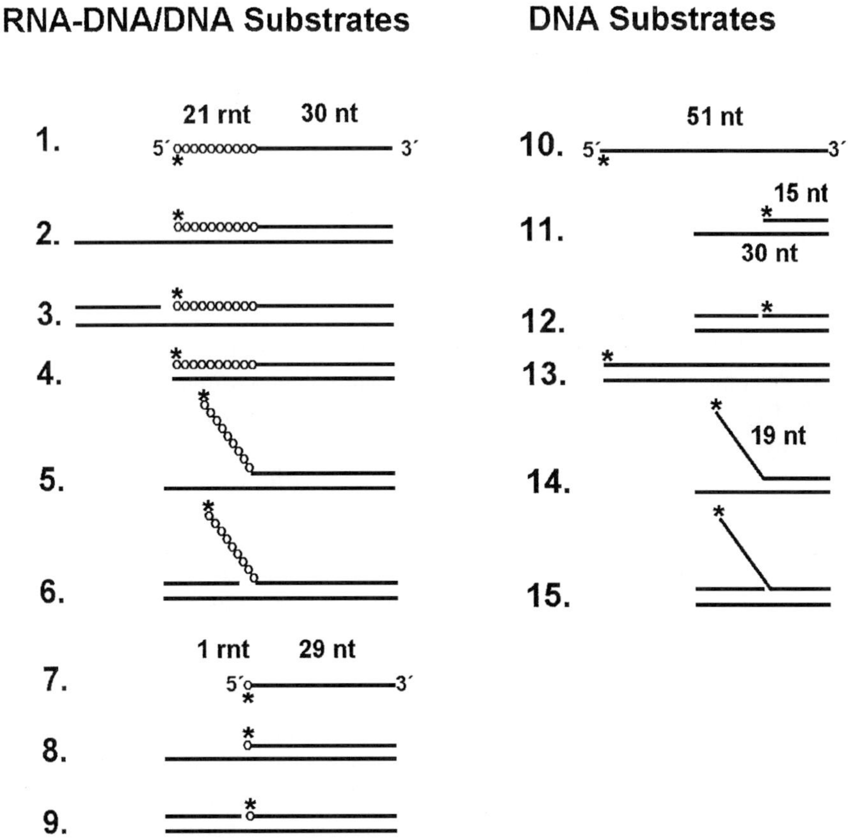

FIG. 1. Schematic representation of substrates for testing specificity of RNases H. Substrates 1–9 are various RNA–DNA/DNA hybrids; control substrates 10–15 are "DNA only" polydeoxyribonucleotides with configurations corresponding to the hybrids. Stretches of circles represent RNA, lines DNA, and asterisks mark the positions of the ^{32}P label; rnt, ribonucleotides, nt, deoxyribonucleotides; the numbers of the substrates correspond to those of Fig. 2.

appropriate custom service center. The labeled 21 ribonucleotide–30 deoxyribonucleotide junction oligo(5′-gggaacaaaagcuugcaugccTGCAGGTCGACTCTAGAGGATCCCCGGGTA-3′) is used as a single-stranded RNA-DNA substrate (substrate 1). It is annealed to the 72-mer DNA strand (5′-TACCCGGGGATCCTCTAGAGTCGACCTGCAGGCATGCAAGCTTTTGTTCCCCATTACGGCTCTCCGAGTTAT-3′) to obtain a 3′ overhang (substrate 2) or a nicked duplex with the upstream complementary DNA sequence (5′-ATAACTCGGAGAGCCGTAATG-3′) (substrate 3). The 51-mer RNA-DNA oligonucleotide also forms a duplex when it is annealed to a complementary DNA strand of identical length (substrate 4). The labeled substrate 1 is also annealed to a partially complementary 51-mer DNA oligo(5′-TACCCGGGGATCCTCTAGAGTCGACCTGCAGTAGACGTCTGAC-ACAGCCGT-3′) to form a pseudo-Y substrate (substrate 5) or a flap substrate (substrate 6) with the presence of an upstream DNA sequence (5′-ACGGCTGT-GTCAGACGTCTA-3′). A single-stranded one ribonucleotide-30 deoxyribonucleotide junction substrate (5′-cTGCAGGTCGACTCTAGAGGATCCCCGGG-TA-3′) serves as substrate 7. It is annealed to a partially complementary 51-mer DNA strand to form a 3′ overhang (substrate 8) or a nicked duplex substrate (substrate 9) with an upstream DNA strand (5′-ACGGCTGTGTCAGACGTCTA-3′). Substrates 10 to 15 are control DNA substrates analogous to substrates 1–6.

^{32}P-Labeled p(A)/p(dT) Substrate (for Renaturation Gel Assay[12]) The synthesis is performed in 3 ml 50 m*M* Tris-HCl (pH 8.0), 5 m*M* $MgCl_2$, 1 m*M* $MnCl_2$, 4 m*M* dithiothreitol, 5% (v/v) glycerol, 0.03 m*M* ATP, 2.5 A_{260} units poly(dT)$_{25\text{-}30}$ (Pharmacia, Uppsala, Sweden; 1 A_{260} unit corresponds to 32 μg single stranded DNA), 150 μCi[α-^{32}P]ATP (3000 Ci/mmol; Amersham, UK) and 53 units of *E. coli* RNA polymerase for 30 min at 37°. The sample is extracted once with an equal volume of phenol and once with an equal volume of chloroform. The aqueous phase is adjusted to 35% ethanol and loaded onto a small cellulose column (e.g., a 5 ml plastic syringe) equilibrated with buffer B/65 (65% Buffer B, see below, 35% ethanol). After washing with buffer B/65, the bound hybrid is eluted into Eppendorf tubes (fractions of 0.5–1.0 ml), with buffer B (10 m*M* Tris-HCl, pH 7.5, 1 m*M* EDTA, 100 m*M* NaCl) and the activity is determined by liquid scintillation counting. For an acrylamide gel of 20 × 15 × 0.1 cm size, hybrid containing around 400,000 cpm has to be included in the separation gel.

Assays

1. Liquid RNase H Assay (using ^{3}H-Labeled Hybrid). One unit of RNase H activity is defined as the amount of enzyme that renders 100 pmol ribonucleotides of a standard RNA/DNA hybrid acid soluble in 10 min at 30°.

Assay conditions (buffer-mix):

50 m*M* Tris-HCl (pH 7.1)
50 m*M* KCl
10 m*M* $MgCl_2$ [or 0.6 m*M* $MnCl_2$]*
0.002% bovine serum albumin (nuclease free)
2 m*M* dithiothreitol

Convenient stock solutions: For the assay using Mg^{2+} as a divalent cation prepare a 10× stock solution of the assay buffer described above. For the assay using Mn^{2+} prepare a 10× stock solution of the buffer described above, but lacking the

*The type of divalent metal ion and its concentration depends on the particular RNase H enzyme, and should be optimized individually.

divalent cation (reason: Mn^{2+} is oxidized if stored for prolonged times). Separately prepare a 2 *M* $MnCl_2$ stock solution. Using both, prepare a 5× stock solution with the appropriate Mn^{2+} concentration immediately before use.

For the determination of RNase H activity (assay volume: 50 μl) approximately 10,000 cpm of the [^{3}H]UTP labeled RNA/DNA hybrid are incubated with 10 μl enzyme solution in an appropriate dilution and 5 μl 10× stock solution (in case of Mg^{2+}) or 10 μl 5× stock solution (in case of Mn^{2+}) for 10 min at 30°. Then 45 μl are spotted on a glassfiber microfilter (Whatman Clifton, NJ GF/C, diameter: 25 mm) and precipitated with ice-cold 10% trichloroacetic acid (TCA) using a specially designed storing chamber, where filters can be placed individually, while soaked with TCA solution. After 10 min the filters are washed extensively with 5% TCA on a sintered glass filter funnel connected to a vacuum, and subsequently washed with ethanol–ether (1 : 1 v/v). Then they are dried in a heat chamber at 60° for 15 min. Dried filters are put in 2 ml LSC vials, and 2 ml of an appropriate scintillator (e.g., Quicksafe A, Zinsser Analytic, Berkshire, UK) is added. The samples are counted in a liquid scintillation counter (LSC), e.g., Tricarb 2100 TR (Canberra-Packard, Meriden, CT). The activity of RNase H is calculated from the portion of RNA in the hybrid remaining acid insoluble. The assay exhibits linearity, with respect to the amount of RNase H present, if less than 40% of the substrate is degraded during the incubation period.

2. *RNase H assay (using defined substrates).*[10] For RNase H assays (total volume: 13 μl), standard reaction mixtures contain 0.8 pmol 5′-^{32}P-labeled substrate, 50 m*M* Tris-HCl (pH 8.0), 10 m*M* $MgCl_2$, and about 0.1 μg of enzyme. The assays are incubated at 37° for 10 min. An equal volume of stop solution (10 m*M* EDTA, 95% deionized formamide, 0.008% bromphenol blue, and 0.008% xylene cyanol) is added to stop the reactions. The samples are mixed, boiled for 3 min, and cooled in ice. Three μl of each reaction product, is run together with appropriate length markers on a sequencing gel (15% denaturing polyacrylamide gel) and exposed overnight to Kodak X-ray film. Figure 2 shows the reaction products obtained with RNase H(35).

3. *Renaturation gel assay*[12]

Solution 1:

25% (v/v) 2-Propanol
50 m*M* Tris-HCl (pH 7.5)
1 m*M* 2-Mercaptoethanol
0.1 m*M* EDTA

Solution 2:

10 m*M* Tris-HCl (pH 7.5)
5 m*M* 2-Mercaptoethanol

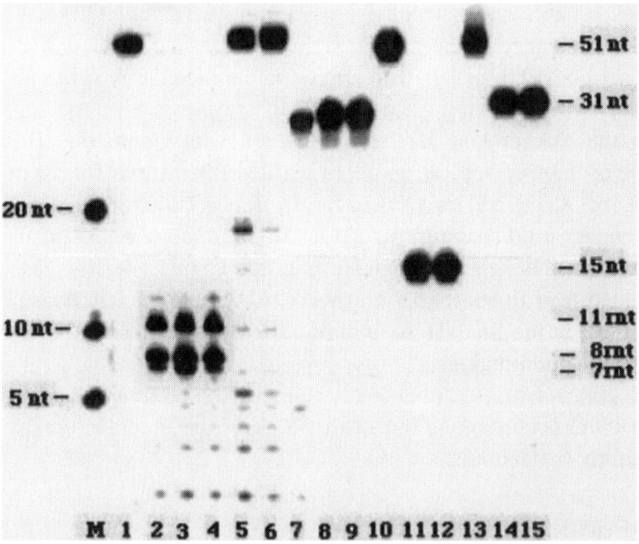

FIG. 2. Products obtained when the substrates of Fig. 1 were incubated with RNase H(35) and separated on a sequencing gel. M, size markers; numbers below the lanes correspond to those of the substrates depicted in Fig. 1; numbers on the right-hand side represent sizes of the labeled oligonucleotides from the substrates as well as from the products.

Solution 3:

50 m*M* Tris-HCl (pH 8.0)
100 m*M* NaCl
0.5 m*M* $MnCl_2$*
1 m*M* 2-Mercaptoethanol
1 m*M* Dithiothreitol
10% (v/v) Glycerol

Solution 4:

2.5% Nonidet P-40 (NP-40) in solution 3

TCA solution:

5% TCA
1% Sodium pyrophosphate

*Although some RNases H are only slightly activated by Mn^{2+} in the liquid assay, we have noticed that using Mn^{2+} instead of Mg^{2+} in the renaturation gel assay greatly enhances the sensitivity (very likely because the renaturation of RNases H is favored by this divalent cation).

Samples for renaturation gel assay are run overnight at 10 mA on a 10% or 12% SDS–polyacrylamide gel ($20 \times 15 \times 0.1$ cm size), containing the labeled substrate (see above), together with appropriate molecular weight markers. After electrophoresis the stacking gel is removed and the part of the separation gel containing the marker lane is cut and stained with Coomassie Brilliant Blue R-250. The remaining separation gel is washed three times for 20 min in solution 1 (150 ml each) and two times for 15 min in solution 2. Afterwards it is incubated (under mild shaking) for 2 h in 200 ml of solution 3 and about 20 h in solution 4. Then this gel, now together with the Coomassie-stained marker lane part, is treated four times for 15 min with TCA solution. Afterward it is stored for another 5 h in the final TCA solution, dried, and exposed to a X-ray film or analyzed by phosphoimaging.

Protein concentrations are determined using the protein dye assay (Bio-Rad, Richmond, CA) according to the instructions of the manufacturer with bovine serum albumin as standard.

Enzyme Purification Procedures

Purification of RNase H(70) from Yeast Cell Extract

The purification procedure is outlined in Fig. 3. It is recommended to use a protease-deficient strain, e.g., strain 20B-12 (Mat α, *pep4-3, trp1*) from the American Type Culture Collection. Proteolytic degradation during the purification procedure is a serious problem!

Materials and Buffers. Polyethylene glycol 6000 (PEG 6000) is from Merck, Darmstadt, Germany. DNA cellulose is prepared according to Litman[13]; in brief, 60 g cellulose (Whatman, fibrous CF11) is washed (2× in 400 ml boiling methanol, 1× in 500 ml 0.1 *M* NaOH, 1× in 250 ml 1 m*M* EDTA, and in bidistilled water until the pH is neutral) and lyophilized. DNA is disolved (2 mg/ml in 1 m*M* NaCl), mixed with 1.5 g lyophilized cellulose (10–16 mg DNA for 1 g cellulose) in a petri dish and dried at 20°. Then it is resuspended in ethanol and UV irradiated for 15 min with 1200 μW/cm^2. Afterward the DNA cellulose is washed 2× with 1 m*M* NaCl and lyophilized. It can be used repeatedly.

Phenyl Sepharose and Mono P are purchased from Amersham Pharmacia; hydroxyapatite and Affi-Gel Blue from Bio-Rad. Restriction enzymes, the protease inhibitors antipain and pepstatin A are from Roche, Basel, Switzerland; the protease inhibitor phenylmethylsulfonyl fluoride (PMSF) is from Merck; and the protease inhibitors *N*-tosyl-L-phenylalanine chloromethyl ketone (TPCK) and N^{α}-tosyl-L-lysine chloromethyl ketone (TLCK) are from Sigma (St. Louis, MO).

For disruption of cells a Braun homogenizer (Braun, Melsungen, Germany) is used. Low-pressure chromatography is performed using a peristaltic pump P1

[13] R. M. Litman, *J. Biol. Chem.* **243,** 6222 (1968).

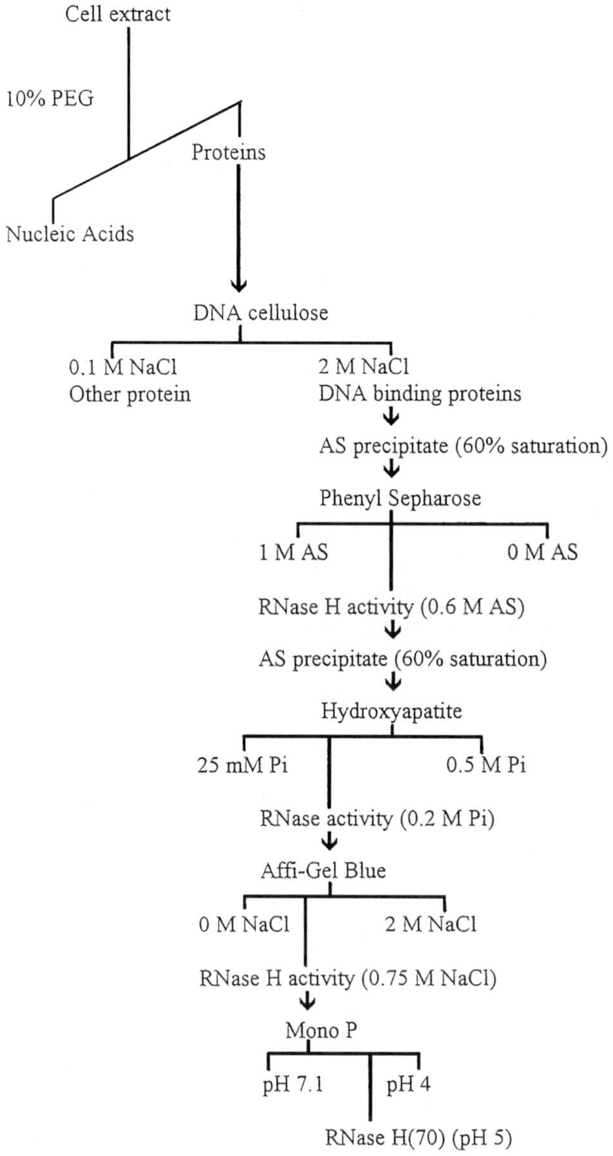

FIG. 3. Flow chart for the purification of RNase H(70). PEG, polyethylene glycol 6000; AS, ammonium sulfate.

(Pharmacia) and a fraction collector. For chromatofocusing, a Pharmacia FPLC system is used.

YPD growth medium (1% yeast extract, 2% peptone, 2% dextrose, w/v)
Buffer A: 20 m*M* Tris-HCl (pH 7.9), 1 m*M* EDTA, 10% glycerol
Buffer K1: 25 m*M* K_2HPO_4/KH_2PO_4 (pH 6.8), 10% glycerol
Buffer K2: 10 m*M* K_2HPO_4/KH_2PO_4 (pH 6.8), 5 m*M* $MgCl_2$, 10% glycerol

Procedure. To minimize proteolysis, all purification steps are carried out at 4°, and protease inhibitors (0.2 m*M* PMSF, 1 m*M* sodium sulfite, pH 8.0; 0.1 m*M* sodium tetrathionate; 1 μM each of TLCK, TPCK, pepstatin A, and antipain) and 0.1% 2-mercaptoethanol are included in all buffers (addition immediately before use). Cells are grown in a 4-liter fermenter as a continous culture at pH 6.2 and 30° using YPD as a medium, and starting with a cell density of 1.6×10^6/ml. The cells are harvested by centrifugation, washed once with buffer A containing 2 *M* NaCl, and frozen in aliquots of 120 g wet weight in liquid N_2. For purification the cell pellet of one aliquot is resuspended in 180 ml of buffer A containing 2 *M* NaCl, and after addition of an equal volume of acid-washed glass beads (diameter 0.45 mm), the suspension is homogenized in 40 ml aliquots in a Braun homogenizer, using special glass bottles. Homogenization takes place under cooling with liquid N_2 in 6 cycles for 30 sec. Between the cycles the bottles are stored on ice. After homogenization and microscopic control of cell disruption the suspension is added to 50 ml Falcon tubes, glass beads are washed with 50 ml buffer A containing 2 *M* NaCl, and the washing buffer is added to the Falcon tubes as well. Then the cell extract is separated from glass beads, undisrupted cells, and cell fragments by centrifugation (Heraeus centrifuge, 3000 rpm, 10 min, 4°). The supernatant is recovered and the pellet is washed with 30 ml of buffer A (2 *M* NaCl) and centrifuged again [Heraeus centrifuge (Heraeus, Hanau, Germany)], 3000 rpm, 10 min, 4°). All supernatants are pooled. The cell extract is then centrifuged for 120 min at 43,000 rpm in a Beckman 45 Ti rotor and the supernatant is filtered through a paper filter.

To remove nucleic acids, the solution is treated with polyethylene glycol 6000 [PEG, 10% (w/v) final concentration] for 30 min on ice followed by centrifugation for 35 min at 22,000 rpm (Beckman 45 Ti rotor). After dialysis of the supernatant against buffer A containing 0.1 *M* NaCl (2 × 2 h) and centrifugation for 25 min at 24,000 rpm (Beckman 45 Ti rotor), the clear supernatant is loaded on a 360 ml (15.2 cm height, 5.5 cm diameter) DNA cellulose column (flow rate: 60 ml/h). After extensive washing (until no protein is detectable in the flow-through by monitoring absorbance at 280 nm), the bound protein is completely eluted with buffer A containing 2 *M* NaCl. Protein containing fractions (10 ml each), as determined by UV monitoring, are collected and analyzed for RNase H activity using assay 1, and active fractions are pooled. The pool is adjusted with solid ammonium sulfate to 60% saturation and, after 20 h of storage in the cold, centrifuged (35,000 rpm, 45 min, 45 Ti rotor). The precipitated proteins are dissolved in buffer A containing 1 *M* ammonium sulfate and loaded onto a 100 ml (13 cm height, 3 cm diameter) phenyl Sepharose column (flow rate 30 ml/h), equilibrated in buffer A containing 1 *M* ammonium sulfate. After washing with 200 ml buffer A containing 1 *M* ammonium sulfate, protein is eluted with 500 ml of a decreasing gradient (1 *M*–0 *M* ammonium sulfate in buffer A) and fractions of 6 ml are collected. Fractions containing RNase H activity are pooled, precipitated with solid ammonium sulfate (60% saturation), and centrifuged (35,000 rpm, 45 min, 45 Ti rotor). The precipitated protein is dissolved in 40 ml buffer K1, dialyzed against buffer K1 (2×2 h), and loaded onto a 40 ml (5.2 cm height, 3 cm diameter) hydroxyapatite column (flow rate 24 ml/h). After washing with buffer K1 (until the baseline of absorbance at 280 nm is reached), the bound protein is eluted with a linear increasing phosphate gradient of 300 ml (25–500 m*M* K_2HPO_4/KH_2PO_4, pH 6.8, 10% glycerol) and fractions of 6 ml are collected. RNase H containing fractions are pooled. The solution is dialyzed two times against buffer K2 and loaded onto a 40 ml (5.2 cm height, 3 cm diameter) Affi-Gel Blue column (flow rate 30 ml/h). After washing with buffer K2 (until the baseline of absorbance at 280 nm is reached), the elution (6 ml fractions) takes place using a 300 ml linear increasing gradient (0–2 *M* NaCl in buffer K2). RNase H containing fractions are pooled. The pool is desalted with a PD10 column (Amersham Pharmacia Biotech, Uppsala, Sweden) and applied to a Mono P column (20 cm height, 0.5 cm diameter), which is equilibrated with 25 m*M* BisTris-HCl (pH 7.1), 10% glycerol, 0.2 m*M* PMSF, and 0.1% 2-mercaptoethanol. Elution is done with Polybuffer 74 (Pharmacia), pH 4.0, 10% glycerol, 0.2 mM PMSF and 0.1% 2-mercaptoethanol. Fractions (1 ml each) are collected and analyzed for the presence of RNase H using assay 3. An example of a result is shown in Fig. 4. Fractions are frozen in liquid N_2 and stored at −70°. If interruption of the procedure is necessary, storage of ammonium sulfate precipitate is recommended.

Overexpression and Purification of RNase H(35)[10]

Growth medium:	1% Peptone 0.5% Yeast extract
Lysis solution I:	100 m*M* NaCl 1 m*M* EDTA 50 m*M* Tris-HCl (pH 8.0) 0.5 mg/ml Lysozyme
Lysis solution II:	100 m*M* NaCl 1 m*M* EDTA 0.1% Sodium deoxycholate 50 m*M* Tris-HCl (pH 8.0)

Wash buffer I:	1% NP-40 100 m*M* NaCl 1 m*M* EDTA 50 m*M* Tris-HCl (pH 8.0)
Wash buffer II:	Wash buffer I without NP-40
Buffer B:	10 m*M* Tris-HCl (pH 7.9) 0.5 *M* NaCl 5 m*M* Imidazole 6 *M* Urea
Desalting buffer:	10 m*M* Tris-HCl (pH 8.0) 150 m*M* NaCl

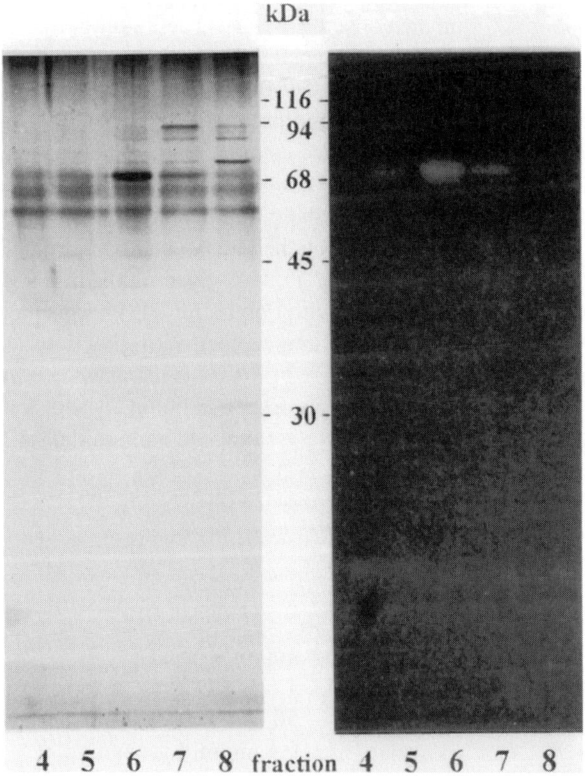

FIG. 4. Visualization of RNase H activity at the expected molecular weight in several fractions from the Mono P column that were separated by 12% SDS–PAGE. *(Left panel)* Silver staining (the bands identical in all fractions result from a staining artifact). *(Right panel) In situ* renaturation assay. Numbers between the two panels: Protein size markers.

For the overexpression of *S. cerevisiae* RNase H(35) in *E. coli,* a His-tag construct carrying the RNH35 coding sequence cloned into pET-28b vector (Novagen, Madison, WI), as described,[10] is used. This construct is transformed into GJ1158, an *E. coli* strain harboring a chromosomal T7 RNA polymerase gene inducible by high salt concentrations.[14] Colonies are inoculated to 1 liter of growth medium with 30 μg/ml kanamycin. Cultures are grown at 37° to OD 0.6 followed by induction at 30° with 0.3 *M* NaCl for 3 h. The cells are harvested by centrifugation (Beckman JA-14 rotor, 6500 rpm, 15 min at 4°) and stored at −80° until use. The RNase H(35) protein is highly expressed. However, the major portion of the protein stays in inclusion bodies.

The protein is purified from inclusion bodies using the built in C-terminal His tag. Purification on a chelating Ni^{2+} matrix is performed under denaturing conditions as described below. Afterward the protein is renatured to regain the activity. All of the purification steps are performed at 4°. The stored cells are thawed in a mixture of ice and water and resuspended in 80 ml lysis solution I. The sample is then incubated at room temperature for 1 h. It is centrifuged at 10,000 rpm for 20 min. The pellet is resuspended in 40 ml ice-cold lysis solution II and incubated on ice with occasional mixing for 10 min. Then the lysate is pushed through a 18-gauge needle for 10 times and centrifuged at 15,000 rpm for 20 min to reduce the viscosity. The pellet is washed by resuspending it in 40 ml of wash buffer I and in 40 ml of wash buffer II. After being centrifuged at 15,000 rpm for 20 min, the pellet is suspended in 25 ml of buffer B and incubated on ice for 1 h to dissolve the inclusion bodies. Then the sample is centrifuged at 14,000 rpm for 45 min to remove the debris. A 5 ml HiTrap chelating Ni^{2+} column (Pharmacia) is equilibrated with buffer B using a fast purification liquid chromatography (FPLC) system (Pharmacia). Then the sample is loaded onto the column. After loading, the column is washed with 25 ml of buffer B, 25 ml of buffer B with 30 m*M* imidazole, and eluted with a 50 ml linear gradient from 30 to 300 m*M* imidazole in buffer B at 2 ml/min. Fractions of 1 ml are collected. Aliquots of the fractions are run on a 10% SDS–polyacrylamide gel and stained with Coomassie Brilliant Blue R-250 to control the purity. The protein is renatured to recover enzymatic activity by exchanging the buffer with decreasing concentration of urea (1× buffer B with 3 *M* urea, 1× buffer B with 1 *M* urea, and 2× buffer B without urea). Finally the protein fractions are desalted with the desalting buffer by a HiTrap desalting column at a flow rate of 2 ml/min (Pharmacia). Fractions are collected (1 ml) and desalted samples are identified by measuring the conductivity. RNase H activity is measured by assay 1 and specific activity is determined as units/mg protein.

[14] P. Bhandari and J. Gowrishankar, *J. Bacteriol.* **179,** 4403 (1997).

```
H. sapiens            VLGVDEAGRGPVLGPMVYAIC   LKVADSKTLLESERERLFA   VNVTQVFVDTVGMPET
C. elegans            VLGIDEAGRGPVLGPMVYAAA   LGVDDSKALNEAKREEIFN   VNVVEIKVDTVGPKAT
S. pombe              RLGVDEAGRGPVLGPMVYAVA   YGFADSKTLASLKREELLK   INVTEIYVDTVGPPIS
S. cerevisiae         IMGIDEAGRGPVLGPMVYAVA   YEFDDSKKLTDPIRRMLFS   VKLSHVYVDTVGPPAS
M. jannaschii         IIGIDEAGRGPVLGPMVVCAF   LGVKDSKELTKNKRAYLKK   STNTKKFEDSF..KDK
B. subtilis           VIGSDEVGTGDYFGPMTVVCA   LGVKDSKDLKDPQIIEIAR   VKPEAILIDQFAEPGV
S. pneumoniae         LIGTDEVGNGSYFGGLAVVAA   LGVGDSKTLTDQKIRQIAP   VQPEKIVIDAFTSAKN
Synechocystis sp.     MAGVDEVGRGALFGPVVAAAV   LGVKDSKQLSPQHRSQLAQ   VMPAWVLVDGRDTLPH
M. tuberculosis       VAGVDEVGRGACAGPLVVAAC   AALDDSKKLSEQAREKLFP   VRPGYVLSDGFRVPG.
M. leprae             VAGVDEVGRGACAGPLVVAAC   .SLDDSKKLSAKGREMLFP   VRPGYVLSDGFRVPG.
Magnetospirillum sp.  VCGIDEVGRGPLAGPVVAAAV   ERLDDSKKLSKRNREELAE   RECAHALVDGNRPPG.
H. influenzae         IAGVDEVGRGPLVGAVVTAAV   EGLADSKKLSEKKRLALAE   IQPHFVLIDGNKIPKD
E. coli               VAGVDEVGRGPLVGAVVTAAV   AGLNDSKKLSEKRRLALYE   IAPEYVLIDGNRCPK.
V. cholerae           .....................   MGLNDSKKLSEKKRLALFP   IQPDLVLIDGNKIPK.
consensus             ..G.DEvGrGp..Gp.v.aa.   .g..DSK.Ls...r..l..   v.p..vl.Dg...p..
                        ---- I -----------     --- II ---------     ------ III----


H. sapiens            KAKADALYPVVSAASICAKVARDQAV   QDLDTDYGSGYPND
C. elegans            TEKADSLFPIVSAASIAAKVTRDSRL   KVPDAGYGSGYPGD
S. pombe              TKKADSLFPIVSLASICAKVTRDIQL   SIRTENWGSGYSSD
S. cerevisiae         AKKADSLYCMVSVASVVAKVTRDILV   D.PDEILGSGYPSD
M. jannaschii         EHKADAKYPVVSAASIIAKAERDEII   .KIYGDIGSGYPSD
B. subtilis           STKAEGIHLAVAAASIIARYSFLMEM   AGMTLPKGAGPHVD
S. pneumoniae         EEKAEGKYLAVAVSSVIARDLFLENL   LGYQLPSGAGTASD
Synechocystis sp.     VTGGDRKSLLISAASILAKVWRDTLI   YDLASNKGYGTAKH
M. tuberculosis       VIGGDAAAACIAAASVLAKVSRDRVM   YGFAEHKGYSTPAH
M. leprae             VVGGDAVVACIAAASVLAKVSRDRLM   YGFAAHKGYCTRAH
Magnetospirillum sp.  VVGGDGISLSIAAASVVAKVRRDAMM   FGWERNAGYGTAEH
H. influenzae         VVKGDSLVAEISAASILAKVARDQEM   YAFAQHKGYPTKLH
E. coli               VVKGDSRVPEISAASILAKVTRDAEM   YGFAQHKGYPTAFH
V. cholerae           VVKGDLRVAQISAASIIAKVIRDQEM   FGFANHKGYPTAAH
consensus             v.kg*.....*saAS*.Akv.rd..m   .g....kGygt..h
                      ------------ IV ----------    ----- V -----
```

FIG. 5. Domains conserved between yeast RNase H(35) and human RNase HI, as well as examples of polypeptides and open reading frames (ORFs) from other organisms, as revealed by alignment, using the "multiple sequence alignment with hierarchical clustering" method of Corpet.[16] The alignment is arranged according to decreasing relationship starting with the domains of human RNase HI. The upper four organisms are eukaryotes, *M. jannaschii* is an archaebacterium, the rest are eubacteria. Accession numbers can be found in Ref. 5. In the deduced amino acid sequence of yeast RNase H(35) (human RNase HI) the domains correspond to the following amino acids: I, 35–55 (30–50); II, 70–88 (63–81); III, 147–162 (133–148); IV, 179–204 (165–190); V, 210–222 (200–213). Amino acids identical or similar in all sequences are marked in bold letters; in the consensus sequence bold uppercase letters indicate identical and bold asterisks similar amino acids; amino acids identical in 50% or more sequences are marked with lowercase letters.

Evolutionarily Related Proteins from Organisms Other than *S. cerevisiae* by Alignments of Predicted Amino Acid Sequences

As originally expected for all proteins exhibiting RNase H activity, the deduced amino acid sequence of yeast RNase H1 is evolutionarily related to *E. coli* RNase HI and to the RNase H domain of viral reverse transcriptases (RT).[15] Human RNase HI and yeast RNase H(35) possess sequences related to each other and to bacterial

[15] M. S. Johnson, M. A. McClure, D. F. Feng, J. Gray, and R. F. Doolittle, *Proc. Natl. Acad. Sci. U.S.A.* **83,** 7648 (1986).

[16] F. Corpet, *Nucleic Acids Res.* **16,** 10881 (1988).

```
RNase H(70)/Rex1p   IFALDCEMC    LTRISLVNFD    EVIYEELVKPDVPIVDYLTRYSGITE
              GOR   IYALDCEMC    LTRVTVVDAD    RVVYDTFVKPDNEIVDYNTRFSGVTE
         caeno594   MFSVDCEMC    LTRISIVDEF    NTILDTLVKPEGRITDYVTRWSGITP
            Rex3p   VLSLDCEMA    MIRLTIVDFF    KTLFDHVIQPIGDIVDLNSDFSGVHE
            XPMC2   TVAMDCEMV    LARVSIVNLF    KCVYDKYVKPTERVTDYRTAVSGIRP
            Rex4p   YIAMDCEFV    LARISIVNYF    HVVLDEFVKPREKVVEWRTWVSGIKP
         caeno271   VYALDCEMV    LARLTMVDMQ    NRVLDVFVKPPTDVLDPNTEFSGLTM
            ISG20   VVAMDCEMV    LARCSLVNVH    AVLYDKFIRPEGEITDYRTRVSGVTP
        consensus   ..a*DCEmv    *.R.siV*.f    ..*y*.f*kP...*v*y.tr.SG*tp
                     -- I --     -- II---       ---------- III -------

RNase H(70)/Rex1p   ILIGHSLQNDLKVMKLKHPLVV..DTA    SLKYLSETFLNKSIQ....NG.EHDSVEDARACLEL
              GOR   ILIGHSLESDLLALKLIHSTVL..DTA    SLRNLAADYLGQIIQ...DSQDGHNSSEDANACLQL
         caeno594   ILVGHSLEHDLQAMKMTHPFCL..DVG    SLKNLTELFLGAQIQ....SEFGHCSYEDAWAAMRL
            Rex3p   ILIGHGLENDLNVMRLFHNKVI..DTA    SLKNLAFEVLSRKIQ...NGE..HDSSQDAIATMDV
            XPMC2   TLVGHAVHNDLKILFLDHPKKAIRDTQ    SLKLLCEKILNVKVQ...TGE..HCSVQDAQAAMRL
            Rex4p   ILVGHALKHDLEALMLSHPKSLLRDTS    SLKKLTREVLKISIQ...EGE..HSSVEDARATMLL
         caeno271   ILIGHSLESDLKAMRVVHKNVI..DTA    ALKVLSAKLLHKNIQGDNEDAIGHDSMEDALTCVDL
            ISG20   LVVGHDLKHDFQALKEDMSGYTIYDTS    SLRVLSERLLHKSIQ...P..LGHSSVEDARATMEL
        consensus   il*GHsle.Dl.a*kl.hp.v...Dta    sLk.L.e..L...*Q.....e.gH.Sv*DA.a.m.*
                     ------ IV--------------        -------------- V -----------------
```

FIG. 6. Domains conserved between yeast RNase H(70) and various eukaryotic polypeptides and ORFs. The related domains originate from the following sequences: GOR, the GOR antigen of the chimpanzee *P. troglodytes* (EMBL a.n. D10017); caeno594, a *C. elegans* ORF (GenBank a.n. AF016430); Rex3p, an *S. cerevisiae* 3′ → 5′-exonuclease (EMBL a.n. Z73280); XPMC2, a *Xenopus laevis* protein (a.n. GenBank U10185); Rex4p, an *S. cerevisiae* 3′ → 5′-exonuclease (a.n. EMBO Z74822); caeno271, a *C. elegans* ORF (Swiss-Prot a.n. Q101024); ISG20, a human protein (a.n. X89773). In the deduced amino acid sequence of RNase H(70) the domains correspond to the following amino acids: I, 225–233; II, 241–250; III, 252–277; IV, 304–328; and V, 342–372. Markings as in Fig. 5; a.n., accession number.

RNases HII. These sequences do not show similarity to *E. coli* RNase HI and to the RNase H domain of RTs, but form a separate group of related polypeptides, which all possess five highly conserved motifs (Fig. 5), most probably responsible for enzyme activity. RNase H(70)/Rex1p and evolutionarily related proteins also possess five highly conserved motifs which are, however, different from those of the RNase H(35) group (Fig. 6).

Acknowledgments

We thank Gabriele Operenyi for her patience during her help in the preparation of the manuscript. The work in the authors laboratory was supported by the Fonds zur Förderung der wissenschaftlichen Forschung in Österreich (Grant S 5806-MOB) to U.W., the Anton Dreher Gedächtnisschenkung für medizinische Forschung (Grant 272/95) to P.F., and the Hochschuljubiläumsstiftung der Stadt Wien (Grant H-00097/96) to P.F.

[27] Human RNases H

By WALT F. LIMA, HONGJIANG WU, and STANLEY T. CROOKE

Introduction

RNase H hydrolyzes RNA in RNA–DNA hybrids.[1] RNase H activity appears to be ubiquitous in eukaryotes and bacteria.[2–7] Although RNases H constitute a family of proteins of varying molecular weight, the nucleolytic activity and substrate requirements appear to be similar for the various isotypes. For example, all RNases H studied to date function as endonucleases exhibiting limited sequence specificity and requiring divalent cations (e.g., Mg^{2+}, Mn^{2+}) to produce cleavage products with 5′-phosphate and 3′-hydroxyl termini.[8]

Two classes of RNase H enzymes have been identified in mammalian cells.[5,9,10] These enzymes were shown to differ with respect to cofactor requirements. For example, RNase H1 is activated by both Mg^{2+} and Mn^{2+}, whereas RNase H2 was shown to be activated by only Mg^{2+} and inhibited by Mn^{2+}. Furthermore, both RNase H1 and H2 were shown to be inhibited by sulfhydryl reagents.[10,11] Although the biological roles of the mammalian enzymes are not fully understood, it has been suggested that mammalian RNase H1 may be involved in replication and that the type 2 enzyme may be involved in transcription.[12,13]

Recently, both human RNase H genes have been cloned and expressed.[11,14,15] The type 1 enzyme is a 286 amino acid protein with a calculated mass of 32 kDa.[11] The enzyme is encoded by a single gene that is at least 10 kb in length and is expressed ubiquitously in human cells and tissues. The amino acid sequence of

[1] H. Stein and P. Hausen, *Science* **166,** 393 (1969).

[2] M. Itaya and K. Kondo, *Nucleic Acids Res.* **19,** 4443 (1991).

[3] M. Itaya, D. McKelvin, S. K. Chatterjie, and R. J. Crouch, *Mol. Gen. Genet.* **227,** 438 (1991).

[4] S. Kanaya and M. Itaya, *J. Biol. Chem.* **267,** 10184 (1992).

[5] W. Busen, *J. Biol. Chem.* **255,** 9434 (1980).

[6] Y. W. Rong and P. L. Carl, *Biochemistry* **29,** 383 (1990).

[7] P. S. Eder, R. T. Walder, and J. A. Walder, *Biochimie* **75,** 123 (1993).

[8] R. J. Crouch and M. L. Dirksen, *in* "Nucleases" (S. M. Linn and R. J. Roberts, eds.), p. 211. Cold Spring Harbor Laboratory Press, Plainview, NY, 1982.

[9] P. S. Eder and J. A. Walder, *J. Biol. Chem.* **266,** 6472 (1991).

[10] P. Frank, S. Albert, C. Cazenave, and J. J. Toulme, *Nucleic Acids Res.* **22,** 5247 (1994).

[11] H. Wu, W. F. Lima, and S. T. Crooke, *Antisense Nucleic Acid Drug Dev.* **8,** 53 (1998).

[12] W. Busen, J. H. Peters, and P. Hausen, *Eur. J. Biochem.* **74,** 203 (1977).

[13] J. J. Turchi, L. Huang, R. S. Murante, Y. Kim, and R. A. Bambara, *Proc. Natl. Acad. Sci. U.S.A.* **91,** 9803 (1994).

[14] P. Frank, C. Braunshofer-Reiter, U. Wintersberger, R. Grimm, and W. Busen, *Proc. Natl. Acad. Sci. U.S.A.* **95,** 12872 (1998).

[15] S. M. Cerritelli and R. J. Crouch, *Genomics* **53,** 307 (1998).

0076-6879/01 $35.00

human RNase H1 displays strong homology with RNase H1 from yeast (21.8% amino acid identity), chicken (59%), *Escherichia coli* (33.6%), and mouse (84.3%). The type 1 enzymes are all small proteins (<40 kDa) and their estimated p*I* values are all 8.7 and greater. The amino acid residues in *E. coli* RNase HI thought to be involved in the Mg^{2+} binding site, catalytic center, and substrate binding region[16–18] are well conserved in the cloned human RNase H1 sequence.[11] The human RNase H2 enzyme is a 299 amino acid protein with a calculated mass of 33.4 kDa and has also been shown to be ubiquitously expressed in human cells and tissues (see Ref. 14; Wu, unpublished data, 1998). Human RNase H2 shares strong amino acid sequence homology with RNase H2 from *Caenorhabditis elegans* (45.5% amino acid identity), yeast (25.7%), and *E. coli* (14.4%). Unlike the RNase H1 isotype, the type 2 enzyme is an acidic protein exhibiting a p*I* of 4.94.

The properties of the cloned and expressed human RNase H1 have been characterized.[19] The activity of the type 1 enzyme is Mg^{+2}-dependent and inhibited by Mn^{2+} and the sulfhydryl blocking agent *N*-ethylmaleimide. Human RNase H1 was also inhibited by increasing ionic strength with optimal activity for both KCl and NaCl observed at 10–20 m*M*. The enzyme exhibited a bell-shaped response to divalent cations and pH, with the optimum conditions for catalysis observed to be, respectively, 1 m*M* Mg^{2+} and pH 7 to 8. The protein was shown to be reversibly denatured under the influence of temperature and destabilizing agents such as urea. Renaturation of human RNase H1 was observed to be highly cooperative and did not require divalent cations. Human RNase H1 was shown to bind selectively to "A-form" duplexes with approximately 10- to 20-fold greater affinity than that observed for *E. coli* RNase H1.[19,20] Finally, human RNase H1 displays a strong positional preference for cleavage, i.e., the enzyme cleaves between 8 and 12 nucleotides from the 5′-RNA–3′-DNA terminus of the duplex.

Molecular Cloning and Expression of Human RNases H

Procedural Comments

The full-length cDNAs of human RNases H were cloned from 5′ and 3′ cDNAs generated using the rapid amplification of cDNA ends (RACE) system (Clontech, Palo Alto, CA).[11] The RACE system combines long-distance PCR (polymerase chain reaction), avion myeloblastosis virus (AMV) reverse transcriptase and a

[16] S. Kanaya, C. Katsuda-Kakai, and M. Ikehara, *J. Biol. Chem.* **266,** 11621 (1991).

[17] K. Katayanagi, M. Okumura, and K. Morikawa, *Proteins: Struct., Funct., Genet.* **17,** 337 (1993).

[18] A. W. Nicholson, *in* "Ribonucleases: Structures and Functions" (G. D'Alessio and J. F. Riordan, eds.), p. 2. Academic Press, New York, 1997.

[19] H. Wu, W. L. Lima, and S. T. Crooke, *J. Biol. Chem.* **274,** 28270 (1999).

[20] W. F. Lima and S. T. Crooke, *Biochemistry* **36,** 390 (1997).

modified lock-docking oligo(dT) primer.[21,22] The advantage of this system is that the same double strand cDNA template can be used for both 3′ and 5′ cDNA amplification and 3′ heterogeneity is eliminated. A minimum of 23 to 28 nucleotides of sequence information is required to design the sense and antisense primers for the RACE amplification system. The oligonucleotide primers used for the RACE amplification of the 5′ and 3′ cDNAs of human RNase H1 were derived from the 361 base pair human RNase H expressed sequence tag (EST) identified in the XREF database of the National Center of Biotechnology Information (NCBI). The human RNase H1 EST is homologous to both the yeast RNase H1 (RNH1) protein sequence (GeneBank accession #Q04740) and chicken homolog (GeneBank accession #D26340) (23). The oligonucleotide primers used for the 5′ and 3′ RACE amplification of human RNase H2 were derived from two overlapping human ESTs which were found to be homologous to yeast RNase H2 (RNH2) protein sequence (GeneBank accession #Z71348) and *C. elegans* (GeneBank accession #Z66524). A liver cDNA library (Stratagene, La Jolla, CA) was then screened using the 1 kb 3′ RACE cDNA product of both RNase H1 and H2 and the 5′ RACE cDNA was used to identify the 5′-most translation initiation codon.

Human RNase H1

Three sets of oligonucleotide primers encoding the human RNase H1 expressed sequence tag (GeneBank accession #H28861) are synthesized. The sense primers are

ACGCTGGCCGGGAGTCGAAATGCTTC (H1)
CTGTTCCTGGCCCACAGAGTCGCCTTGG (H3)
GGTCTTTCTGACCTGGAATGAGTGCAGAG (H5)

The antisense primers are

CTTGCCTGGTTTCGCCCTCCGATTCTTGT (H2)
TTGATTTTCATGCCCTTCTGAAACTTCCG (H4)
CCTCATCCTCTATGGCAAACTTCTTAAATCTGGC (H6)

The human RNase H1 3′ and 5′ cDNAs are amplified by RACE (Clontech) according to the manufacturer's instructions. The cDNA templates used for amplification are from either human liver or leukemia (lymphoblastic Molt-4) Marathon ready cell lines (Clontech). The primers used for the first and second run amplification are, respectively, H1 or H3/AP1 and H4 or H6/AP2. The fragments are subjected

[21] W. M. Barnes, *Proc. Natl. Acad. Sci. U.S.A.* **91,** 2216 (1994).
[22] M. A. Frothman, *Proc. Natl. Acad. Sci. U.S.A.* **218,** 340 (1991).
[23] A. R. Mushegian, H. K. Edskes, and E. V. Koonin, *Nucleic Acids Res.* **22,** 4163 (1994).

to agarose gel electrophoresis[24] and transferred to a nitrocellulose membrane (Bio-Rad, Hercules, CA) for confirmation by Southern blot, using ^{32}P-labeled H2 probe for the 3′ RACE products and H1 probe for the 5′ RACE products.[24] The confirmed fragments are excised from the agarose gel and purified by gel extraction (Qiagen, Germany). The purified RACE products are subcloned into a Zero-blunt vector (Invitrogen, Carlsbad, CA) and subjected to DNA sequencing with an automated DNA sequencer (Retrogen, San Diego, CA). The overlapping sequences are aligned and then combined by the assembling program MacDNASIS v3.0 (Hitachi Software Engineering Co, CA). The program Mac Vector v6.0 (Oxford Molecular Group, UK) is used to analyze the structure of the protein.

The 3′ RACE product corresponding to the carboxy terminal portion of the protein was used to screen the human liver cDNA lambda phage Uni-zap library (Stratagene). The positive cDNA clones are subsequently screened with the 5′ RACE product in order to identify full-length cDNAs. The full-length cDNA clones are sequenced using an automated DNA sequencer. The cDNA fragment coding the full-length RNase H protein sequence is amplified by PCR using a primer that contains the restriction enzyme *NdeI* site adapter, six histidine (His_6 tag) codons, and 22 base pairs corresponding to the N-terminal coding sequence of the protein, and a second primer that contains an XhoI site and 24 base pairs corresponding to the C-terminal coding sequence including the stop codon. The PCR product is cloned into the expression vector pET17b (Novagen, Madison, WI) and confirmed by DNA sequencing. The plasmid is transfected into *E. coli* BL21 (DE3) (Novagen). The bacteria are grown in M9ZB medium[24] at 32° and harvested at OD_{600} of 0.8. The cells are induced with 0.5 m*M* isopropylthiogalactoside (IPTG) at 32° for 2 h. The cells are lysed in 8 *M* urea solution and the recombinant protein is partially purified with Ni-NTA agarose (Qiagen, Valencia, CA).

Human RNase H2

The cloning of the cDNA and expression of the human RNase H2 protein is performed as described for the human RNase H1 enzyme with the following exceptions. The sense and antisense primers are derived from two overlapping human ESTs (GenBank accession #W05602 and H43540). The sense primers are

AGCAGGCGCCGCTTCGAGGC (H1A).
CCCGCTCCTGCAGTATTAGTTCTTGC (H1B)

The antisense primers are

TTGCAGCTGGTGGTGGCGGCTGAGG (H1C)
TCCAATAGGGTCTTTGAGTCTGCCAC (H1D)

[24] F. M. Ausubel, R. Brent, R. E. Kingston, D. D. Moore, J. A. Smith, J. G. Seidman, and K. Struhl, eds., *in* "Current Protocols in Molecular Biology." Wiley and Sons, New York, 1988.

CACTTTCAGCGCCTCCAGATCTGCC (H1E)
GCGAGGCAGGGGACAATAACAGATGG (H1F)

The primers for amplification of the 3′ and 5′ RACE products are H1A or H1B/AP1 (for first-run PCR) and H1B or H1C/Ap2 (for second-run PCR). Finally, the protein is expressed using *E. coli* BL21 (DE3) (Novagen) grown in LB medium at 37° and harvested when the OD_{600} of the culture reaches 0.8.[24]

Purification of Human RNase H Proteins

The RNase H proteins are purified by C_4 reversed-phase chromatography (Beckman, System Gold, Fullerton, CA) using a 0% to 80% gradient of acetonitrile in 0.1% trifluoroacetic acid/distilled water (%v/v) over 40 min.[25] The recombinant protein is collected, lyophilized, and analyzed by 12% SDS–PAGE.[24] The purified protein and control samples are resuspended in 6 *M* urea solution containing 20 m*M* Tris-HCl, pH 7.4, 400 m*M* NaCl, 20% glycerol, 0.2 m*M* phenylmethylsulfonyl fluoride (PMSF), 5 m*M* dithiothreitol (DTT), and 10 μg/ml each aprotinin and leupeptin (Sigma, St. Louis MO). The protein is refolded by dialysis with decreasing urea concentration from 6 *M* to 0.5 *M* and DTT concentration from 5 m*M* to 0.5 m*M*.[25] The refolded protein is concentrated 10-fold using a Centricon apparatus (Amicon, Danvers, MA).

Preparation of Heteroduplex Substrate

Procedural Comments

The RNA–DNA substrate is heat-denatured at 90° and subsequently annealed by slowly reducing the temperature of the hybridization reaction to 37°. It is recommended that the hybrids be denatured in the absence of $MgCl_2$, as the combination of Mg^{2+} ions and high temperature will cause RNA cleavage. The kinetics of hybrid formation is dependent on the concentration and composition of the oligonucleotides used to prepare the heteroduplex substrate. For example, in order to ensure that the RNA is fully duplexed at hybrid concentrations <100 n*M*, the heteroduplex substrate is prepared with the DNA concentration at a minimum 2-fold excess over the RNA concentration. Human RNase H1 has been shown to bind ssDNA with a K_d of 1.5 μM.[19] For that reason, heteroduplex substrates at concentrations >1 μM are prepared with equimolar DNA and RNA in order to avoid the potential inhibition of the enzyme activity by the presence of free ssDNA. Finally, phosphorothioate oligonucleotides have been shown to be strong competitive inhibitors of human RNase H1 (K_i = 36 n*M*), and as a result, heteroduplex substrates containing phosphorothioate oligonucleotides should be prepared with

[25] M. P. Deutscher, *Methods Enzymol.* **182,** 311 (1990).

the heteroduplex concentration below the K_j for the single-strand phosphorothioate oligonucleotides.

The oligoribonucleotide is either 5′-end labeled with [γ-^{32}P]ATP using T4 polynucleotide kinase or 3′-end labeled with [^{32}P]pCp using T4 RNA ligase. Each labeling method offers certain advantages depending on the respective labeling efficiencies and the method used to analyze the activity of the RNase H1 enzyme. The criteria for choosing the appropriate labeling method is discussed in detail elsewhere in this volume.[25a] The recommended procedure for the purification of the ^{32}P-labeled oligoribonucleotide is by electrophoresis on a polyacrylamide gel. This procedure routinely yields purity levels of >95%. Achieving this level of purity is critical as the presence of contaminating deletion products introduced during the synthesis of the oligoribonucleotide will greatly interfere with the interpretation of the enzymatic assay. Although duplexed RNA is fairly impervious to nonspecific degradation resulting from contaminants, RNase inhibitors [e.g., Prime RNase Inhibitor (5 Prime → 3 Prime, Boulder, CO), rRNasin (Promega, Madison, WI] are included as an added precaution.

Synthesis of Oligonucleotides

The oligoribonucleotides are synthesized on a PE-ABI 380B synthesizer using 5′-*O*-silyl-2′-*O*-bis)(2-acetoxyethoxy)methyl ribonucleoside phosphoramidites and procedures described elsewhere.[25a] The oligoribonucleotides are purified by reversed-phase HPLC. The DNA oligonucleotides are synthesized on a PE-ABI 380B automated DNA synthesizer and standard phosphoramidite chemistry. The DNA oligonucleotides were purified by precipitation 2 times out of 0.5 *M* NaCl with 2.5 volumes of ethanol.

Preparation of ^{32}P-Labeled Substrate

The sense strand is 5′-end labeled with ^{32}P using [γ-^{32}P]ATP, T4 polynucleotide kinase or 3′-end labeled with [^{32}P]pCp using T4 RNA ligase and is purified by electrophoresis on 12% denaturing PAGE as described elsewhere (Lima and Crooke, this volume). The specific activity of the labeled oligonucleotide is approximately 3000 to 8000 cpm/fmol.

Preparation of the Heteroduplex

The heteroduplex substrate is prepared in 100 μL containing 50 n*M* unlabeled oligoribonucleotide, 10^5 cpm of ^{32}P-labeled oligoribonucleotide, 100 n*M* complementary oligodeoxynucleotide, and hybridization buffer [20 m*M* Tris, pH 7.5, 20 m*M* KCl]. Reactions are heated at 90° for 5 min, then cooled to 37°, and 60 U of Prime RNase Inhibitor (5 Prime → 3 Prime, CO) and $MgCl_2$ at

[25a] W. F. Lima and S. T. Crooke, *Methods Enzymol.* **341,** [32] 2001 (this volume).

a final concentration of 1 mM are added. Hybridization reactions are incubated 2–16 h at 37° and 2-mercaptoethanol (2-ME) is added at final concentration of 20 mM.

RNase H1 Cleavage Assay

Procedural Comments

The cleavage rate of human RNase H1 is dependent on the state of ionization of the enzyme, concentration of the $MgCl_2$ cofactor, and ionic strength of the reaction. The optimal cleavage rate was observed at pH 7–8, 1–10 mM $MgCl_2$, 20 mM KC1, and 20 mM 2-ME. The cleavage products are analyzed by polyacrylamide gel electrophoresis or acid precipitation and the merits of these methods are discussed in detail elsewhere in this volume.[25a]

Determinations of Initial Rates (V_0)

The background control is prepared by incubating a 10 μl aliquot of the heteroduplex substrate without human RNase H1 at 37° for the duration of the assay. The heteroduplex substrate is digested with 0.5 ng human RNase H1 at 37°. A 10-μl aliquot of the cleavage reaction is removed at time points ranging from 2 to 120 min and quenched by adding 5 μL of stop solution (8 M urea and 120 mM EDTA) and snap-freezing on dry ice. The aliquots are heated at 90° for 2 min and resolved in a 12% denaturing polyacrylamide gel, and the substrate and product bands are quantitated on a Molecular Dynamics PhosphorImager.

For acid precipitation the 10 μl aliquot of the cleavage reaction is quenched with 90 μl of 0.6 mg/ml yeast tRNA and then precipitated on ice with 100 μl 10% trichloroacetic acid (Sigma) for 5 min. The sample is centrifuged at 15,000g, for 5 min at 4°. A 150-μl aliquot of the supernatant is removed and added to 2 ml of scintillation cocktail and the solubilized radioactivity counted in a scintillation counter.

The concentration of converted substrate is calculated by measuring the fraction of substrate converted to product (acid-soluble counts or counts for cleavage product bands/total counts) for each time point, multiplying by the substrate concentration and correcting for background {[fraction product × (total substrate)] – background}. The background values represent the fraction corresponding to the degradation products (counts for nonspecific degradation products/total counts). The concentration of the converted product is plotted as a function of time. The initial cleavage rate is obtained from the slope (mol RNA cleaved/min) of the best-fit line for the linear portion of the plot, which comprises, in general, <10% of the total reaction. For accuracy, the line should represent data from at least five time points. The time points are selected through iterative testing to obtain a sufficient number of data points within the linear portion of the rate curve. Alternatively, the concentration of the RNase H1 enzyme is adjusted.

Multiple Turnover Kinetics

Cleavage reactions are performed using excess heteroduplex substrate and with the human RNase H1 concentration below the K_m of the enzyme/substrate complex. The heteroduplex substrate is prepared in 100 μl containing unlabeled oligoribonucleotide at concentrations ranging from 10 to 1000 nM, 10^5 cpm ^{32}P-labeled oligoribonucleotide, twofold excess complementary oligodeoxynucleotide, and hybridization buffer (20 mM Tris, pH 7.5, 20 mM KCl). Reactions are heated at 90° for 5 min, cooled to 37°, and 60 U of Prime RNase Inhibitor (5 Prime → 3 Prime) and $MgCl_2$ at a final concentration of 1 mM are added. Hybridization reactions are incubated 2–16 h at 37° and 2-mercaptoethanol (2-ME) is added at final concentration of 20 mM. The background control is prepared as described for the initial rate determinations. The heteroduplex substrate is digested with 0.5 ng human RNase H1 at 37°. The cleavage reactions are quenched and analyzed by either the PAGE or acid precipitation procedures as described for the initial rate determinations. The product concentration is calculated as described for the initial rate determinations and plotted as a function of time. The multiple- turnover reaction rates (k_{obs}) are calculated from the slope (mol RNA cleaved/min) of a linear least-squares fit from the linear portion of the curve, normally the first 5–10% of the reaction. The (k_{obs}) values are plotted as function of the substrate concentration and the data is fit to the Michaelis–Menten equation using the program Ultrafit (Biosoft, NJ). The multiple-turnover K_m corresponds to the heteroduplex substrate concentration at half-maximum rate and the $k_{cat} = V_{max}/[\text{total peptide}]$, where V_{max} corresponds to the horizontal asymptote of the hyperbolic curve. Alternatively, the multiple-turnover K_m and k_{cat} are determined by plotting k_{obs} vs $k_{obs}/[\text{heteroduplex substrate}]$. The slope of the line is equal to $-K_m$ and the y intercept is equal to the V_{max}. The heteroduplex substrate concentrations used to determine the multiple-turnover K_m and k_{cat} are selected by trial and error, and the Michaelis–Menten binding curve is fit using a minimum of four heteroduplex substrate concentrations.

Determination of Binding Affinity

Procedural Comments

The dissociation constants (K_d) of human RNase H1 for single- and double-strand oligonucleotides are determined using a competition assay.[20] Here, the cleavage rate is determined for the RNA/DNA heteroduplex at variety of concentrations in both the presence and absence of competing noncleavable single- and double-strand oligonucleotides. Lineweaver–Burk and Augustinsson analysis of the data is used to determine the inhibitory constant (K_i) for the competing oligonucleotides. Alternatively, the IC_{50} (concentration of inhibitor required to achieve a half-maximal degree of inhibition) is determined by measuring the cleavage rate for the heteroduplex substrate at a single substrate concentration in both the presence and absence of noncleavable inhibitor.[26] The equations of Cheng and Prusoff are used to convert IC_{50} values to K_i values.[27] In both cases, the value for K_i is equivalent to dissociation constant (K_d) when noncleavable inhibitors are used.

Determination of Dissociation Constants (K_d)

The wild-type RNA/DNA hybrids are prepared as described for multiple turnover kinetics with the exception that the final volume is 50 μl. The competing oligonucleotide reactions are prepared in 50 μl containing hydridization buffer and unlabeled oligonucleotide in excess of the wild-type RNA/DNA hybrid concentration. Competing double-strand oligonucleotides are prepared with equimolar concentration of the complementary strands. Reactions are heated at 90° for 5 min, then cooled to 37°, and 60 U of Prime RNase Inhibitor (5 Prime → 3 Prime) and 1 mM $MgCl_2$ (final concentration) are added. Reactions are incubated 2–16 h at 37° and 2-mercaptoethanol (2-ME) is added at final concentration of 20 mM. The competing oligonucleotide reaction is added to the wild-type RNA/DNA hybrid and the background control is prepared as described for the initial rate determinations. For the wild-type RNA/DNA only control, 50 μl of hybridization buffer is substituted for the competing oligonucleotide reaction. The hybrid is digested with 0.5 ng human RNase H1 at 37°. The cleavage reactions are quenched and analyzed by either the PAGE or acid precipitation procedures as described for the initial rate determinations. The product concentration is calculated as described for the initial rate determinations and plotted as a function of time. The initial cleavage rate (V_0) is obtained from the slope (mol RNA cleaved/min) of the best-fit line for the linear portion of the plot, which comprises, in general, <10% of the total reaction. For accuracy, the line should represent data from at least five time points. The data is analyzed by the Lineweaver–Burk method (plot of V_0 vs V_0/[wild-type RNA/DNA hybrid]) and the K_i is calculated from the slope of a linear least-squares fit of the data, where K_i = [competing oligonucleotide]/(slope with competing oligonucleotide/slope without competing oligonucleotide). Ideally, the K_i is determined using a minimum of three competing oligonucleotide concentrations. Alternatively, the IC_{50} is determined by measuring the initial rate (V_0) for the wild-type RNA/DNA hybrid and the initial

[26] R. A. Copland, ed., "Enzymes: A Practical Introduction to Structure, Mechanism, and Data Analysis." VCH Publishers, Inc., New York, 1996.

[27] Y.-C. Cheng and W. H. Prusoff, *Biochem. Pharmacol.* **22,** 3099 (1973).

rate (V_i) for wild-type RNA/DNA hybrid with competing oligonucleotide, where IC_{50} = [competing oligonucleotide]/((V_0/V_i) − 1). The K_m is determined for the wild-type RNA/DNA hybrid as described for multiple turnover kinetics and the K_i is calculated from equations of Cheng and Prusoff; $K_i = IC_{50}/(([\text{RNA/DNA hybrid}/K_m) + 1)$.

Conclusions

RNase H1 from *E. coli* is the best-characterized member of the RNase H family, and many of the properties observed for the bacterial enzyme are consistent with the human isotype (e.g., the cofactor requirements, substrate specificity, and binding specificity).[19,20] In fact, the carboxy-terminal portion of human RNase H1 is highly conserved with the amino acid sequence of the *E. coli* enzyme. The glutamic acid and two aspartic acid residues of the catalytic site, as well as the histidine and aspartic acid residues of the proposed second metal binding site of the *E. coli* enzyme, are conserved in human RNase H1.[16,28–30] In addition, the lysine residues within the α-helical basic substrate-binding region of *E. coli* RNase H1 are also conserved in the human enzyme. Whether these conserved amino acid residues serve the same function in the human enzyme remains to be seen.

Aside from these similarities, the structures of the two enzymes differ in several important ways. For example, the amino acid sequence of the human enzyme is approximately twofold larger than the *E. coli* enzyme. The human enzyme contains an extra amino-terminal region homologous with a double-strand RNA binding motif, and human RNase H1 contains two additional cysteine residues. The role of the double-strand RNA-binding region is unclear. The strong positional preference for cleavage displayed by human RNase H1, as well as the enhanced binding affinity of the enzyme for various nucleic acids, may be due to the presence of this additional binding region. Although the involvement of the additional cysteine residues also remains unclear, the inhibition of the human enzyme with sulfhydryl blocking agents suggest that at least some of these residues are essential for the function of the enzyme.

Compared with human RNase H1, significantly more work is needed to support the characterization of the RNase H2 enzyme. Preliminary studies involving human RNase H2 suggest that the protein exhibits little to no ribonuclease activity (Wu, unpublished data). This is surprising considering that the RNase H2

[28] H. Nakamura, Y. Oda, S. Iwai, H. Inoue, E. Ohtsuka, S. Kanaya, S. Kimura, C. Katsuda, K. Katayanagi, K. Morikawa, H. Miyashiro, and M. Ikehara, *Proc. Natl. Acad. Sci. U.S.A.* **88,** 11535 (1991).

[29] K. Katanagi, M. Miyagawa, M. Matsushima, M. Ishikawa, S. Kanaya, M. Ikehara, T. Matsuzaki, and K. Morikawa, *Nature* **347,** 306 (1990).

[30] W. Yang, W. A. Hendrickson, R. J. Crouch, and Y. Satow, *Science* **249,** 1398 (1990).

enzyme is ubiquitously expressed in human cells and tissues and has been shown to exhibit immunological cross-reactivity with the RNase H active protein from calf thymus.[14] Perhaps the enzyme requires significantly different conditions for activity. Alternatively, the acidic nature of human RNase H2 (p*I* of 4.94) is unusual in that the majority of nucleic acid binding proteins are basic (p*I* > 7) and perhaps the enzyme requires an additional protein subunit for activity.

[28] Assays for Retroviral RNase H

By CHRISTINE SMITH SNYDER and MONICA J. ROTH

Assays for Retroviral RNase H

The retroviral reverse transcriptase (RT) is a multifunctional enzyme consisting of RNA- and DNA-dependent DNA polymerase and RNase H activities. The polymerase functions are localized to the N terminus of the protein; the RNase H domain is at the C terminus. Both polymerase and RNase H functions are required for successful replication of the viral RNA into double-stranded DNA. Synthesis of DNA using RNA as a template generates an RNA/DNA hybrid. RNase H activity selectively degrades the RNA strand within an RNA/DNA hybrid. The RNase H activity is therefore required to remove the viral RNA throughout the replication process. In addition, more specific RNase H cleavages are required.

Retroviral replication (for a review, see Ref. 1) is initiated using a cellular tRNA as a primer, which hybridizes near the 5′ end of the plus-strand viral RNA (vRNA) at the primer binding sequence (PBS). Reverse transcription to the capped RNA terminus generates the (−) strand strong stop DNA. Degradation of the RNA within the minus-strand strong-stop DNA exposes the repeated sequence (R region). Completion of minus-strand synthesis requires the transfer or "jump" of the DNA to the 3′ terminus of the vRNA, through base-pairing of the R regions. The primer for plus-strand synthesis is generated by an RNase H cleavage at the polypurine tract (PPT) within the viral RNA. The generation and subsequent removal of the (+) strand primer is critical for the virus because the position of cleavage defines the U3 terminus of the viral long terminal repeats (LTRs). More complex retroviruses, such as the human immunodeficiency virus (HIV), contain a second PPT within the center of the viral genome.[2,3] For

[1] A. Telesnitsky and S. P. Goff, *in* "Retroviruses" (J. M. Coffin, S. H. Hughes, and H. E. Varmus, eds.), p. 121. Cold Spring Harbor Laboratory Press, Plainview, NY, 1997.

[2] P. Charneau and F. Clavel, *J. Virol.* **65,** 2415 (1991).

[3] P. Charneau, M. Alizon, and F. Clavel, *J. Virol.* **66,** 2814 (1992).

0076-6879/01 $35.00

all retroviruses, the tRNA remains covalently linked to the minus-strand DNA after the first strand-transfer reaction. Plus-strand synthesis utilizes the minus-strand DNA as template and proceeds to copy the first 18 nucleotides of the tRNA primer.[4] Synthesis utilizing the tRNA as template creates another substrate for RNase H. Removal of the tRNA primer must be precise, since incomplete removal subsequently affects the U5 terminus of the LTRs. Removal of the tRNA exposes the sequences that can hybridize with the PBS sequences at the 5′ terminus of the complete minus-strand DNA. The second-strand transfer event can proceed and results in the completion of the double-strand DNA synthesis.

With the long half-life of cells bearing integrated retroviral genomes and the rapid selection of retroviruses resistant to drug therapy, there is a great need to identify new and effective drugs that target divergent aspects of the viral life cycle. For HIV-1, drugs that inhibit the polymerase activity of RT have proven effective. Potent inhibitors of the RNase H activity of reverse transcriptase remain to be developed. These could target the specific or the nonspecific RNase H cleavages required for successful viral replication. Screens for such compounds indicate the potential of RNase H as a target.[5,6] In this chapter, we summarize the assays used for retroviral RNase Hs in general, as well as the assays that define the unique cleavages of retroviral replicative intermediates.

Nonspecific RNase H Assays

RNase H cleaves the phosphodiester bond of an RNA within an RNA/DNA hybrid, yielding a 3′-OH and a 5′-P terminus. For nonspecific cleavages, the RNA is usually degraded to small products that are acid-soluble. Although there are many possible assays to measure RNase H activity, a major consideration regarding the assay of choice depends on the source and purity of the enzyme. Contamination by host cellular RNase Hs in preparations of reverse transcriptase must be considered. Of particular note, the *Escherichia coli* RNase H is a highly active enzyme and trace contamination can obscure the retroviral RNase H activity in RT preparations from bacteria. To address this, *in situ* RNase H assays are performed, where the RNA/DNA labeled substrate is cross-linked into a sodium dodecyl sulfate (SDS)–polyacrylamide gel. Protein fractions are electrophoresed and renatured within the gel, and the RNase H activity is visualized by the loss of the radioactive substrate corresponding with the position of the protein in the gel. This protocol has been previously described (see Ref. 7), as well as in a chapter in this book. Although

[4] M. J. Roth, P. Schwartzberg, and S. P. Goff, *Cell* **58,** 47 (1989).

[5] S. Gabbara, W. R. Davis, L. Hupe, D. Hupe, and J. A. Peliska, *Biochemistry* **38,** 13070 (1999).

[6] G. Borkow, R. S. Fletcher, J. Bernard, D. Arion, D. Motakis, G. I. Dmitrienko, and M. A. Parniak, *Biochemistry* **36,** 3179 (1997).

[7] A. Telesnitsky, S. Blain, and S. P. Goff, *Methods Enzymol.* **262,** 347 (1995).

this assay definitively assigns the RNase H activity with the reverse transcriptase (or subdomains[8]), the protocol can take days and is not suitable for rapid analysis.

Purified reverse transcriptase can be assayed for nonspecific RNase H cleavages by measuring the release of acid-soluble RNA products. From our experience, the purity of His_6-tagged reverse transcriptase subdomains encoding RNase H purified through two nickel-affinity (NTA-Qiagen) columns and eluted with pH 4.5 steps have been suitable for analysis with this assay.[8,9]

Method

Preparation of Labeled RNA : DNA Substrate. Nonspecific RNA/DNA hybrids are prepared in which the RNA is uniformly labeled. Synthesis by *E. coli* RNA polymerase on single-stranded (ss) templates yields RNA/DNA hybrids. Commercial preparations of *E. coli* RNA polymerase core or holoenzyme have proven suitable for substrate preparation. A convenient and suitable source of template DNA is the single-stranded circular DNA of bacteriophage M13. This can be purchased (M13+ strand DNA, Pharmacia, Piscataway, NJ) or isolated using standard procedures.[10] Briefly, the bacteria are separated from the phage by centrifugation. Bacteriophage are then precipitated by the addition of 0.25 v/v of 20% PEG (8000), 3.5 *M* ammonium acetate, pelleted at 17,000 *g*, resuspended in TE and extracted 4× with equal volumes of phenol : chloroform and 3× with chloroform. The ssDNA is ethanol precipitated in the presence of 1.5 *M* ammonium acetate. In this assay, the substrate is labeled by incorporation of the nucleotide [α-^{35}S]UTP. However, ^{3}H-NTP or [α-^{32}P]NTP may be substituted. The isotope selected influences the half-life of the RNA/DNA hybrid substrate and the frequency with which the substrate must be synthesized.

Reaction mixtures (0.5 ml) containing 15 μg single-stranded M13 DNA, 68 μg *E. coli* RNA polymerase (Promega, Madison, WI), 40 m*M* Tris-HCl, pH 8.0, 8 m*M* $MgCl_2$, 2 m*M* dithiothreitol (DTT), 100 m*M* KCl, 115 μM each of CTP, GTP, ATP, 14 μM UTP, and 250 μCi [α-^{35}S]UTP are incubated at 37°. The extent of incorporation is monitored by spotting an aliquot of the reaction (1 μl) on DE81 Whatman (Clifton, NJ) paper, washing 2 times with 2× SSC, once with ethanol, and counting the dried filter in a scintillation counter. Reactions are stopped by the addition of EDTA to 10 m*M*, when the incorporation reaches a plateau (generally 1.5 hr). Reactions are extracted once with phenol/chloroform and ethanol precipitated in the presence of 1.5 *M* ammonium acetate.

In order to ensure the integrity of the substrate as an RNA/DNA hybrid, single-stranded radiolabeled RNA is removed by digestion with RNase A in the presence

[8] J. S. Smith and M. J. Roth, *J. Virol.* **67,** 4037 (1993).

[9] J. S. Smith, K. Gritsman, and M. J. Roth, *J. Virol.* **68,** 5721 (1994).

[10] J. Sambrook, E. F. Fritsch, and T. Maniatis, "Molecular Cloning: A Laboratory Manual," 2nd Ed., p. 4.31. Cold Spring Harbor Laboratory Press, Cold Spring Harbor, NY, 1989.

of high salt. After ethanol precipitation (see above), the substrate is resuspended and treated with 50 mM sodium acetate, pH 5.0, 300 mM NaCl, 10 μg RNase A (100 μl), at 25° for 2 h. The substrate is extracted with phenol/chloroform and chloroform. Any remaining free nucleotides are removed by size fractionation on a Sephadex G-50 column. The substrate preparation is loaded onto a 1 cm × 22 cm Sephadex G-50 column (made in a disposable 5 ml pipette) equilibrated in TE and 100 μl fractions are collected. The radioactive fractions within the void volume of the column are pooled, ethanol precipitated, and resuspended in RNase-free H_2O [diethyl pyrocarbonate (DepC) treated].

Nonspecific RNase H Assay. Standard reaction mixtures (50 μl) contain 3 pmol of RNA/DNA hybrid substrate. Reaction conditions can be modified based on the specific RT/RNase H enzyme being studied. RNase H assays for HIV-1 RT are performed in 20 mM Tris-HCl, pH 7.5, 50 mM KCl, 0.6 mM to 8 mM $MgCl_2$, and 2 mM DTT. Although the wild-type HIV-1 RT is active in either $MgCl_2$ or $MnCl_2$, specific HIV-1 RT mutants, such as E478Q, are only active in $MnCl_2$.[11] Similarly, an isolated HIV-1 RNase H domain, NY427,[8] requires 8 mM $MnCl_2$ and is assayed in the presence of 20 mM 2-[*N*-morpholino]ethanesulfonic acid (MES), pH 6.0 and 2 mM DTT. Reaction mixtures for MuLV RT/RNase H contain 50 mM Tris-HCl, pH 7.5, 2 mM DTT, 40 mM KCl, and 1 mM $MnCl_2$.[12] In reaction mixtures with $MnCl_2$ and DTT, care should be taken that the DTT is fresh and added to the buffer last. If the $MnCl_2$ is oxidized, the buffer will turn brown and can affect the activity of the enzyme. Reactions are incubated for 30 min at 37° and are stopped by the addition of 50 μl of 100 mM $NaPP_i$ (pH 6), followed by precipitation with 8% trichloroacetic acid (TCA). The mixture is centrifuged for 10 min, and the radioactivity in the supernatant is determined by liquid scintillation counting. Carrier BSA or sonicated DNA can be added to assist in the TCA precipitation. The percent UMP released within the 30-min reaction can be determined by dividing the number of counts released by the number of input counts.[8,9]

Generation of Alternative RNase H Substrates. RNA/DNA hybrids can be generated through the direct hybridization of RNA to single-stranded DNA. Two sources for RNA are frequently used. The first is to chemically synthesize the RNAs, defining the sequences that are to be studied. The cost of synthesis of RNA oligonucleotides is higher than that for DNA and therefore the size of the oligoribonucleotide is a consideration. Even under optimal synthesis conditions, the oligoribonucleotide should be gel purified from a polyacrylamide gel prior to use.

Alternatively, larger RNAs can be synthesized using T7 RNA polymerase cassettes. The sequences of interest are cloned downstream of a bacteriophage

[11] N. M. Cirino, C. E. Cameron, J. S. Smith, J. W. Rausch, M. J. Roth, S. J. Benkovic, and S. F. J. Le Grice, *Biochemistry* **34,** 9936 (1995).

[12] M. J. Roth, N. Tanese, and S. P. Goff, *J. Biol. Chem.* **260,** 9326 (1985).

T7 promoter. Restriction enzyme digestion of the plasmid followed by transcription using T7 RNA polymerase yields a runoff transcript of a defined length. The RNA is purified by gel electrophoresis and eluted. The use of the T7 RNA polymerase allows for the generation of RNA/DNA hybrid substrates varying in length.[13]

To monitor the RNA cleavages, the RNA termini are 5′-end labeled with T4 polynucleotide kinase and [γ-^{32}P]ATP. RNA molecules produced by T7 RNA polymerase require treatment with calf intestine alkaline phosphatase prior to the kinase reaction. Reaction mixtures (15 μl) for efficient 5′-labeling of RNA contain 40 pmol of RNA or RNA–DNA oligonucleotide, RNasin (20 units; Promega), T4 polynucleotide kinase (20 units), 320 μCi [γ-^{32}P]ATP (7000 Ci/mmol), 50 m*M* Tris-HCl, pH 7.6, 10 m*M* $MgCl_2$, and 10 m*M* DTT. Reactions are incubated for 1.5 h at room temperature.

The isolated RNA (either the oligoribonucleotide or T7 RNA transcript) is then hybridized to single-stranded DNA. The DNA can be a chemically synthesized oligodeoxynucleotide or, if a long template is necessary, single-stranded M13 phage DNA containing the complementary sequences (see above, nonspecific substrate). Annealing of the RNA and DNA is performed in 10 m*M* Tris-HCl (pH 8.0), 1 m*M* EDTA, and 80 m*M* KCl. The sample is heated to 60° for 10 min and slowly cooled for 90 min. Excess unannealed primer can be removed by gel filtration spin chromatography. To ensure the complete removal of any unannealed primer, the sample can be subjected to native polyacrylamide gel electrophoresis and purified from the gel.[13]

The use of the 5′-end-labeled RNA allows for visualization of particular reaction products rather than determining the percent of RNA released. Reaction mixtures are specific for the retroviral reverse transcriptase/RNase H studied (see nonspecific RNase H assays, above). Reactions are terminated by the addition of 2× termination mix [90% formamide (v/v), 10 m*M* EDTA (pH 8.0), and 0.1% each of xylene cyanol and bromphenol blue]. Samples can then be subjected to denaturing gel electrophoresis to resolve reaction products. These reaction products can be quantified in order to determine the percentage of substrate being cleaved.

Specific RNase H Cleavages

Cleavages at the PPT

RNase H is responsible for generation and removal of the PPT primer. Assays have been developed to analyze both of these steps during HIV-1 and Moloney

[13] C. Palaniappan, G. M. Fuentes, L. Rodriguez-Rodriguez, P. J. Fay, and R. A. Bambara, *J. Biol. Chem.* **271,** 2063 (1996).

murine leukemia virus (M-MuLV) reverse transcription processes.[14,15] The PPT sequence is relatively resistant to cleavage by RNase H. Directed RNase H cleavages at either side of the PPT produce a stable RNA/DNA hybrid that is extended by RT. Once plus-strand priming has occurred from the PPT, RNase H must now function to remove this sequence. Assays to look at both the generation and removal of the PPT will be described herein.

Generation of (+) Strand Primer. Substrates encoding the PPT sequences are generated through annealing of sequence specific RNAs and DNA. This system can be modified to analyze any retroviral system by changing the sequence of the input RNA/DNA hybrid to sequences representative of the particular retroviral PPT. Substrates can be long RNA/DNA hybrid duplexes or short oligonucleotides (see alternative RNA/DNA substrates). For M-MuLV, the PPT consists of 15 nucleotides with the sequence 5′AGAAAAAGGGGGAA 3′. Oligoribonucleotides as short as 16-mers containing the sequence 5′AGAAAAAGGGGGGAAU 3′ within an RNA/DNA hybrid are recognized by MuLV RT and are cleaved to yield the fragment 5′AGAAAAAGGGGG 3′OH,[14] where cleavage occurs between a G and an A nucleotide at positions 7815 and 7816.[16]

The RNA oligonucleotide is 5′-labeled with [γ-^{32}P]ATP (see Alternative RNA/DNA Substrates, above, for conditions). The 5′ labeled RNA oligonucleotide (5 pmol) is annealed with excess template strand (10 pmol), to drive the reaction in a 50 μl reaction containing 0.2 *M* NaCl, 25 m*M* Tris-HCl, pH 8.0. The sample is heated at 90° for 3 min and slowly cooled to room temperature. The sample is then heated to 37° for 15 min, followed by precipitation in the presence of 2 μg glycogen and 2 volumes 100% ethanol. After centrifugation, the pellet is redissolved in 10 m*M* Tris-HCl, pH 8.0, 1 m*M* EDTA.[14]

The resulting RNA/DNA hybrid substrate mimicking the PPT intermediate can be assayed for cleavage by a viral RT/RNase H enzyme. For MuLV, reaction mixtures (20 μl) contain 0.1 pmol (5 m*M* final concentration) of hybrid oligonucleotide, 1 pmol of RT or RNase H enzyme, 50 m*M* Tris-HCl, pH 8.0, 6 m*M* $MgCl_2$, 1 m*M* DTT, 100 μg/ml bovine serum albumin (BSA). Reactions are incubated at 37° and can be terminated by the addition of an equal volume of formamide stop mix (2×: 80% deionized formamide, 1 m*M* EDTA, 0.2% bromphenol blue, 0.2% xylene cyanol). Samples can then be analyzed on a 20% sequencing gel.[14]

Polypurine Tract Primer Removal Assay. For HIV-1 RT, an assay for PPT primer removal has been developed that analyzes the ability of an RNase H to remove the PPT primer once plus-strand initiation has occurred. An RNA

[14] S. J. Schultz, M. Zhang, C. D. Kelleher, and J. J. Champoux, *J. Biol. Chem.* **274,** 34537 (1999).

[15] M. D. Powell, M. Ghosh, P. S. Jacques, K. J. Howard, S. F. Le Grice, and J. G. Levin, *J. Biol. Chem.* **272,** 13262 (1997).

[16] T. M. Shinnick, R. A. Lerner, and J. G. Sutcliffe, *Nature* **293,** 543 (1981).

oligonucleotide mimicking the 15-nucleotide sequence of the HIV-1 PPT (5′ AAAAGAAAAGGGGGG 3′) is annealed to a 35-mer template oligonucleotide (5′-AGTGAATTAGCCCTTCCAGTCCCCCCTTTTCTTTT-3′) and extended using T4 DNA polymerase in the presence of [α-^{32}P]dATP. The resulting RNA–DNA hybrid will be 35 nucleotides in length, and removal of the PPT primer will result in the release of 20-mer DNA oligonucleotide.[15]

Reactions are performed at 37° with 1 pmol of the RNA/DNA hybrid substrate and 10 pmol of enzyme in a total reaction volume of 15 μl in RT buffer (50 m*M* Tris-HCl, pH 7.9, 50 m*M* KCl, 10 m*M* $MgCl_2$, 6 m*M* DTT).[17] Reactions are carried out for 20 min and stopped by the addition of formamide stop mix (see above). The reaction products are fractionated on polyacrylamide gels and analyzed by autoradiography.[15]

Removal of the tRNA Primer

Reverse transcriptase associated RNase H enzymes function to specifically remove the tRNA primer after plus-strand strong stop DNA synthesis has occurred. This step can be analyzed in an *in vitro* assay by using a substrate that mimics the intermediate. Mimics of HIV-1, M-MuLV, or ASLV intermediate can be generated by encoding the cognate U5 and tRNA sequences. HIV-1 uses the $tRNA^{Lys,3}$, MuLV uses $tRNA^{Pro}$, and ASLV uses $tRNA^{Trp}$. Using defined U5-tRNA substrates *in vitro,* the initial cleavage site by both the M-MuLV and HIV-1 RT/RNase H has been determined to occur between the C and A within the CCA sequence at the 3′ terminus of the tRNA. For MuLV, the terminal A nucleotide of the tRNA is rapidly removed.[18] In contrast, in HIV-1, the terminal ribonucleotide of the tRNAs remains associated with the DNA and is copied upon completion of plus-strand DNA synthesis.[19,20]

Preparation of Substrate. The RNA/DNA hybrid substrate must first be constructed to mimic the intermediate found during viral replication. At this stage of replication, the tRNA is covalently linked to the minus-strand DNA and is hybridized with DNA through the first 18 nucleotides of the tRNA. An RNA oligonucleotide is synthesized to mimic the first 18 nucleotides of the tRNA primer being analyzed (for HIV-1, $tRNA^{Lys,3}$), and a DNA oligonucleotide is synthesized complementary to the tRNA sequence that mimics the newly synthesized (+) strand of the U5/PBS, approximately 38 nucleotides in length.

To produce the substrate, 40 pmol of RNA oligonucleotide encoding the 3′ terminus of the tRNA is 5′ end-labeled with [γ-^{32}P]ATP (see Alternative RNA/DNA Substrates, above, for conditions). The 5′-end-labeled RNA oligonucleotide is isolated from a 20% sequencing gel and eluted overnight in 0.5 *M* NH_4Ac and 0.01 *M*

[17] M. D. Powell and J. G. Levin, *J. Virol.* **70,** 5288 (1996).
[18] C. M. Smith, W. B. Potts III, J. S. Smith, and M. J. Roth, *Virology* **229,** 437 (1997).
[19] J. S. Smith, S. Kim, and M. J. Roth, *J. Virol.* **64,** 6286 (1990).
[20] J. S. Smith and M. J. Roth, *J. Biol. Chem.* **267,** 15071 (1992).

magnesium acetate (900 μl). The eluate is concentrated sevenfold by lyophilization, followed by precipitation of the sample with 100% ethanol at $-80°$, overnight. The pellets are carefully washed three times with 100% ethanol and resuspended in 20 μl DepC-treated water. The recovery of the sample should be constantly measured by monitoring the radioactivity present in the washes, as well as that remaining in the pellet. The labeled RNA is annealed to 40 pmol of the template strand DNA encoding the 5′ U5 DNA-PBS-3′ sequence in a 20 μl reaction containing 50 m*M* Tris-HCl pH 8.0, 50 m*M* KCl, 8 m*M* $MgCl_2$, and 2 m*M* DTT. Extension reactions are performed to synthesize the U5 sequence using the 3′ OH of the tRNA as a primer. The extension reaction contains 0.4 m*M* dCTP, 0.4 m*M* dATP, 0.4 m*M* dGTP, 0.4 m*M* TTP, and 8 units Klenow Exo(−) polymerase, and is performed for 5 min on ice, followed by room temperature for 5 min, then 1 h at 37°. The extension reactions are stopped with addition of 22 μl formamide stop buffer, separated on a 15% denaturing polyacrylamide gel, isolated (in the same manner as described above), and resuspended in 20 μl RNase-free water. The RNA–DNA molecule is annealed to a complementary DNA oligonucleotide in 20 m*M* Tris-HCl (pH 7.5), 50 m*M* KCl buffer by heating at 70° for 4 min followed by slow cooling to room temperature.

For tRNA removal, the specificity of cleavage can be monitored with substrates labeled either at the 5′ terminus of the RNA primer or within the DNA linked to the RNA. Radiolabel within the DNA can be incorporated either during synthesis or through ligation of a 5′-labeled DNA oligonucleotide to the RNA oligonucleotide in the presence of a complementary bridging oligonucleotide.[21] For substrates in which the DNA portion is internally labeled, 20 pmol RNA is annealed to 20 pmol DNA using the same conditions as described above. The extension reactions are performed with the final reaction volume containing 0.4 m*M* dGTP, 0.4 m*M* dCTP, 0.4 m*M* TTP, 0.2 m*M* dATP, 50 m*M* Tris-HCl pH 8.0, 8 m*M* $MgCl_2$, 2 m*M* DTT, 8 units Exo(−) Klenow polymerase, and 20 μCi of [α-^{32}P]dATP. The reactions are performed like those described above.

Substrate preparation can be simplified by chemically synthesizing an oligonucleotide containing RNA joined to DNA. This eliminates the need to perform extension reactions to synthesize the DNA using Exo(−) Klenow polymerase. In this case, the RNA–DNA oligonucleotide is simply 5′ labeled and annealed to the complementary DNA strand. Minimally, substrates containing 5 deoxynucleotides covalently linked to the RNA are specifically cleaved in the tRNA removal assays using HIV-1 model substrates and the HIV-1 RT/RNase H.[22]

In addition to using RNA oligonucleotides to mimic the tRNA primer, authentic tRNA primer as well as *in vitro* transcribed tRNA primer can be used to construct the model substrate. The benefit of using authentic tRNA primer versus the *in vitro* transcribed primer is that it possesses the necessary modifications (m1A

[21] M. Moore and P. Sharp, *Science* **256,** 992 (1992).
[22] C. M. Smith, O. Leon, J. S. Smith, and M. J. Roth, *J. Virol.* **72,** 6805 (1998).

methylation at position 19 within the tRNA primer) for termination of synthesis to occur.[4,23] This is an important factor if RNase H activity is being investigated along with the production of plus-strand strong-stop DNA.

Because of the high degree of secondary structure within the tRNA molecule, it is necessary to modify the substrate preparation protocol when constructing the RNA/DNA hybrid. The authentic $tRNA^{Lys,3}$ substrate is ligated to a DNA oligonucleotide mimicking the U5 sequences of the viral genome, in the presence of a complementary bridging oligonucleotide.[21] The bridging oligonucleotide can be the same oligonucleotide used as the annealing template strand, indicated above. In this case, it is necessary to synthesize an oligonucleotide complementary to the U5 sequences in the annealing strand such that it may be ligated to the tRNA. This oligonucleotide will be 5′ labeled using T4 polynucleotide kinase and [γ-^{32}P]ATP. The radiolabeled DNA is isolated utilizing Sephadex G-25 spin columns (Boehringer Mannheim). The 5′ DNA oligonucleotide and authentic tRNA are annealed to the bridging oligonucleotide in a reaction containing 200 m*M* KCl and 50 m*M* Tris-HCl, pH 7.5. The sample should be heated at 95° for 5 min, shifted to 58° for 30 min, then allowed to slowly cool to room temperature. After annealing, the authentic $tRNA^{Lys,3}$ and DNA oligonucleotide are ligated using T4 DNA ligase in a reaction containing 50 m*M* Tris-HCl, pH 7.5, 10 m*M* $MgCl_2$, 10 m*M* dithiothreitol, 1 m*M* ATP, and 25 μg/ml bovine serum albumin for 4 h at room temperature. The tRNA/DNA oligonucleotide is gel isolated and eluted overnight as described above. The tRNA/DNA oligonucleotide is annealed to the complementary annealing strand, using the same modified annealing conditions as those described above.

tRNA Removal Assay. Any of the annealed, extended RNA/DNA substrates described above can be utilized in the RNase H assay. The substrate can be incubated with wild-type HIV-1 or MuLV RT, mutant RTs or an isolated RNase H domain. Reactions (20 μl) contain 4 pmol of substrate, in 50 m*M* Tris-HCl, pH 7.5, 50 m*M* KCl, 2 m*M* DTT, and 8 m*M* $MnCl_2$ or 8 m*M* $MgCl_2$. The reactions are incubated at 37°. Time course reactions can monitor the extent of degradation of the RNA. Aliquots (3 μl) are taken at various internals (time points varying between 10 sec and 30 min) and are stopped by the addition of formamide stop buffer. Reactions are separated on 20% denaturing polyacrylamide gel. An example of a tRNA removal assay is shown in Fig. 1. The substrate is a 38-mer RNA/DNA hybrid encoding the 3′-terminal 18 nucleotides of the $tRNA^{Lys,3}$ attached to 20 deoxynucleotides. Since the RNA is 5′-end-labeled, specific cleavage of the tRNA would release a 17-mer, as the terminal A residue of the tRNA remains associated with the DNA. With the mutant HIV-1 RT, this initial cleavage is detected (center panel). In this assay, a low level of −2 product is detected; however, on long incubation, products smaller than this are not observed. With the wild-type HIV-1 RT, the initial cleavage occurs and

[23] S. Auxilien, G. Keith, S. F. J. Le Grice, and J.-L. Darlix, *J. Biol. Chem.* **274,** 4412 (1999).

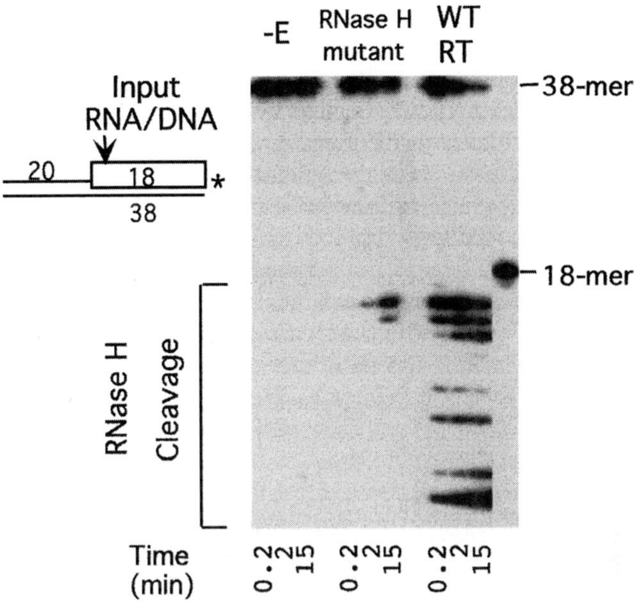

FIG. 1. tRNA removal assay. Time course of cleavage of a $tRNA^{Lys,3}$ mimic substrate catalyzed by wild-type HIV-1 RT and an RNase H mutant is shown. Aliquots were analyzed at the time (in minutes) indicated at the bottom. The substrate is diagrammed on the left and consists of a 5′-end-labeled 18-mer RNA linked to 20 nucleotides of DNA hybridized to a 38-mer oligonucleotide. Reactions were performed as described in the text for the tRNA removal assay in the presence of 8 m*M* $MnCl_2$.

is followed by rapid further cleavages to yield smaller products (Fig. 1, right-hand side). This further degradation of the RNA assists in the release of the RNA and is termed directional processing, or polymerase-independent RNase H cleavage (for review, see Ref. 24). The mutation, therefore, affects the directional processing of the RNase H. Interestingly, the wild-type avian Rous sarcoma virus RT/RNase H is reported to lack the RNase H 3′->5′ directed processing activity.[25] Therefore, retroviral RTs differ in their subunit composition and catalytic activities.

Coupled Polymerase and RNase H Activities

In many instances, the activities of the polymerase and the RNase H domains of RT are coupled. RNase H cleavages that occur between 18 and 20 nucleotides behind the site of active polymerization are termed polymerase-dependent RNase H cleavages. Assays have been developed to monitor these coupled reactions (for review, see Ref. 24).

[24] E. J. Arts and S. F. Le Grice, *Prog. Nucleic Acid Res. Mol. Biol.* **58,** 339 (1998).
[25] S. Werner and B. M. Wohrl, *J. Biol. Chem.* **274,** 26329 (1999).

Quite fortuitously, the virus has evolved such that during plus-strand synthesis, when the enzyme is stopped at the modified nucleotide of the tRNA, the cleavage site required for tRNA removal is optimally positioned at the RNase H active site. Removal of the tRNA allows the strand-transfer reaction to proceed. The assay outlined in Fig. 2 highlights the coordination between these activities allowing strand transfer to occur. The substrate is a modification of that used in the tRNA removal assay, in which the RNA is attached to a 32-mer DNA and hybridized to a DNA primer (Fig. 2A, Step 1). The input substrate and primer are both 5′-end labeled and can be visualized in Fig. 2B (minus enzyme panel). Extension by RT to the terminus of the RNA yields a 58-mer DNA product (Fig 2A, Step 2; and Fig. 2B, +dNTP panel). The 5′ tail of the primer distinguishes the input 50-mer from the 58-mer extension product. Synthesis utilizing the RNA as template creates an RNA/DNA hybrid. RNase H activity removes the tRNA mimic at the −1 position, releasing a 17-mer (Fig. 2A, Step 3). For HIV-1 RT, the RNA within the hybrid is subsequently degraded to smaller products by RNase H directional processing (Fig. 2B, HIV-1 RT panels). Removal of the RNA sequences allows entry of an acceptor molecule that serves as a template for synthesis. Strand-transfer yields synthesis until the 5′ terminus of the acceptor is reached, producing a 70-mer product (Fig. 2A, Step 4 and Fig. 2B, HIV-1 RT +dNTP + acceptor).

HIV-1 RT bearing RNase H active site mutations are capable of synthesis of 58-mer extension product; however, no RNase H cleavages or subsequent strand-transfer reactions can be produced in the presence of Mg^{2+} (Fig. 2B, RNase H RT panel). For HIV-1 RT, manipulation of the metal ion can permit the initial RNase H cleavage (see above, tRNA removal assay), to produce the 17-mer product. However, this cleavage is not sufficient for entry of the acceptor.[26]

Preparation of Substrate. The substrate for the plus-strand transfer reaction is diagrammed in Fig. 2, Step 1. It consists of a 50-mer 5′-end labeled RNA–DNA molecule hybridized to a 26-mer DNA primer. The 50-mer RNA-DNA substrate is prepared similarly to the preparation of tRNA removal substrates. Twenty pmol of an 18-mer RNA mimic encoding the 3′ terminus of the tRNA is 5′ labeled with T4 polynucleotide kinase and [γ-^{32}P]ATP (see Alternative RNase H Substrates) and isolated using a Sephadex G-25 spin column. The labeled RNA is annealed to 40 pmol of a 60-mer oligonucleotide. The DNA oligonucleotide has a 10 base extension at its 3′ terminus from the region complementary to the RNA. Hybridization and extension of the RNA would therefore yield a 50-mer product. In a reaction mixture (25 μl) containing 50 m*M* Tris-HCl (pH 8.0), 50 m*M* KCl, 8 m*M* $MgCl_2$, and 2 m*M* dithiothreitol, the substrate is extended with Klenow exonuclease (−) polymerase, and the 50-mer product is isolated on a 15% sequencing polyacrylamide gel. Extreme care should be taken in excision of this species to avoid copurification of the template strand, which will yield a substrate for RNase H.

[26] C. M. Smith, J. S. Smith, and M. J. Roth, *J. Virol.* **73,** 6573 (1999).

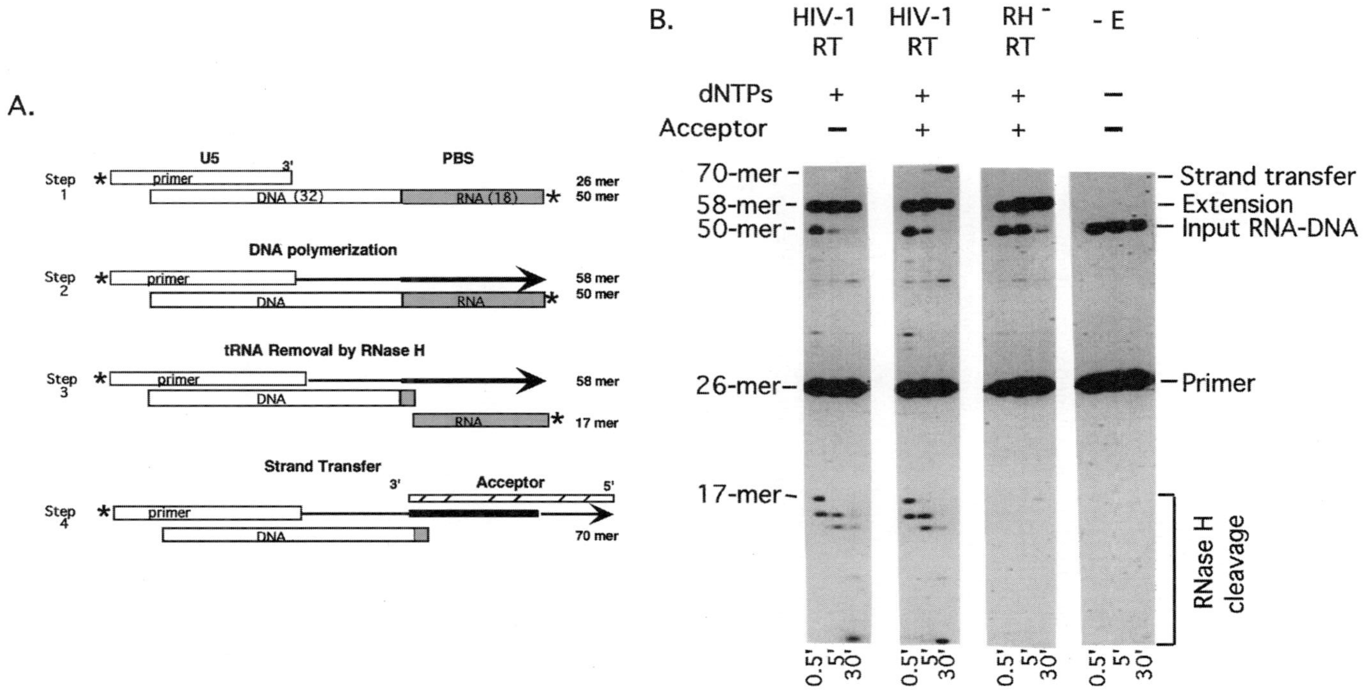

FIG. 2. (A) Model of plus-strand transfer reaction. The four steps to the strand transfer reaction are outlined. The four steps are: (1) input substrate, (2) DNA polymerization, (3) RNase H cleavage of the RNA : DNA hybrid, and (4) entry of acceptor and strand transfer. The RNA corresponding to the first 18 nucleotides of the primer tRNA is in gray. The sizes of the DNA products are indicated in parentheses as well as on the right-hand side. The newly synthesized DNA is shown as an arrow. The radiolabel is shown by a star. The acceptor molecule is the hatched box. (B) Plus-strand transfer assay. A time course of a strand-transfer reaction is shown; the aliquots were analyzed at the time indicated at the bottom of the panel. The size of the nucleic acids are indicated on the left-hand side and their identities are indicated on the right-hand side. The grid on top indicates the presence or absence of dNTPs, acceptor molecules, or enzyme in the reactions. Enzymes used are either wild-type HIV-1 RT and RNase H minus mutant (RH RT) or no enzyme (−E).

If the template strand is copurified, RNase H cleavage of the substrate will be detected in the subsequent strand transfer reaction in the absence of addition of dNTPs in the reaction. The RNA–DNA hybrid is eluted from the gel overnight as described above (tRNA removal assay). The 26-mer DNA primer is similarly 5′-end labeled with T4 polynucleotide kinase and a twofold molar excess is annealed to the RNA–DNA molecule in 20 m*M* Tris-HCl, pH 7.5, 50 m*M* KCl buffer by heating at 70° for 4 min followed by slow cooling to room temperature.

(+) Strand-Transfer Reactions. The substrate for the plus-strand transfer reaction is the 50-mer RNA–DNA hybrid annealed to the 26-mer primer. Both molecules are end-labeled to allow simultaneous monitoring of DNA synthesis and RNase H cleavage. The acceptor molecule encodes the sequence of the tRNA at its 3′ terminus plus additional nucleotides of defined length at the 5′ terminus. In the experiments outlined below, the acceptor molecule is a 30-mer. Incorporation of a ddNTP at the 3′ terminus of the acceptor molecule can eliminate the use of the acceptor molecule as a primer. In this assay, the extension product from the acceptor is not radioactively labeled and therefore cannot be detected. For HIV-1 RT, strand-transfer reactions (20 μl) contain approximately 4 pmol of substrate (50-mer RNA-DNA hybrid annealed to 26-mer primer), 4 pmol of acceptor, 2 pmol of RT, in reaction buffer containing 20 m*M* Tris-HCl, pH 7.5, 50 m*M* KCl, 8 m*M* $MgCl_2$, or 8 m*M* $MnCl_2$, and 0.25 m*M* each deoxynucleotide (dATP, dCTP, dGTP, and TTP). The reactions are initiated by addition of enzyme at 37°. For time-course reactions, aliquots are removed and analyzed by electrophoresis on a 15% denaturing gel followed by autoradiography. The viral nucleocapsid (NC) protein can be included in the reactions.[27]

Acknowledgment

This article was written with the support of award RO1 CA90174 from the National Institutes of Health.

[27] T. Wu, J. Guo, J. Bess, L. E. Henderson, and J. G. Levin, *J. Virol.* **73,** 4794 (1999).

Section V

Synthetic Ribonucleases

[29] Sequence-Selective Artificial Ribonucleases

By MAKOTO KOMIYAMA, JUN SUMAOKA, AKINORI KUZUYA, and YOJI YAMAMOTO

Significance of Artificial Ribonucleases

In current biotechnology and molecular biology, naturally occurring enzymes are being used to transform biomolecules (DNA, RNA, proteins, and others) into desired forms. When new types of transformations are necessary, pertinent enzymes are hunted for in nature. These attempts have been fruitful hitherto. For further developments, however, we have to create nonnatural tools and accomplish what natural enzymes cannot do. Artifical enzymes for sequence-selective RNA scission—the object of this chapter—are typical examples. If only one RNA can be chosen from many others and cleaved at the desired site, it opens the way to new RNA science (regulation of expression of a specific gene in cells, advanced therapy, RNA manipulation, and others). None of the naturally occurring ribonucleases shows such a high sequence-selectivity.

Ribozymes are eminently suited for these purposes.[1] However, these "semiartificial" enzymes, which are composed of an RNA framework, do not necessarily fulfill all the requirements for versatile applications. The following factors sometimes impose limitations: (1) ribozymes are easily destroyed by the ribonucleases in living cells, (2) the scission occurs only at the specific sequences in RNA, and (3) RNAs having complicated tertiary structures are difficult to cleave (ribozymes are highly ordered macromolecules and cause significant steric hindrance). In contrast, "completely artificial ribonucleases," presented here, are designed and synthesized according to the functions and properties we need. The factors (1)–(3) can be minimized. Furthermore, in these man-made catalysts, even nonnatural functional groups can be easily incorporated.

Molecular Design

There are two strategies for sequence-selective RNA scission. In the first one, some catalysts for RNA hydrolysis are covalently attached to the molecules that bind the substrate RNA near the target scission site (Fig. 1a). The molecular scissors are placed at the target phosphodiester linkage and cleave it. A number of molecular scissors, either organic or inorganic, are now available (see Tables I

[1] R. Flores, C. Hernandez, M. de la Pena, A. Vera, and J-A. Daros, *Methods Enzymol.* **341,** [34]; K. J. Hampel, R. Pinard, and J. M. Burke, *Methods Enzymol.* **341,** [36]; A. Michienzi and J. J. Rossi, *Methods Enzymol.* **341,** [37] 2001 (this volume).

METHODS IN ENZYMOLOGY, VOL. 341
0076-6879/01 $35.00

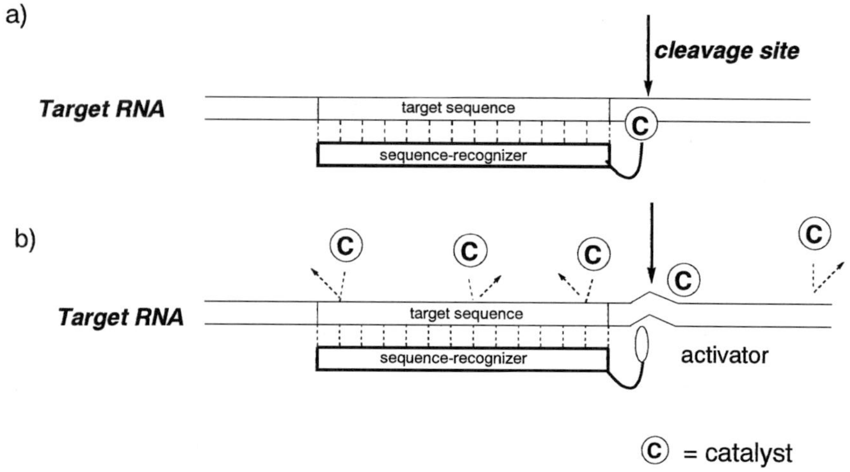

FIG. 1. Two strategies for site-selective RNA scission: (a) hybridization of catalyst and sequence recognizer and (b) noncovalent activation of the target phosphodiester linkage.

and II). As the sequence-recognizing moieties, DNA oligomers, which are complementary with the substrate RNA, are widely used. If necessary, they are chemically modified to provide desired physicochemical and/or biological properties (e.g., increased permeability through cell membranes and stability against enzymatic cleavage). Protein nucleic acids (PNAs), RNA-binding proteins, and other organic molecules can be also used. There is no limitation in their molecular design.

Another strategy is to activate the target phosphodiester linkage in the substrate RNA by using some noncovalent interactions (Fig. 1b). For example, the conformation of the ribose residue at the target site is perturbed so that the intramolecular attack by the 2′-OH toward the phosphorus atom is facilitated. Alternatively, some strain is induced to the target phosphodiester linkage. When catalysts for RNA hydrolysis are added to the system (without being bound to any sequence-recognizing moiety), the activated phosphodiester linkage at the target site is cleaved preferentially over the other linkages.

Methods

Catalysts for RNA Hydrolysis

In some experiments the target RNA is cleaved leading the researcher to conclude that RNA is an unstable molecule. It is an incorrect assumption as RNA is intrinsically very stable (the apparent instability is simply due to contamination of ribonucleases in the specimens). The half-life for spontaneous hydrolysis of one phosphodiester linkage in RNA at pH 7 and 30° is 100–1000 years. For many years in spite of a number of attempts, nonenzymatic RNA hydrolysis was unsuccessful.

TABLE I
TYPICAL CATALYSTS FOR RNA CLEAVAGE

Species	Catalyst form	Activity ($k_{obs} \times 10^3/s^{-1}$) [a]	Reference
Lanthanide(III)	Tm(III), Yb(III), and Lu(III) ions	3 (pH 7, 30°)	3, 4
	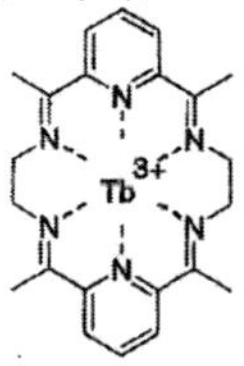 1	0.06 (pH 7, 37°)	5
Copper(II)	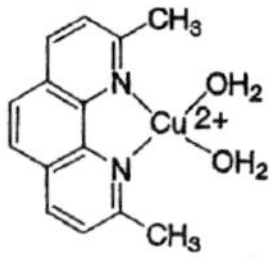 2	4 (pH 7, 25°)	6
	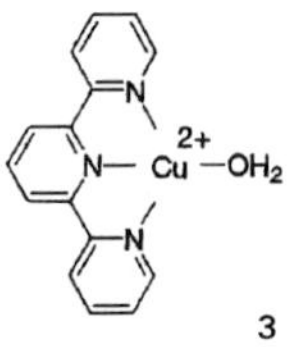 3	0.005 (pH 8, 25°)	7, 8
Cobalt(III)	$[Co((H_2NCH_2CH_2)_3N)(OH_2)_2]^{3+}$	0.004 (pH 7, 50°)	10
Oligoamine	*N,N′*-Dimethylethylenediamine	0.02 (pH 8, 50°)	11, 12

[a] Approximate value for dinucleotide cleavage.

In the last 10 years, however, various useful catalysis for RNA hydrolysis have been found.[1a,2]

Metal Ions and Their Complexes. Lanthanide(III) series of ions, especially Tm(III), Yb(III), and Lu(III) are enormously active catalysts (see Table I). Under

[1a] B. N. Trawick, A. T. Daniher, and J. K. Bashkin, *Chem. Rev.* **98,** 939 (1998).
[2] M. Oivanen, S. Kuusela, and H. Lönnberg, *Chem. Rev.* **98,** 961 (1998).
[3] M. Komiyama, K. Matsumura, and Y. Matsumoto, *J. Chem. Soc., Chem. Commun.,* 640 (1992).
[4] K. Matsumura and M. Komiyama, *J. Biochem.* **122,** 387 (1997).
[5] J. R. Morrow, L. A. Buttrey, V. M. Shelton, and K. A. Berback, *J. Am. Chem. Soc.* **114,** 1903 (1992).
[6] B. Linkletter and J. Chin, *Angew. Chem., Int. Ed. Engl.* **34,** 472 (1995).
[7] M. K. Stern, J. K. Bashkin, and E. D. Sall, *J. Am. Chem. Soc.* **112,** 5357 (1990).
[8] S. Liu and A. D. Hamilton, *J. Chem. Soc., Chem. Commun.,* 587 (1999).
[9] E. L. Hegg, K. A. Deal, L. L. Kiessling, and J. N. Burstyn, *Inorg. Chem.* **36,** 1715 (1997).
[10] M. Komiyama, Y. Matsumoto, H. Takahashi, T. Shiiba, H. Tsuzuki, H. Yajima, M. Yashiro, and J. Sumaoka, *J. Chem. Soc., Perkin Trans.* **2,** 691 (1998).

TABLE II
MULTINUCLEAR COMPLEXES FOR RNA CLEAVAGE

Metal ion	Structure of complex	Activity ($k_{obs} \times 10^3/s^{-1}$)[a]	Reference
Zinc(II)	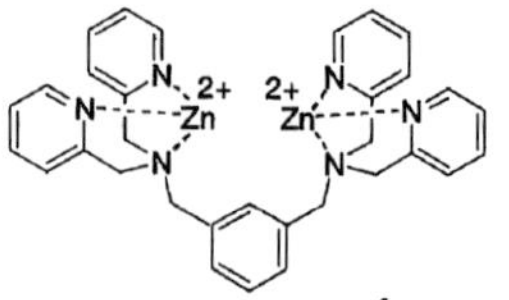 4	0.1 (pH 7, 50°)	13
	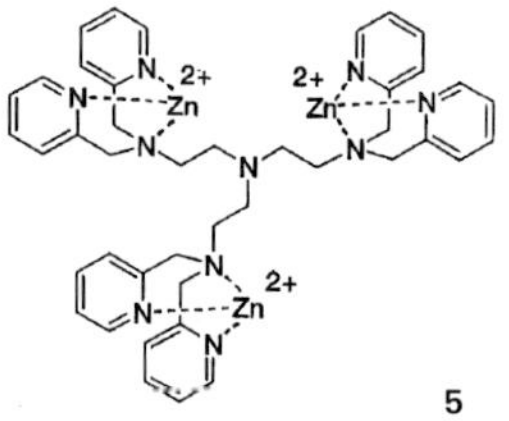 5	0.7 (pH 7, 50°)	14
Copper(II)	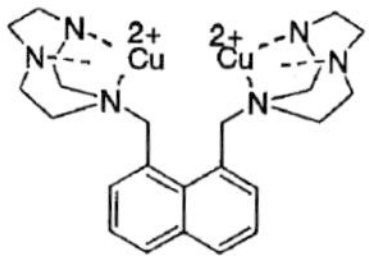	0.2 (pH 7, 50°)	15
Zinc(II) + copper(II)	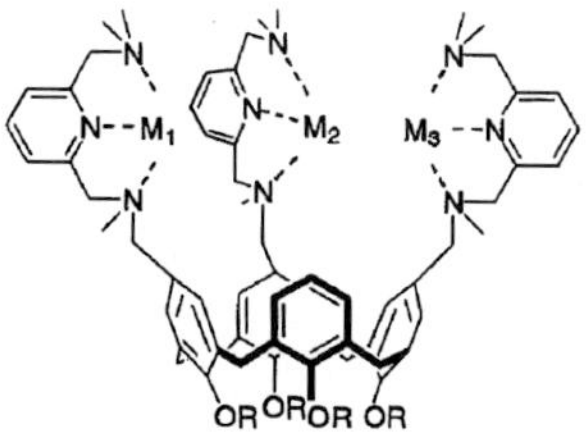 R = C_2H_4OEt M_1, M_2, M_3 = Zn(II), Zn(II), Cu(II)	0.9 (pH 8, 50°, 35% ethanol)	16

[a]Approximate value for dinucleotide cleavage.

physiological conditions, diribonucleotides are completely hydrolyzed in less than 30 min. The acceleration is almost 10^8-fold. Lanthanide(III) chlorides are the most widely and conveniently used (the nitrates and perchlorates show similar activities). RNA hydrolysis by Lu(III) ion (the last lanthanide ion) is carried out as follows. To buffer solutions (HEPES or Tris), $LuCl_3$ is added and the pH is adjusted to the desired value[7,8] by using NaOH. Then, the substrate RNA in water is added. After a predetermined time (e.g., 5 min), the reaction solution is analyzed by either reversed-phase high-performance liquid chromatography (HPLC) or polyacrylamide gel electrophoresis. When $[LuCl_3]_0 = 5$ and $[ApA]_0 = 0.1$ m*M* at pH 7 and 30°, more than half of the ApA disappears in 5 min, and 1 : 1 mixture of A and (A2′p + A3′p) is formed. The reaction intermediate, A > p, is rapidly hydrolyzed by the Lu(III) to A2′p or A3′p, and not much is accumulated. The pseudo-first-order rate constant for the disappearance of ApA is $3 \times 10^{-3}\ s^{-1}$ (the half-life < 4 min). Other diribonucleotides are hydrolyzed at similar rates. The hydrolysis rate is proportional to $[LuCl_3]_0$.

Oligoribonucleotides are also efficiently hydrolyzed by lanthanide(III) ions. The scission occurs randomly without any specific base preference. When $[LuCl_3]_0 = 10$ m*M* at pH 7.5 and 30°, substrate RNAs are almost completely degraded into small fragments within 1 h. The termini of the RNA fragments are either the 2′- or 3′-monophosphates, and the 2′,3′-cyclic monophosphate termini are hardly formed (they are rapidly hydrolyzed to the monophosphates). It is noteworthy that the rate of RNA hydrolysis by lanthanide ions is drastically dependent on the pH of soluton. With $LuCl_3$, the RNA hydrolysis at pH 8 is 300-fold faster than that at pH 7 (the slope of the logarithm of rate constant vs pH is around 2.5, corresponding to the formation of $[Lu_2(OH)_2]^{4+}$ as the active species). In order to obtain reproducible results, the pH must be carefully controlled.

Several metal complexes for RNA hydrolysis are available. Lanthanide(III) complexes of Schiff-base macrocycles (**1**)[5] are prepared by refluxing the corresponding diamine and dialdehyde in the presence of metal salts. The $[Cu(neocuproine)(H_2O)_2]^{2+}$ complex (**2**)[6] is obtained as follows. First, $[Cu(neocuproine)Cl_2]$ is prepared by mixing the ligand (commercially available) with $CuCl_2$ in methanol (the ligand and $CuCl_2$ are soluble in methanol, whereas the chloride complex precipitates out of the solvent). Then, this complex is dissolved in water, and the chloride ligands are exchanged with water. When the concentration of the complex is at 10 m*M* at pH 7 and 25°, the pseudo-first-order rate constants for diribonucleotide hydrolysis are around $4 \times 10^{-3}\ s^{-1}$. The Cu(II)/terpyridine complex (**3**)[7,8] is also active. In contrast, Cu(II) ion itself is virtually inactive, probably because the gel of the metal hydroxide is formed in the solutions and the effective concentration of the active species is too small. The Cu(II) complex of a macrocyclic amine ligand hydrolyzes the double-stranded portion in hairpin-structured RNA, although this portion is less susceptible to hydrolysis than the single-stranded one.[9]

Organic Catalysts. Oligoamines such as ethylenediamine and diethylenetriamine effectively hydrolyze RNA at about pH 7 (Table I).[11,12] The presence of two or more amino residues in a molecule is essential (monoamines are inactive).

[11] K. Yoshinari, K. Yamazaki, and M. Komiyama, *J. Am. Chem. Soc.* **113,** 5899 (1991).
[12] M. Komiyama and K. Yoshinari, *J. Org. Chem.* **62,** 2155 (1997).

When [ethylenediamine]$_0$ is 1.0 *M* at pH 8 and 50°, more than 50% of the phosphodiester linkages in RNA oligomers are hydrolyzed in 2 days. At physiological pH, one of the two amino residues in ethylenediamine exists as a neutral amine, and the other takes a protonated form (the pK_a values are 9.2 and 6.5). Other oligoamines also have neutral amines and ammonium cations on one molecule. In RNA hydrolysis, the neutral amine (base catalyst) and the ammonium ion (acid catalyst) show intramolecular cooperation. The RNA hydrolysis by diazabicyclo [2.2.2]octane derivatives proceeds in a similar fashion.[12a] These catalytic mechanisms are reminiscent of that of ribonuclease A, in which two imidazoles (one is neutral, and the other is protonated) cooperate. The simple molecular structures of oligoamines, as well as their physicochemical and biological stability, are advantageous for practical applications. Organic scissors bearing two or more imidazole molecules are synthesized.

Bimetallic and Trimetallic Complexes. Except for lanthanide(III) ions, no metal ions hydrolyze RNA at reasonable rates. However, some of dinuclear and trinuclear complexes are sufficiently active (see Table II[13–16]). For example, the dinuclear Zn(II) complex (**4**) promptly hydrolyzes RNA, although Zn(II) ion and monomeric Zn(II) complexes are inactive.[13] The pseudo-first-order rate constant for diribonucleotide hydrolysis at pH 7 and 50° is 10^{-4} s^{-1} (concentration of compound **4** is 5 m*M*). Here, one of the Zn(II) ions increases the electrophilicity of the phosphodiester moiety (acid catalysis), whereas the other activates the 2′-OH for the nucleophilic attack (base catalysis). The trinuclear complex **5** is still more active.[14] An interesting substrate specificity is achieved by the trinuclear Cu(II) complex, in which three *N,N*-bis (2-pyridylmethy) amino residues are bound to the benzene moiety. Nonnatural substrate 2′-5′ UpU is efficiently hydrolyzed by this complex, whereas natural RNA having 3′-5′ phosphodiester linkages is not.

Sequence-Selective Artificial Ribonucleases

Various artificial enzymes are obtained by tethering molecular scissors to sequence-recognizing moieties (Fig. 2, structures **6–16**). In most of the cases, DNA oligomers are used as the sequence-recognizing moieties. The design of linker portion, which connects the scissors and the sequence-recognizing moieties, is important for accomplishing both high selectivity and high efficiency.

[12a] M. Zenkova, N. Beloglazova, V. Sil'nikov, V. Vlassov, and R. Giegé, *Methods Enzymol.* **341,** [31] 2001 (this volume).

[13] S. Matsuda, A. Ishikubo, A. Kuzuya, M. Yashiro, and M. Komiyama, *Angew. Chem., Int. Ed. Eng.* **37,** 3284 (1998).

[14] M. Yashiro, A. Ishikubo, and M. Komiyama, *J. Chem. Soc., Chem. Commun.,* 83 (1997).

[15] M. J. Young and J. Chin, *J. Am. Chem. Soc.* **117,** 10577 (1995).

[16] P. Molenveld, J. F. J. Engbersen, and D. N. Reinhoudt, *Angew. Chem., Int. Ed. Eng.* **38,** 3189 (1999).

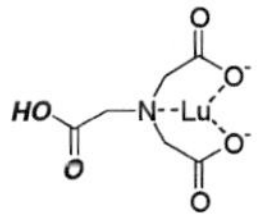

6 (ref. 17)

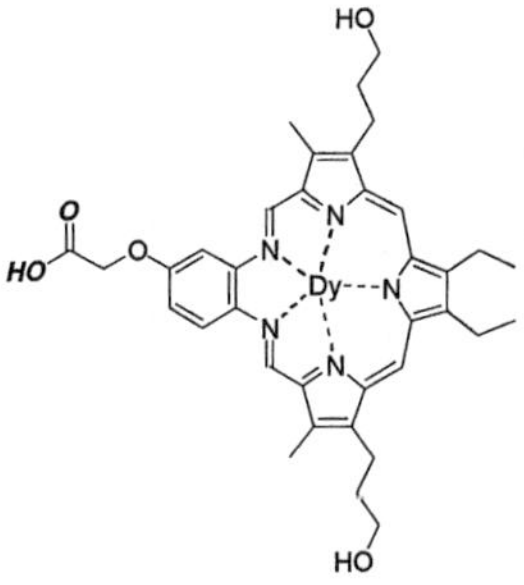

7 (ref. 18)

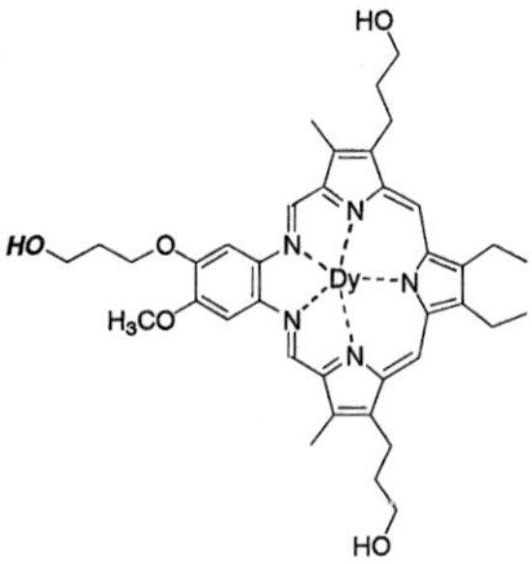

8 (ref. 19)

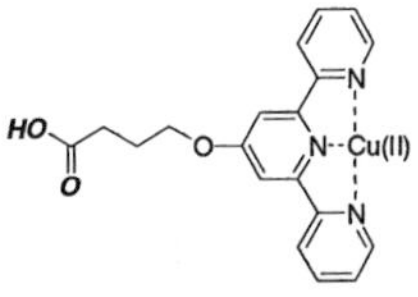

9 (ref. 22)

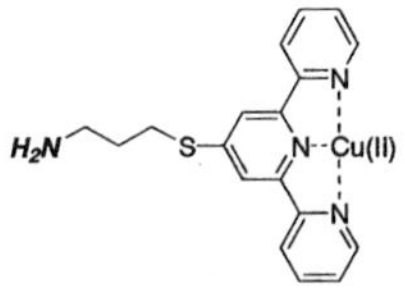

10 (ref. 23)

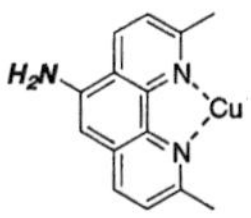

11 (ref. 24)

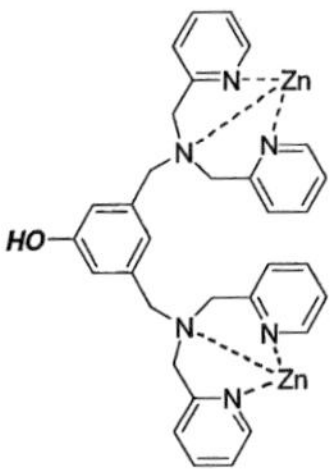

12 (ref. 13)

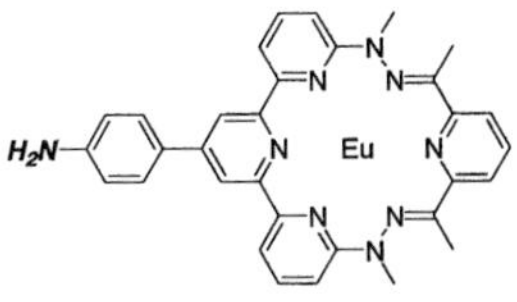

13 (ref. 20)

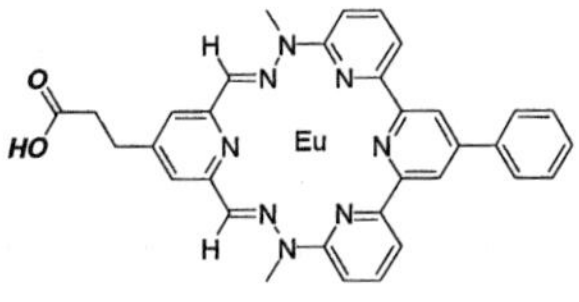

14 (ref. 20)

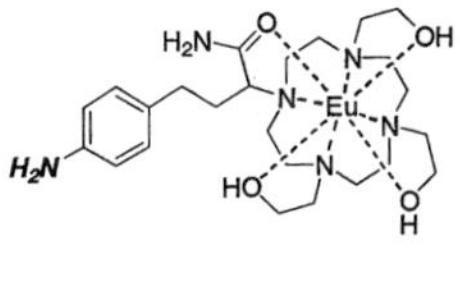

15 (ref. 21)

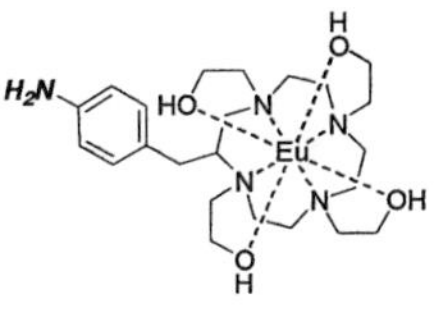

16 (ref. 21)

FIG. 2. Sequence-selective artificial ribonucleases (**6–16**). The catalysts presented here are attached to DNA oligomers (sequence-recognizing moieties) at the groups indicated by the bold characters.

[Lanthanide Complex]/DNA Hybrids. The conjugates of an iminodiacetate (the ligand to bind metal ions) and DNA oligomers are synthesized as shown in Fig. 3a.[17] First, DNA oligomer, which is complementary with the substrate RNA (near the target scission site), is prepared on an automated synthesizer. At the final stage of synthesis, an amino-linker [$H_2N–(CH_2)_n$-; $n = 3, 6$, or 12] is attached to the 5′ end of DNA by using the commercially available phosphoramidite monomer. This modified DNA is reacted with *p*-nitrophenyl ester of *N,N*-bis (ethoxycarbonylmethyl)glycine, and then the ethyl esters are hydrolyzed under alkaline conditions. By these procedures, an iminodiacetate group is bound to the DNA via an amide linkage. The artificial enzymes (**6** in Fig. 2) are prepared *in situ* by mixing these conjugates with lanthanide(III) chloride in water. According to the polyacrylamide gel electrophoresis, the RNA scission occurs selectively at the target site where the lanthanide complex is placed (Fig. 3b). When [the DNA–iminodiacetate conjugate]$_0 = 10$, [Lu(III)]$_0 = 10$, and [RNA]$_0 = 0.3$ μM at pH 8 (Tris buffer) and 37°, the conversion for the sequence-selective RNA scission is 17 mol% at 8 h. In order to achieve the selective scission, the [Lu(III)]/[iminodiacetate] ratio must be 1.0 or smaller. Otherwise, nonselective RNA scission by free Lu(III) ions concurrently occurs.

In place of the iminodiacetate group, so-called "texaphyrins" (**7** and **8:** expanded porphyrins involving five nitrogen atoms) can be also used.[18,19] In a typical example, the Dy(III)/texaphyrin complex is attached to the 5′ end of 20-mer DNA, and a substrate RNA (2 n*M*) is treated with this hybrid DNA (50 n*M*) at pH 7.5 and 37°. The sequence selectivity for the scission is satisfactorily high. The half-life of RNA cleavage varies from 2 to 10 h, depending on the length of linker between the Dy(III)/texaphyrin complex and the DNA. Shorter linkers, in general, provide better results.

[Copper(II) Complex]/DNA Hybrids. The phosphoramidite monomers bearing a terpyridine are synthesized in a straightforward fashion, and the DNA–terpyridine conjugates are obtained on an automated synthesizer (**9** and **10**).[22–24] The conditions for the sequence-selective RNA scission is as follows: [the conjugate]$_0 = 5$, $[CuCl_2]_0 = 10$, and [RNA]$_0 = 10^{-3}$ μM at pH 7.5 (HEPES buffer) and 45°. The conversion of RNA scission after 3 days is 28 mol%.

[17] K. Matsumura, M. Endo, and M. Komiyama, *J. Chem. Soc., Chem. Commun.*, 2019 (1994).

[18] D. Magda, R. A. Miller, J. L. Sessler, and B. L. Iverson, *J. Am. Chem. Soc.* **116,** 7439 (1994).

[19] D. Magda, S. Crofts, A. Lin, D. Miles, M. Wright, and J. L. Sessler, *J. Am. Chem. Soc.* **119,** 2293 (1997).

[20] J. Hall, D. Hüsken, and R. Häner, *Nucleic Acid Res.* **24,** 3522 (1996).

[21] L. Huang, L. L. Chappell, O. Iranzo, B. F. Baker, and J. R. Morrow, *J. Biol. Inorg. Chem.* **5,** 58 (2000).

[22] J. K. Bashkin, E. K. Frolova, and U. Sampath, *J. Am. Chem. Soc.* **116,** 5981 (1994).

[23] A. T. Daniher and J. K. Bashkin, *Chem. Commun.*, 1077 (1998).

[24] W. C. Putnam and J. K. Bashkin, *Chem. Commun.*, 767 (2000).

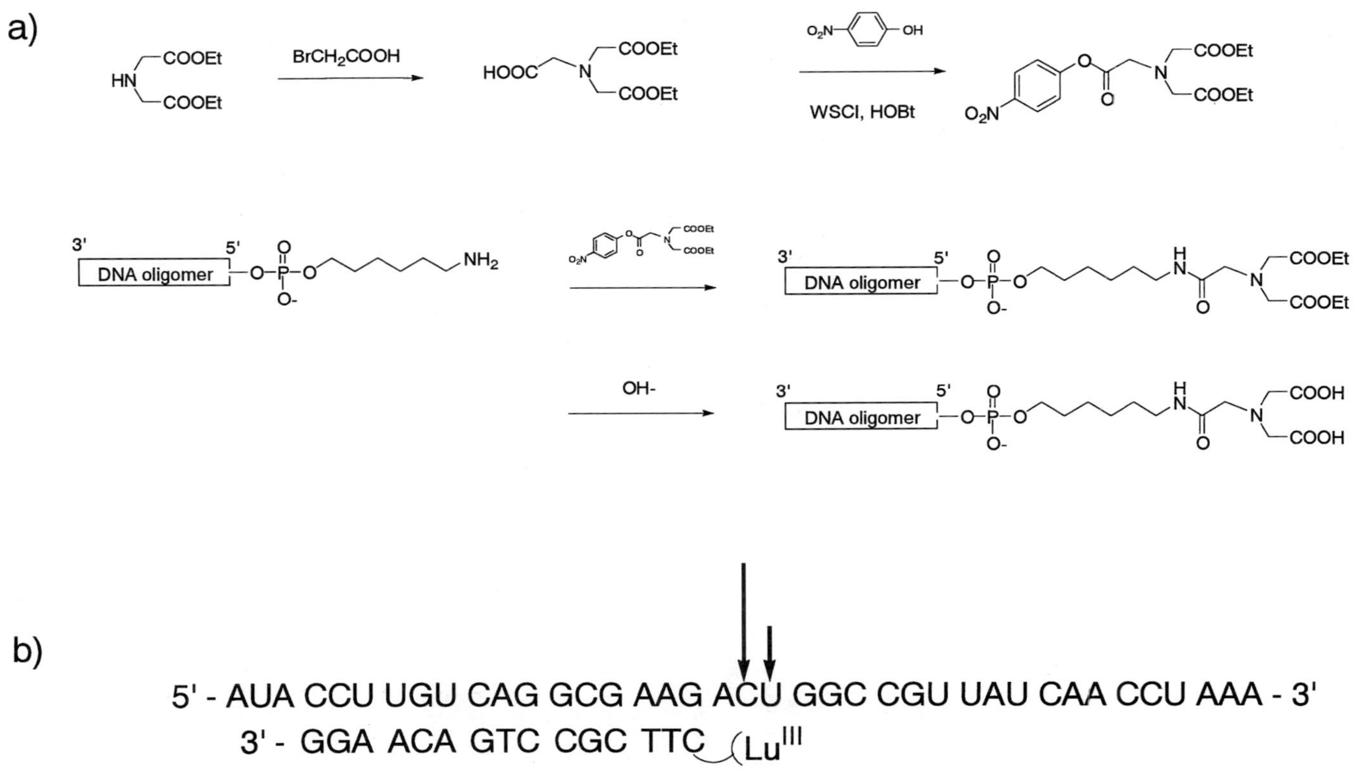

FIG. 3. Lu(III)/DNA hybrid as sequence-selective ribonuclease: (a) synthetic procedure for DNA–iminodiacetate hybrid preparation and (b) the RNA scission pattern.

Oligoamine/DNA Hybrids. Oligoamines are the simplest mimics of the active site of ribonuclease A (see above). Their hybrids with DNA are prepared as depicted in Fig. 4a.[25] First, the DNA oligomer is prepared on an automated synthesizer. In 1,4-dioxane, the CPG column carrying the DNA is directly treated with 1,1′-carbonyldiimidazole (0.3 *M*) at 30°, and then with diethylenetriamine (0.2 *M*) at 50° (these on-column procedures facilitate the purification process). An ethylenediamine group is bound to the 5′-end of DNA via urethane linkage (one of the primary amines of diethylenetriamine is used to form the linkage). Then, the hybrid is detached from the CPG column and purified. Exactly as designed, the RNA scission by this hybrid occurs only at the target site (Fig. 4b). These simple artificial ribonucleases involve no inorganic moieties and thus are easily incorporated into cells. By using PNAs in place of the DNA oligomers, the stability against enzymatic degradation can be improved.[26]

[Dinuclear Metal Complex]/DNA Hybrids. Another attractive tactics is to use a multinuclear metal complex as the molecular scissors. The ligand to form the dinuclear Zn(II) complex is attached to DNA oligomers. The [Zn(II) complex]/DNA hybrid (**12**) is prepared simply by adding $ZnCl_2$ to the DNA derivative.[13] The sequence selectivity remains 100%, even in the presence of large amounts of free Zn(II) ions [note that free Zn(II) ion is inactive for RNA hydrolysis, and thus the scission occurs only at the site where the dinuclear ligand is placed]. The applications *in vivo* are easy. In order to regulate the expression of a specific gene, for example, only the ligand–DNA conjugate [without Zn(II) ions] can be incorporated into cells. The conjugate is activated by using the Zn(II) ions in the cells and cleaves the target RNA. The greatest obstacle in these approaches (how to penetrate metal complexes through cell membranes) has been solved.

More Advanced Artificial Ribonucleases

Catalytic Turnover. All the artificial ribonucleases described above possess the molecular scissors at the termini (mostly the 5′ ends) of DNA oligomers. Here, the RNA is cleaved outside the portion that is complementary with the DNA oligomer. Even after the scission, this portion is kept intact and strongly binds to the DNA. Catalytic turnover is hardly accomplished.

In the second-generation artificial ribonucleases, the molecular scissors are incorporated into the inside of DNA strand.[22–24,27,28] Upon scission, the RNA

[25] M. Komiyama, T. Inokawa, and K. Yoshinari, *J. Chem. Soc., Chem. Commun.,* 77 (1995).

[26] J. C. Verheijin, B. A. L. M. Deiman, E. Yeheskiely, G. A. van der Marel, and J. H. van Boom, *Angew. Chem., Int. Ed. Engl.* **39,** 369 (2000).

[27] M. Endo, Y. Azuma, Y. Saga, A. Kuzuya, G. Kawai, and M. Komiyama, *J. Org. Chem.* **62,** 846 (1997).

[28] D. Magda, M. Wright, S. Crofts, A. Lin, and J. L. Sessler, *J. Am. Chem. Soc.* **119,** 6947 (1997).

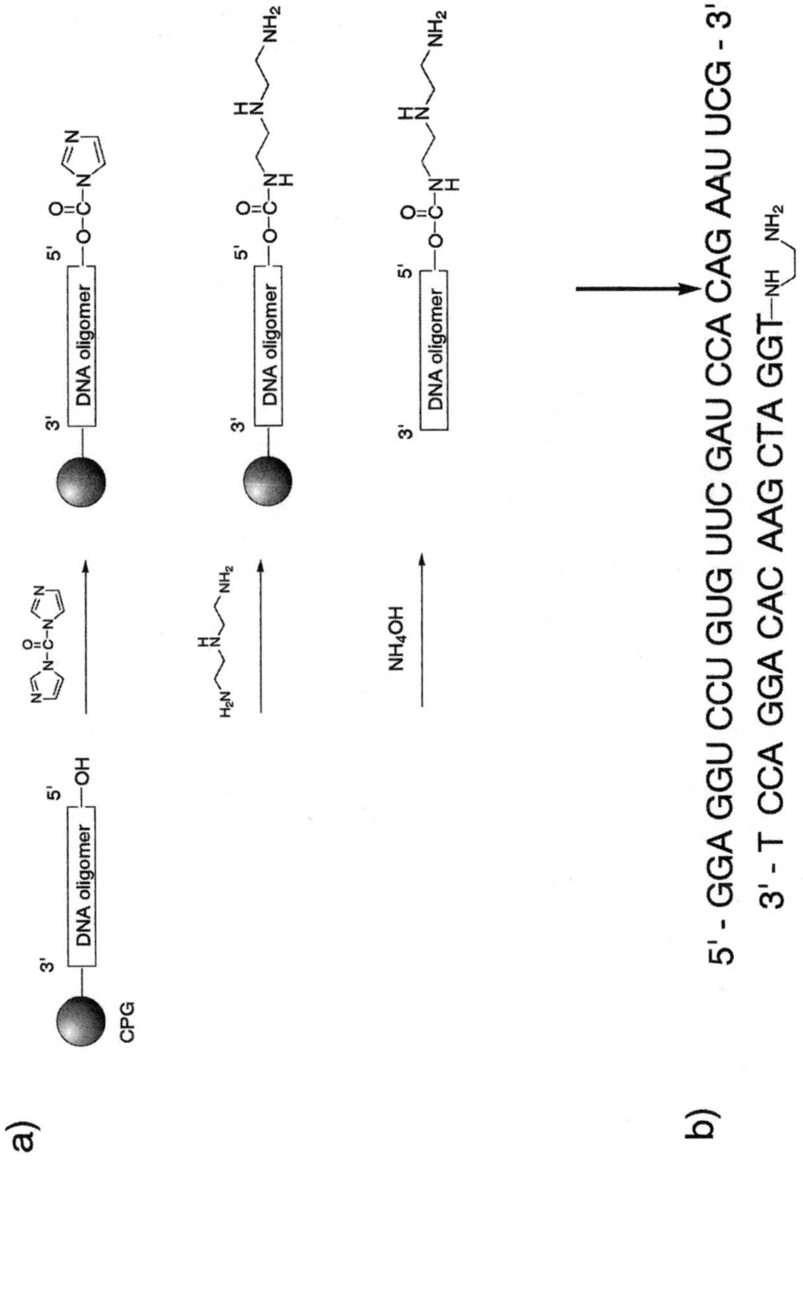

FIG. 4. (a) Synthesis of ethylenediamine/DNA hybrid and (b) its sequence-selective RNA cleavage.

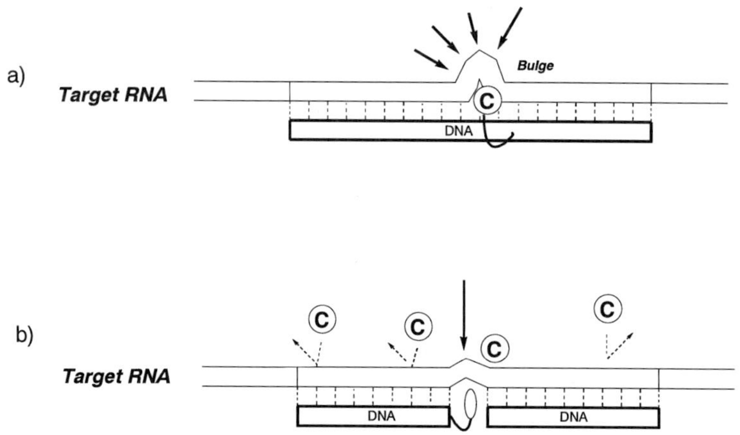

FIG. 5. Sequence-selective RNA scission by (a) the artificial ribonucleases involving bulge-forming DNA and (b) the ternary noncovalent systems composed of intercalator–DNA hybrid, another DNA, and lanthanide(III) ion.

portion (used for the binding to the DNA) is divided into two small fragments. The stability of the RNA/DNA heteroduplex decreases as the RNA becomes shorter, so that the small RNA fragments are spontaneously removed from the artificial enzymes. Efficient turnover takes place.

Increase of Scission Efficiency by Bulge-Structure Formation. A bulge structure is intentionally formed in the substrate RNA, and the target phosphodiester linkage is activated (Fig. 5a).[20,29] The DNA oligomer, used as the sequence-recognizing moiety in these attempts, is complementary with the target RNA, but a few nucleotides in the middle are absent. When the molecular scissors (the lanthanide complexes of Schiff-base macrocyles: **13** and **14**) are bound to this DNA oligomer, the phosphodiester linkages in the bulge are selectively hydrolyzed. The scission efficiency is increased by the bulge formation. In a proposed mechanism, the conformations of the ribose residues in the bulge structures are more suitable for the intramolecular attack by the 2′-OH toward the phosphorus atom.

Noncovalent Systems for Sequence-Selective RNA Scission

In these systems, the target phosphodiester linkage is activated by some noncovalent interactions and is selectively hydrolyzed. The scissors do not have to be bound to sequence-recognizing moieties, and thus complicated organic synthesis

[29] D. Hüsken, G. Goodall, M. J. J. Blommers, W. Jahnke, J. Hall, R. Häner, and H. E. Moser, *Biochemistry* **35,** 16591 (1996).

is unnecessary. All we need are DNA oligomers (slightly modified when necessary) and molecular scissors.

The ternary system composed of (1) DNA bearing an intercalator (e.g., acridine), (2) another DNA, and (3) lanthanide(III) ion is a typical example (Fig. 5b). These two DNAs form Watson–Crick base pairs with the substrate RNA, except for one ribonucleotide. The DNA bearing an intercalator can be synthesized on an automated synthesizer by using the commercially available agents. RNA scission selectively occurs at the ribonucleotide that is not base-paired with either of two DNAs. One of the important characteristics of these noncovalent systems is that the RNA hydrolysis comes up to high conversion under physiological conditions. At pH 8 and 37° ($[LuCl_3]_0 = 100\ \mu M$), more than 85% of the RNA is hydrolyzed in 2 days. When two unmodified DNAs are used, however, scission is virtually nil. The intercalators are essential to activate the target phosphodiester linkage.

Artificial Nucleases for Sequence-Selective DNA Scission: Artificial Restriction Enzymes

Artificial enzymes for site-selective scission of DNA are also valuable tools for biotechnology. They are prepared by hybridizing the Ce(IV)/iminodiacetate complex with DNA oligomers.[30,31] The Ce(IV) complex, placed near the target phosphodiester linkage, hydrolyzes this linkage. The length of the DNA oligomer and its sequence are freely chosen so that the sequence specificity can be increased at will. Thus, even huge DNA of higher animals and higher plants can be manipulated with these artificial enzymes. The reader is referred to Refs. 30 and 31 for details, since they are beyond the scope of this article.

Prospect

A number of sequence-selective artificial ribonucleases are now available. Their sequence specificity is far greater than those of naturally occurring ribonucleases and can be increased still more if necessary. RNA scission is achieved under physiological conditions. These artificial enzymes can pick up the target RNA from a family of RNAs and cut it at the desired site. Furthermore, their potential should be enormously increased by tethering various functional molecules. For example, the covalent conjugates of these artificial ribonucleases and appropriate signal peptides can be transported to the desired part in cells and cleave the target RNA therein. By attaching photoresponsive molecules to them, the

[30] M. Komiyama, N. Takeda, T. Shiiba, Y. Takahashi, Y. Matsumoto, and M. Yashiro, *Nucleosides Nucleotides* **13,** 1297 (1994).

[31] M. Komiyama, T. Shiiba, Y. Takahashi, N. Takeda, K. Matsumura, and T. Kodama, *Supramol. Chem.* **4,** 31 (1994).

RNA-hydrolyzing activity can be photocontrolled. The scope of these kinds of approaches is unlimited. These "completely artificial ribonucleases" have many different characteristics from those of ribozymes. The combination of these two types of artificial ribonucleases will be essential for the future of RNA science.

Acknowledgments

This study was partially supported by the grant from "Research for the Future" Program of the Japan Society for the Promotion of Science (JSPS-RFTF97I00301) and Grants-in-Aid for Scientific Research from the Ministry of Education, Science, and Culture, Japan.

[30] RNA Cleavage by 1,4-Diazabicyclo[2.2.2]octane–Imidazole Conjugates

By MARINA ZENKOVA, NATALIA BELOGLAZOVA, VLADIMIR SIL'NIKOV, VALENTIN VLASSOV, and RICHARD GIEGÉ

Introduction

In recent years considerable effort has been expended in designing compounds capable of cleaving RNA.[1,2] These chemical constructs have been referred to as "artificial" or "chemical" ribonucleases (RNases). Artificial ribonucleases are of great current interest because of their potential applications in molecular biology and in drug design. Small molecules that interact with and cleave RNA can be used for probing RNA structure.[3] Conjugation of small RNA cleaving groups to antisense oligonucleotides can improve the efficacy of these specific RNA targeted agents, which may be of therapeutic value, and also serve as valuable tools for manipulating RNA.[4–6]

Previously we found that conjugation of specific groups found in active centers of ribonucleases to compounds capable of binding to nucleic acids yields catalysts that cleave phosphodiester bonds in RNA.[7–10] In this chapter we describe new artificial ribonucleases, consisting of conjugates of 1,4-diazabicyclo[2.2.2]octane

[1] E. L. Hegg, J. N. Burstyn, *Coordination Chem. Rev.* **173,** 133 (1998).

[2] V. V. Vlassov, V. N. Sil'nikov, and M. A. Zenkova, *Mol. Biol. (Russ.)* **32,** 50 (1998).

[3] R. Giege, B. Felden, V. N. Sil'nikov, M. A. Zenkova, and V. V. Vlassov, *Methods Enzymol.* **318,** 165.

[4] R. Häner and J. Hall, *Antisense Nucleic Acid Drug Dev.* **7,** 423 (1997).

[5] N. G. Beloglazova, V. N. Sil'nikov, M. A. Zenkova, and V. V. Vlassov, *FEBS Lett.,* in press (2000).

[6] M. A. Reynolds, T. A. Beck, P. B. Say, D. A. Schwartz, B. P. Dwyer, W. J. Daily, M. M. Vagheti, M. D. Metzler, R. E. Klem, and L. J. Arnold, Jr., *Nucleic Acids Res.* **24,** 760 (1996).

0076-6879/01 $35.00

and imidazole-containing molecules (histamine, histidine, or methyl ester of histidine). The compounds cleave RNA under physiological conditions and display RNase A-like specificity.

Design of RNA Cleaving Conjugates

The conjugates consist of four domains A, B, C, and L (Fig. 1). In the conjugates ABLkCm, catalytic domain C contains functional groups found in active centers of ribonucleases (imidazole and carboxylic groups). The cationic domain—bisquaternary salt of 1,4-diazabicyclo[2.2.2]octane—is responsible for binding to RNA. The distance between the RNA binding center and the catalytic part is dependent on the number of methylene groups in the linker structure L. The hydrophobic domain A is expected to prevent cooperative binding of the cationic structures to RNA. Compounds AC, AB, and B were synthesized as control structures for elucidation of the functional role of the domains.

The strategy of synthesis of these conjugates was designed so as to provide easy variation of the structural parameters of the constructs (n, m, k), and involved a universal synthesis procedure that preserved the main structural element of the RNA-binding site (Scheme 1). To this end, a bisquaternary salt of 1,4-di(azonia)bicyclo[2.2.2]octane was obtained using monoquaternary salt of 1,4-diazabicyclo[2.2.2]octane, containing 3-azidopropyl and alkyl residues as the starting compounds and activated ethers of ω-Br-carbonic acids (Scheme 1, step a). After condensation with the catalytic part of a construct (Scheme 1, step b), the azido group can be reduced to an amino group, which confers an additional positive charge of the constructs at neutral pH (Scheme 1, step c). Condensation of the amine obtained in step (c) of Scheme 1 with activated ethers of derivatives of 1,4-diazabicyclo[2.2.2]octane yields conjugates with different numbers of positive charges (combination of steps c and d in Scheme 1).

Comparison of Ribonuclease Activities of the Designed Conjugates

The ribonuclease activity of conjugates was studied using a 5′-^{32}P-labeled *in vitro* transcript of yeast tRNA^{Asp} as substrate. The RNA was incubated with the

[7] D. A. Konevetz, I. E. Beck, N. G. Beloglazova, I. V. Sulemenkov, V. N. Sil'nikov, M. A. Zenkova, G. V. Shishkin, and V. V. Vlassov, *Tetrahedron* **55,** 503 (1999).

[8] N. Beloglazova, A. Vlassov, D. Konevetz, V. Sil'nikov, M. Zenkova, R. Giege, and V. Vlassov, *Nucleosides Nucleotides* **18,** 1463 (1999).

[9] M. A. Zenkova, N. L. Chumakova, A. V. Vlassov, N. I. Komarova, A. G. Venyaminova, V. V. Vlassov, and V. N. Sil'nikov, *Mol. Biol. (Russ.)* **34,** 456 (2000).

[10] D. A. Konevetz, M. A. Zenkova, V. N. Sil'nikov, and V. V. Vlassov, *Dokl. Akad. Nauk (Russ.)* **360,** 554 (1998).

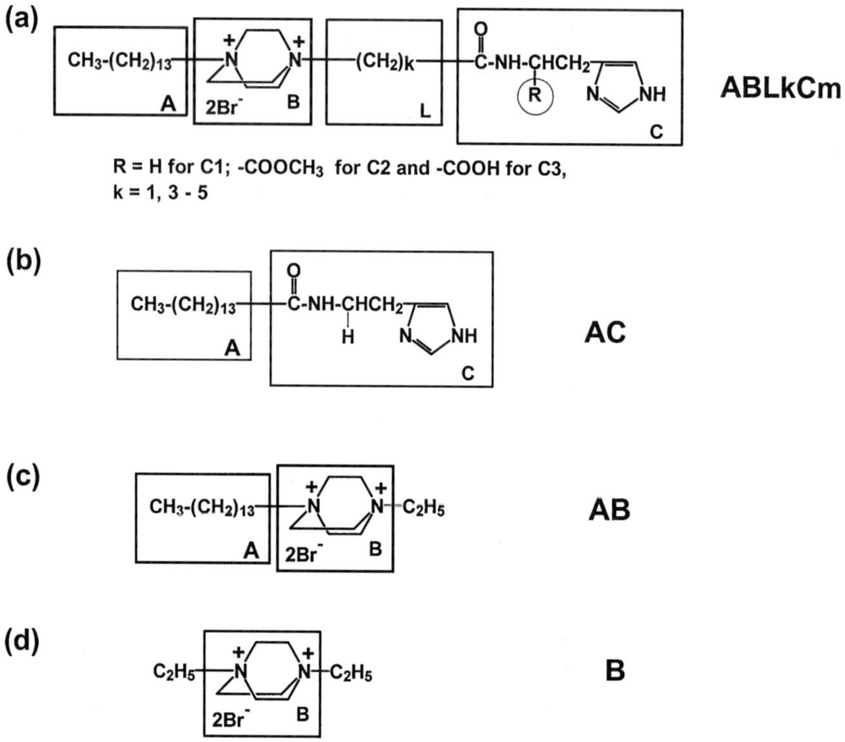

FIG. 1. Ribonuclease mimics: conjugates of 1,4-diazabicyclo[2.2.2]octane and imidazole containing compounds. A, Lipophilic domain; B, cationic RNA binding domain; C, catalytic domain; L, linker group. Cm refers to the type of catalytic group: C1 for histamine; C2 for methyl ester of histidine; C3 for histidine. (a) Ribonuclease mimics ABLkCm are built of the four functional domains A, B, C, and L. Linker length k is 3, 4, or 5 methylene groups, for ribonuclease mimics ABL3Cm, ABL4Cm, and ABL5Cm, respectively. (b) The conjugate AC contains only catalytic (C) and lipophilic (A) parts. (c) Conjugate AB is built of lipophilic (A) and RNA-binding (B) domains. (d) Compound B is cationic 1,4-diazabicyclo[2.2.2]octane only.

compounds under study at 37° in 50 m*M* imidazole buffer at neutral pH. Some of the compounds efficiently cleave the RNA and exhibit a similar sequence specificity, with strong preference for CA and UA sequences. About 90% of the RNA degradation products result from cleavages occurring at these motifs. The main cleavage sites within $tRNA^{Asp}$ were phosphodiester bonds U8-A, U13-A, C20-A, C43-A, and C56-A, all present in the single-stranded regions of the RNA. Table I summarizes the activities of conjugates ABLkCm with total positive charge of +2 and linkers L of different length, and the control constructs AB, AC, and B.

All three catalytic domains used in the compounds—histamine (ABLkC1), the methyl ester of histidine (ABLkC2), and histidine (ABLkC3)—provide similar specificity and high RNA cleavage activity: 20%, 70%, and 95% of tRNA was

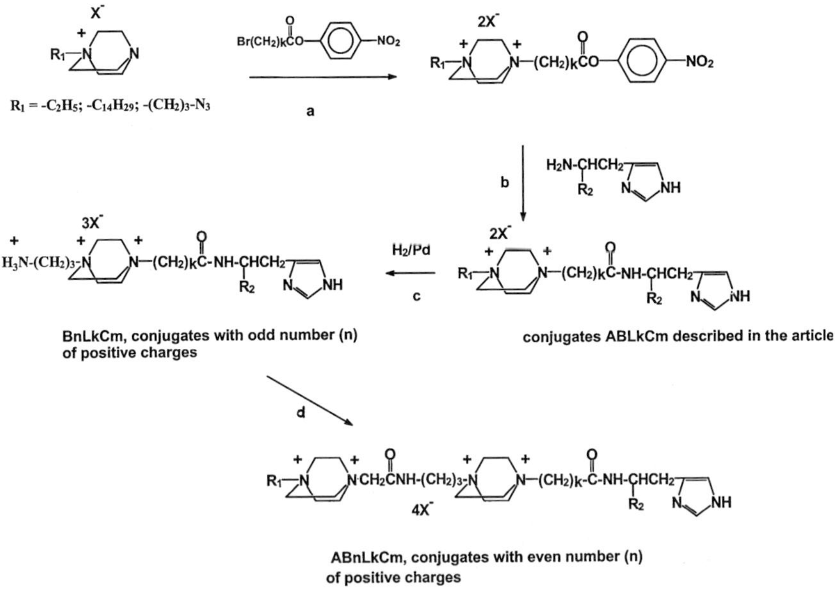

SCHEME 1. Synthesis of conjugates of 1,4-diazabicyclo[2.2.2]octane and imidazole-containing compounds.

cleaved during 18 h incubation at 37° with the compounds ABL3C1, ABL3C2, and ABL3C3, respectively. The rate of RNA cleavage with constructs ABLkC1 with one imidazole residue in the catalytic domain was lower then the rates observed in experiments with constructs ABLkC2 and ABLkC3, which contains both imidazole and carboxylic groups in domain C.

In the case of conjugates ABLkC1, an increase of the spacer length caused an increase in cleavage rate. The compound ABL1C1 displayed negligible cleavage activity at any concentration (data not shown), whereas the conjugates ABL5C1 cleaved the tRNA under identical conditions as efficiently as the most active conjugate ABL3C3. Among the ABLkC3 conjugates maximal activity was exhibited by the compound ABL3C3 with a linker built of three methylene groups. The conjugates ABLkC2 with 3, 4, and 5 methylene groups in the linker structure cleaved RNA with similar high efficiency.

The data presented in Table I indicate that the cleavage activity of the conjugates is considerably affected by their architecture. The control constructs AC and B containing an aliphatic fragment connected only to a histamine residue or RNA binding domain do not cleave RNA. Apparently, both the catalytic histamine residue and the RNA binding dicationic structure are needed for catalytic activity. Linker part L provides an optimal arrangement of functional groups of conjugates participating in the catalysis of RNA cleavage and strongly affects cleavage. The cationic RNA binding domain B has an affinity for the ribose–phosphate backbone

TABLE I
COMPARISON OF RIBONUCLEASE ACTIVITIES OF COMPOUNDS ABLKCM, AC, AB, AND B

Conjugate	Cm[a]	k[b]	RNA cleaving activity[c]
ABL3C1	C1	3	20
ABL4C1	C1	4	26
ABL5C1	C1	5	76
ABL3C2	C2	3	70
ABL4C2	C2	4	45
ABL5C2	C2	5	90
ABL3C3	C3	3	91
ABL4C3	C3	4	85
ABL5C3	C3	5	64
AC	C1	—	0
AB	—	—	20
B	—	—	0

[a] Cm refers to the type of catalytic group: C1 for histamine; C2 for methyl ester of histidine; C3 for histidine.
[b] k, The number of CH_2 groups in the linker domain L; total positive charge of the conjugates is +2.
[c] 5′-^{32}P-labeled *in vitro* transcript of $tRNA^{Asp}$ was incubated in 50 m*M* Imidazole buffer, pH 7.0, containing 200 m*M* KCl, 0.5 m*M* EDTA, 0.1 mg/ml of tRNA carrier with each of the RNase mimics at their optimal concentrations (see Experimental part) at 37° for different time intervals (up to 24 h); RNA cleaving activity was calculated as percentage of RNA depolymerization during incubation at 37° for 10 h.

because of its positive charge; this domain also may be expected to contribute to the cleavage process in a similar manner as the positively charged lysine in the active center of ribonuclease.[11] Surprisingly, construct AB demonstrates some cleavage activity generating a pattern similar to that of the other conjugates. It has been demonstrated[12,13] that some nonionic or zwitterionic detergents cause cleavage of RNA. The mechanism of this process is not understood yet.

Mechanism of RNA Cleavage by Imidazole Conjugates

The "classic" ribonuclease mechanism requires the simultaneous participation of a proton acceptor (histidine residue) and a proton donor.[14,15] There is little

[11] S. Vishveshwara, R. Jacob, G. Nadig, and J. V. Maizel, Jr., *J. Mol. Struct.* **471,** 1 (1998).
[12] A. Riepe, H. Beier, and H. J. Gross, *FEBS Lett.* **457,** 193 (1999).
[13] R. Kierzek, *Nucleic Acids Res.* **20,** 5079 (1992).
[14] A. Fersht, "Enzyme Structure and Mechanism." W. H. Freeman, New York, 1985.
[15] C. A. Deakyne and L. C. Allen, *J. Am. Chem. Soc.* **101,** 3951 (1979).

doubt that cleavage of phosphodiester bonds by RNase A proceeds by the linear mechanism.[15] Figure 2A shows a hypothetical mechanism of interaction of conjugates ABLkCm with RNA and emphasizes the role of the histidine residue and the cationic structure in catalysis. The mechanism suggests participation of RNA-binding fragment in catalysis, eliminating the need of a long linker between RNA-binding and the catalytic domains of the constructs. In contrast, the linker is indispensable in the case of synthetic RNases based on intercalators.[16,17]

The mechanism of RNA cleavage by the conjugates has been investigated using diribonucleoside monophosphate CpA and the decaribonucleotide UUCAUGUA-AA as substrates. The minimal substrate CpA is incubated with compound ABL3C3 and with RNase A. The reactions are monitored using reversed phase ion-pair HPLC. The results of the experiments, shown in Fig. 2B, indicate that adenosine and cytidine 2′,3′-cyclophosphate are formed at the initial step of the process in all the reactions. In the case of RNase A, formation of the cyclophosphate is accompanied by further hydrolysis to Cp(3′). Cp(3′) or Cp(2′) are not formed with catalyst ABL3C3. In contrast to the native enzyme, the conjugate cannot efficiently bind cytidine 2′,3′-cyclophosphate; therefore, the reaction stops at this step. Acid treatment of C > p(2′,3′) isolated from all the reaction mixtures leads to an equimolar mixture of Cp(2′) and Cp(3′).

The cleavage of the 5′- or 3′-^{32}P-labeled oligoribonucleotide UUCAUGUAAA, carried out in standard conditions either with RNase A or with conjugate ABL3C3, occurred at two sites: C(3)pA(4) and U(7)pA(8). Both the conjugate and RNase A cleave the phosphodiester bond in the CpA sequence several times faster than the bond in the UpA sequence (Fig. 2C). Similarly to the CpA cleavage, the initial step of the reaction generates products containing a 2′,3′-cyclophosphate group. This is inferred from the fact that treatment of the reaction products with 0.1 *N* HCl results in an increase in the electrophoretic mobility of the oligonucleotides. These changes in the electrophoretic mobilities were equal for the cleavage products of the decaribonucleotide generated by RNase A and by ABL3C3. The results of the experiments suggest that cleavage with synthetic catalysts occurs by a mechanism similar to that of RNase A.

It is assumed that phosphodiester bonds may be cleaved in the presence of diamines according to two mechanisms, based on nucleophilic or general acid–base catalysis.[18,19] We have compared the cleavage of CpA by the conjugate ABC3L3 in water and in D_2O, and we have noted an isotope effect ($K_H/K_D = 2.28$) of a magnitude expected for the acid–base catalysis as the rate-limiting step of the

[16] V. N. Sil'nikov, N. P. Lukjanchuk, G. V. Shishkin, R. Gige, and V. V. Vlassov, *Dokl. Akad. Nauk (Russ.)* **364,** 690 (1999).

[17] M. A. Podyminogin, V. V. Vlassov, and R. Giege, *Nucleic Acids Res.* **21,** 5950 (1993).

[18] J. W. Bats and G. Durner, *Angew. Chem. Int. Ed. Engl.* **31,** 207 (1992).

[19] M. Komiyama and K. Yoschinari, *J. Chem. Soc. Chem. Commun.,* 1880 (1989).

(A)

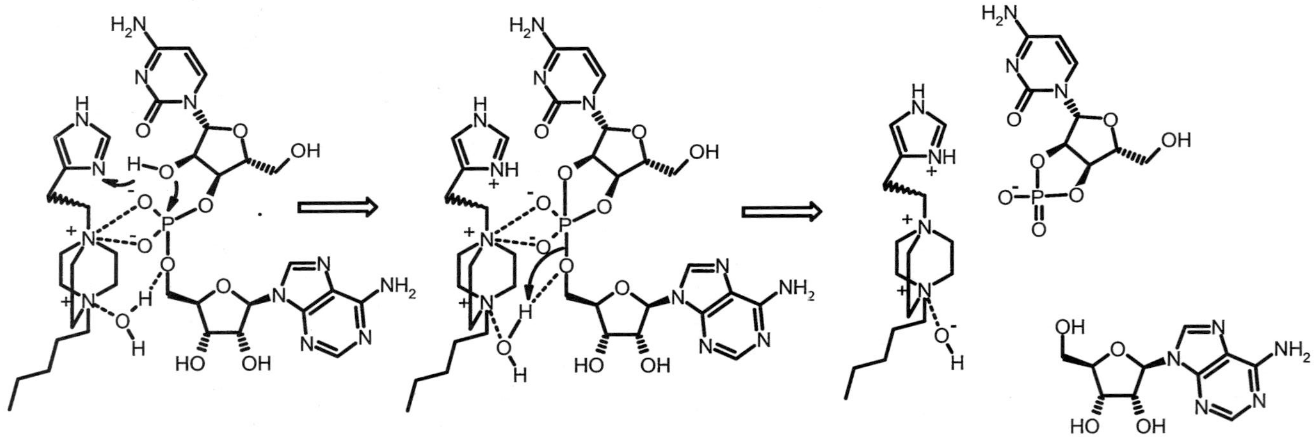

FIG. 2. Cleavage of RNA by the synthetic ribonuclease mimics. (A) Proposed scheme for RNA cleavage. In the first step, transesterification occurs to form a 2′,3′-cyclophosphate at the scission site. The presence of positively charged fragments provides a high affinity of synthetic RNases for internucleotide phosphates. The compact structure of the synthetic RNases allows them access to most of the phosphodiester bonds in the RNA substrates. (B) Typical profiles of chromatographic separation of the products formed by the initial step of cleavage of CpA with ABL3C3 and RNase A. 5×10^{-5} *M* CpA was incubated at 37° in 50 m*M* imidazole buffer, pH 7.0, containing 200 m*M* KCl, 0.5 m*M* EDTA in the presence of 1×10^{-4} *M* ABL3C3 for 40 h or 1×10^{-5} *M* RNase A for 1 h. The reaction mixtures were analyzed by ion-pair RP-HPLC using a Lichrospher RP-18 column (volume 50 μl) and 0–10% acetonitrile gradient in the presence of 2 m*M* *tert*-butylammonium phosphate. 1, CpA cleaved by RNase A; 2, CpA cleaved with ABL3C3; 3, control mixture of cytidine 2′,3′-cyclophosphate and adenine. (C) Electrophoretic analysis of cleavage products of 5′-^{32}P-labeled oligonucleotide pUUCAUGUAAA generated by conjugate ABL3C3 and by RNase A. Autoradiograph of 18% polyacrylamide/8 *M* urea gel. Lanes 1 and 2, intact oligonucleotide; lane 3, oligonucleotide cleaved by ABL3C3 and treated with 0.1 *M* HCl; lane 4, oligonucleotide cleaved by RNase A and treated with 0.1 *M* HCl; lane 5, oligonucleotide cleaved by ABL3C3; lanes 6 and 7, oligonucleotide cleaved by RNase A at 20° and 50°, respectively; lane 8, oligonucleotide cleaved by RNase T1. The oligonucleotide was incubated in 50 m*M* imidazole buffer at pH 7.0, containing 200 m*M* KCl, 0.5 m*M* EDTA, 0.1 mg/ml of tRNA carrier with 5×10^{-4} *M* ABL3C3 for 1.5 h at 37° or with 10^{-6} *M* of RNase A for 15 min at 20° or for 5 min at 50°.

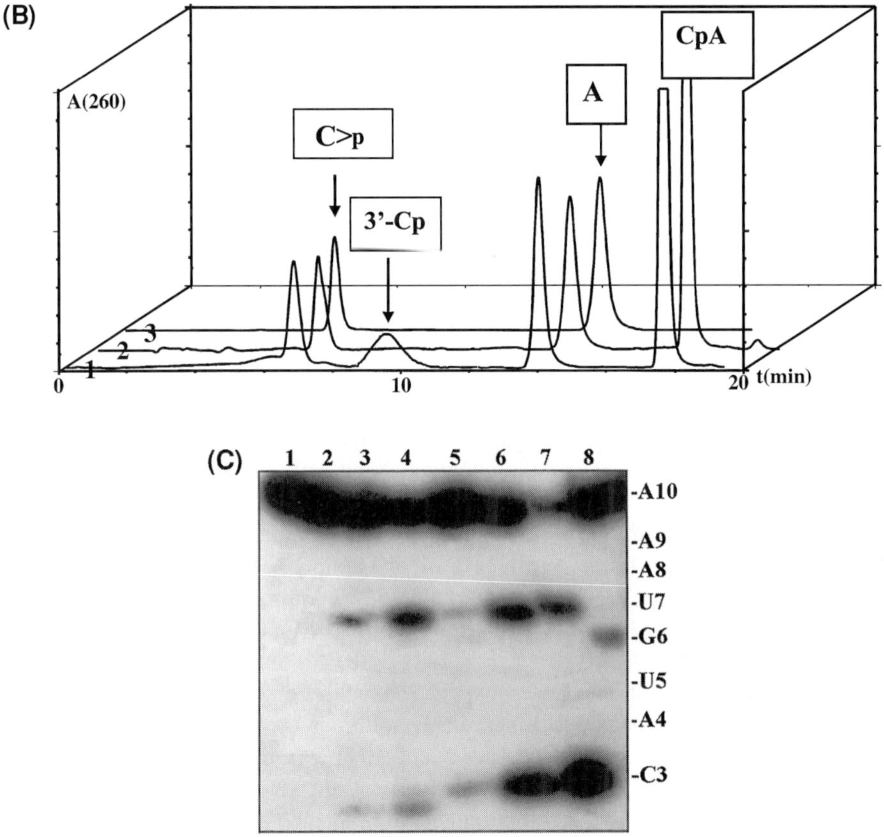

FIG. 2 (*continued*)

reaction.[19] Apparently, in this case as well as in the case of RNase A, the imidazole group of the catalyst functions as a base by abstracting a proton from the ribose 2′-hydroxyl.[14,20] The quaternary ammonium groups of 1,4-diazabicyclo[2.2.2] octane therefore participate in one or several steps of the cleavage process: to interact with the negatively charged oxygen of the phosphate group to increase the electrophilicity of the phosphorus atom; to be a proton donor for the leaving group 5′-oxygen; or to enhance the stability of the pentacoordinated phosphorus atom through electrostatic interactions. Apparently, the efficiency of binding of the conjugates to RNA strongly depends on the substrate structure, and the efficiency of cleavage of structured substrates does not accurately reflect the efficiency of a catalytic construct.

[20] A. Wlodawer, R. Bott, and L. Sjölin, *J. Biol. Chem.* **257,** 1325 (1982).

RNA Cleavage by Conjugates ABL3Cm

The cleavage activity of conjugates ABL3Cm can be assayed using labeled $tRNA^{Asp}$ as a substrate. The reactions are performed in 50 m*M* imidazole buffer at neutral pH and physiological ionic strength. The cleavage products are resolved in 12% polyacrylamide/8 *M* urea gel and quantified. The concentration dependence of the cleavage efficacy of conjugates ABL3Cm shows a bell-shaped curve with a maximum at a conjugate concentration of 0.1 m*M* (Fig. 3A). The decrease in cleavage rate at high conjugate concentrations can be explained by the limited solubility of the constructs in water, and micelle formation. An alternative explanation is that the saturation of the negatively charged centers of RNA by the cationic conjugate molecules can interfere with each other in establishing the complexes necessary for catalysis.

The kinetics of $tRNA^{Asp}$ cleavage by conjugates ABL3C1–ABL3C3 has been investigated at an optimal conjugate concentration 0.1 m*M*. Figure 3B shows the kinetics of RNA cleavage by the conjugates and by RNase A. Within the first 4 h for ABL3C3 (curve 3, Fig. 3B) and 18 h for ABL3C2 (curve 2, Fig. 3B) the percentage of RNA that is cleaved reaches 90%. Complete RNA cleavage with these conjugates was observed after 10, 24, and almost 40 h for mimics ABL3C3, ABL3C2, and ABL3C1, respectively. Under these conditions efficient cleavage of RNA is achieved by RNase A at concentrations of 1 and 5 n*M* (curves 5 and 6, Fig. 3B). To prove that the cleavage activity of the conjugates was not a result of contamination with ribonucleases, experiments with Centricon-3 filters (Amicon, Danvers, MA) have been performed, using solutions containing 10 times higher concentrations of conjugates ABL3Cm and RNase A than in experiments described above (1 m*M* and 50 n*M*, respectively). The solutions are filtered through membrane Centricon-3 with cutoff equal to M_r 3000 and assayed under the above conditions. Centricon-3 efficiently removes RNase A with M_r 13,799 from the solution, since no cleavage of RNA is detected by the filtrate (curves 9 and 10, Fig. 3B). Pancreatic RNase is one of the smallest RNases; therefore the filtration procedure could be expected to remove other putative contaminating RNases. As is seen from curves 6–8 (Fig. 3B), the kinetics of RNA cleavage with conjugates ABL3Cm is not affected by the filtration procedure (in Fig. 3B compare curves 1 and 5 for ABL3C1, curves 2 and 6 for ABL3C2, and curves 3 and 8 for ABL3C3), thus demonstrating that the RNA cleavage activity is associated with these small molecules.

The goal of the next set of experiments was to determine whether the main cuts produced by conjugates ABL3Cm within the $tRNA^{Asp}$ transcript are the cuts of the primary reaction, or are secondary cleavages within the structure promoted by the preceding cleavages. To this end we have compared the kinetics of cleavage, and the cleavage patterns of tRNA transcripts labeled at either the 3′-end or the 5′-end. The experiments reveal that the same cuts at phosphodiester bonds U8-A, U13-A,

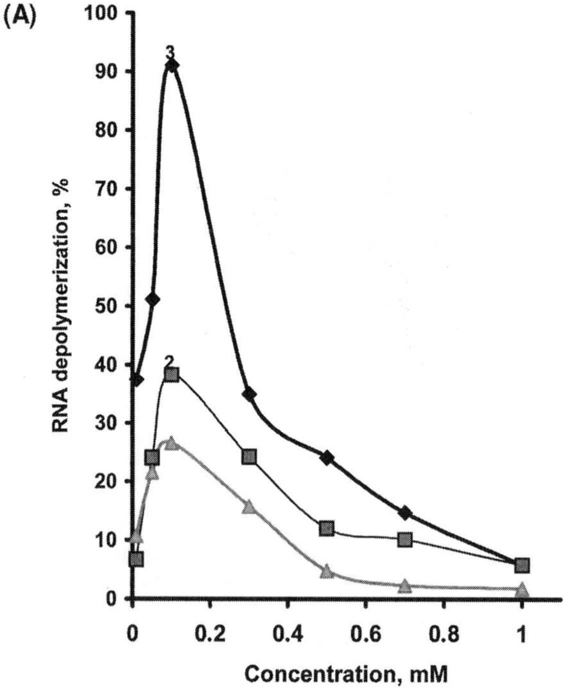

FIG. 3. The effect of reaction conditions on the rate of RNA cleavage with conjugates ABL3Cm. (A) Effect of the conjugates concentration. 5′-^{32}P-labeled *in vitro* transcript of tRNA^{Asp} was incubated at 37° for 18 h in 50 m*M* imidazole buffer, pH 7.0, containing 200 m*M* KCl, 0.5 m*M* EDTA, 0.1 mg/ml of carrier tRNA, and conjugates ABL3C1, ABL3C2, and ABL3C3 (curves 1, 2, and 3, respectively) in concentrations ranging from 1×10^{-5} to 1×10^{-3} *M*. (B) Kinetics of cleavage of the tRNA^{Asp} by conjugates ABL3Cm and RNase A. 5′-End labeled RNA was incubated at 37° in 50 m*M* imidazole buffer, pH 7.0, containing 200 m*M* KCl, 0.5 m*M* EDTA, 0.1 mg/ml tRNA carrier in the presence of mimics ABL3C1, ABL3C2, ABL3C3, and RNase A. Curves 1, 2, 3, 4, 5 show kinetics of RNA cleavage with 1×10^{-4} *M* ABL3C1, ABL3C2, ABL3C3, 5 n*M* RNase A, and 1 n*M* RNase A, respectively. Curves 6, 7, 8, 9, 10 correspond to RNA treatment with the same samples of mimics ABL3C1, ABL3C2, ABL3C3, and RNase A, respectively, that were filtered through the Centricon-3 membrane before addition to the solution of tRNA^{Asp}. (C) Graphs demonstrating the cleavage of the *in vitro* transcript of tRNA^{Asp} with ABL3C3 occurs as parallel reactions at each of the sensitive phosphodiester bonds. Kinetics of accumulation of cleavage products of 5′-^{32}P- and 3′-^{32}P-labeled tRNA^{Asp}. The experiments were performed in parallel with 5′-[^{32}P]tRNA^{Asp} (curves 1, 2, 3) and 3′-[^{32}P]tRNA^{Asp} (curves 4, 5, 6). Curves 1 and 4—for cleavage at U8-A9 phosphodiester bond; curves 2 and 5—for cleavage at C20-A21; and curves 3 and 6—for cleavage at C56-A57 phosphodiester bonds, respectively. Reaction conditions were the same as in Fig. 3B. Cleavage products were resolved in 12% PAAM/8 *M* urea gel and quantified as described in the experimental methods.

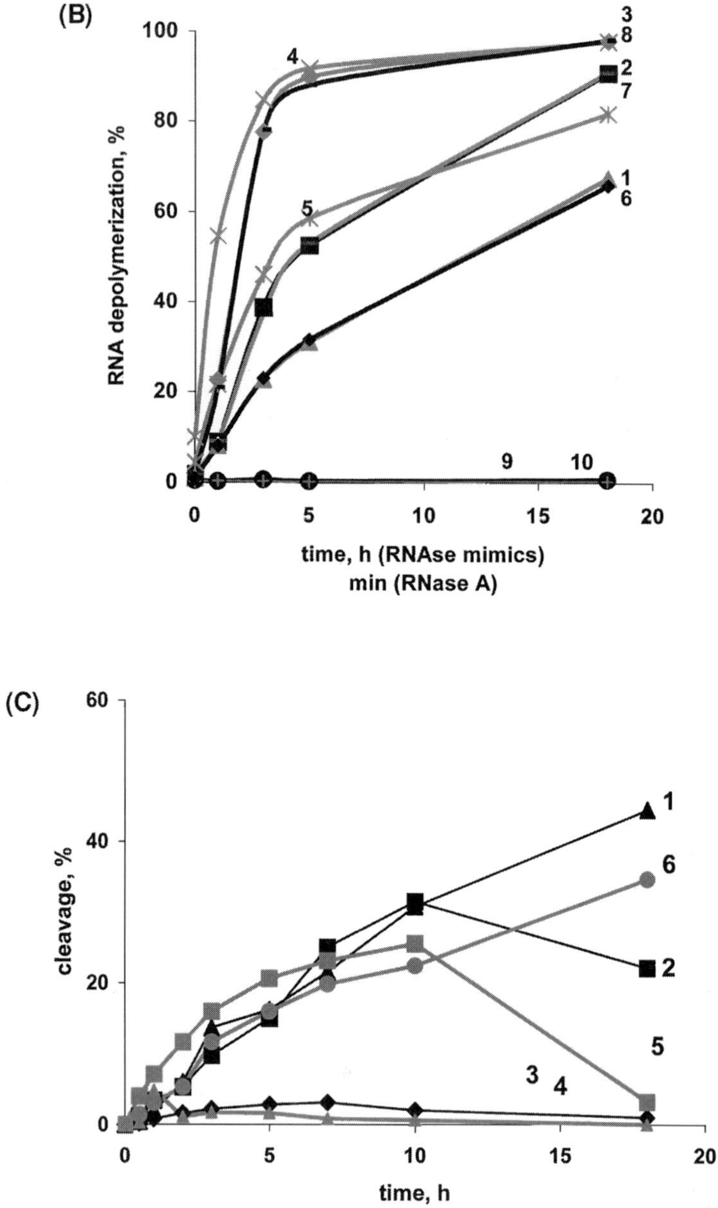

FIG. 3 (*continued*)

TABLE II
CLEAVAGE OF 5′-^{32}P-LABELED tRNA^{Asp} BY ABL3C3 AND RNase A IN DIFFERENT BUFFERS

Buffer[a]	Cleavage extent[b]	
	ABL3C3	RNase A
Imidazole	70	98
HEPES	49	34
Tris	67	73
Cacodylate	60	65
Phosphate	45	13
6 *M* Urea/HEPES[c]	38	98
4 *M* Guan/HEPES[c]	42	44

[a] tRNA was incubated in indicated buffers (buffer concentration was 50 m*M*), pH 7.0, containing 200 m*M* KCl, 0.5 m*M* EDTA, 0.1 mg/ml tRNA carrier in the presence of 1×10^{-4} *M* ABL3C3 at 37° for 18 h or in the presence of 5×10^{-9} *M* RNase A at 37° for 30 min.

[b] The extent of cleavage was calculated as the percentage of tRNA depolymerization after incubation at 37° for 10 h and 10 min with ABL3C3 and RNase A, respectively.

[c] Solutions of 6 *M* urea and 4 *M* guanidine isothiocyanate contained 50 m*M* HEPES–KOH, pH 7.0; 200 m*M* KCl, 0.5 m*M* EDTA and were supplemented with 0.1 mg/ml tRNA carrier.

C20-A, C43-A, C56-A are observed in both cases. The accumulation of cleavage products of tRNA labeled either at the 3′-end or at the 5′-end shows similar kinetic curves (Fig. 3C), demonstrating an independent cleavage of the tRNA transcript at each accessible site.

Table II displays the effect of different buffers on RNA cleavage by the conjugate ABL3C3 and RNase A. The maximal rate of RNA cleavage by the conjugate is observed in imidazole buffer, perhaps since the imidazole of the buffer may directly participate in the cleavage step. It was satisfying to observe that the efficiencies of tRNA cleavage with construct ABL3C3 is only slightly affected by different buffers and denaturing agents. The rate of RNA cleavage in HEPES, phosphate buffer, 6 *M* urea with HEPES, or 4 *M* guanidine isothiocyanate buffered with HEPES at pH 7.0 is only 1.5 times lower than the reaction in imidazole buffer, whereas the rate of RNA cleavage in Tris and cacodylate buffers are almost equal to the rate in imidazole buffer. Cationic conjugate molecules can bind to any site in negatively charged RNA and can form catalytically active complexes with substrate. From the data shown in Table II it is seen that the activity of RNase A is strongly affected by buffer type and is inhibited by phosphate buffer and in the presence of 4 *M* guanidine.

Specificity of RNA Cleavage with Conjugates ABL3Cm; Comparison with RNase A

The RNA cleaving conjugates described in this section represent a new family of tools potentially useful for the investigation of RNA conformation. The conjugates with imidazole groups representing simple mimics of RNase A demonstrate a substrate sequence and structure specificity similar to that of RNase A. Figure 4 shows the results of the experiments where conjugate ABL3C3 is used in parallel with RNase A and RNase T1, to probe the structure of *in vitro* transcripts of wild-type $tRNA^{Lys}$ under native conditions (Fig. 4A). ABL3C3 and RNase A cleave the tRNA mainly in the single-stranded regions and junctions (Fig. 4B). Some weak cuts observed in the D stem (U12 and U17) and the anticodon stem (U23 and U36) are located opposite to each other and indicate tRNA structure instability in these regions. Interestingly, the two probes ABL3C3 and RNase A cut $tRNA^{Lys}$ only at phosphodiester bonds in CA and UA sequences, except for a slight cleavage after C26 (C-C motive) in the anticodon loop.

Figure 5 summarizes the probing data on an *in vitro* transcript of influenza virus M2 RNA using conjugates ABL3C1, ABL3C2, and ABL3C3, RNase T1, and RNase ONE. It is seen that cleavages with the conjugates occur exclusively within CA sequences. It should be mentioned that all the CA sequences present in M2 RNA are cleaved by the conjugates. In some cases additional cuts were observed at phosphodiester bonds located on the 5′ side of the cleaved CA sequence: in these cases the cleavage occurred at G–C (G69, G74, G167), U–C (U82, U161, U285), and A–C (A186, A281) phosphodiester bonds. These cuts are likely secondary cuts occurring at the flexible fragile dangling ends formed by the primary cuts. Strong RNase T1 and RNase ONE cuts are located in the same M2 RNA regions. The positions of M2 RNA cleavage by the conjugates correlate well with RNase T1 cuts and position of RNA cleavage by single-strand specific RNase ONE. All together the results demonstrate that the new conjugates cleave RNA preferably at phosphodiester bonds within CA and UA sequences located in single stranded regions, in junctions or regions with unstable structure.

Specific Features of RNA Cleavage by Conjugates of 1,4-Diazabicyclo[2.2.2]octane and Imidazole

A straightforward approach to the design of RNA cleaving molecules would consist of mimicking the active center of ribonuclease A, using as catalysts the imidazole groups of two essential histidine residues.[21] The sole sufficiency of imidazole groups to imitate the ribonuclease active center is suggested by the

[21] C. A. Deakyne and L. C. A. Allen, *J. Am. Chem. Soc.* **101,** 3951 (1979).

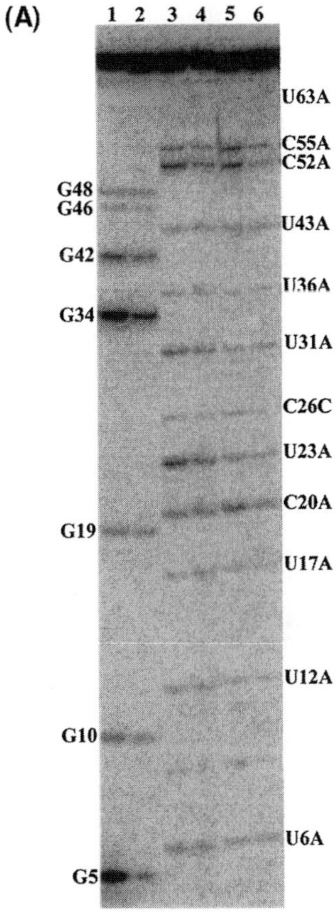

FIG. 4. Probing of 5′-end-labeled *in vitro* transcript of $tRNA^{Lys}$ with conjugate ABL3C3, RNase T1, and RNase A. (A) Autoradiograph of 12% polyacrylamide/8 *M* urea gel. Lanes 1 and 2, tRNA incubated at 37° in 50 m*M* Tris-HCl, pH 7.0, containing 200 m*M* KCl, 10 m*M* $MgCl_2$, 0.5 m*M* EDTA, 0.1 mg/ml tRNA carrier for 15 min with 0.5 or 0.1 units of RNase T1, respectively; lanes 3 and 4, tRNA incubated for 10 min in the same buffer with 10^{-8} *M* or 3×10^{-9} *M* RNase A, respectively; lanes 5 and 6, tRNA incubated in the same buffer at 37° with 1×10^{-4} *M* ABL3C3 for 1 h and 3 h, respectively. (B) Hairpin structure of the *in vitro* transcript of wild-type $tRNA^{Lys}$. Arrows (black, ABL3C3; gray RNase A) indicate the residues after which phosphodiester bonds are easily attacked by cleaving agents. The sizes of the arrows are proportional to the intensity of the cuts.

cleavage of RNA in concentrated imidazole buffers.[22–24] Several attempts to mimic the ribonuclease active center by conjugating imidazole residues to intercalating

[22] R. Breslow and M. Labelle, *J. Am. Chem. Soc.* **108,** 2655 (1986).
[23] R. Breslow and R. Xu, *Proc. Natl. Acad. Sci. U.S.A.* **90,** 1201 (1993).
[24] A. V. Vlassov, V. V. Vlassov, and R. Giege, *Dokl. Akad. Nauk (Russ.)* **349,** 411 (1996).

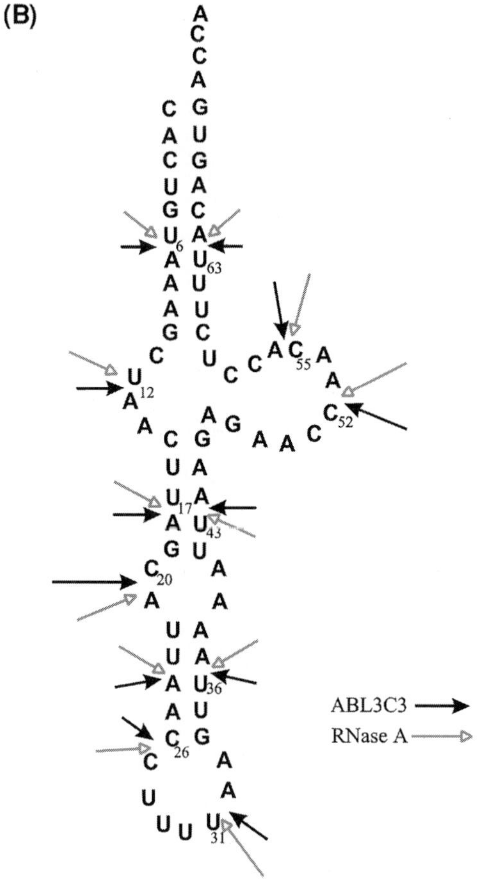

FIG. 4 (*continued*)

dyes,[16,25] spermidine,[26] or peptide molecules[27–29] have been reported. It was shown that these conjugates at millimolar concentrations cleave RNA substrates at an appreciate rate. Conjugates of acridine with histidine are able to cleave ribosomal

[25] A. Lorente, J. F. Espinosa, M. Fernandez-Saiz, J.-M. Lehn, D. Wilson, and Yi. Yi. Zhong, *Tetrahedron Lett.* **37,** 4417 (1996).

[26] V. V. Vlassov, G. Zuber, B. Felden, J.-P. Behr, and R. Giege, *Nucleic Acids Res.* **23,** 3161 (1995).

[27] C.-H. Tung, Z. Wei, M. J. Leibowitz, and S. Stern, *Proc. Natl. Acad. Sci. U.S.A.* **89,** 7114 (1992).

[28] N. S. Zhdan, I. L. Kuznetsova, A. V. Vlassov, V. N. Sil'nikov, M. A. Zenkova, and V. V. Vlassov, *Nucleosides Nucleotides* **18,** 1491 (1999).

[29] N. S. Zhdan, I. L. Kuznetsova, M. A. Zenkova, A. V. Vlassov, V. N. Sil'nikov, and V. V. Vlassov, *Russian J. Bioorgan. Chem.* **25,** 639 (1999).

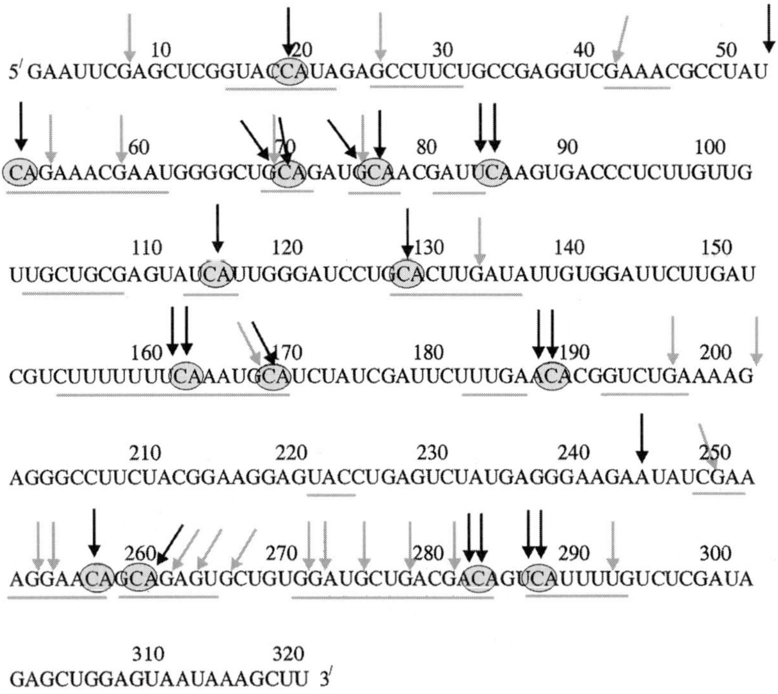

FIG. 5. Probing of the *in vitro* transcript of influenza virus M2 mRNA with conjugates ABL3C1, ABL3C2, ABL3C3, RNase T1, and RNase ONE. CA sequences are marked by ovals; phosphodiester bonds sensitive to cleavage with conjugates ABL3C1–ABL3C3 are shown by black arrows; gray arrows show RNase T1 cleavage sites; single-stranded RNA regions detected by RNase ONE probing are underlined. Experimental conditions are described in Materials and Methods.

RNA, although very slowly.[25] The efficacy of the cleavage increases when a second histidine residue or a free α-amino group is introduced in the conjugate structure.[27] We have designed conjugates of cationic 1,4-diazabicyclo[2.2.2] octane with constructs containing imidazole, or imidazole and ester, or imidazole and carboxylic group, which can at low concentrations (0.01–0.1 m*M*) efficiently cleave RNA under physiological conditions.

Surprisingly, the conjugates that mimic the ribonuclease A active center preferentially cleave RNA at the same motifs as the ribonuclease itself. The specificity of RNase A is assumed to result from the recognition of a pyrimidine nucleotide through formation of hydrogen bonds between the functional groups of the pyrimidine base and the groups in the active center of the enzyme.[30] The conjugates do not possess structures that recognize pyrimidine bases. However

[30] S. DelCardayre and R. Raines, *Biochemistry* **33,** 6031 (1994).

artificial ribonucleases obtained by conjugation of imidazole residues to RNA binding structures exhibit pronounced sequence specificity. The major cleavage sites of $tRNA^{Asp}$ by the spermine–imidazole (Sp–Im) conjugate are two CpA sequences in positions 55–56 and 20–21.[26] When the same $tRNA^{Asp}$ is incubated with the phenazine–imidazole conjugate (Phen–Im), cleavage occurs mainly at the C55-A56 phosphodiester bond.[16] The high rate of cleavage of this particular bond by the Sp–Im and Phen–Im conjugates may be explained by the high intrinsic susceptibility of this bond to cleavage.

The cleavage specificities of the conjugates ABLkCm are somewhat different from that observed for spermine–imidazole and phenazine–imidazole conjugates.[16,23] The ABLkCm conjugates demonstrate a pronounced preference for pyrimidine–purine sequences located in single-stranded regions. A source of the difference may be that these conjugates have a high affinity for phosphate anion and can access riboses and phosphates of RNA easily from different directions and cleave phosphodiester bonds, which are the most sensitive to cleavage. Apparently, a favorable situation is realized for Py-Pu sequences. The instability of such sequences in RNA is a known phenomenon,[12,31–34] the exact nature of which remains to be elucidated. In the case of the conjugates Sp–Im or Phen–Im, their access to phosphodiester bonds is governed by the nature of the RNA binding moiety. For example, the pattern of RNA cleavage by Sp–Im depends on two factors: the presence of susceptible phosphodiester bonds and the presence of specific spermine binding sites.

The conjugates ABLkCm cleave RNAs at pyrimidine–purine phosphodiester bonds located in the single-stranded regions (under conditions that stabilize RNA structure). In the $tRNA^{Asp}$ the bonds U8-A, U13-A, C20-A, C43-A, C56-A are the major cleavage sites and are cleaved by conjugates ABLkCm at similar rates. The *in vitro* transcript of wild-type $tRNA^{Lys}$ with a loose structure[35] is cleaved by these conjugates at each pyrimidine–adenine sequence (CpA and UpA), whereas in the influenza virus M2 RNA only CA bonds are cleaved by these conjugates. Comparison of the rates of oligonucleotide UUCAUGUAAA cleavage at the C3-A4 and U7-A8 phosphodiester bonds by ABL3C3 give the ratio $V_{CpA}/V_{UpA} = 3.2$. Interestingly, a similar ratio of the rates of cleavage at these bonds was measured for the reaction catalyzed by RNase A.[36] According to published data, the k_{cat} of cleavage of phosphodiester bond by RNase A in the dinucleotide CpA and UpA are

[31] C. Florentz, J.-P. Briand, P. Romby, L. Hirth, J.-P. Ebel, and R. Giegé, *EMBO J.* **1,** 269 (1982).

[32] A.-C. Dock-Bregeon and D. Moras, *Cold Spring Harbor Symp. Quant. Biol.* **52,** 113 (1987).

[33] R. Kierzek, *Nucleic Acids Res.* **19,** 5073 (1992).

[34] H. Hosaka, I. Sakabe, K. Sakamoto, S. Yokoyama, and H. Takaku, *J. Biol. Chem.* **269,** 20090 (1994).

[35] M. Helm, R. Giege, and C. Florentz, *Biochemistry* **38,** 13338 (1999).

[36] N. Beloglazova, D. Konevetz, V. Petyuk, V. Sil'nikov, V. Vlassov, and M. Zenkova, *Biochem. Biophys. Res. Commun.,* in press (2001).

$8.4 \times 10^6\ s^{-1}$ and $5.0 \times 10^6\ s^{-1}$, respectively.[37,38] From our data we estimated that cleavage of RNA by the conjugate is approximately 5000 times slower compared to the reaction catalyzed by RNase A, and that differences in reactivities of CA and UA sequences toward the chemical catalyst are more pronounced than in the case of the enzymatic reaction.

The high catalytic efficiency and ability of the conjugates ABLkCm to cleave RNA in different buffers even in denaturing conditions suggests the possibility of using these compounds as probes of RNA structure. Our initial studies support this possibility.[3,8] This simple structure can be used as reactive group for design of the next generation of antisense oligonucleotide conjugates capable of cleaving RNA at specific sequences[5,39] and for design of other RNA cleaving affinity reagents.

Materials and Methods

Materials

RNase Mimics. Syntheses of conjugates ABL3Cm followed published procedures.[7] The compounds ABLkCm were synthesized as described in Ref. 40. Synthesis of compounds AC and AB will be published elsewhere. The conjugates ABLkC3 and ABLkC2 are available on request to Dr. V. Sil'nikov.

Miscellaneous Chemicals and Enzymes. Nuclease- and metal-free imidazole solutions should be used. Stock imidazole solutions should be prepared and tested as described elsewhere.[3] Chemicals for electrophoresis can be obtained from Sigma (St. Louis, MO). [γ-^{32}P]ATP (3000 Ci/mmol) and [α-^{32}P]pCp (3000 Ci/mmol) were from Amersham (Les Ulis, France). All solutions for RNA handling should be prepared using Milli-Q water containing 0.5 m*M* EDTA. Solutions should be filtered through membranes with 0.2 μm pore size (e.g., Millex GS filters from Millipore, Bedford, MA). T4 polynucleotide kinase is from Amersham. Restriction nucleases, T4 RNA Ligase, RNase ONE, and RNasin are from Promega (Madison, WI). RNase T1 is from Boehringer Mannheim (Germany). T7 RNA polymerase is prepared according to established procedures.[41] Enzymes and chemicals used for end-labeling and electrophoresis are essentially as described.[42] Oligonucleotides are prepared by standard phosphoramidite chemistry and purified by HPLC.

[37] S. Zhenodarova, *Mol. Biol. (Russia)* **27,** 245 (1993).

[38] J. Thompson and R. Raines, *J. Am. Chem. Soc.* **116,** 5467 (1994).

[39] V. Vlassov, T. Abramova, T. Godovikova, R. Giege, and V. Silnikov, *Antisense Res. Dev.* **7,** 39 (1997).

[40] D. A. Konevetz, M. A. Zenkova, I. E. Beck, V. N. Sil'nikov, R. Giege, and V. V. Vlassov, *Tetrahedron,* in press (2000).

[41] B. Rether, J. Bonnet, and J.-P. Ebel, *Eur. J. Biochem.* **50,** 281 (1974).

[42] P. Romby, D. Moras, M. Bergdoll, P. Dumas, V. V. Vlassov, E. Westhof, J.-P. Ebel, and R. Giegé, *J. Mol. Biol.* **184,** 455 (1985).

RNAs. Wild-type aspartate tRNA transcript is synthesized enzymatically *in vitro* using T7 RNA polymerase and linearized plasmid essentially as described earlier.[43] Influenza virus M2 mRNA is synthesized *in vitro* using T7 transcription from *Hin*dIII linearized plasmid pSVK3M2. RNA concentrations are determined spectrophotometrically, assuming 1 A_{260} unit corresponds to 40 μg/ml RNA in a 1 cm path-length cell. For unambiguous analysis of cleavage products, RNA samples must be free of background nicks. Bulk yeast RNA, used as carrier to supplement labeled RNAs, is from Boehringer-Mannheim.

General Procedures for RNA Cleavage by RNase Mimics

End-Labeling of RNA. RNAs are 5′-end-labeled by dephosphorylation with alkaline phosphatase followed by phosphorylation with [γ-^{32}P]ATP and T4 polynucleotide kinase as described.[42,44] Labeling at the 3′ end of the $tRNA^{Asp}$ is performed by ligation of 5′-[^{32}P]pCp with T4 RNA ligase.[44] After labeling, RNAs are purified by electrophoresis in 12% denaturing polyacrylamide gels. The labeled RNAs are eluted from gels by 0.5 *M* ammonium acetate, pH 6.0, containing 0.5 m*M* EDTA and 0.1% sodium dodecyl (SDS). After ethanol precipitation, RNAs are dissolved in water and stored at −20°.

Cleavage Experiments. These experiments are usually performed with 50,000 to 100,000 Cerenkov cpm of 5′- or 3′-end-labeled RNA. These experiments include a sequencing ladders to assign cleavage sites (here, an imidazole ladder, and a G-ladder produced by RNase T1), the cleavage experiments, and controls, including incubations without reagents in order to visualize unspecific cuts. The standard assay volume is 10 μl. All reaction mixtures are supplemented by 1 μg of carrier RNA to facilitate precipitation, and to control the stoichiometry between cleavage reagents and RNA. Ladders and cleavage reaction are done according to the following protocols:

Imidazole ladders

Mix in an Eppendorf tube, 5 μl 4 *M* Im stock solution and 5 μl of solution of end-labeled RNA in water. Cap the tube tightly, vortex to mix, and place in a hot water bath at 90° for 10 min.

Transfer the tube to ice, remove the cap, and precipitate the RNA by adding 200 μl of 2% lithium perchlorate in acetone. Cap the tube, vortex, and centrifuge for 15 min at 15,000 rpm at 4°.

Carefully remove the liquid, wash the precipitate with acetone (200 μl), carefully remove the acetone, dry the pellet in desiccator, and dissolve in

[43] T. E. England and O. C. Uhlenbeck, *Biochemistry* **17,** 2069 (1978).

[44] P. Romby, D. Moras, P. Dumas, J.-P. Ebel, and R. Giegé, *J. Mol. Biol.* **195,** 193 (1987).

6 μl of loading buffer (7 *M* urea containing 0.02% bromphenol blue and 0.02% xylene cyanol).

G-ladder

Mix in an Eppendorf tube, 8 μl of buffer consisting of 7 *M* urea, 25 m*M* sodium citrate, pH 4.5, 0.1 m*M* EDTA, 0.25 mg/ml total tRNA, 0.01% bromphenol blue, 0.01% xylene cyanol, and 1 μl of solution of end-labeled RNA in water. Cap the tube tightly, vortex to mix, and place in a water bath at 50° for 10 min.

Transfer the tube to ice, remove the cap, and add 0.5 or 0.02 unit of RNase T1. Cap the tube tightly, vortex to mix, and place in a water bath at 50° for 10 min again.

Add 1 μl of 1 *M* Tris-HCl, pH 8.0 to stop the reaction, vortex, and keep frozen at −20° before use.

Cleavage of tRNA^{Asp} with conjugates ABLkCm. The procedures involving the ABLkCm, AC, AB, and B conjugates are essentially the same, except for the incubation times. The concentrations of the probes have to be optimal for each particular conjugate. We perform the cleavage experiments using the following buffer: 50 m*M* imidazole, pH 7.0, (pH adjusted by HCl), 200 m*M* KCl, 0.1 m*M* EDTA, 0.1 mg/ml total tRNA *Escherichia coli.* The stock solution is 5× Im buffer: 250 m*M* imidazole, pH 7.0, 1 *M* KCl, 0.5 m*M* EDTA. If needed, 50 m*M* $MgCl_2$ is added. We perform tRNA^{Asp} cleavage under the following conditions: the concentrations of the conjugates vary from 0.01 to 1 m*M*, with the optimum at 0.1 m*M* for ABLkCm. The incubation time at 37° varies from 30 min up to 48 h.

Solutions of RNase mimics are prepared as follows:

Dissolve an appropriate quantity of compound in dimethyl sulfoxide (DMSO) to provide a 50 m*M* stock solution. Store the solution at 4°.

Mix 20 μl of the stock solution with 480 μl of water to produce 2 m*M* solution.

Centrifuge the solution through a Centricon-3 membrane at 12,000 rpm at room temperature for 40 min. This solution can be stored at 4° for 1 month without loss of activity.

tRNA^{Asp} cleavage with conjugates ABLkCm

Mix 1 μl of the labeled tRNA solution, 1 μl of the carrier RNA solution (1 mg/ml), 5 μl of water, and 2 μl of stock 5× Im buffer. Incubate for 20 min at 20°.

Add 1 μl of 1 m*M* conjugate solution, vortex to mix, and incubate for 10 h at 37°.

Add 200 μl of 2% lithium perchlorate in acetone. Cap the tube, vortex, and centrifuge at 14,000 rpm for 15 min at 4°.

Carefully remove the liquid, wash the precipitate with acetone (200 μl), carefully remove the acetone, dry the pellet in desiccator, and dissolve in 5 μl of loading buffer (7 *M* urea containing 0.02% bromphenol blue and 0.02% xylene cyanol).

Structural Probing of RNA with Conjugates ABL3Cm

The procedures with conjugates ABL3C1, ABL3C2, and ABL3C3 are essentially the same as for RNA cleavage by enzymes, except that the incubation time has to be adjusted for each particular RNA and each conjugate. The reason is that the susceptibility of RNAs to the conjugates is considerably affected by RNA structure. Tightly folded molecules are more resistant to the RNA cleavers than are RNAs with loose structure. For instance, reasonable cleavage of *in vitro* transcript of $tRNA^{Asp}$ by ABL3C3 is achieved after 8 h incubation in the presence of 0.1 m*M* ABL3C3 while only 2 h incubation in 0.05 m*M* ABL3C3 is sufficient for cleavage of influenza virus M2 RNA. Only under optimal concentration of the cleaving compound and incubation time can a "one-hit" cleavage pattern be obtained.

We usually perform the cleavage experiments using the following buffer: 50 m*M* imidazole pH 7.0 (pH adjusted by HCl), 0.2 *M* KCl, 10 m*M* $MgCl_2$, 0.1 m*M* EDTA, 0.1 mg/ml total tRNA (*E. coli*). The stock solution is 5× Im buffer: 250 m*M* imidazole pH 7.0; 1 *M* KCl; 50 m*M* $MgCl_2$, 0.5 m*M* EDTA. Prior to cleavage the RNA is renatured by incubation at 70° for 1 min, slowly cooled down for 15 min, then incubated for 10 min at 25°. 5× Im buffer is then added, and RNA incubated 10 min at 25°. A standard assay is performed with 0.15 μg M2 mRNA supplemented with 1 μg of total tRNA from *E. coli* as carrier. Cuts in the M2 RNA are identified by primer extension. The primers are 5′-end-labeled oligodeoxyribonucleotides. Primer labeling is performed as described.[45] Elongation controls are run in parallel in order to detect nicks in the template or reverse transcription pauses. The RNA structure is probed with RNase T1 (5×10^{-3}–2×10^{-2} units) and RNase ONE (2×10^{-3}–8×10^{-3} units) at 37° for 15 min to identify structure specificity of the conjugates ABLkCm. Reactions with RNase T1 and RNase ONE are performed in the same buffer, except 50 m*M* imidazole, pH 7.0, is replaced by 50 m*M* HEPES, pH 7.0.

Primer extension analysis is according to published protocol.[46] Standard reaction mixture (15 μl) contains 50 m*M* Tris-HCl buffer, pH 8.3, 60 n*M* 5′-^{32}P-labeled

[45] A. M. Maxam and W. Gilbert, *Proc. Natl. Acad. Sci. U.S.A.* **74,** 560 (1977).

[46] L. Lempereur, M. Nicoloso, N. Riehl, C. Ehresmann, B. Ehresmann, and J.-P. Bachellerie, *Nucleic Acids Res.* **13,** 8339 (1985).

primer, 140 μg/ml M2 mRNA, 100 m*M* KCl, 4 m*M* dithiothreitol (DTT), 10 m*M* $MgCl_2$, 250 μ*M* of each dNTP, and 130 U/ml AMV RT (Life Science). Elongation is performed at 37° for 60 min. The products are ethanol precipitated and analyzed using 8% denaturing PAAG.

Influenza Virus M2 mRNA Probing with RNase Mimics ABL3Cm

Mix 1 μl of the *in vitro* synthesized M2 mRNA (0.15 μg), 1 μl of the carrier RNA solution (1 mg/ml), 5 μl of water, and 2 μl of 5× Im buffer. Incubate for 20 min at 20°.

Add 1 μl of 0.5 m*M* solution of conjugate ABL3C1, ABL3C2, or ABL3C3. Incubate for 2 h at 37°.

Add 200 μl of ethanol, containing 0.3 *M* NaAc, pH 5.2, and precipitate RNA at −20° for 1 h. Centrifuge at 14,000 rpm for 15 min at 4°.

Carefully remove the liquid, wash the precipitate with 70% (v/v) ethanol (200 μl), carefully remove the ethanol, dry the pellet in desiccator and dissolve in 5 μl of water. Analyze by primer extension reaction as described above.

Probing $tRNA^{Lys}$ with Conjugate ABL3C3:

Mix 1 μl of the 5′-end-labeled transcript of $tRNA^{Lys}$ (10000 cpm), 1 μl of the carrier RNA solution (1 mg/ml), 5 μl of water, and 2 μl of 5× Tris buffer. Vortex to mix, incubate for 20 min at 20°. (5× Tris buffer is 250 m*M* Tris-HCl, pH 7.0; 1 *M* KCl, 50 m*M* $MgCl_2$, 2.5 m*M* EDTA.

Add 1 μl of 1 m*M* solution of conjugate ABL3C3, mix, incubate at 37° for 1, 3, or 8 h.

Add 200 μl of 2% lithium perchlorate in acetone. Cap the tube, vortex, and centrifuge at 14,000 rpm for 15 min at 4°.

Carefully remove the liquid, wash the precipitate with acetone (200 μl), carefully remove the acetone, dry the pellet in desiccator and dissolve in 5 μl of loading buffer (7 *M* urea containing 0.02% bromphenol blue and 0.02% xylene cyanol).

In parallel one can probe $tRNA^{Lys}$ in the same buffer with RNase A (concentration in the reaction mix 0.01 and 0.05 μg/ml) at 37° for 10 min and with RNase T1 (0.1–0.5 units per mix) for 15 min at 37°.

Analysis of Cleaved RNA and Quantitation of Data

The RNA samples are dissolved in 7 *M* urea containing 0.02% bromphenol blue and 0.02% xylene cyanol and subjected to electrophoresis in a denaturing polyacrylamide gel (12% or 8% acrylamide, 7 *M* urea, 30 × 40 × 0.04 cm^3). Gels are electrophoresed in 1× TBE running buffer (100 m*M* Tris, 100 m*M* borate, and 2.8 m*M* EDTA). For calibration of cleavage patterns, end-labeled RNA is statistically cleaved at G residues by digestion with RNase T1 or cleaved by imidazole at 90° as described above. To detect cleavage positions in M2 mRNA, primer extension reactions are performed in the presence of ddATP, ddTTP, ddCTP, ddGTP. Gels must be run at 45°–50° to keep RNA denatured. After separation, the gels are transferred onto Whatman 3 m*M* filter paper, covered with plastic wrap, and dried. Dried gels are exposed to X-ray film (Kodak X-Omat, Rochester, NY) either at room temperature, or at −70° with intensifying screens.

Cleavage patterns are quantitated using a Bio-Imaging Analyzer, e.g., the FUJIX Bio-Imaging Analyzer BAS 2000 system. Photostimulatable imaging plates (type BAS-III from Fuji Photo Film Co, Ltd, Japan) are pressed on gels and exposed at room temperature for 30 min. Imaging plates are analyzed by performing volume integration of specific cleavage sites and reference blocks using the FUJIX BAS 2000 Work Station Software (version 1.1).

Acknowledgment

We thank Dr. D. Konevets who designed and prepared the ABLkCm conjugate. We thank Dr. A. Vlassov who did some original work with conjugates; we acknowledge N. Tamkovich, N. Komarova, and N. Chumakova for the experiments on oligonucleotide and CpA cleavage with described conjugates. We thank Prof. C. Florenz for helpful discussion and constant interest. This work was supported by INTAS Grant 96-1418, RFBR Grants 99-04-49538 and 00-15-97969, CNRS and University Louis Pasteur, Strasbourg.

[31] Preparation and Use of ZFY-6 Zinc Finger Ribonuclease

By WALT F. LIMA and STANLEY T. CROOKE

Introduction

The 30-amino acid (aa) peptide (KTYQCQYCEYRSADSSNLKTHIKTKHS-KEK) of the ZFY zinc finger protein has been shown to exhibit single-strand specific endoribonuclease activity.[1] The ZFY protein is a human male-associated transcription factor and contains zinc finger domains arranged in a distinctive two-finger repeat motif.[2,3] The sequence of the peptide corresponds to the

[1] W. F. Lima and S. T. Crooke, *Proc. Natl. Sci. U.S.A.* **96,** 10010 (1999).

[2] J. M. Berg, *Science* **232,** 485 (1987).

[3] T. J. Gibson, J. P. M. Postama, R. S. Brown, and P. Argos, *Protein Eng.* **2,** 209 (1988).

0076-6879/01 $35.00

even-numbered zinc finger domain ZFY-6 and has been shown to form an independent zinc finger structural unit through the tetrahedral coordination of the zinc ion with both the cysteine thiolates and the histidine imidazoles.[2,4]

The ribonuclease active structure of the ZFY-6 peptide consists of a homodimer, and the formation of the homodimer was shown to occur through a single intermolecular disulfide bridge.[1] Furthermore, the ribonuclease active structure of the ZFY-6 peptide was observed to form only in the absence of zinc. The coordination of zinc ion with the cysteine thiolates of the ZFY-6 peptide inhibited the formation of the homodimer and resulted in the complete ablation of the ribonuclease activity. In addition, the mutant sequence of the ZFY-6 peptide in which both the cysteines had been substituted with alanines as well as the ZFY-6 peptide treated with the thiol blocking agent *N*-ethylmaleimide were also shown not to form homodimers and not to exhibit RNase activity.

The ribonuclease activity of the ZFY-6 peptide was observed to be dependent on the state of ionization of the peptide as well as the ionic strength of the reaction buffer.[1] For example, the ZFY-6 peptide exhibited a bell-shaped response with respect to changes in KCl concentration. Likewise, the pH profile curve for the ZFY-6 peptide was determined to be bell-shaped. The observed pH profile of the ZFY-6 peptide is characteristic of the general acid–base catalytic mechanism described for the single-strand specific RNase A family of enzymes.[5] Also consistent with this catalytic mechanism were the observations that (i) cleavage of the RNA by the ZFY-6 peptide involved a metal-independent mechanism; (ii) the termini of the RNA cleavage products consisted of 3′-phosphate and 5′-hydroxyl; and (iii) the ZFY-6 peptide cleaved specifically at the 3′ side of pyrimidines.

The cleavage rates observed for the ZFY-6 peptide were many orders of magnitude faster than the rates observed for other synthetic ribonuclease peptides.[6,7] The rate of cleavage was dependent on the sequence of the RNA substrate with RNA sequences containing 5′-pyrimidine-adenosine-3′ (5′-pyr-A-3′) cleaved 3- to 8-fold faster than sequences without the dinucleotide. Moreover, RNA sequences containing three 5′-pyr-A-3′ dimers were degraded 40-fold faster than sequences without the preferred dinucleotide. Kinetic analysis of the cleavage activity under multiple- and single-turnover conditions revealed, respectively, values for k_{cat} of 0.26 and 0.20 min^{-1} and values for K_m of 494 and 209 n*M*. Furthermore, the values for k_{cat} were comparable to those observed for the group II intron ribozymes.[8]

[4] A. D. Frankel, J. M. Berg, and C. O. Pabo, *Proc. Natl. Acad. Sci. U.S.A.* **84,** 4841 (1987).

[5] J. J. Beintema, H. J. Breukelman, A. Carsana, and A. Furia, *in* "Ribonucleases: Structures and Function" (G. D'Alessio and J. F. Riordan, eds.), p. 245. Academic Press, New York, 1997.

[6] M. A. Podyminogin, V. V. Vlassov, and R. Giege, *Nucleic Acids Res.* **21,** 5950 (1993).

[7] T. Ching-Hsun, Z. Wei, M. J. Leibowitz, and S. Srein, *Proc. Natl. Acad. Sci. U.S.A.* **89,** 7114 (1992).

[8] W. J. Michels and A. M. Pyle, *Biochemistry* **34,** 2965 (1995).

Synthesis and Purification of ZFY-6 Zinc Finger Peptide

The peptide is synthesized by the solid-phase procedure using the Applied Biosystems automated synthesizer (Foster City, CA) and *tert*-butyloxycarbonyl (*t*-BOC) amino acids.[9] The peptide is synthesized with a free amine at the amino terminus and with an amide group at the carboxyl terminus. The crude peptide is purified to >90% by reversed-phase high-performance liquid chromatography (HPLC) on a C_{18} column (250 × 4.6 mm) using a 5.0–65.0% gradient of acetonitrile in 0.1% trifluoroacetic acid/distilled water (%, v/v). The peptide is lyophilized and the pellet is resuspended in distilled water. Peptide preparations found to exhibit poor solubility in water are dissolved by adding acetic acid to a final concentration of 0.1% acetic acid (%, v/v). Peptide purity is checked by mass spectrometry and analytical reversed-phase HPLC. The concentration of the peptide is determined by amino acid analysis.

Preparation of ZFY-6 Zinc Finger Homodimer

Procedural Comments

Formation of the ZFY-6 homodimer is critical as only the homodimeric form of the peptide exhibits endoribonuclease activity. The amount of homodimer formed is dependent on both the reduced state of the cysteines and the concentration of the peptide. Consequently, the complete reduction of the peptide prior to dimerization is essential in order to ensure that potential previously formed inactive disulfide isomers are denatured. The formation of the active conformer is best accomplished with the rapid one-step oxidation of the fully reduced peptide, with a peptide concentration greater than 1 m*M*. While these conditions favor the formation of the homodimer (e.g., dimerization >90%), lower yields were sometimes improved by increasing the oxidation pressure through the addition of the glutathione redox pair (GSSG/GSH). Finally, the formation of ZFY-6 homodimer requires the use of nonionic detergents [e.g., Nonidet P-40 (NP-40, CalBiochem, La Jolla, CA) and Triton X-100 (Sigma, St. Louis, MO)].

Reduction of Peptide

The ZFY-6 peptide is reduced under denaturing conditions by incubating 0.25–0.5 μmol of peptide in 200 m*M* dithiothreitol (DTT) and 6 *M* urea for 16 h at 60°. The reduced peptide is purified by gel filtration using a Sephadex G-25 column (1.5 × 4.9 cm) in 0.1 *M* acetic acid. Reduction of the peptide is monitored by polyacrylamide gel electrophoresis using a gradient (10–20%) Tricine–SDS gel (Novex, San Diego, CA) and by thiol assay using the Ellman method. The

[9] G. Barany and R. B. Merrifield, *in* "The Peptides" (E. Gross and J. Meienhofer, eds.), Vol. 2, p. 18. Academic Press, New York, 1979.

determination of free sulfhydryl by the Ellman method is carried out in 5.2 ml containing approximately 4 nmol purified peptide (in 0.1 *M* acidic acid), 0.4 mg Ellman reagent [5,5′-dithiobis(2-nitrobenzoic acid), Sigma], and 100 m*M* sodium phosphate, pH 8.0. After 15 min, the absorbance is measured at 412 nm. Here, 10 μM sulfhydryl solution gives an absorbance of 0.136 (1 cm light path), or a molar absorbance for sulfhydryl of 13,600 (1 cm light path) at 412 nm.

Formation of Homodimer

The eluent from the gel filtration column is lyophilized, and the pellet is resuspended in 50 μl of oxidation buffer (20 m*M* sodium phosphate, pH 8.0 and 0.1% NP-40). Alternatively, the oxidation strength of the buffer is increased with the addition of 0.5 m*M* GSSG and 5 m*M* GSH (Sigma) and incubating for 16 h at 4°. The glutathione redox pair is removed by dialysis using a 3500 molecular weight cutoff Slide-Lyser cassette (Pierce, Rockford, IL) and two 1 liter changes of the oxidation buffer at 4° for 4 h. Homodimer formation is monitored by polyacrylamide gel electrophoresis using gradient (10–20%) Tricine–SDS gel (Novex) and/or liquid chromatography–mass spectrometry (LC-MS).

Preparation and Purification of Oligoribonucleotide Substrate

Procedural Comments

The cleavage activity of the ZFY-6 peptide is dependent on the sequence and structure of the oligoribonucleotide substrate. The ZFY-6 zinc finger peptide specifically cleaves at pyrimidines and preferentially hydrolyzes the dinucleotide sequence 5′-pyr-A-3′.[1] Therefore, the RNA substrate must contain pyrimidines and ideally, the 5′-pyr-A-3′ dinucleotide. Furthermore, RNA sequences capable of forming secondary structures should be avoided as the ZFY-6 peptide has been shown not to digest double-strand RNA.[1] The potential for forming secondary structure increases with increasing oligonucleotide length and therefore the use of short oligoribonucleotides (e.g., 15–20 nucleotides in length) is recommended.

Two methods are provided for the synthesis of the oligoribonucleotide substrate. The first involves the *in vitro* transcription of the RNA substrate using T7 RNA polymerase and the method of Milligan and Uhlenbeck, which has been modified for the synthesis of short oligoribonucleotides.[10] The second involves the synthesis of the oligoribonucleotide substrate on an automated DNA synthesizer.

T7 Transcription

The synthetic DNA template for transcription is prepared in 100 μl containing 100 pmol of the template strand (a fragment containing the 17 nucleotide (nt) T7

[10] J. F. Milligan and O. C. Uhlenbeck, *Methods Enzymol.* **180,** 51 (1989).

promoter sequence followed by the complement of the desired RNA sequence), 1 nmol of the 17-mer complement of the promoter sequence, 10 m*M* Tris, pH 8.0, and 1 m*M* EDTA. The template is heated at 90° for 5 min and incubated at 37° for 2 h. The T7 transcription reaction is prepared in 100 μl containing 100 n*M* synthetic DNA template, 1 m*M* each NTP, 0.1 mg/ml T7 RNA polymerase (Promega, Madison, WI), 40 m*M* Tris, pH 8.1, 6 m*M* $MgCl_2$, 5 m*M* DTT, 1 m*M* spermidine, 0.01% Triton X-100 (%v/v), 50 mg/ml bovine serum albumin (BSA), and 80 mg/ml polyethylene glycol (PEG) 8000. The reaction is incubated at 37° for 2 h and the DNA template is removed by adding 1 unit of RQ1 RNase-free DNase (Promega), and incubating the reaction for an additional 15 min. The RNA transcript is extracted two times with (1 : 1) phenol : chloroform (Amresco, Cleveland, OH) and two times with (24 : 1) (v/v) chloroform : iosamyl alcohol (Amresco, OH). The extracted RNA is precipitated with 300 m*M* sodium acetate, pH 4.6, or 1 *M* ammonium acetate, 40 μg glycogen (Boehringer Mannheim, Indianapolis, IN), and 2.5 volumes of ethanol at −20° for 16 h, and centrifuged in a microfuge at 30,000 rpm for 30 min. The pellet is lyophilized and resuspended in double-distilled water. The RNA is purified using a Quick Spin Sephadex G-25 column according to the manufacturer's instructions (Boehringer Mannheim, Indianapolis, IN).

Automated Oligoribonucleotide Synthesis

The oligonucleotide is synthesized on a PE-ABI 380B synthesizer using 5′-*O*-silyl-2′-*O*-bis(2-acetoxyethoxy)methylribonucleoside phosphoramidites.[11] The phosphate methyl group is cleaved in 1 *M* disodium 2-carbamoyl-2-cyanoethylene-1,1-dithiolate in dimethylformamide (DMF) for 30 min. The oligonucleotide is cleaved from the support and the acyl and acetyl groups on, respectively, the exocyclic amines and 2-*O*-orthoester are removed in 150 m*M* sodium acetate buffer, pH 3.0, at 55° for 10 min. The pH is raised to pH 7.8 with an equal volume of 300 m*M* Tris buffer, pH 8.7 and the deprotected oligonucleotide is then incubated at 55° for an additional 10 min. The oligoribonucleotide is purified by reversed-phase HPLC.

^{32}P Labeling of Oligoribonucleotide Substrate

Procedural Comments

The oligoribonucleotide is either 5′-end-labeled with [γ-^{32}P]ATP using T4 polynucleotide kinase or 3′-end-labeled with [^{32}P]pCp using T4 RNA ligase. Selecting the appropriate labeling method is dependent on several factors. These include (i) the method used to analyze the ribonuclease activity of the peptide and

[11] S. A. Scaringe, F. E. Wincott, and M. H. Caruthers, *J. Am. Chem. Soc.* **120,** 11820 (1998).

the corresponding size of the ^{32}P-labeled cleavage products; (ii) the reagents used to prepare the RNA substrate, and (iii) the respective labeling efficiencies of the two methods. For example, the sensitivity of the acid precipitation method used to analyze the ribonuclease activity of the peptide is significantly affected by the size differential between cleavage products and the intact oligoribonucleotide substrate. In this case, shorter cleavage products result in a greater size differential and ultimately improved sensitivity. Consequently, if the predominant cleavage sites are positioned on the 3′-side of the oligoribonucleotide, then placing the ^{32}P label on the 3′-end of the substrate would result in shorter cleavage products. Similarly, if the predominant cleavage sites are positioned on the 5′-side of the oligoribonucleotide, then the 5′-end labeling method is recommended. The choice of labeling method is also influenced by the reagents used in the preparation of the RNA substrate. For instance, ammonium salts are often used to increase the precipitation efficiency of the oligoribonucleotide in ethanol. However, ammonium salts are not compatible with the 5′-end labeling method as these ions are strong inhibitors of T4 polynucleotide kinase. Finally, the 5′-end labeling method is generally 2 to 3 times more efficient than the 3′-end labeling method. The expected specific activity for 40 pmol preparation of 5′-end-labeled RNA oligonucleotide is approximately 3000–8000 cpm/fmol, whereas the specific activity for 3′-end-labeled RNA is approximately 1000–3000 cpm/fmol.

The recommended procedure for the purification of the labeled substrate is by electrophoresis on a polyacrylamide gel. Although other less labor-intensive procedures are available (e.g., ethanol precipitation or gel filtration), this procedure consistently provides the level of purity (e.g., >95%) required for the cleavage assay. Achieving this level of purity is crucial, as the presence of contaminating deletion products introduced during the synthesis of the RNA will greatly interfere with the interpretation of the enzymatic assay.

Dephosphorylation of T7 Transcript

The 5′-triphosphate of the enzymatically synthesized T7 transcript is hydrolyzed by incubating 40 pmol RNA, 1 U calf intestine alkaline phosphatase (Boehringer Mannheim, GmbH), 50 m*M* Tris, pH 8.5, and 0.1 m*M* EDTA at 37° for 30 min. The alkaline phosphatase is removed by extracting two times with (1 : 1) phenol : chloroform (Amresco) and two times with (24 : 1) chloroform : iosamyl alcohol (Amresco) and the extracted RNA is precipitated as described above. The RNA is resuspended in double-distilled water.

^{32}P Labeling

The RNA substrate is 5′-end-labeled with ^{32}P using 20 U of T4 polynucleotide kinase (Promega, Madison, WI), 120 pmol (7000 Ci/mmol) [γ-^{32}P]ATP (ICN, Costa Mesa, CA), 40 pmol RNA, 70 m*M* Tris, pH 7.6, 10 m*M* $MgCl_2$, and

50 m*M* DTT. The kinase reaction is incubated at 37° for 30 min. Alternatively the RNA substrate is 3′-end-labeled with ^{32}P using 50 U T4 RNA ligase (Boehringer Mannheim), 80 pmol (3000 Ci/mmol) [^{32}P]pCp (ICN), 40 pmol RNA, 50 m*M* Tris, pH 7.5, 10 m*M* $MgCl_2$, 10 m*M* DTT, 10 m*M* ATP, and 5 μg bovine serum albumin (BSA). The ligation reaction is incubated at 4° for 16 h.

Purification of the ^{32}P Labeled RNA

The labeled RNA is purified by electrophoresis. An equal volume of loading buffer (8 *M* urea, 0.25% xylene cyanol, and 0.25% bromphenol blue) is added to the labeled oligoribonucleotide and the RNA is separated on 12% denaturing polyacrylamide gel. The gel fragment corresponding to the full-length labeled RNA is localized by autoradiography, excised, crushed, and soaked in 0.5 *M* NaCl at 4° for 16 h. The gel slurry is centrifuged at 3000 rpm for 5 min and the supernatant is extracted two times with phenol : chloroform, two times with chloroform : isoamyl alcohol, and ethanol precipitated as described above. The RNA pellet is resuspended in double-distilled water.

Cleavage Assay

Procedural Comments

The cleavage rate of the ZFY-6 peptide is greatly influenced by the solubility and state of ionization of the peptide. The optimal cleavage rate was observed at pH 7.0, 100 m*M* KCl, and 0.1% NP-40. The cleavage rates are obtained from the initial 10% of the reaction. Consequently, even very low levels of nonspecific degradation of the RNA substrate resulting from contaminants within the reaction can greatly affect the kinetic analysis of the cleavage reaction. Therefore it is particularly important to follow the common practices for working with RNA. These include the use of gloves, double-distilled water, RNase-free certified tubes (Sorenson Bioscience, UT), and pipette tips (Rainin, MA), and RNase inhibitors (e.g., Prime RNase Inhibitor (5 Prime → 3 Prime, Boulder, CO), rRNasin (Promaga, Madison, WI). Finally, with the precautions listed here, the treatment of buffers and water with diethyl pyrocarbonate (DEPC) is not necessary.

Analysis of the cleavage products is performed by either polyacrylamide gel electrophoresis or acid precipitation. The acid precipitation method is advantageous for quantitating large sample numbers, but this method should be reserved for longer substrates as the precipitation efficiencies of oligoribonucleotides substrates <15-mer is poor. On the other hand, analysis by PAGE is amenable to a significantly broader range of oligoribonucleotide lengths and offers the added advantage of providing information about the position and number of cleavage site on the RNA.

Initial Cleavage Rate (v_0)

The cleavage reaction is prepared in 100 μl containing 500 n*M*–1 μM unlabeled RNA substrate, 10^5 cpm ^{32}P-labeled RNA, digestion buffer (20 m*M* sodium phosphate, pH 7.0, 100 m*M* KCl, 1 m*M* EDTA, 0.1% NP-40), and 60 U of Prime RNase Inhibitor (5 Prime → 3 Prime). The background control is prepared by incubating a 10 μl aliquot of the cleavage reaction without ZFY-6 peptide at 37° for the duration of the assay. The peptide is diluted to a final concentration of 4.5 μM in sodium phosphate buffer, pH 8.0, and 2 μl of the diluted ZFY-6 peptide homodimer is added to the digestion reaction (resulting in a final peptide concentration of 100 n*M*). A 10 μl aliquot of the cleavage reaction is removed at time points ranging from 2 to 120 min and quenched by adding 10 μl of 8 *M* urea and snap-freezing on dry ice. The aliquots are heated at 90° for 2 min and resolved in a 12% denaturing polyacrylamide gel, and the substrate and product bands are quantitated on a Molecular Dynamics PhosphorImager.

For acid precipitation the 10 μl aliquot of the cleavage reaction is quenched with 90 μl of 0.6 mg/ml yeast tRNA and then precipitated on ice with 100 μl 10% trichloroacetic acid (Sigma) for 5 min. The sample is centrifuged at 15,000*g*, for 5 min at 4°. A 150 μl aliquot of the supernatant is removed and added to 2 ml of scintillation cocktail and the solubilized radioactivity counted in a scintillation counter.

The concentration of converted substrate is calculated by measuring the fraction of substrate converted to product (acid-soluble counts or counts for cleavage product bands/total counts) for each time point, multiplying by the substrate concentration, and correcting for background ((fraction product × [total substrate]) − background). The background values represent the fraction corresponding to the degradation products (counts for nonspecific degradation products/total counts). The concentration of the converted product is plotted as a function of time. The initial cleavage rate is obtained from the slope (mol RNA cleaved/min) of the best-fit line for the linear portion of the plot, which comprises, in general, <10% of the total reaction. For accuracy, the line should represent data from at least five time points. The time points are selected through iterative testing to obtain a sufficient number of data points within the linear portion of the rate curve. Alternatively, the concentration of ZFY-6 peptide homodimer is adjusted.

Single-Turnover Kinetics

Cleavage reactions are performed using excess ZFY-6 peptide and the RNA substrate concentration well below the K_m. The cleavage reaction is prepared in 100 μl containing 50 n*M* unlabeled RNA substrate, 10^5 cpm ^{32}P-labeled RNA, digestion buffer (20 m*M* sodium phosphate, pH 7.0, 100 m*M* KCl, 1 m*M* EDTA, 0.1% NP-40) and 60 U of Prime RNase Inhibitor. The background control is prepared as described for the initial rate determinations. The ZFY-6 peptide is diluted in sodium phosphate buffer, pH 8.0, and added to the cleavage reactions at a final concentration ranging from 50 nM to 1 μM. The cleavage reactions are quenched and analyzed by either the PAGE or acid precipitation procedures as described for the initial rate determinations. The product fraction is calculated as described for the initial rate determinations. The first-order reaction rates (k_{obs}) for each peptide concentration are determined by plotting the natural logarithm of the intact substrate fraction (1 − fraction product) vs time. The $t_{1/2}$ of the reaction is obtained from the slope of the best-fit line, and values for $k_{obs} = (ln\,2)/t_{1/2}$. The plots should be linear for more than five half-lives based on an arbitrary reaction end point of 95%. The values for k_{obs} are plotted as a function of peptide concentration and fit to a Michaelis–Menten binding curve using the program Ultrafit (Biosoft, Ferguson, MO). The single-turnover K_m and k_{cat} correspond to, respectively, the ZFY-6 peptide concentration at half-maximum rate and the horizontal asymptote of the hyperbolic curve. The enzyme concentrations used to determine the single-turnover K_m and k_{cat} are selected by trial and error, and the Michaelis–Menten binding curve is fit using a minimum of four peptide concentrations.

Multiple-Turnover Kinetics

Cleavage reactions are performed using excess RNA substrate and with a ZFY-6 peptide concentration below the K_m of the complex. The cleavage reaction is prepared in 100 μl containing an unlabeled RNA substrate at concentrations ranging from 100 to 1000 nM, 10^5 cpm ^{32}P-labeled RNA, digestion buffer (20 mM sodium phosphate, pH 7.0, 100 mM KCl, 1 mM EDTA, 0.1% NP-40) and 60 U of Prime RNase Inhibitor. The background control is prepared as described for the initial rate determinations. The peptide is diluted to a final concentration of 4.5 μM in sodium phosphate buffer, pH 8.0, and 2 μl of the diluted ZFY-6 peptide homodimer is added to the digestion reaction (resulting in a final peptide concentration of 100 nM). The cleavage reactions are quenched and analyzed by either the PAGE or acid precipitation procedures as described for the initial rate determinations. The product concentration is calculated as described for the initial rate determinations and plotted as a function of time. The multiple-turnover reaction rates (k_{obs}) are calculated from the slope (mole RNA cleaved/min) of a linear least-squares fit from the linear portion of the curve, normally the first 5–10% of the reaction. The (k_{obs}) values are plotted as function of the substrate concentration and the data is fit to the Michaelis–Menten equation using the program Ultrafit (Biosoft, NJ). The multiple-turnover K_m corresponds to the RNA substrate concentration at half-maximum rate and the $k_{cat} = V_{max}$/[total peptide], where V_{max} corresponds to the horizontal asymptote of the hyperbolic curve. Alternatively, the multiple-turnover K_m and k_{cat} are determined by plotting k_{obs} vs k_{obs}/[RNA substrate]. The slope of the line is equal to $-K_m$ and the y intercept is equal to the V_{max}. The RNA substrate concentrations used to determine the multiple-turnover K_m and k_{cat} are selected by trial and error and the Michaelis–Menten binding curve is fit using a minimum of four RNA substrate concentrations.

pH Dependence

Procedural Comments

The state of ionization of the free substrate and peptide has a profound effect on the affinity of the complex (K_m). Therefore, in order to eliminate the effect of pH on (K_m), the reaction rates (k_{obs}) are determined with sufficiently high RNA substrate concentration to saturate the ZFY-6 peptide. Multiple buffers are used to measure the activity of the peptide over a broad range of pH values. In order to ensure that changes in cleavage activities are not due to buffer-specific effects, duplicate measurements are made at pH values in which the buffering ranges overlap (e.g., sodium acetate and sodium phosphate for pH 5.0, sodium phosphate and sodium carbonate for pH 8.0).

Procedure

The cleavage reactions are prepared in 100 μl containing 1 μM unlabeled RNA substrate, 10^5 cpm ^{32}P-labeled RNA, 100 mM KCl, 1 mM EDTA, 0.1% NP-40, and 60 U of Prime RNase Inhibitor. The pH of the reaction is adjusted to pH 4.0 with 20 mM sodium acetate, pH 5.0 to 8.0 with sodium phosphate, and pH 9.0 with sodium carbonate buffer. The background control is prepared as described for the initial rate determinations. The peptide is diluted to a final concentration of 9.0 μM in sodium phosphate buffer, pH 8.0, and 2 μl of the diluted ZFY-6 peptide homodimer is added to the digestion reaction (resulting in a final peptide concentration of 200 nM). The cleavage reactions are quenched and analyzed by either the PAGE or acid precipitation procedures as described for the initial rate determinations. The reaction rates (k_{obs}) are calculated as described for multiple-turnover kinetics and the logarithm of k_{obs} is plotted as a function of pH. A nonlinear least-squares fit of the data is generated using the equations of Dixon and Webb[12] and implemented with the program Winnonlin (Scientific Consulting, CA).

Conclusions

The previously reported observations together with the procedures outlined here set the framework for further characterization of the ZFY-6 peptide. Clearly, numerous questions pertaining to the active structure, the catalytic mechanism,

[12] M. Dixon and E. C. Webb, *in* "Enzymes," 3rd Ed. (M. Dixon and E. C. Webb, eds.), p. 148. Academic Press, New York, 1979.

and the possible biological roles of the zinc finger ribonuclease remain. For example, the sequence of the ZFY-6 zinc finger peptide contains the classical Cys_2His_2 zinc finger motif ($CX_{2,4}CX_9LX_2HX_{3-5}H$), and it is unclear whether other members of this zinc finger family exhibit ribonuclease activity. Nor is it clear whether the five conserved amino acids of the zinc finger domain are required for ribonuclease activity. Whereas it has been shown that the cysteine residues are required for the formation of the ribonuclease active structure, the histidine residues as well as the glutamic acid residue were found not to be required for catalysis.[1] Furthermore, the ZFY-6 peptide homodimer contains four cysteine residues and as a result several disulfide isomers are possible. As to which of the disulfide isomers is responsible for forming the active conformation, this remains unclear. Finally, it is also not known whether the conserved leucine residue is required for RNase activity.

Section VI

Ribonucleolytic Nucleic Acids

[32] RNA Cleavage by the 10-23 DNA Enzyme

By GERALD F. JOYCE

Introduction

An idealized reagent for the sequence-specific cleavage of RNA would be a compound that is easily synthesized, chemically stable, generalizeable to any target sequence, and highly efficient in cleaving the target RNA. If the reagent is to be utilized *in vivo,* then it must be delivered readily to the cells of interest, remain active in those cells, encounter the target RNA within the appropriate cellular compartment, and not produce cellular toxicity or other undesirable side effects. Clearly this list of requirements is formidable and cannot be met by any known class of compounds. Antisense agents, broadly defined, perhaps come closest. Their most important attribute is that, by controlling the sequence of the antisense oligodeoxynucleotide, one can direct these compounds to almost any target RNA in a sequence-specific manner. Antisense agents themselves do not have RNA-cleavage activity, but instead rely on the protein enzyme RNase H to degrade the RNA component of the heteroduplex formed between the antisense DNA and target RNA.

Shortly after the discovery of catalytic RNA (ribozymes), efforts began to apply these compounds as "catalytic antisense" agents that could be made to cleave target RNAs with high sequence specificity.[1,2] Rather than depend on RNase H, the ribozyme itself is responsible for cleaving the RNA following recognition through Watson–Crick pairing. Compared to antisense oligodeoxynucleotides, however, ribozymes are more difficult to synthesize and must be significantly modified to prevent their degradation by cellular nucleases.[3] Ribozymes are more sequence-specific than conventional antisense agents, but less generalizeable with respect to the target sequence.

In recent years a new class of catalytic antisense agents has emerged, based on DNA enzymes that have been obtained by *in vitro* selection. In one study, designed to elicit Mg^{2+}-dependent RNA-cleaving DNA enzymes, dozens of different catalytic motifs were produced.[4] The best characterized of these and the one that comes closest to meeting the criteria listed above is the "10-23" DNA enzyme, derived from the 23rd clone obtained after the 10th round of *in vitro* selection. This enzyme has been utilized by many investigators to cleave a broad

[1] J. Haseloff and W. L. Gerlach, *Nature* **334,** 585 (1988).

[2] T. R. Cech, *JAMA* **260,** 3030 (1988).

[3] L. Beigelman, J. A. McSwiggen, K. G. Draper, C. Gonzalez, K. Jensen, A. M. Karpeisky, A. S. Modak, J. Matulic-Adams, A. B. DiRenzo, P. Haeberli, D. Sweedler, D. Tracz, S. Grimm, F. E. Wincott, V. G. Thackray, and N. Usman, *J. Biol. Chem.* **270,** 25702 (1995).

[4] S. W. Santoro and G. F. Joyce, *Proc. Natl. Acad. Sci. U.S.A.* **94,** 4262 (1997).

0076-6879/01 $35.00

range of RNA targets, both *in vitro* and *in vivo*. Applications of the 10-23 enzyme include processing of *in vitro* transcribed RNAs for biophysical studies,[5] mapping of accessible sites within structured RNAs,[6] cleavage of reporter molecules in quantitative PCR-based diagnostic assays,[7] analysis of single-nucleotide polymorphisms in RNA,[8] and inactivation of target RNAs both in cultured cells and in whole animals.[9–11]

This chapter focuses on the biochemical properties of the 10-23 DNA enzyme and provides guidelines for its "best use" as a reagent for the cleavage of target RNAs in the test tube and in cells. The enzyme contains only ~30 deoxynucleotides and thus can be synthesized easily and inexpensively. It is generalizeable to almost any target sequence. When optimized for the cleavage of a particular target RNA, the enzyme exhibits a potent combination of high sequence specificity and high catalytic efficiency. Issues of cellular uptake, subcellular localization, and stability *in vivo* are similar to those encountered with traditional antisense oligodeoxynucleotides. The possibility of cellular toxicity, although not yet fully explored, is likely to be less of a problem for DNA enzymes compared to RNase-H-dependent antisense agents because the former are considerably more sequence specific and do not contain nucleoside phosphorothioates, which are prone to nonspecific interactions with cationic proteins.

10-23 Motif

The 10-23 DNA enzyme consists of a catalytic core of 15 nucleotides, surrounded by substrate-binding domains of 6–12 nucleotides each (Fig. 1). The sequence of the catalytic core was discovered by *in vitro* selection and is very narrowly specified.[4] Almost any single-nucleotide change within the catalytic core results in the complete loss of catalytic activity. The only known exceptions are a T → A or T → C change at position 8, a G → A change at position 14, and an A → G change at position 15, each of which results in about a 10-fold decrease in catalytic rate. Certain point mutations within the catalytic core have been used as standard controls to provide a molecule that lacks catalytic activity but is otherwise

[5] A. M. Pyle, V. T. Chu, E. Jankowsky, and M. Boudvillain, *Methods Enzymol.* **317,** 140 (2000).

[6] M. J. Cairns, T. M. Hopkins, C. Witherington, L. Wang, and L.-Q. Sun, *Nature Biotechnol.* **17,** 480 (1999).

[7] A. V. Todd, C. J. Fuery, H. L. Impey, T. L. Applegate, and M. A. Haughton, *Clin. Chem.* **46,** 625 (2000).

[8] M. J. Cairns, A. King, and L.-Q. Sun, *Nucleic Acids Res.* **28,** e9 (2000).

[9] M. Warashina, T. Kuwabara, Y. Nakamatsu, and K. Taira, *Chem. Biol.* **6,** 237 (1999).

[10] Y. Wu, L. Yu, R. McMahon, J. J. Rossi, S. J. Forman, and D. S. Snyder, *Human Gene Therapy* **10,** 2847 (1999).

[11] F. S. Santiago, H. C. Lowe, M. M. Kavurma, C. N. Chesterman, A. Baker, D. G. Atkins, and L. M. Khachigian, *Nature Med.* **5,** 1264 (1999).

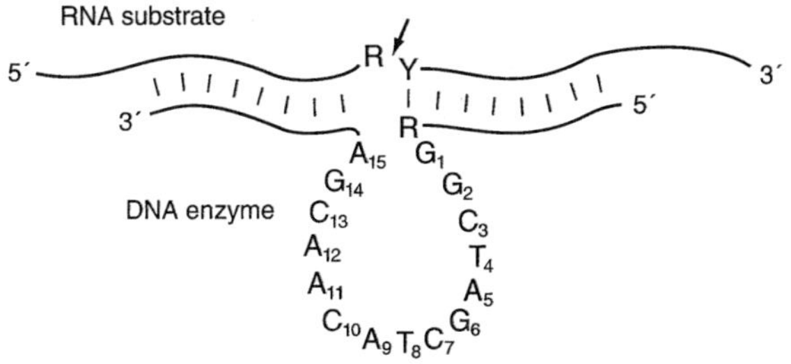

FIG. 1. Nucleotide composition of the 10-23 DNA enzyme, shown bound to a complementary RNA substrate. Arrow indicates the cleavage site, located between an unpaired purine (R) and paired pyrimidine (Y) residue.

comparable to the active compound. These "knockout" mutations are either a T → G change at position 4 or a G → C change at position 6. The inactivated molecule serves as an important control in determining whether the biological response to a DNA enzyme is due to its catalytic activity.

Little is known about the structural properties of the catalytic core of the 10-23 DNA enzyme. NMR studies of the enzyme–substrate complex revealed base pairing of the two substrate-binding domains, but could not resolve features of the catalytic core on the NMR time scale. Two different high-resolution crystal structures of the enzyme–substrate complex have been reported,[12,13] but in both cases the complex rearranged in the crystal to form a structure that is not relevant to catalytic function. The lack of structural information regarding the 10-23 DNA enzyme has not prevented its application as an RNA cleavage agent, but structural information would be useful in interpreting the biochemical properties of the enzyme and modifying its chemical composition.

The substrate requirements of the DNA enzyme are very simple: any target RNA sequence that is accessible to Watson–Crick pairing and contains a purine–pyrimidine junction can be cleaved at the phosphodiester located between the purine and pyrimidine residues.[4] The substrate-binding domains of the enzyme must be complementary to nucleotides that lie both upstream and downstream from the cleavage site. The target purine must be unpaired and all of the flanking nucleotides must be paired in a Watson–Crick fashion (see section on Substrate Specificity). Summarizing over the large number of substrate sequences that have been examined to date, there appears to be a preference for R · U compared to R · C sequences at the cleavage site (R = A or G). Substitution of the unpaired

[12] J. Nowakowski, P. J. Shim, G. S. Prasad, C. D. Stout, and G. F. Joyce, *Nature Struct. Biol.* **6,** 151 (1999).

[13] J. Nowakowski, P. J. Shim, C. D. Stout, and G. F. Joyce, *J. Mol. Biol.* **300,** 93 (2000).

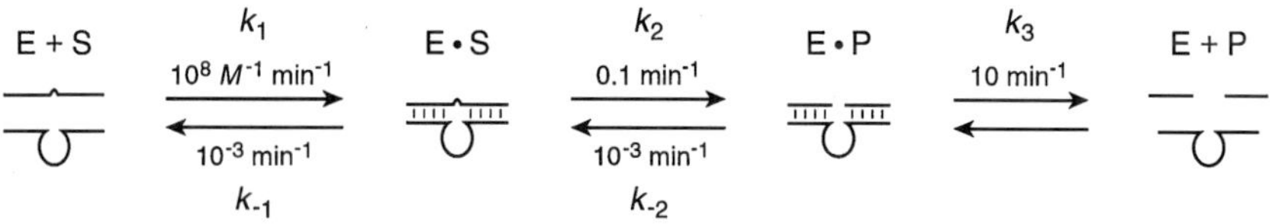

FIG. 2. Minimal kinetic scheme for RNA cleavage catalyzed by the 10-23 DNA enzyme. The enzyme (E) binds the substrate (S) to form an enzyme–substrate complex (E · S). Following cleavage, the two products (P) are released from the enzyme. The rate constant k_3 refers to release of the product that dissociates most slowly from the enzyme. Values for individual rate constants are typical for a well-behaved substrate and an enzyme with substrate-binding domains of optimal length, measured in the presence of 1 mM $MgCl_2$ and 150 mM KCl at pH 7.5 and 37°.[14]

purine residue by uridine reduces activity by about 10-fold, whereas substitution by cytidine eliminates activity altogether.[14]

The catalytic mechanism of the 10-23 DNA enzyme is thought to involve positioning of a divalent metal cation to assist in deprotonation of the 2-hydroxyl that lies adjacent to the cleavage site. The resulting 2′-oxyanion attacks the neighboring phosphate to form a 2′,3′-cyclic phosphate at the 3′ end of the upstream product and a 5′-hydroxyl at the 5′ end of the downstream product. This mechanism is supported by the log-linear dependence of catalytic rate on pH, the saturable dependence of catalytic rate on divalent metal cation concentration, and the rough correspondence between catalytic rate and the pK_a of the corresponding metal hydroxide for a variety of divalent metal cofactors.[14] The DNA enzyme is active in the presence of Mn^{2+}, Pb^{2+}, Mg^{2+}, Ca^{2+}, Cd^{2+}, Sr^{2+}, Ba^{2+}, Zn^{2+}, and Co^{2+}, in order of decreasing reactivity. The enzyme is not affected by the presence of monovalent cations.

Kinetic Properties

Optimal use of the 10-23 DNA enzyme requires an understanding of its kinetic properties and how these properties are influenced by the reaction conditions, target sequence, and length of the two substrate-binding domains. A minimal kinetic scheme for DNA-catalyzed RNA cleavage is shown in Fig. 2. The enzyme binds the substrate with second-order rate constant k_1 to form the enzyme–substrate complex. This complex can either dissociate, with rate k_{-1}, or proceed through the chemical step of the reaction, with rate k_2, to yield the enzyme–product complex. The products can undergo the reverse reaction, with rate k_{-2}, to regenerate the substrate, or dissociate from the enzyme, with rate k_3, to give the free products and allow the enzyme to turn over. For simplicity, k_3 refers to the rate of dissociation of the product that is the slowest to release. In order to progress optimally through the

[14] S. W. Santoro and G. F. Joyce, *Biochemistry* **37,** 13330 (1998).

reaction, it is important that k_1 be as close as possible to the rate of duplex formation, that k_2 substantially exceed k_{-1} for matched but not mismatched substrates, that k_2 substantially exceed k_{-2}, and that k_3 not be rate limiting for catalytic turnover. These are the guiding principles in tailoring the DNA enzyme to cleave a particular target RNA.

Enzyme–Substrate Complex

For unstructured RNA substrates, under a broad range of reaction conditions, the rate of formation of the enzyme–substrate complex is roughly equal to the rate of duplex formation: $4.7 \times 10^8 M^{-1}\text{min}^{-1}$ in the presence of 2 mM Mg^{2+} and $6.0 \times 10^9 M^{-1}\text{min}^{-1}$ in the presence of 100 mM Mg^{2+}.[14] This is in contrast to the hammerhead, hairpin, and group I ribozymes, which have roughly a 10-fold lower value for k_1 under the same reaction conditions. The 10-23 DNA enzyme takes full advantage of the inherent rate of duplex formation, without imposing additional barriers to formation of a productive enzyme–substrate complex. Often, however, secondary and tertiary structure within the RNA substrate or, less likely, involving the substrate-binding domains of the enzyme, reduces k_1 by impeding access to Watson–Crick pairing.

Once bound, the RNA substrate is held by extensive base pairing and dissociates slowly from the DNA enzyme. The rate of dissociation is determined by the number of base pairs, their GC content, and, to a lesser extent, their detailed sequence. If there is insufficient binding between the enzyme and substrate, then k_{-1} will be as fast or faster than k_2 and catalytic efficiency will be compromised. On the other hand, if enzyme–substrate binding is too strong, then there will be two undesirable consequences. First, for applications where substrate specificity is important, there will be less discrimination between matched and mismatched substrates, especially when the mismatch occurs far away from the cleavage site. Second, if the substrate binds very tightly to the enzyme, then this will tend to be true for the products as well, which could reduce the rate of catalytic turnover by decreasing k_3 and causing product release to become rate limiting.

Based on past experience, a general rule has been formulated to guide the design of substrate-binding domains with optimal stability. Referring to the data of Sugimoto and colleagues for the thermodynamic stability of DNA–RNA heteroduplexes,[15] one should aim for a predicted ΔG°_{37} of -8 to -10 kcal/mole for *each* substrate-binding domain (Table I). This value includes the helical nucleation penalty of 3.1 kcal/mol for each domain, but does not consider possible cooperativity between the two substrate-binding domains or contributions from the unpaired purine of the substrate and catalytic core of the enzyme.

[15] N. Sugimoto, S.-I. Nakano, M. Katoh, A. Matsumura, H. Nakamuta, T. Ohmichi, M. Yoneyama, and M. Sasaki, *Biochemistry* **34,** 11211 (1995).

TABLE I
THERMODYNAMIC PARAMETERS FOR RNA–DNA HETERODUPLEXES[a]

RNA sequence[b]	$\Delta G°_{37}$ (kcal/mol)[c]
AA	−1.0
AC	−2.1
AG	−1.8
AU	−0.9
CA	−0.9
CC	−2.1
CG	−1.7
CU	−0.9
GA	−1.3
GC	−2.7
GG	−2.9
GU	−1.1
UA	−0.6
UC	−1.5
UG	−1.6
UU	−0.2

[a] Data from Sugimoto *et al.*,[15] obtained in 1 *M* NaCl at pH 7.0.
[b] Nearest-neighbor pairs reading 5′ to 3′.
[c] Overall duplex stability is calculated by summing the $\Delta G°_{37}$ values for each nearest-neighbor pair and adding a helix initiation penalty of 3.1 kcal/mol. There are $n - 1$ pairs for a sequence of length n.

Chemical Step of Reaction

Under simulated physiological conditions (1 m*M* $MgCl_2$, 150 m*M* KCl, pH 7.5, 37°), the rate of the chemical step of the reaction is ~0.1 min^{-1}.[14] If the DNA enzyme is to be applied to cells, then it should be evaluated in the laboratory under conditions such as these. If it is to be applied in the test tube, then the reaction conditions can be chosen to maximize both the catalytic rate and catalytic rate enhancement of the enzyme. In this case a good choice of reaction conditions is 25 m*M* $CaCl_2$ at pH 7.5 and 37°, which gives a catalytic rate of ~2 min^{-1} and an uncatalyzed rate of ~10^{-6} min^{-1}. Compared to physiological conditions, the increased concentration of divalent metal cation results in a 20-fold increase in catalytic rate and the replacement of Mg^{2+} by Ca^{2+} results in both a 2-fold increase in catalytic rate and a 30-fold decrease in the uncatalyzed rate of reaction.[14] For applications where catalytic rate but not catalytic rate enhancement is important, one could employ either 25 mM $CaCl_2$ at pH 8.5 or 10 mM $MnCl_2$ at pH 7.5 to obtain a rate of ~10 min^{-1}.

The catalytic rate of the 10-23 DNA enzyme is 0.1–10 min^{-1}, depending on reaction conditions, which is considerably slower than that of most protein ribonucleases. On the other hand, the K_m of the DNA enzyme for an ideal substrate is <1 nM, providing a catalytic efficiency, k_{cat}/K_m, of 10^8–10^{10} M^{-1} min^{-1}. The very low K_m is important for cellular applications because the concentration of target RNA may be low and it may be difficult to deliver high concentrations of the DNA enzyme. In the test tube it usually is more convenient to work at higher concentrations. Different situations call for operating under either single-turnover, enzyme-excess or multiple-turnover, substrate-excess conditions. The former is preferred when it is important to drive the reaction quickly to completion. The latter may be preferred for preparative work in which large amounts of RNA are to be cleaved.

Once RNA cleavage has occurred, the products dissociate rapidly from the DNA enzyme. There does not appear to be significant interaction between the two bound products, each dissociating at a rate similar to that predicted for the corresponding RNA-DNA heteroduplex.[14] If the substrate-binding domains are chosen so that each has a predicted ΔG°_{37} of −8 to −10 kcal/mole (see above), then the rate of product release will be at least 100 times faster than the rate of cleavage. Under subsaturating conditions catalytic turnover is limited by the rate of formation of the enzyme–substrate complex, while under saturating conditions it is limited by the rate of the chemical step. The rate of the reverse reaction, k_{-2}, is 450-fold slower than the rate of the forward reaction.[14] Thus, the enzyme continues to turn over until it reaches completion with essentially all of the substrate being cleaved.

Measurement of Kinetic Parameters

DNA-catalyzed RNA cleavage can be measured under either multiple-turnover (excess substrate) or single-turnover (excess enzyme) conditions. Multiple-turnover kinetics provide values for k_{cat} and K_m that are relevant to most applications. Single-turnover kinetics can be used to determine whether product release is rate-limiting, in which case the observed rate will be faster under single-turnover compared to multiple-turnover conditions. The protocol for multiple-turnover kinetic analysis is described here, whereas that for single-turnover kinetic analysis is described elsewhere.[14]

1. Prepare the DNA enzyme using an automated DNA synthesizer (or purchase from a commercial supplier). The DNA must be deprotected, then purified by either HPLC or denaturing polyacrylamide gel electrophoresis, desalted, and dissolved in

a solution containing 10 m*M* *N*-(2-hydroxyethyl)piperazine-*N*′-3-propanesulfonic acid (EPPS, pH 7.5) and 0.1 m*M* Na_2EDTA.

2. Prepare the RNA substrate either synthetically or by *in vitro* transcription. Long substrates can be body labeled during transcription by including [α-^{32}P]ATP in the reaction mixture (final specific activity ~0.1 μCi/nmol). Purify as above and dissolve in a solution containing 10 m*M* EPPS (pH 7.5) and 0.1 m*M* Na_2EDTA.

3. Determine the concentration of DNA enzyme and RNA substrate spectrophotometrically. For the enzyme and short substrates calculate the molar extinction coefficient based on published values for nearest-neighbor pairs.[16,17] For long substrates assume a molar extinction coefficient of 8.5 OD/μmol/nucleotide at 260 nm and 25°.

4. Substrates that were not labeled during transcription must be kinased, employing 20–100 pmoles of the purified RNA, T4 polynucleotide kinase, and [γ-^{32}P]ATP (specific activity ~5 μCi/pmol). Purify the kinased material by denaturing polyacrylamide gel electrophoresis. The labeled and unlabeled material are then mixed to obtain the desired concentration and specific activity of substrate in the final reaction mixture.

5. Choose a range of substrate concentrations that span the expected K_m, typically 0.1–100 n*M*. Choose an enzyme concentration that is at least 10-fold lower than the concentration of substrate and at least 10-fold lower than the expected K_m.

6. Premix the enzyme and substrate in separate solutions, each containing 2 m*M* $MgCl_2$, 150 m*M* NaCl, 30 m*M* EPPS (pH 7.5), and 0.01% (w/v) sodium dodecyl sulfate (SDS) at 37°. The SDS helps to prevent material from sticking to the inside of the reaction vessel.

7. Start the reaction by mixing appropriate amounts of the enzyme and substrate solutions, then maintain the combined mixture at 37°.

8. Withdraw aliquots at various times (typically 1–30 min) and quench by adding an ice-cold mixture containing 8 *M* urea, 20% sucrose, 90 m*M* Tris–borate (pH 8.3), Na_2EDTA in excess over the Mg^{2+}, 0.05% xylene cyanol, 0.05% bromphenol blue, and 0.1% (w/v) SDS. At least 5 aliquots should be taken over the first 10–15% of the reaction.

9. Separate the reaction products by denaturing polyacrylamide gel electrophoresis and determine the fraction cleaved.

10. For each substrate concentration obtain a value for k_{obs} based on a best-fit line to a plot of fraction cleaved versus time. Generate a modified Eadie–Hofstee plot of k_{obs} versus k_{obs}/[S] and determine k_{cat} and k_m from the *y* intercept and negative slope, respectively, of a best-fit line through the data.

[16] M. M. Warshaw and I. Tinoco, Jr., *J. Mol. Biol.* **20,** 29 (1966).
[17] C. R. Cantor, M. M. Warshaw, and H. Shapiro, *Biopolymers* **9,** 1059 (1970).

Substrate Specificity

Substrate *selectivity* refers to the ability of the DNA enzyme to recognize a particular target RNA from among many possible targets. It is determined by the summed length of the two substrate-binding domains, typically 16–20 base pairs, which allow the enzyme to recognize a particular target RNA from among 4^{16}–4^{20} (10^9–10^{12}) possibilities. Substrate *specificity* refers to the ability of the enzyme to discriminate between different RNAs. It is defined by the relative efficiency of the enzyme in cleaving a fully matched substrate compared to those that contain one or more base mismatches relative to the enzyme.

The 10-23 DNA enzyme is highly sensitive to base mismatches, more so than antisense oligodeoxynucleotides that recognize RNAs of similar length. The two substrate-binding domains of the DNA enzyme appear to operate independently in reading the target sequence. A single-base mismatch within either domain results in a significant decrease in catalytic efficiency.[14] Not surprisingly, discrimination is greatest for mismatches that significantly disrupt Watson–Crick pairing, such as purine–purine pairs, and least for dG · rU or dT · rG wobble pairs. Mismatches that occur close to the cleavage site are more disruptive than those that lie far away.

In choosing an optimal length for the two substrate-binding domains, one should aim for the minimum length that provides full catalytic activity. A further increase in length may reduce catalytic turnover by slowing the rate of product release (see section on Kinetic Properties). It may also reduce substrate specificity by slowing the rate of release of a mismatched substrate. For applications in which substrate specificity is not critical, such as the *in vitro* cleavage of RNA for preparative work, it is preferable to use substrate-binding domains that are slightly too long for optimal specificity but provide maximum catalytic efficiency. For other applications, such as the discrimination of single-nucleotide polymorphisms in RNA, it is preferable to accept slightly reduced catalytic activity in order to maximize substrate specificity.

Target Site Selection

If the target RNA is devoid of secondary and tertiary structure, then a DNA enzyme designed based on the principles discussed above will cleave the substrate in a highly efficient and sequence-specific manner. In the real world, however, RNA seldom exists as a random coil. Most RNAs of biological interest tend to adopt a complex structure that largely precludes access to hybridization by a complementary oligonucleotide. This problem is well known in the antisense field and typically is addressed, especially in the commercial arena, by testing many different antisense oligodeoxynucleotides in order to find the small subset that hybridize readily to the target.[18,19] A similar approach has been taken with the 10-23 DNA enzyme in targeting mRNAs.[6] A multiplexing technique has been developed that allows dozens of DNA enzymes to be tested simultaneously, although each must still be synthesized individually.

Many investigators have chosen simply to target the A · U site within the AUG start codon of an mRNA. They reason that because this site is accessible to the translation machinery it must be accessible to the DNA enzyme. However, the start codon region of some mRNAs is not accessible, and even when accessible may not allow hybridization of a DNA enzyme at surrounding nucleotide positions. The main consequence of poor target site accessibility is a significantly elevated K_m for DNA-catalyzed RNA cleavage, negating an important feature of the DNA enzyme. Catalytic efficiency can sometimes be regained by slightly altering the length of one or both substrate-binding domains. Shortening a domain by 1–2 nucleotides or lengthening it by 1–4 nucleotides may have a significant benefit, depending on the detailed local structure of the target RNA.[20]

If one chooses to employ a screening approach to finding accessible sites within a target RNA, it is important that the RNA be folded into its native structure. RNAs that are prepared by *in vitro* transcription should have an intact 5′ end and contain at least 100 nucleotides downstream from the last potential target site, preferably all the way to their natural 3′ end. Assuming a "first-come, first-folded" model for the adoption of RNA secondary structure, a construct of this length will place the target site within a reasonable approximation of its native context. There are two important caveats to this model: first, downstream elements that are omitted from the transcript might otherwise produce structural alterations far upstream; second, the environment of an *in vitro* transcription mixture can hardly be called "native." In the cell, RNA folds in the presence of various cellular proteins, organic and inorganic salts, polyamines, and other small molecules, conditions that cannot be replicated in the test tube. A method has been devised that allows the target RNA to be expressed *in vivo,* then harvested as part of a cell extract and analyzed *ex vivo.*[21] Another approach, described below, involves *in vitro* transcription under simulated physiological conditions and subsequent analysis of the target RNA in its folded state without ever having exposed it to denaturing conditions.

Determining Target Site Accessibility within in Vitro Transcribed RNA

1. Prepare a template DNA, either as a PCR product or linearized plasmid, that encodes the RNA of interest and extends >100 nucleotides downstream from the last target site.

[18] W. F. Lima, V. Brown-Driver, M. Fox, R. Hanecak, and T. W. Bruice, *J. Biol. Chem.* **272,** 626 (1997).
[19] N. Milner, K. U. Mir, and E. M. Southern, *Nature Biotechnol.* **15,** 537 (1997).
[20] K. U. Mir and E. M. Southern, *Nature Biotechnol.* **17,** 788 (1999).
[21] M. Scherr and J. J. Rossi, *Nucleic Acids Res.* **26,** 5079 (1998).

2. Perform *in vitro* transcription in a 100-μl reaction mixture containing 1 μM (100 pmol) DNA, 10 mM $MgCl_2$, 150 mM NaCl, 1 mM spermidine, 5 mM dithiothreitol, 30 mM EPPS (pH 7.5), 2 mM of each NTP, 20 μCi [α-^{32}P]ATP (final specific activity ~0.2 μCi/nmol), and 2000 U T7 RNA polymerase, which is incubated at 37° for 2 h. Add all reagents at room temperature to avoid precipitation of the DNA.

3. Quench the reaction by adding 10 μl 100 mM Na_2EDTA and chilling on ice.

4. Extract twice with an equal volume of phenol, then once with an equal volume of chloroform/isoamyl alcohol (24 : 1, v/v). Do not ethanol precipitate.

5. Dilute with H_2O to 500 μl volume. If the RNA contains >1000 nucleotides, the amount of nonspecific cleavage that occurs during *in vitro* transcription may obscure the results of DNA-catalyzed RNA cleavage. In this case it may be beneficial to purify the RNA in a nondenaturing 4% polyacrylamide gel, with 2 mM $MgCl_2$ included in both the gel and running buffer. The isolated RNA should be eluted into a 500-μl volume containing 10 mM EPPS (pH 7.5) and 0.1 mM Na_2EDTA.

6. Carry out each RNA-cleavage reaction in a 10-μl volume containing 1 μl of the diluted *in vitro* transcription mixture (~5 pmol RNA, ~750 kcpm for a 1 kb RNA), 0.1 μM (1 pmol) DNA enzyme, 2 mM $MgCl_2$, 150 mM NaCl, and 30 mM EPPS (pH 7.5), which is incubated at 37° for 10–60 min. It may be useful to test three different concentrations of DNA enzyme (e.g., 0.01, 0.1, and 1 μM) in order to stratify the performance of the various constructs.

7. Quench the reaction by adding 2 μl 100 mM Na_2EDTA and chilling on ice.

8. Separate the products by electrophoresis in a denaturing 5% polyacrylamide gel and determine the fraction cleaved. Under these conditions >50% of the RNA will be cleaved after 60 min, provided that the target site is readily accessible.

Another consideration in target site selection is the avoidance of secondary structure within the corresponding DNA enzyme. It is important that the two substrate binding arms of the enzyme not be complementary to each other or to the catalytic core. Several nucleic acid folding programs are available that can be used to predict whether inter- or intramolecular secondary structure will be a problem (see, for example, http://mfold.wustl.edu/~folder/dna/form1.cgi). Note that the 10-23 DNA enzyme contains two short palindromic sequences within its catalytic core: 5′-GCTAGC-3′ at positions 2–7 and 5′-TAGCTA-3′ at positions 4–9. Neither is stable enough to perturb DNA-catalyzed RNA cleavage.

For some applications one is free to choose the sequence of the target RNA, for example, in generating a precise terminus at the 3′ end of an *in vitro* transcribed RNA. In such cases one can either design a well-behaved sequence or choose one from the literature. An especially well-behaved RNA sequence that has been studied extensively is 5′-GGAGAGAGA·UGGGUGCG-3′,[14] which corresponds to the start codon region of human immunodeficiency virus type 1 (HIV-1) *gag-pol* mRNA.

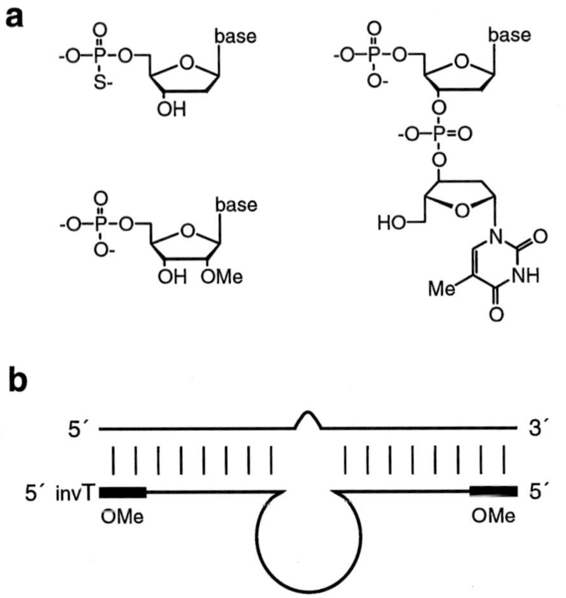

FIG. 3. Nucleotide analogs used to protect the DNA enzyme against degradation by nucleases. (a) *Top left,* deoxynucleoside phosphorothioate; *bottom left,* 2′-*O*-methyl nucleotide; *right,* inverted thymidylate. (b) Stabilization of the DNA enzyme using two 2′-*O*-methyl nucleotides at the end of each substrate-binding domain and an inverted thymidylate at the 3′ end.

Delivery to Cells

Protection against Nucleases

DNA is stable in the test tube under a broad range of reaction conditions. When applied in a biological context, however, it must be protected against degradation by cellular nucleases.[22] The greatest vulnerability of linear, single-stranded DNA is to 5′- and 3′-exonucleases.[23,24] Thus the DNA enzyme must be protected at both ends with nuclease-resistant nucleotide analogs. Although it is possible to replace all but the most proximal residue of each substrate-binding domain by a variety of DNA analogs, it usually is sufficient to replace just 2–3 residues at each end of the enzyme (Fig. 3).

The DNA analog employed most commonly in antisense studies is a deoxynucleoside phosphorothioate, chosen because it confers protection against nucleases

[22] T. M. Woolf, C. G. B. Jennings, M. Rebagliati, and D. A. Melton, *Nucleic Acids Res.* **18,** 1763 (1990).

[23] D. M. Tidd and H. M. Warenius, *Br. J. Cancer* **60,** 343 (1989).

[24] J.-P. Shaw, K. Kent, J. Bird, J. Fishback, and B. Froehler, *Nucleic Acids Res.* **19,** 747 (1991).

yet retains the ability to induce cleavage of the target RNA by RNase H. However, as mentioned previously, oligophosphorothioates are prone to nonspecific interactions with cationic proteins, complicating their use in a biological setting.[25] Most of the "second-generation" antisense agents contain none or only a few phosphorothioate residues and rely on 2′-*O*-methyl nucleotides or other nucleotide analogs to protect the 5′ and 3′ termini.[26,27] 2′-*O*-Methyl nucleotides do not induce cleavage by RNase H and thus must be used sparingly in antisense oligodeoxynucleotides.

For the DNA enzyme there is no need to maintain RNase H activity, allowing either deoxynucleoside phosphorothioates or 2′-*O*-methyl nucleotides to be used to protect against nuclease degradation. The latter analog is preferred because it avoids the nonspecific effects of phosphorothioates, although it may require slight adjustment of the length of the substrate-binding domains because base pairs involving 2′-*O*-methyl nucleotides are about 0.2 kcal/mol more stable than those involving standard deoxynucleotides.[28] Surprisingly, a DNA enzyme that contains substrate-binding domains of 8 nucleotides each and has just two 2′-*O*-methyl nucleotides at each end does not induce appreciable cleavage of the target RNA by RNase H.[9] Apparently the catalytic core of the enzyme interferes with RNase H activity at proximal positions while the 2′-*O*-methyl nucleotides prevent cleavage at the distal positions.

Another analog that has been employed in protecting the DNA enzyme against nuclease digestion is an inverted (3′,3′-linked) thymidylate, located at the 3′ end of the molecule.[11] The 3′ terminus is the most vulnerable position and modifying it in this way avoids a free 3′ end altogether (Fig. 3). A generally useful approach for stabilizing the DNA enzyme is to include two 2′-*O*-methyl nucleotides at the distal end of each substrate-binding domain and an inverted thymidylate at the 3′ end. The four 2′-*O*-methyl nucleoside phosphoramidites and the thymidine 5′-phosphoramidite CPG support all are commercially available (e.g., Glen Research, Sterling, VA) and can be included in custom syntheses performed by most commercial suppliers of synthetic DNA (e.g., Operon Technologies, Alameda, CA).

Cellular Uptake

A critical issue when applying the DNA enzyme *in vivo* is the need to achieve good cellular uptake without incurring cellular toxicity. It is difficult to deliver

[25] C. A. Stein, *Trends Biotechnol.* **14,** 147 (1996).

[26] B. P. Monia, E. A. Lesnik, C. Gonzalez, W. F. Lima, D. McGee, C. J. Guinosso, A. M. Kawasaki, P. D. Cook, and S. M. Freier, *J. Biol. Chem.* **268,** 14514 (1993).

[27] R. Zhang, Z. Lu, H. Zhao, X. Zhang, R. B. Diasio, I. Habus, Z. Jiang, R. P. Iyer, D. Yu, and S. Agrawal, *Biochem. Pharmacol.* **50,** 545 (1995).

[28] E. A. Lesnik, C. J. Guinosso, A. M. Kawasaki, H. Sasmor, M. Zounes, L. L. Cummins, D. J. Ecker, P. D. Cook, and S. M. Freier, *Biochemistry* **32,** 7832 (1993).

significant concentrations of an oligodeoxynucleotide to cultured cells without the assistance of a delivery vehicle, such as cationic lipids, a membrane-penetrating peptide, or an attached hydrophobic group. The situation is very different in the whole animal, where such assistance does not appear to be necessary and may even be deleterious.

The most common approach for the delivery of oligodeoxynucleotides to cultured cells is to employ a cationic lipid. Most formulations consist of a mixture of a cationic lipid and a neutral lipid, such as Lipofectin[29] (a 1 : 1 mixture of *N*-[1-(2,3-dioleyloxy)propyl]-*N,N,N*-trimethylammonium chloride and dioleoylphosphatidylethanolamine) or LipofectAMINE (a 3 : 1 mixture of 2,3-dioleyloxy-*N*-[2-sperminecarboxamido)ethyl]-*N,N*-dimethyl-1-propanaminium trifluoroacetate and dioleoylphosphatidylethanolamine; Life Technologies, Rockville, MD). Other formulations include an additive, such as cholesterol, polylysine, a fusogenic peptide, or a covalently attached spermine group. The cationic lipid interacts with the DNA to form a compacted complex that enters the cell through endocytosis.[30–32] The accompanying neutral lipid is thought to promote release of the DNA from the endosome.[33] The various additives are used to stabilize the lipid–DNA complex, assist in cellular uptake, or facilitate endosomal release.

It is difficult to choose a cationic lipid formulation because its performance is strongly influenced by cell type, degree of cell confluency, absolute concentration of both lipid and DNA, and the charge ratio of lipid relative to DNA. A reasonable strategy is to screen several different lipid formulations against the target cells, relying on the manufacturer's recommendations in choosing a concentration range for the lipid and DNA and testing several different charge ratios.[34,35] The number of passages and the degree of confluency of the cells should be kept constant. Delivery of the DNA enzyme can be assessed by attaching a fluorescein label to its 5′ end and monitoring cellular uptake using either fluorescence microscopy or fluorescence-activated cell sorting (FACS).[36] Propidium iodide should be added to the growth medium (1 μM final concentration) to distinguish nonviable cells that are unable to exclude the dye.

Once the DNA enzyme has been delivered to cells, it must escape from the endosomes and become localized to the nucleus. Fluorescently labeled DNA that

[29] P. L. Felgner, T. R. Gadek, M. Holm, R. Roman, H. W. Chan, M. Wenz, J. P. Northrop, G. M. Ringold, and M. Danielsen, *Proc. Natl. Acad. Sci. U.S.A.* **84,** 7413 (1987).

[30] X. Zhou and L. Huang, *Biochim. Biophys. Acta* **1189,** 195 (1994).

[31] J. Zabner, A. J. Fasbender, T. Moninger, K. A. Poellinger, and M. J. Welsh, *J. Biol. Chem.* **270,** 18997 (1995).

[32] O. Zelphati and F. C. Szoka, Jr., *Pharm. Res.* **13,** 1366 (1996).

[33] O. Zelphati and F. C. Szoka, Jr., *Proc. Natl. Acad. Sci. U.S.A.* **93,** 11493 (1996).

[34] M. A. Reynolds, *in* "Intracellular Ribozyme Applications: Principles and Protocols" (J. J. Rossi and L. A. Couture, eds.) p. 125. Horizon Scientific Press, Wymondham, U.K., 1999.

[35] S. A. Williams and J. S. Buzby, *Methods Enzymol.* **313,** 388 (1999).

[36] L. Benimetskaya, J. Tonkinson, and C. A. Stein, *Methods Enzymol.* **313,** 287 (1999).

remains trapped in endosomes gives the cells a punctate cytoplasmic fluorescence, whereas that which escapes produces diffuse or punctate nuclear fluorescence.[33] Because the DNA enzyme operates with a very low K_m, the observation of even modest nuclear fluorescence may be sufficient to allow one to conclude that adequate amounts have been delivered. Further enhancement of delivery may come at the cost of reduced cell viability, as evidenced by a larger proportion cells that become stained with propidium iodide. Cationic lipids themselves are toxic to cells, especially when used in high concentrations or without an accompanying noncationic lipid. A dose range of 10–500 n*M* DNA enzyme and a charge ratio of lipid : DNA of 1 : 1 to 4 : 1 is likely to give the best results.

The efficacy of the DNA enzyme can be assessed in a variety of ways, many of which are specific to the particular application. It usually is possible to measure the amount of the target RNA, for example, by RT-PCR (reverse transcriptase-polymerase chain reaction) or an RNase protection assay. In most cases it will not be possible to detect the cleavage products, which tend to degrade rapidly once they have been cleaved by the DNA enzyme. Another quantitation strategy relies on reporter assays, such as luciferase expression, which is coupled to expression of the target RNA.[9] In the best of circumstances, Western blot analysis can be used to measure the corresponding protein product directly.[10,11]

Conclusions

The 10-23 DNA enzyme is a simple and powerful tool for the site-specific cleavage of RNA. When directed to cleave an unstructured RNA it exhibits high catalytic efficiency and high sequence specificity. The difficulty comes in trying to apply the DNA enzyme to RNAs of biological interest, which adopt secondary and tertiary structures that tend to exclude hybridization by a complementary oligonucleotide. By screening several different DNA enzymes, each designed according to the principles described in this chapter, it usually is possible to find one that efficiently cleaves the target RNA. Another screen must be carried out to define a suitable formulation for delivering the DNA enzyme to the cells of interest. Based on prior experience with antisense oligonucleotides and RNA-cleaving ribozymes, and the rapidly growing literature concerning cellular applications of the 10-23 DNA enzyme, there is reason to be optimistic concerning the broad utility of RNA-cleaving DNA enzymes.

Acknowledgments

I am grateful to Stephen Santoro for helpful comments on the manuscript. This work was supported by the National Institutes of Health Grant AI30882 and Johnson & Johnson Research.

[33] Leadzyme

By LAURENT DAVID, DOMINIC LAMBERT, PATRICK GENDRON, and FRANCOIS MAJOR

Introduction

Since the finding of catalytic RNAs in the 1980s,[1–3] other RNA enzymes have been discovered in a wide variety of organisms. These new classes of enzymes, which are also called *ribozymes,* considered distinct from protein metalloenzymes.[4] Earlier studied systems include the hammerhead[5] and the hairpin[6] ribozymes that work through specific self-cleavage of a phosphodiester bond. The hammerhead and hairpin ribozymes were found during RNA processing consisting of self-cleavage and splicing.

Currently, there are two classes of ribozymes. The first class comprises the hammerhead, hairpin, HDV (hepatitis delta virus), and tRNA ribozymes, in which transesterification is accomplished using an internal nucleophile O2′. The attack of the 2′-OH group proceeds by an in-line S_N2 nucleophilic displacement with a configuration inversion of the phosphorus atom of the next nucleotide. The second class contains group I and group II introns, and RNase P RNA ribozymes, where transesterification is performed through an external nucleophile,[7] such as a water molecule, which attacks the phosphorus center by means of S_N2 displacement.

It was long thought that all catalytic RNAs required divalent ions to be effective,[8,9] e.g., Mg^{2+} for the hammerhead, or both Mg^{2+} and Pb^{2+} for the yeast phenylalanine tRNA.[10] However, this has recently been challenged in the hairpin ribozyme by using $Co(III)NH_3^{3+}$ in place of $Mg(H_2O)_6^{2+}$.[11] Although different roles have been assigned to these ions, the most important one is structural. The hammerhead ribozyme is unfolded in the absence of divalent metal ions,[12] and the

[1] T. Cech, A. Zaug, and P. Grabowvski, *Cell* **27,** 487 (1981).
[2] K. Kruger, P. Grabowski, A. Zaug, J. Sands, D. Gottschling, and T. Cech, *Cell* **31,** 147 (1982).
[3] C. Guerrier-Takada, K. Gardiner, T. Marsh, N. Pace, and S. Altman, *Cell* **3,** 849 (1983).
[4] A. Pyle, *Science* **261,** 709 (1993).
[5] A. Forster and R. Symons, *Cell* **49,** 211 (1987).
[6] J. Buzayan and W. Gerlach, *Nature* **323,** 349 (1986).
[7] R. Kuimelis and L. McLaughlin, *Chem. Rev.* **98,** 1027 (1998).
[8] T. Pan, D. Long, and O. Uhlenbeck, *in* "RNA Folding and Catalysis." Cold Spring Harbor Laboratory Press, Cold Spring Harbor, NY, 1993.
[9] D. Lilley, *Curr. Opin. Struct. Biol.* **9,** 330 (1999).
[10] R. Brown, J. Dewan, and A. Klug, *Biochemistry* **24,** 4785 (1985).
[11] A. Hampel and J. Cowan, *Chem. Biol.* **4,** 513 (1997).
[12] G. Bassi, N. Mollegaard, A. Murchie, E. vonKitzing, and D. Lilley, *Nat. Struct. Biol.* **2,** 45 (1995).

METHODS IN ENZYMOLOGY, VOL. 341
0076-6879/01 $35.00

addition of Mg^{2+} induces its folding.[13] The hairpin ribozyme needs two divalent ions, such as Mg^{2+} or Ca^{2+}, to fold either in the hinged form[14] or in the junction form.[15] The yeast tRNAs contain at least four site-specific Mg^{2+} binding sites[16,17] and require Pb^{2+} ions to be activated.

A more controversial role of divalent metal ions is their action during catalytic reaction, where they could act as general base catalysts to deprotonate a nucleophile, for instance using a coordinated hydroxide, or as a Lewis acid to stabilize the transition state by neutralizing negative charges of the phosphate oxygens. In the hammerhead ribozyme, the cleavage rate is linearly pH-dependent and the metal ions are located close to the active site. These two observations suggest a direct involvement of the ions, suspected to act as a general base in catalysis.[18] Another role that could explain the linear pH dependency would be a direct interaction between the ion and the O2′ atom, which decreases the pK_a.[19,20]

The size of catalytic RNA systems varies considerably from hundreds of nucleotides, for group I and II introns, to fewer than a hundred, for the hammerhead and hairpin ribozymes. Ribozymes have been minimized in size to simplify their biochemical studies. Several active variants have been created. The smallest characterized ribozyme is the UUU trinucleotide, which catalyzes the cleavage of GAAA between G and A in the presence of Mn^{2+} ions.[21] This ribozyme is still active with Cd^{2+}, but to a lesser extent. The 2′-OH groups are not essential in this catalytic reaction, as it was demonstrated by producing an active dUdUdU trinucleotide.

The leadzyme is a small RNA molecule that was isolated by *in vitro* selection under autolytic cleavage with Pb^{2+} using a library of random yeast $tRNA^{Phe}$ fragments.[22,23] It does not fold in the same manner as the tRNA. It consists of an asymmetric internal loop of six nucleotides flanked by two stems as shown in Fig. 1a. A two-step reaction mechanism occurs. First a catalysis of transesterification gives 5′- and 2′,3′-cyclic phosphodiester products, and then a hydrolysis reaction of the 2′,3′-cyclic phosphodiester yields a 3′-phosphate. The reaction is highly specific to Pb^{2+}, and the enhancement of the cleavage rate at the specific site is estimated to be equal to 1100 for the variant LZ2 of Pan and

[13] G. Bassi, A. Murchie, F. Walter, R. Clegg, and D. Lilley, *EMBO J.* **16,** 7481 (1997).
[14] N. Walter, K. Hampel, K. Brown, and J. Burke, *EMBO J.* **17,** 2378 (1998).
[15] A. Murchie, J. Thompson, F. Walter, and D. Lilley, *Mol. Cell.* **1,** 873 (1998).
[16] S. Holbrook, J. Sussman, R. Warrant, G. Church, and S. Kim, *Nucl. Acids Res.* **4,** 2811 (1977).
[17] A. Jack, J. Ladner, D. Rhodes, R. Brown, and A. Klug, *J. Mol. Biol.* **111,** 315 (1977).
[18] W. Scott, J. Murray, J. Arnold, B. Stoddard, and A. Klug, *Science,* **274,** 2065 (1996).
[19] B. Pontius, W. Lott, and P. von Hippel, *Proc. Natl. Acad. Sci. U.S.A.* **94,** 2290 (1997).
[20] W. Scott, *Curr. Opin. Struct. Biol.* **8,** 720 (1998).
[21] S. Kazakov and S. Altman, *Proc. Natl. Acad. Sci. U.S.A.* **89,** 7939 (1992).
[22] T. Pan and O. Uhlenbeck, *Nature* **358,** 560 (1992).
[23] T. Pan and O. Uhlenbeck, *Biochemistry* **31,** 3887 (1992).

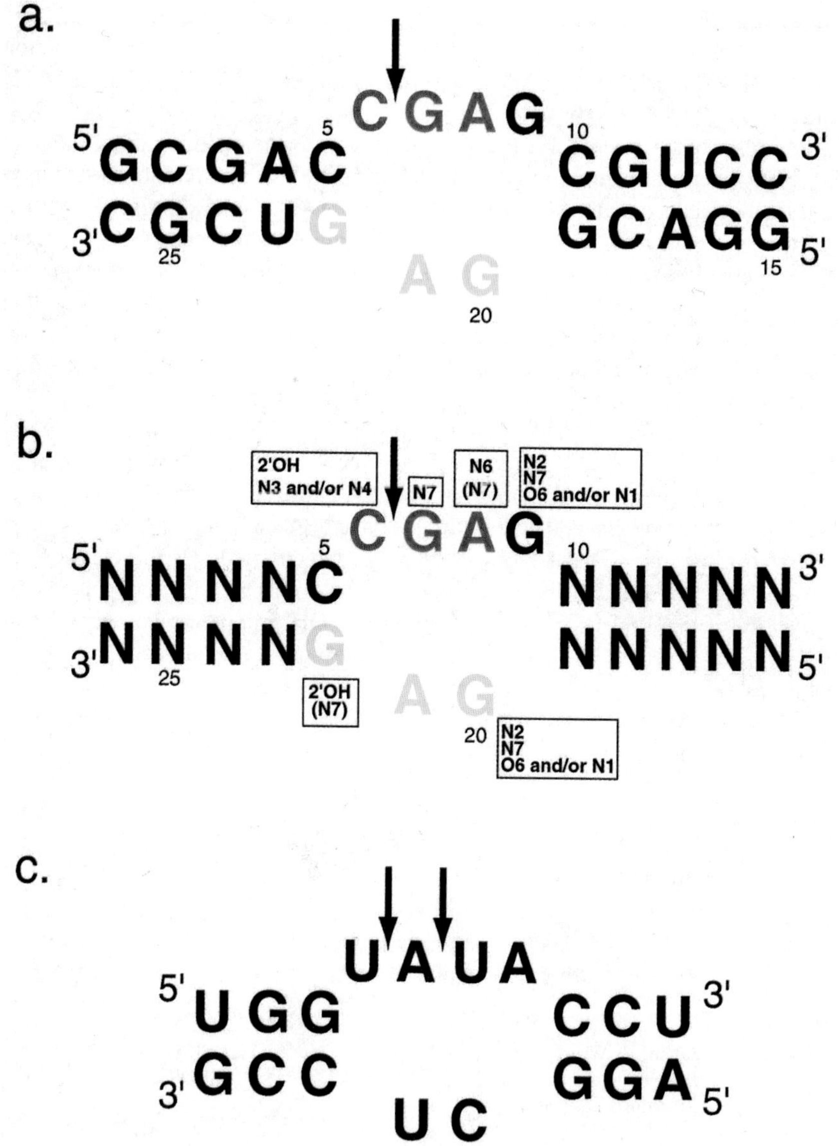

FIG. 1. Secondary structure and cleavage sites of the leadzyme. (a) Cleavage occurs at the phosphodiester bond indicated by the arrow between C6 (in red) and G7 (in green). The colors black (stems and G9), red (C6), green (G7 and A8), and yellow (G20, A21, and G22) apply in Fig. 4. (b) Biochemical modification data. The chemical groups indicated in the boxes were determined as necessary for the catalytic activity. (c) Secondary structure of a leadzyme variant with two adjacent cleavage sites, as indicated by the arrows.

Uhlenbeck[22] compared to a random sequence. This ribozyme can be formed by a single strand or by two strands commonly called leadzyme and substrate, containing the scissile bond.[22] The small size of this molecule and its high specificity to the lead ions make this system very suitable for structural studies such as nuclear magnetic resonance (NMR), circular dichroism (CD), and X-ray diffraction, as well as mutagenesis studies.

Composed of fewer than 30 nucleotides, the leadzyme is one of the smallest catalytic RNAs. The most important advantages of studying small ribozyme systems, such as the leadzyme, are found in experiments, where they prove to be easily synthesized, crystallized, or concentrated in solution. Small ribozymes can be used for developing theoretical calculations on RNA, by giving access to computational tools such as molecular dynamics and MM/QM (a combination of molecular mechanics/quantum mechanics) studies. In particular, the specificity of the leadzyme for Pb^{2+} ions has led to new research. The determination of different structural and catalytic models, using NMR,[24]crystallography,[25] and molecular modeling[26] has shed some light on the inner mechnism of the leadzyme. Contradictions between interpretations of some experimental results warrant new experimental and theoretical investigations.

In the first part of this article we discuss the leadzyme and characterize its catalytic mechanism. Structural and catalytic models derived from NMR, X-ray crystallography, and computer modeling studies are summarized and compared. The second part deals with the importance and implication of metal ions, and in particular of lead ions, in RNA catalytic reactions. We describe computational methods that are currently applied to help characterize the leadzyme catalytic reaction. Finally, we also present the development of empirical parameters for performing reliable simulations that include lead ions, and that are directed to the elucidation of the catalytic reaction of the leadzyme.

Active Structure Determination

Biochemical Modifications

An analysis of sequence variants in the asymmetric internal loop showed that only a C at position 6, and a G at position 9 (see Fig. 1a) were allowable for active leadzymes, such as LZ2.[27] By fixing these two bases, Pan *et al.* found 36 active variants with a preferred G at position 7. The single variants G7U and A8G (in green in Fig. 1a) have a decreased catalytic activity when compared to the wild-type

[24] P. Legault and A. Pardi, *J. Am. Chem. Soc.* **116,** 8390 (1994).
[25] J. E. Wedekind and D. B. McKay, *Nature Struct. Biol.* **6,** 261 (1999).
[26] S. Lemieux, P. Chartrand, R. Cedergren, and F. Major, *RNA* **4,** 739 (1998).
[27] T. Pan, B. Dichtl, and C. Uhlenbeck, *Biochemistry* **33,** 9561 (1994).

(LZ2), as well as the double mutant G2U, A3G, which suggests a cooperation between these two nucleotides.

Biochemical modifications were applied to all internal loop residues (see Fig. 1b).[28] For most positions, when the nucleotide is substituted by deoxynucleotide, the activity is diminished by a factor of about 3, except for G22 where a factor of about 16 was measured, and for C6 which suppressed the activity. These results are in agreement with those of Ohmichi and Sugimoto.[29] The substitutions of C6, G9, and G20 reduced the activity by factors of at least 10. The replacement of A21 with several adenosine substituents increased the leadzyme activity, except for ethenoadenosine, for which the activity was decreased by 40%. This latter result suggested that A21 could be bulged out of the asymmetric loop, in contrast to the NMR results of Legault and Pardi who observed a wobble base pair between A21 and C6. Figure 1b summarizes the important functional groups of the leadzyme and its substrate. The fact that the binding free energy of one hydrogen bond is estimated to be about 1–2 kcal/mol[30] provides a basis for interpreting the effect of the molecular replacements. Using nondenaturing gel electrophoresis, Chartrand *et al.* showed that the reduction in activity was not due to a reduction in the stability of the leadzyme–substrate complex.[28] Therefore, the reduced activity must be a consequence of the mechanism itself.

An interesting result was obtained by Ohmichi *et al.*,[31] who examined various substrates. A variant where C6 was deleted was active with an increased cleavage rate of almost twice that of the wild type. This result suggests a reorganization of the active site structure due to the absence of C6 since Chartrand *et al.* determined C1 : N3 to be involved in catalysis.[28] This also is in contrast, but to a lesser extent, to the finding of Legault and Pardi, who observed, at pH 5.5, a protonated A21 forming a wobble base-pair with C6.[24] All variants lacking G9 were inactive, which suggests that either stacking or hydrogen bonding involving G9 could be a major factor for the stability of the leadzyme active structure.

Two variants, one containing a dG at position 20 and the other a dA at position 21, were built to assess the role of 2′-hydroxyls in the binding of the leadzyme with its substrate, and to verify their participation in the catalytic reaction.[29] On the one hand, the k_{obs} for the dG variant was found to be twice as small as that for the wild-type, and 10 times lower for the dA variant. The binding constant was affected by the dG variant, but not by the dA variant. This led to a model involving an interaction between the 2′-OH of G20 and the G9 : O6. On the other hand, the 2′-hydroxyl group of A21 was found to be important for the cleavage rate, in

[28] P. Chartrand, N. Usman, and R. Cedergren, *Biochemistry* **36,** 3145 (1997).

[29] T. Ohmichi and N. Sugimoto, *Bull. Chem. Soc. Jpn.* **70,** 2743 (1997).

[30] D. Turner and P. C. Bevilacqua, "The RNA World." Cold Spring Harbor Laboratory Press, Cold Spring Harbor, NY, 1993.

[31] T. Ohmichi, O. Okumoto, and N. Sugimoto, *Nucl. Acids Res.* **26,** 5655 (1998).

contrast to G20. The weak variation of the cleavage rate, compared to the one obtained for the group I ribozyme (about 1700), suggests that the hydroxyl group of A21 could act indirectly in the catalytic reaction by means of a metal ion Pb^{2+}.[29]

NMR

Legault and Pardi determined the pK_a of each N^1-adenine using heteronuclear NMR spectroscopy. At 25°, without lead ions, they determined the pK_a of A21 : N1 to be 6.5, which represents a shift of about 2.6 units from the normal value.[24,32,33] A21 was determined to lie close to the C6pG7 cleavage site and was thought to act as a general acid catalyst in the cleavage reaction, or to participate in a wobble base-pair interaction with C6.

Further NMR data allowed the team to propose a three-dimensional (3-D) model of the leadzyme.[34–36] The data were collected at pH 5.5, in the absence of lead ion. In their model, the internal loop is roughly helical, with G7 and G9 bulged out of the helix. The Watson–Crick groups of the protonated nucleotide A21 form two hydrogen bonds with the Watson–Crick groups of C6. Nucleotides G7 and A8 are flexible. G9 is clearly bulging out of the helix, as interpreted from the spectra. Because of the absence of Pb^{2+}, none among the ensemble of structures consistent with the NMR data exhibits an in-line conformation of the 2′-OH with the phosphate group of G7. The NMR structure of the leadzyme is inactive and strongly affected by the absence of divalent ions. This raises the question whether the NMR determined conformation is close to the active form of the leadzyme.

NMR studies were also performed by Katahira *et al.*, with the solution containing Pb^{2+} ions. To prevent cleavage, they incorporated a 2′-*O*-methylcytidine at position 6. They performed the same experiments in solution without Pb^{2+} and found no evidence for a stable CA base pair in the internal loop at neutral pH. However, when the pH was lowered to 5.5, some signals in the spectra appeared, showing the possibility of a sheared GA interaction. The stem structure was mainly in the A-form, except for G19. A right-handed helical structure is maintained not only in the stem region, but also in the loop with the exception of G9. The bases of A8 and C10 were found to be close to each other, leading to the hypothesis that G9 could be flipped out of the helix. This is underscored by the fact that no cross peak involving G9 was observed. Nucleotides G7, A8, G20, and A21 adopt an *anti*-conformation at neutral pH. A21 : H2 has a strong nuclear overhauser effect (NOE) with G22 : H1′, and a weak NOE with G7 : H1′, A8 : H1′, and G9 : H8. These NOEs support the strong dependence of the leadzyme structure on the pH

[32] P. Legault and A. Pardi, *J. Am. Chem. Soc.* **119,** 6621 (1997).

[33] W. Saenger, "Principles of Nucleic Acid Structure." Springer-Verlag, New York, 1984.

[34] C. Hoogstraten, P. Legault, and A. Pardi, *J. Mol. Biol.* **284,** 337 (1998).

[35] L. Mueller, P. Legault, and A. Pardi, *J. Am. Chem. Soc.* **117,** 11043 (1995).

[36] P. Legault, J. Li, J. Mogridge, L. Kay, and J. Greenblatt, *Cell* **93,** 289 (1998).

of the solution. The results obtained with the noncleavable analog led to similar conclusions. Structural changes were observed for G22, suggesting a binding of the Pb^{2+} near this residue.[37]

Using these NMR spectra, it was shown that the variation of the pH from 7.0 to 5.5 brings a chemical shift for A21 : H8, which could be due to its protonation,[38] as seen by Legault and Pardi,[32] where a pK_a of about 6.5 was determined for A21 : N1. A temperature increase from 37° to 45° at pH 5.5 brings back the A21 : H8 chemical shift to its value at pH 7.0.

X-Ray Diffraction

Wedekind and McKay obtained X-ray diffraction data for the leadzyme–substrate complex[25] with Mg^{2+} and Ba^{2+} ions. Their final structure was refined at 2.7 Å resolution, with two independent molecules per asymmetric unit. A non-Watson–Crick interaction between C6 and A21 was found with one hydrogen bond between the exocyclic amine of A21 and the C6 : O2. No protonation state for A21 : N1 was observed. The X-ray leadzyme structure is helical, whereas the substrate has three bulged-out nucleotides: G7, A8, and G9. The two X-ray determined structures share a significant 2.0 Å root mean square deviation (RMSD). In the first structure, the three bulged-out nucleotides are stacked, whereas in the second structure only A8 and G9 are stacked, and G7 does not create a kink at the phosphodiester link of the scissile bond. C6 adopts a C2′-endo pucker mode in molecule 2, and a C3′-endo in molecule 1. It is important to note that the three bulged-out nucleotides in molecule 1 form two intermolecular hexanucleotide stacks and non-Watson–Crick interactions with another copy of molecule 1 in the crystal. In molecule 2, G7 is not stacked and contributes to the formation of leadzyme multimers by interdigitating between G7 of one leadzyme and A21 of another leadzyme molecule.[25] It shows clearly that the crystal packing stabilizes these conformations.

A working model of the catalytic reaction was built using molecule 2 that was thought to be close to the active form. To obtain a possible active conformation, the ribose pucker of C6 was changed from C2′- to C3′-endo, and the phosphodiester bond was slightly twisted. The experimental Ba^{2+} was replaced by a $Pb(OH)^+$ in the active site, leading to a plausible catalytic mechanism, which, nonetheless, contrasts with experimental activity data.[39] The main inconsistency concerns the involvement of A21 : N1 in the reaction. Where the model of Wedekind and McKay includes an interaction between the lead ion and A21 : N1, Chartrand *et al.*,[28] as well as Pan *et al.*,[27] determined that A21 : N1, which was shown to have a high

[37] M. Katahira, T. Sugiyama, M. Kanagawa, M. Kim, S. Uesugi, and T. Kohno, *Nucleosides Nucleotides* **15,** 489 (1996).

[38] M. Katahira, M. Kim, T. Sugiyama, Y. Nishimura, and S. Uesugi, *Eur. J. Biochem.* **255,** 727 (1998).

[39] E. Westhof and T. Hermann, *Nature Struct. Biol.* **6,** 208 (1999).

pK_a of about 6.5 by NMR,[24] is not well suited to be a catalytic metal ligand and to act as a Lewis base.

Molecular Modeling

As a first step toward the establishment of the leadzyme active structure using molecular modeling, we considered every possible arrangement exhibiting at least one hydrogen bond between all nucleotides in the internal asymmetric loop. A set of base pairing hypotheses was defined from the biochemical modification results of Chartrand *et al.* A consistent structure was proposed by Lemieux *et al.*[26] that included a base-triple between C6-G20-G9. The model, obtained using the molecular modeling program *MC-Sym,* includes a base-pair interaction between G9 and G20 that provides the possibility for the Watson–Crick face of G20 to be exposed to the internal loop, and to form hydrogen bonds with the Watson–Crick groups of C6 previously shown to be crucial for the catalytic reaction.[28] In the *MC-Sym* model, nucleotides G20, A21, and G22, bearing unfamiliar conformations, form a sharp turn as described in Table I and shown in Fig. 2c. This organization of the backbone bulges A21 out of the internal loop, which is the main discrepancy between this model and the NMR and crystallographic ones.

Lemieux *et al.* also proposed positions for the lead ions. One is near the catalytic site, and one is in the neighborhood of G7 and A8. The latter is necessary to stabilize the sharp turn structure to accommodate formation of the base triple. Katahira and co-workers have shown that these two sites were important for the positioning of the lead ions.[37] The *MC-Sym* model also indicates an in-line attack conformation of the scissile bond with an angle of 149.9° of the atoms C6 : O2′, G7 : P and G7 : O5′, and a distance of 3.6 Å between the atoms C6 : O2′ and G7 : P.

Structure Analysis and Comparison

Using an analysis procedure developed in our laboratory,[39a] we compared the three different models on the basis of their various nucleotide conformations and interactions. Table I summarizes the results of this comparison for the core structures. We also performed an analysis of the statistical deviation of each base relations from the normal, in order to shed light on the structural specificity of the models. In Fig. 2, the nucleotides colored in red are involved in relations that are close to the normal, whereas nucleotides colored in yellow are involved in nonstandard interactions. This description generally indicates regions that adopt crucial conformations to maintain a particular fold.

The X-ray structure has a poorly ordered internal asymmetric loop 5′ strand. This can be observed in Fig. 2a, where nucleotides C6 to G9 are colored in yellow. That portion of the molecule displays unusual relationships between its nucleotides

[39a] P. Gendron, S. Lemieux, and F. Major, *J. Mol. Biol.* **308,** 919 (2001).

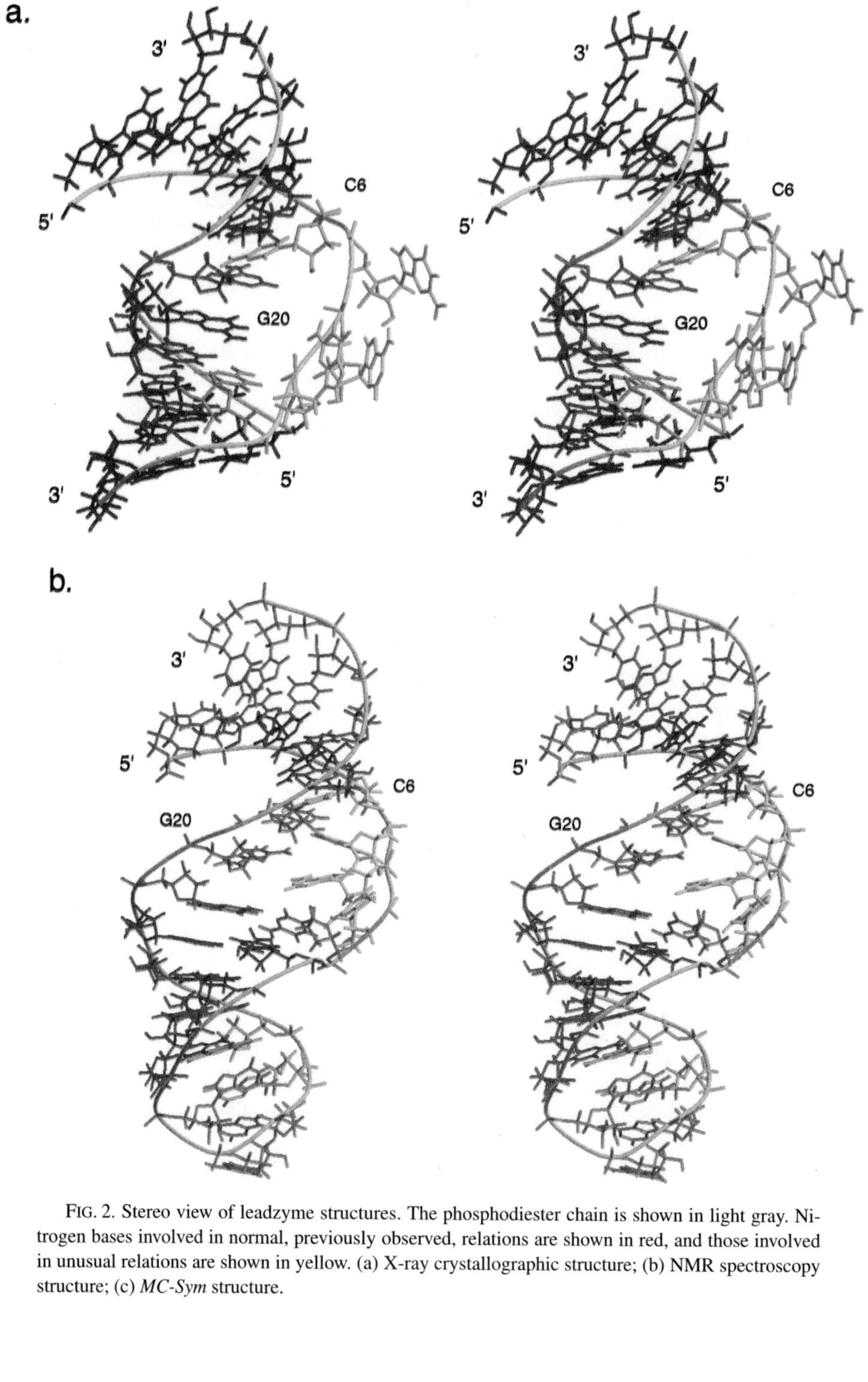

FIG. 2. Stereo view of leadzyme structures. The phosphodiester chain is shown in light gray. Nitrogen bases involved in normal, previously observed, relations are shown in red, and those involved in unusual relations are shown in yellow. (a) X-ray crystallographic structure; (b) NMR spectroscopy structure; (c) *MC-Sym* structure.

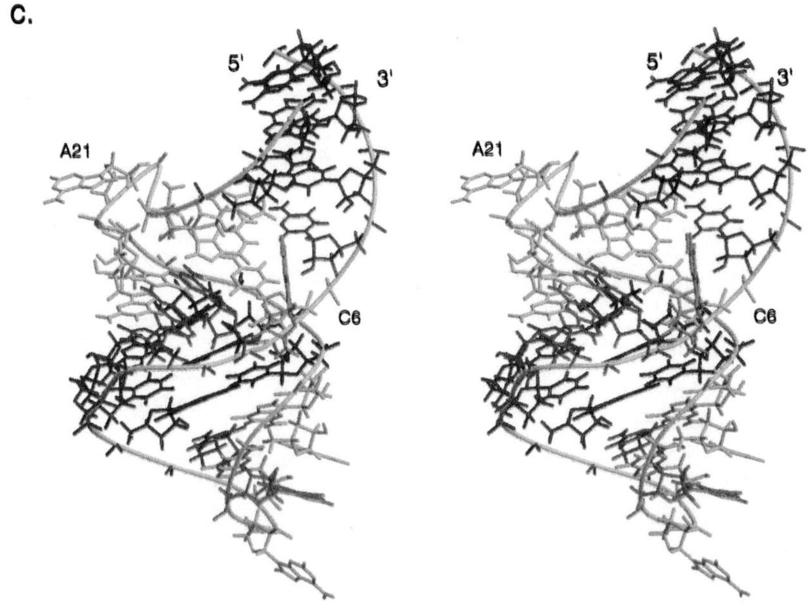

FIG. 2 (*Continued*)

(see Table I) since nucleotides A7 to G9 are bulging out of the helix, and therefore do not participate in internal hydrogen bonding. Interestingly, these bases interact with another unit in the crystal (described in the previous section), which provides a rationale for the particular arrangement. The same region is also depicted in yellow in the NMR-generated model, as can be seen in Fig. 2b. However, in this model the internal asymmetric loop 5′ strand looks more ordered because of the stacking between C6 and G7, which is not the case in the crystal structure. Nevertheless, the other nucleotides in this region are either stacked inside the helix or slightly bulging out of it and are not involved in hydrogen bonding interactions. The *MC-Sym* model suggests that this same region of the leadzyme internal loop is in helical form. Nucleotides C6 through G9 are tinged with orange, which represents more conventional base relations (see Fig. 2a).

Likewise, the internal asymmetric loop 3′ strand of the X-ray diffraction structure is composed completely of typical helix-like RNA strand and displays a reddish color (see Fig. 2a). This segment is also similarly structured in the NMR structure and is exposed in shades of orange. In contrast, the *MC-Sym* structure shows a shiny yellow throughout this region, indicating peculiar base relations. A hydrogen bond is suggested between the keto group of C6 and the Watson–Crick face hydrogen of the exocyclic amino group of A21 in the crystal structure. The

TABLE I
COMPARATIVE ANALYSIS OF X-RAY NMR, AND *MC-Sym* STRUCTURES

Source	Crystal	NMR	*MC-Sym* model
Residues			
5	C3′_endo anti	C3′_endo anti	C3′_endo anti
6	C2′_endo anti	C3′_endo anti	C2′_endo anti
7	C3′_endo anti	C2′_endo syn	C3′_endo anti
8	C3′_endo anti	C2′_exo anti	C3′_endo anti
9	C1′_exo syn	C1′_exo anti	C3′_endo anti
10	C3′_endo anti	C3′_endo anti	C3′_endo anti
19	C3′_endo anti	C3′_endo anti	C3′_endo anti
20	C3′_endo anti	C3′_endo anti	C4′_exo syn
21	C3′_endo anti	C3′_endo anti	C4′_endo syn
22	C3′_endo anti	C2′_exo anti	C2′_exo syn
Relations			
5–6	Stack	Stack	No stack
6–7	No stack	Stack	No stack
7–8	No stack	No stack	No stack
8–9	Stack	No stack	Stack
9–10	No stack	No stack	Stack
19–20	Stack	Stack	Stack
20–21	Stack	Stack	No stack
21–22	Stack	Stack	No stack
5–22	Pairing XIX	Pairing XIX	Pairing XXII
6–20	—	—	Pairing XXII
6–21	Pairing A:N6-C:O2	Pairing A:N6-C:N3	—
6–22	—	—	Stack
9–20	—	—	Pairing VI
10–19	Pairing XIX	Pairing XIX	Pairing XIX

distance measured between the two participating groups is 2.6 Å. Analogously, in the NMR structure, the cyclic amino group of C6 and the Watson–Crick face hydrogen of the exocyclic amino group of A21 form a hydrogen bond. The distance between these two groups is 3.2 Å. Another hydrogen bond could possibly be established from the observation of this structure. In fact, the keto group of C6 is at 3.1 Å from the cyclic amino of A21, but the angle between those two groups is too obtuse. In contrast, the *MC-Sym* structure infers the formation of a base-triple in which many hydrogen bonds are implicated.

Catalysis and Ions

The leadzyme structure provides the basis for its catalytic activity. The absence of an experimental structure in the active conformation raises the question of how similar the proposed experimental and theoretical models discussed above are to the active conformations? In this section, we discuss how kinetic studies were applied to quantify the relationships between structure and activity, and how kinetic data were used to build an active conformation of the leadzyme.

Kinetic Parameters

Under physiological conditions, Li and Breaker[40] have determined the kinetic parameters that are needed to estimate the transesterification rate of RNAs. In order to simplify the system and to put the focus on only one RNA phosphodiester cleavage, a 22-nucleotide DNA with a single ribonucleotide was used. Here is a summary of their observations. The cleavage rate constant increases exponentially when the pH is between 9.0 and 13.0 and reaches a plateau after 13.0, which is the pK_a value for the 2′-hydroxyl group. The cleavage rate is proportional to the fraction of 2′-oxyanion versus 2′-hydroxyl groups. They determined that 2′-oxyanion is about 10^8 times more powerful than 2′-hydroxyl group in a nucleophilic attack. The pK_a value of the 2′-hydroxyl group is also linearly dependent with the monovalent (potassium) ion concentration. An increase in potassium concentration, from 0.25 *M* to 3.16 *M*, decreases the pK_a value by 0.6 unit, and thus increases the catalytic rate. The temperature is also an important factor for catalysis. When the temperature is increased from 4° to 50°, the rate constant increases by a factor of 3. Another important parameter for catalysis is the nucleotide sequence. Four different factors were determined for four different variants. The rate constants for the pairs UA and CA, not tested by Li and Breaker, were found to be 100-fold greater than AA.[41] Hence, catalysis is strongly dependent on various factors that influence RNA structure.

Two Metal-Binding Sites

Two metal-binding sites are located within or close to the asymmetric internal loop. Using chemical shifts, Kim *et al.* localized the two sites. Site A is located at nucleotides G20-A21-G22, and site B at nucleotides G7-A8.[42,38] By combining different metal ions with the leadzyme and its substrate, Sugimoto and Ohmichi clearly showed that the leadzyme/substrate complex possesses at least two different ion binding sites.[43] Pb^{2+} binds to site A (near G22) 50 to 100 times more strongly than Mg^{2+}, and five times more strongly to site B than Mg^{2+}.[42] The addition of Nd^{3+} affects the cleavage step, as well as the stability of the leadzyme–substrate complex.[44] Nd^{3+} could induce a conformational change in the complex that increases the cleavage rate. Furthermore, Nd^{3+} is a better acid than Pb^{2+}. Hence,

[40] Y. Li and R. Breaker, *J. Am. Chem. Soc.* **121,** 5364 (1999).
[41] K. Williams, S. Ciafre, and G. T. Valentini, *EMBO J.* **14,** 4558 (1995).
[42] M. Kim, M. Katahira, T. Sugiyama, and S. Uesugi, *J. Biochem.* **122,** 1062 (1997).
[43] N. Sugimoto and T. Ohmichi, *FEBS Lett.* **393,** 97 (1996).
[44] T. Ohmichi and N. Sugimoto, *Biochemistry* **36,** 3514 (1997).

a dual ion mechanism would be more efficient with Nd^{3+} than using only Pb^{2+}. In this mechanism Pb^{2+} would activate the deprotonation of the 2′-OH, and Nd^{3+} would act as an acid to coordinate the 5′-oxygen leaving group.[44] A Hill coefficient of 3.3 was determined by Chartrand *et al.*,[28] showing that at least three metal ions interact cooperatively during catalysis. This coefficient, also computed by Kim *et al.*[42] was found to lie between 3.9 and 4.4 for Pb^{2+}, and between 0.7 and 0.9 for Mg^{2+}. These results show that there are four lead ions binding the leadzyme, but since they all do not necessarily participate directly in catalysis, they would still be in agreement with the dual Pb^{2+} binding site.

Two structures were built based on the X-ray diffraction maps.[25] In the first structure, two Mg^{2+} were located, one in contact with G3 in stem I and another with G19, and a Ba^{2+} ion was found to interact with G22. In the second structure, one Ba^{2+} was located at the same position as found in the first structure, that is, close to G22, and another was found near the scissile bond. A Pb^{2+} ion was found to interact with U17-G18 in stem II. Mg^{2+} was not found in the second structure, suggesting that Mg^{2+} could lead to a noncatalytic conformation of the leadzyme.

Pb^{2+} versus Mg^{2+}

Pan *et al.* showed that in the absence of Mg^{2+}, the cleavage rate of the leadzyme increases with Pb^{2+} concentration up to a certain point, and then decreases abruptly. This can be explained by the variation of Pb^{2+} polyhydroxide concentration, and polyhydrates present in a Pb^{2+} solution at pH 7.0.[27] In the presence of Mg^{2+}, the cleavage is slower at low Pb^{2+} concentration, but increases to approximately its maximal value at higher Pb^{2+} concentrations. This suggests a competition between both ions. Pan *et al.*[27] also found that the cleavage rate increases exponentially with pH between 5.5 and 7.0, corresponding to a 10-fold increase per pH unit. At a higher pH, the Pb^{2+} ions tend to form polyhydrates or polyhydroxides. A linear slope of the pH rate profile is consistent with the involvement of a basic group in catalysis. This group is likely to be the $Pb(OH)^{+}$, as in the cleavage of yeast $tRNA^{Phe}$.[10] Because of the different pK_a values of the metal ions, their action in RNA catalysis is dependent on the pH of the solution. The Mg^{2+} ion is effective at high pH values, whereas the Pb^{2+} ion is more effective at neutral pH. These two metals coordinate hydroxide ions, which act as a base to deprotonate the 2′-OH at the cleavage site. The phosphodiester attack may occur anywhere, but as observed in RNA 3-D structures that contain ion-binding pockets, it becomes more specific.[4] In the case of leadzyme, the specificity for Pb^{2+} suggests a direct participation, or a very important structural role.

Some experiments were performed to understand the action of the lead ion in the catalytic reaction. In the continuation of the work of Pan *et al.*,[27] CD was used to study the competition between Mg^{2+} and Pb^{2+}. Kim *et al.*[42] showed that the leadzyme is cleaved at two adjacent cleavage sites, with either enhanced catalysis or repression, depending on the ion concentration. The asymmetric loop in the leadzyme was conserved, but a base pair, C2-G25, was added in the 3′ stem, and the first three base pairs in the 5′ stem were inverted. This inversion was shown to be responsible for the appearance of the second cleavage site adjacent to the original one. CD spectroscopic measurement of the ellipticity of the leadzyme showed that Pb^{2+} and Mg^{2+} have different effects on the overall structure.[42] However, these differences have not been quantified. Using a 2′-*O*-methyl group at the cleavage site, and lowering the pH value of the solution to 5.5 to decrease the nonspecific RNA degradation, Katahira *et al.*[38] did not observe any change in the overall structure of the complex, with Mg^{2+} or Pb^{2+} at pH 7.0 and pH 5.5. Nonetheless, this latter experiment did not give any indication about local changes in the asymmetric loop, which may be responsible for the slight variation in the spectra. It was found that Mg^{2+} and Pb^{2+} induce similar as well as different effects on the structure. Specifically, the shift of the spectrum at 210 nm is the same when the ions are added, but they are different at 270 nm.

The CD spectra of the wild type, and of the variant without C6,[31] with 100 mM Na^+ and 50 μM Pb^{2+} show a normal A-form structure with both cations.[45] However, the differences between the spectra show that the active complexes in presence of Pb^{2+} and Na^+ are different. Furthermore Pb^{2+} induces slow association and dissociation rate constants, as compared to Na^+.[31]

Other Leadzyme Variants?

The major surface proteins of procyclic trypanosomes, EP1, EP2, EP3, and GPEET, are not expressed in forms found in the bloodstream. Domain I of the 3′ untranslated region (UTR) of EP1 mRNA contains three leadzyme-like regions that do not conform to the previously described ones.[46] Two sites contain an asymmetric internal loop of the same size, but sharing low sequence similarity, with that of the consensus leadzyme. The third site contains a smaller internal loop formed by a U-U mismatch, flanked by wobble G-U and standard A-U base pairs. It has been shown that the catalytic activity of the 3′ UTR of EP1 mRNA decreases with an increase of temperature. This shows that loss of secondary structure reduces the activity, which is an important characteristic of the leadzyme. We extracted the RNA sequence of these sites and used mFold[47] to predict their secondary structure. One site, shown in Fig. 1c, possesses a leadzyme-like secondary structure, even though the sequence is different.

[45] K. Hall and L. McLaughlin, *Biochemistry* **30,** 10606 (1991).

[46] M. Drozdz and C. Clayton, *RNA* **5,** 1632 (1999).

[47] M. Zuker, D. Mathews, and D. Turner, *in* "RNA Biochemistry and Biotechnology" (J. Barciszewski and B. F. C. Clark, eds.), pp. 11–43. Kluwer Academic Publishers, Dordrecht, The Netherlands, 1999.

Computational Studies

Activity Quantification

The intrinsic structure of an RNA is a major factor in cleavage activity. The nucleophilic oxygen O2′ creates RNA folds that are less stable than DNA by a factor of about 10^5.[40] The main factors contributing to an increase of the transesterification reaction are a deprotonation of the O2′, or the protonation of a nonbridging phosphate oxygen, and a protonation of the O5′.[48] Another important factor is the in-line positioning between the nucleophile O2′, the phosphorus atom, and the leaving 5′-oxyanion group. Li and Breaker characterized the 3-D structure by the angle formed between O2′-P-O5′, a fitness coefficient depending on the previous angle, and the distance between O2′ and P. An angle of 180° and a distance of 3 Å represent a structure with an optimal rate of transesterification. Using different systems, they showed that the internucleotide linkage geometry plays a major role in the kinetics of catalysis. A low fitness coefficient correlates with a weak cleavage rate. They found that the in-line geometry provides a minimum of about a 10- to 20-fold rate enhancement for the cleavage of RNA by intramolecular transesterification catalysis. Different studies on systems with metal ions (see, for instance, Ref. 49) show that metal ions promote new cleavage sites. Nonetheless, they found that the extent of cleavage with or without metal ions is nearly the same. It was proposed that metal ions may alter the tertiary structure and may generate linkage conformations that favor in-line attacks. The ions may not be involved as general base catalysts, or as Lewis acid catalysts.

In the remainder of this chapter, we present the docking of Pb^{2+} ions in different models. Coordination bonds and bond angles between the lead ion and the leadzyme are defined. The coordination parameters are then determined, and the results of molecular dynamics simulations applied to the X-ray and *MC-Sym* structures are presented. The simulations were analyzed in order to understand and characterize the leadzyme catalytic mechanism.

Lead Ion Docking

The absence of lead ions in solution renders leadzyme inactive. This implies that determined structures may vary from the active one. Computational simulations are necessary to derive an active conformation, with lead ions explicitly represented. Hence, docking ions into the determined structures was required. This important initial step was performed by only specifying nonbonded interactions between the lead ion and the leadzyme, using the program Mining Minima.[50]

[48] M. Oivanen, S. Kuusela, and H. Lonnberg, *Chem. Rev.* **98,** 961 (1998).

[49] S. Kazakov and S. Altman, *Proc. Natl. Acad. Sci. U.S.A.* **88,** 9193 (1991).

[50] M. S. Head, J. A. Given, and M. K. Gilson, *J. Comp. Aid. Mol. Des.* **101,** 1609 (1997).

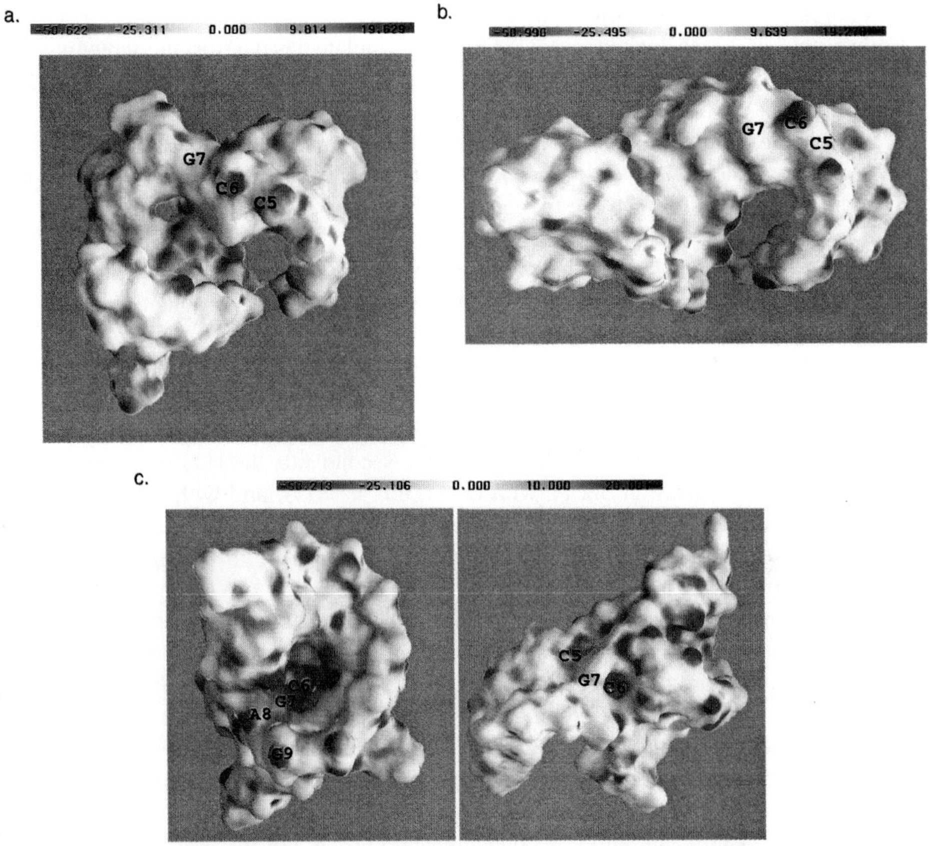

FIG. 3. Molecular surface of leadzyme structures. Areas of electronegative potential values are indicated in red, whereas electropositive areas are indicated in blue. (a) X-ray crystallographic structure; (b) NMR structure; (c) *MC-Sym* structure.

Figure 3 shows the molecular surfaces of the leadzyme by NMR, X-ray, or *MC-Sym*. The regions on the surfaces are colored according to electrostatic potentials in solution.[51] The electronegative binding pocket shown in the *MC-Sym* structure surface (in red) clearly indicates a high affinity for cations, as is also the case for the X-ray structure, but to a slightly lesser extent. In the NMR model, there is no lead-ion attractive region or pocket next to the catalytic site. To dock the ion, an 18 Å cubic box centered on the binding pocket was defined.

[51] A. Nicholls, K. Sharp, and B. Honig, *Proteins Struct. Funct. Gen.* **11,** 281 (1991).

A 2.0 Å van der Waals radius sphere with a 0.3 kcal/mol energy depth and a two-electron charge was docked inside the box. To simulate the electrostatic screening due to solvent, a distance dependent dielectric was used, with the dielectric set to 4*r*.[50] The mesh of the nonbonded grids was set to 0.2 Å and 100 docking sites were tested using a maximum of 220,000 generated conformations per docking.

Figure 4 shows 100 lead ions docked onto the different leadzyme structures. In the X-ray structure, we see two main clusters of ions. The blue ion is the ion of lowest binding energy in the first site, whereas the orange ion has the lowest binding energy near the second proposed lead binding site. The X-ray and the *MC-Sym* structures have the same binding sites, even though they do not share the same energies.

In the case of the *MC-Sym* structure, four different global minima were obtained. Using the experimental data, only two sites were retained. In the first site, the lead ion interacts directly with G9 : O6 at 3.1 Å, with C6 : O2 at 4.6 Å, and with G7 : O2P at 3.2 Å (in orange in Fig. 4). In the second site, the lead ion interacts directly with G7 : O2P at 3.2 Å, with G19 : O2P at 3.0 Å, and with A8 : O2P at 3.2 Å (in blue in Fig. 4). In the X-ray structure that was considered to be the closest to the active conformation (second structure in the PDB file), Pb^{2+} interacts in the first site directly with A8 : O1P at 3.2 Å, with C6 : O2′ at 3.4 Å, with C6 : O1 at 3.4 Å, with G20 : O6 at 3.9 Å, and with A21 : N1 at 3.2 Å (in blue in Fig. 4). The main difference between this model and the one proposed by Wedekind and McKay is the coordination of Pb^{2+} with A21 : N1, instead of with G7 : P in their model. For the second site (in orange in Fig. 4), there are three main interactions. The lead ion interacts with G20 : O2P at 3.2 Å, with A21 : N7 at 3.3 Å, and with A21 : O2P at 2.9 Å. These results correlate well with the NMR spectroscopic data obtained by Katahira *et al.*,[38] where two binding sites were located, in the first site near G20-A21-G22, and in the second site near G7-A8. For the NMR model, no ions were placed close to the catalytic site, and thus we abandoned this model in further computer simulations.

Computer Simulations

Only nonbonded parameters are used to dock the lead ions into the leadzyme structures. This implies that coordination bonds and bond angles between the ions and the leadzyme are not explicitly represented. In order to get a better representation of the lead ion during molecular dynamics simulations, the bonds and bond angles are parameterized, and hence we define a more realistic representation of the Pb^{2+} ions. The parameterization is performed using small compound structures that are extracted from the Cambridge Structural Database,[52] and using a

[52] F. Allen, J. Davies, J. Galloy, O. Johnson, O. Kennard, C. Macrae, E. Mitchell, G. Mitchell, J. Smith, and D. Watson, *J. Chem. Inf. Comput. Sci.* **31,** 187 (1991).

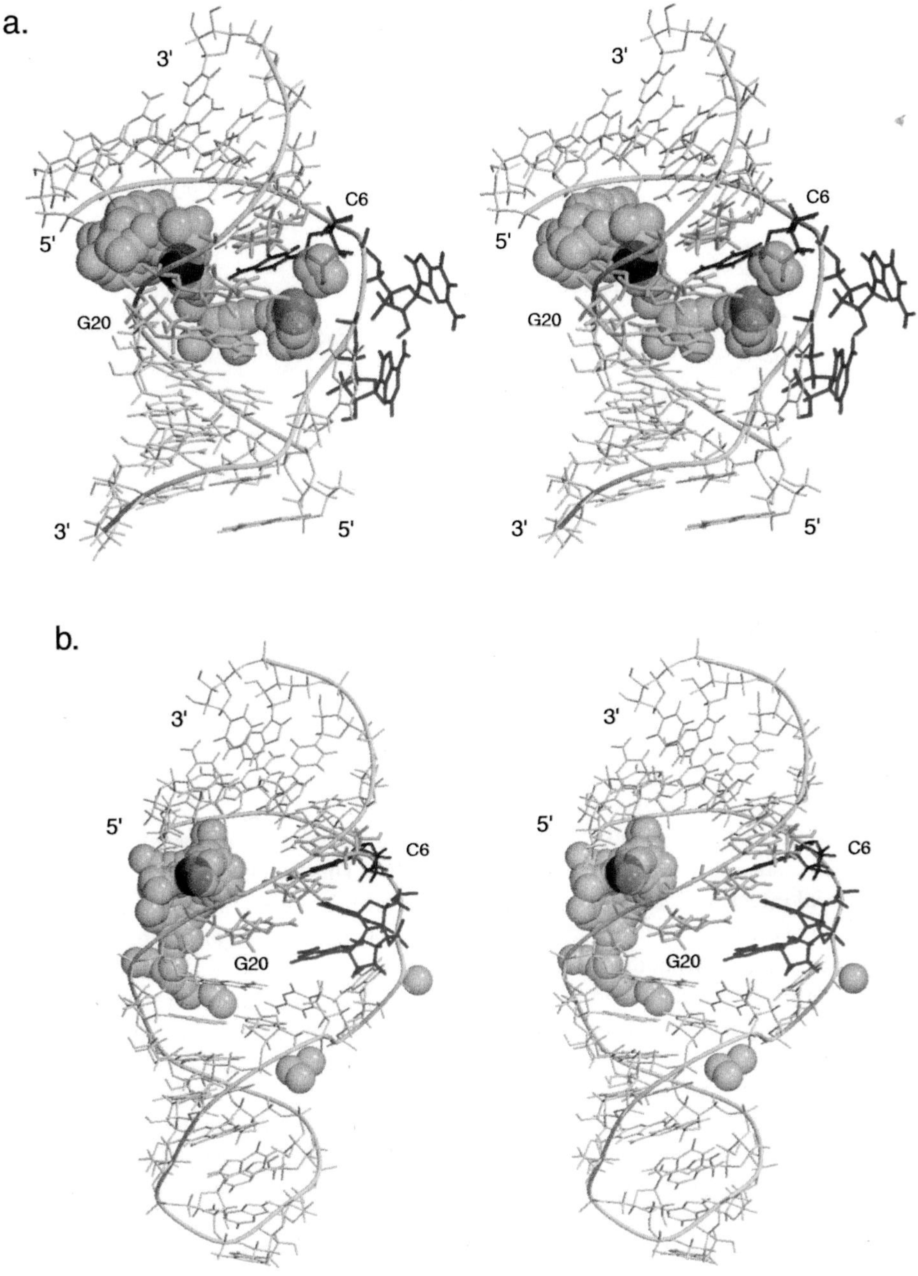

FIG. 4. Lead ions in leadzyme structures. The ion location of lowest binding energies near the first binding site are shown in blue. The ion location of lowest binding energies near the second binding site are shown in orange. The colors defined in Fig. 1 apply here. (a) X-ray crystallographic structure; (b) NMR structure; (c) *MC-Sym* structure.

C.

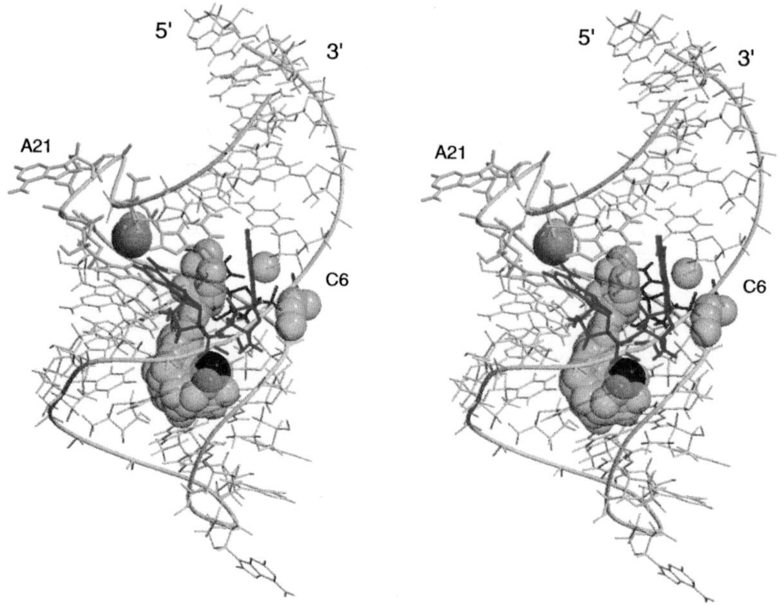

FIG. 4 (*Continued*)

Boltzmann distribution to fit the parameters, such as the force constants and equilibrium distances and angles (unpublished results). This step is important to study the local effects of lead ions on leadzyme structure. The coordination bonds and angles between the leadzyme atoms and Pb^{2+} are obtained directly from the docked structures described in the previous section.

The force constants and the equilibrium distances and angles are shown in Table II. The charges of the lead ion are shared by all RNA ligand atoms: 0.3 electron per ligand for up to three ligand atoms, or 0.2 electron for more than three (Table III). The Pb^{2+} van der Waals parameters are set to $\epsilon = -0.5$ kcal/mol, and $R = 2.5$ Å. Two simulations are performed with the X-ray and *MC-Sym* structures. All hydrogen atoms are added to the structures. To neutralize the system, Na^+ ions are placed next to the phosphates, 25 in the *MC-Sym* structure and 19 in the X-ray structure. Then, the complex leadzyme–Pb^{2+}–Na is placed into a cubic water box containing 3375 water molecules. All water molecules overlapping the RNA–Pb^{2+}–Na complex with a distance of 2.8 Å are removed. The nonbonded list is updated every 20 fsec. An energy minimization is performed on the complex prior to the molecular dynamics simulations. Keeping all RNA atoms

TABLE II
Pb^{2+} BOND AND BOND ANGLE PARAMETERS

Atoms			Force constant (kcal/mol/A^2)	Force constant (kcal/mol/rad^2)	Equilibrium constant (Å)	Equilibrium constant (○)
Ox	Pb^{2+}		8.0		2.62	
Nx	Pb^{2+}		44.0		2.60	
Ox	Pb^{2+}	Ox		5.0		77.0
Ox	Pb^{2+}	Nx		40.0		80.0
P	Ox	Pb^{2+}		7.0		130.0

frozen, 1000 steps of steepest descent followed by 1000 steps of the Powell conjugate gradient method[53] are applied. With all the free atoms set, 1000 further steps of steepest descent, 5000 steps of Powell conjugate gradient, and 1000 steps of the Adopted Basis Newton–Raphson method are applied.[53] The final gradient RMSD (GRMSD) is equal to 0.008 kcal/mol/Å^2 for the X-ray structure, and to 0.075 kcal/mol/Å^2 for the *MC-Sym* structure. A 1 fs step is used

TABLE III
Pb^{2+}–LIGAND ATOMS AND CHARGES

X-ray first site			X-ray second site		
Atom	Residue	Charge (*e*)	Atom	Residue	Charge (*e*)
O2	Cyt6	−0.46	N7	Ade21	−0.33
O2′	Cyt6	−0.28	O2P	Ade21	−0.50
O1P	Ade8	−0.60	O2P	Gua20	−0.50
O6	Gua20	−0.27	PB		1.10
N1	Ade21	−0.54			
PB		1.00			
MC-Sym first site			*MC-Sym* second site		
Atom	Residue	Charge (*e*)	Atom	Residue	Charge (*e*)
O1P	Gua7	−0.50	O2P	Gua19	−0.60
O6	Gua9	−0.17	O1P	Gua20	−0.60
O3′	Gua7	−0.25	O5′	Gua20	−0.35
PB		1.10	N3	Gua20	−0.46
			PB		1.20

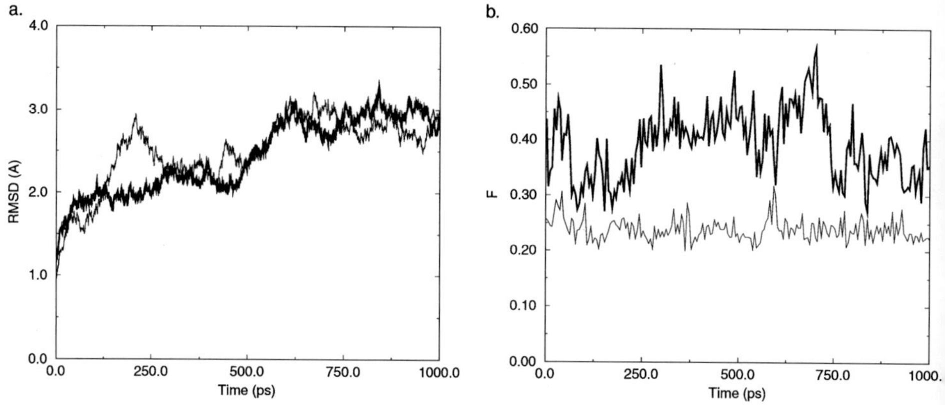

FIG. 5. Molecular dynamics results. (a) RMSD over time. (b) S_N2 Fitness coefficients over time. The dark curves represent the *MC-Sym* structure simulation, whereas light curves represent the X-ray crystallographic structure simulation.

to integrate the Langevin equations with a friction coefficient of 10 sec^{-1} assigned to every atoms. A 10-Å cutoff value is used for the nonbonded interactions with the Particle Mesh Ewald summation.[53] The simulations are performed over 1 nsec.

The molecular dynamics simulations have given us information about the motion and evolution of the structures as a function of time. Hence, molecular dynamics simulations return a quantification of specific system properties, such as the stability of the system over time. By calculating the Fitness coefficient [Eq. (1)], given by Soukup and Breaker,[54] an assessment of the likelihood of having an S_N2 attack can be assessed. One way to quantify the evolution of the structure as a function of time is given by the RMSD between the generated structures and the initial one (at time 0). The RMSD oscillates within a range of 3 Å for the *MC-Sym* structure, and within a range of 2.8 Å for the X-ray structure. In both simulations, a plateau is reached (Fig. 5), showing convergence to an equilibrium state. The stems for both simulations are well conserved, and therefore only the internal loop structure has been analyzed. For the *MC-Sym* simulation, the interaction C5-G22 has never been observed, and both nitrogen bases remain stacked. The internal loop base pair C6-G24(XXII) was conserved over time, whereas base pair G9-G20(VI) is changed to a A8-G20(64), or to a G9-G20(110) in the last 500 psec. For the

[53] B. Brooks, R. Bruccoleri, B. Olafson, D. States, S. Swaminathan, and M. Karplus, *J. Comp. Chem.* **4,** 187 (1983).

[54] G. Soukup and R. Breaker, *RNA* **5,** 1308 (1999).

X-ray structure, an interaction between C6-A21 (type 75 or 78) appears during the simulation, but only with a probability of 0.5. The base pairs C5-G22 and C10-G19 were very well conserved. Figure 5 shows both distributions of F over time. The curve has been smoothed by taking the maximum F values over a window of 5 ps, and by sliding the window by increments of 5 ps. In the X-ray simulation, this coefficient remains about 0.2, showing a low probability for catalytic activity. In the *MC-Sym* simulation, the factor F remains about 0.4, with a peak at around 250 ps. This is a better value than with the X-ray simulation, which increases the likelihood of a S_N2 catalytic mechanism in this conformation. As C6 : O2′ was kept protonated, the value of F can never be closer to 1; however, the tendency of increasing values supports an increasing probability of catalysis.

$$F = \frac{(\theta - 45)}{(180 - 45)} \frac{3^3}{d^3_{O2',P}} \tag{1}$$

The simulations did not bring any new insights about the assessment of the current structures, but nonetheless they were shown to be stable. In the case of the *MC-Sym* structure, the dynamics showed a tendency to remain close to the catalytic optimum geometry, whereas this was not observed in the X-ray structure. It is clear at this time that more experimental and activity data, that is, in the presence of lead ions, are needed to determine a more precise structure, and to precisely observe the catalytic reaction of the leadzyme.

Conclusion

State-of-the-art studies on the leadzyme were presented in this review. Experimental and computational work were described in detail, and new theoretical approaches to obtain further information about the leadzyme structure and catalytic mechanism were introduced. Experimental data shed light on understanding the catalytic process by identifying, for instance, which residues are important for catalysis. Two experimental structures, using X-ray crystallography and NMR spectroscopy, as well as an *MC-Sym* theoretical structure based on experimental activity studies including leadzyme analogs[28] have been thoroughly compared. Both experimental structures exhibit differences in the internal loop, and both contain interactions between C6 and A21, while the other loop residues are unpaired. In contrast, in the *MC-Sym* structure the main difference remains in A21, which was found unimportant for catalysis in activity studies, and hence this base was left bulging out of the internal loop. The theoretical structure contains a base-triple composed of the C6-G20(XXII) and G9-G20(VI) base pairs, which optimize the free energy of the internal loop. Interestingly, the NMR model does not show any binding pocket in the vicinity of the active site, whereas for both X-ray and *MC-Sym* structures two binding pockets can be observed, as described from the results of NMR experiments.[38] Molecular dynamics simulations were performed using the X-ray and *MC-Sym* structures in the presence of lead ions. Both simulations showed very stable structures, and the geometry of the catalytic phosphodiester link was studied. The X-ray structure does not exhibit a geometry suitable to an S_N2 reaction, whereas the *MC-Sym* structure remains close to it. At this time, more experimental data are needed to discriminate between those two conformations. Although the *MC-Sym* structure is the most different, it is able to explain the catalytic activity of the leadzyme.

Acknowledgment

We thank Allen W. Nicholson for useful comments and suggestions. We thank Michael Gilson for providing an access to the Cambridge Structural Database at CARB/NIST, and to Mining Minima. This work was supported by a grant from the Medical Research Council of Canada (MT-14604) to FM. PG holds a M.Sc. scholarship from the Fonds pour la Formation de Chercheurs et l'Aide a la Recherche du Quebec.

[34] Hammerhead Ribozyme Structure and Function in Plant RNA Replication

By RICARDO FLORES, CARMEN HERNANDEZ, MARCOS DE LA PEÑA, ANTONIO VERA, and JOSE-ANTONIO DAROS

Introduction

Evidence that there are biological entities with a lower complexity than viruses has been accumulating for the past 30 years. The most conspicuous representatives of this subviral world are viroid and viroid-like satellite RNAs from plants, which are solely composed of a small [250–450 nucleotide (nt)] single-stranded circular RNA with a high secondary structure content and without any apparent messenger capacity.[1] This makes the replication cycle of these molecular parasites, proposed to occur through a rolling circle mechanism, extremely dependent on enzymes provided by the host in the case of viroids, or by the host and the helper virus in the case of viroid-like satellite RNAs. However, three viroids and all viroid-like satellite RNAs code, in an ample sense, ribozymes catalyzing self-cleavage of their oligomeric strands generated in replication and, in some cases, self-ligation

[1] T. O. Diener, *FASEB J.* **5,** 2808 (1991).

0076-6879/01 $35.00

of the resulting linear monomers. The hammerhead structure[2–4] is the most widely distributed ribozyme in these small plant RNAs,[5] and it will be the subject of the present review. To provide a more general outlook we will also consider, in addition to plant viroid and viroid-like satellite RNAs, other small plant and animal RNAs with less-defined biological roles but also exhibiting catalytic activity through hammerhead ribozymes.

Structure of Hammerhead Ribozymes

Catalog of Natural Hammerhead Structures and Conserved Sequence and Structural Motifs

The hammerhead, ribozyme is a small RNA motif that self-cleaves in the presence of a divalent metal ion, generally Mg^{2+}, at a specific phophodiester bond producing 2′,3′-cyclic phosphate and 5′-hydroxyl termini.[2,4,6] Figure 1 shows the known natural hammerhead structures in panels corresponding to viroid-like satellite RNAs grouped according to their helper virus (B to D),[2,4,7–11] viroids (E),[3,12,13] one viroid-like RNA from cherry whose biological nature remains undetermined (F),[14] and three small RNAs, one from carnation (G)[15] and two of animal origin (H),[16,17] which have a homologous DNA counterpart; the consensus hammerhead structure (A) is displayed as originally proposed[6] (Fig. 1, right), and in a more recent representation derived from crystallography[18,19] (Fig. 1, left). It can be observed that these RNAs have hammerhead structures in their plus or in both polarity strands.

[2] G. A. Prody, J. T. Bakos, J. M. Buzayan, I. R. Schneider, and G. Bruening, *Science* **231,** 1577 (1986).
[3] C. J. Hutchins, P. D. Rathjen, A. C. Forster, and R. H. Symons, *Nucleic Acids Res.* **14,** 3627 (1986).
[4] A. C. Forster and R. H. Symons, *Cell* **49,** 211 (1987).
[5] G. Bruening, *Methods Enzymol.* **180,** 546 (1989).
[6] K. J. Hertel, A. Pardi, O. K. Uhlenbeck, M. Koizumi, E. Ohtsuka, S. Uesugi, R. Cedergren, F. Eckstein, W. L. Gerlach, R. Hodgson, and R. H. Symons, *Nucleic Acids Res.* **20,** 3252 (1992).
[7] J. M. Kaper, M. E. Tousignant, and G. Steger, *Biochem. Biophys. Res. Commun.* **154,** 318 (1988).
[8] L. Rubino, M. E. Tousignant, G. Steger, and J. M. Kaper, *J. Gen. Virol.* **71,** 1897 (1990).
[9] P. Keese and R. H. Symons, *in* "Viroids and Viroid-like Pathogens" (J. S. Semancik, ed.), p. 1. CRC Press, Boca Raton, FL, 1987.
[10] R. F. Collins, D. L. Gellatly, O. P. Seghal, and M. G. Abouhaidar, *Virology* **241,** 269 (1998).
[11] W. A. Miller, T. Hercus, P. M. Waterhouse, and W. L. Gerlach, *Virology* **183,** 711 (1991).
[12] C. Hernández and R. Flores, *Proc. Natl. Acad. Sci. U.S.A.* **89,** 3711 (1992).
[13] B. Navarro and R. Flores, *Proc. Natl. Acad. Sci. U.S.A.* **94,** 11262 (1997).
[14] F. Di Serio, J. A. Daròs, A. Ragozzino, and R. Flores, *J. Virol.* **71,** 6603 (1997).
[15] C. Hernández, J. A. Daròs, S. F. Elena, A. Moya, and R. Flores, *Nucleic Acids Res.* **20,** 6323 (1992).
[16] L. M. Epstein and J. G. Gall, *Cell* **48,** 535 (1987).
[17] G. Ferbeyre, J. M. Smith, and R. Cedergren, *Mol. Cell. Biol.* **18,** 3880 (1998).
[18] H. W. Pley, K. M. Flaherty, and D. B. McKay, *Nature* **372,** 68 (1994).
[19] W. G. Scott, J. T. Finch, and A. Klug, *Cell* **81,** 991 (1995).

Inspection of natural hammerhead structures shows that they are characterized by a central core with a cluster of strictly conserved nucleotide residues flanked by three double-helix regions (I, II, and III) with loose sequence conservation except positions 15.2 and 16.2, which in most cases form a C-G pair, and positions 10.1 and 11.1, which in most cases form a G-C pair (Fig. 1). X-ray crystallography of two model hammerhead structures has revealed a complex array of noncanonical interactions between the residues forming the central core, thus providing clues as to why they are strictly conserved in all natural hammerhead structures.[18,19] Prominent in this array are three non-Watson–Crick pairs (involving A9 and G12, G8 and A13, and U7 and A14 that extend helix II) and a uridine turn motif, the tetranucleotide CUGA (positions 3–6), which along with the active site residue forms the catalytic pocket.

Unusual Features of Some Hammerhead Structures

Deviations from the consensus model have been observed in some natural hammerhead structures (Fig. 1). Apart from some intrinsic flexibility these deviations are most probably the consequence that because of their smallness, the genetic information in viroid and viroid-like RNAs is very compressed and the involvement of specific nucleotide residues in additional functions other than self-cleavage can be presumed. In such cases, a compromise must be reached that may not be optimal for the individual functions.

The pair between C15.2 and G16.2 is substituted by U15.2 and A16.2, respectively, in the plus and minus hammerhead structures of sRPV and csc RNA1 (for abbreviations see Fig. 1). The common C17 residue preceding the self-cleavage site is A in both polarity hammerhead structures of csc RNA1, as well as in those corresponding to the plus strand of sRPV, the minus strand of sLTSV, and in a sequence variant of the minus strand of ASBVd. The other two possible substitutions are excluded, most likely because an U17 could base pair with the conserved A14 extending helix III, and a G17 could base pair with the conserved C3 extending helix I; as mentioned above, conserved residues of the central core are involved in a series of noncanonical interactions that are crucial for the catalytic activity. The C7 or U7 is A in the plus hammerhead structures of sLTSV and csc RNA1, and in the minus hammerhead structures of ASBVd and CarSV RNA; a G at position 7 has never been found. It has already been indicated that the base pair of helix II closest to the central core is generally G10.1 and C11.1. However, there are two exceptions to this rule. First, the plus and minus hammerhead structures of sRPV and CarSV RNA, respectively, have two extra residues, a C between A9 and G10.1, and an A between C11.1 and G12 (Fig. 1); whether or not these two extra residues form a noncanonical base pair that lengthens helix II remains to be determined. And second, between A9 and G10.1, there is an extra U in the plus hammerhead structures of sLTSV and sArMV, and an extra A in the plus hammerhead

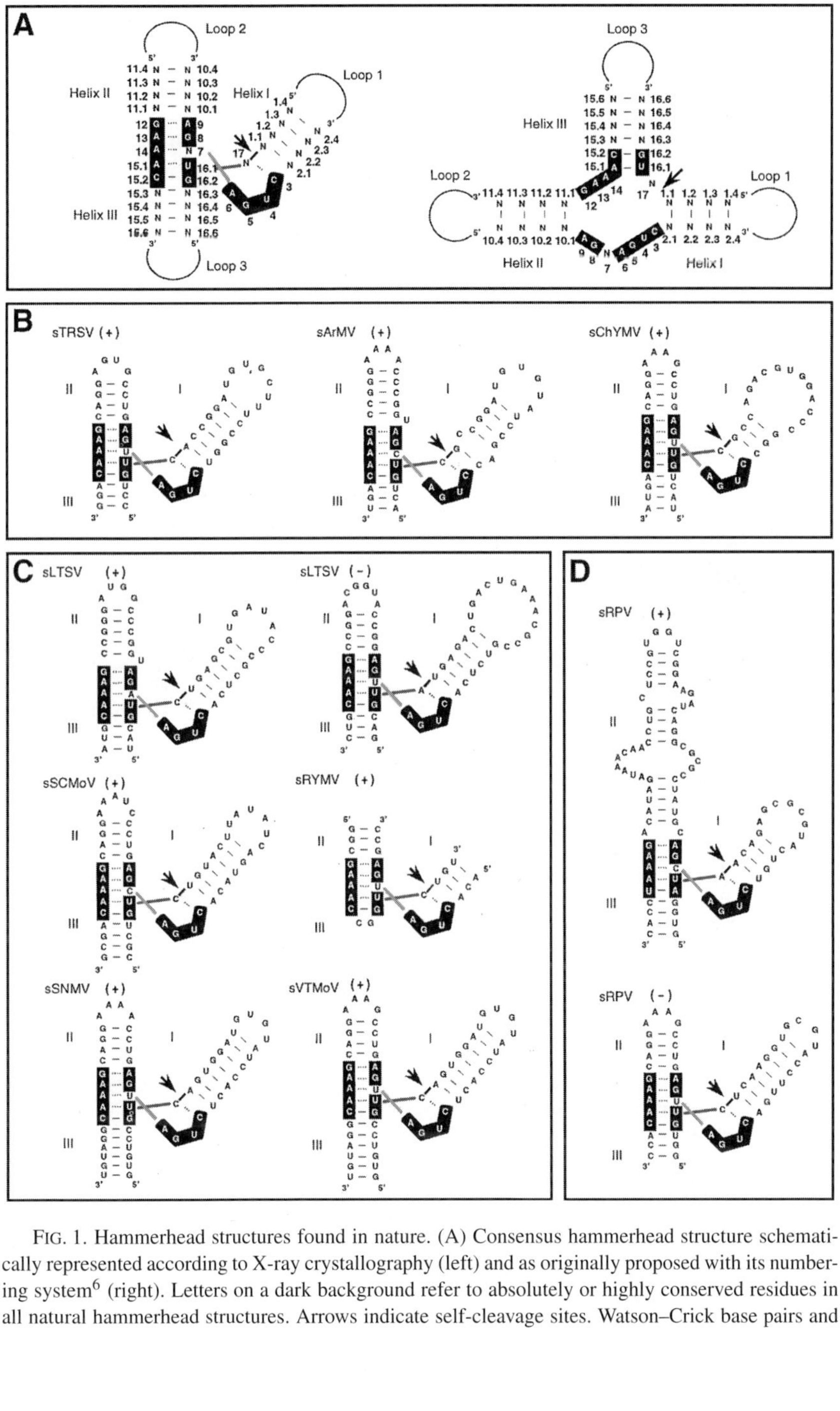

FIG. 1. Hammerhead structures found in nature. (A) Consensus hammerhead structure schematically represented according to X-ray crystallography (left) and as originally proposed with its numbering system[6] (right). Letters on a dark background refer to absolutely or highly conserved residues in all natural hammerhead structures. Arrows indicate self-cleavage sites. Watson–Crick base pairs and

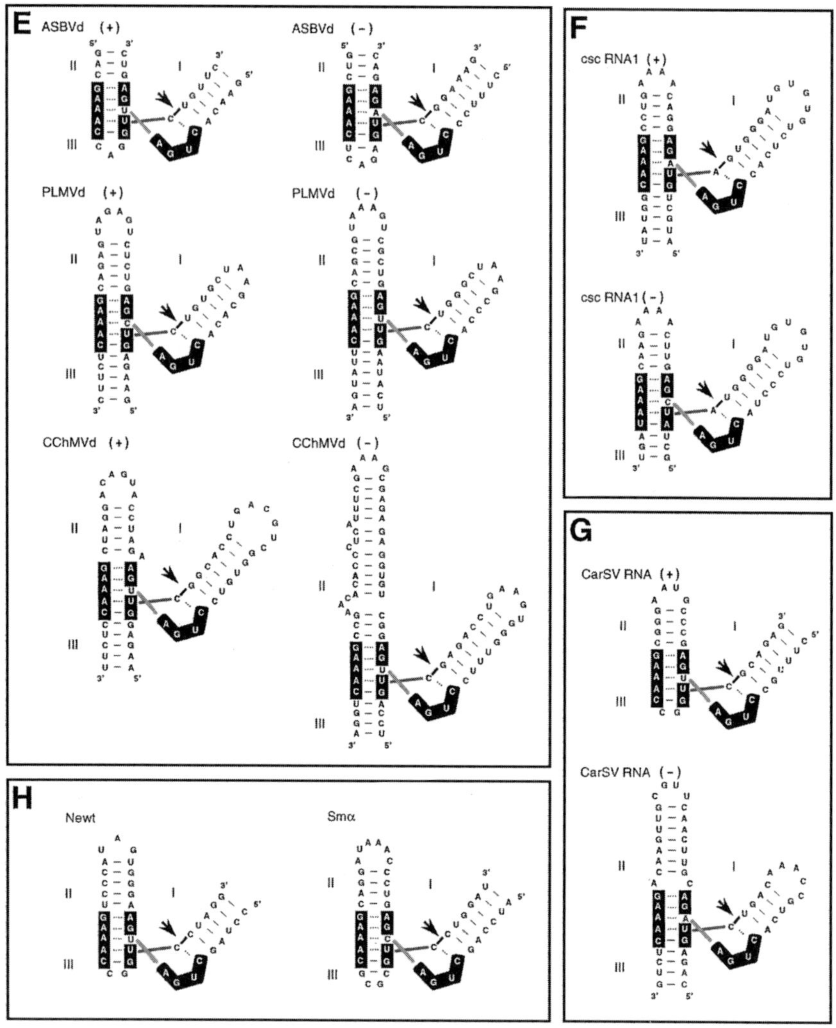

noncanonical interactions are denoted with continuous and broken lines, respectively. (B) Viroid-like satellite RNAs from nepoviruses: sTRSV (tobacco ringspot virus), sArMV (*Arabis* mosaic virus) and sChYMoV (chicory yellow mottle virus).[2,7,8] (C) Viroid-like satellite RNAs from sobemoviruses (also called virusoids): sLTSV (lucerne transient streak), sSNMV (*Solanum nodiflorum* mottle), sSCMoV (subterranean clover mottle), sVTMoV (velvet tobacco mottle), and sRYMV (rice yellow mottle).[4,9,10] (D) Viroid-like satellite RNAs from luteoviruses: sRPV (cereal yellow dwarf virus-RPV).[11] (E) Viroids: ASBVd (avocado sunblotch), PLMVd (peach latent mosaic), CChMVd (chrysanthemum chlorotic mottle).[3,12,13] (F–H) Other small RNAs with hammerhead structures: csc RNA1 (cherry small circular RNA1),[14] CarSV RNA (carnation small viroid-like RNA),[15] and Newt and Smα (transcripts of the newt satellite DNA II and of the schistosome satellite DNA).[16,17]

structure of CChMVd. The extra U or A, which are compatible with extensive *in vitro* self-cleavage, could either induce a rearrangement of the junction between helix II and the three adjacent noncanonical interactions of the central core, or be accommodated as a bulging residue.

Similarities between Hammerhead Structures of Some Catalytic RNAs

The plus and minus hammerhead structures of sLTSV are more closely related to each other than to any of the other known hammerhead structures.[4] This is a relatively frequent situation as illustrated by the cases of both polarity hammerhead structures of PLMVd,[12] CChMVd,[3] and csc RNA1,[14] with the sequence similarity being particularly noticeable in the domain formed by helix I and loop 1 (Fig. 1). These intramolecular similarities may have resulted from template switching by an RNA polymerase in the course of evolution.[4] More specifically, the sequences forming the hammerhead structures of one polarity in these RNAs could have served as the templates for synthesis of the corresponding hammerhead structure of the opposite polarity.[14]

There are also limited sequence similarities between the hammerhead structures of different catalytic RNAs. This is the situation found between the residues of helices I and II close to the central core of both PLMVd hammerhead structures, and the corresponding residues of the hammerhead structures of sSCMoV, sLTSV (plus and minus), and ASBVd (plus).[12] Also in this line, the octanucleotide GGAUGUGU forming part of helix I and loop 1, and the hexanucleotide CAAAAG forming part of helix II and loop 2 in both csc RNA1 hammerhead structures, are found in equivalent positions in the hammerhead structures of sArMV and sSNMV.[14] This can be regarded as support for a common phylogenetic origin of these RNAs, although other possibilities cannot be dismissed.

Computed-Assisted Search for Novel Hammerhead Structures in Databases

On the basis of the consensus structural and sequence features of the hammerhead (Fig. 1), a program was developed to look for domains of this kind in databases. This search not only retrieved the hammerhead structures described previously in plant viroid and viroid-like RNAs as well as in the satellite DNA II sequences of several newt species but, additionally, it uncovered a new hammerhead structure in the satellite DNA sequences of the human blood fluke *Schistosoma mansoni.*[17] PCR (polymerase chain reaction) amplification with primers flanking the repeated unit led mainly to a 335-bp fragment that was found to contain a hammerhead domain preceded by a region with RNA polymerase III promoter elements and followed by a 3′-terminal region. The hammerhead domains of some of the clones fulfill the criteria for activity and self-cleave *in cis* during *in vitro* transcription. The presence of the RNA polymerase III promoter elements suggested that the schistosome repeat was also transcribed *in vivo;*

analysis of total cellular RNA by Northern blotting, RT-PCR, and primer extension confirmed this prediction and showed that self-cleavage occurs also *in vivo*. Moreover, sequence analysis identified in schistosome DNA a potential *trans* cleavage target within a gene coding for a synaptobrevin-like protein, and the corresponding RNA transcript was efficiently cleaved *in vitro*. The discovery of this novel hammerhead structure underscores the potential of bioinformatics for mining in databases functional RNA domains once their key structural motifs have been identified.[17]

In Vitro *Selection of Hammerhead Ribozymes*

The importance of the conserved nucleotide residues forming the central core of the hammerhead structure was initially analyzed by introducing single mutations by site-directed mutagenesis.[20] With the advent of *in vitro* selection methods it has been possible to apply this new tool to isolate (i) hammerhead ribozymes with increased AUA cleavage activity from an RNA pool randomized at three positions 7, 10.1 and 11.1,[21] (ii) *trans*-acting hammerhead ribozymes from an RNA pool in which the helix II–loop 2 sequence and most of the central core were randomized,[22] and (iii) *cis*-acting hammerhead ribozymes from an RNA pool in which positions 10.1 and 11.1 and most of the central core were randomized.[23] Interestingly, the active ribozymes recovered in all cases predominantly contain the central core of the natural consensus hammerhead structure, thus showing that it offers the optimal catalytic efficiency for this motif.

The hammerhead ribozyme has been optimized by natural selection for cleavage after GUC (Fig. 1), and it is assumed that the inability to cleave after a G17 results from forming a pair with C3 that impairs the participation of these two residues in alternative interactions required for catalysis. *In vitro* selection has been used to investigate the nature of hammerhead-like domains capable of cleaving after AUG. Such catalytic domains do exist in the sequence space, and one of the ribozymes selected cleaves *in cis* and *trans* after this triplet with rates comparable to that of the GUC-cleaving hammerhead.[24] Nuclease probing indicates that the selected ribozyme has an overall secondary structure similar to that of the consensus hammerhead ribozyme, although there are differences that include substitutions and two deletions (positions 3 and 9) in the central core, and an altered helix II–loop 2 sequence. These results expand the application of hammerhead ribozymes for specific RNA cleavage and underline the power of *in vitro* selection.

[20] D. E. Ruffner, G. D. Stormo, and O. C. Uhlenbeck, *Biochemistry* **29,** 10695 (1990).

[21] K. L. Nakamaye and F. Eckstein, *Biochemistry* **33,** 1271 (1994).

[22] M. Ishizaka, Y. Ohshima, and T. Tani, *Biochem. Biophys. Res. Commun.* **214,** 403 (1995).

[23] J. Tang and R. R. Breaker, *RNA* **3,** 914 (1997).

[24] N. K. Vaish, P. A. Heaton, O. Fedorova, and F. Eckstein, *Proc. Natl. Acad. Sci. U.S.A.* **95,** 2158 (1998).

Function of Hammerhead Ribozymes in Their Natural Habitat

Rolling Circle Mechanism for Replication of Viroid and Viroid-like Satellite RNAs

The available data indicate that replication of these small RNAs occurs through a rolling-circle mechanism with exclusively RNA intermediates.[25,26] Two variants of the mechanism, symmetric and asymmetric, have been proposed, their main difference being the existence and absence, respectively, of the circular monomeric minus RNA acting as a replicative intermediate. Detection of this RNA species in tissue infected by ASBVd[26–28] and by PLMVd[29] is taken as evidence that these viroids, and most likely CChMVd and the viroid-like satellite RNAs with ribozymes in both polarity strands, follow the symmetric pathway.[30,31] In contrast, nonhammerhead viroids and other viroid-like satellite RNAs with ribozymes only in one polarity strand seem to replicate via the asymmetric alternative.[32,33] In this asymmetric variant of the rolling circle mechanism, the infecting monomeric plus circular RNA is transcribed into linear multimeric minus strands that are directly used as a template for the generation of the linear multimeric plus strands. In the symmetric variant, the linear multimeric minus strands are processed and ligated to the minus circular monomer that in the second half of the cycle, symmetric to the first half, serves as the template for synthesis of the multimeric plus linear RNAs. In both cases, the plus multimeric forms are cleaved to unit-length molecules that are then ligated to yield the circular progeny. The rolling circle mechanism, therefore, requires a highly specific processing activity to excise precisely the monomeric forms from the oligomeric intermediates. Such specificity is achieved in some viroid and viroid-like satellite RNAs by ribozymes embedded in one or in both polarity strands, thus circumventing the need to rely on host factors for this critical replication step. The ribozymes are of the hammerhead type, except in the case of the minus polarity strands of the viroid-like satellite RNAs from nepoviruses, which self-cleave through hairpin ribozymes.[30]

[25] A. D. Branch and H. D. Robertson, *Science* **223,** 450 (1984).

[26] C. J. Hutchins, P. Keese, J. E. Visvader, P. D. Rathjen, J. L. McInnes, and R. H. Symons, *Plant Mol. Biol.* **4,** 293 (1985).

[27] J. A. Daròs, J. F. Marcos, C. Hernández, and R. Flores, *Proc. Natl. Acad. Sci. U.S.A.* **91,** 12813 (1994).

[28] J. A. Navarro, J. A. Daròs, and R. Flores, *Virology* **253,** 77 (1999).

[29] F. Bussière, J. Lehoux, D. A. Thompson, L. J. Skrzeczkowski, and J. P. Perrault, *J. Virol.* **73,** 6353 (1999).

[30] G. Bruening, B. P. Passmore, H. Van Tol, J. Buzayan, and P. E. Feldstein, *Mol. Plant–Microbe Interact.* **4,** 219 (1991).

[31] R. H. Symons, *Nucleic Acids Res.* **25,** 2683 (1997).

[32] A. D. Branch, B. J. Benenfeld, and H. D. Robertson, *Proc. Natl. Acad. Sci. U.S.A.* **85,** 9128 (1988).

[33] P. A. Feldstein, Y. Hu, and R. A. Owens, *Proc. Natl. Acad. Sci. U.S.A.* **95,** 6560 (1998).

In Vivo *Functional Significance of Hammerhead Ribozymes*

Distinct lines of evidence support the involvement of hammerhead structures in the *in vivo* processing of oligomeric viroid and viroid-like RNAs containing these catalytic domains. For sTRSV,[2] sRPV,[11] ASBVd,[27,34] CChMVd,[13] PLMVd (Hernández, unpublished data), and CarSV RNA (Daròs, unpublished data), linear monomeric RNAs of one or both polarities with 5′ termini identical to those produced in the corresponding *in vitro* self-cleavage reactions have been isolated from infected tissue. Moreover, compensatory mutations or covariations that preserve the stability of the hammerhead structures are frequently found in sequence variants of PLMVd[12,35] and CChMVd,[13,36] and the observed *in vivo* reversion of mutations introduced *in vitro* to eliminate self-cleavage of the hammerhead structure of a viroid-like satellite RNA[37] further supports the *in vivo* role of these ribozymes, as does the correlation existing between the infectivity of different PLMVd and CChMVd variants and the extent of their self-cleavage during *in vitro* transcription.[35,36] On the other hand, the presence of a 2′-phosphomonoester, 3′,5′-phospho diester bond at a unique position of the plus circular sVTMoV and sSNMV RNAs has been interpreted as the signature of an RNA ligase at the ligation site.[38] Since such a position is coincidental with the self-cleavage sites predicted by the hammerhead structures contained in these RNAs, this can be considered as more indirect proof in favor of their *in vivo* significance. Data suggesting the existence of an extra 2′-phospho monoester at the nucleotide preceding the predicted self-cleavage/ligation site have also been obtained for other viroids[35,36] and viroid-like satellite RNAs.[11,37]

With the exception of PLMVd, for which certain degree of self-ligation of the linear monomers resulting from self-cleavage of the dimeric transcripts has been reported,[39] no significant reversibility of the cleavage reaction mediated by hammerhead structures has been detected.[2] Most likely this is because the *in vitro* folding of PLMVd linear molecules into a conformation in which the close proximity of the 5′ and 3′ termini favors their spontaneous circularization. However, since the phosphodiester bonds produced are mostly 2′,5′ instead of the 3′,5′ usually found in RNA,[39] the *in vivo* significance of this reaction is unclear. Intriguingly, the only report that disagrees with the *in vivo* functioning of a hammerhead structure concerns a nonpathogenic RNA, the newt transcript, for which the *in vitro* self-cleavage site has been mapped 46–47 residues downstream of the *in vivo* cleavage site in ovarian tissue.[16] Because of the very distinct biological nature of

[34] J. A. Navarro and R. Flores, *EMBO J.* **19,** 2662 (2000).

[35] S. Ambrós, C. Hernàndez, J. C. Desvignes, and R. Flores, *J. Virol.* **72,** 7397 (1998).

[36] M. de la Peña, B. Navarro, and R. Flores, *Proc. Natl. Acad. Sci. U.S.A.* **96,** 9960 (1999).

[37] C. C. Sheldon and R. H. Symons, *Virology* **194,** 463 (1993).

[38] P. A. Kibertis, J. Haseloff, and D. Zimmern, *EMBO J.* **4,** 817 (1985).

[39] F. Côte and J. P. Perrault, *J. Mol. Biol.* **273,** 533 (1997).

this molecule with respect to most hammerhead RNAs, this discrepancy in their processing mechanisms should not be surprising.

Regulation of Hammerhead Ribozymes in Vivo

The activity of these catalytic domains must be finely tuned during replication of viroid and viroid-like RNAs in which they are embedded; hammerhead ribozymes must catalyze processing of oligomeric RNAs to monomers but at the same time their activity should be regulated to preserve a certain level of monomeric circular RNAs to act as templates in the rolling circle replication. Two different strategies have been proposed to achieve this. Some of the hammerhead structures, such as those found in both ASBVd RNAs,[3] in the plus sRYMV and CarSV RNAs,[10,15] and in the newt and schsistosome transcripts,[16,17] are thermodynamically unstable, having a stem III of only two or three base pairs closed by a small loop of two or three residues (Fig. 1). In line with this instability, *in vitro* self-cleavage of these monomeric RNAs is very inefficient. However, in the corresponding dimeric or multimeric replicative intermediates, the sequences of two single-hammerhead structures can form a stable double-hammerhead structure with an extended helix III that promotes efficient self-cleavage[40]. A different situation within this same scheme has been observed in the plus sRPV RNA in which adoption of an active single-hammerhead structure in the monomeric form is prevented by a pseudoknot between residues in loop 1 and in a G+C-rich bulge of helix II; in a multimeric context, a double-hammerhead structure that lacks the pseudoknot can be formed and mediate efficient self-cleavage.[41]

A distinct strategy has been proposed for sLTSV, PLMVd, CChMVd, and csc RNA1, in which stable single-hammerhead structures can be potentially adopted by the monomeric plus and minus strands. However, the formation of these hammerhead structures is impeded in their predicted conformations of lowest free energy because the conserved sequences of both polarity hammerhead structures, because of their extensive complementarity, are involved in an alternative folding that does not promote self-cleavage of the monomeric RNAs. The catalytically active hammerhead structures are probably only formed transiently during transcription, promoting self-cleavage of the multimeric RNAs.[4,12] Therefore, there seems to be an interplay between two conformations, one with the hammerhead structure and promoting self-cleavage, and another favoring circularization, mediated probably by a host RNA ligase.

[40] A. C. Forster, C. Davies, C. C. Sheldon, A. C. Jeffries, and R. H. Symons, *Nature* **334,** 265 (1988).

[41] S. I. Song, S. L. Silver, M. A. Aulik, L. Rasochova, B. R. Mohan, and W. A. Miller, *J. Mol. Biol.* **293,** 781 (1999).

In Vitro Assays to Test Hammerhead-Mediated RNA Cleavage

Kinetic Pathway for Hammerhead Ribozyme

A major effort has been made over the past few years to understand the mechanism that governs the hammerhead cleavage reaction, which can occur *in cis* and *in trans*. In their natural context, hammerhead ribozymes act *in cis* and, therefore, they catalyze a single turnover intramolecular cleavage. Initial studies were performed in the *cis* format, and the extent of RNA self-cleavage under standard conditions (see below) was found to be extremely rapid (completed in less than 1 min) but no further kinetic analysis was done.[4,42] To gain a deeper insight, the hammerhead was divided into two separate fragments, the ribozyme itself that remains unchanged after the reaction, and the substrate that is cleaved *in trans* when combined with the ribozyme in the presence of Mg^{2+}. This intermolecular format, in which the ribozyme may proceed through multiple rounds of substrate binding, cleavage, and product release, has permitted a detailed kinetic dissection.[43,44] A minimal kinetic pathway has been proposed for the intermolecular reaction:

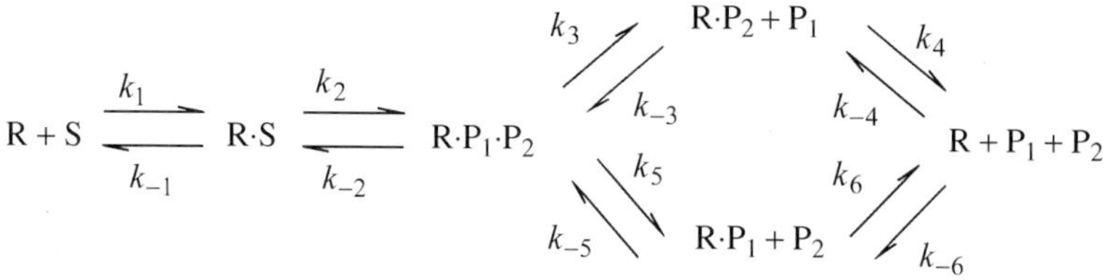

in which the five main species are the ribozyme (R), substrate (S), ribozyme–substrate complex (R · S), ribozyme–products complex ($R \cdot P_1 \cdot P_2$), and products (P_i).[45] The reaction starts with the binding between ribozyme and substrate to form the R · S complex, a step governed by the rate constant k_1. At this point, if Mg^{2+} or other divalent metal ion is present, cleavage occurs, giving the $R \cdot P_1 \cdot P_2$ complex with a rate constant k_2. These steps are reversible and elemental dissociation rate constants for the complexes (k_{-1}, k_{-3}, k_{-4}, k_{-5} and k_{-6}), and even a rate constant for ligation (k_{-2}), can be defined.

Methods for Testing RNA Cleavage Using a Cis *Format*

The hammerhead RNA is obtained by *in vitro* transcription of a linearized recombinant plasmid (0.1 μg/μl) containing, under the control of a phage RNA polymerase promoter, either a minimal hammerhead motif[42] or a larger sequence in which the hammerhead motif is embedded.[46] Alternatively, the template for

[42] C. C. Sheldon and R. H. Symons, *Nucleic Acids Res.* **17,** 5679 (1989).
[43] O. C. Uhlenbeck, *Nature* **328,** 596 (1987).
[44] T. K. Stage-Zimmermann and O. C. Uhlenbeck, *RNA* **4,** 875 (1998).
[45] M. J. Fedor and O. C. Uhlenbeck, *Biochemistry* **31,** 12042 (1992).

hammerhead RNA synthesis can also be prepared by hybridization of a pair of oligodeoxyribonucleotides (5 ng/μl), one containing the transcription promoter followed by the hammerhead motif, and the other complementary to the promoter sequence in order to obtain a partially duplex DNA.[47] In both cases, *in vitro* transcription reactions containing 1 U/μl T7 RNA polymerase, 40 m*M* Tris-HCl (pH 7.5), 6 m*M* Mg^{2+}, 0.1 μg/μl bovine serum albumin (BSA), 10 m*M* dithiothreitol (DTT), 0.5 m*M* each of ATP, CTP, and GTP, 0.025 m*M* UTP, and 1 μCi/μl [α-^{32}P]UTP are incubated at 37° for 1 h. The extent of self-cleavage during transcription (fraction of cleaved versus total RNA) is determined by electrophoresis of the transcription products on 1× TBE polyacrylamide gels (5–15% w/v) containing 7 *M* urea and, occasionally, 40% formamide to improve the separation of RNAs with a high secondary structure content. A method to obtain intramolecular cleavage rates during *in vitro* transcription has been devised.[48] The fraction of transcribed RNA that remains uncleaved (F) is independent of the transcription rate as long as the latter stays constant. Therefore, F can be plotted versus time (t) and the experimental data fitted to the equation

$$F = (1 - e^{-kt})/(kt)$$

in which k is the rate constant for intramolecular cleavage. Values obtained for this constant (around 1 min^{-1}) agree well with those derived from assays *in trans,* indicating that no fundamental kinetic distinction exists between the intra- and the intermolecular reactions.[48]

To study the self-cleavage reaction in a protein-free environment, the uncleaved transcript is eluted in the absence of divalent metal ions and then is incubated under standard self-cleavage conditions: 50 m*M* Tris-HCl (pH 8), 5 m*M* $MgCl_2$, and 0.5 m*M* EDTA at 25° or 40° for 1 h.[46]

Methods for Testing RNA Cleavage Using a Trans *Format*

RNAs are obtained by *in vitro* transcription (see above) or chemically synthesized. The substrate is 5′-end-labeled with T4 polynucleotide kinase and [γ-^{32}P]ATP[46]. The reaction can be studied under two alternative conditions: single turnover in which the ribozyme is in excess (μ*M* to n*M*) over the substrate (n*M* to p*M*) and a single cleavage event occurs, and multiple turnover in which consecutive rounds of substrate binding (in excess over the ribozyme), cleavage, and product release take place. The ribozyme and substrate are heated together at 95° for 2 min in 50 m*M* Tris-HCl (pH 7.5) without Mg^{2+} and allowed to cool to 25°. A zero time point is taken and $MgCl_2$ (10 m*M* final concentration) is then

[46] A. C. Forster, C. Davies, C. J. Hutchins, and R. H. Symons, *Methods Enzymol.* **181,** 583 (1990).
[47] J. P. Milligan and O. C. Uhlenbeck, *Methods Enzymol.* **180,** 51 (1989).
[48] D. M. Long and O. C. Uhlenbeck, *Proc. Natl. Acad. Sci. U.S.A.* **91,** 6977 (1994).

added to initiate the cleavage reaction. Aliquots are removed at different times and quenched immediately with a 5-fold excess of stop solution (9 *M* urea in 50 m*M* EDTA), and the substrate and products from each aliquot are separated on denaturing polyacrylamide gels (15–20%) and quantitated by an image analyzer. Under single-turnover conditions data can be fitted to the equation

$$F_t = F_\infty(1 - e^{-kt})$$

where F_t and F_∞ are the fractions of product at time t and at the end point, respectively, and k the rate constant of cleavage. For a hammerhead behaving ideally, every substrate molecule is bound to a ribozyme molecule, and the observed rate of cleavage is determined by the balance between the forward (k_2) and the reverse rate constants (k_{-2}). For the hammerhead ribozyme $k_2 >>> k_{-2}$ and, therefore, the observed rate of cleavage corresponds to k_2. For the multiple turnover reaction, steady-state rates are measured over a range of substrate concentrations, and the rate of cleavage, normalized to the ribozyme concentration, is plotted versus substrate concentration. The resulting data are fitted to the Eadie–Hofstee equation to obtain the rate constant of cleavage and the Michaelis constant.[44]

Conclusions and Perspectives

Discovery of ribozymes has produced a scientific revolution with deep functional, biotechnological, and even evolutionary implications. We know now that processing of some cellular and some invading RNAs is mediated by ribozymes, which likely represents a relic of the RNA world presumed to have existed on Earth before the advent of DNA and proteins. Most of the ribozymes have been found in viroid and viroid-like satellite RNAs from plants as well as in other small RNAs of animal and fungal origin, suggesting that further research on these RNAs may lead to discovery of new ribozymes. All ribozymes found in nature behave as RNases (and some, additionally, as RNA ligases). Certain ribozymes, including the hammerhead, have been manipulated to act *in trans* against specific target RNAs. This has opened the possibility of engineering "restriction RNases," a tool with a great potential impact in medicine and industry. The hammerhead ribozyme, because of its extreme structural simplicity, has served as a model system for extensive structural and functional analysis, and it is expected to continue playing a capital role in future developments of ribozymology.

Acknowledgments

Work in the laboratory of R.F. has been partially supported by Grants PB95-0139 and PB98-0500 from the Comisión Interministerial de Ciencia y Tecnología de España. C.H., J.-A.D., and A.V. were recipients of postdoctoral contracts and M. de la P. of a predoctoral fellowship from the Ministerio de Educación y Cultura de España.

[35] Kinetic Analysis of Bimolecular Hepatitis *delta* Ribozyme

By SIRINART ANANVORANICH, KARINE FIOLA, JONATHAN OUELLET, PATRICK DESCHÊNES, and JEAN-PIERRE PERREAULT

Hepatitis *delta* ribozymes are derived from the self-cleaving motifs found in the genome of defective hepatitis *delta* virus (HDV). In the presence of hepatitis B virus, infectious HDV particles are assembled as a ribonucleoprotein complex that includes multiple copies of *delta* antigen, hepatitis B surface antigen, and the circular 1.7-kb RNA single-stranded HDV genome. The circular HDV genome is highly self-complementary with a secondary structure that can be depicted as a rod shape composed of two domains, referred to as the viroid-like and protein encoding domains (Fig. 1A). During viral replication multimeric strands of both the genomic and antigenomic polarities are generated. Self-cleaving motifs located in the viroid-like domain on both the genomic and antigenomic strands subsequently cleave these multimeric strands, releasing unit-length monomers of both polarities. The resulting monomers are then ligated to produce circular conformers.

Following molecular cloning of the HDV genome,[1] the minimum contiguous sequence and secondary structure of this catalytic RNA were thoroughly characterized using a series of nested deletion mutants and nuclease mapping assays.[2] Genomic and antigenomic self-cleaving motifs exhibit similar secondary structures and catalytic activities[3] (Fig. 1B). Several different secondary structures have been predicted[4,5]; however, the pseudoknot structure proposed by Perrotta and Been[3] is the most commonly used. According to this secondary structure, four stems, named P1, P2, P3, and P4, are formed (Fig. 1B). The stems are joined by single-stranded regions called junctions named J1/2, J1/4, and J4/2, respectively. Crystallographic data have shown *cis*-acting genomic *delta* ribozyme to possess an additional stem, referred to as the P1.1 stem, generated by base pair interactions between the L3 loop and the J1/4 helix.[6] The cleavage of a phosphodiester bond located at the 5′-end of stem P1 gives rise to reaction products containing a 5′-hydroxyl and a 2′,3′-cyclic phosphate terminus. This reaction requires divalent

[1] S. Makino, M. F. Chang, C. K. Shieh, T. Kamahora, D. M. Vannier, V. Govindarajan, and M. M. C. Lai, *Nature* **329,** 343 (1987).

[2] H. N. Wu, Y. J. Lin, F. P. Lin, S. Makino, M. F. Chang, and M. M. C. Lai, *Proc. Natl. Acad. Sci. USA* **86,** 1831 (1989).

[3] A. T. Perrotta and M. D. Been, *Nature* **350,** 434 (1991).

[4] H. N. Wu, Y. J. Wang, C. F. Hung, H. J. Lee, and M. M. Lai, *J. Mol. Biol.* **223,** 233 (1992).

[5] A. D. Branch and H. D. Robertson, *Proc. Natl. Acad. Sci. USA* **88,** 10163 (1991).

[6] A. R. Ferre-D'Amare, K. Zhou, and J. A. Doudna, *Nature* **395,** 567 (1998).

0076-6879/01 $35.00

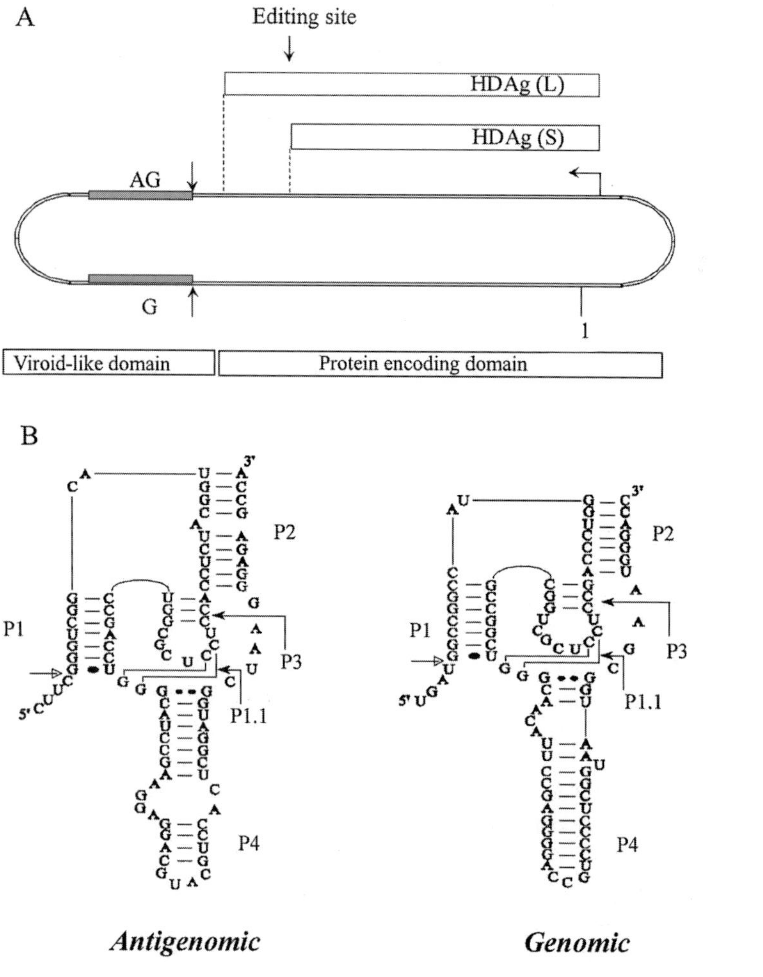

FIG. 1. (A) Schematic representation of the hepatitis *delta* virus (HDV) genome. Shaded boxes indicate the HDV self-cleaving motifs on genomic (G) and antigenomic (AG) strand. Clear boxes indicate the open reading frame of *delta* antigens (HDAg). The antigens are produced in two forms, either small (S) or large (L), as a consequence of an editing event occurring at the site indicated by an arrow. (B) Identification of the self-cleaving motifs on both the antigenomic and genomic single-stranded RNAs.

metal such as Mg^{2+}, Mn^{2+}, and Ca^{2+} and is referred to as *cis*-cleavage, in which the motif acts as a pseudoenzyme.

Based on the pseudoknot structure, a true enzyme, or *trans*-acting *delta* ribozyme, can be generated by eliminating a single-stranded junction so that one molecule acts as a substrate, while the other resulting molecule contains both the substrate binding site and catalytic domain (Fig. 2A). In our previous study we

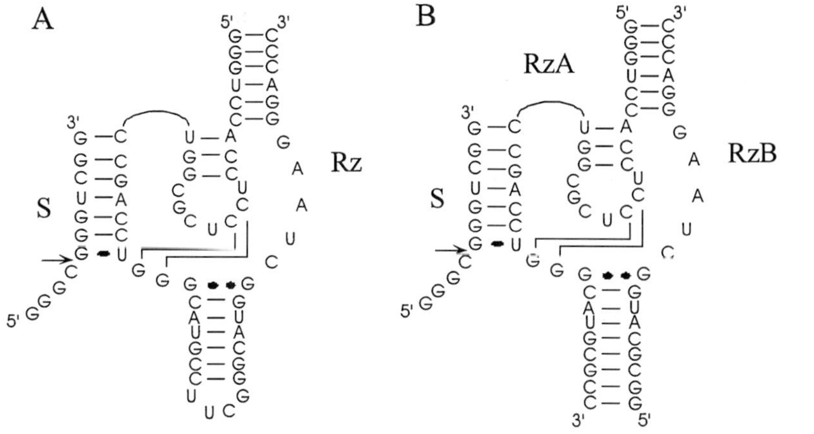

FIG. 2. (A) *Trans*-acting *delta* ribozyme derived from the antigenomic self-cleaving motif. (B) Bimolecular *delta* ribozyme system.

used a version derived from the antigenomic self-cleaving motif, with its J1/2 junction removed. In addition, stems P2 and P4 were shortened so that only minimum sequences remained, and the 5′-end sequence of the ribozyme began with GGG to ensure an efficient RNA production in *in vitro* transcription reactions. This *trans*-acting ribozyme derived from the *delta* ribozyme exhibits a unique characteristic in its substrate binding, namely that only the 3′-portion of the substrate is required for binding to the ribozyme. Specifically, a short stretch of nucleotides (7 nt) located on the substrate, which can be referred to as the ribozyme recognition site, is all that is required for binding. We view this characteristic as an advantage for the future development of a therapeutic means of controlling, for example, a viral infection. Moreover, when we thoroughly characterized the effect of neighboring sequences located on both ends of the recognition site, we found that the substrate specificity of *delta* ribozyme can be altered to accommodate the cleavage of a long mRNA molecule,[7] and can also be enhanced.[8,9]

This article was designed so that several techniques used in enzymatic studies on hepatitis *delta* ribozymes can be presented. In order not to repeat previous studies on *trans*-acting antigenomic *delta* ribozyme,[10,11] a bimolecular *delta* ribozyme is used in this study to illustrate techniques that include RNA production, ribozyme assembly, and both cleavage and kinetic studies. This bimolecular *delta*

[7] G. Roy, S. Ananvoranich, and J. P. Perreault, *Nucleic Acids Res.* **27,** 942 (1999).

[8] S. Ananvoranich, D. A. Lafontaine, and J. P. Perreault, *Nucleic Acids Res.* **27,** 1473 (1999).

[9] P. Deschênes, D. A. Lafontaine, S. Charland, and J. P. Perreault, *Antisense Nucleic Acid Drug Dev.* **10,** 53 (2000).

[10] S. Ananvoranich and J. P. Perreault, *J. Biol. Chem.* **273,** 13182 (1998).

[11] S. Mercure, D. A. Lafontaine, S. Ananvoranich, and J. P. Perreault, *Biochemistry* **37,** 16975 (1998).

ribozyme (Fig. 2B) is derived from the HDV antigenomic self-cleaving motif and can reconstitute its catalytic activity. Although an analogous system has previously been reported by Sakamoto *et al.*,[12] it was a mixture of both antigenomic and genomic sequences and was used for different purposes. Our findings on the bimolecular *delta* ribozyme derived from the antigenomic sequence are therefore worth presenting and comparing to the former system. Ultimately, we are striving to reach two goals: (i) to describe the methodology used; and (ii) to provide insight into some of the characteristics of the bimolecular *delta* ribozyme system.

Design of Bimolecular *delta* Ribozyme System

Based on the *trans*-acting version of *delta* ribozyme, a bimolecular version has been designed so that the ribozyme can be assembled using two molecules, referred to as RzA and RzB (Fig. 2B). RzA can act as a substrate binding moiety so that the antisense effect can be explored; RzB is required to reconstitute the enzymatic property of the ribozyme complex. This system might be worthwhile for further investigation of its therapeutic potential regarding to both antisense and ribozyme effects. In addition, because of their relatively small size, the transcripts can either be chemically synthesized at an affordable cost, or enzymatically produced using T7 RNA polymerase in an *in vitro* transcription reaction. For *in vitro* transcription reactions, individual pairs of deoxyoligonucleotides for RzA, RzB, or substrate are designed so that the sequence immediately downstream of the T7 RNA promoter is either CCC or CCG in order to produce the best yield as reported by Milligan *et al.*[13] As a result, the transcripts of RzA, RzB, and substrate contain either GGG or GGC at their 5′-ends.

Materials and Methods

Deoxyoligonucleotides

The oligonucleotides Oligo-T7 promoter, 5′-GTA ATA CGA CTC ACT ATA GGG-3′; Oligo-substrate, 5′-CCG ACC CGC CCT ATA GTG AGT CGT ATT AC-3′; Oligo-RzA,5′-GGC GCA TGC CCA GGT CGG ACC GCG AGG AGG TGG ACC CTA TAG TGA GTC GTA TTA C-3;′ Oligo-T7RzB, GTA ATA CGA CTC ACT ATA GGC-3′; Oligo-RzB, GGG TCC CTT AGC CAT GCG CCT ATA GTG AGT CGT ATT AC-3′ are synthesized, deprotected, and purified commercially (Gibco-BRL, Gaithersburg, MD). The oligonucleotide pair, Oligo-T7 promoter and Oligo-substrate, Oligo-T7 promoter and Oligo-RzA, and Oligo-T7RzB

[12] T. Sakamoto, Y. Tanaka, T. Kuwabara, M. H. Kim, Y. Kurihara, M. Katahira, and S. Uesugi, *J. Biochem.* **121,** 1123 (1997).

[13] J. F. Milligan, D. R. Groebe, G. W. Witherell, and O. C. Uhlenbeck, *Nucleic Acids Res.* **15,** 8783 (1987).

and Oligo-RzB are used to produce the 11-nt substrate, the 37-nt RzA and the 20-nt RzB, respectively.

Enzymatic RNA Synthesis

Substrate is prepared by denaturing the deoxyoligonucleotides (Oligo-substrate and Oligo-T7 promoter, 500 pmol) at 95° for 5 min in a 20 μl mixture containing 10 m*M* Tris-HCl pH 7.5, 10 m*M* $MgCl_2$, 50 m*M* KCl, and then slowly cooling to 37°. *In vitro* transcription reactions are carried out using the resulting partial duplex as the template in a solution containing 27 units RNAguard RNase inhibitor (Amersham Pharmacia, Piscataway, NJ), 4 m*M* of each rNTP (Amersham Pharmacia), 80 m*M* HEPES–KOH pH 7.5, 24 m*M* $MgCl_2$, 2 m*M* spermidine, 40 m*M* dithiothreitol (DTT), 0.01 unit pyrophosphatase (Roche Boehringer Mannheim), and 25 μg purified T7 RNA polymerase[14] in a final volume of 50 μl at 37° for 4 h. After incubation, the reaction mixtures are fractionated by denaturing 20% polyacrylamide gel electrophoresis (PAGE, 19 : 1 ratio of acrylamide to bisacrylamide) containing 45 m*M* Tris–borate pH 7.5, 7 *M* urea, and 1 m*M* EDTA. RzA and RzB are synthesized under conditions similar to those described above for substrate synthesis. Following electrophoresis, the reaction products are visualized by UV shadowing, the bands corresponding to the correct sizes of substrate, RzA or RzB cut out, and the transcripts eluted overnight at 4° in a solution containing 0.1% sodium dodecyl sulfate (SDS) and 0.5 *M* ammonium acetate, pH 7.0 (w/v). The eluted transcripts are precipitated by the addition of 0.1 volume of 3 *M* sodium acetate, pH 5.2 and 2.2 volumes of ethanol. Transcript yields are determined by UV spectroscopy at 260 nm.

Chemical RNA Synthesis

RzB transcripts containing individual deoxyribose residues in place of the ribose ones at positions 9 to 14, which span a single-stranded portion (J4/2) in the ribozyme complex, are chemically synthesized (Fig. 5, inset). The synthesis has been carried out by the Keck Oligonucleotide Synthesis Facility (Yale University, New Haven, CT) using β-cyanoethyl chemistry. The resulting RNA/DNA mixed oligonucleotides are deprotected as described by Perreault and Altman,[15] and purified on a PAGE gel.

End-Labeling of RNA with [γ-^{32}P]ATP

Purified transcripts (10 pmol) are dephosphorylated in a 20 μl reaction mixture containing 200 m*M* Tris-HCl pH 8.0, 10 units RNA Guard, and 0.2 units calf

[14] P. Davanloo, A. H. Rosenberg, J. J. Dunn, and F. W. Studier, *Proc. Natl. Acad. Sci. U.S.A.* **81,** 2035 (1984).

[15] J. P. Perreault and S. Altman, *J. Mol. Biol.* **226,** 399 (1992).

intestine alkaline phosphatase (Amersham Pharmacia). The mixture is incubated at 37° for 30 min, and then extracted twice with an equal volume of phenol : chloroform (1 : 1). Dephosphorylated transcripts (1 pmol) are 5′-end-labeled in a mixture containing 1.6 pmol [γ-^{32}P]ATP (6000 Ci/mmol), 10 m*M* Tris-HCl pH 7.5, 10 m*M* $MgCl_2$, 50 m*M* KCl, and 3 units T4 polynucleotide kinase (Amersham Pharmacia) at 37° for 30 min. Excess [γ-^{32}P]ATP is removed by spin chromatography through a Sephadex G-50 gel matrix (Amersham Pharmacia). The concentration of labeled transcripts is adjusted to 0.01 pmol/μl by the addition of water.

Nuclease Digestion and Alkaline Hydrolysis

The lengths and partial sequences of the RNAs are verified using specific cleavage by nucleases that digest ribose–phosphate backbones. Trace amounts of 5′-end-labeled RNA (<1 n*M*, *ca* 50,000 cpm) are dissolved in 4 μl of buffer containing 50 m*M* Tris-HCl, pH 7.5, 10 m*M* $MgCl_2$, and 100 m*M* NH_4Cl, and are then incubated at 37° in the presence of the ribonucleases. Ribonucleases T1 (Amersham Pharmacia), which digests Gp↓N in the single-stranded RNA, and U2 (Amersham Pharmacia), which preferably digests Ap↓N in the single-stranded RNA, are routinely used at final concentrations of 1 unit/μl. The reaction mixtures are incubated for 1 to 5 min at 37°. The cleavage reaction mixtures are fractionated on denaturing PAGE gels along with the corresponding RNA ladder. The ladders are generated by mixing, in a final volume of 5 μl, 5′-end-labeled RNA (50,000 cpm) in a solution containing 50 m*M* $NaHCO_3$ and 5 m*M* EDTA and incubating at 95° for 5 min. The reaction is quenched by the addition of 5 μl of loading buffer [95% formamide, 0.05% bromphenol blue, and 0.05% xylene cyanol (v/v)]. The reaction products and the ladders on the electrophoresed gels are visualized following the exposure of the gels to a phosphorimaging screen. The screens are then scanned and analyzed using ImageQuant version 5.0 (Molecular Dynamics, Sunnyvale, CA).

Determination of the Equilibrium Dissociation Constants (K_d)

In order to evaluate the assembly of the ribozyme complex, we have determined K_d values of both RzA : substrate and RzA : RzB complexes. The RzA : substrate equilibrium dissociation constant is determined by mixing individual RzA concentrations ranging from 0.1 to 300 n*M* with trace amounts of end-labeled substrate (<1 n*M*) in a 9 μl solution containing 50 m*M* Tris-HCl, pH 7.5. The mixtures are heated at 95° for 2 min, and then cooled to 37° for 5 min prior to the addition of $MgCl_2$ to 10 m*M* final, in a manner analogous to that of a regular cleavage reaction. The samples are incubated at 37° for 1.5 h, at which point 2 μl of sample loading solution [50% (v/v) glycerol, 0.025% of each bromphenol blue, and xylene cyanol] is added, and the resulting mixtures electrophoresed through a nondenaturing polyacrylamide gel [12% acrylamide with a 29 : 1 ratio of acrylamide to bisacrylamide, 45 m*M* Tris–borate buffer, pH 7.5, containing 10 m*M* $MgCl_2$ and 10% (v/v) glycerol]. Polyacrylamide gels are prerun at 20 W for 1 h prior to sample loading, and are then electrophoresed at 15 W for 4.5 h at room temperature. Quantification of the amount of bound and free substrate is performed following the exposure of the gels to a phosphorimaging screen, and the screen is scanned and analysed using ImageQuant version 5.0 (Molecular Dynamics). The RzA : RzB complex K_d is determined by mixing either RzA or RzB concentrations that ranged from 0.1 to 300 n*M* with trace amounts of either end-labeled RzB or RzA, respectively, and following a procedure similar to that described above in order to determine K_d^{RzA} and K_d^{RzB}, respectively.

Cleavage Reactions

Optimal ratios of substrate to RzA, and RzA to RzB, are determined by cleavage activity assessment following measurement of the equilibrium dissociation constants. Various concentrations of ribozymes are mixed with trace amounts of substrate (final concentration <1 n*M*) in an 18 μl reaction mixture containing 50 m*M* Tris-HCl, pH 7.5, and subjected to denaturation by heating at 95° for 2 min. The mixtures are quickly placed on ice for 2 min, and then equilibrated to 37° for 5 min prior to the initiation of the reaction. Unless stated otherwise, cleavage is initiated by the addition of $MgCl_2$ to 50 m*M* final concentration. The cleavage reactions are incubated at 37° and are followed for 3.5 h or until the end point of cleavage is reached. The reaction mixtures are periodically sampled (2–3 μl), and these samples are quenched by the addition of 5 μl stop solution containing 95% formamide, 10 m*M* EDTA, 0.05% bromphenol blue, and 0.05% xylene cyanol. The resulting samples are analyzed on a 20% PAGE as described above. Both the substrate (11 nt) and the reaction product (4 nt) bands are detected using a Molecular Dynamic phosphorimager after exposure of the gels to a phosphorimaging screen.

Measurement of Cleavage Rate (k_{obs})

Unlabeled substrate is mixed with trace amounts of end-labeled substrate (<1 n*M*) to a final concentration of 50 n*M*, and then incubated with various ribozyme concentrations (50 to 500 n*M*). The fraction cleaved is determined, and the rate of cleavage (k_{obs}) obtained by fitting the data to the equation $A_t = A_\infty(1 - e^{-kt})$, where A_t is the percentage of cleavage at time t, A_∞ is the maximum percent cleavage (or the end point of cleavage), and k is the rate constant (k_{obs}). Each observed rate is calculated from at least two measurements. Values obtained from independent experiments vary less than 15%.

Mg^{2+} Requirement of Bimolecular Ribozyme System

The ribozyme system is studied by incubating the reaction mixtures with various concentrations of $MgCl_2$ (1 to 500 m*M*) in the presence of RzA and RzB

(1 : 3, 50 : 150 n*M*) with substrate (50 n*M*). The concentrations of Mg^{2+} at the half-maximal velocity are calculated.

Results and Discussion

RNA Production

Approximately 25 to 100 μg of RNA was recovered from a 50-μl *in vitro* transcription reaction and gel purification. When higher amounts of RNA were required, the reaction volume was proportionally increased. We observed that the sequence downstrem of the T7 RNA polymerase promoter has a drastic influence on the RNA yield.[13] The sequence of the RNA products were verified using RNase T1 and U2 digestions.

Ribozyme Assembly

Since the design of the bimolecular ribozyme system was based solely on the predicted secondary structure of the HDV self-cleaving motif, it is possible that the transcripts might adopt an alternative conformation in solution. In order to test for the presence of any alternative conformers, RzA, RzB, and a mixture of RzA and RzB were fractionated on a native gel. When the individual transcripts were tested, a single band was observed for each RNA species. The RzA–RzB complex was observed to migrate more slowly than the individual RzA and RzB transcripts. We also observed that approximately 2% of the end-labeled RzB behaved like another retarded complex migrating at the same level as the unbound RzA and lower than the RzA–RzB complex (Fig. 3A, indicated by an arrowhead). The amount of this unexpected band could not be accurately measured due to the elevated background level in the lanes.

Two independent experiments were performed using 11 different concentrations of either RzA or RzB against trace amounts of either RzB* or RzA* (the asterisk denotes 5′-end-labeled transcripts). The amount of expected complex formed was used to calculate the equilibrium dissociation constants that indicate the mutual affinity of the two components of the complex. The migration of end-labeled RzA is proportionally retarded in the presence of increasing amounts of RzB. The amounts of retarded complex (RzA : RzB) and unbound labeled RzA are determined and used to calculate the percentage of complex formed as a function of the RzB concentration. The data are then fit to the simple binding equation, RzA : RzB complex formed (%) = $[\mathrm{RzB}]/K_d + [\mathrm{RzB}]$, in order to obtain the equilibrium dissociation constant (K_d^{RzB}), which indicates the concentration of RzB required to retard 50% of the end-labeled RzA. The values of K_d^{RzA} for RzB* and S* were determined in a similar fashion (Fig. 3B). The observed K_d values are similar (4 to 7 n*M*), suggesting that individual transcripts have similar affinities for the complex formation.

A

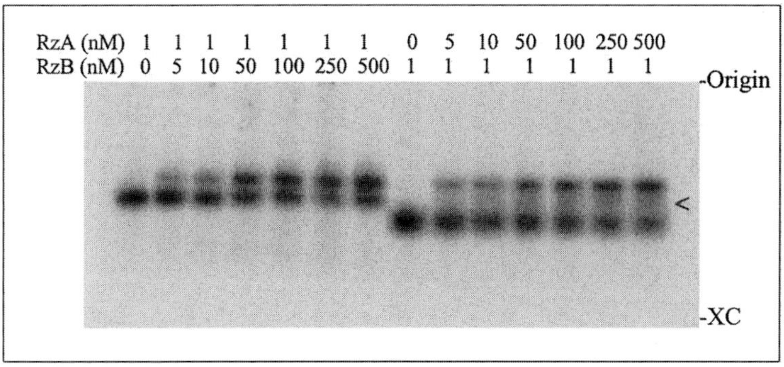

B

Equilibrium dissociation constants (K_d)	
K_d^{RzB} (RzA* and RzB)	6.7 ± 0.9 nM
K_d^{RzA} (RzB* and RzA)	4.4 ± 0.9 nM
K_d^{RzA} (S* and RzA)	6.6 ± 0.8 nM

FIG. 3. (A) Autoradiogram of a gel retardation assay used in the measurement of equilibrium dissociation contants. (B) Calculated values of K_d.

Ribozyme Cleavage

Native gel electrophoresis indicated that there is a low level of an alternative RzA–RzB complex formed in solution. Prior to performing the kinetic study, an optimal molar ratio of RzA and RzB was assayed using a cleavage assay. Time course experiments in which the amounts of substrate and RzA were kept equal (S : RzA, 1 : 1, 50 n*M*) and were incubated with increasing amounts of RzB (50 to 400 n*M*) were then performed. The fractions of product formed and substrate remaining were measured and plotted as a function of time as shown in Fig. 4A. Observed cleavage rates of $0.16 \pm 0.02 \text{min}^{-1}$ were observed for all mixtures, whereas maximal cleavage product levels were detected when the RzB concentration was higher than 150 n*M*, or when the ratio of S : RzA : RzB was 1 : 1 : 3. At lower RzB concentrations (50 to 100 n*M*), a suboptimal level of complex (S : RzA : RzB) was formed, resulting in the formation of lower amounts of product. In comparison to the

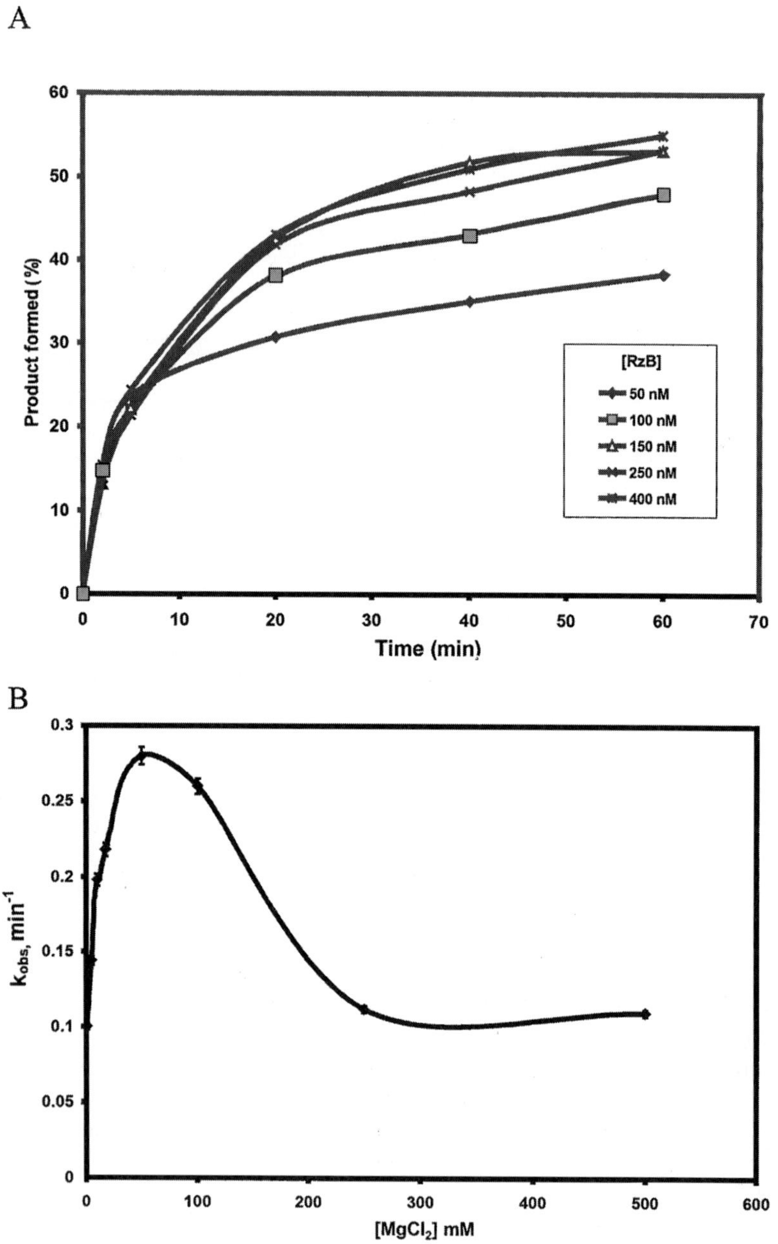

FIG. 4. (A) Time course experiments on the cleavage activity of assembled bimolecular systems containing 50 n*M* substrate and 50 n*M* RzA and various concentrations of RzB. (B) Determination of the $MgCl_2$ requirement of the bimolecular *delta* ribozyme in the presence of 50 n*M* of both substrate and RzA, and 150 n*M* RzB.

analogous system described by Sakamoto *et al.*,[12] higher ribozyme concentrations (RzA–RzB, 0.1 to 100 μM) were required in order to obtain a similar level of activity in the cleavage of substrate (0.1 μM). Since there was no detailed analysis on the previously reported ribozyme system, we assumed that alternative conformers are also present in the mixture and are responsible for the higher ribozyme concentration requirement. In comparison to the *trans*-acting *delta* ribozyme (whose structure is shown in Fig. 2A), the bimolecular ribozyme system has a similar observed rate of cleavage (i.e., $0.29 \pm 0.2\ \text{min}^{1}$) and $0.16 \pm 0.2\ \text{min}^{-1}$, respectively. Although the maximum extent of cleavage of the bimolecular ribozyme is lower (60% of the substrate was cleaved), it has several features that encourage the further development of other ribozyme systems based on *delta* ribozyme structures (see below).

Magnesium Requirements of Bimolecular Ribozyme System

Similar to *cis*- and *trans*-acting *delta* ribozymes, the bimolecular ribozyme system requires divalent metal ions for its cleavage activity. More importantly, the bimolecular ribozyme might require higher concentration of metal ion for its assembly. For this reason we characterized its magnesium requirement. The optimal ratio of S : RzA : RzB (1 : 1 : 3) described earlier was used in cleavage reactions in the presence of various $MgCl_2$ concentrations (1 to 400 m*M*). When the values of k_{obs} were plotted as a function of magnesium concentration (Fig. 4B), the maximum cleavage rate was observed at 50 m*M* $MgCl_2$. These results are different from those of Sakamoto's system.[12] This might be due to the differences in the ratios of RzA, RzB, and total ribozyme used in the two studies.We suspect that the bell-shaped curve is due to the fact that higher $MgCl_2$ concentrations might promote the formation of an alternative conformation that yields no cleavage product. In order to determine the $MgCl_2$ concentration at which this occurs, other structural analyses are required.

Perspectives

The bimolecular *delta* ribozyme system possesses several interesting features that warrant further studies, including ones directed toward determining the structure–function relationship and developing an antisense/ribozyme dual system. For example, the construction of a miniribozyme with a higher stability (i.e., one resistant to nucleases) could be achieved using a similar strategy and introducing deoxyribonucleotides in place of the nuclease-sensitive riboses. With the generous contribution of a T7 RNA polymerase mutant by Dr. Rui Sousa (University of Texas Health Science Center, San Antonio, CA),[16] we have begun preliminary tests of a

[16] R. Padilla and R. Sousa, *Nucleic Acids Res.* **27,** 1561 (1999).

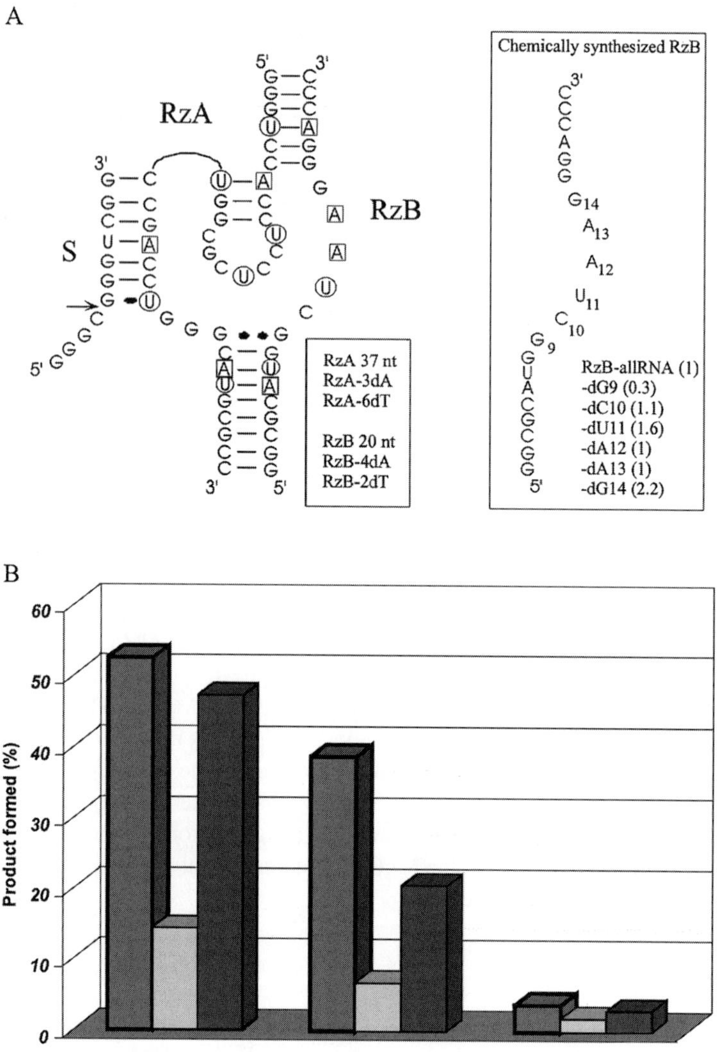

RzB \ RzA	all RNA	dA	dT
all RNA	52.5	38.7	3.8
dA	14.4	6.8	1.9
dT	47.3	20.6	3

FIG. 5. (A) Schematic representation of the RzA and RzB transcripts containing either single or multiple positions replaced with deoxyribose residues. Circled and boxed nucleotides represent the positions of replaced deoxyriboses. The extent of cleavage of singly substituted RzB transcripts are compared to chemically synthesized all RNA RzB (shown in the inset). The numbers in the brackets represent the relative extents of cleavage. (B) Histogram showing the extents of cleavage of RzA and RzB transcripts containing multiple substitutions.

mixed DNA/RNA bimolecular ribozyme system. This was achieved by substituting individual deoxyribonucleotides for ribonucleotides (at the same concentration) in *in vitro* transcription reactions using this mutant T7 RNA polymerase to produce RzA and RzB. The reactions using dATP and dTTP, to replace rATP and rUTP, respectively, produced RzA and RzB transcripts in comparable amounts as those using rATP and rUTP and wild-type T7 RNA polymerase. However, when either dGTP or dCTP was used, no RzA and RzB transcripts were produced. Most likely this is due to the greater number of guanidine and cytosine nucleotides in RzA and RzB. We then used the RzA and RzB transcripts containing dA and dT in a cleavage assay (Fig. 5). It should be noted that several positions on both RzA and RzB can be substituted by deoxyribose moieties. When the three adenine nucleotides of RzA are deoxyribose residues (RzA-3dA), RzB was observed to have a cleavage extent (38.7%) as compared to that of all RNA RzA and RzB (52.5%, Fig. 5B). The replacement of uracil by thymidine in the RzA transcript is detrimental to cleavage, most likely because the hydrogen bonds formed between these bases are impaired, specifically the wobble G-U base pair adjacent to the scissile phosphodiester bond. When the riboses of RzB were replaced by deoxyriboses (RzB-2dT), the ribozyme complexes of all RNA RzA and RzB-2dT result in 47.3% product formation, whereas RzB-4dA gave a lower cleavage extent (Fig. 5B).

In order to have a better understanding of the effect of the 2′-hydroxyl groups on the cleavage and assembly of the bimolecular ribozyme complex, single substitutions are required. According to the secondary structure, we suspect that the ribose residues of J4/2 might influence the formation of the RzA and RzB complex and its subsequent cleavage. Seven RzB transcripts possessing either all ribose residues or those with individual positions (9 to 14) replaced by deoxyribose residues were chemically synthesized (Fig. 5, inset). It should be noted that the chemically synthesized RzB exhibited slightly lower affinity for all RNA RzA, as well as a slightly lower activity than the enzymatically synthesized RzB. In order to have comparable values, all chemically synthesized transcripts including all RNA RzB were assayed with enzymatically synthesized RzA and substrate. When the extents of cleavage were compared, the presence of deoxyriboses at positions 10, 12, and 13 (RzB-dC10, -dA12, and -dA13) had no significant effect. RzB-dU11 and -dG14 promoted cleavage, whereas RzB-dG9 had a detrimental effect on it. Although we could not distinguish whether or not the 2′-hydroxyl group is important in the ability of RzB to form the complex, or in other interactions, we have initiated several studies on the structure–function relationship.

Conclusion

We described here the methodology used in the enzymatic study of a bimolecular ribozyme system derived from hepatitis *delta* ribozyme. The bimolecular ribozyme system is very promising for the further development of a mixed

RNA/DNA ribozyme system that is likely to be more stable to nuclease degradation than the all-RNA one.

Acknowledgment

This work was supported by a grant from the Medical Research Council (MRC) of Canada to J-P.P.

[36] Catalytic and Structural Assays for the Hairpin Ribozyme

By KEN J. HAMPEL, ROBERT PINARD, and JOHN M. BURKE

Introduction

The hairpin ribozyme is a catalytic RNA that functions in nature as a site-specific ribonuclease and an RNA ligase.[1,2] The catalytic motif originally was identified in the processing of tobacco ringspot virus negative strand satellite RNA [(−)TRSV], and *in vivo,* it is responsible for processing rolling-circle replication intermediates.[1] The minimal catalytic fragment was excised from the context of the 359 nucleotide satellite RNA and was shown to cleave exogenous substrates and to possess the capacity of multiple turnover of substrates, thus fulfilling the requirements of a biological catalyst (Fig. 1).[3] The capacity to cleave exogenous substrates has generated interest in converting the hairpin ribozyme into an RNA-inactivating therapeutic agent with potential application to viral and genetic diseases.[4] In this chapter we will outline the known catalytic and structural properties of the hairpin ribozyme through a brief introduction followed by a more detailed description of catalytic assays and structural methods. The emphasis will be placed on the unique characteristics of the hairpin ribozyme system, especially its strengths and weaknesses as an experimental model.

Ribozyme Reaction Chemistry

The kinetic pathway for the reversible transesterification reaction catalyzed by the ribozyme is shown in Fig. 2. The forward (cleavage) reaction is initiated

[1] J. M. Buzayan, W. L. Gerlach, and G. Bruening, *Nature* **323,** 348 (1986).
[2] M. J. Fedor, *J. Mol. Biol.* **297,** 269 (2000).
[3] A. Hampel and R. Tris, *Biochemistry* **28,** 4929 (1989).
[4] D. J. Earnshaw and M. J. Gait, *Antisense Nucleic Drug Dev.* **7,** 403 (1997).

0076-6879/01 $35.00

by base-mediated deprotonation of the 2′OH, which allows the resulting alkoxide ionic group (RO−) to carry out an in-line nucleophilic attack of the adjacent 3′-phosphate group. An acid function is required to supply a proton to the 3′ cleavage product-leaving group. The reaction results in formation of 5′ and 3′ cleavage products with 2′,3′-cyclic phosphate and 5′-hydroxyl termini, respectively.[1] The same general reaction pathway is shared by several other ribozymes and protein RNases described in this volume. The chemical participants at the active site of the hairpin ribozyme have not been determined. However, it is clear that metal ions do not play any direct roles in the chemistry.[5–7] Thus, in the hairpin ribozyme, nucleotide functional groups on the RNA provide both acid and base functions at the active site, and specific cations serve to direct the correct folding of the tertiary structure.

Structure

The principal structural requirement for cleavage and ligation is a direct interaction between the two domains of the ribozyme, A and B, shown in Fig. 3.[8] Each domain is dominated by an internal loop flanked by two short double helices. The majority of evolutionarily conserved bases and important nucleotide functional groups are located in the two internal loop structures, loops A and B.[9–13] The wild-type secondary structure for the ribozyme and the cleavage site is shown in Fig. 1. Fluorescence resonance energy transfer (FRET) methods have been used to show that the sequence at the native four-way interdomain helical junction directs the orientation of the four helices, and in doing so, brings the internal loops A and B into close proximity.[14] In addition, time-resolved FRET studies demonstrate that the interdomain interaction is more stable in the native form of the ribozyme than in the minimal construct containing a two-way helical junction.[15] The only known

[5] S. Nesbitt, L. A. Hegg, and M. J. Fedor, *Chem. Biol.* **4,** 619 (1997).

[6] A. Hampel and J. A. Cowan, *Chem. Biol.* **4,** 513 (1997).

[7] K. J. Young, F. Gill, and J. A. Grasby, *Nucl. Acids. Res.* **25,** 3760 (1997).

[8] S. E. Butcher, J. E. Hechman, and J. M. Burke, *J. Biol. Chem.* **270,** 29648 (1995).

[9] P. Anderson, J. Montforte, R. Tris, S. Nesbitt, J. Hearst, and A. Hampel, *Nucleic Acids* **22,** 1096 (1994).

[10] A Berzal-Herranz, S. Joseph, B. M. Chowrira, S. E. Butcher, and J. M. Burke, *EMBO J.* **13,** 2567 (1993).

[11] S. Schmidt, L. Beigelman, A. Karpeisky, N. Usman, U. S. Sorensen, and M. J. Gait, *Nucleic Acids Res.* **24,** 573 (1996).

[12] J. A. Grasby, K. Mersmann, M. Singh, and M. J. Gait, *Biochemistry* **34,** 4068 (1995).

[13] B. M. Chowrira, A Berzal-Herranz, C. F. Keller, and J. M. Burke, *J. Biol. Chem.* **268,** 19458 (1993).

[14] A. I. Murchie, J. B. Thomson, F. Walter, and D. M. Lilley, *Mol. Cell.* **1,** 873 (1998).

[15] N. G. Walter, J. M. Burke, and D. P. Millar, *Nat. Struct. Biol.* **6,** 544 (1999).

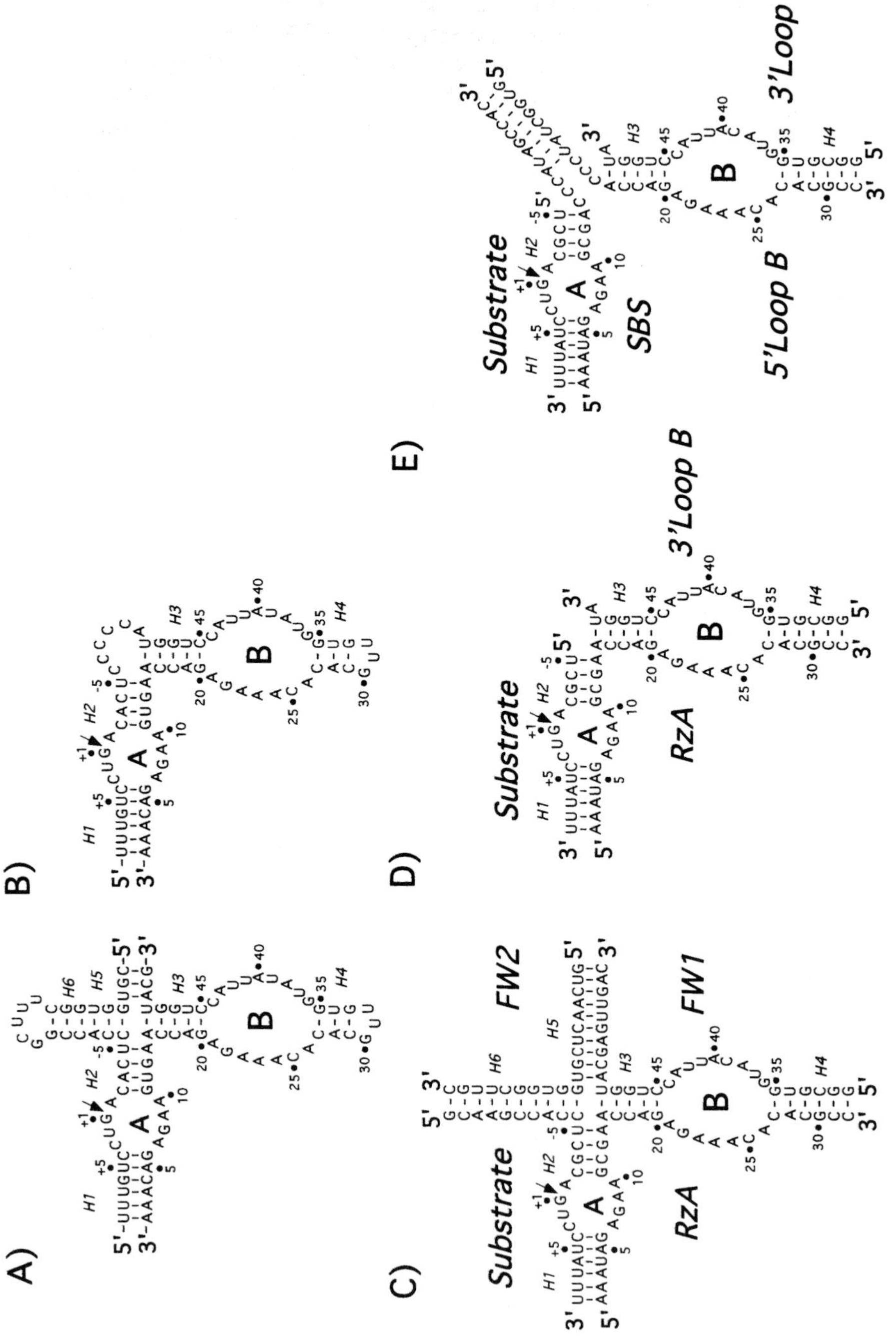
A)
B)
C)
D)
E)
Substrate
RzA
FW1
FW2
Substrate
RzA
3'Loop B
Substrate
SBS
5'Loop B
3'Loop

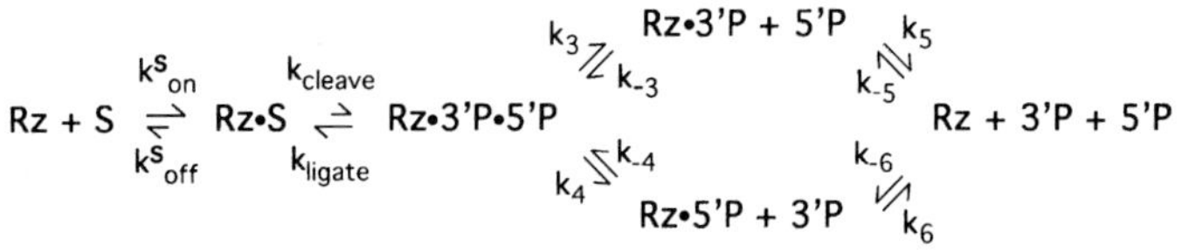

FIG. 2. Kinetic pathway for the hairpin ribozyme.

tertiary interaction that links the A and B domains is a Watson–Crick base pair between G+1 in the substrate and C25.[16]

Enzymatic and structural studies of the hairpin ribozyme are greatly aided by the short minimal sequence requirements for catalysis. The minimal ribozyme motif that retains catalytic activity is 50 nucleotides. This means that the ribozyme can be synthesized by solid-phase chemistry. By utilizing solid-phase RNA synthesis, any of a large number of modified nucleotides can be inserted into the ribozyme in order to address specific questions about structure and function.[17]

Catalytic Assays

Preliminary Considerations

One of the most difficult problems with studying cleavage and ligation kinetics of the hairpin ribozyme is the fact that these activities oppose one another (Fig. 2). The measured ribozyme cleavage rates are, thus, a function of the intrinsic cleavage rate and the ligation rate under the specific conditions employed. This was shown to be especially true for single-turnover cleavage assays of the native ribozyme form.[18] Initial *trans*-cleavage assays performed on the native hairpin motif showed very slow single turnover kinetics relative to the minimal construct.[19] A more recent kinetic study demonstrated that the relatively poor apparent kinetic performance of the native form is due to slow product disassociation, with subsequent ligation.[18]

[16] R. Pinard, D. Lambert, N. G. Walter, J. E. Heckman, F. Major, and J. M. Burke, *Biochemistry* **38,** 16036 (1999).

[17] R. A. Zimmermann, M. J. Gait, and M. J. Moore, *in* "Modification and Editing of RNA" (E. Grosjean and R. Benne, eds.), p. 59. ASM Press, Washington, D.C., 1998.

[18] M. J. Fedor, *Biochemistry* **38,** 11040 (1999).

[19] J. B. Thomson and D. M. J. Lilley, *RNA* **5,** 180 (1998).

FIG. 1. The secondary structures of self-splicing and *trans*-activing hairpin ribozymes. Numbering is according to Hampel and Tris[3]; arrowheads indicate the site of cleavage and ligation. (A) Native self-cleaving motif; (B) minimal self-cleaving motif; (C) native *trans*-active SV5 ribozyme and substrate; (D) minimal *trans*-acting SV5 ribozyme–substrate complex; (E) SV5 ribozyme–substrate complex for photoaffinity cross-linking.

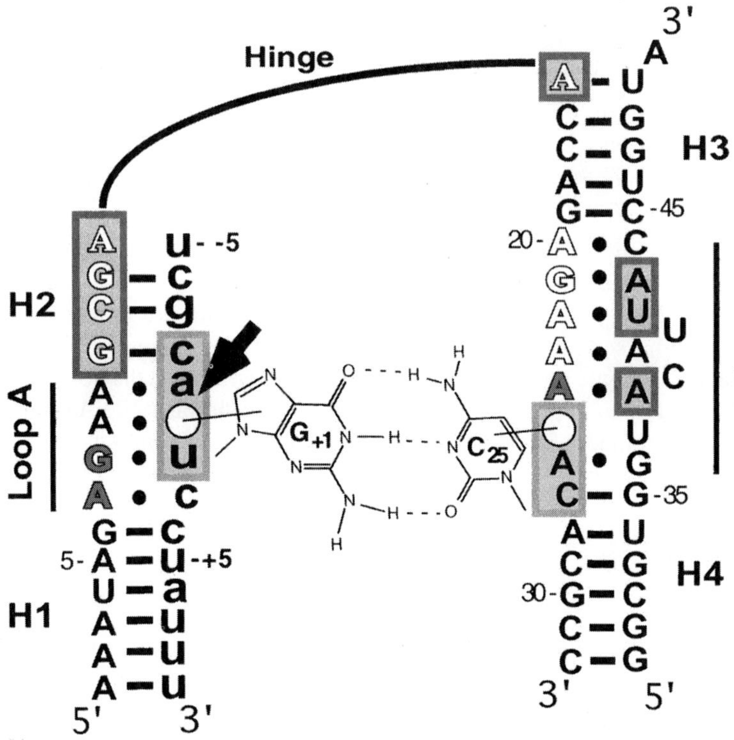

FIG. 3. Interactions between domain A and B in the docked complex. Rectangles indicate two pairs of regions protected from hydroxyl radical attacks in the docked complex. White and gray letters indicate regions that have been photochemically cross-linked. There is a strong correlation between the regions identified by these two methods. These regions potentially are involved in the formation of tertiary contacts. The interdomain interaction between position +1 in the substrate and C_{25} in the ribozyme also is indicated.

The extended pairing of the 5′ product to the ribozyme decreases the dissociation, whereas strengthened tertiary structures in the native form contribute to increased affinity of the 3′ cleavage product for the ribozyme. Shortening the base pairing in helix 1 increases the 3′ product K_D, thus yielding a ribozyme with very similar cleavage rates to the minimal ribozyme motif. The ligation rate for the ribozyme will similarly be affected by the opposing cleavage activity. The intrinsic rate constant for ligation is, therefore, equal to the sum of the measured ligation and intrinsic cleavage rates.[19,20]

Sequence changes to the ribozyme have been made to improve the structural and kinetic homogeneity of ribozyme–substrate complexes. First, our laboratory

[20] S. M. Nesbitt, H. A. Erlacher, and M. J. Fedor, *J. Mol. Biol.* **286,** 1009 (1999).

has noted that the wild-type (−)TRSV sequence for the substrate is partially self-complementary. This complementarity poses problems in *trans*-cleavage assays and prohibits the use of the wild-type sequence in multiple-turnover experiments. In our laboratory we use a sequence referred to as SV5 (sequence variant 5).[8,21] Because the sequences in helices 1 and 2 can be varied as long as substrate binding is not dramatically impaired, a wide variety of sequences in H1 and H2 can be employed. This characteristic makes the ribozyme an especially attractive candidate for RNA inactivation applications. SV5 represents an attempt to limit substrate self-complentarity while maintaining conserved elements of helix 2. The other major difference between the wild-type ribozyme and laboratory model systems is the number of base pairs in helix 4 and the helix 4-capping loop sequence. Evidence from our laboratory shows that increasing the length of helix 4 in the wild-type ribozyme, from 3 to 6 base pairs, enhances the cleavage rate by 6-fold; however, this cleavage rate increase is not observed in ribozymes with SV5 sequences.[21] One interpretation of this result is that these improvements to the ribozyme structure are at least partially redundant.

Trans Cleavage Assays

Cleavage and ligation require concentrations of cations sufficient to promote docking between the A and B domains.[22,23] Thus, cleavage assays can be initiated by the addition of specific cations to the preformed substrate–ribozyme complex and terminated using denaturing agents. Cleavage reactions can, alternatively, be initiated by the addition of substrate to the ribozyme. In this case both RNAs are preincubated in a buffered solution at the final folding cation concentration.

Chemical Stocks and Buffers

0.5 *M* HEPES–NaOH, pH 7.5
1 *M* $MgCl_2$
0.5 *M* EDTA, pH 8
Reaction quenching buffer (20 m*M* Tris-HCl, pH 7.5; 0.02% (w/v) bromphenol blue; 0.02% (w/v) xylene cyanol; 15 m*M* EDTA, pH 8; prepared in formamide)

All solutions are prepared from molecular biology grade reagents. Our minimum precautions against RNase contamination include wearing gloves at all times during solution preparation, and using only distilled, deionized water to dilute chemicals. We have found that if gloves are worn while dispensing chemicals, using instruments that are cleaned by rinsing with distilled, deionized water, then RNase contamination is very rarely a problem.

[21] J. A. Esteban, A. R. Banerjee, and J. M. Burke, *J. Biol. Chem.* **272,** 13629 (1997).
[22] K. J. Hampel, N. G. Walter, and J. M. Burke, *Biochemistry* **37,** 14672 (1998).
[23] N. G. Walter, K. J. Hampel, K. M. Brown, and J. M. Burke, *EMBO J.* **17,** 2378 (1998).

We will describe a typical cleavage reaction where ribozyme and substrate are premixed, and the reaction initiated by dilution of the RNA complex into cation solution. Reactions are set up by first mixing ribozyme (0.5 μM) with ^{32}P-end-labeled substrate RNA (1–2 nM) in a buffer containing 50 mM HEPES–NaOH, pH 7.5, 0.1 mM EDTA. The cation solution contains 50 mM HEPES–NaOH, pH 7.5, 12 mM $MgCl_2$. Both mixes are equilibrated at 25° for 5 min. The reaction is initiated by diluting 40 μl of RNA mix into an equal volume of the cation solution. Aliquots of 3 μl are removed at 7–12 time points over the course of the reaction, and quenched with 10 μl of the quench buffer on ice.

A zero time point is an essential control used to demonstrate that the quench stops the reaction. To perform a zero time point 1.5 μl of the RNA solution, and cation are separately added to the same 10 μl of quench solution. Reactions initiated with cations that are not chelated by EDTA must be quenched in a larger volume of quench solution. For example, in reactions where $Co(NH_3)_6^{3+}$ is the folding cation, we typically dilute the reactions in at least 5 volumes of quench buffer on ice.

Reaction products are separated by electrophoresis using 20% polyacrylamide-8 M urea gels. A typical autoradiogram is shown in Fig. 4. The fractional substrate cleavage is then quantified by phosphorimager analysis. After subtracting the background, the counts of product are divided by the sum of the product and substrate counts in each lane. The resulting plot of fraction substrate reacted versus time can be fit to the exponential growth equation $y = y_0 + A(e^{-t/\tau})$, where A represents the amplitude, and the rate equals $1/\tau$. Our laboratory has shown that the minimal *trans*-acting ribozyme reacts under single-turnover condition with double exponential kinetics, which can be fit to the equation $y = y_0 + A_1(e^{-t/\tau_1}) + A_2(e^{-t/\tau_2})$.[20] The slow phase of this reaction was subsequently found to be due to a slow

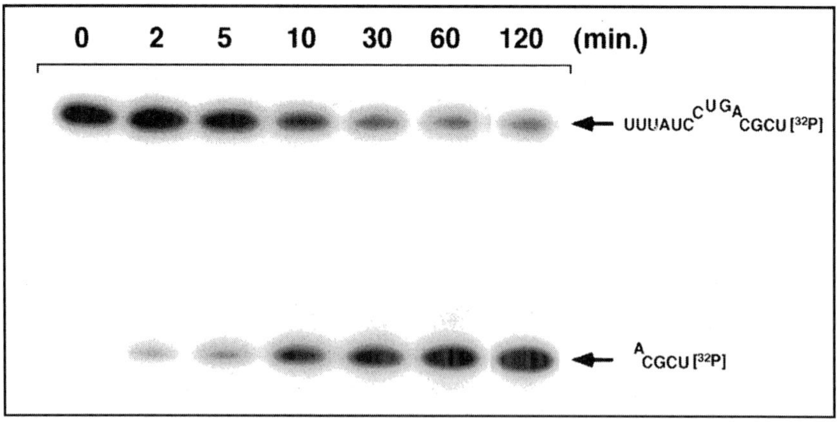

FIG. 4. Cleavage activity or ribozymes. Assays performed with construct D (Figure 1), using 5′-^{32}P-end-labeled substrate. The full-length and cleaved products are indicated by arrows on the right.

rearrangement from an inactive ribozyme–substrate complex, where helices 2 and 3 are stacked, to the active, docked form.[24]

Ligation Assays

The hairpin ribozyme ligates RNAs with 5′-OH and 2′,3′-cyclic phosphate termini. RNAs containing 5′-OH termini can be generated by solid-phase synthesis or by dephosphorylation of transcription products. The only way to generate an RNA with a 2′,3′-cyclic phosphate, however, is to cleave a hairpin ribozyme substrate and gel-purify the 5′ product. For kinetic studies, the strategy most often followed is to transcribe a self-cleaving hairpin construct in a reaction that contains [α-^{32}P]CTP.[20,21] The internally labeled 5′-self-cleavage product that accumulates during transcription can be easily gel purified and quantified. If site-specific nucleotide changes or modifications are to be tested for their ligation effects, then an unmodified active ribozyme can be used to cleave a 5′-^{32}P-labeled substrate, from which the 5′ product is purified. The mutant ribozyme can then be used to study the ligation of the purified 5′ product to a synthetic 3′ product.

The general methods used to determine cleavage and ligation rates are essentially identical. In ligation reactions, however, great care should be taken to ensure that the ribozyme is saturated with cleavage products. In other words, the point at which the ligation rate no longer depends on the concentrations of cleavage products should be clearly defined.

Structural Analysis

In this section we will detail two methods for probing the structure of the hairpin ribozyme: hydroxyl-radical footprinting and photoaffinity cross-linking. Both methods have been invaluable tools used in our laboratory for defining the global structure of the ribozyme, and for diagnosing the structural effects of mutations or functional group substitutions.[22,25]

Hydroxyl-Radical Footprinting

Chemical modification or cleavage footprinting has been used extensively to define the internalized sites in globular RNA molecules at structural equilibrium.[26,27] We have observed that, in the presence of cations that are required for activity, the hairpin ribozyme–substrate complex folds into a compact tertiary structure with five discrete sites that are protected from hydroxyl radical (·OH) attack (Fig. 3).[22] In contrast to base-modifying probes such as dimethyl sulfate,

[24] J. A. Esteban, N. G. Walter, G. Kotzorek, J. E. Heckman, and J. M. Burke, *Proc. Natl. Acad. Sci. USA* **95,** 6091 (1998).

[25] R. Pinard, J. E. Heckman, and J. M. Burke, *J. Mol. Biol.* **287,** 239 (1999).

[26] J. A. Latham and T. R. Cech, *Science* **245,** 276 (1989).

[27] S. Stern, D. Moazed, and H. F. Noller, *Methods Enzymol.* **164,** 481 (1988).

hydroxyl radicals (·OH) do not discriminate between single and double-stranded sites on the RNA.[26] This makes ·OH footprinting a particularly useful method for defining long-range tertiary contacts involving helical regions of RNA. Furthermore, the ·OH cleavage of *Tetrahymena* ribozyme P4-P6 domain yields results that correlate very well with the solvent accessibility calculated from its tertiary structure, as determined by X-ray crystallography.[28]

Tullius and co-workers developed a method of footprinting protein binding sites on DNA by taking advantage of the oxidation of Fe(II) to Fe(III) by hydrogen peroxide in order to produce ·OH, which represents the Fenton reaction.[29] In order to prevent Fe(II) from binding to the DNA directly, it is chelated with EDTA. Fe(II)-EDTA is regenerated from Fe(III)-EDTA by ascorbate. In previous experiments using this reagent, exposure of RNA to the mix of ·OH-generating reagents was on the order of minutes.[29] We have observed that only a few seconds of exposure are required to generate sufficient RNA cleavage to visualize solvent-protected sites.[30] A slightly modified method is presented in detail below, in addition to general considerations of RNA handling, ·OH scavengers, and specific application of the method to the hairpin ribozyme.

Preliminary Considerations

RNA preparation. Since background RNA cleavage can easily obscure the appearance of protected sites in the folded RNA, great care must be taken during the preparation and storage of labeled RNA employed in footprinting analysis. We typically gel purify RNA prepared by *in vitro* transcription or solid-phase synthesis in order to remove products that are shorter or longer than the desired "full-length" product. Depending on the source of the RNA, however, one or more enzymatic steps are required to radioactively end-label the purified target RNA for use in footprinting experiments. We have observed, however, that any enzymatic modification of the RNA introduces some nonspecific degradation, which increases the noise on footprinting gels. Thus, freshly end-labeled RNA should be gel purified prior to use in footprinting experiments.

Radiolysis of labeled RNAs is another source of background noise in footprinting experiments. Typically we store labeled RNAs at a specific activity $\sim 2 \times 10^5$ dpm/μl in 10 m*M* sodium cacodylate, pH 7, 0.1 m*M* EDTA at $-20°$. Labeled RNA shows only minimal degradation within 1 week of gel purification under these conditions.

Free-radical scavengers. Some buffers, notably Tris, reduce RNA and DNA cleavage by hydroxyl radicals, presumably by acting as free radical scavengers.[29]

[28] J. H. Cate, A. R. Gooding, E. Podell, K. Zhou, B. Golden, L. Kundrot, C. E. Cech, and J. A. Doudna, *Science* **272,** 1678 (1996).

[29] T. D. Tullius, B. A Dombroski, M. E. A. Churchhill, and L. Kam, *Methods Enzymol.* **155,** 537 (1987).

[30] K. J. Hampel and J. M. Burke, *Methods,* in press.

Our footprinting buffers contain 25 m*M* sodium cacodylate, pH 7, which does not quench free radicals. In addition, hydroxylated compounds such as glycerol, alcohols, and sugars are known ·OH scavengers and can inhibit ·OH-mediated cleavage.[29]

Application to hairpin ribozyme. There are five sites of hydroxyl-radical protection in the docked form of the hairpin ribozyme–substrate complex (Fig. 3). In the ribozyme, positions 12–15, 25–27, 38, and 42–43 are protected, and positions −2 to +2 are protected in the substrate.[22] In order to assay ·OH protection of the substrate, cleavage must be blocked by 2′-deoxy or 2′-*O*-methyl modification of the −1 nucleotide. The RNA that carries the cleavage site must, therefore, be made by solid-phase synthesis. This modification does not negatively affect interdomain docking.[22,23] Another important consideration to note is that interdomain docking is much more stable in the native catalytic motif.[15] In the minimal catalytic motif, we have found that the SV5 sequences contribute significantly to the stability of the docked structure and, thus, the formation of observable ·OH-protected sites.[22]

Chemical Stocks and Buffers

- 0.25 *M* Cacodylate–NaOH, pH 7
- 10 m*M* $Co(NH_3)_6^{3+}$
- 0.5 *M* EDTA, pH 8
- 30% (v/v) H_2O_2
- 60 m*M* sodium ascorbate
- $Fe(NH_4)_2(SO_4)_2$ (solid, Aldrich, Madison, WI)
- Quench buffer (50 m*M* Tris-HCl, pH 7.5; 10 m*M* thiourea; 0.02% (w/v) bromphenol blue; 0.02% (w/v) xylene cyanol, 0.2 μg/μl tRNA; prepared in formamide)

We will describe a typical footprinting assay carried out on the native ribozyme construct with 5′-^{32}P-labeled FW1 RNA. The RNAs in the assay are always mixed first by adding 1 μl of each cold RNA (2.5 μM, stock concentration) and 1 μl of labeled RNA (2×10^5 dpm/μl, stock concentration) to each reaction tube. The final volume is adjusted to 5 μl with distilled, deionized water. Preliminary mixing of the RNAs in the absence of folding cations may prevent the formation of stable self-structures by the RNAs and allows for prefolding of the secondary structure. To this mixture add 1 μl of 0.25 *M* cacodylate–NaOH, pH 7, 1 μl 10 m*M* $Co(NH_3)_6^{3+}$, and distilled, deionized water to a final volume of 8 μl. The contents of the tube are then mixed by a quick microcentrifuge spin.

Interdomain folding of the native form of the ribozyme–substrate complex is very rapid ($k_{obs} > 10 s^{-1}$)[31]; thus, ·OH-generating reagents can be applied to the folding mix almost immediately. Equal volumes (0.7 μl) of the three

[31] K. J. Hampel, C. Ralston, J. M. Burke, and M. Brenowitz, unpublished observations (1999).

·OH-generating reagents are added, in separate droplets, to the interior wall of the micro-test tube. The reagents are 0.35% (v/v) H_2O_2 [freshly prepared from a 30% (v/v) stock solution], 60 m*M* sodium ascorbate (60 m*M* stocks can be stored at −20°), and Fe(II)-EDTA [25 m*M* EDTA, 20 m*M* $Fe(NH_4)_2(SO_4)_2$, mixed immediately prior to experiment]. These three droplets are introduced into the RNA folding mix by a quick microcentrifuge spin, and the reaction is terminated by the addition of 20 μl of quench buffer. The time between initiation of the reaction and its termination can be very short (<5 s), since no further RNA cleavage occurs beyond this time.[30] Samples can be loaded directly onto a 20% polyacrylamide–8 *M* urea gel.

Fractional protection at specific sites on the RNA is calculated from phosphorimager analysis. A critical control in this experiment is a sample where the RNA is mixed with the folding cation, but not treated with the ·OH-generating solution (lane − ·OH, Fig. 5). The counts from sites in this lane represent background degradation and are subtracted from the corresponding bands of interest in the experimental lanes. In addition, the number of counts at two sites, where the accessibility to ·OH does not change on formation of the docked complex, are counted to correct for the lane-to-lane variation in gel loading and RNA cleavage. In the FW1 fragment of the native ribozyme these sites are 36 and 47–48. Fractional protection at specific sites is calculated by comparing the number of counts in the + and − cation lanes using the formula Fractional protection = 1 − [Counts in + cation lane/Counts in − cation lane].

Photoaffinity Cross-Linking

Few structural constraints are available to guide modeling efforts of the hairpin ribozyme–substrate complex. We have therefore developed a photoaffinity cross-linking system for the hairpin ribozyme that allows the efficient incorporation of the cross-linking agent, azidophenacyl, at specific and defined sites within an RNA oligonucleotide.[25] This approach can be used for other small RNA molecules for which there are no X-ray crystal or complete nuclear magnetic resonance (NMR) structures and few topographical constraints available to guide modeling efforts.

The cross-linking approach provides specific distance constraints and indicates sites that represent attractive candidates for the formation of tertiary interactions. This approach is also an excellent complement to the chemical and hydroxyl-radical probing which, as described above, provides information concerning the regions protected on the folding of an RNA complex and indicate regions that are potentially involved in tertiary contacts.

A variant of the hairpin ribozyme has been designed where three short synthetic RNA oligonucleotides assemble into a fully active and properly folded ribozyme complex has been designed (see Fig. 1E).[24] In this construct the normal hinge region between positions 14 and 15, which linked the two domains together, is replaced by six unpaired cytidines and a seven base-pair helix (the zipper helix).

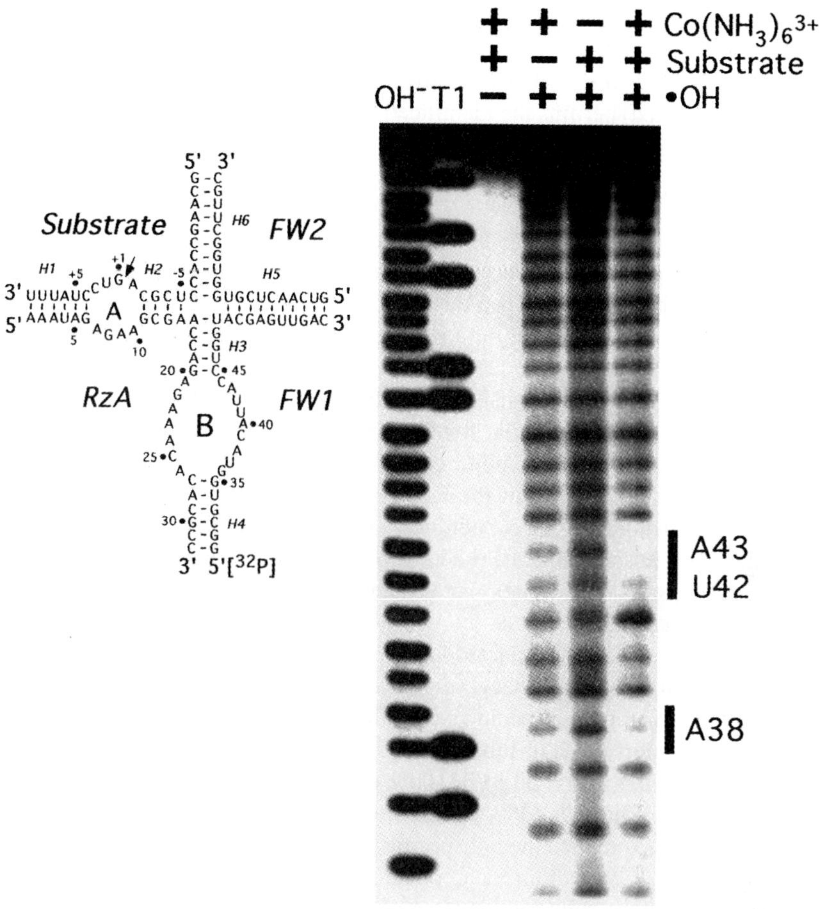

FIG. 5. Hydroxyl-radical footprinting of the native *trans*-acting SV5 ribozyme–substrate complex. The solvent accessibility of 5′-^{32}P-labeled FW1 was probed according to the described method. *Note:* The assignment of protections is one base below the corresponding alkali hydrolysis band because ·OH-mediated cleavage results in removal of the nucleotide at the site of attack.

The four-way junction variant (Fig. 1C), which also is assembled from four short strands, also has been used successfully with this particular method. The principle of the cross-linking method is simple. One of the ribozyme RNA strands is produced by solid phase synthesis and made to contain a single phosphorothioate linkage at a chosen attachment site. The modified oligonucleotide and the azidophenacyl bromide are combined in solution, and the azidophencyl moiety is transferred to the sulfur.[32] The photoaffinity probe-containing strand is incubated with the other strands to assemble the ribozyme complex. The substrate strand then

[32] A. Burgin and N. R. Pace, *EMBO J.* **9,** 4111 (1990).

is added to complete the assembly. Upon UV irradiation using a 312 nm light, the azido group is converted to a nitrene, which reacts with its surroundings (effective radius of 9 Å) and forms a covalent link. Selective labeling of one of the strand (5′-^{32}P-end-labeled) permits the identification and analysis of the cross-link products by polyacrylamide gel electrophoresis.

Chemicals

Azidophenacyl bromide (Sigma, St. Louis, MO)
O-Methyl and regular phosphoramidites for solid-phase chemistry (Glen Research, Inc.) Sulfurization agent (3*H*-1, 2-benzodithiol-3-one 1,1-dioxide) (Glen Research, Inc.) RNase T1 (Boehringer Mannheim)

The cross-linking approach consists of the following four steps.

Oligonucleotide Preparation. Synthetic RNA oligonucleotides are synthesized using solid-phase phosphoramidite chemistry and purified as described previously.[25,33] The sulfurization of the RNA oligonucleotides is carried out during their synthesis using 3*H*-1,2-benzodithiol-3-one 1,1-dioxide. A 2′-*O*-methyl group adjacent to the phosphorothioate is added in order to avoid degradation of the synthesized RNA due to the formation of unstable triesters following alkylation of the phosphorothioate group.

Coupling Reaction. As depicted in Fig. 6, the cross-linking agent, in this case the commonly used azidophenacyl agent is coupled to a sulfur that replaces a nonbridging oxygen atom at a specific site in the ribose–phosphate backbone.[25,32,34] In amber Eppendorf tubes or tubes protected from light (the azido group is light sensitive) add in this order: 60 μl of 100% methanol, 10 μl of sulfur/2′-*O*-methyl synthetic oligonucleotides (300–1500 pmol), 60 μl of H_2O, 2μl of a 100 m*M* solution of ATP (to reduce the nonspecific incorporation of azidophenacyl group to adenosine), 4μl of sodium bicarbonate (from a 1 *M* stock solution at pH 9), and finally 60μl of a solution containing 250 m*M* (60 mg/ml) of azidophenacyl bromide or related cross-linking agents in 100% methanol. Complete with water to a final volume of 200 μl. Incubate 4 hr at room temperature in the dark. Lyophilize the samples in the dark (cover the Speed-Vac lid with aluminum foil). Add 100 μl of H_2O to the dry pellet and extract three times with 2 volumes of diethyl ether. The lower phase is the aqueous phase. Purify the coupled oligonucleotides by high-performance liquid chromatography (HPLC). Because of the stereospecificity of the oxygen atoms, the sulfurization of the oligonucleotides gives rise to two diastereoisomers (R_p and S_p). Although difficult to separate before coupling, the two isomers can be discriminated and isolated efficiently by HPLC using a reversed-phase C_8 column (see Fig. 5A for a typical HPLC pattern).

[33] F. Wincott, A. Di Renzo, C. Shaffer, S. Grimm, D. Tracz, C. Workman, D. Sweedler, C. Gonzalez, S. Scaringe, and N. Usmann, *Nucl. Acids Res.* **25,** 2677 (1995).

[34] J. M. Nolan, D. H. Burke, and N. R. Pace, *Science* **262,** 762 (1993).

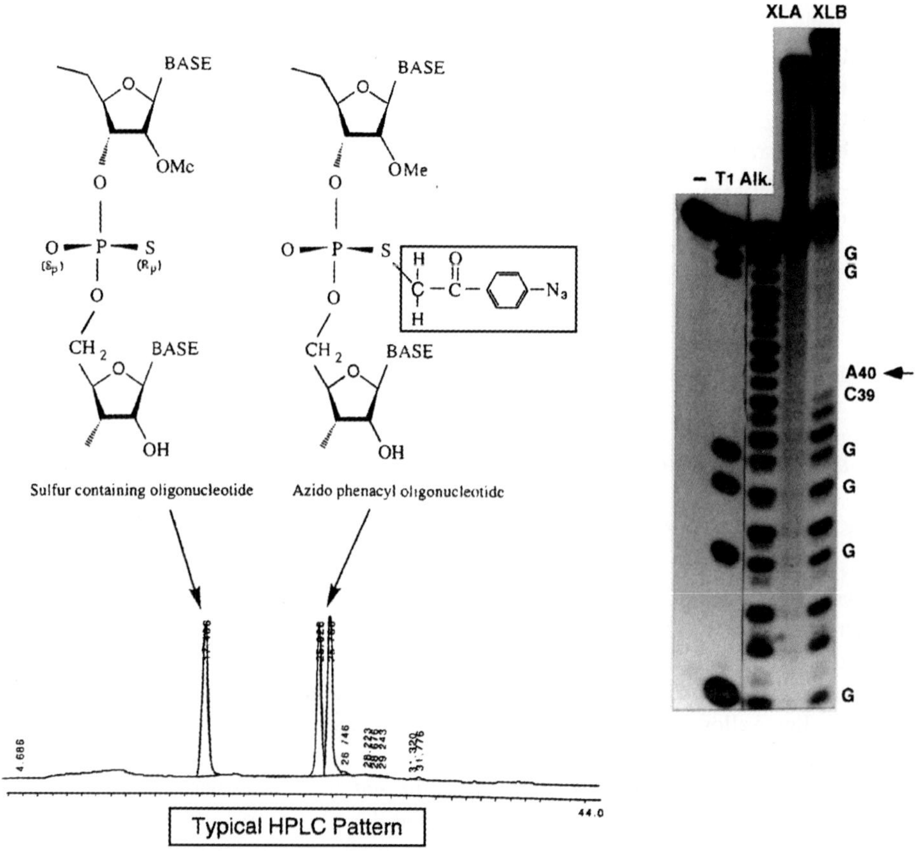

FIG. 6. Sulfur containing oligonucleotide modification and mapping of the cross-linked sites. The azidophenacyl moiety is transferred to the sulfur and the diastereoisomers of the azidophenacyl oligonucleotides are separated and purified by HPLC. The cross-linked sites are mapped by limited alkaline cleavage and/or partial digestion with RNase T1. Interruption in the pattern is indicated by an arrow on the right-hand side and permits the deduction of the cross-linked site.

Assembly of Complexes and Formation of Cross-Link Species. In a typical cross-linking experiment, one stand of the ribozyme contains the cross-linking agent and another strand is 5′-^{32}P-end-labeled and constitutes the "target" strand. The assembly of the ribozyme complex or the ribozyme–substrate complex is performed in 50 m*M* Tris-HCl (pH 7.5), 12 m*M* $MgCl_2$. In a tube protected from light exposure, incubate the azidophenacyl-containing oligonucleotide (100–200 n*M*) with a 5- to 10-fold molar excess of the appropriate synthetic ribozyme RNA fragments in the reaction buffer for 15 min at 37°. When ribozyme–substrate complexes are analyzed, use a noncleavable version of the substrate that contains a 2′-deoxy-1-nucleotide as described above. In each reaction one of the fragments contains the cross-linking agent and one is ^{32}P-end-labeled at either the 5′ or 3′ end. Allow the assembled complex to equilibrate for 10 min at room temperature by transferring the mixture into a well of a Falcon 96-well U-bottom flexible assay plate. Screen the assay plate with a polystyrene filter (use the lid of a large petri dish) and irradiate with 312 nm ultraviolet light (hand-held model VL-6M, IBI Inc. or equivalent) for 10 to 20 min at room temperature. The cross-linked species can then be separated on a 20% polyacrylamide, 8 *M* urea gel. Cut out the band from the gel and elute overnight in 0.5 *M* ammonium acetate, 1 m*M* EDTA, and 0.1% SDS. The cross-linked species are precipitated with ethanol and passed through a small column to remove salts (Centrisep column, Princeton Separation Inc.).

Mapping of Cross-Linking Sites. Identification of the cross-linked site in the target strand is determined by partial digestion with ribonuclease T1 and by limited alkaline hydrolysis.[32] For partial ribonuclease digestion, cross-linked species are incubated in 33 m*M* sodium citrate, pH 3.5, 9.4 *M* urea, 1.5 m*M* EDTA, and 0.04% (w/v) xylene cyanol for 10 min at 50° in the presence of 0.1 unit of RNase T1. For partial alkaline cleavage, the species are incubated in 50 m*M* sodium phosphate pH 12 for 10 min at 50°C. In alkaline buffer, however, the cross-linked species are less stable; 3 to 5% of the labeled strand can be released (the two strands come apart). We have noticed that the urea buffer used for the ribonuclease digestion greatly reduces the dissociation of the cross-linked species. Partially digested cross-linked species are loaded on a 20% denaturing polyacrylamide gel. Non-cross-linked 5′-end-labeled "target" strands are subjected to the same treatment and run in parallel to allow the identification of the cross-linked site. The digested fragments resulting from cleavage occurring 5′ from the linkage site will comigrate with the fragements generated from the non-cross-link segments. Alternatively, cleavage to the 3′ side of the cross-linked site (beyond the site of linkage) will generate fragments that will migrate with a lower electrophoretic mobility compared to the non-cross-linked strand since the 5′-end-labeled RNA and the azidophenacyl-containing strand still are linked together. The distinct gap created in the digestion or hydrolysis pattern of the conjugated species allows the identification of the cross-link site (see Fig. 5B).

Controls and Other Useful Information. In order to determine if the addition of the cross-linking agent perturbs the cleavage activity and to establish that the complex assembled reflects an active conformation, a single turnover cleavage assay should be performed as described above. Only cross-links that are formed reproducibly should be selected for analysis. In addition, for each set of cross-link experiments, uncoupled phosphorothioate-containing oligonucleotides should be used to verify that all the cross-linked species were azidophenacyl-dependent. Some low-efficiency cross-links can be detected occasionally with non-photoaffinity probe-containing RNA. These background cross-links are not analyzed. The efficiency of formation of the cross-link analyzed varies. One should expect 1–5% of the molecules to be cross-linked.

[37] Intracellular Applications of Ribozymes

By ALESSANDRO MICHIENZI and JOHN J. ROSSI

The discovery of ribozymes (RNA with catalytic activity) in 1981 by T. Cech and colleagues[1] is a milestone in the comprehension of the origin of life ("RNA world" theory) and also provided a new tool for studying gene expression as well as a potential new therapeutic agent.

Biochemical, structural, and mutational studies have probed deeply into the structural–functional bases of ribozyme activity and have also provided the tools for modifying these RNAs to bind and cleave any given RNA target. Ribozymes have been successfully used to inactivate cellular, viral, and reporter RNAs in different model organisms including *Xenopus laevis,*[2,3] *Drosophila,*[4] zebrafish,[5] yeast,[6] mice,[7,8] and human cells.[9–11] Progress in the development of these RNA molecules has led to tests of function in human clinical trials as well.

The main goal of this chapter is to offer to readers a general view of different strategies that can be exploited, from ribozyme design to *in vivo* expression applications.

Catalytic RNAs

Catalytic RNAs discovered thus far have been divided into seven classes: (1) self-splicing group I introns, (2) self-splicing group II introns, (3) RNA component of RNase P, (4) hepatitis δ virus ribozyme, (5) hammerhead ribozyme (plant viroids and virusoids), (6) hairpin ribozyme (plant viroids and virusoids), and (7) *Neurospora* VS RNA ribozyme.

[1] T. R. Cech, A. J. Zaug, and P. J. Grabowski, *Cell* **27,** 487 (1981).
[2] M. Cotten and M. L. Birnstiel., *EMBO J.* **8,** 3861 (1989).
[3] A. Michienzi, S. Prislei, and I. Bozzoni, *Proc. Natl. Acad. Sci. U.S.A.* **93,** 7219 (1996).
[4] J. J. Zhao and L. Pick, *Nature* **365,** 448 (1993).
[5] Y. Xie, X. Chen, and T. E. Wagner, *Proc. Natl. Acad. Sci. U.S.A.* **94,** 13777 (1997).
[6] D. A. Samarsky, G. Ferbeyre, E. Bertrand, R. H. Singer, R. Cedergren, and M. J. Fournier, *Proc. Natl. Acad. Sci. U.S.A.* **96,** 6609 (1999).
[7] S. Larsson, G. Hotchkiss, M. Andang, T. Nyholm, J. Inzunza, I. Jansson, and L. Ahrlund-Richter, *Nucleic Acids Res.* **22,** 2242 (1994).
[8] S. Effrat, M. Leiser, Y. J. Wu, D. Fusco-DeMane, O. A. Emran, M. Surana, T. L. Jetton, M. A. Magnuson, G. Weir, and N. Fleischer, *Proc. Natl. Acad. Sci. U.S.A.* **91,** 2051 (1994).
[9] J. J. Rossi, *Curr. Biol.* **4,** 469 (1994).
[10] J. D. Thompson, *Methods Enzymol.* **306,** 241 (1999).
[11] N. Sarver, E. M. Cantin, P. S. Chang, J. A. Zaia, P. A. Ladne, D. A. Stephens, and J. J. Rossi, *Science* **247,** 1222 (1990).

METHODS IN ENZYMOLOGY, VOL. 341
0076-6879/01 $35.00

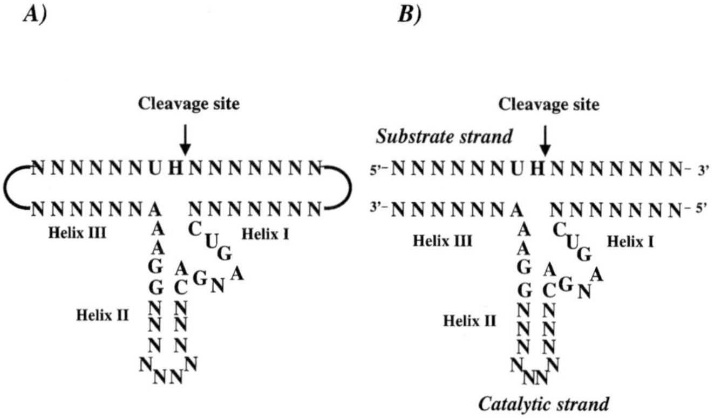

FIG. 1. Schematic representation of the hammerhead ribozyme. N represents any nucleotide; H is A, C or U. (A) *Intramolecular* cleaving ribozyme, (B) *intermolecular* cleaving ribozyme.

The hammerhead is the simplest of the ribozyme motifs in terms of size and structure.[12] This catalytic RNA was originally discovered in plant viroids and virusoids as part of their RNA replication process. The hammerhead ribozyme catalyzes the self-cleavage reaction of genomic RNA concatamers produced by a rolling circle replication, yielding monomers of the pathogenic genome.[13–15] Comparative sequence analyses of a variety of different hammerhead RNA self-cleavage domains as well as biochemical and mutational analyses have defined the structure and sequence elements critical for catalysis. The hammerhead ribozyme has 11 conserved nucleotides that form the catalytic core and three base-paired helixes (I–III) (Fig. 1A). The *intramolecular* self-cleaving reaction mediated by the natural hammerhead ribozyme can be converted to an *intermolecular, trans*-cleaving reaction by dividing the ribozyme into two strands, the substrate and the catalytic strand, that interact to form an active hammerhead structure[16] (Fig. 1B). Since helixes I and III do not contain conserved nucleotides, it is possible to design a *trans*-cleaving ribozyme containing the conserved catalytic core sequence, helix II, and flanking regions that are complementary to the substrate strand containing the cleavage site. A hammerhead ribozyme can potentially be directed against any given RNA target. Although *in vitro* mutagenesis experiments have shown that the cleavage can occur at any UH (where H is any nucleotide but G), the efficiency

[12] K. R. Birikh, P. A. Heaton, and F. Eckstein, *Eur. J. Biochem.* **245,** 1 (1997).
[13] A. C. Forster and R. H. Symons, *Cell* **49,** 211 (1987).
[14] J. Haseloff and W. L. Gerlach, *Nature* **334,** 585 (1988).
[15] R. H. Symons, *Annu. Rev. Biochem.* **61,** 641 (1992).
[16] O. C. Uhlenbeck, *Nature* **328,** 596 (1987).

of cleavage varies with the RNA target sequence.[17–19] Nevertheless, there are potentially dozens to hundreds of hammerhead ribozyme cleavage sites in any given messenger RNA.

Because the parameters affecting the hammerhead catalytic activity have been determined from *in vitro* studies, they are not always useful for the designed ribozymes intended for *in vivo* (intracellular) use. Several factors affect the intracellular activity of the hammerhead ribozyme, such as stability, intracellular level of expression, colocalization with the target RNA, folding, and target RNA accessibility.

Target Accessibility

Because the ribozyme must bind the target RNA to mediate the cleavage reaction, the accessibility of the cleavage sites is a major starting point in ribozyme design. In a cellular environment, the RNA is complexed with sequence specific binding proteins as well as with nonsequence specific heterogeneous ribonucleoproteins (hnRNPs). These protein–RNA interactions can dramatically change the secondary and tertiary structure of RNA. Therefore, in a cell, some of the potential cleavage sites may not be accessible for ribozyme interaction or differentially accessible when compared to the naked RNA *in vitro*. In the past, researchers have most often used RNA secondary structure prediction programs (such as Mfold)[20] to find potential single-stranded regions of the RNA, that are considered to be the most accessible sites. Based on the variable successes of this approach, other strategies have been developed. Various approaches are based on random libraries of ribozymes or antisense oligonucleotide (ODNs) and RNase H cleavage. The efficiency of the cleavage reaction on the RNA target by the ODN is directly correlated with the accessibility of the complementary RNA sequence. Using the antisense oligonucleotide technique Birikh *et al.* identified accessible sites on human acetylcholinesterase RNA.[21] Ribozymes designed to cleave the identified sites were much more active *in vitro* than the ones designed to bind sites predicted based on a computer RNA folding program. Lieber and Strauss screened the entire sequence of the human growth hormone RNA using a library of hammerhead ribozymes with randomized arms to find accessible cleavage sites on this target RNA.[22] This analysis was performed using total cellular RNA or

[17] D. E. Ruffner, G. D. Stormo, and O. C. Uhlenbeck, *Biochemistry* **29,** 10695 (1990).

[18] R. Perriman, A. Delves, and W. L. Gerlach, *Gene* **113,** 157 (1992).

[19] T. Shimayama, S. Nishikawa, and K. Taira, *Biochemistry* **34,** 3649 (1995).

[20] A. B. Jacobson and M. Zuker, *J. Mol. Biol.* **233,** 261 (1993).

[21] K. R. Birikh, Y. A. Berlin, H. Soreq, and F. Eckstein, *RNA* **3,** 429 (1997).

[22] A. Lieber and M. Strauss, *Mol. Cell. Biol.* **15,** 540 (1995).

cytoplasmic extract. The cleavage products were detected using RACE PCR and sequencing. Ribozymes designed to cleave the identified accessible sites were highly active *in vitro* and *in vivo*. One of these ribozymes was also used successfully in an animal experiment.[23] In an effort to mimic the intracellular environment, we have established a simple and rapid method to identify RNA-accessible sites using ODNs in cellular extracts or native mRNA.

As an example, ODNs targeted to three different regions of the murine DNA methyltransferase mRNA (MTase) were tested for their ability to direct RNase H-mediated cleavage in a cellular extract prepared from NIH 3T3 cells.[24] The correlation between ODNs and ribozyme accessibility on the MTase message has been demonstrated in extracts and cells.[25] The initial screening with ODNs in extracts is both fast and cost effective.[26] The most accessible sites then can be used for ribozyme design.

Intracellular Ribozyme Expression

Once ribozyme accessible sites on the target RNA have been identified, the next step is to design a ribozyme that can specifically bind one of them. The catalytic activity of the ribozyme is first tested in the cell extract against the native RNA target, and then in cell culture. For intracellular expression, the ribozyme sequence is inserted in an appropriate expression cassette to achieve high levels of expression, specific subcellular localization, and stability.

Pol II and Pol III promoters have both been used for ribozyme transcription (Figs. 2 and 3). The use of Pol II promoters leads to the addition of a 5′ cap and 3′ poly(A) tail to the transcribed RNA, conferring both stability and cytoplasmic localization (Fig. 2A). An advantage of Pol II promoters is tissue specificity. A potential disadvantage is the requirement for splicing of the primary transcript to ensure nuclear export. Alternatively, the ribozyme can be inserted within small nuclear or cytoplasmic RNAs transcribed from Pol II or Pol III promoters (such as U1 or U6 snRNAs). These RNAs are well characterized, ubiquitously expressed and can accommodate small ribozyme inserts. The U1 snRNA (small nuclear RNA) promoter is very useful for expression of ribozymes or small therapeutic RNAs. Ribozymes have been inserted within the coding sequence of U1 RNA or between the promoter sequence and the U1 transcriptional termination signal[3,27] (Fig. 2B). The presence of the U1 coding sequence confers stability to the ribozyme as a

[23] A. Lieber and M. A. Kay, *J. Virol.* **70,** 3153 (1996).

[24] M. Scherr and J. J. Rossi, *Nucleic Acids Res.* **26,** 5079 (1998).

[25] M. Scherr, M. Reed, C.-F. Huang, A. D. Riggs, and J. J. Rossi, "Molecular Therapy," **2,** 26 (2000).

[26] D. Castanotto, M. Scherr, and J. J. Rossi, *Methods Enzymol.* **313,** 401 (1999).

[27] E. Bertrand, D. Castanotto, C. Zhou, C. Carbonnelle, N. S. Lee, P. Good, S. Chatterjee, T. Grange, R. Pictet, D. Kohn, D. Engelke, and J. J. Rossi, *RNA* **3,** 75 (1997).

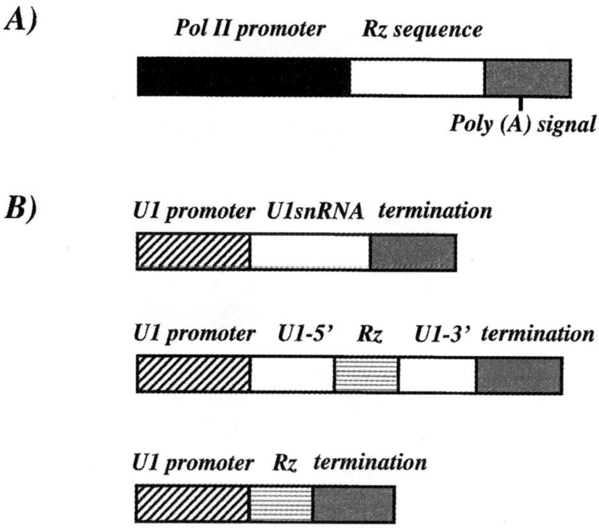

FIG. 2. Pol II expression cassettes. (A) Ribozymes (Rz) are inserted downstream of the Pol II promoter sequence and upstream of the transcription termination sequence. The resulting transcribed ribozymes contain a 5′ cap and a 3′ poly(A) tail that confer intracellular stability and cytoplasmic localization. (B) The U1 snRNA gene can be used as an expression cassette for ribozymes. Ribozymes can be inserted within the U1 snRNA coding sequence and the resulting transcribed ribozymes are generally localized in the nucleus. If the ribozyme replaces the U1 coding sequence, intracellular localization will depend on the appended flanking sequences.

result of the binding of U1 specific proteins, which serve to protect the ribozyme from degradation.

Pol III promoters naturally drive the expression of short highly structured RNAs such as tRNA, U6 snRNA, adenoviral VA1 RNA, 7SL RNA, Y RNAs, and 5S rRNA. The promoter elements are either contained in the RNA coding sequence (box A and B in 5S, VA1 RNA and tRNA) or in an upstream nontranscribed region (U6 snRNA and 7SL RNA) (Fig. 3). Transcription normally terminates within a stretch of 4–6 uridines. VA1 is a highly structured RNA into which ribozymes, antisense RNAs, or RNA decoys have been inserted in a specific stem–loop region called "the central domain"[28,29] (Fig. 3B). The VA1 chimeric RNAs accumulate predominantly within the cytoplasm.

tRNA driven transcriptional units have been also designed to express ribozymes via insertions in the anticodon loop,[2,30,31] in the aminoacyl acceptor stem,[32,33] or

[28] L. Cagnon, M. Cucchiarini, J. C. Lefebvre, and A. Doglio, *J. Acquir. Immune. Defic. Syndr. Hum. Retrovirol.* **9,** 349 (1995).

[29] S. Prislei, S. B. Buonomo, A. Michienzi, and I. Bozzoni, *RNA* **3,** 677 (1997).

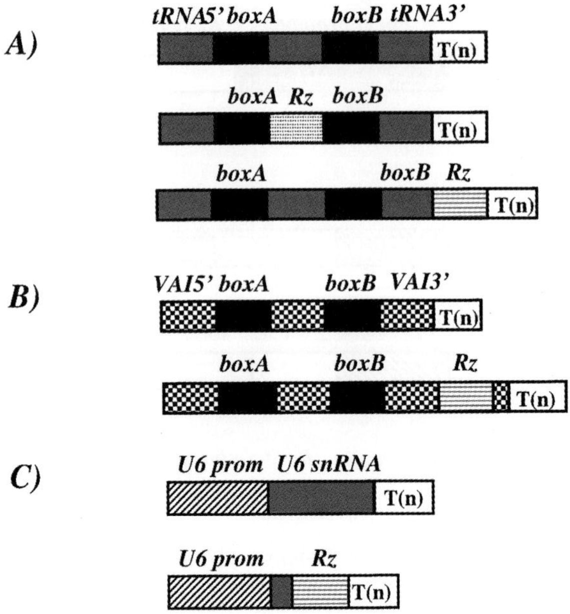

FIG. 3. Pol III expression cassettes. Ribozymes (Rz) can be inserted within the coding sequence of tRNA or VAI genes (A and B) or downstream of the U6 snRNA external promoter with or without additional U6 coding sequence (C).

downstream of the aminoacyl acceptor stem[34,35] (Fig. 3A). The resulting chimeric tRNAs can accumulate in the cytoplasm and/or in the nucleus, depending on how much the tertiary structure has been perturbed. In yeast and in *X. laevis* it has been demonstrated that the tRNA is exported to the cytoplasm only after splicing and 5′ and 3′ end processing have taken place. Thus altering the structure can block processing and transport.[36]

[30] J. Ohkawa, N. Yuyama, Y. Takebe, S. Nishikawa, and K. Taira, *Proc. Natl. Acad. Sci. U.S.A.* **90,** 11302 (1993).
[31] N. Yuyama, J. Ohkawa, T. Koguma, M. Shirai, and K. Taira, *Nucleic Acids Res.* **22,** 5060 (1994).
[32] M. Yu, J. Ojwang, O. Yamada, A. Hampel, J. Rappaport, D. Looney, and F. Wong-Staal, *Proc. Natl. Acad. Sci. U.S.A.* **90,** 6340 (1993).
[33] J. D. Thompson, D. F. Ayers, T. A. Malmstrom, T. L. McKenzie, L. Ganousis, B. M. Chowrira, L. Couture, and D. T. Stinchcomb, *Nucleic Acids Res.* **23,** 2259 (1995).
[34] S. K. Westaway, L. Cagnon, Z. Chang, S. Li, H. Li, G. P. Larson, J. A. Zaia, and J. J. Rossi, *Antisense Nucleic Acid Drug Dev.* **8,** 185 (1998).
[35] A. Hampel, *Prog. Nucleic Acids Res. Mol. Biol.* **58,** 1 (1998).
[36] G. Simos and E. Hurt, *Curr. Biol.* **9,** 238 (1999).

Several different expression cassettes have been constructed from the U6 snRNA gene. These all contain the 5′ promoter element and various amounts of the U6 snRNA coding sequence (Fig. 3C). In some instances, an artificial stem and loop immediately preceding the transcriptional termination sequence have been constructed to help to stabilize the transcript from degradation.[37] Therapeutic RNAs cloned in the U6 expression cassettes accumulate primarily in the nucleus. The presence of the first 27 nucleotides of the U6 RNA sequence signals the phosphomethyl γ cap of the chimeric RNAs, which may serve as a nuclear retention signal.[37]

Colocalization of Ribozyme and RNA Target

A critical factor for intracellular ribozyme effectiveness is colocalization with the target RNA. Because specific RNAs may localize within different cellular compartments (such as the nucleus, nucleolus, or cytoplasm), it is important that the ribozyme reach the same location.

Sullenger and Cech first demonstrated the importance of colocalization. They inserted a hammerhead ribozyme and the LacZ target sequence inside two different retroviral vectors.[38] Following coexpression of the two viral RNAs in a packaging cell line, the hammerhead ribozyme would be expected to bind and cleave the LacZ sequence in the cytoplasm where translation occurs.

In this work the authors demonstrated that the cleavage of LacZ RNA and inhibition of β-gal activity took place during the copackaging in the virions rather than within the cytoplasm during translation. This was a result of the ribozyme colocalizing with the LacZ target via the packaging signals contained in the two different viral RNAs. Following this work, several papers have been published using diverse strategies, which further demonstrate the importance of the ribozyme–target colocalization within the same subcellular compartments. For example, a hammerhead ribozyme was appended to the U1 small nuclear RNA to deliver it into the nucleus where the target pre-mRNA of human immunodeficiency virus type 1 (HIV-1) Rev undergoes splicing. The U1-hammerhead ribozyme (U1-Rz) was demonstrated to be highly active against the Rev pre-mRNA in two different experimental systems: in oocytes of *Xenopus laevis* coinjected in the nucleus with the plasmids coding for Rev pre-mRNA and U1-Rz,[3] and in a Jurkat cell line stably expressing the U1-Rz infected by HIV-1.[39]

[37] P. D. Good, A. J. Krikos, S. X. Li, E. Bertrand, N. S. Lee, L. Giver, A. Ellington, J. A. Zaia, J. J. Rossi, and D. R. Engelke, *Gene Ther.* **4,** 45 (1997).

[38] B. A. Sullenger and T. R. Cech, *Science* **262,** 1566 (1993).

[39] A. Michienzi, L. Conti, B. Varano, S. Prislei, S. Gessani, and I. Bozzoni, *Hum. Gene Ther.* **9,** 621 (1998).

The 3′ UTR of several mRNAs have been found to have subcellular localization signals. These signals have been given the designation "zipcode."[40] The zipcode is sufficient to deliver a reporter molecule to the appropriate subregion of the cytoplasm in embryonic myoblasts and fibroblasts. We expressed ribozymes and a LacZ target sequence containing either the α- or β-actin 3′UTR zipcode. When the localization signals in the ribozyme and in the LacZ target RNA were matched (α–α or β–β) there was a statistically significant increase in the percentage of cellular colocalization between the two RNAs. Consequently, the ribozyme mediated a reduction in β-galactosidase (3-fold) consistent with the extent of observed colocalization.[41]

Even more discrete colocalization of a hammerhead ribozyme and a target RNA was achieved by appending both sequences to the U3 small nucleolar RNA, which colocalized the transcripts to the nucleolus, resulting in nearly complete ribozyme-mediated destruction of the target in a yeast system.[6]

In this article we describe experiments for subcellular colocalization of a hammerhead ribozyme with its HIV-1 RNA target.

On the basis of the controversial results obtained using *in situ* hybridization to detect HIV RNA within the nucleoli of human cells,[42–46] we tried to address the same problem using an alternative approach based on the design of an anti-HIV-1 hammerhead ribozyme with nucleolar localization.[47] If HIV-1 RNAs pass through this compartment, the ribozyme would be colocalized with HIV-1 RNA, thereby maximizing the potential for interaction with the target. To deliver the ribozyme into the nucleolus, we appended it to the U16 small nucleolar RNA (snoRNA)[48] (Fig. 4). U16 is a member of a large family of snoRNAs that accumulate in the nucleoli of eukaryotic cells.[49] SnoRNAs are primarily involved in posttranscriptional modifications of rRNA (ribose methylation and pseudouridylation).[49]

The expression of the U16-ribozyme (U16Rz) was achieved by cloning it downstream of the human U6 promoter (Fig. 5A, top). A functionally disabled control ribozyme was also constructed by mutating the C_3 nucleotide of the hammerhead catalytic core to a G.[50]

[40] E. Kislauskis, X. Zhu, and R. H. Singer, *J. Cell. Biol.* **127,** 441 (1994).

[41] N. S. Lee, E. Bertrand, and J. J. Rossi, *RNA* **5,** 1200 (1999).

[42] G. Zhang, M. L. Zapp, G. Yan, and M. R. Green, *J. Cell Biol.* **135,** 9 (1996).

[43] S. O. Boe, B. Bjorndal, B. Rosok, A. M. Szilvay, and K. H. Kalland, *Virology* **244,** 473 (1998).

[44] J. P. Favaro, K. T. Borg, S. J. Arrigo, and M. G. Schmidt, *Virology* **249,** 286 (1998).

[45] J. P. Favaro, F. Maldarelli, S. J. Arrigo, and M. G. Schmidt, *Virology* **255,** 237 (1999).

[46] V. I. Romanov, A. S. Zolotukhin, N. A. Aleksandroff, P. P. Da Silva, and B. K. Felber, *Virology* **228,** 360 (1997).

[47] A. Michienzi, L. Cagnon, I. Bahner, and J. J. Rossi, *Proc. Natl. Acad. Sci. U.S.A.* **97,** 8955 (2000).

[48] P. Fragapane, S. Prislei, A. Michienzi, E. Caffarelli, and I. Bozzoni, *EMBO J.* **12,** 2921 (1993).

[49] L. B. Weinstein and J. A. Steitz, *Curr. Opin. Cell Biol.* **11,** 378 (1999).

[50] K. J. Hertel, A. Pardi, O. C. Uhlenbeck, E. Ohtsuka, S. Uesugi, R. Cedergren, F. Eckstein, W. L. Gerlach, R. Hodgson, and R. H. Symons, *Nucleic Acids Res.* **20,** 3252 (1992).

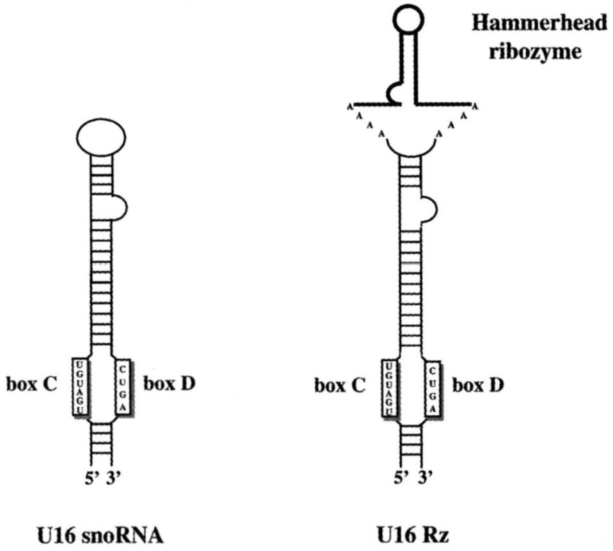

FIG. 4. Schematic representation of U16Rz. To investigate whether the HIV RNAs traffic through the nucleolus, we designed a nucleolar localizing anti-HIV hammerhead ribozyme. The hammerhead ribozyme is targeted against a highly conserved sequence located in the 5′ nontranslated region of HIV-1 and designed to cleave at position +115 relative to the transcription initiation site.[56] The ribozyme was inserted in the apical loop of the U16 snoRNA (clone U16Rz).

Ribozyme Delivery

Ribozymes can be expressed endogenously or delivered exogeneously. For endogenous expression, the ribozyme genes are inserted in an appropriate vector and then delivered to cells by transient transfection or transduction, thereby effecting either transient or long term expression of the ribozymes. In an exogenous system, presynthesized ribozymes are delivered to the target cells using various carriers such as liposomes.

Endogenous Delivery

Viral Delivery. Several viral vectors have been exploited for gene delivery (retroviral, lentiviral, adenoviral, adeno-associated viral, herpes). The murine-based retroviral vectors are the most well characterized and widely used.[51] The genomes of murine retroviruses contain three genes: *gag* (capsid-group antigen), *pol* (polymerase), and *env* (envelope). These encode all the viral proteins necessary for replication, reverse transcription, encapsidation, and infection. The retroviral system is based on the use of a retroviral vector containing the *cis* sequences

[51] R. A. Morgan and W. F. Anderson, *Annu. Rev. Biochem.* **62,** 191 (1993).

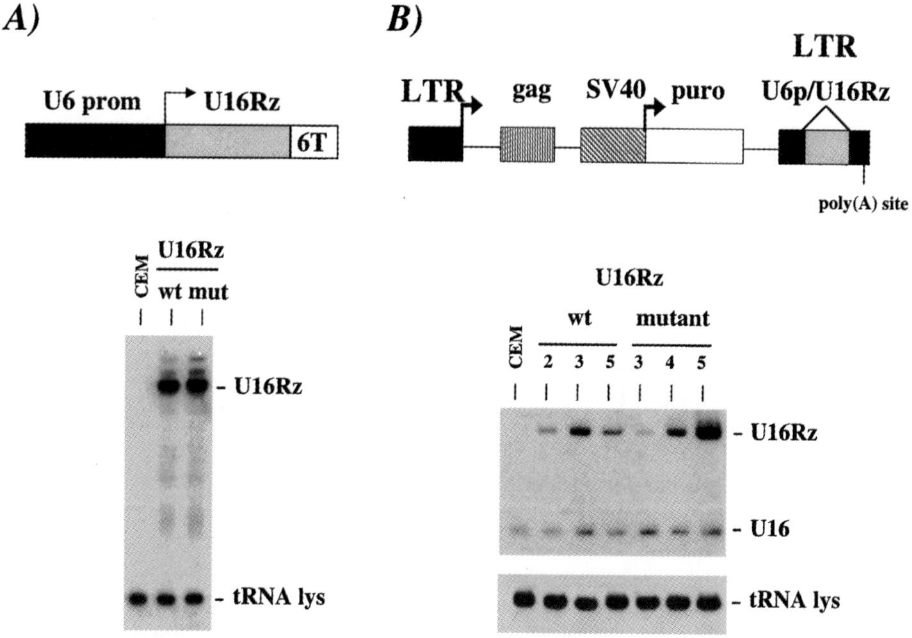

FIG. 5. Intracellular expression of the U16Rz. To test intracellular expression of the U16Rz we inserted the sequence within a U6 expression cassette (Fig. 3C). (A) *Top:* Schematic representation of the U6-U16Rz expression cassette *Bottom:* Northern blot analysis of 10 μg of total RNAs extracted from 293 cells transiently transfected with the U6-U16 wild-type and mutant plasmids. The RNAs were electrophoresed in a 6% polyacrylamide–7 *M* urea gel. After blotting onto a nylon filter, hybridizations with specific probes were carried out to detect the U16Rz RNA and the $tRNA_3^{Lys}$ used as loading control. (B) To establish human cell lines stably expressing the U16Rz, we inserted the U6-U16Rz wt and mutant expression cassettes within the U3 region of the 3′LTR of the pBabe retroviral vector (*top*). CEM human T-lymphoblastoid cells were then transduced with the pBabe constructs and single stable clones were isolated by puromicin selection. Northern blot analysis was performed following electrophoresis of 5 μg total RNA in a 6% polyacrylamide–7*M* urea gel. RNAs from cells expressing the U16Rz wild type (clones 2, 3, and 5) and U16Rz mutant (3, 4, and 5) were analyzed. After blotting onto a nylon filter, we performed hybridization with specific probes to detect the U16Rz (upper) and the $tRNA_3^{Lys}$ used as loading control (*bottom*).

necessary for the viral RNA genome encapsidation and a packaging cell line that stably expresses *in trans* all the viral proteins required to assemble infectious virions. The retroviral vector contains only minimal viral sequences such as the 5′ and 3′ long terminal repeats (LTRs) the packaging (psi) signal, splice donor and acceptor sites, an antibiotic-resistance gene for positive selection, and often multiple cloning sites. Using transient or stable transfection of the packaging cell line with the retroviral vector, infectious virions are produced and collected. These virions can be then used to infect a suitable cell line. The tropism of the vector depends on the type of envelope used for packaging. The most popular envelopes are amphotropic (infect murine and human cells) or pseudo-typed envelopes such as vesicular stomatitis virus, which has a broad host range.[52] The retroviral system offers the advantage of efficient transduction of a wide variety of cells and stable integration into the host genome. A potential drawback is epigenetic silencing of the transgene in the absence of selection.

Nonviral Delivery. An increasing number of transfection reagents and methods are available for vector delivery. These include procedures such as calcium phosphate, electroporation, and liposome-mediated transfection.

Exogenous Delivery

This alternative method is based on the utilization of chemically or biochemically synthesized ribozymes. For chemically synthesized ribozymes, several modifications have been developed to increase the intracellular stability of the ribozyme. These include 2′-OH modifications using methylation, amino, allyl, and fluoro derivatives. These modifications are often combined with terminal phosphorothioate linkages and inverted 5′- and 3′-terminal linkages.[53] The chemically synthesized ribozymes can be either directly delivered to the cell culture media or transfected using carriers such as cationic liposomes. The development of chemical modifications that increase the stability of ribozymes without compromising catalytic activity and sequence specificity make this an attractive approach for ribozyme delivery. A disadvantage of this approach is that for most applications, multiple deliveries of the ribozyme must be made to maintain the desired target down-regulation.

As an example of endogenous ribozyme expression we describe the use of our U6-U16Rz constructs (Fig. 5A, top). Initially we tested the intracellular expression of the U16Rz by transient transfection of human 293 cells with the U6-U16Rz plasmid using calcium phosphate (GIBCO-BRL, Gaithersburg, MD) (Fig. 5A, bottom).

We next delivered the ribozyme gene into the human T-lymphoblastoid CEM1 (provided by J. Zaia, City of Hope) cell line using the murine pBabe retroviral vector.[54]

The ribozyme gene was inserted in the U3 region of the 3′LTR to take advantage of mechanism of duplication mediated by the viral reverse transcriptase (Fig. 5B). This enzyme duplicates the U3 region of the 3′LTR into the U3 region of the 5′LTR,[55] thus generating a double copy of the ribozyme gene (DCT).

[52] J. K. Yee, T. Friedmann, and J. C. Burns, *Methods Cell Biol.* **43,** 99 (1994).

[53] F. Eckstein, *Ciba Found. Symp.* **209,** 212 (1997).

[54] J. P. Morgestern and H. Land, *Nucleic Acids Res.* **18,** 3587 (1990).

[55] B. A. Sullenger, T. C. Lee, C. A. Smith, G. E. Ungers, and E. Gilboa, *Mol. Cell. Biol* **10,** 6512 (1990).

The Phoenix packaging cell line (http//www.stanford.edu/nolan/NL-phnxr.html) was chosen to produce infectious virions for CEM cell transductions. This cell line is a derivative of 293T cells and was designed to express the *gag-pol* and the *env* genes from separate promoters. To monitor the production of gag-pol proteins an IRES-CD8 was fused downstream of the gag-pol sequence. The CD8 level and therefore the amount of gag-pol produced can be easily monitored by flow cytometry.

Transduction of CEM Cells

The Phoenix Packaging cell line was cultured in Dulbecco's modified Eagle's medium (DMEM) (Irvine Scientific, Santa Ana, CA) containing 10% fetal calf serum (FCS, Irvine Scientific), penicillin (10 units/ml, Irvine Scientific), and streptomycin (100 μg/ml, Irvine Scientific) and plated at 1.5×10^6 cells per 60 mm dish, one day prior to transfection. Five minutes before the transfection, 25 μM chloroquine was added to each plate. The cells were then transiently transfected with 6 μg of the different pBabe puro constructs (pBabe Puro/U16Rz wt and mutant) using a calcium phosphate kit (GIBCO-BRL). Eight hours after transfection the precipitate was washed and replaced by fresh media. Thirty-two hours after transfection fresh media was exchanged for the spent media. Forty-eight hours after transfection a pellet of 1×10^6 CEM cells was resuspended with 2 ml of virus (Phoenix cell supernatants) and 28 μl of protamine sulfate (400 μg/ml, ELKINS-SINN, Inc., Cherry Hill, NJ) and spun for 90 min at 2500 rpm at 32°. After the spin, the virally infected CEM cells were incubated for 150 min at 37°. The supernatant was then removed and the CEM cells resuspended in 5 ml of RPMI-1640 supplemented with 10% FCS (Irvine Scientific), penicillin (10 units/ml, Irvine Scientific), and streptomycin (100μg/ml, Irvine Scientific) and incubated for 48 hr at 37° under 5% (v/v) CO_2.

Stable Clone Selection

For puromycin-resistant clonal selection, 1.5μg/ml puromycin was added to the medium and cells were incubated in the presence of this drug for 3 weeks to obtain pooled, drug-resistant populations of cells. Single stable clones were obtained from the pools by limiting dilution. A Northern blot analysis was performed on total RNA extracted from the selected, CEM single, stable clones to monitor the levels of expression of the U16Rz (Fig. 5B).

After analyzing the U16Rz expression in 293 cells and CEM cells, we next tested the intracellular localization of the U16Rz. Two methods are generally used to study RNA localization: cellular fractionation and *in situ* hybridizations. Here

[56] L. Ratner, W. Haseltine, R. Patarca, K. J. Livak, B. Starcich, S. F. Josephs, E. R. Doran, J. A. Rafalski, E. A. Whitehorn, and K. Baumeister, *Nature (London)* **313,** 277 (1985).

we describe *in situ* hybridization which is based on the protocol from the R. H. Singer laboratory (http://singerlab.aecom.yu.edu/protocols).

In Situ *Hybridization*

Probe Design. DNA oligonucleotides or *in vitro* transcribed RNAs can be used as probes for *in situ* hybridization. The DNA oligonucleotides are preferable because their size allows more efficient penetration into fixed cells.

DNA oligonucleotides are designed to be complementary to the sequence of the RNA to be detected and contain amino-allyl T modified nucleotides. The amino-allyl T nucleotides (amino-modifier C6-dT, Glen Research, Sterling, VA) are incorporated during the primer synthesis at approximately one every 10 bases. The primers should be around 50 bases in length, designed to contain about 5 amino-modified nucleotides with a G-C content of about 50%. The primers should be gel purified after synthesis.

Chemical Conjugation. The purified amino-allyl-T modified primers are conjugated with a specific fluorophore (Fluorolink Cy-3, Amersham Pharmacia Biotech, Buckinghamshire, UK and Oregon Green 488, Molecular Probes, Eugene, OR). The labeling reaction is performed by adding 5 μg of amino-allyl T primer (resuspended in 70 μl of 0.1 *M* $NaHCO_3$, pH 8.8) to 30 μl of dimethylsulfoxide (DMSO) containing the activating fluorophore (1 vial of Cy-3 or 1 mg of Oregon Green). After 48 h of labeling conducted in the dark at room temperature, the probes are purified by either gel filtration through Sephadex G-50 columns or by gel purification.

Cell Fixation

Adherent cells. The cells are grown on coverslips that have been previously fixed on the bottom of a 100 mm dish treated with 0.5% gelatin, and incubated overnight. The cells are next transiently transfected with a plasmid containing the ribozyme expression cassette, and after 48 h are washed once with phosphate-buffered saline (PBS) then fixed for 30 min with 4% paraformaldehyde, in PBS. The cells are then washed twice with PBS, permeabilized with 70% (v/v) ethanol, and maintained at 4° before the hybridization (the cells can be maintained at 4° for weeks before *in situ* hybridization analysis).

Suspension cells. Ten to 40 μl of cells (2×10^7 cells/ml density) stably transfected or transduced with a plasmid containing the ribozyme expression cassette is dropped onto a polylysine-coated glass slide. After 30 min the glass slides are transferred to 100 mm dishes containing 4% paraformaldeheyde in PBS and incubated for 30 min at room temperature. The cells are washed twice with PBS, permeabilized with 70% ethanol, and stored at 4° before the hybridization. The polylysine-coated slides are prepared by dropping a 0.01% poly(L-lysine) solution (Sigma St. Louis, MO) under the tissue culture hood onto a glass slide previously immersed in a solution of detergent. The slides are washed first with tap water and then with deionized water, sterilized by autoclaving, and dried at 80°. The glass slides with the Polylysine solution are maintained for 20 min at room temperature. After removal of excess solution, the glass slides are dried at 60° for 1 h. The coated glass slides are stored at 4° for no more than 1 week prior to use.

Hybridization. After fixation, the cells stored in 70% ethanol are rehydrated in 2× SSC, 50% formamide. Hybridization is performed overnight at 37° in a solution containing 10% dextran sulfate, 2 m*M* vanadyl–ribonucleoside complex (Sigma), 0.02% RNAse-free BSA, 40 μg *Escherichia coli* tRNA, 2 × SSC, 50% formamide, and about 30 ng of conjugated probe.

Washing and Mounting. Following hybridization the cells are washed twice at 37° in 2× SSC, 50% formamide and then in PBS prior to mounting in 90% glycerol, PBS, and 0.1 μg/ml 4′,6 diamidino-2-phenylindole (DAPI) (for the nuclear staining).

Utilizing the above protocol we were able to detect the U16Rz in the nucleoli of transiently transfected 293 cells (Fig. 6) and in stable CEM clones (data not shown). For the U16Rz probe we used an amino-allyl T specific primer conjugated with the Oregon Green 488 fluorophore (green fluorescence). To visualize the nucleoli we designed an amino-allyl T primer specific for the U3 snoRNA, the most abundant small nucleolar RNA. The U3 primer was conjugated with the Cy-3 fluorophore (red fluorescence).

The pattern obtained demonstrated that the U16Rz and the U3 snoRNA colocalize within the nucleoli (Fig. 6).

Intracellular *in Vivo* Ribozyme Activity

To test the *in vivo* activity of the designed ribozyme on the target RNA, a variety of techniques can be used. These include assays to determine the intracellular level of the target RNA, such as primer extension, Northern blot analysis, RNase protection, RT-PCR (reverse transcriptase-polymerase chain reaction), and specific assays to monitor the level of the protein encoded by the target RNA. To test the *in vivo* inhibitory effect of the U16Rz, we infected the parental CEM as well as the CEM clones expressing the U16Rz wild-type and mutant with the HIV-1 NL4-3. We monitored the extent of inhibition of HIV-1 replication using an ELISA assay. Using this assay, we detect the amount of HIV-1 p24 viral protein released from the infected CEM cells. In Fig. 7 we show that CEM cells expressing the U16Rz wt dramatically inhibit HIV-1 replication, as determined by the undetectable levels of p24 protein in comparison to the levels in the CEM cells expressing the U16Rz mutant and in the parental CEM cells. Therefore a nucleolar anti-HIV ribozyme can dramatically inhibit HIV-1 replication in a T cell line. These results support our hypothesis that the HIV RNAs might pass through the nucleolus prior to cytoplasmic export.

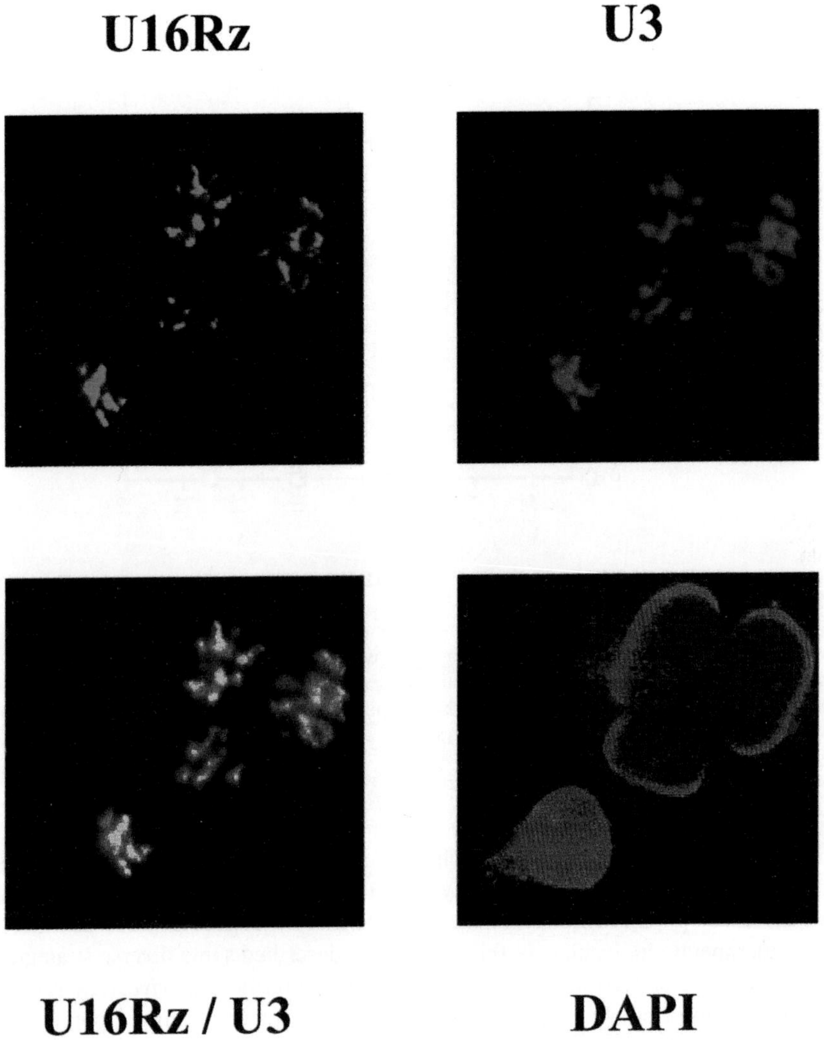

FIG. 6. *In situ* hybridization analysis. 293 cells were transiently transfected with the U6-U16Rz wt and mutant constructs. After 48 h the cells were fixed with 4% paraformaldeyde and *in situ* hybridization was performed as described in the text. As probes, we used amino-allyl T modified primers specific for the U16Rz RNA and for the endogenous U3 snoRNA as a nucleolar control. The U3 primer was conjugated with the Cy3 fluorophore (CyTM3 monofunctional dye, Amersham Pharmacia Biotech), while the U16Rz primer was conjugated with the Oregon Green 488 fluorophore (Molecular Probes). Images were collected using an Olympus BX50 microscope and DEI-50 video camera (optronics). Cy3 and FITC filters were used to detect respectively the U3 snoRNA and U16Rz RNA signals. Using a dual filter FITC+Cy3, the overlapping signals are observed (yellow). The DAPI filter was used to identify the nucleus. *Upper left:* Hybridization with the U16Rz probe. *Upper right:* Hybridization with the U3 probe. *Bottom left:* Overlapping of the U3 and U16Rz signals. *Bottom right:* DAPI staining.

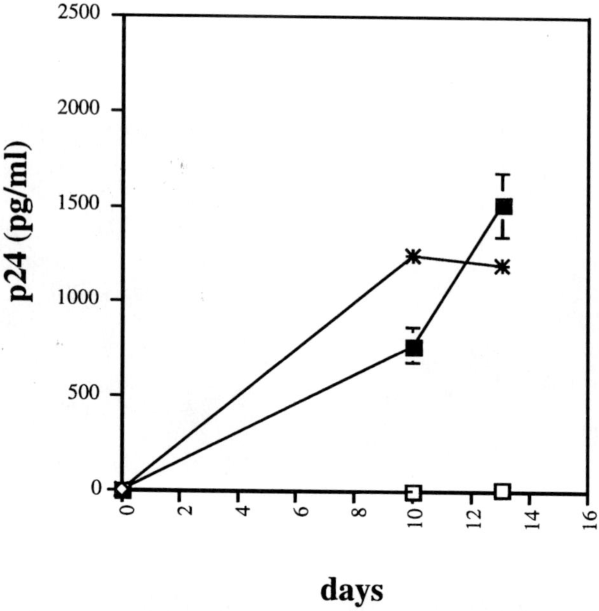

FIG. 7. *In vivo* activity of the U16Rz. A single, transduced CEM clone expressing the U16Rz wt (clone 2) (□), and a clone expressing the mutant version of the U16Rz (clone 4) (■) along with the parental CEM cells (∗) were infected with HIV-1 $_{NL4-3}$at moi (0.0002). Ten and 13 days after the supernatants were collected, the HIV-1 p24 protein antigen produced during viral infection was determined using a p24 antigen capture assay (SAIC Frederick, Frederick, MD).

Conclusions

During the past two decades, much effort has been expended to study the rules governing catalytic activity of ribozymes *in vitro*. These studies gave the basis for the development of ribozymes designed to inactivate gene expression *in vivo* or as therapeutic molecules. In this chapter we described some diverse strategies to increase the effectiveness of ribozymes intracellularly or *in vivo*. The results obtained using ribozymes in cells are promising and lend strong support to the idea that effective ribozyme utilization will involve combined knowledge of the target RNA structure, intracellular trafficking, and localization as well as the use of optimal ribozyme expression and delivery systems.

Acknowledgments

The authors acknowledge Laurence Cagnon for conducting the HIV-1 assays. J.J.R. was supported by NIH Grants AI 29329, AI 42552, and AI 46030.

Section VII

Ribonuclease Inhibitors

[38] Barnase–Barstar Interaction

By ROBERT W. HARTLEY

Introduction

Barnase and barstar are two small soluble proteins produced by the bacterium *Bacillus amyloliquefaciens*. Barnase is a secreted ribonuclease that is specifically inhibited by the intracellular barstar. With 110 and 89 amino acid residues, respectively, neither protein is cross-linked by disulfide bonds, nor do either require any metal ions or other cofactors to fold or function. Inhibition of barnase by barstar involves the formation of a bimolecular complex with a dissociation constant on the order of 10^{-14}–10^{-13} *M*, in which the active site of the enzyme is covered and access to substrate effectively denied. Both proteins can be reversibly unfolded in solution and have been widely used in studies of protein unfolding and refolding and as a model pair in protein–protein investigations.[1] As the genes are available and can be expressed in *Escherichia coli*, much of this work has involved protein engineering and the effects of directed mutagenesis. Variations on barnase and barstar are also available naturally from other strains of *Bacillus*,[2–6] with sequence identities ranging down to 60% (in *B. polymyxa*[6]). More distantly related homologs, with sequence identities below 25%, occur in *Streptomyces* strains.[7,8] In spite of the great difference in sequence, the enzymes of both genera have essentially the same fold and the similarity of their active sites is indicated by the fact that the enzymes of each group are inhibited by the inhibitors of the other. Comparative studies with all of these proteins and with the even more distantly related fungal ribonucleases related to RNase T1 will be useful in determining just what elements of sequence are most important in establishing the folds. All of the proteins mentioned here are available from genes carried on plasmids in *E. coli*.

[1] R. W. Hartley, *in* "Ribonucleases: Structures and Functions" (G. D'Alessio and J. F. Riordan, eds.), p. 51. Academic Press, New York, 1997.

[2] G. A. Aphanasenko, S. M. Dudkin, L. B. Kaminir, I. B. Leschinskaya, and E. S. Severin, *FEBS Lett.* **97,** 77 (1979).

[3] A. A. Dementiev, G. P. Moiseyev, and S. V. Shlyapnikov, *FEBS Lett.* **334,** 247 (1993).

[4] S. V. Shlyapnikov and A. A. Dementiev, *Dokl. Akad. Nauk (Transl.)* **332,** 150 (1993).

[5] L. V. Znamenskaya, L. A. Gabdrakhmanova, E. B. Chernokalskaya, I. B. Leschinskaya, and R. W. Hartley, *FEBS Lett.* **357,** 16 (1995).

[6] A. A. Dementiev, O. A. Mirgorodskaya, G. P. Moiseyev, G. I. Yakovlev, S. V. Shlyapnikov, and M. P. Kirpichnikov, *Biochem. Mol. Biol. Int.* **39,** 158 (1996).

[7] R. W. Hartley, V. Both, E. J. Hebert, D Homerova, M. Jucovic, V. Nazarov, I. Rybajlak, and J. Sevcik, *Protein Peptide Lett.* **3,** 225 (1996).

[8] D. Krajcikova, R. W. Hartley, and J. Sevcik, *J. Bacteriol.* **180,** 1582 (1996).

In this article I will first present a method using only a simple fluorescence assay to measure the dissociation constants (K_d) for various pairs of these ribonucleases and inhibitors and their mutants. Second, I will describe an application of combinatorial methods to produce useful variations in the sequence of barstar.

Barnase–Inhibitor Equilibria

Two very sophisticated (and expensive) instruments currently available for investigating protein–protein interactions are an isothermal titration calorimeter (ITC, Microcal) and a plasmon resonance biosensor (Biacore). The former measures in-solution enthalpy directly and in good cases can provide theoretically satisfying values of free energy and entropy as well. The biosensor, which requires much less material, measures on-rates and off-rates from which the free energy can be derived. One of the proteins must be attached to a rigid or flexible substrate, however, always leaving the question as to whether this attachment affects the reaction under study. Although calorimetric measurements can provide reaction enthalpy (ΔH), neither of these methods can yield free energy (ΔF) for reactants that bind as tightly as wild-type barnase and barstar. Schreiber and Fersht[9] were able to determine ΔF for the barnase–barstar reaction by directly measuring the on and off rates in solution. The very fast on rates were obtained by stopped-flow measurements of a change in fluorescence on complex formation, the very slow off-rate by measuring the displacement of either tritium-labeled component from the complex by a large excess of that component unlabeled. The procedure presented here, a simple titration of enzyme activity by the inhibitor, although less direct, is also technically much less demanding. It can be applied to derivatives or homologs of barnase and barstar in combinations with dissociation constants in the range of 10^{-8} to$10^{-14}M$. The upper limit is due to competition by substrate binding, which is on the order of $10^{-6}M$.

Storage of Barstar, Its Mutants, and Its Homologs

Barnase is very stable and will maintain full activity for years at 5° in a pH 8 buffer with only EDTA to inhibit microorganisms. Barstar, on the other hand, slowly loses activity under the same conditions. One sample, stored in this way for 6 years, was found to have lost more than 70% of its barnase-inhibiting titer. Full activity was recovered on the addition of 10 m*M* of 2-mercaptoethanol and incubation at 37° overnight. Loss of activity was clearly due to oxidation of cysteines, presumably during the rare periods when the molecules are fully or partly unfolded. Barstar homologs from *Streptomyces aureofaciens* and *Bacillus subtilis*

[9] G. Schreiber and A. R. Fersht, *Biochemistry* **32,** 5145 (1993).

168 and various mutants of barstar also lose activity without protection by a reducing agent.

Assay

For the published work using this procedure,[10,11] the assay involved the fluorogenic barnase substrate polyethenoadenosine. A superior fluorogenic substrate fluorescein-dArGdAdA-TAMRA is now available (Integrated DNA Technologies) and its use is described here. It is a modification of a substrate for pancreatic ribonuclease first described by Zelenko *et al.*[12] and reported on elsewhere in this volume.[12a] Barnase, and its homologs from *Streptomyces,* have a strong preference for cutting 5′ of guanine bases and rapidly separate the fluorescent part of this compound from its inhibitory partner.

The description here refers to the use of a Perkin-Elmer (Norwalk, CT) LS-50B fluorescence spectrophotometer using FL Winlab Version 2.00 software and a thermostatted and stirred four-cuvette accessory. Adaptation to other instruments should be straightforward.

Instrument settings are as follows: Time Drive program; Excitation wavelength 489 nm; excitation slit 10 nm; Emission wavelength 519 nm; emission slit 10 nm; duration 20 s; data interval 0.1 s.

Substrate is dissolved in water at a nominal concentration of 3.5 μg/ml (A_{260} of 0.105). Aliquots are stored at -20° or -80°.

A standard assay buffer is 0.2 *M* ammonium acetate, 0.01 *M* EDTA, 0.01 *M* 2-mercaptoethanol, pH 8.0. For particular experiments, other salts and buffers may be appropriate.

A pure barnase standard is used to calibrate the instrument, its concentration being based on an A_{280} of 2.21 for a 1.0 mg/ml solution.[13] A typical assay begins with 1.8 ml of buffer in a 1 cm × 1 cm stirred cuvette in the thermostatted instrument at 25°. Ten μl of barnase solution at a concentration of 0.1 to 2 μg/ml is added, followed, just after starting the time drive, by 10 μl substrate. The Data Handling program can be used to fit a straight line to the linear portion of the resulting fluorescence versus time plot and provides a number for the slope, which is proportional to the ribonuclease activity. Figure 1 illustrates a plot of this slope

[10] R. W. Hartley, *Biochemistry* **32,** 5978 (1993).

[11] R. C. Fitzgerald and R. W. Hartley, *Anal. Biochem.* **214,** 544 (1993).

[12] O. Zelenko, U. Neumann, W. Brill, U. Pieles, H. E. Moser, and J. Hofsteenge, *Nucleic Acids Res.* **22,** 2731 (1994).

[12a] C. Park, B. R. Keleman, T. A. Klink, R. Y. Sweeney, M. A. Behlke, S. R. Eubanks, and R. T. Raines, *Methods Enzymol.* **341,** [6] 2001 (this volume).

[13] R. Loewenthal, J. Sancho, and A. R. Fersht, *Biochemistry* **30,** 6775 (1991).

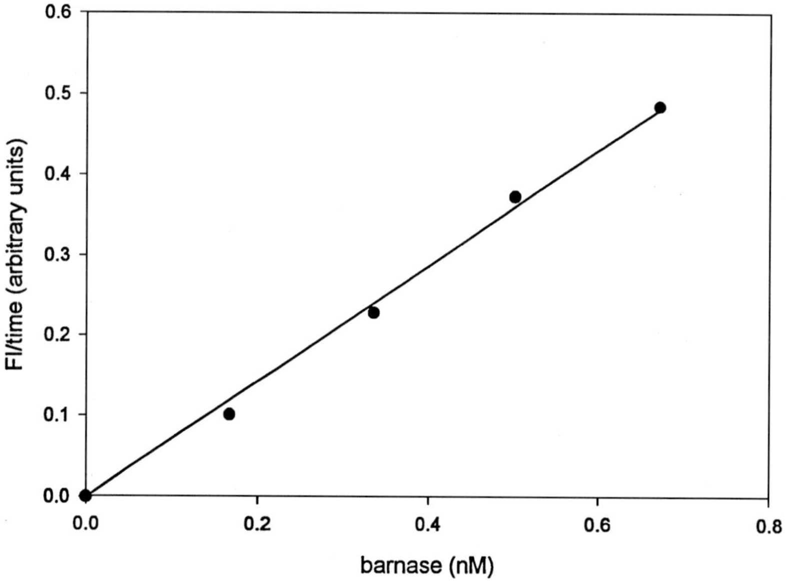

FIG. 1. Slope of fluorescence versus time plotted against barnase concentration.

versus barnase added. Measured activities of unknown samples are then expressed as equivalent concentrations of pure barnase.

Determination of Barnase–Inhibitor Dissociation Constants

Assays are performed as above, except that 10 μl of various dilutions of the inhibitor titrant is added before adding the substrate. It is best to allow 2–3 min to allow equilibrium between barnase and inhibitor before adding substrate, especially near barnase–inhibitor equivalence.

According to the law of mass action, the barnase–inhibitor dissociation constant K_d is given by:

$$K_d = bb^*/(bb^*) \tag{1}$$

where b, b^*, and (bb^*) are the molar concentrations of barnase, the inhibitor, and their complex, respectively. Also:

$$(bb^*) = b_0 - b = b_0^* - b^* \tag{2}$$

where b_0 and b_0^* are the total amount of each protein added. Combining these relations we can get:

$$b_0^* = K_d b_0/b + b_0 - b - K_d \tag{3}$$

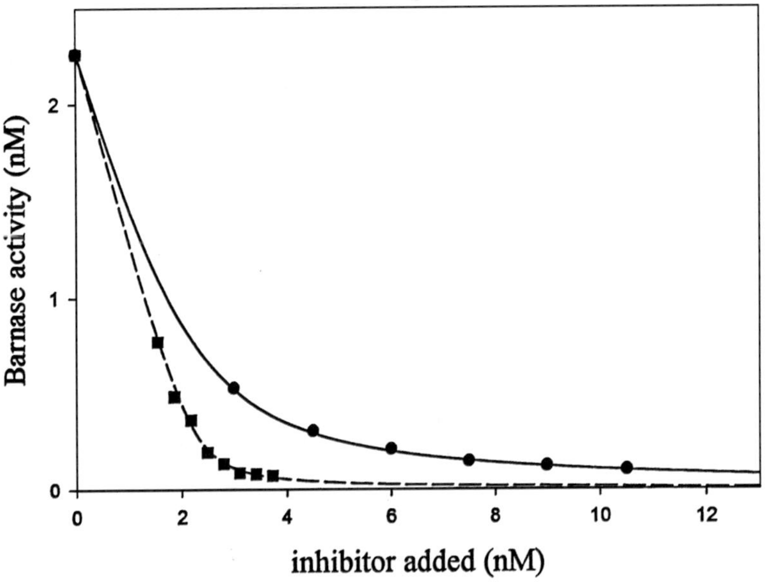

FIG. 2. Titrations of barnase with two inhibitors from *Streptomyces aureus* with dissociation constants derived from fitted curves. (■) SaI20, $K_d = 4.4 \times 10^{-11}$; (●) SaI14, $K_d = 3.8 \times 10^{-10}$.

b_0 is known (2–3 n*M*) and b is measured as a function of inhibitor, b_0^*, added. Note that it is not necessary to know the concentration of the barstar titrant in advance. It can be determined from the initial slope ($= -1$) of the plot of barnase activity (b) versus added inhibitor (b_0^*) and refined in fitting Eq. (3) to the data. This refinement also provides K_d. The inhibitor preparation can be very crude. Extraction of the cells by acetone (with 10 m*M* 2-mercaptoethanol), followed by extraction of inhibitor into the assay buffer provides adequate purity. There appears to be nothing else in *E. coli* extracts to affect the barnase assay.

For some barnase–inhibitor combinations, those with a K_d of about 10^{-12} *M* or above, a series of assays with increasing inhibitor, starting with none, as illustrated in Fig. 2, provides data from which K_d can be derived by curve fitting. For pairs that bind more tightly, such as the wild-type barnase and barstar, the result of this operation is a straight line with little or no curvature at the equivalence point where the line approaches the abscissa. It is still possible to get adequate curvature simply by greatly increasing the barnase concentration from $2\text{–}3 \times 10^{-9}\,M$ to $2\text{–}3 \times 10^{-7}\,M$. A stock solution is prepared at this high concentration (this is still only micrograms per ml) with a slightly lower amount of inhibitor, the concentration of which has been determined by the titration at low levels. This stock can then be distributed to cuvettes and titrated through the equivalence point. If the stock as

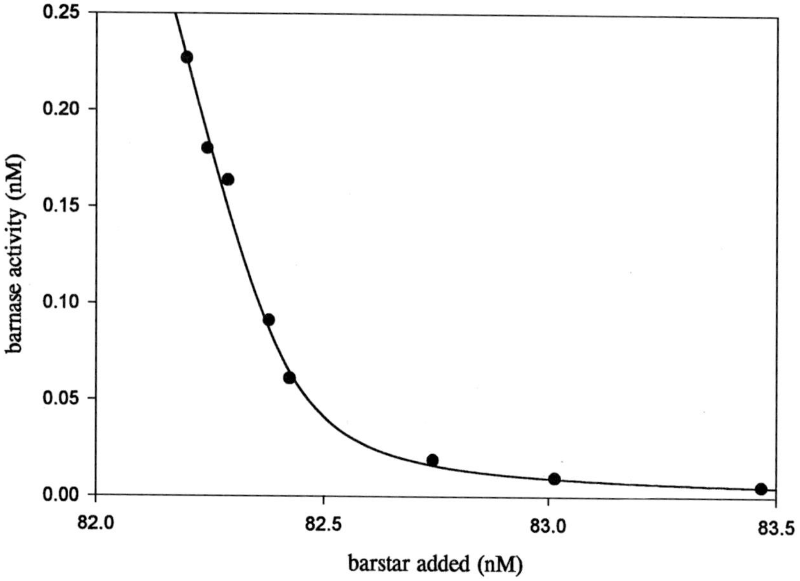

FIG. 3. Titration of barnase by wild-type barstar. $K_d = 7 \times 10^{-14}$.

first prepared is too far from equivalence, it can be adjusted by addition of small amounts of barnase or inhibitor as long as one keeps track of the concentrations. The fitting of such a titration of barnase by barstar is shown in Fig. 3. The value of K_d obtained, 7×10^{-14} *M*, compares with the value of 7×10^{-13} *M* from direct measurements of on and off rates[9] in 0.2 *M* NaCl, pH 8.

Simultaneous Randomization of Groups of Structurally Related Residues of Barstar and Selection for Activity

Gene Synthesis

There are now many procedures available for directing the mutation of particular codons of a cloned gene or simultaneous mutation of several such codons if they are closely spaced. In such cases, each mutation may code for a particular new amino acid or for any of a selected group, depending on the mix of bases used at each position on the synthetic oligodeoxynucleotide used in the operation. For examples, a mixture of all four bases (acgt) at each position of a codon randomizes the residue at that position to all 20 amino acids, while the codon (acgt)(t)(cg) yields only the hydrophobic residues Leu, Ile, Val, Met, and Phe. Examination of the structural requirements for folding or function of a protein, however, may require mutation of several spatially adjacent residues that are widely scattered along the peptide chain. If specific mutations are needed at each site, they may be introduced stepwise. This is clearly unsatisfactory, however, if one wants each of the positions to be randomized independently. The simplest solution to this problem is a total synthesis of the gene with each of the desired codon positions randomized.

The methods discussed here were developed in the course of a study of the hydrophobic core of barstar but can be adapted to other aspects of its structure and to other small soluble proteins with functions that allow some sort of genetic selection. In experiments of this nature we are seeking to determine what combinations of mutations are compatible with the basic fold of the native protein and what effect such substitutions have on the detailed structure, stability, and function of the molecule. This work was inspired by reports that for two enzymes, T4 lysozyme[14] and barnase,[15] some activity survived drastic modification of their hydrophobic cores. In the latter case, 20% of the genes tested after essentially complete randomization of all hydrophobic core residues (to Leu, Ile, Val, Met, or Phe) coded for proteins retaining some ribonuclease activity.

In contrast to the two proteins mentioned, whose hydrophobic residues are distributed in several distinctly separate cores, barstar has a single large and well-defined core, consisting of the side chains of 22 residues. Our initial experiments involved randomization of a group of eight residues, which, in the folded structure, formed a compact group well away from the active site. As none of some 200 such clones tested produced any functional barstar, it became clear that we needed methods for selecting clones with functional barstar. As the proportion of functional barstars would be expected to decrease as we increased the number of residues randomized, it also seemed advisable to increase the multiplicity of our randomized gene, that is, the number of independently randomized genes in our library.

Our first preparations involved a PCR-type of stitching together and then amplifying an appropriate collection of small (25–40 bases) overlapping oligonucleotides. Codons for the eight residues mentioned above were replaced by (ACGT)T(CG), providing eight possible codons at each of these sites coding for Leu, Ile, Val, Met, or Phe. Restriction sites were included on the ends to allow facile introduction into a plasmid expression vector. The stitching phase of this operation was clearly very inefficient, as subsequent semiquantitative PCR indicated multiplicities of only about 10^5.

To have a library large enough to have several molecules of each possible combination ($8^8 = 1.68 \times 10^7$) we need a multiplicity on at least the order of 10^8. This is also approximately the library size that can be conveniently packaged in our phage system for display selection as described below. We have tried a number of possible schemes for achieving such libraries, two of which will be outlined.

a. Ligation of a Set of Oligonucleotides Making Up the Complete Sequence in the Forward Direction. The joints between oligonucleotides are placed midway

[14] D. D. Axe and A. R. Fersht, *Proc. Natl. Acad. Sci. U.S.A.* **93,** 5590 (1996).

[15] N. C. Gassner, W. A. Baase, and B. W. Matthews, *Proc. Natl. Acad. Sci. U.S.A.* **93,** 12155 (1996).

between randomized codons that are at least 16–20 bases apart. Shorter complementary oligonucleotides can then hybridize to and hold together the forward string without overlapping the randomized codons. Ligation, using T4 ligase, then produces a full-length single strand that can be amplified by PCR (polymerase chain reaction). In practice, the yield of this procedure for a fragment of 270–300 base length is inadequate. Using this procedure to synthesize the gene in two overlapping 150 base fragments, however, with a unique restriction site engineered into the overlap, allows us to synthesize the whole gene with multiplicity to spare. The two-part synthesis provides two advantages. (1) The shorter fragments are produced in greater number, with multiplicities in our hands on the order of 10^9–10^{10}. (2) Since the randomized codons are distributed over both fragments, the required multiplicity for each is less. In our first case, for example, there are four randomized codons on each fragment, so that only $8^4 = 4096$ combinations are possible for each. Since each fragment presumably maintains its multiplicity on amplification by PCR, ligation of the amplified fragments after cutting at the overlap restriction site can yield multiplicities as high as 10^{18}–10^{20}. This method has been used to construct a complete library of molecules with the eight residues randomized and a large but necessarily incomplete library with the entire 22-residue core similarly randomized. A complete library of the latter would require a multiplicity of at least an order of magnitude greater than 8^{22}, or about 300 g of DNA. As we are limited by our phage packaging system to about 10^9 or 10^{10} plaque-forming units (pfu) we have not tried to push our DNA libraries beyond that level. Our 22 codon randomization scheme is detailed in Fig. 4.

b. Use of Very Long Oligonucleotides. The simplest method for synthesizing a gene would be to extend a synthesis in a commercial oligonucleotide synthesizer to include a single strand of the entire gene. The manufacturers of these machines generally claim coupling efficiencies of better than 97% per nucleotide. We use a Beckman Oligo 1000M DNA Synthesizer. For a nominal 50 n*M* synthesis of a 270 bp gene such as barstar, 97% efficiency would provide a multiplicity of 8×10^{12}. A great advantage of such a synthesis is that it puts no restriction on the distribution or limit on the number of randomized codons. A single large-scale PCR would then provide ample double-stranded DNA with something close to this multiplicity. Sequences, both 3′ and 5′, with restriction sites for cloning in appropriate phage or plasmid vectors are added in the PCR step. Again, we have found it better to use this method to produce overlapping half gene fragments, which could then be amplified and joined.

Selection of Genes Coding for Functional Barstar

Two methods of selection will be described, one based on phage display and the other a two-plasmid *in vivo* system using the lethal effect of barnase expression in the absence of a functional barstar. The latter is a simpler derivative of a method

```
5' Fragment
           EcoR1                    Ile5          Ile10
CAGCCCGTGCGAATTCGAGCAAAAAAGCAGTCXXXAACGGGAACAAXXXAGAAGTATCAGCGACX
     GCACGCTTAAGCTCGTTTTTTCGTCAG                      TCTTCATAGTCGCTG

Leu16      Leu20
XXCACCAGACAXXXAAAAAGGAGCTTGCCCTTCCGGAATACTACGGTGAAAACCTGGACGCTTTAT
              TTTTTCCTCGAACGGGAAGGCCTTATG      CTTTTGGACCTGCGAAATA

            Age1
GGGATTGTCTGACCGGTTGGGTGGAGT
CCCT

PCR oligos for 5' Fragment:

          forward  CAGCCCGTGCGAATTCGAGCAAAAAAGCAGTC
          reverse  ACTCCACCCAACCGGTCAGAC


3' Fragment
         Age1                        Leu51
ATTGTCTGACCGGTTGGGTGGAGTACCCGCTCGTTXXXGAATGGAGGCAGTTTGAACAAAGCAAGC
   CAGACTGGCCTACCCACCTCATGGCGAGCAA                    CTTGTTTCGTTCG

                         Val70      Phe74
AGCTGACTGAAAATGGCGCCGAGAGTXXXCTTCAGGTTXXXCGTGAAGCGAAAGCGGAAGGCTGCG
TCGACTGACTTTTACCG                                     CGCCTTCCGACGC

       Ile86         HindIII
ACATCACCXXXATACTTTCTTAAGCTTCATGTCAGGC
TGTAGTGG

PCR oligos for 3' Fragment:

          forward  ATTGTCTGACCGGTTGGGTGGAGT
          reverse  GCCTGACATGAAGCTTAAGAAAGTAT

XXX = (AGCT)T(GC)
```

FIG. 4. Scheme for synthesis of barstar with eight residues of the hydrophobic core randomized to Leu, Ile, Val, Met, or Phe. Every other forward oligonucleotide is underlined. Ligation of each forward strand is followed by PCR amplification. Amplified fragments are joined through their *Age*1 sites and amplified again. Wild-type residues at the randomized positions are indicated.

reported previously[16] and is now used as a secondary screen for phage selected libraries.

Phage Display. A commercially available derivative of the *E. coli* phage T7 (T7Select 10-3b, Novagen, Inc.) is well suited for this purpose. Its genome includes

[16] M. Jucovic and R. W. Hartley, *Proc. Natl. Acad. Sci. U.S.A.* **93,** 2343 (1996).

a multi-restriction-site linker positioned so that the gene for a peptide or small protein can be inserted in phase and expressed as a replacement for the unnecessary C-terminal portion of the phage coat protein. Each resulting phage carries approximately 10 copies of the peptide or protein on its surface. As T7 is a lytic phage, there is no problem obtaining large numbers of phage particles. General information and detailed protocols are available with the phage or from the supplier's Web site (www.novagen.com). A kit includes, in addition to the phage DNA, all that is needed for packaging the DNA into infective phage and amplifying them in *E. coli* strain BLT5403.

For insertion between the *Eco*RI and *Hin*dIII sites of the phage, the synthetic barstar gene is terminated in the following fashion:

```
          K   S   S   K   K          S  end
GGGATCCG AAT TCG AGC AAA AAA-------TCT TAA GCTTGCGGCCGCAC
          EcoRI                          HindIII
```

The phage DNA is available (Novagen) cut by *Eco*RI and *Hin*dIII as a purified mixture of the two arms. When this gene is transferred from the phage to an expression vector, the N terminus is modified to:

```
 M   K   S   S   K   K
ATG AAT TCG AGC AAA AAA -----
```

preceded by a *tac* promoter and ribosome binding site. The additional K-S-S tripeptide, in addition to allowing the *Eco*RI site, is designed to provide a flexible hinge between the phage coat protein and barstar. It is well away from the barnase-binding region of barstar and seems to have no effect on its function. Mutant barstars selected for detailed study can have their N termini corrected to the wild-type by directed mutagenesis.

Phage Selection. Phage bearing functional barstar on their surface are selected by binding to immobilized barnase. The barnase, in turn, had been bound to a metal chelate adsorbent through a 10-histidine peptide attached to its N terminus. A His_6-barnase did not provide adequate selection. Recovery of bound phage is based on the fact that barnase can be reversibly denatured by SDS (sodium dodecyl sulfate).[17]

His_{10}-Barnase. The His_{10}-barnase is prepared from *E. coli* carrying a plasmid, pMT4001, modified from plasmid pMT1002 by the insertion of the underlined portion of the sequence:

```
 K  S  A  N  H  H  H  H  H  H  H  H  H  H  S  K  A  N  V
AAAAGCCGCACAGCATCACCATCATCATCACCATCACCATCACAGCAAGGCACAGGTT
```

[17] R. W. Hartley, *Biochemistry* **14,** 2367 (1975).

The KSAN upstream of the histidines is designed to reconstruct the signal protease recognition site of pMT1002. These plasmids, derived from pTN441,[18] have the structural barnase gene on a *phoA* signal sequence and a lambda P_R promoter, controlled by a temperature-sensitive lambda repressor. Grown in rich medium such as Super Broth at 37°, *E. coli* XL1-blue carrying pMT4001 produces and secretes about 25 mg/liter of His_{10}-barnase. The growth medium, centrifuged and passed through a sterilizing filter, can be applied directly to the chelate column.

Chelate Column. A cobalt-loaded TALON Resin (Clontech Laboratories, Inc) has been used, but nickel chelate columns appear to work as well. The column is washed with binding buffer, followed by His_{10}-barnase in the growth medium. Loading is complete when barnase activity rises in the effluent. Except where noted otherwise, all of the following is performed at room temperature.

Buffers

Wash buffer: 0.5 *M* NaCl, 20 m*M* Tris-HCl, 5 m*M* imidazole, pH 7.9
Phage elution buffer: 0.5 *M* NaCl, 20 m*M* Tris-HCl, 0.5% SDS, pH 7.9

Column. An 0.8 ml packed volume of resin in a column of 6 mm diameter. For the loading and wash operations, flow rate is not critical. Gravity flow with a head of about 100 cm is convenient. A loop of the input tubing that falls below the effluent level stops the flow when the input reservoir is empty. Barnase capacity is about 300 μg.

Phage Selection Cycles. The packaged and amplified phage need only be clarified by centrifugation and sterile filtration before application to the column. For the first cycle, where the targets (phage carrying functional inhibitors) are at a minimum, we apply at least 50 ml of the cleared lysate, containing on the order of 10^{13} pfu (plaque-forming units). For subsequent cycles, 2–5 ml is plenty. The phage application is followed by at least 50 ml of wash buffer. For elution of bound phage, the top of the column and the effluent tube at the bottom are removed and the column allowed to drain. Starting with no liquid at the top of the column, aliquots of 0.6 ml and 1.0 ml of elution buffer are added and their effluents collected. The second (1 ml) effluent contains essentially all of the bound phage. The column is washed with 5 ml more of elution buffer and 10 ml of wash buffer. It is then ready for the next cycle.

Next, 0.1 ml of 10 *M* KAcetate is added to the 1 ml effluent, which is then placed on ice for one-half hour to allow precipitation of most of the dodecyl sulfate. After a 2-min spin in a cold microfuge, the supernate can be used to infect an appropriately growing culture of *E. coli* (see Novagen kit protocol) of at least 20 ml. A portion of the cleared lysate from this culture is then applied to the barnase column and

[18] A. L. Okorokov, R. W. Hartley, and K. I. Panov, *Protein Express. Purif.* **5,** 547 (1994).

the process repeated. One or two cycles can be carried out per day, and if there are any phage carrying functional barstar they will dominate the population after eight or nine cycles.

A new barnase column must be prepared for each different library screened.

Secondary Screening and Testing of Barstar Genes. Gene libraries from the selected phage can be prepared directly by PCR, using a small sample of the clarified phage lysate as template. After cutting with *Eco*RI and *Hin*dIII, the library is ligated into a similarly cut vector (pMT3020) derived from pMT3018 by deletion of most of the barstar gene and addition of a 5′ *Eco*RI site. pMT3018 is a barstar expression vector with a *tac* promoter and confers chloramphenicol resistance. An *E. coli* strain such as HB101, lacking the *lacI*^{Q} gene, provides high barstar expression without induction by IPTG. Cells transformed by this plasmid library can be plated out on LB agar with chloramphenicol for the isolation of clones, or grown up in liquid LB with chloramphenicol to be tested by transformation by pMT816. pMT816 is based on pOU61,[19] which is compatible with the pUC origin of pMT3018 and carries ampicillin resistance and the barnase gene on a lambda P_L promoter. Both the copy number and the barnase gene are controlled by a temperature-sensitive lambda repressor. Even at 30°, where the copy number is near 1 and the barnase gene is largely repressed, pMT816 is lethal in *E. coli* not carrying a functional barstar gene. Sharing *E. coli* with pMT3018, however, it is not lethal and, at 37°, directs the secretion of substantial amounts of barnase. When cells carrying our plasmid library of barstar genes are transformed by pMT816 and selected on LB agar with both ampicillin and chloramphenicol, only those cells with plasmids carrying genes for a functional barstar will form colonies at 30°. Theoretically, survival at higher temperatures should correlate with inhibitor effectiveness and stability, but this has not been adequately explored.

Example

Our synthetic barstar library, with codons randomized to Leu, Ile, Val, Met, or Phe for eight structurally contiguous core residues, was screened by nine cycles of phage display. The genes were transferred to the expression plasmid and transformed into *E. coli*. Clones that grew in the presence of chloramphenicol were further selected only by their binding an appropriate barstar oligonucleotide probe, i.e., one based on a part of the barstar gene not containing a randomized codon.

Cells of each of 71 clones were extracted for comparison of *in vitro* barstar activity with that of cells carrying a comparable wild-type barstar gene. Twenty-eight clones yielded barstar titers comparable to wild type. In 22, no barstar activity could be detected. The rest were intermediate. Barstar genes from 20 of the high-yielding and 10 of the intermediate clones were sequenced. No two coded for

[19] J. Larsen, K. Gerdes, J. Light, and S. Molin, *Gene* **28,** 45 (1984).

the same set of residues in the eight randomized positions. At one such position, Phe-74 in the wild type, phenylalanine was found in all 30 clones, indicating the importance of this residue in the context of the eight residues randomized. One of the 30 translated to wild-type barstar. The probability of finding wild-type barstar in 30 clones from our unselected library is only about 10^4, indicating that either sequences coding for functional barnase are rare or that the wild-type sequence is more strongly selected. When challenged with the barnase plasmid pMT816, all 71 clones survived and secreted barnase even at 41°, indicating function *in vivo*.

Further studies on some of these mutant barstars will be reported elsewhere.

[39] Cytoplasmic Ribonuclease Inhibitor

By ROBERT SHAPIRO

Introduction

Mammalian tissues contain a potent proteinaceous inhibitor of pancreatic RNase superfamily enzymes. The existence of such an RNase inhibitor (RI) was first inferred in 1952 from the increase in RNase activity in liver supernatants after acidification.[1] Over the next 25 years, the nature and distribution of RI and the changes in its levels under various physiological conditions were extensively investigated. The utility of RI as a reagent for protecting RNA from the action of adventitious RNases was also demonstrated during this time. The "modern" era of RI research began in the late 1970s with the development of methods to isolate quantities of the homogeneous protein sufficient for physicochemical and functional analysis.[2] The following decade then saw detailed kinetic studies on the interactions of RI with various RNases, the initial exploration of the physical basis for tight binding, the determination of the primary structure of RI, and the construction of systems for recombinant production of this protein. During the 1990s, the three-dimensional (3D) structures of RI and two of its complexes were determined; these structures, together with mapping of the interfaces by mutagenesis, have provided considerable insight into how RI recognizes its various ligands. These and other developments in RI research (up to 1997) are discussed in several reviews.[3-6] This chapter will focus primarily on methods for isolating and functionally

[1] M. Pirotte and V. Desreux, *Bull. Soc. Chim. Belg.* **61,** 167 (1952).

[2] P. Blackburn, G. Wilson, and S. Moore, *J. Biol. Chem.* **252,** 5904 (1977).

[3] J. S. Roth, *Methods Cancer Res.* **3,** 153 (1967).

[4] P. Blackburn and S. Moore, "The Enzymes," 3rd Ed., Vol. 15, p. 317. Academic Press, New York, 1982.

0076-6879/01 $35.00

characterizing RI, while briefly summarizing the available information on the physicochemical properties, structure, specificity, and biological role of RI.

Physicochemical Properties and Primary Structure

RI is an acidic protein (pI ~4.7) of M_r ~50,000 with an amino acid composition characterized by unusually high contents of leucine and cysteine (Table I).[2] It is readily inactivated by heat, acid, and oxidation, and is highly sensitive to treatment with sulfhydryl reagents, some of which (e.g., *p*-hydroxymercuribenzoate) can rapidly release active RNases from their tight complexes with RI. The primary structures of porcine,[7] human,[8,9] and rat[10] RI (pRI, hRI, and rRI, respectively) have been determined; the three proteins are 74–77% identical in sequence. pRI and rRI contain 456 amino acids, whereas hRI has 460 residues owing to a four-residue N-terminal extension. pRI[7] and hRI[8] have been shown to contain no posttranslational modifications other than acetylation of the N terminus, and the hRI,[8] pRI,[11] and rRI[10] cDNAs lack signal peptide sequences, in agreement with the proposed cytosolic localization of the inhibitor. All of the cysteines in hRI and pRI are present in the reduced form.[7,8] Each animal species appears to contain only a single RI (i.e., RIs isolated from different tissues are identical). Although RNase inhibitory activity has been detected in amphibian,[12] avian,[13] and insect[14] species, none of the proteins responsible have been characterized and their relationships to the mammalian inhibitor are unknown.

RI is constructed almost entirely of leucine-rich repeats (LRRs). Such repeats have been identified in more than 100 proteins that exhibit a wide range of functions, including cell-cycle regulation, DNA repair, extracellular matrix interactions, and enzyme inhibition.[15,16] The bulk of RI is formed by 14 alternating

[5] F. S. Lee and B. L. Vallee, *Prog. Nucleic Acid. Res.* **44,** 1 (1993).

[6] J. Hofsteenge, *in* "Ribonucleases: Structures and Functions" (G. D'Alessio and J. F. Riordan, eds.), p. 621. Academic Press, New York, 1997.

[7] J. Hofsteenge, B. Kiefer, R. Matthies, B. A. Hemmings, and S. R. Stone, *Biochemistry* **27,** 8537 (1988).

[8] F. S. Lee, E. A. Fox, H.-M. Zhou, D. J. Strydom, and B. L. Vallee, *Biochemistry* **27,** 8545 (1988).

[9] R. Schneider, E. Schneider-Scherzer, M. Thurnher, B. Auer, and M. Schweiger, *EMBO J.* **7,** 4151 (1988).

[10] M. Kawanomoto, K. Motojima, M. Sasaki, H. Hattori, and S. Goto, *Biochim. Biophys. Acta* **1129,** 335 (1992).

[11] A. M. Vicentini, B. Kieffer, R. Matthies, B. Meyhack, B. A. Hemmings, S. R. Stone, and J. Hofsteenge, *Biochemistry* **29,** 8827 (1990).

[12] H. Nagano, H. Kiuchi, Y. Abe, and R. Shukuya, *J. Biochem.* **80,** 19 (1976).

[13] N. Kraft and K. Shortman, *Aust. J. Biol. Sci.* **23,** 175 (1970).

[14] Y. Aoki and S. Natori, *Biochem. J.* **196,** 699 (1981).

[15] B. Kobe and J. Deisenhofer, *Trends Biochem. Sci.* **19,** 415 (1994).

[16] S. G. S. Buchanan and N. J. Gay, *Prog. Biophys. Molec. Biol.* **65,** 1 (1996).

TABLE I
AMINO ACID COMPOSITIONS OF HUMAN, PORCINE, AND RAT RNase INHIBITORS[a]

	Residues/mol (mol %) in		
Amino acid	Human RI	Porcine RI	Rat RI
Asp	27 (5.9)	23 (5.0)	30 (6.6)
Asn	17 (3.7)	17 (3.7)	20 (4.4)
Glu	35 (7.6)	35 (7.7)	33 (7.2)
Gln	24 (5.2)	25 (5.5)	24 (5.3)
Ser	45 (9.8)	38 (8.3)	40 (8.8)
Gly	32 (7.0)	37 (8.1)	28 (6.1)
His	5 (1.1)	7 (1.5)	4 (0.9)
Arg	24 (5.2)	22 (4.8)	20 (4.4)
Thr	13 (2.8)	23 (5.0)	21 (4.6)
Ala	32 (7.0)	32 (7.0)	25 (5.5)
Pro	14 (3.0)	16 (3.5)	15 (3.3)
Tyr	3 (0.7)	4 (0.9)	5 (1.1)
Val	25 (5.4)	19 (4.2)	25 (5.5)
Met	2 (0.4)	2 (0.4)	3 (0.7)
Ile	12 (2.6)	9 (2.0)	12 (2.6)
Leu	92 (20.0)	98 (21.5)	91 (20.0)
Phe	4 (0.9)	0 (0)	4 (0.9)
Lys	16 (3.4)	14 (3.1)	20 (4.4)
Cys	32 (7.0)	30 (6.6)	30 (6.6)
Trp	6 (1.3)	5 (1.1)	6 (1.3)

[a] Compositions are from the amino acid sequences.[7,8,10]

tandem homologous A type (28-residue) and B type (29-residue) LRRs; these are flanked by single N- and C-terminal repeats that are less closely related. The average sequence identity among the repeats of each subtype is ~40%. Interestingly, the hRI gene contains 8 exons, with exons 2–7 each comprising a single 57-amino-acid A + B unit [as evident from comparison of the cDNA (GenBank M22414)[8] with the gene sequences in GenBank U73630]. However, in structural terms (see below) it is clear that RI is constructed from the smaller units.

Specificity

RI has extraordinarily high affinity for widely divergent ~14 kDa proteins of the mammalian pancreatic RNase superfamily, all of which it binds with 1 : 1 stoichiometry. As shown in Table II, K_i values for RI with pancreatic RNase,[11,17,18] angiogenin (Ang),[17,19] RNase-2 (also known as eosinophil-derived neurotoxin, RNase U_s, and placental RNase),[20] and RNase-4[21]—proteins that typically share only ~25–35% sequence identity—are all extremely low, spanning less than 3

TABLE II
INHIBITION CONSTANTS FOR RNase INHIBITOR COMPLEXES[a]

RI	Enzyme	K_i (M)	Ref.
Human	Pancreatic RNase, bovine	4.4×10^{-14}	17
Porcine	Pancreatic RNase, bovine	5.9×10^{-14}	11
Human	Pancreatic RNase, human	2.0×10^{-13}	18
Human	RNase-2, human	9.4×10^{-16}	20
Human	Angiogenin, human	7.1×10^{-16}	17
Human	Angiogenin, rabbit	1.5×10^{-16}	19
Human	Angiogenin, bovine	3.4×10^{-15}	19
Human	Angiogenin, porcine	2.2×10^{-15}	19
Human	Onconase, frog	$\geq 1.0 \times 10^{-6}$	18
Porcine	RNase-4, porcine	4.0×10^{-15}	21

[a] K_i values were measured at pH 6.0 (100 mM MES–NaOH, 100 mM NaCl, 1 mM EDTA[17,19,20]; 50 mM HEPES–NaOH, 125 mM NaCl, 1 mM EDTA[11,18,21]), 25°.

orders of magnitude. Variations in affinity across species are small (e.g., K_i values for bovine pancreatic RNase A with hRI and pRI are similar). The RNases examined in detail include representatives of each of the four families within the mammalian superfamily, and others (e.g., RNases-3 and -6[22,23]) are also known to be inhibited effectively by RI. Indeed, the only superfamily protein that has been shown to escape the action of RI is bovine seminal RNase, a close homolog of RNase A whose native form is a dimer.[24] (The monomeric form is inhibited by RI.[24]) All of these RNases are noncytoplasmic and are either secreted or packaged into secretory granules. RI is much less effective against homologous RNases of nonmammalian species such as the frog oocyte enzyme onconase ($K_i \sim 1\ \mu M$[18]) and turtle pancreatic RNase,[23] and in some cases no inhibition whatsoever has been observed.[13,25] The cytoplasmic RNases involved in catabolism of intracellular RNA that have been investigated thus far, which are unrelated to the pancreatic superfamily proteins, appear to be completely insensitive to RI (see citations listed

[17] F. S. Lee, R. Shapiro, and B. L. Vallee, *Biochemistry* **28,** 225 (1989).
[18] E. Boix, Y. Wu, V. M. Vasandani, S. K. Saxena, W. Ardelt, J. Ladner, and R. J. Youle, *J. Mol. Biol.* **257,** 992 (1996).
[19] M. D. Bond, D. J. Strydom, and B. L. Vallee, *Biochim. Biophys. Acta* **1162,** 177 (1993).
[20] R. Shapiro and B. L. Vallee, *Biochemistry* **30,** 2246 (1991).
[21] J. Hofsteenge, A. Vicentini, and O. Zelenko, *Cell. Mol. Life. Sci.* **54,** 804 (1998).
[22] E. Boix, Z. Nikolovski, G. P. Moiseyev, H. F. Rosenberg, C. M. Cuchillo, and M. V. Nogues, *J. Biol. Chem.* **274,** 15605 (1999).
[23] R. Shapiro, unpublished results, 1992.
[24] B. S. Murthy and R. Sirdeshmukh, *Biochem. J.* **281,** 343 (1992).
[25] J. S. Roth, *Biochim. Biophys. Acta* **61,** 903 (1962).

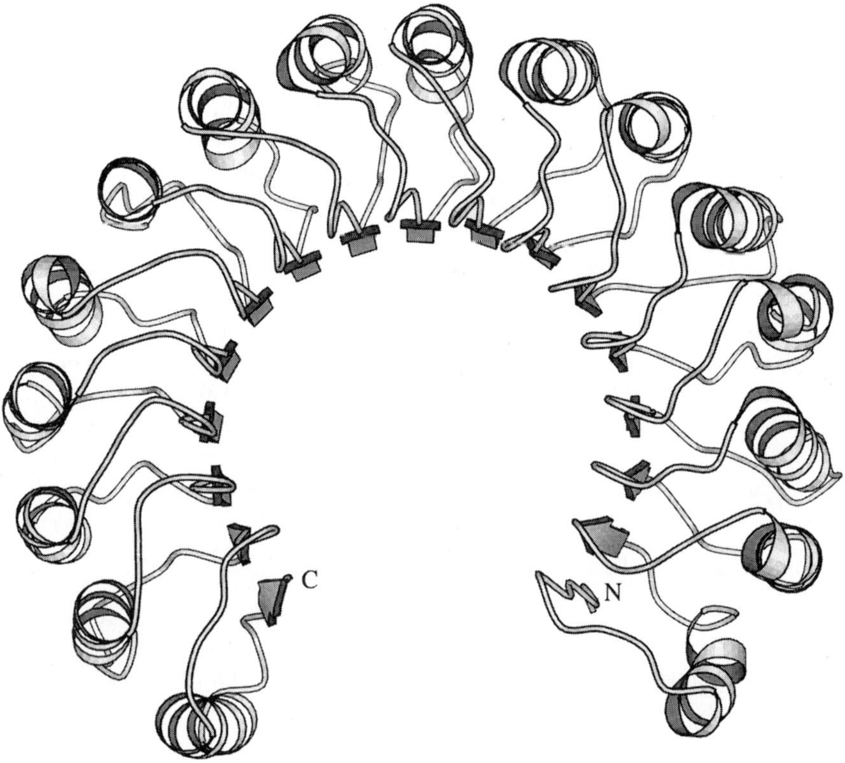

FIG. 1. Ribbon diagram of the crystal structure of porcine RI. From Ref. 15.

in Ref. 6). The enzymatic activities of *E. coli* RNase I and the fungal RNases N1, T1, and U1 are also unaffected by RI (see citations listed in Ref. 5).

Structural Basis for Tight Binding

Crystal structures have been determined for pRI[26] and its complex with RNase A,[27,28] and for the complex of hRI with human Ang[29]; in addition, the hRI–RNase A and hRI–Ang complexes have been studied extensively by mutagenesis.[5,6,30–32] The pRI structure was the first determined for any LRR protein. It revealed a novel nonglobular fold in which the LRR units are arranged symmetrically into a horseshoe-like shape with inner diameter ~21 Å, outer diameter 67 Å, and thickness 32 Å (Fig. 1). Each unit contains a short β strand and a longer α helix, connected by β/α and α/β loops; the β strands line the inner surface of the horseshoe and form a parallel β sheet, whereas the α helices decorate the outside surface.

[26] B. Kobe and J. Deisenhofer, *Nature* **366,** 751 (1993).

When RNase A binds to pRI, about one-third of it sits inside the central cavity of the horseshoe, with the remainder extending above the cavity and the C-terminal region of the inhibitor (orientation as in Fig. 1). Twenty-eight residues distributed over 12 different repeat units of pRI form direct interactions with 24 residues located in the active site and several other regions of RNase A, burying 2551 $Å^2$ of solvent-accessible surface area. The effects of hRI mutations suggest that one small part of this large contact surface—containing the C-terminal LRR of RI and the active site of RNase A—contributes much of the binding energy.[30]

The overall docking of Ang to RI is similar to that of RNase A, with 26 RI residues contacting 24 amino acids on Ang. However, most of the specific interactions in the two complexes are distinctive, even when the same RI residue is involved. The C-terminal segment of RI and the active site of Ang are again most important, although the functioning of the key region in the Ang and RNase A complexes is strikingly different in many respects.[30–32] These findings and comparisons of the sequences of the various RNases suggest that the broad specificity of RI is based largely on its capacity to recognize features unique to each of its ligands.[29,33]

Biological Role of RI

No direct investigations of the *in vivo* function of RI have been reported as of this writing. The earliest hypothesis was that RI is involved in the regulation of intracellular levels of RNA (discussed in Refs. 3, 5, and 6). This proposal now seems largely untenable in view of the RI-insensitivity of all cytoplasmic RNases tested. A second possible role for RI is to serve as a "sentry" to protect cells against noncytosolic RNases that gain entry to the cytosol, e.g., through breakdown of secretory granules carrying them, endocytosis, or targeting errors.[34] All of the RI ligands examined are highly toxic if injected into *Xenopus* oocytes, which contain no effective inhibitor of mammalian RNases.[35] Moreover, noncytosolic RNases that are naturally RI-resistant (e.g., bovine seminal RNase and onconase) or have been engineered to be relatively insensitive to RI are toxic to RI-containing cells.[36–39] Another function of RI might be to regulate the potent biological effects

[27] B. Kobe and J. Deisenhofer, *Nature* **374,** 183 (1995).
[28] B. Kobe and J. Deisenhofer, *J. Mol. Biol.* **264,** 1028 (1996).
[29] A. C. Papageorgiou, R. Shapiro, and K. R. Acharya, *EMBO J.* **16,** 5162 (1997).
[30] C.-Z. Chen and R. Shapiro, *Proc. Natl. Acad. Sci. U.S.A.* **94,** 1761 (1997).
[31] C.-Z. Chen and R. Shapiro, *Biochemistry* **38,** 9273 (1999).
[32] R. Shapiro, M. Ruiz-Gutierrez, and C.-Z. Chen, *J. Mol. Biol.* **302,** 497 (2000).
[33] R. Shapiro, J. F. Riordan, and B. L. Vallee, *Nat. Struct. Biol.* **2,** 350 (1995).
[34] J. J. Beintema, C. Schuller, M. Irie, and A. Carsana, *Prog. Biophys. Molec. Biol.* **51,** 165 (1988).
[35] S. K. Saxena, S. M. Rybak, G. Winkler, H. M. Meade, P. McGray, R. J. Youle, and E. J. Ackerman, *J. Biol. Chem.* **266,** 21208 (1991).
[36] Y. Wu, S. M. Mikulski, W. Ardelt, S. M. Rybak, and R. J. Youle, *J. Biol. Chem.* **268,** 10686 (1993).

of some RNase family members.[5,40] Finally, the presence of appreciable amounts of RI in mature erythrocytes and platelets, which lack any functional RNA, indicates that this protein may play some additional role(s) unrelated to inhibition of RNases.[41,42]

Isolation of RI

RI has been purified to homogeneity from human placenta,[2,43] brain,[44] and blood cells,[41,42] porcine liver,[45] testis,[46] and brain,[47] bovine liver[44,7] and brain,[48] rat liver[45] and testis,[49] mouse liver,[45] and sheep liver[45]; its production in three recombinant systems has also been reported.[50,11,37] Virtually all isolation procedures are based on that developed by Blackburn and co-workers[2,43] for RI from human placenta; the key step is affinity chromatography on RNase A-Sepharose, which can achieve a purification of >1000-fold. The basic Blackburn protocol is simple and robust and was used routinely in this laboratory for over a decade without a single failure. This method (with minor modifications) is described below, including recommendations for increasing the scale of preparation. Adaptations for obtaining homogeneous RI from other natural and recombinant sources are also outlined. Additional details can be found in the references cited.

Human Placental RI

Human placenta is a rich, accessible source of RI: ~2.5 mg of RI can be obtained from a 400 g placenta in 2 days, and three placentas can be readily worked up together. Moreover, the procedure can be scaled up to obtain as much as ~60 mg of RI in 2 weeks. For smaller quantities of hRI, blood may be a more convenient

[37] S. Vescia, D. Tramontano, G. Augusti-Tocco, and G. D'Alessio, *Cancer Res.* **40,** 3740 (1980).

[38] P. A. Leland, L. W. Schultz, B.-M. Kim, and R. T. Raines, *Proc. Natl. Acad. Sci. U.S.A.* **95,** 10407 (1998).

[39] M. Suzuki, S. K. Saxena, E. Boix, R. J. Prill, V. M. Vasandani, J. E. Ladner, C. Sung, and R. J. Youle, *Nat. Biotechnol.* **17,** 265 (1999).

[40] R. Shapiro and B. L. Vallee, *Proc. Natl. Acad. Sci. U.S.A.* **84,** 2238 (1987).

[41] D. Nadano, T. Yasuda, H. Takeshita, and K. Kishi, *Int. J. Biochem. Cell Biol.* **27,** 971 (1995).

[42] M. Moenner, M. Vosoghi, S. Ryazantsev, and D. G. Glitz, *Blood Cells. Mol. Dis.* 24, 149 (1998).

[43] P. Blackburn, *J. Biol. Chem.* **254,** 12484 (1979).

[44] D. Nadano, T. Yasuda, H. Takeshita, K. Uchide, and K. Kishi, *Arch. Biochem. Biophys.* **312,** 421 (1994).

[45] L. E. Burton and N. P. Fucci, *Int. J. Peptide Protein Res.* **19,** 372 (1982).

[46] M. Ferreras, J. G. Gavilanes, C. Lopez-Otin, and J. M. Garcia-Segura, *J. Biol. Chem.* **270,** 28570 (1995).

[47] S. Cho and J. G. Joshi, *Anal. Biochem.* **176,** 175 (1989).

[48] L. E. Burton, P. Blackburn, and S. Moore, *Int. J. Peptide Protein Res.* **16,** 359 (1980).

[49] J. M. Fominaya, J. M. Garcia-Segura, and J. G. Gavilanes, *Biochim. Biophys. Acta* **954,** 216 (1988).

[50] F. S. Lee and B. L. Vallee, *Biochem. Biophys. Res. Commun.* **160,** 115 (1989).

source: Moenner *et al.*[42] has obtained 430 μg of hRI from the erythrocyte fraction in 100 ml of whole blood using the Blackburn method, modified only with respect to the homogenization procedure.

General Considerations. All steps are performed at 0–4°. All buffers contain 1 m*M* EDTA and 5 m*M* dithiothreitol (DTT), unless stated otherwise, to maintain RI stability, and all purification procedures should be performed quickly. Our experience has been that it is neither necessary nor productive to assay RI activity[2] at each stage during routine isolations.

Homogenization. Placentas should be placed at 4° within ~1 hr after delivery and used within 8 hr. Swirl in TS buffer (20 m*M* Tris-HCl, pH 7.5, containing 250 m*M* sucrose) and blot most of the liquid on a paper towel. After removing the thick membrane, cut each placenta into small pieces, rinse and blot as above, weigh, and add to TS (the usual volume is 2.0 ml/g, but as little as 1.0 ml/g can be used). Homogenize with a Polytron instrument (Brinkmann, Westbury, NY), a blender, or a similar device. Centrifuge the homogenate at 16,000*g* for 30 min and measure the volume of the supernatant.

Ammonium Sulfate Fractionation. Over the course of 20–30 min, add ammonium sulfate to gently stirred supernatant from the previous step to reach 35% saturation. After another 40 min of stirring, centrifuge as above, collect the supernatant, and repeat the ammonium sulfate precipitation procedure, in this case bringing the solution to 60% saturation. RI will be in the pellets. These should be drained to remove free liquid and then processed for affinity chromatography (see next step) or stored at −20° (the RI is stable for months under these conditions). For large-scale preparations, it is convenient to accumulate pellets from multiple placentas (up to 20) before proceeding further.

Affinity Chromatography. RNase A-Sepharose is made from CNBr-activated Sepharose (Pharmacia, Piscataway, NJ) by standard methods,[2,43] using 2 mg of enzyme per ml of swollen resin. The resin can be stored for years in 0.02% sodium azide at 4° and is conveniently prepared in batches of ~50 ml. Use ~1.3 ml per mg of RI; it can generally be assumed that the ammonium sulfate pellet from 1 kg of placenta contains ~8 mg of RI. Elution of inhibitor is more efficient if the resin volume is not appreciably greater than that calculated by these guidelines. The resin should be equilibrated in loading buffer (45 m*M* potassium phosphate, pH 6.4).

Dissolve the ammonium sulfate pellet in a minimum volume of loading buffer (typically ~400 ml per 100 g of pellet) and dialyze overnight against 20 volumes of the same buffer. Centrifuge at 40,000–48,000*g* for 1 hr and then load onto the column; a flow rate of 3 ml/min is used for a 1.6-cm diameter column. Wash the column with additional loading buffer until the OD_{280} of the eluting material is <0.1, and then wash with loading buffer that has been supplemented with 0.5 *M* NaCl to remove additional non-RI proteins. When the absorbance returns to baseline levels (~0.01), elute the RI with 0.1 *M* sodium acetate, pH 5.0, containing 3.0 *M* NaCl and 15% (v/v) glycerol at one-half the original flow rate. Collect fractions in 1.5- or 2.0-ml screw-cap polypropylene tubes. Measure the OD_{280} and activity of each fraction (see below); assess purity by SDS–PAGE (note that some silver-staining procedures are not effective for visualizing RI). In general, material in the early ascending portion of the peak is impure, but the remainder is >95% homogeneous. This level of purity is sufficient for use in physical and enzymological studies. However, these preparations are contaminated with trace amounts of RNase A (≪1%) from the affinity column, and additional purification by anion-exchange chromatography is recommended if the RI is to be used for protection of RNA.

Mono Q Chromatography. This step was originally introduced for final purification of porcine liver RI[7]; its use for hRI is as follows.[44] The active fractions from the RNase column are dialyzed into 20 m*M* Tris-HCl, pH 7.5, containing 15% (v/v) glycerol, 5 m*M* DTT, and 1 m*M* EDTA, concentrated by ultrafiltration, and applied to a Mono Q column (HR 5/5; Pharmacia) that is equilibrated in 20 m*M* Tris-HCl, pH 7.5, containing 0.1% (w/v) polyethylene glycol 6000, 5 m*M* DTT, and 1 m*M* EDTA. A linear 50-ml gradient of 0–0.5 *M* NaCl in the same buffer is then applied at a flow rate of 0.7 ml/min; hRI elutes with ~0.25 *M* NaCl.

Storage. hRI begins to precipitate over the course of several weeks at 4° in the affinity column elution buffer. Long-term (>4 year) stability can be achieved by storing the protein instead at −20° in 20 m*M* HEPES, pH 7.6, containing 50 m*M* KCl, 8 m*M* DTT, 1 m*M* EDTA, and 50% (v/v) glycerol. hRI is also stable for at least 2 months at 4° in various other neutral pH buffers containing at least 5 m*M* DTT and 15% glycerol.[2,51] Buffer changes and sample concentration are conveniently performed with a Centricon-30 ultrafiltration device (Amicon, Danvers, MA), with recoveries generally >80%.

Porcine Liver RI

Porcine liver has been the other major natural source of RI for biochemical and structural studies; yields ranging from 3 to 8 mg/kg of wet tissue have been reported.[45,7] The purification procedure most commonly used differs from that for hRI only with respect to the conditions for affinity chromatography. pRI remains on the column if the standard hRI elution conditions are applied, and elution is performed instead with 50 m*M* boric acid, 59 m*M* sodium borate, pH 6.4, containing 4 *M* NaCl, 15% glycerol, 5 m*M* DTT, and 1 m*M* EDTA at room temperature. (Loading and washing are at 4°.) An additional change is that the sample is applied to the affinity column in 50 m*M* potassium phosphate, pH 6.4, plus 0.75 *M* NaCl (and the usual amounts of DTT and EDTA). The RI obtained usually contains significant impurities and is purified to homogeneity by Mono Q chromatography as described above, except that the gradient is 0–300 m*M* NaCl in 45 min at a flow rate of 1 ml/min.[7] The same method (through the affinity step) has been used to isolate RIs from bovine, rat, mouse, and sheep liver.[45] All of the organs were obtained

[51] F. S. Lee and B. L. Vallee, *Biochemistry* **29,** 6633 (1990).

shortly after slaughter, stored on wet ice during transit, and processed the same day. However, we have obtained high yields of ovine RI from placental cotyledons that had been quick-frozen in liquid nitrogen, stored at -70° for 3 months, and then thawed overnight at 4°. It seems likely that RI in other tissues can also survive this procedure, and that it may not be necessary to use only fresh tissue.

Recombinant RI

Three recombinant systems for production of RI have been reported in the literature: two for hRI in *Escherichia coli*[50,38] and one for pRI in *Saccharomyces cerevisiae.*[11] The yields from the bacteria are ~0.2–0.3 mg/liter of culture, whereas 0.2 mg/g of wet cells is obtained from yeast. The RIs isolated from one of the bacterial sources[50] and from yeast have been shown to be functionally indistinguishable from the natural products. The first of these systems has been used routinely in this laboratory for over a decade to obtain wild-type and variant hRIs. The inhibitor is expressed under the control of the *E. coli trp* promoter (plasmid pTRP-PRI) in *E. coli* strain W3110. Most of the RI is soluble after cell disruption, and no attempt is made to recover any additional protein residing in inclusion bodies. The cell lysate is subjected to ultracentrifugation, followed by treatment with streptomycin sulfate to precipitate nucleic acids. In the original protocol,[50] the supernatant is diluted and loaded directly onto RNase A-Sepharose. However, the product obtained sometimes contains up to 30% impurity as judged by SDS–PAGE. For this reason, an ammonium sulfate fractionation is now performed prior to affinity chromatography. This step had been added previously, along with a change in affinity column buffers, to improve binding of some RI variants to RNase A-Sepharose.[51] The modified buffers can also be used for wild-type hRI. The binding of some RI variants to RNase A-Sepharose is too weak to allow purification even by the revised procedure, and in these cases Ang-Sepharose has been used.[30]

Stability of RI

RI is generally considered to be an extremely unstable protein. However, this reputation is only partially warranted. Clearly, it is difficult to maintain RI activity in many complex mixtures, particularly tissue extracts, and most or all of the precautions described above for isolating RI are indeed necessary to prevent sulfhydryl oxidation, which causes denaturation,[52] and possibly proteolysis. RI administered intravenously to mice or incubated in mouse serum at 37° has been shown to be completely inactivated with 1 hour.[53] Moreover, RI loses >99% of its activity when incubated for 30 min at 37° with H_2O_2 concentrations as low as 0.01% if

[52] J. M. Fominaya and J. Hofsteenge, *J. Biol. Chem.* **267,** 24655 (1992).

[53] I. J. Polakowski, M. K. Lewis, V. Muthukkaruppan, B. Erdman, L. Kubai, and R. Auerbach, *Am. J. Pathol.* **143,** 507 (1993).

DTT and glycerol are decreased well below the recommended levels.[54] [Glycerol ($\geq$15%, v/v) affords essentially complete protection against even 0.8% H_2O_2.]

Despite these indications, pure RI is in fact relatively stable in the absence of both DTT and glycerol under conditions of the type commonly used for biochemical and functional studies *in vitro.* hRI (~8 μM) dialyzed into 20 m*M* HEPES, pH 7.6, containing 50 m*M* KCl (using a Centricon-30 device; DTT and glycerol reduced to <40 μM and <0.25%, respectively) and stored at 4° retains full activity against RNase A for at least 1 week. Dilution of this inhibitor preparation (after 1 week) to ~0.5 μM in water, followed by 1 min of vigorous vortexing, destroys only 15% of activity, with 50% activity lost after a full 5 min of vortexing. These findings suggest that researchers need not automatically limit their experimental systems for studying or using RI to those where high concentrations of reducing and stabilizing agents can be maintained. If additional precautions are nonetheless desired, oxidation can be minimized by storing the RI solution in a sealed container under nitrogen or argon.

Quantification of RI

Activity-Based Methods

RI for functional studies can be accurately quantified based on inhibitory activity toward RNase A. The concentration of the RNase A stock solution itself is determined by absorbance at 278 nm ($\varepsilon = 9800\ M^{-1}\text{cm}^{-1}$).[55] Inhibition is then measured in any of several convenient assay systems. In all cases, the RI concentration is many orders of magnitude above the K_i value; thus, essentially all active RI molecules will be bound to the enzyme and a titration plot measuring inhibition with 3 or 4 amounts of RI will provide a reliable assessment of inhibitor concentration. In the standard assay,[43] cytidine 2′,3′-cyclic phosphate (C > p) is the substrate, and the increase in absorbance at 286 nm that accompanies hydrolysis is measured continuously; 73 n*M* RNase is used with 1 m*M* substrate in 0.1 *M* Tris–acetate, pH 6.5. Advantages of this system are that aliquots from RI stocks and column fractions can be assayed without dilution and that the substrate is inexpensive; a disadvantage is that the product, 3′-CMP, binds much more tightly than the substrate, so that initial rates must be derived from only the first 1–2% of the reaction. An alternative is to use the dinucleotide CpG (60–100 μM) as substrate, in this case measuring the decrease in absorbance at 286 nm as C > p is formed.[56] In the buffer usually employed (0.1 *M* MES–NaOH, 0.1 *M* NaCl, pH 6.0), a suitable initial velocity is achieved with 3–5 n*M* RNase A and 60 μM CpG,

[54] B.-M. Kim, L. W. Schultz, and R. T. Raines, *Prot. Sci.* **8,** 430 (1999).

[55] M. Sela and C. B. Anfinsen, *Biochim. Biophys. Acta* **24,** 229 (1957).

[56] H. Witzel and E. A. Barnard, *Biochem. Biophys. Res. Commun.* **7,** 295 (1962).

and product inhibition is insignificant.[57] The use of poly(C) as substrate for determination of RI concentrations has also been described.[38] All of these continuous assays are generally faster and more accurate than the precipitation methods that have been employed for this purpose.[2–4]

Quantities of RI are frequently reported in terms of units of activity rather than molar amounts. One unit is defined as the amount of RI needed to produce 50% inhibition of 5 ng of RNase A.[58] Thus one unit corresponds to 0.183 pmol of RI (9.1 μg for hRI).

Molar Absorptivity

The ε_{280} values for hRI and pRI calculated[59] based on tryptophan and tyrosine contents (no cystines are present in either RI) are 37,470 M^{-1} cm^{-1} and 33,460 M^{-1} cm^{-1}, respectively. The experimental value reported by Burton *et al.*[48] for hRI (37,000 M^{-1} cm^{-1}) is similar, but in the author's laboratory a much higher value, 45,000 M^{-1} cm^{-1} (at 281 nm, the absorbance maximum), has been measured, and Ferreras *et al.*[46] have determined a value for pRI ($\varepsilon_{280} = 43{,}200\ M^{-1}$ cm^{-1}) that is also well above that calculated. This question requires further investigation. Irrespective of the actual values, the use of UV absorbance to measure RI concentrations in standard buffer solutions can be problematic because of the presence of DTT. Although DTT itself has no significant absorbance at 280 nm, its oxidized form, which is rapidly generated upon exposure to air, is a moderately strong chromophore with $\varepsilon_{283} = 273\ M^{-1}$ cm^{-1}.[60] It should be noted that the glycerol in RI stock solutions in incompatible with common amino acid analysis procedures, ruling out this method for routine quantification of RI.

Kinetic Characterization of RI Complexes

The dissociation constants of most RI complexes are extremely low (Table II) and cannot be determined by standard kinetic approaches such as Dixon or Lineweaver–Burk plots, which require that there be no significant depletion of inhibitor by enzyme. The use of these methods in early studies of RI–RNase A interactions resulted in gross underestimations of binding strength and the misidentification of the (competitive) inhibition mode as noncompetitive. Several procedures have now been devised for characterizing the kinetics of RI–ligand interactions,[17,57,11,20,21] and are outlined below. (Additional details are described in the references cited. For a general discussion of tight-binding and slow-binding

[57] F. S. Lee, D. S. Auld, and B. L. Vallee, *Biochemistry* **28,** 219 (1989).

[58] K. Shortman, *Biochim. Biophys. Acta* **51,** 37 (1961).

[59] C. N. Pace, F. Vajdos, L. Fee, G. Grimsley, and T. Gray, *Prot. Sci.* **4,** 2411 (1995).

[60] W. W. Cleland, *Biochemistry* **3,** 480 (1964).

inhibition, see Morrison,[61] Cha,[62,63] and Williams and Morrison.[64]) Thus far, these methods have been applied to complexes of hRI and/or pRI with Ang, pancreatic RNase, RNase-2, and RNase-4. It is likely that any additional RI complexes of interest can be analyzed by adapting at least one of these procedures. Indeed, similar techniques have been applied to the study of many other tight protein–protein complexes.[65,66]

General Considerations

The approaches followed fall into two broad categories: enzymatic and physical. Enzymatic methods are applicable only when activity can be measured accurately at enzyme concentrations that are no more than ~2 orders of magnitude greater than K_i' [i.e., $K_i(1+[S]/K_m)$]. This has proved feasible for the complexes of pRI with RNase A and RNase-4, and that of hRI with bovine and human pancreatic RNase A. For the complexes of Ang, a very weak RNase[67] whose activity is usually measured at concentrations of 0.1–10 μM, purely physical techniques are required: apparent rate constants for association and dissociation (k_a and k_d, respectively) are measured and then used to calculate K_i. A largely physical method has also been used for investigating the complexes of hRI with RNase A and human RNase-2. It should be noted that the slow dissociation of RI complexes precludes the use of surface plasmon resonance methods for assessing K_i values and that calorimetric methods,[68] although theoretically applicable to very tight complexes, are based on assumptions that would be difficult to validate in this particular case.

Kinetic studies have been performed mostly in Mes buffer (pH 6) at ionic strength ~0.15, i.e., standard conditions for assaying RNase A activity. K_i values for the complexes of hRI and pRI with RNase A rise substantially with increasing ionic strength[11,17]: e.g., by 900-fold at $I = 0.55$ and 22,000-fold at $I = 1.05$ (for hRI). Lowering the pH to 5 weakens binding of hRI to RNase A by 430-fold (i.e., $K_i = 19$ pM).[17] At pH values above 6, only upper limits for K_i have been set: 0.9 pM, 3 pM, and 4 pM for pH 7, 8, and 9, respectively. For the hRI–Ang complex, the k_a value is markedly affected by both pH and ionic strength: optima are at pH 5.5 and $I = 0.125$; the k_a values at pH 9 ($I = 0.15$) and $I = 1.05$ (pH 6) are 13-fold and 140-fold, respectively, lower than under the standard conditions.[57] The effects of ionic strength and pH on k_d values have not been determined.

[61] J. F. Morrison, *Biochim. Biophys. Acta* **185,** 269 (1969).
[62] S. Cha, *Biochem. Pharmacol.* **24,** 2177 (1975).
[63] S. Cha, *Biochem. Pharmacol.* **25,** 2695 (1976).
[64] J. W. Williams and J. F. Morrison, *Methods Enzymol.* **63,** 437 (1979).
[65] S. R. Stone and J. Hofsteenge, *Biochemistry* **25,** 4622 (1986).
[66] G. Schreiber and A. R. Fersht, *Biochemistry* **32,** 5145 (1993).
[67] R. Shapiro, J. F. Riordan, and B. L. Vallee, *Biochemistry* **25,** 3527 (1986).
[68] J. F. Brandts and L.-N. Lin, *Biochemistry* **29,** 6927 (1990).

RI–Ang Complexes

Formation of the complex of hRI with human Ang is accompanied by a 50% increase in tryptophan fluorescence. The kinetics of association can be analyzed by monitoring this increase under stopped-flow conditions.[57] Pseudo-first-order rate constants, k_{obs}, are measured as a function of the concentration of one component (usually Ang), which is always at >4-fold excess over the other. A plot of k_{obs} vs concentration is hyperbolic, consistent with a two-step mechanism in which formation of a loose complex EI is followed by isomerization to the tight complex EI*[69]:

$$\mathrm{E} + \mathrm{I} \underset{k_{-1}}{\overset{k_1}{\rightleftharpoons}} \mathrm{EI} \underset{k_{-2}}{\overset{k_2}{\rightleftharpoons}} \mathrm{EI}^*$$

where E is Ang and I is RI. In this mechanism, the relationship of k_{obs} to the rate constants is:

$$k_{obs} = k_{-2} + k_2[\mathrm{E}]/(K_1 + [\mathrm{E}]) \qquad ([\mathrm{E}] \gg [\mathrm{I}]) \tag{1}$$

where K_1 equals k_{-1}/k_1, the dissociation constant for the EI complex, if the first step is fast compared to the second (as has been assumed to be the case). Because k_{-2} is many orders of magnitude smaller than k_2 (see below), this equation simplifies to

$$k_{obs} = k_2[\mathrm{E}]/(K_1 + [\mathrm{E}]) \tag{2}$$

Values for k_2 and K_1 can be obtained from a nonlinear fit of the data to this equation, or from a double-reciprocal plot of $1/k_{obs}$ vs $1/[\mathrm{E}]$, which is linear with intercepts at $1/k_2$ and $-1/[K_1]$. In general, k_{obs} values in these experiments become less reliable as [E] increases; therefore, a standard unweighted linear regression for the double-reciprocal plot will automatically rely most heavily on the more accurate data. If a direct nonlinear fit is performed, appropriate weighting of data points [e.g., $(1/k_{obs})^2$] is recommended. The apparent second-order rate constant for association (i.e., k_a) is closely approximated by k_2/K_1. The values of K_1, k_2, and k_a obtained for the wild-type hRI–Ang complex by this procedure are 0.53 μM, 97 s^{-1}, and $1.8 \times 10^8\ M^{-1}\ \mathrm{s}^{-1}$, respectively (Table III). A value for k_a alone can also be obtained by measuring k_{obs} at [Ang]$\ll K_1$, so that Eq. (2) reduces to

$$k_{obs} = k_2[\mathrm{E}]/K_1 = k_a[\mathrm{E}] \tag{3}$$

Dissociation rate constants for RI–Ang complexes are determined by first forming the complex, then adding a large molar excess of a scavenger for free RI, and finally measuring the amount of free Ang present at various times. This can be performed using unlabeled Ang to form the initial complex and RNase A (at 250-fold excess over Ang) as scavenger[17]; alternatively, [^{14}C]Ang can be used in the first step, with unlabeled Ang (at 10- to 20-fold excess) as the scavenger.[30] Free Ang is measured by cation-exchange HPLC, using peak area (214 nm) for

[69] An alternative mechanism, which is kinetically indistinguishable, is discussed in Ref. 57.

TABLE III
KINETIC PARAMETERS FOR VARIOUS RI COMPLEXES[a]

RI	Enzyme[b]	k_a ($M^{-1} \cdot s^{-1}$)	k_d (s^{-1})	K_i (fM)	Method	Ref.
Human	Ang	1.8×10^8	1.3×10^{-7}	0.71	Physical; RNase A scavenger for k_d	17,57
Human	Ang	2.0×10^8	1.1×10^{-7}	0.54	Physical; [^{14}C]Ang used for k_d	32
Human	RNase A	3.4×10^8	1.5×10^{-5}	44	Physical–enzymatic; HPLC detection of free RNase for k_d	17,57
Human	RNase A	3.4×10^8	1.2×10^{-5}	35	Physical–enzymatic; free RNase for k_d measured enzymatically	17,57
Porcine	RNase A	—	—	74	Enzymatic; initial rates	11
Porcine	RNase A	1.7×10^8	9.8×10^{-6}	59	Enzymatic; progress curves, UpA substrate	11
Porcine	RNase A	1.3×10^8	1.5×10^{-5}	113	Enzymatic; progress curves, fluorescent substrate	72
Human	RNase-2	1.9×10^8	1.8×10^{-7}	0.94	Physical–enzymatic; free RNase for k_d measured enzymatically	20
Porcine	RNase-4	1.5×10^8	6.1×10^{-7}	4.0	Enzymatic; progress curves, fluorescent substrate	21

[a] See legend to Table II for conditions.
[b] Ang and RNase-2 are human, RNase A is bovine, and RNase-4 is porcine.

the complex of unlabeled Ang and scintillation counting for that of labeled Ang. Incubation mixtures contain 0.1 M Mes (pH 6.0), 0.1 M NaCl, 100 μg/ml bovine serum albumin (BSA; RNase-free, Worthington, Lakewood, NJ), 120 μM DTT at 25°, 0.7–1.4 μM Ang, and 1.5 equivalents of RI. All of the ingredients are sterilized with Centrex cellulose–acetate filters (0.45 μm; Schleicher and Schuell, Keene, NH). Aliquots of 70–140 μl are injected onto a Mono S column (HR 5/5; Pharmacia) equilibrated in 10 mM Tris-HCl buffer (pH 8.0) containing 0.2 M NaCl, and a 15-min linear gradient of 0.2–0.6 M NaCl in the same buffer is applied at 1.5 ml/min. Free Ang elutes at 10–12 min, whereas RNase A, RI, and the RI complexes appear in the breakthrough. Control mixtures that lack either RI or scavenger are also incubated. [^{14}C]Ang is produced in *E. coli*[30]: BL21(DE3) cells harboring the plasmid pET-Ang (i.e., pET11 with a synthetic gene for Ang cloned into the *Nde*I and *Bam*HI sites) are grown in M9 medium containing [^{14}C]glucose as the sole carbon source, and [^{14}C]Ang (typically 2700 cpm/μg) is purified to homogeneity from inclusion bodies by standard methods.[70,71]

A few percent of the Ang is released from the hRI complex within several minutes after addition of scavenger, and an additional ~8% dissociates over the next 10 days.[17] (After 10 days, dissociation gradually becomes somewhat faster,

[70] R. Shapiro, J. W. Harper, E. A. Fox, H.-W. Jansen, F. Hein, and E. Uhlmann, *Anal. Biochem.* **175,** 450 (1988).
[71] R. Shapiro and B. L. Vallee, *Biochemistry* **31,** 12477 (1992).

reflecting a decrease in the strength of the interaction perhaps due to chemical changes in one or both proteins over the long course of the experiment.) The dissociation rate constant, k_d, is calculated from the slower process, fitting the data to the equation:

$$[A] = [A]_{eq} - ([A]_{eq} - [A]_0)e^{-k_d t} \tag{4}$$

where $[A]_{eq}$ is the free Ang concentration at equilibrium and $[A]_0$ is that measured at the first time point after addition of scavenger. [A] is normalized with respect to the value obtained in the control sample lacking RI (which typically declines by ~10% over 10 days). When the procedure involving [^{14}C]Ang is used, $[A]_{eq}$ can be assumed to be the same as the total Ang concentration. However, when unlabeled Ang is used with RNase A as scavenger, the $[A]_{eq}$ is ~80% of the total[17]; this reflects the 60-fold greater affinity of hRI for Ang than for RNase A. The values obtained by the first and second methods are similar: $1.3 \times 10^{-7} s^{-1}$ and $1.1 \times 10^{-7} s^{-1}$, respectively, corresponding to half-lives of 62 days and 73 days.

The K_i value is related to the individual kinetic parameters according to the equation

$$K_i = K_1 k_{-2}/(k_{-2} + k_2) \tag{5}$$

Because $k_2 \gg k_{-2}$, this reduces to

$$K_i = K_1 k_{-2}/k_2 \tag{6}$$

K_1 and k_2 (or K_1/k_2) are from the association experiments described above, and k_{-2} is essentially equivalent to the k_d value measured in the dissociation experiment.[17] The K_i value for the hRI–Ang complex originally obtained by this method is 0.71 f*M* (Table III). Repeated determinations in this laboratory since that time have generally given somewhat lower values (0.39–0.55 f*M*, with standard errors of ~20%).

RI–RNase A Complexes

Method 1: Physical–Enzymatic. Rate constants for dissociation of RI–RNase A complexes can be determined by procedures similar to those used for RI–Ang complexes.[17] Ang (in 100-fold molar excess) is the scavenger and the concentration of free RNase A is measured by HPLC or, more conveniently, by assaying for enzymatic assay toward CpA, UpA, or CpG; Ang has no detectable activity in these assays. Incubations are typically performed with 20–40 n*M* RNase A in 0.1 *M* Mes (pH 6), 0.1 *M* NaCl, 1 m*M* EDTA, 100 μg/ml BSA, and 5 m*M* DTT. Again, there is an initial "burst" of free ligand (in this case up to ~20% of the total), with the remainder dissociating much more slowly. Analysis of the slower process yields a k_d value of 1.2×10^{-5} s^{-1}($t_{1/2}$ = 16 h) (Table III).

Formation of the RI–RNase A complex has only a small effect on Trp fluorescence, and k_a is measured by competition in an enzymatic assay rather than by physical techniques.[57] RNase A (R) and Ang (A) are mixed, RI is added, and the partitioning of RI between the two ligands is then determined by measuring RNase A activity. Because dissociation is slow, partitioning depends entirely on the relative rates of association of the RI · R and RI · A complexes. (Another RI ligand can also be used as a competitor if it has low enzymatic activity and the k_a value for its complex is known.) The association rate constant ($k_{a,R}$) is calculated from the equation

$$k_{a,R} = k_{a,A}\{\ln([\mathrm{R}]_\mathrm{T}/[\mathrm{R}]_\mathrm{F})/\ln([\mathrm{A}]_\mathrm{T}/[\mathrm{A}]_\mathrm{F})\} \tag{7}$$

where $k_{a,A}$ is the rate constant for association of A, and T and F refer to the total and free ligand concentrations, respectively. Equation (7) is valid for both two-step and single-step mechanisms, but in the former case, $[\mathrm{R}]_\mathrm{F}$ and $[\mathrm{A}]_\mathrm{F}$ must always be well below both K_1 values (this can be established by demonstrating that $k_{a,R}$ values are independent of the protein concentrations). The competition is performed in the assay cuvette, with $[\mathrm{R}]_\mathrm{T}$ and $[\mathrm{I}]_\mathrm{T}$ kept fixed and equal, and Ang varied to give ~25–75% inhibition. The value for $k_{a,R}$ obtained by this procedure is 3.4×10^8 M^{-1} s^{-1}. Application of Eq. (6) then gives $K_i = 44$ fM.

Method 2 (Enzymatic): Effect of RI on Progress Curves. The values of k_a, k_d, and K_i for the RI · RNase A complex can also be determined by measuring the time course of inhibition of RNase A in the presence of substrate at very low concentrations of RNase and RI so that the onset of inhibition is slow.[11] Progress curves are measured with multiple concentrations of RI and fit to the equation:

$$P = v_s t + \frac{(v_0 - v_s)(1-\gamma)}{\gamma\lambda}\ln\left[\frac{1-\gamma e^{-\lambda t}}{1-\gamma}\right] \tag{8}$$

where P is the amount of product at time t, v_0, and v_s are the reaction velocities at time zero and at steady state, respectively, γ is a function of $[\mathrm{E}]_\mathrm{T}$, $[\mathrm{I}]_\mathrm{T}$, and K_i' (defined as above), and λ is a function of these parameters and k_{obs}, the observed second-order rate constant of association. This gives values for K_i' and k_{obs}, which are corrected for substrate concentration to obtain K_i and k_a (the inhibition mode is competitive; see below). k_d is then calculated as $K_i k_a$. (The full details are provided in Ref. 11.) When this procedure was first applied to the analysis of the pRI · RNase A complex, the substrate was UpA (at $\sim 3 \times K_m$) and the enzymatic reaction was followed by HPLC. In a subsequent improvement,[72] a fluorescent substrate was used that allowed the reaction to be monitored continuously at an RNase A concentration only ~20-fold above K_i. The kinetic parameters obtained with the two assay systems differ by <2-fold (Table III).

Method 3 (Enzymatic): Effect of RI on Initial Velocities. A simpler way to determine the K_i value for the RI · RNase A complex (which, however, does not yield any information on the individual rate constants) is to measure initial

[72] O. Zelenko, U. Neumann, W. Brill, U. Pieles, H. E. Moser, and J. Hofsteenge, *Nucl. Acids Res.* **22,** 2731 (1994).

steady-state velocities (v_i) for the enzymatic reaction at various concentrations of RI. These data are then fit to the equation

$$v_i = (v_0/2[\mathrm{E}]_\mathrm{T}) \cdot \{[(K_i' + [\mathrm{I}]_\mathrm{T} - [\mathrm{E}]_\mathrm{T})^2 + 4K_i'[\mathrm{E}]_\mathrm{T}]^{1/2} - (K_i' + [\mathrm{I}]_\mathrm{T} - [\mathrm{E}]_\mathrm{T})\} \quad (9)$$

where v_0 is the velocity in the absence of inhibitor. This yields the value of K_i', which is again corrected for substrate concentration to give K_i. This method was applied to the pRI · RNase A complex, using UpA as substrate (at $2 \times K_m$) in a discontinuous, HPLC-based assay.[11] The concentration of RNase A was 9.2 pM, i.e., 40-fold above K_i', which approaches the upper limit for this procedure.

Other RI Complexes

The complexes of hRI with human RNase-2,[20] hRI with human pancreatic RNase,[18] and pRI with porcine RNase-4[21] have been analyzed by modifications of the methods (1, 3, and 2, respectively) used for the RI · RNase A complex. The values obtained are listed in Tables II and III. These procedures, and those described for the RI · Ang complex, have also been adapted for characterizing complexes of RI, Ang, and RNase A variants.[30–32,38,51,71,73] In addition, k_d values for some rapidly dissociating (i.e., $t_{1/2} < 40$ min) variant RI · Ang complexes have been determined by monitoring Trp fluorescence after addition of W89M-Ang as a scavenger for free RI: the emission intensity of the hRI · W89M–Ang complex is approximately half that of the wild-type Ang complexes.[31,32]

Mode of Inhibition

In a competitive mechanism, the apparent k_a and K_i values (k_a' and K_i') vary with the concentration of either substrate S or a known competitive inhibitor C; k_a' decreases by the factor $(1 + [\mathrm{S}]/K_m)$ or $(1 + [\mathrm{C}]/K_{i,c})$ and K_i' increases by the same factor. This type of dependence has been demonstrated for k_a' with the complex of hRI with RNase A[17] (using the inhibitor 2′-CMP as a competitor) and for K_i' with the complex of pRI with RNase A.[11] The competitive inhibition mode indicated by these kinetic experiments is strongly supported by the crystal structure of the RI · RNase A complex, as discussed above. Although kinetic studies of this type have not been performed for other RI complexes, it is clear from the RI · Ang crystal structure that inhibition is competitive in this case as well, and homology considerations suggest this will apply generally.

Acknowledgments

I thank Matthew Crawford and Daniel P. Teufel for skilled technical assistance, and Dr. Bostjan Kobe for kindly supplying Fig. 1.

[73] U. Neumann and J. Hofsteenge, *Prot. Sci.* **3,** 248 (1994).

[40] Small Molecule Inhibitors of RNase A and Related Enzymes

By ANIELLO RUSSO, K. RAVI ACHARYA, and ROBERT SHAPIRO

Introduction

The interactions of bovine pancreatic RNase A (EC 3.1.27.5) with small nucleotide and oligonucleotide inhibitors have been studied extensively over the past four decades by kinetic and physical methods (reviewed by Richards and Wyckoff [1] and Eftink and Biltonen[2]), and crystal or NMR structures for at least 20 complexes have been determined (summarized by Gilliland[3]). Nonetheless, it is only during the past several years that efforts have been launched to develop tight-binding low molecular weight inhibitors of this enzyme. Interest is this area has been stimulated largely by research demonstrating that several pancreatic RNase homologs, including angiogenin (Ang), eosinophil-derived neurotoxin (EDN), and bovine seminal RNase, utilize their enzymatic activities to elicit potent physiological effects.[4–6] Some of these proteins are known or suspected to play important roles in human disease processes, and inhibitors are therefore potential drug candidates. Specific antagonists can also be used to illuminate the biological mechanisms of these molecules. The impetus to develop improved small-molecule inhibitors also derives in part from the need to neutralize adventitious mammalian RNases for *in vitro* applications utilizing RNA. The proteinaceous cytosolic RNase inhibitor (discussed elsewhere in this volume[7]) is often extremely effective for this purpose, but is irreversibly inactivated under the denaturing or oxidizing conditions used in some systems.

In this chapter, we summarize the current state of knowledge of low molecular weight inhibitors of RNase A and related enzymes, with particular emphasis on the most potent class of inhibitors developed thus far, the adenosine 5′-pyrophosphate derivatives. First, however, we provide some background on the active site

[1] F. M. Richards and H. W. Wyckoff, "The Enzymes," 3rd Ed., Vol. 4, p. 647. Academic Press, New York, 1971.

[2] M. R. Eftink and R. L. Biltonen, *in* "Hydrolytic Enzymes" (A. Neuberger and K. Brocklehurst, eds.), p. 333. Elsevier, Amsterdam, 1987.

[3] G. L. Gilliland, *in* "Ribonucleases: Structures and Functions" (G. D' Alessio and J. F. Riordan, eds.), p. 305. Academic Press, New York, 1997.

[4] R. Shapiro, E. A. Fox, and J. F. Riordan, *Biochemistry* **28,** 1726 (1989).

[5] S. Sorrentino, D. G. Glitz, K. J. Hamann, D. A. Loegering, J. L. Chekel, and G. J. Gleich, *J. Biol. Chem.* **267,** 14859 (1992).

[6] G. D' Alessio, A. Di Donato, A. Parente, and R. Piccoli, *Trends Biochem. Sci.* **16,** 104 (1991).

[7] R. Shapiro, *Methods Enzymol.* **341,** [40] 2001 (this volume).

0076-6879/01 $35.00

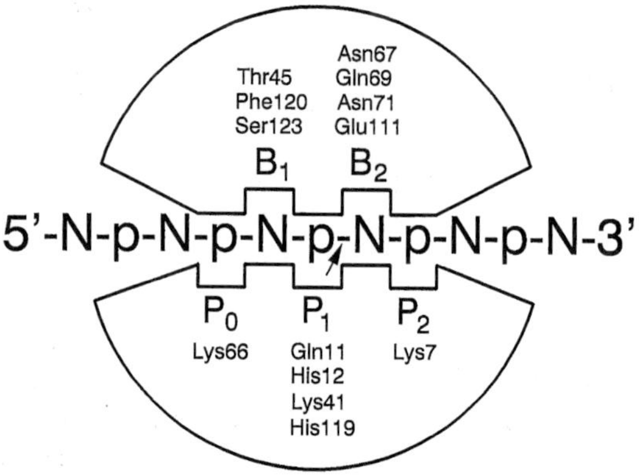

FIG. 1. Subsites of the ribonucleolytic center of RNase A. 5′-N-p . . . N-3′ represents the RNA substrate, and the arrow indicates the site at which it is cleaved. B_1 and B_2 are base-binding sites. P_0, P_1, and P_2 are phosphate-binding sites. Additional subsites B_3 and P_{-1} have also been shown to exist. The major residues whose side chains interact with RNA at each subsite, as indicated by crystal and NMR structures of RNase A–inhibitor complexes, are indicated. From Russo and Shapiro.[36]

architecture and functional properties of RNase A and the other enzymes targeted by these compounds.

Binding Subsites on RNase A

The active site of RNase A is conventionally described in terms of multiple subsites for binding the various phosphate, base, and ribose moieties of RNA substrates (Fig. 1) (see reviews by Richards and Wyckoff,[1] Nogues *et al.*,[8] and Raines[9] and references therein). The phosphate-binding subsite where $P–O^{5'}$ bond cleavage occurs is designated P_1; it contains the catalytic residues His-12, Lys-41, and His-119, as well as Gln-11 and the main-chain NH of Phe-120. The B_1 site binds the base whose ribose donates its 3′ oxygen to the P_1 phosphate; residues contributing directly or indirectly to this site are Thr-45, Val-43, Asn-44, Asp-83, Phe-120, and Ser-123. This site has a nearly absolute specificity for pyrimidines. The B_2 site binds the base, preferably a purine, whose ribose provides the 5′ oxygen of the scissile bond; its major component is Asn-71,[10] which hydrogen bonds to the base, but crystal and NMR structures have also shown inhibitors to interact with Asn-67, Gln-69, Glu-111, and His-119. In addition, four peripheral

[8] M. V. Nogues, M. Moussaoui, E. Boix, M. Vilanova, M. Ribo, and C. M. Cuchillo, *Cell. Mol. Life Sci.* **54,** 766 (1998).

[9] R. T. Raines, *Chem. Rev.* **98,** 1045 (1998).

[10] A. Tarragona-Fiol, H. J. Eggelte, S. Harbron, E. Sanchez, C. J. Taylorson, J. M. Ward, and B. R. Rabin, *Protein Eng.* **6,** 901 (1993).

subsites make significant contributions: P_0 (Lys-66), P_2 (mainly Lys-7, but Arg-10 may also participate), P_{-1} (Arg-85[11]), and B_3 (Gln-69 and the B_2 purine[12]). The crystal structure of the complex of RNase A with d(ApTpApApG) shows no specific B_0 and B_4 subsites on the enzyme.[12]

Pancreatic RNase Superfamily

Four major families of pancreatic RNase superfamily proteins have been identified in mammals (reviewed by Sorrentino[13] and Beintema and Kleineidam[14]; also see chapters by Rosenberg,[15] Boix,[16] and Riordan[17] in this volume): (i) pancreatic RNase (RNase-1) and the seminal and brain RNases found only in ruminants; (ii) liver–spleen RNase (RNase-2, EDN), RNase-6, and several RNases limited to certain species (e.g., RNase-3 or eosinophil cationic protein); (iii) a second liver enzyme termed RNase-4 that apparently has no close relatives; and (iv) angiogenin (RNase-5) and similar proteins present only in nonhuman species. Each of the four families is represented in blood plasma, but their distributions in different tissues and cell types are diverse. All of the RNases appear to be synthesized with signal peptides and are either secreted directly or packaged into secretory granules; thus, the early classification of the second family of enzymes as "nonsecretory," although still frequently seen in the literature, is incorrect.

Pancreatic RNase superfamily proteins share certain general enzymatic characteristics: they cleave RNA on the 3′ side of pyrimidine nucleotides via a transphosphorylation mechanism to form products terminating with 2′,3′-cyclic phosphates, and they universally prefer purines in the B_2 position. Most (perhaps all) can also catalyze the hydrolysis of the $P–O^{2'}$ bond of the 2′,3′-cyclic phosphate. Beyond these common features, however, the enzymes vary widely in their actions. The pancreatic group is the most effective toward both polynucleotides and small substrates. The activities of RNase-2 from humans (the only species examined in detail) and RNase-4 with RNA are similar to or somewhat lower than that of RNase-1, depending on the assay, but k_{cat}/K_m values with their best dinucleotide substrates are at least 10-fold smaller than for the pancreatic enzyme. Human RNase-3 and RNase-6 are ~100-fold less efficient than RNase A with polynucleotides, and RNase-3 has been shown to have even lower activity toward dinucleotides. The

[11] B. M. Fisher, J. E. Grilley, and R. T. Raines, *J. Biol. Chem.* **273,** 34134 (1998).

[12] J. C. Fontecilla-Camps, R. de Llorens, M. H. le Du, and C. M. Cuchillo, *J. Biol. Chem.* **269,** 21526 (1994).

[13] S. Sorrentino, *Cell. Mol. Life Sci.* **54,** 785 (1998).

[14] J. J. Beintema and R. G. Kleineidam, *Cell. Mol. Life Sci.* **54,** 825 (1988).

[15] H. F. Rosenberg and J. B. Domachowske, *Methods Enzymol.* **341,** [18] 2001 (this volume).

[16] E. Boix, *Methods Enzymol.* **341,** [19] 2001 (this volume).

[17] J. F. Riordan, *Methods Enzymol.* **341,** [17] 2001 (this volume).

angiogenins are the weakest RNases yet described, with potencies toward standard substrates typically 5–6 orders of magnitude lower than for RNase A. Substrate specificity is also quite variable among the different homologs: RNase-1, RNase-2, and RNase-3 cleave CpN dinucleotides slightly more rapidly than UpN substrates,[16,18,19] whereas Ang has a 10- to 20-fold preference for cytidine,[20] and RNase-4 has a unique several-hundred-fold preference for uridine.[21] With only one reported exception (bovine Ang[22]), the RNases under consideration are more active toward NpA than NpG substrates, but the magnitude of this preference differs markedly among the subtypes: ~3-fold for Ang,[23] ~10-fold for RNase-1[18] and RNase-4,[21] and up to 1000-fold for RNase-2[19] and RNase- 3.[16]

Consistent with the enormous variability in potency and specificity among the RNase subtypes, most of the base- and phosphate-binding subsites are not well conserved (Table I). Comparisons of the sequences and 3D structures of RNase A,[12,24] RNase-2,[25] RNase-4,[26] and Ang[27] reveal that only P_1 is strictly maintained; even here, residues that indirectly influence the function of this site in RNase A (e.g., Lys-7[28] and Asp-121[29]) are either replaced or positioned differently in some of the nonpancreatic RNases. Although the primary B_1 residue Thr-45 is conserved in all of the enzymes, other architectural features of this subsite are altered in RNase-4 to impose uridine specificity[30,26] and this subsite is obstructed in free Ang.[27] The regions of RNase-2 and Ang that correspond to the B_2, P_2, and P_0 subsites of RNase A are also structurally quite distinct. The variability in the structures of subsites other than P_1 among the different RNase families might allow the design of compounds that are specific for each of the RNase families.

Finally, the available evidence indicates that the enzymatic properties of orthologous RNases from different species are similar. Thus, compounds that inhibit

[18] H. Witzel and E. A. Barnard, *Biochem. Biophys. Res. Commum.* **7,** 295 (1962).

[19] R. Shapiro and B. L. Vallee, *Biochemistry* **30,** 2246 (1991).

[20] R. Shapiro, J. W. Harper, E. A. Fox, H.-W. Jansen, F. Hein, and E. Uhlmann, *Anal. Biochem.* **175,** 450 (1988).

[21] R. Shapiro, J. W. Fett, D. J. Strydom, and B. L. Vallee, *Biochemistry* **25,** 7255 (1986).

[22] D. J. Strydom, M. D. Bond, and B. L. Vallee, *Eur. J. Biochem.* **247,** 535 (1997).

[23] J. W. Harper and B. L. Vallee, *Biochemistry* **28,** 1875 (1989).

[24] B. Borah, C.-W. Chen, W. Egan, M. Miller, A. Wlodawer, and J. S. Cohen, *Biochemistry* **24,** 2058 (1985).

[25] S. C. Mosimann, D. L. Newton, R. J. Youle, and M. N. G. James, *J. Mol. Biol.* **260,** 540 (1996).

[26] S. S. Terzyan, R. Peracaula, R. de Llorens, Y. Tsushima, H. Yamada, M. Seno, F. X. Gomis-Ruth, and M. Coll, *J. Mol. Biol.* **285,** 205 (1999).

[27] K. R. Acharya, R. Shapiro, S. C. Allen, J. F. Riordan, and B. L. Vallee, *Proc. Natl. Acad. Sci. U.S.A.* **91,** 2915 (1994).

[28] B. M. Fisher, L. W. Schultz, and R. T. Raines, *Biochemistry* **37,** 17386 (1998).

[29] L. W. Schultz, D. J. Quirk, and R. T. Raines, *Biochemistry* **37,** 8886 (1998).

[30] J. Hofsteenge, C. Moldow, A. M. Vicentini, O. Zelenko, Z. Jarai-Kote, and U. Neumann, *Biochemistry* **37,** 9250 (1998).

TABLE I
OBSERVED OR PREDICTED NUCLEOTIDE-BINDING RESIDUES ON RNase A, RNase-2, RNase-4, AND ANGIOGENIN[a]

Subsite	RNase A	RNase-2	RNase-4	Angiogenin[b]
P_0	Lys-66 (c)	Ser-64	Lys-65	—
		Arg-132		
B_1	Thr-45 (h)	Thr-42	Thr-44 (h)	Thr-44
			Arg-101 (h)	
	Val-43 (v)	Gln-40	Phe-42 (v)	Ile-42
	Asn-44 (v)	Asn-41	Asn-43 (v)	Asn-43
	Phe-120 (v)	Leu-130	Phe-117 (v)	Leu-115
	Ser-123 (h-w)	Ile-133		
P_1	Gln-11 (h)	Gln-14	Gln-11 (h)	Gln-12
	His-12 (h)	His-15	His-12 (h)	His-13
	Lys-41 (h)	Lys-38	Lys-40 (h)	Lys-40
	His-119 (h)	His-129	His-116 (h)	His-114
B_2	Asn-67 (h)	Asn-65	Asn-66	Leu-69
	Gln-69 (h)	Arg-68	Lys-68	
	Asn-71 (h)	Asn-70	Asn-70	
	Glu-111 (h)	Asp-112	Glu-108	Glu-108
	His-119 (s)	His-129	His-116	His-114
P_2	Lys-7 (h)	Trp-7	Arg-7	Arg-5
	Arg-10 (c)		Arg-10	

[a] Subsite residues listed for RNase A are those observed to form hydrogen bonds (h), water-mediated hydrogen bonds (h-w), van der Waals contacts (v), or stacking interactions (s) with nucleotides in crystal structures of complexes [see G. L. Gilliland *in* "Ribonucleases: Structures and Functions" (G. D' Alessio and J. F. Riordan, eds.), p. 306. Academic Press, New York, 1997]. RNase A residues listed as forming only Coulombic interactions (c) were identified by site-specific mutagenesis [E. Boix, M. V. Nogues, C. H. Schein, S. A. Benner, and C. M. Cuchillo, *J. Biol. Chem.* **269,** 2529 (1994); B. M. Fisher, J.-H. Ha, and R. T. Raines, *Biochemistry* **37,** 12121 (1998)]. RNase-4 residues listed as forming hydrogen bonds or van der Waals contacts are those interacting with 2′-deoxy 3′-UMP in the crystal structure of the complex [S. S. Terzyan, R. Peracaula, R. de Llorens, Y. Tsushima, H. Yamada, M. Seno, F. X. Gomis-Ruth, and M. Coll, *J. Mol. Biol.* **285,** 205 (1999)]. The other residues listed for RNase-4, RNase-2, and angiogenin are based on modeling of nucleotide complexes.

[b] The B_1 site of angiogenin is obstructed by Gln-117; it has been postulated that prior to or during catalysis this site opens and becomes similar to the B_1 site of RNase A [K. R. Acharya, R. Shapiro, S. C. Allen, J. F. Riordan, and B. L. Vallee, *Proc. Natl. Acad. Sci. USA.* **91,** 2915 (1994); N. Russo, R. Shapiro, K. R. Acharya, J. F. Riordan, and B. L. Vallee, *Proc. Natl. Acad. Sci. USA* **91,** 2920 (1994)].

TABLE II
K_i VALUES FOR LOW MOLECULAR WEIGHT INHIBITORS OF RNase A

Inhibitor[a]	K_i^b (μM)	pH[c] (buffer, salts)	Ref.[d]
Ordinary Mononucleotides and Dinucleotides			
pT-3′-p	1.0	5.0 (A)	1
pU-2′(3′)-p	1.6	5.0 (B)	2
4-Thio-U-3′-p	2.7	5.0 (A)	1
pA-3′-p	5.0	5.9 (C)[e]	3
C-2′-p	7.0	6.0 (D)	4
U-2′-p	7.1	5.5 (D′)	4
T-3′-p	13.2	5.0 (A)	1
pTpdA-3′-p	14.2	5.0 (A)	5
dU-3′-p	25	6.0 (D′)	6
pA	80	5.9 (C)[e]	3
U-3′-p	82	6.0 (D′)	4
C-3′-p	103	6.0 (D)	4
TpdA-3′-p	137	5.0 (A)	5
pT	257	5.0 (A)	1
TpdA	1200	5.9 (C)[e]	7
Oligonucleotides			
Fluorescein-d(pApUpApA)[f,g]	88	6.0 (E)	8
Fluorescein-d(pUpApA)[f]	130	6.0 (E)	8
Nonnucleotides			
Uridine vanadate[h]	0.45	7.0 (F)	9
Pyrophosphate	172	6.0 (D″)	4
Orthophosphate	4600	6.0 (D″)	4
Adenosine 5′-Pyrophosphate Derivatives			
pdUppA-3′-p	0.027	5.9 (C)	7
pTppA-3′-p	0.041	5.9 (C)	7
dUppA-3′-p	0.12	5.9 (C)	7
dU2′ppA-3′-p	0.13	5.9 (C)	7
ppA-3′-p	0.24	5.9 (C)	3
ppA-2′-p	0.52	5.9 (C)	3
ppA	1.2	5.9 (C)	3
TppdA	4.0	5.9 (C)	7
pppA	14.3	5.5 (G)	10

[a] The letter "p" preceding a nucleoside denotes a 5′-linked phosphate; "pp" is diphosphate; "ppp" is triphosphate; "d" preceding a nucleoside indicates that the ribose is 2′-deoxy. All dinucleotide linkages are 3′→5′, except for that in dU2′ppA-3′-p, which is 2′→5′.

[b] All K_i values from Refs. 1–3, 5, 7, 9, and 10 and the values for nonnucleotide inhibitors from Ref. 4 are based on inhibition of enzymatic activity. Other values from Ref. 4 and those from Ref. 6 were determined by UV difference spectroscopy, and the values from Ref. 8 were measured by fluorescence anisotropy.

[c] Conditions at the indicated pH values are: (A) 0.1 *M* acetate; (B) 6.7 m*M* acetate; (C) 0.2 *M* MES–NaOH; (D) 50 m*M* Tris–acetate, 50 m*M* sodium acetate, 100 m*M* KNO_3; (D′) as for D but with KCl in place of KNO_3; (D″) as for D but with NaCl replacing KNO_3; (E) 20 m*M*

any particular RNase effectively will most likely have comparable activity against its orthologs.

Inhibitors of RNase A

Table II lists K_i values for many of the nucleotide inhibitors of RNase A that have been characterized to date, as well as for selected nonnucleotide inhibitors. Early work in this field focused largely on ordinary mononucleotide and dinucleotide substrate or product analogs, with one foray into the realm of transition state analogs. The most effective inhibitors identified [thymidine 3′,5′-diphosphate (pT-3′-p), pU-2′(3′)-p, adenosine 3′,5′-diphosphate (pA-3′-p), 2-thiouridine-3′-p, and cytidine and uridine 2′-monophosphates (C-2′-p and U-2′-p)] exhibited low micromolar dissociation constants in the optimum pH range of 5–6. The pyrimidine nucleotides occupy the B_1, P_1, and in some cases P_0 subsites, whereas pA-3′-p is expected to bind at P_1, B_2, and P_2. As discussed below, K_i values are strongly dependent on pH and ionic strength. Hence, the potencies of the inhibitors listed in the table can be compared rigorously only when measured under identical conditions. In general, K_i values at a given ionic strength vary by less than a factor of 2 within the pH range 5–6 (all conditions except for F), and binding is expected to be tightest at low ionic strength (conditions A and B).

Some of the major findings of these studies that have potential relevance for the development of more effective inhibitors are: (i) cytidine, uridine, and thymidine bind to B_1 with similar affinities; (ii) removal of the 2′-hydroxyl group of the B_1

MES-NaOH, 100 m*M* NaCl; (F) 5 m*M* Tris-HCl, 0.35 *M* KCl; (G) not specified in the reference cited. All measurements were performed at 25°.

[d] *Key to References:* (1) K. Iwahashi, K. Nakamura, Y. Mitsui, K. Ohgi, and M. Irie, *J. Biochem.* **90,** 1685 (1981); (2) F. Sawada and *M* Irie, *J. Biochem.* **66,** 415 (1969); (3) N. Russo, R. Shapiro, and B. Vallee, *Biochem. Biophys. Res. Commun.* **231,** 671 (1997); (4) D. G. Anderson, G. C. Hammes, and F. G. Walz, *Biochemistry* **7,** 1637 (1968); (5) M. Irie, H. Watanabe, K. Ohgi, M. Tobe, G. Matsumura, Y. Arata, T. Hirose, and S. Inayama, *J. Biochem.* **95,** 751 (1984); (6) F. G. Walz, *Biochemistry* **10,** 2156 (1971); (7) N. Russo and R. Shapiro, *J. Biol. Chem.* **274,** 14902 (1999); (8) B. M. Fisher, J. E. Grilley, and R. T. Raines, *J. Biol. Chem.* **273,** 34134 (1998); (9) C. H. Leon-Lai, M. J. Gresser, and A. S. Tracey, *Can. J. Chem.* **74,** 38 (1996); (10) H. Katoh, M. Yoshinaga, T. Yanagita, K. Ohgi, M. Irie, J. J. Beintema, and D. Meinsma, *Biochim. Biophys. Acta* **873,** 367 (1986).

[e] K_i values for pA-3′-p, pA, and TpdA were originally reported in Ref. 5 (pH 5.0, condition A); the values obtained were 5.0 μM, 46 μM, and 1300 μM, respectively.

[f] Fluorescein is coupled to the 5′-terminal phosphate of the oligonucleotide through a six-carbon spacer.

[g] The dissociation constant for this oligonucleotide in 20 m*M* MES–NaOH (pH 6.0) containing 50 m*M* NaCl (i.e., conditions more similar to condition C) is 11 μM [B. M. Fisher, J.-H. Ha, and R. T. Raines, *Biochemistry* **37,** 12121 (1998)].

[h] Uridine vanadate is unstable and only a small fraction of added uridine and vanadate is present as the inhibitory complex (see text).

nucleotide improves binding a fewfold; (iii) 2′ nucleotides at B_1/P_1 bind>10-fold more tightly than the corresponding 3′ nucleotides; (iv) introduction of a phosphate to occupy the P_0 or P_2 subsite can enhance affinity by ~1 order of magnitude; and (v) dinucleotides bind much less strongly than their mononucleotide substituents. The last observation in part appears to reflect the preference of the P_1 site for a phosphate dianion vs monoanion,[31] but may also be due to difficulty optimizing the interactions of the individual components at B_1, P_1, and B_2 when the two nucleotides are linked by a standard phosphodiester bond.

During the early 1970s, Lindquist *et al.*[32] investigated the complexes of uridine with oxovanadium (IV) and vanadium (V) ions as possible transition state analogs for RNase A–catalyzed hydrolysis of uridine 2′,3′-cyclic phosphate (U>p), and measured K_i values of 10 μM and 12 μM, respectively. Although both vanadate complexes are covalent, they are unstable and the calculations of the inhibition constants took into account the fact that relatively small amounts of the complexes are actually present in mixtures of uridine and either vanadate form. Later studies revealed the existence of additional equilibria in uridine vanadate (U>v) solutions, and when the K_i was reexamined on this basis, a considerably lower value (0.45 μM) was obtained.[33] This is ~10,000-fold below the K_m of 4.2 mM measured for U>p under identical conditions, but at least five orders of magnitude higher than that predicted for the transition state of the U>p reaction.[33] In practical terms, it should be noted that U>v, despite its relatively impressive K_i value, is not very useful for inhibiting RNase A in experimental systems: e.g., a mixture of 10 mM uridine and 0.1 mM V(V) at pH 7 contains only sufficient U>v to bind ~80% of the RNase A present.[32] Moreover, vanadate itself is an inhibitor of numerous other enzymes, and its use may be problematic in many situations.

Over the past several years, a new class of nucleotide inhibitors with high affinity for RNase A has been identified (Table II).[34–36] These compounds are mononucleotide and dinucleotide derivatives of adenosine 5′-pyrophosphate; their development was based on the initial observation that 5′-ADP (ppA; K_i = 1.2 μM) binds to RNase A 67-fold more tightly than does 5′-AMP (pA). It was known from earlier studies[37] that the K_i value for pAp is also much lower than that for pA. Thus, 5′-diphosphoadenosine 3′-phosphate (ppA-3′-p; Fig. 2) was synthesized and tested to determine the combined effect of the two phosphate additions; the isomer

[31] M. Flogel and R. L. Biltonen, *Biochemistry* **14,** 2610 (1975).

[32] R. N. Lindquist, J. L. Lynn, and G. E. Lienhard, *J. Am. Chem. Soc.* **95,** 8762 (1973).

[33] C. H. Leon-Lai, M. J. Gresser, and A. S. Tracey, *Can. J. Chem.* **74,** 38 (1996).

[34] N. Russo, K. R. Acharya, B. L. Vallee, and R. Shapiro, *Proc. Natl. Acad. Sci. U.S.A.* **93,** 804 (1996).

[35] N. Russo, R. Shapiro, and B. L. Vallee, *Biochem. Biophys. Res. Commun.* **231,** 671 (1997).

[36] N. Russo and R. Shapiro, *J. Biol. Chem.* **274,** 14902 (1999).

[37] M. Irie, H. Watanabe, K. Ohgi, M. Tobe, G. Matsumura, Y. Arata, T. Hirose, and S. Inayama, *J. Biochem.* **95,** 751 (1984).

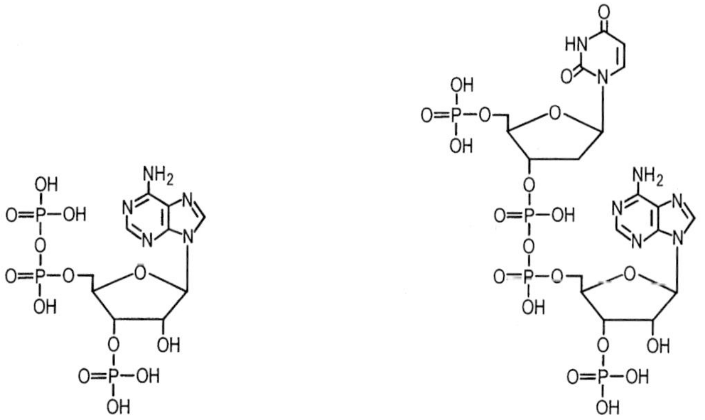

5'-diphosphoadenosine 3'-phosphate (ppA-3'-p)

5'-phospho-2'-deoxyuridine 3'-pyrophosphate, P'→5'-ester with adenosine 3'-phosphate (pdUppA-3'-p)

FIG. 2. Structures of ppA-3′-p (5′-diphosphoadenosine 3′-phosphate) and pdUppA-3′-p (5′-phospho-2′-deoxyuridine 3′-pyrophosphate P′→5′ ester with adenosine 3′-phosphate).

diphosphoadenosine 2′-phosphate (ppA-2′-p) was investigated as well. The K_i values for the two compounds were found to be 0.24 μM and 0.52 μM, respectively (i.e., 5-fold and 2.5-fold lower than for ppA).[35]

Crystal structures of the complexes of RNase A with ppA-3′-p and ppA-2′-p were then determined and used as the basis for design of tighter binding nucleotides.[38,36] Modeling suggested that linkage of a pyrimidine nucleotide to the β-phosphate would extend the inhibitor into the B_1 site where additional interactions could be formed. Therefore, dUppA-3′-p (2′-deoxyuridine 3′-pyrophosphate P′→5′ ester with adenosine 3′-phosphate) and the corresponding uridine 2′-pyrophosphate ester were synthesized and tested. The affinities of dUppA-3′-p and U2′ppA-3′-p for RNase A were both 2-fold greater than that of ppA-3′-p at the standard assay pH of 5.9. However, the pH dependence for binding of the 3′→5′ dinucleotide was more favorable, and only this compound was used for further design. An RNase A-dUppA-3′-p model suggested that extension of the inhibitor by a 5′-phosphate [yielding pdUppA-3′-p (5′-phospho-2′-deoxyuridine 3′-pyrophosphate P′→5′ ester with adenosine 3′-phosphate); Fig. 2] would allow additional interactions at P_0. The K_i value for pdUppA-3′-p (27 nM) was indeed severalfold lower than for dUppA-3′-p. The effect of replacing the 2′-deoxyuridine moiety with thymidine, which is more readily available and less expensive, was also tested; the affinity of pTppA-3′-p for RNase A was found to be similar to that

[38] D. D. Leonidas, R. Shapiro, L. I. Irons, N. Russo, and K. R. Acharya, *Biochemistry* **36,** 5578 (1997).

of the uridine nucleotide (K_i = 41 nM). These two novel 3′,5′-pyrophosphate-linked dinucleotides are the most potent stable low-molecular-weight inhibitors of RNase A described thus far. The specific energetic contribution of the pyrophosphate linkage as compared to an ordinary phosphate linkage was shown to be 3.4 kcal/mol: i.e., the K_i for TppdA is 300-fold lower than that of TpdA (Table II). Indeed, TppdA binds more tightly than a tetranucleotide [fluorescein-d(pApUpApA)] with standard phosphate linkages that occupies four additional functional subsites (P_{-1}, P_0, P_2, and B_3). The recent determination of a high-resolution crystal structure for the RNase A–pdUppA-3′-p complex[39] may now facilitate the design of additional members of this class of nucleotides that are even more effective.

Inhibitors of Human RNase-2, RNase-4, and Angiogenin

Relatively little information is available on low molecular weight inhibitors of the pancreatic superfamily RNases other than RNase A. To our knowledge, dissociation constants for ordinary nucleotide inhibitors have been reported with only three of these enzymes: bovine seminal RNase, human RNase-4, and human Ang. The interactions of the dimeric seminal RNase with small inhibitors are cooperative[40] and will not be considered here. Human RNase-4 was shown to be inhibited by 2′-UMP with a K_i of 40 μM (in 0.1 M Mes, 0.1 M NaCl, pH 6.0),[21] a value that is in the same range as that measured with RNase A. The 2′- and 3′-phosphates of uridine and cytidine, pA, and pA-3′-p bind to Ang 90- to 1000-fold less tightly than to RNase A.[34,41] The generally low affinity of Ang for nucleotides is also evident from the >60 mM K_m value measured for this enzyme with the substrate CpA (~100-fold higher than for RNase A).[41]

Inhibition of human RNase-2, RNase-4, and Ang by the new class of adenosine 5′-pyrophosphate derivatives has been investigated (Table III). ppA-3′-p is an effective inhibitor of both RNase-2 and RNase-4, with K_i values of 0.25 and 0.54 μM, respectively, similar to that measured for RNase A. However, no substantial improvement is seen for the dinucleotide pyrophosphates pdUppA-3′-p and pTppA-3′-p, which bind 5- to 11-fold less tightly than to the pancreatic enzyme. The adenosine 5′-pyrophosphates have higher affinity for Ang than do ordinary nucleotides, but binding is still weak compared to that measured with the three more active RNases. The most effective of these compounds against Ang is ppA-2′-p, with a K_i value of 110 μM. The value for the 3′-phosphate is 3-fold higher, consistent with the preference of Ang for pA-2′-p over pA-3′-p.[34] Extending these

[39] D. D. Leonidas, R. Shapiro, L. I. Irons, N. Russo, and K. R. Acharya, *Biochemistry* **38,** 10287 (1999).
[40] R. Piccoli, A. Di Donato, and G. D' Alessio, *Biochem. J.* **253,** 329 (1988).
[41] N. Russo, R. Shapiro, K. R. Acharya, J. F. Riordan, and B. L. Vallee, *Proc. Natl. Acad. Sci. U.S.A.* **91,** 2920 (1994).

TABLE III
INHIBITION OF HUMAN RNase-2, RNase-4, AND ANGIOGENIN BY ADENOSINE 5′-PYROPHOSPHATE DERIVATIVES

	$K_i(\mu M)$		
Inhibitor	RNase-2[a]	RNase-4[a]	Angiogenin[b]
ppA-3′-p	0.25	0.54	300
ppA-2′-p	—	—	110
dUppA-2′-p	—	—	150
pdUppA-3′-p	0.18	0.26	360
pTppA-3′-p	0.46	0.21	—

[a] Values for pdUppA-3′-p and pTppA-3′-p are from Ref. 36; values for ppA-3′-p were determined by the procedure described therein.

[b] The value for ppA-2′-p listed is from R. Shapiro, *Biochemistry* **37,** 6847 (1998), and differs slightly from that originally reported in Ref. 34. Values for the other compounds were measured as described in these references.

compounds (ppA-2′-p to dUppA-2′-p; ppA-3′-p to pdUppA-3′-p) has only a minor effect on affinity.

Practical Considerations for Use or Assay of RNase Inhibitors

Binding of nucleotide inhibitors to RNase A is markedly influenced by pH and ionic strength; the magnitude of this dependence varies with the size and other structural features of the inhibitors. Ordinary nucleotides such as 2′-CMP, 3′-CMP, and 3′-UMP exhibit maximum affinity at a pH of ~5.5[42]. With 2′-CMP, the K_i is slightly (less than 2-fold) higher at pH 5 and 6, and increases much more dramatically as the pH is lowered or raised further: e.g., K_i values at pH 7 and 7.5 are 26-fold and 200-fold higher, respectively, than at pH 5.5. Binding of 3′-CMP falls off somewhat less rapidly at higher pH, by 8-fold at pH 7, 16-fold at pH 7.5, and 32-fold at pH 8.[31,42] (3′-UMP behaves similarly.[42]) Biltonen and co-workers[31,43] have proposed that the pH dependence arises because (i) the favored state of the inhibitor phosphate group (pK_a 5.7–6.0) is the dianion and (ii) the inhibitor binds preferentially to the form of RNase A in which the P_1 residues His-12 and His-119 (pK_a values in free enzyme 5.8–6.2) and perhaps His-48 are protonated.

Ionic strength is a less important consideration for binding of mononucleotide inhibitors such as 3′-CMP and 2′-CMP. The affinity of 3′-CMP for RNase A

[42] D. G. Anderson, G. C. Hammes, and F. G. Walz, *Biochemistry* **7,** 1637 (1968).

[43] M. Flogel and R. L. Biltonen, *Biochemistry* **14,** 2610 (1975).

TABLE IV
INHIBITION OF RNase A BY pdUppA-3′-p UNDER VARIOUS CONDITIONS

Condition[a]	K_i^b (n*M*)
0.2 *M* MES, pH 5.9	27
0.1 *M* MES, pH 5.9	13
0.05 *M* MES, pH 5.9	6
0.1 *M* MES, 0.1 *M* NaCl, pH 5.9	132
0.1 *M* MES, 0.2 *M* NaCl, pH 5.9	1040
0.2 *M* HEPES, pH 7.0	220
0.2 *M* HEPES, pH 8.0	22,000

[a] Measurements were at 25°. The first condition listed is that used for the studies of adenosine 5′-pyrophosphate inhibitors in refs. 35 and 36. Solutions were titrated to the desired pH with NaOH.

[b] Values were determined as described in Ref. 36; those listed for 0.2 *M* MES, pH 5.9 and 0.2 *M* HEPES, pH 7.0 were reported in this reference.

is only 14-fold lower in 1.2 *M* sodium acetate (pH 5.5) than in 0.05 *M*, and is essentially constant in the range 0.05 to 0.15 *M*.[44] The K_i value for 2′-CMP is only 70% higher at an ionic strength of 0.75 M as compared to 0.10.[45] However, as the number of nucleotides in the inhibitor molecule increases, the dependence on ionic strength becomes much larger. Raines and co-workers have measured the affinity of the oligonucleotides fluorescein-d(pUpApA) and fluorescein-d(pApUpApA) for RNase A (Table II) as a function of sodium ion concentration at pH 6.[11,46] They find that the trinucleotide binds 5.4-fold less tightly at 142 m*M* [Na^+] than at 33 m*M*. The effect of [Na^+] is even greater with the tetranucleotide: the dissociation constant at 142 m*M* [Na^+] is 28-fold larger than at 33 m*M*, and 107-fold higher than at 18 m*M* [Na^+]. They show that the strong dependence of affinity on ionic strength for oligonucleotides reflects the weakening of interactions of the inhibitor phosphate groups at the P_2, P_0, and P_{-1} subsites.

The effectiveness of pdUppA-3′-p for inhibiting RNase A has been measured at various pH values and ionic strengths (Table IV). Under the standard assay conditions used for studying the adenosine 5′-pyrophosphate derivatives (0.2 *M* MES.NaOH, pH 5.9), [Na^+] (which approximates the ionic strength) is 78 m*M*

[44] D. W. Bolen, M. Flogel, and R. Biltonen, *Biochemistry* **10,** 4136 (1971).

[45] M. Irie, *J. Biochem.* **57,** 355 (1965).

[46] B. M. Fisher, J.-H. Ha, and R. T. Raines, *Biochemistry* **37,** 12121 (1998).

and the K_i value is 27 nM. Decreasing [Na^+] to 39 mM and 19.5 mM improves affinity by factors of 2 and 4.5, respectively; raising [Na^+] to 139 mM and 239 mM weakens binding by ~5-fold and 38-fold, respectively. In this comparison, we have considered ionic strength provided by MES-NaOH and by NaCl to be equivalent. This is supported by evidence that there are no significant specific ion effects on binding of other types of inhibitors (e.g., see Ref. 46).

The K_i value for pdUppA-3′-p in 0.2 M HEPES, pH 7.0 is 8-fold higher than under the standard conditions at pH 5.9. However, [Na^+] in the pH 7.0 buffer (50 mM) is significantly lower, and we estimate that the factor attributable to pH alone is ~13-fold, compared to factors of 21 and 6 for 2′-CMP and 3′-CMP, respectively.[42] In 0.2 M HEPES at pH 8.0, the K_i value is 100 times higher than under the pH 7 conditions and 800 times greater than under the standard conditions at pH 5.9. In this case, the effect of pH alone is less extensive because [Na^+] in this buffer (154 mM) is well above that in the pH 5.9 and pH 7.0 buffers. This effect is estimated to be 100- to 150-fold for pH 8.0 vs. pH 5.9, a dependence somewhat weaker than for 2′-CMP but more marked than for 3′-CMP.

In light of the preceding discussion, it is clear that whenever nucleotide inhibitors are used to protect RNA from the action of RNase A and other pancreatic-type RNases, maximum efficacy can be achieved by maintaining the pH as close to the range 5–6 as possible, and by keeping the ionic strength as low as possible. The effects of the conditions on inhibition of other groups of RNases has not been examined, but the same considerations no doubt apply, at least qualitatively. Finally, it should be borne in mind that all of the inhibitors listed in Table II are expected to act competitively. This has been demonstrated kinetically in many instances, and crystal or NMR structures of numerous RNase–inhibitor complexes have directly shown that these compounds occupy the active site such that RNA substrates cannot bind productively. Therefore, the effectiveness of inhibitors will also depend on the level of RNA present, and in those situations where [RNA] approaches or exceeds K_m, the relative ratio of $[I]/K_i$ vs $[S]/K_m$ becomes a critical parameter to consider when judging how much inhibitor to use.

Preparation and Characterization of ppA-3′-p, ppA-2′-p, pdUppA-3′-p, and pTppA-3′-p

The four compounds are synthesized in solution by combined chemical and enzymatic procedures (see Fig. 3, **I–IX**), in part based on the methods reported by Moffat and Khorana[47,48] for the preparation of ADP and coenzyme A. All of the inhibitors are made from a common precursor, adenosine 2′,3′-cyclic phosphate 5′-phosphomorpholidate (**II**), which is prepared by reacting a mixture of

[47] J. G. Moffat and H. G. Khorana, *J. Am. Chem. Soc.* **83,** 649 (1961).

[48] J. G. Moffat and H. G. Khorana, *J. Am. Chem. Soc.* **83,** 663 (1961).

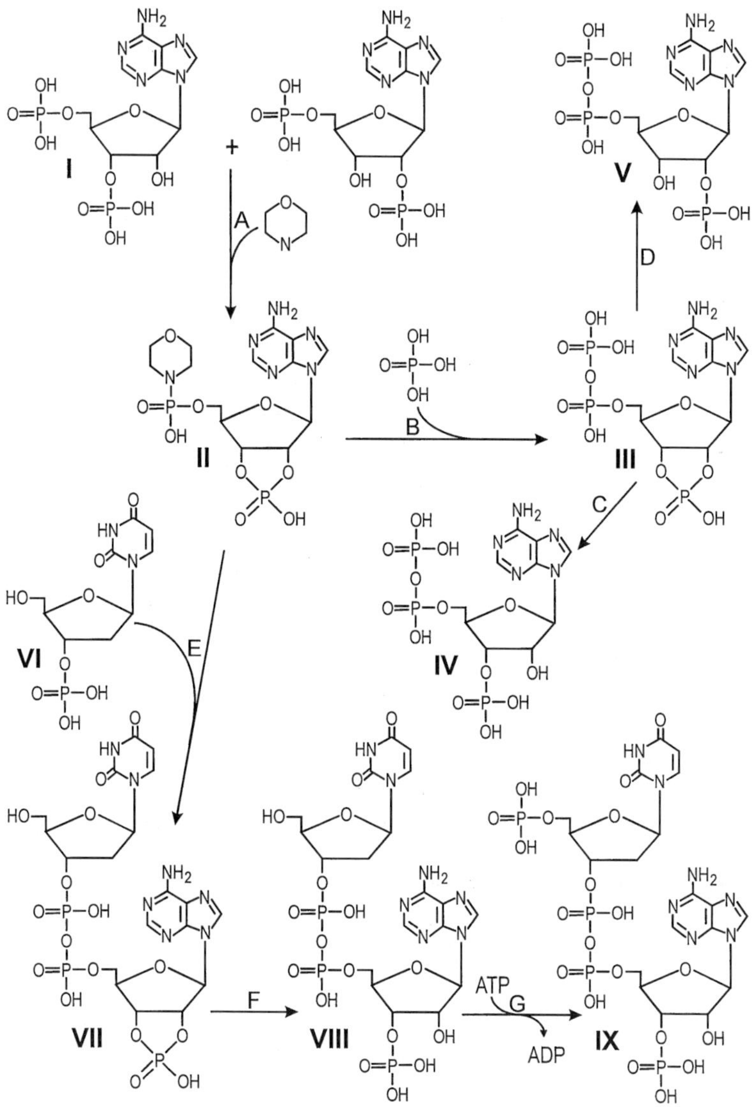

FIG. 3. Synthesis of ppA-3′-p, ppA-2′-p, and pdUppA-3′-p. The various compounds are:(**I**) a mixture of adenosine 2′,5′- and 3′,5′-diphosphate; (**II**) adenosine 2′,3′-cyclic phosphate 5′-phosphomorpholidate; (**III**) 5′-diphosphoadenosine 2′,3′-cyclic phosphate; (**IV**) 5′-diphosphoadenosine 3′-phosphate (ppA-3′-p); (**V**) 5′-diphosphoadenosine 2′-phosphate (ppA-2′-p); (**VI**) 2′-deoxyuridine 3′-phosphate; (**VII**) 2′-deoxyuridine 3′-pyrophosphate, P′→5′-ester with adenosine 2′,3′-cyclic phosphate; (**VIII**) 2′-deoxyuridine 3′-pyrophosphate, P′→5′-ester with adenosine 3′-phosphate; (**IX**) 5′-phospho-2′-deoxyuridine 3′-pyrophosphate, P′→5′-ester with adenosine 3′-phosphate (pdUppA-3′-p). The reaction conditions are as follows: (A) dicyclohexylcarbodiimide in 70% *tert*-butyl alcohol

adenosine 2′,5′- and 3′,5′-diphosphate (Fig. 3, **I**) with morpholine and carbodiimide. This reaction simultaneously protects the 2′(3′)-phosphate by converting it to a 2′,3′-cyclic phosphate and activates the 5′-phosphate for subsequent addition of phosphate, 2′-deoxyuridine 3′-phosphate, or thymidine 3′-phosphate. The products of these couplings are then subjected to enzyme treatments to obtain the final compounds, ppA-3′-p (**IV**), ppA-2′-p (**V**), pdUppA-3′-p (**IX**), and pTppA-3′-p. The overall yields typically range from 40 to 60% (20 to 30 mg for the scale of synthesis described).

Materials and General Procedures

Nucleotides, RNase T2 (type VII) from *Aspergillus oryzae,* inorganic pyrophosphatase from baker's yeast, and nucleotide pyrophosphatase from *Crotalus adamanteus* venom can be obtained from Sigma (St Louis, MO). Sigma was also the supplier of the bovine brain 2′,3′-cyclic nucleotide 3′ phosphodiesterase used in the original syntheses, but at present this enzyme is no longer available from this or any other commercial source to our knowledge; procedures for isolating the phosphodiesterase, which is used only for preparing inhibitors containing adenosine 2′-phosphate, are described by Drummond *et al.*[49] and Suda and Tsukada.[50] We use T4 polynucleotide kinase from Promega (Madison, WI), calf intestinal phosphatase from New England Biolabs (Beverly, MA) triethylamine (sequenal grade) from Pierce (Rockford, IL), dicyclohexylcarbodiimide from Fluka (Ronkonkoma, NY), and morpholine, tri-*n*-butylamine, dry pyridine, and dry ether from Aldrich (Milwaukee, WI). SP-Sepharose, QAE-Sepharose, and the Mono Q column (HR5/5) are products of Amersham Pharmacia Biotech (Piscataway, NJ).

A stock solution of 0.5 *M* triethylammonium bicarbonate (TEAB), pH 7.3, can be prepared by mixing 140 ml (101 g) of triethylamine with 1.7 liters of water, titrating the pH with CO_2 gas (from dry ice) with vigorous stirring, and finally bringing the volume to 2 liters with water. All chromatographic procedures are performed at room temperature. The sodium, morpholinium, or tri-*n*-butylammonium forms of SP-Sepharose are obtained by washing the resin first with 15 volumes of 2 *M* NaCl, 2 *M* morpholinium chloride (pH 7.0), or 0.5 *M* tri-*n*-butylammonium acetate (pH 7.0), respectively, then with 15 volumes of water. Counterion exchanges are performed by loading the samples onto the specified resins and washing with

[49] G. I. Drummond, N. T. Iyer, and J. Keith, *J. Biol. Chem.* **237,** 3535 (1962).
[50] H. Suda and Y. Tsukada, *J. Neurochem.* **34,** 941 (1980).

at boiling temperature; (B and E) dry pyridine at room temperature; (C and F) RNase T2 in 0.1 *M* MES-NaOH (pH 5.9), and 60 m*M* Tris-HCl (pH 7.5), respectively, at 37°; (D) 2′,3′-cyclic nucleotide 3′-phosphodiesterase in 30 m*M* MES (pH 5.9) at 37°; (G) T4 polynucleotide kinase in 80 m*M* Tris-HCl (pH 7.4). Other details are provided in the text.

two column volumes of water (flow by gravity); the entire eluate is collected. Synthetic reactions can be monitored by analytical anion-exchange chromatography on a Mono Q column with a 25-min linear gradient from 25 to 400 m*M* NaCl in 10 m*M* Tris-HCl (pH 8.0) at a flow rate of 1.2 ml/min, recording the absorbance at 254 nm. The purity of the final products can be assessed by the same procedure (suggested loading: 50 nmol). Nucleotides and dinucleotides can be quantitated spectrophotometrically by using the following ε_{260} values (M^{-1} cm^{-1}): 15,000 for adenosine 2′,5′- and 3′,5′-diphosphate, ppA-3′-p, and ppA-2′-p; 10,000 for 2′-deoxyuridine 3′-phosphate; 8,400 for thymidine 3′-phosphate; 25,000 for pdUppA-3′-p; 23,400 for pTppA-3′-p. These molar absorptivities are derived from those for mononucleotides listed by Beaven *et al.*[51]

Adenosine 2′,3′-Cyclic Phosphate 5′-Phosphomorpholidate (**II,** *mpA>p)*

A mixture of adenosine 2′,5′- and 3′,5′-diphosphate (pA-2′-p and pA-3′-p;**I**) (sodium salt, 0.3 mmol total) is dissolved in 4 ml of water and converted to the morpholinium salt by passage through a 1 × 6 cm column of SP-Sepharose (morpholinium form). The sample is then lyophilized and redissolved in 3 ml of water. Morpholine (0.21 ml), *tert*-butyl alcohol (7.5 ml), and dicyclohexylcarbodiimide (618 mg) are added, and the mixture is refluxed for 3 h to yield **II**. The solution is then cooled to room temperature, filtered, and subjected to rotary evaporation until the *tert*-butyl alcohol has been largely removed. The remaining solution is diluted to 15 ml with water, refiltered, extracted 5 times with an equal volume of ether, lyophilized, and stored at −20°.

ppA-3′-p (**IV**)

Two ml of pyridine are combined with 20 μl of 85% (w/w) orthophosphoric acid and 70 μl of tri-*n*-butylamine, and the solvent is evaporated. In separate vessels, the residue from this procedure and compound **II** (from 75 μmol of ADP starting material) are made anhydrous by 5 coevaporations with dry pyridine in a rotary evaporator, and each is redissolved in 2 ml of pyridine. The two samples are combined and incubated at room temperature for 24 h to obtain 5′-diphosphoadenosine 2′,3′-cyclic phosphate (ppA>p; **III**). The reaction mixture is then supplemented with 35 μl of tri-*n*-butylamine, the solvent is evaporated, and residual pyridine is removed by two coevaporations with water. Finally, the material is treated with RNase T2, which cleaves the 2′-O-P bond of ppA>p to produce ppA-3′-p; 40 units of enzyme are used in 4 ml of 0.1 *M* MES–NaOH (pH 5.9) and the incubation is for 5 h at 37°.

Purification of ppA-3′-p is achieved by anion-exchange chromatography. The RNase T2–treated sample is diluted with 16 ml of 0.1 *M* TEAB (pH 7.3) and loaded

[51] G. H. Beaven, E. R. Holiday, and E. A. Johnson, *in* "The Nucleic Acids" (E. Chargaff, and J. N. Davidson, eds.), Vol. 1, p. 493. Academic Press, New York, (1955).

onto a QAE-Sepharose column (1.5 × 10 cm) that has been equilibrated in the same buffer. The resin is washed with 40 ml of this buffer and elution is performed with a 6-h linear gradient from 0.1 to 0.4 *M* TEAB (pH 7.3) at a flow rate of 1.5 ml/min. The main peak of absorbance at 280 nm, which elutes with ~0.25 *M* TEAB, is diluted with an equal volume of water and lyophilized. Residual TEAB and other non-nucleotide impurities are removed by four coevaporations with methanol followed by precipitation of the ppA-3′-p (in 2 ml of methanol) with 10 volumes of dry ether (0°, 20 min). The precipitate is collected by centrifugation, placed under vacuum to evaporate residual ether, and reconstituted in 1 ml of water. Finally, ppA-3′-p is converted from the triethylammonium to the sodium salt by passing it through a 0.5 × 5 cm column of SP-Sepharose (sodium form) in water, and stored at −20°.

The product is expected to be >95% homogeneous, as judged by Mono Q chromatography, with pA-3′-p as the only detectable contaminant. The identity of this material as ppA-3′-p can be confirmed by mass spectrometry and by digesting aliquots with calf intestinal phosphatase [0.5 m*M* nucleotide and 5 units of enzyme in 50 μl of 10 m*M* Tris-HCl (pH 7.9) containing 50 m*M* NaCl, 10 m*M* $MgCl_2$, and 1 m*M* dithiothreitol; 37° for 1 h] and inorganic pyrophosphatase [0.5 m*M* nucleotide and 5 units of enzyme in 50 μl of 50 m*M* Tris-HCl (pH 7.2) containing 50 m*M* NaCl; 25° for 1 h]: the phosphatase yields a single detectable product that coelutes with adenosine by C_{18} HPLC,[34] and the pyrophosphatase produces a single peak of A_{254}-absorbing material that coelutes with pA-3′-p during Mono Q chromatography.

ppA-2′-p

ppA-2′-p (**V**) can be prepared and characterized as described for ppA-3′-p except that ppA>p is treated with 2′,3′-cyclic nucleotide 3′-phosphodiesterase, which cleaves the 3′-O–P bond, instead of with RNase T2. The reaction is performed in 10 ml of 30 m*M* MES–NaOH (pH 5.9) at 37° for 4 h with 2 units of phosphodiesterase.

pdUppA-3′-p

2′-Deoxyuridine 3′-phosphate (dU-3′-p; **VI**)(sodium salt, 85 μmol) is dissolved in 4 ml of water and converted to the tri-*n*-butylammonium salt by passage through a 1 × 6 cm column of SP-Sepharose (tri-*n*-butylammonium form) in water. The sample is then lyophilized, made anhydrous by 5 coevaporations with dry pyridine, and redissolved in 2 ml of pyridine. This material is then combined with mpA>p (from 75 μmol of starting material) that has been made anhydrous by the same procedure, and the mixture is incubated at room temperature for 24 h to obtain 2′-deoxyuridine 3′-pyrophosphate, P′→5′-ester with adenosine 2′,3′-cyclic phosphate (**VII**). The solvent is then evaporated and residual pyridine is removed by two coevaporations with water. Next, the sample is dissolved in 6 ml of 60 m*M* Tris-HCl (pH 7.5) and treated with 60 units of RNase T2 at 37° for 24 h to hydrolyze the

2′,3′-cyclic phosphate to obtain dUppA-3′-p (**VIII**). [The alternative isomer dUppA-2′-p, an inhibitor of Ang, can be produced from **VII** by performing the hydrolysis with 2′,3′-cyclic nucleotide 3′-phosphodiesterase instead of RNase T2.] In the final synthetic step, the sample is supplemented with 9 ml of water, 1.1 ml of 1 *M* Tris-HCl (pH 7.5), 180 μl of 0.5 *M* dithiothreitol, 175 μl of 1 *M* $MgCl_2$, 1.6 ml of 100 m*M* ATP, and 900 units of polynucleotide kinase, and incubated at 37° for 27 h to phosphorylate the 5′ terminal oxygen. This reaction proceeds nearly to completion.

The product, pdUppA-3′-p (**IX**), is purified by QAE-Sepharose chromatography as described above for ppA-3′-p, except that the sample is diluted with 2 volumes of 0.25 *M* TEAB (pH 7.3) prior to loading, 0.25 *M* TEAB is used for column equilibration, and the gradient for elution is from 0.25 to 0.5 *M* TEAB. The last peak of absorbance at 280 nm, which elutes with ~0.5 *M* TEAB, is diluted with 2 volumes of water and lyophilized. Residual TEAB is then removed and pdUppA-3′-p is converted to the sodium salt, both as described for ppA-3′-p.

The final preparation is expected to be $>95\%$ homogeneous, as judged by Mono Q chromatography. The identity of the product can be confirmed by UV spectroscopy (the spectrum is indistinguishable from that of UpA), mass spectrometry, and digestion with nucleotide pyrophosphatase [0.5 m*M* nucleotide, 0.04 units of enzyme in 50 μl of 50 m*M* Tris-HCl (pH 7.5) containing 50 m*M* NaCl and 1 m*M* $MgCl_2$; incubation for 1 h at 25°]. Analysis of the digestion products by Mono Q chromatography reveals two peaks of 254-nm absorbance; one coelutes with pA-3′-p and the other elutes ~3 min later than dU-3′-p and slightly before pA-3′-p, consistent with the expected behavior of pU-3′-p (to our knowledge, this nucleotide is not available commercially).

pTppA-3′-p

pTppA-3′-p can be prepared as for pdUppA-3′-p, substituting T-3′-p for dU-3′-p in the reaction with mpA>p. The final product is purified and characterized by the same procedures used for pdUppA-3′-p; elution times of the inhibitor and pT-3′-p during anion-exchange chromatography are similar to those of the corresponding 2′-deoxyuridine compounds.

Structures of Complexes of RNase A with Adenosine 5′-Pyrophosphate Derivatives

The high-resolution (1.7 Å) crystal structures for the complexes of RNase A with the mononucleotides ppA-3′-p and ppA-2′-p[38] reveal binding modes that differ strikingly from those anticipated on the basis of earlier structures of RNase A complexes (Fig. 4).[3,12] The key difference is that the 5′-β-phosphate rather than the 5′-α-phosphate occupies the P_1 site. As a consequence, the ribose moieties are shifted by ~2 Å from their normal positions and the adenine rings are rotated

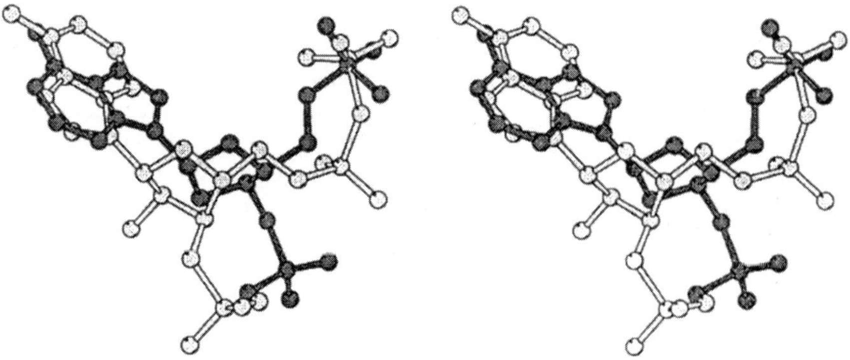

FIG. 4. Stereoview of the superimposed structures of ppA-3′-p (unfilled bonds and atoms) and the corresponding part of d(ApTpApApG)[12] (filled) when bound to RNase A.

by almost 180° into unusual *syn* conformations. Thus, the six-membered and five-membered rings of both adenines are nearly reversed compared to their counterparts in previous complexes. Despite this reversal, the base is able to form a set of interactions with the B_2 site of RNase A that is at least as extensive as those seen in complexes with ordinary nucleotides. The 6-amino group of adenine makes similar hydrogen bonds with Asn-71 in all cases, plus a second hydrogen bond with Asn-67 or Gln-69. The *syn* conformation also allows an additional hydrogen bond, between N1 of adenine and Asn-71 of RNase A. Moreover, the six-membered rings of the adenines in ppA-3′-p and ppA-2′-p engage in stacking interactions with the imidazole group of His-119 that are comparable to those between the five-membered ring of the base and His-119 in earlier complexes. Contacts between the 5′-diphospho inhibitors and RNase A residues outside the B_2 subsite include (i) four hydrogen bonds of the β-phosphate with P_1 residues Gln-11, His-12, His-119, and Phe-120 (main-chain NH) that are nearly identical to those formed by the 5′-α-phosphate in ordinary nucleotide inhibitors and (ii) one or three hydrogen bonds between the 5′-α-phosphate and Lys-7.[52] The 5′-α-phosphate of ppA-3′-p also forms hydrogen bonds with Gln-11 of RNase. Its 3′-phosphate hydrogen bonds to Lys-7; however, no interactions of the 2′-phosphate of ppA-2′-p are observed and the basis for the 2.3-fold tighter binding of this inhibitor vs ppA is unclear.

A 1.7-Å-resolution crystal structure has also been determined for the complex of RNase A with pdUppA-3′-p.[39] This inhibitor occupies the $P_0/B_1/P_1/B_2/P_2$ region of the active site in a manner similar to that predicted by modeling from the ppA-3′-p complex structure.[36] The interactions at P_1 and B_2 are indistinguishable from those in the ppA-3′-p complex, and contacts made by the deoxyuridine

[52] The number of hydrogen bonds between the α-phosphate and Lys-7 is different for the two inhibitors and for the two molecules in the noncrystallographic dimer.[38]

5′-phosphate portion of pdUppA-3′-p appear to be responsible for the 9-fold higher affinity of this compound. The uracil base forms the same two hydrogen bonds with Thr-45 of RNase A found in earlier uridine nucleotide complexes, and the terminal 5′-phosphate is positioned to form medium-range Coulombic interactions with Lys-66. However, some of the hydrogen bonds of Lys-7 and Gln-11 with the terminal 3′-phosphate and the adenylate 5′-phosphate observed in the ppA-3′-p complex are not present. The loss of these contacts, which was not predicted by modeling, may prevent the full potential benefit of the B_1/P_0 interactions from being realized.

Acknowledgments

This work was supported by the Endowment for Research in Human Biology, Inc. (Boston, MA), under a research agreement with Promega Corporation (Madison, WI) and by the National Institutes of Health (HL52096 to R.S.).

[41] Ribonuclease-Resistant RNA Controls and Standards

By DAVID BROWN and BRITTAN L. PASLOSKE

Introduction

Reverse transcription–polymerase chain reaction (RT-PCR),[1] branched DNA assays,[2] and transcription-mediated amplification (TMA)[3] are all technologies that have been developed to detect and quantify specific RNA targets. RT-PCR has been the technology of choice for most researchers because it is easily adapted to different RNA sequences. In research laboratories, RT-PCR is mainly used to study the relative changes in the levels of a particular mRNA from cells or tissue exposed to different conditions over time. However, many researchers are also measuring the absolute concentration of a particular RNA sequence using competitive quantitative RT-PCR.

In diagnostics, all three of the above-mentioned technologies are used for the detection and absolute quantification of viral RNA in the plasma of patients infected

[1] J. Mulder, N. McKinney, C. Christopherson, J. Sninsky, L. Greenfield, and S. Kwok, *J. Clin. Microbiol.* **32,** 292 (1994).

[2] M. L. Collins, I. Irvine, D. Tyner, E. Fine, C. Zayati, C. Chang, T. Horn, D. Ahle, J. Detmer, L.-P. Shen, J. Kolberg, S. Bushnell, M. S. Urdea, and D. D. Ho, *Nucleic Acids Res.* **25,** 2979 (1997).

[3] D. L. Kacian and T. J. Fultz, U.S. Patent 5,399,491 (1995).

0076-6879/01 $35.00

with human immunodeficiency virus (HIV) and hepatitis C virus (HCV). Assays are not only used to diagnose infected patients, but also to monitor the effects of different drug therapies by measuring the absolute concentrations of viral RNA in plasma. If the viral RNA concentration in the plasma of the patient is decreasing, then the drug treatment is effective. If viral RNA concentration is rising, then the virus has probably developed drug resistance and a new drug regimen must be designed and implemented.

RNA controls and standards are a critical component of quantitative assays for a specific RNA. To clarify terms, a "control" is used to demonstrate that the assay is performing as expected. A "standard" has an absolute value (in this case, copies of RNA per ml) that is used to quantify the target RNA in the sample. If the stability or concentration of the RNA control or standard is in question, then the whole assay becomes unreliable.

In vitro transcription is the method that has been most often used to produce RNA controls and standards. A DNA template encoding the RNA sequence of interest is transcribed using a phage RNA polymerase. At Ambion, the transcription reaction has been optimized to produce as much as 150 μg of RNA in a 20 μl reaction from 1 μg of DNA template. The disadvantage of using *in vitro* transcribed RNA for controls or standards is that it is susceptible to both enzymatic and chemical degradation.

Ribonucleases are ubiquitous and extreme care must be used when handling RNA. All reagents must be ribonuclease free; therefore compatible solutions must be treated with DEPC and then autoclaved to inactivate any ribonucleases. The plasticware (tips and tubes) should be purchased from a vendor that certifies them RNase free. However, all of these precautions are moot if the RNA control or standard is inadvertently contaminated with RNase from another source. In addition, RNA should be stored at a low pH (pH 7.0 or less) and in the presence of a chelator that binds divalent metals. RNA will undergo a slow chemical cleavage at higher pH values and more so in the presence of divalent cations.

Ribonuclease-Resistant RNA Controls and Standards

Two different technologies were developed to increase the stability and reliability of RNA for use as RNA controls and standards. Using the Armored RNA technology, an RNA sequence (such as a sequence from HIV) is assembled with bacteriophage coat protein to form pseudo-bacteriophage particles. The coat protein structure protects the RNA from ribonucleases.[4,5] Armored RNA particles are stable for over 11 months at 4° in human plasma. However, once the RNA

[4] B. L. Pasloske, C. R. WalkerPeach, R. D. Obermoeller, M. Winkler, and D. B. DuBois, *J. Clin. Microbiol.* **36,** 3590 (1998).

[5] C. R. WalkerPeach, M. Winkler, D. B. DuBois, and B. L. Pasloske, *Clin. Chem.* **45,** 2079 (1999).

is extracted from the coat protein, it is completely susceptible to ribonuclease degradation. It is this property that makes Armored RNA a good control for RNA extraction. Armored RNA can be added directly to a plasma sample from a patient and then the sample can be processed to isolate total RNA. In this procedure, the viral RNA and the Armored RNA will undergo the same purification process. If there is a deficiency in any of the downstream processes (such as RNA isolation, reverse transcription, cDNA amplification, and detection), then that deficiency will be reflected in the signal obtained from the Armored RNA control.

The construction, production, isolation, and application of Armored RNA in detail is beyond the scope of this section. However, more information can be obtained in Pasloske *et al.*[4] and WalkerPeach *et al.*[5]

The other method used for increasing ribonuclease resistance is to incorporate modified ribonucleotides into RNA during transcription. Ribonucleotides modified at the 2′-ribose position can confer resistance to ribonucleases, depending on the specificity of the ribonuclease. The most useful application for this type of RNA is as standards for RT-PCR or for transcription-mediated amplification (TMA) and nucleic acid sequence-based amplification (NASBA). Modified RNA is inherently more resistant to ribonucleases than nonmodified RNA, and therefore it is not an appropriate control for monitoring the efficiency of an RNA isolation procedure. However, because it is much more stable than nonmodified RNA, it is much more reliable, and therefore is the preferred material to be used to generate a set of external standards for quantifying unknown samples.

The synthesis and properties of RNA produced with modified nucleotides will be described in the following section.

In Vitro Transcription using Modified NTPs

Ribonuclease A (RNase A) cleaves RNA 3′ to pyrimidines. The reaction mechanism employed by the enzyme has been determined and provides a general understanding of how riboncleases degrade RNA.[6] RNase A binds an RNA substrate and localizes a cytidine or uridine to the enzyme active site. The action of two histidines in the active site removes a proton from the 2′-OH of the pyrimidine, causing the formation of a cyclic 2′,3′-phosphate. Phosphate cyclization releases the portion of the RNA chain that is 3′ to the pyrimidine, resulting in cleavage of the RNA strand. The cyclized phosphate is then hydrolyzed creating a 2′-OH and 3′-phosphate on the 3′-terminal ribose of the cleaved RNA. The reaction mechanism requires the 2′-OH on the nucleotide recognized by the nuclease to facilitate the attack and subsequent cleavage of the phosphate backbone. This requirement is common among ribonucleases and has been exploited to generate RNA resistant to enzymatic degradation.

[6] G. Zubay, *in* "Biochemistry" (G. Zubay, ed.), p. 307. Macmillan Publishing Company, New York, 1988.

Modified nucleotides have been incorporated into synthetic ribo-oligomers to enhance the stability of ribozymes.[7] The 2′-hydroxyl groups of cytidines and uridines were replaced with amino or fluoro groups to block formation of the 2′,3′-cyclic phosphate required for nuclease digestion. More recently, 2′ amino- and 2′ fluoro-NTPs were incorporated by T7 RNA polymerase to create 2′ modified transcripts.[8] Theoretically, an RNA must be modified at every nucleotide to be completely resistant to all ribonucleases. This is possible using chemical synthesis, but the size of ribo oligonucleotides is limited by relatively inefficient coupling reactions. *In vitro* transcription is the method of choice for synthesizing RNA longer than 50 bases. However, existing *in vitro* transcription systems provide extremely low yields of completely modified RNA. We have developed transcription conditions that yield partially modified RNA in quantities significantly greater than conventional reactions. These modified RNAs have displayed surprising stability under conditions that rapidly degrade nonmodified RNA.

A standard transcription reaction [40 m*M* Tris pH 8.0, 20 m*M* NaCl, 6 m*M* $MgCl_2$, 2 m*M* spermidine, 10 m*M* dithiothreitol (DTT), 25 ng/μl template, and 2 units/μl T7 RNA polymerase] will generate 1 to 100 ng of modified RNA from 0.5 m*M* ATP, GTP, 2′F- or 2′-NH_2-CTP, and 2′-F- or 2′-NH_2-UTP. This amount can be increased by tenfold or more by adding $MnCl_2$ to the reaction and using specially designed DNA templates. We typically titrate $MnCl_2$ over a range of 0.5–5 m*M* to identify the concentration that provides maximal transcription. The optimal concentration depends largely on the purity of the water and polymerase being used as well as the sequence of the RNA being synthesized. We have also found that templates that do not require incorporation of CTP and UTP in the first 10–12 positions of a transcript provide much greater yields. This is apparently due to an increased likelihood of abortive transcription when modified nucleotides are incorporated prior to the formation of a stable transcription elongation complex.[9]

As expected, RNA with 2′ modifications at cytidines and uridines is not appreciably degraded by the pyrimidine-specific RNase A (Fig. 1). Only when RNase A concentrations approach 10 μg/ml does any degradation occur. We suspect that at these higher concentrations, RNase A loses substrate specificity and thus degrades RNA at the unmodified purines. In contrast, standard RNA is completely degraded by 10 ng/ml of RNase A (Fig. 1). This indicates that RNA possessing pyrimidines with 2′ modifications is more than 1000-fold more stable than unmodified RNA in the presence of RNase A.

The stability of pyrimidine-modified RNA in the presence of RNase A provides important benefits owing to the widespread use of the enzyme in plasmid preparations performed in research laboratories. However, additional RNases tend

[7] W. Piecken, *Science* **253,** 314 (1991).

[8] H. Aurup, D. M. Williams, and F. Eckstein, *Biochemistry* **31,** 9636 (1992).

[9] M.-L. Ling, S. S. Risman, J. F. Klement, N. McGraw, and W. T. McAllister, *Nucleic Acids Res.* **17,** 1605 (1989).

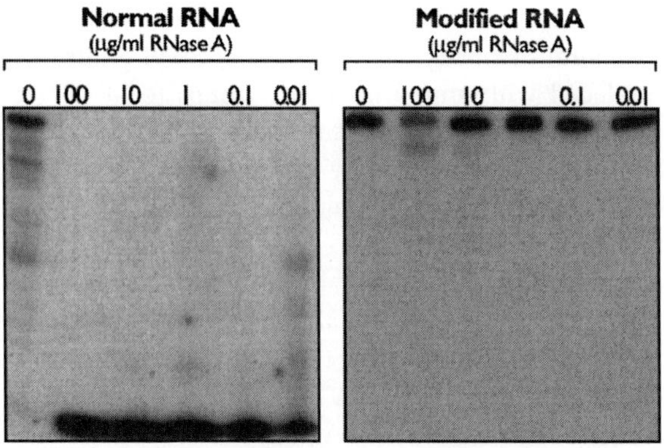

FIG. 1. Nuclease stability of RNA synthesized with the RT-PCR Competitor Construction Kit. RNase A at the concentrations indicated was incubated at room temperature for 1 hr with radiolabeled standard RNA or RNA synthesized by the RT-PCR Competitor Construction Kit in 1× transcription buffer. The samples were assessed on 8% denaturing polyacrylamide gels and products were detected by autoradiography.

to be present in experimental samples or introduced by human contact. The existence of purine-specific ribonucleases such as ribonuclease T1 and U2 suggests that pyrimidine-modified RNA would not be substantially more stable than nonmodified RNA when faced with other sources of ribonuclease. To our surprise, however, pyrimidine-modified RNA was essentially unaffected by serum, saliva, and skin samples, whereas equivalent nonmodified RNA was completely degraded (Fig. 2). These results point either to a paucity of purine-specific ribonucleases in human samples or an inability of nucleases to interact and degrade modified RNA. The fact that purified preparations of the purine-specific ribonuclease T1 degrade pyrimidine-modified RNA and nonmodified RNA with equal efficiency (data not shown) suggests that there were relatively few purine-specific nucleases in the samples tested.

The stability of RNA modified at the 2′ positions of pyrimidines provides significant advantages for research and diagnostic laboratories. Modified RNA can be used in samples that might be contaminated with nuclease without adversely affecting the assay. For instance, modified RNA probes specific to one or more targets of interest can be added to contaminated samples to detect and quantify the target nucleic acids. In contrast, a normal RNA probe would likely be degraded during the hybridization, leading to a false-negative result. More importantly, the modified RNAs can be stored without fear that they will be degraded prior to their incorporation into assays. As above, a nonmodified RNA probe that has been contaminated with nuclease might provide a lower than expected signal or false negative result when used in an assay to detect a specific nucleic acid.

The benefits of modified RNA standards in competitive RT-PCR assays are equally significant. Using nonmodified RNA standards, any nuclease that is

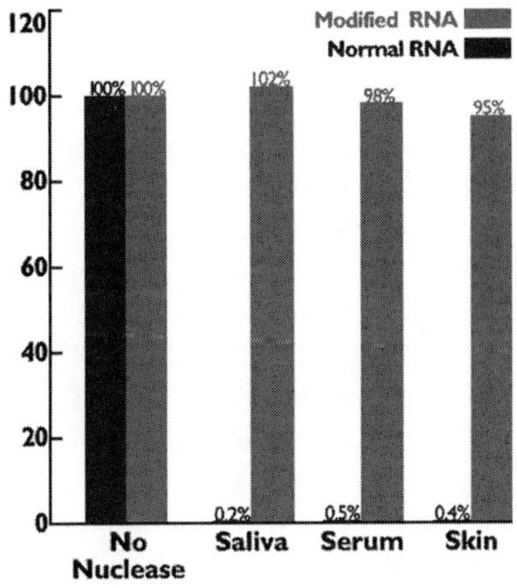

FIG. 2. Stability of RNA synthesized with the RT-PCR Competitor Construction Kit in various nuclease-contaminated samples. Saliva, serum, and skin samples were added to microfuge tubes containing 10 μl of 1× transcription buffer and either radiolabeled standard RNA or RNA synthesized with the RT-PCR Competitor Construction Kit. The samples were incubated for 1 hr at room temperature and then assessed on denaturing polyacrylamide gels. Full-length products were excised from the gel and quantified by scintillation counting.

introduced during preparation of the standard or while removing aliquots for use in RT-PCR experiments will decrease the concentration of RNA competitor being used and thus affect the accuracy of the data being generated (Fig. 3). In cases where nonmodified RNA standards are contaminated with nuclease, the best case is actually one where the RNA competitor is completely degraded. This way, the researcher will not waste time or effort on additional RT-PCR reactions. The worse case is one where a slight contamination occurs. The data being generated become progressively less accurate as the concentration of the competitor decreases. Comparing data generated on different days will result in inaccurate conclusions.

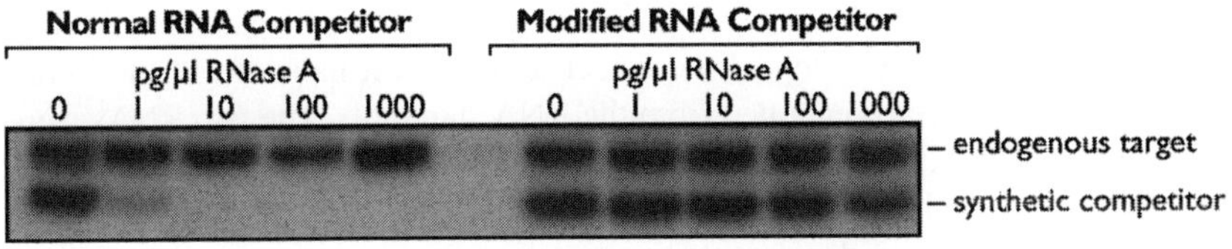

FIG. 3. Effects of nuclease contamination on competitive RT-PCR. RNase A, at the concentrations indicated, was stored at −20° in aliquots of 1 nM cyclophilin RNA competitor synthesized using either standard transcription or Ambion's RT-PCR Competitor Construction Kit. The various aliquots were diluted into mouse liver total RNA and the resulting mixtures were reverse transcribed using Ambion's RETROscript Kit and PCR amplified using radiolabeled, cyclophilin specific primers and Ambion's SuperTaq polymerase. The PCR products were assessed by gel electrophoresis and autoradiography.

Modified RNA is far less susceptible to nuclease degradation; therefore, the risks associated with using RNA standards are greatly reduced (Fig. 3).

We have used modified RNA in a variety of assays to compare its functional characteristics to those of nonmodified RNA. When radiolabeled, modified RNA can be used as a probe for Northern, Southern, and dot-blot analyses without altering the hybridization and wash conditions used for the identical, nonmodified probe. Pyrimidine-modified RNA can be used in ribonuclease protection assays provided that the nuclease digestion step is performed with a purine-specific nuclease such as RNase T1. The mobility of modified RNA when assessed by gel electrophoresis is reduced compared to nonmodified RNA. We therefore do not substitute modified RNA for standard molecular weight markers.

A particularly important application for modified RNA is in RT-PCR assays. Using modified RNA as on RT-PCR standards requires that the modified nucleotides meet two criteria. First, modified RNA must not act as a template for the thermostable DNA polymerases used for PCR. If modified RNA could serve as templates for primer extension by DNA-dependent DNA polymerases, then reverse transcription would not be required for amplification. This would keep the modified RNA from controlling for the reverse transcription efficiency of the sample. Second, modified RNA must be reverse transcribed with the same efficiency as standard RNA. This ensures that competitive RT-PCR experiments are quantitative and that other RT-PCR experiments are properly controlled. Our experimental data indicate that modified RNA must be reverse transcribed in order for it to be amplified by PCR and that it is reverse transcribed at the same rate as nonmodified RNA (data not shown).

The ultimate test of the modified RNA for RT-PCR diagnostics applications is to show that it functions in an existing commercial diagnostic product. A template encoding the sequence of the standard for the Amplicor HIV Monitor assay was transcribed with 2′ F-CTP and 2′ F-UTP as above. A series of dilutions of the 2′ F modified RNA was applied to the Amplicor HIV Monitor test. The modified RNA standard produced a signal in a concentration dependent fashion, indicating that the modified RNA was being reverse transcribed and that the resulting cDNA was being PCR amplified. The modified RNA standard was then compared to an unmodified RNA. Both standards produced approximately equal signals when introduced after the lysis step of the detection protocol, indicating that similar amounts of cDNA were generated from the RNA standards. The two RNAs were then incubated in human plasma for 15 min prior to being reverse transcribed. The cDNA was amplified by the Amplicor procedure. The signal produced by the unmodified RNA was indistinguishabe from background whereas the modified RNA produced signals were nearly equivalent to that observed when the RNA was not preincubated in plasma. This latter observation points to one of the advantages of the modified RNA standard, namely that the RNA can be added earlier in the protocol (prior to sample lysis) providing a better control for the overall experiment.

Section VIII

Nonenzymatic Cleavage of RNA

[42] Nonenzymatic Cleavage of Oligoribonucleotides

By RYSZARD KIERZEK

Introduction

For many years it has been noted that RNAs are unstable, and the selective cleavage of these molecules has been observed in many procedures.[1–4] In the absence of direct experimental evidence, this instability has been commonly attributed to ribonuclease contamination.

The discovery that some RNAs have catalytic properties changed this point of view and provided a new explanation for cleavage. The catalytic capacity of RNA has been studied with ribozymes such as group I introns, hepatitis delta virus (HDV), hammerhead, hairpin, and M1 RNA.[5] In addition to cleavage of phosphodiester bonds, some ribozymes demonstrate RNA ligase activity, nucleotide transferase, phosphotransferase, monoester transferase, restriction endonuclease, and poly(C) polymerase. Ribozymes are polynucleotides containing 40–500 nucleotides that form complex structures with specific secondary and tertiary interactions. The size of ribozymes and variety of interactions have made them a challenge to study.

The observation that short oligoribonucleotides can be selectively and quantitatively cleaved in the absence of a ribonuclease has allowed the detailed study of the chemistry of this process. The cleavage of oligoribonucleotides without protein enzymes has been studied by several groups. Hecht and co-workers observed that 31- and 47-nucleotide long RNAs can be selectively cleaved in presence of Mn^{2+}.[6] Both oligomers can fold into hairpin structures, and the cleavage occurs at a single-stranded fragment, suggesting the importance of a secondary structure of RNA for cleavage. Altman and Kozakov reduced the size of this RNA[7] and demonstrated that the hairpin structure of RNA is not required and the sequence corresponding to the stem of hairpin is sufficient for cleavage. However, the cleaved diester bond has to be in a G-A linkage and placed as a 5′-dangling end to the helical fragment.

[1] N. Watson, M. Gurevitz, J. Ford, and D. Apirion, *J. Mol. Biol.* **172,** 301 (1984).

[2] D. E. Kennell, *in* "Maximizing Gene Expression" (W. Reznikoff and L. Gold, eds.), p. 101. Butterworths, Boston, 1986.

[3] A. C. Dock-Bregeon and D. Moras, *in* "Cold Spring Harbor Symposia on Quantitative Biology," Vol. LII, p. 113. Cold Spring Harbor Laboratory Press, Cold Spring Harbor, NY, 1987.

[4] N. Stange and H. Beier, *EMBO J.* **6,** 2811 (1987).

[5] F. Eckstein and D. M. J. Lilley (eds.), "Catalytic RNA." Springer, New York, 1997.

[6] V. Dange, R. B. Van Atta, and S. M. Hecht, *Science* **248,** 585 (1990).

[7] S. Kozakov and S. Altman, *Proc. Natl. Acad. Sci. U.S.A.* **89,** 7939 (1992).

0076-6879/01 $35.00

Watson and co-workers[1] and Takaku and co-workers[8] analyzed the stability of oligoribonucleotides containing a sequence corresponding to a fragment of a pre-RNA (p2Sp1 RNA) of bacteriophage T4. They observed that the model RNA was cleaved at a C-A diester bond, forming a 5′ fragment containing 2′,3′-cyclic phosphate and a 3′ fragment bearing a 5′-hydroxyl group. No divalent metal ion is required for cleavage, and the presence of a nonionic detergent such as Brij 58, Nonidet P-40, or Triton X-100 promotes this process.

Materials and Methods

Synthesis and Purification of Oligoribonucleotides

Oligoribonucleotides are synthesized on an Applied Biosystems DNA/RNA synthesizer, using β-cyanoethyl phosphoramidite chemistry.[9,10] Removal of the oligoribonucleotide from CPG support and the base-labile protecting groups is accomplished by treatment with ethanol/ammonia (1 : 3, v/v) for 16 h at 55°. Removal of the 2′ *tert*-butyldimethylsilyl group is accomplished by incubation with freshly made 1 *M* triethylammonium fluoride in pyridine for 48 h at 55° (50 equivalents of fluoride per silyl protecting group in the oligomer). The excess fluoride is removed on a Sep-Pak C_{18} column. Purification of oligoribonucleotides up to 10 nucleotides is performed on silica gel plates (0.5 mm, Merck) in 1-propanol/ammonia/water (55 : 35 : 10, v/v/v). The position of the oligomer on the silica gel plate is identified by a UV lamp, and the proper band is scratched out. The oligomer is eluted from the silica gel with water. To remove any impurities, the aqueous solution is passed through a Sep-Pak column. After loading of material on the column, it is washed with 5 m*M* ammonium acetate. The oligomer is eluted with a solution of acetonitrile/5 m*M* ammonium acetate (3 : 7, v/v). Longer oligoribonucleotides are purified by preparative 20% polyacrylamide gel electrophoresis. The purity of the oligoribonucleotides is analyzed by C_8 high-performance liquid chromatography (HPLC) or by 20% polyacrylamide gel electrophoresis (PAGE).

Materials

T4 polynucleotide kinase and T4 RNA ligase are obtained from US Biochemicals, and [γ-^{32}P]ATP and [5′-^{32}P]pCp are from Amersham. Polyvinylpyrrolidone (PVP, molecular mass 360 kDa) is from Sigma (St. Louis, MO). The

[8] H. Hosaka, H. Haruta, K. Takai, K. Sakamoto, S. Yokoyama, and H. Takaku, *Nucleic Acids Symp. Ser.* **31,** 169 (1994).

[9] L. J. McBride and M. H. Caruthers, *Tetrahedron Lett.* **24,** 245 (1983).

[10] F. Wincott, A. DiRenzo, C. Shaffer, S. Grimm, D. Tracz, C. Workman, D. Sweedler, C. Gonzalez, S. Scaringe, and N. Usman, *Nucleic Acids Res.* **23,** 2677 (1995).

polyamines, diamines, methylated diamines, and amino alcohols used in the experiments are from Aldrich (Milwaukee, WI) or Fluka (Ronkonkoma, NY) at the highest available purity, and further purified by vacuum distillation or crystallization. The buffers and glassware are autoclaved or treated with diethyl pyrocarbonate (DEPC) prior to the experiments.

Labeling of Oligoribonucleotides and Standard Conditions for Cleavage.

Oligoribonucleotides are labeled at the 5′ termini with [γ-^{32}P]ATP and T4 polynucleotide kinase or at the 3′ termini with [5′-^{32}P]pCp and T4 RNA ligase. The labeled strands are purified by TLC on silica gel plates (0.5 mm, Merck) in 1-propanol/ammonia/water (55 : 35 : 10, v/v/v), and after exposure of the X-ray film, the proper spot on the silica gel plate is identified, removed, and the oligoribonucleotide eluted with water. Oligomers longer than 10 nucleotides are purified by 20% polyacrylamide gel electrophoresis and isolated by standard methods. The $tRNA_i^{Met}$ sequences are labeled at the 5′ or 3′ termini and purified by 14% polyacrylamide gel electrophoresis, according to standard procedures.

Under standard conditions for cleavage of oligoribonucleotides, ca. 0.1 pmol of 5′- or 3′-^{32}P-labeled oligoribonucleotide is incubated in 50 m*M* Tris-HCl (pH 7.5), 1 m*M* spermidine (or other polyamines at concentrations described below), and 1 m*M* EDTA in the presence of 0.1% PVP solution at 37° for up to 24 h.[11,12] For some experiments, 50 m*M* sodium chloride is added to the reaction mixture, or cleavage is performed at lower temperature for a suitably longer time. Addition of 3′-terminal pCp does not affect the reaction. Aliquots are quenched with formamide and analyzed by 20% or 14% (for tRNA) polyacrylamide gel electrophoresis.

Features of Oligoribonucleotide Cleavage

Products of Cleavage

The products of the nonenzymatic cleavage of a oligoribonucleotide are two shorter oligomers.[11] The 5′ fragment carries a 2′,3′-cyclic phosphate and the 3′ fragment has a 5′-hydroxyl. The products of the cleavage were identified by comparison with the products of alkaline cleavage. Moreover, incubation of the product containing 2′,3′-cyclic phosphate in HCl (pH 3) for 3 h at 50° partially converts this fragment to oligomers terminated with 2′- and 3′-phosphates.[13] The identity of the 3′-fragment was confirmed by comparison with the appropriate marker. For example, the expected 3′-product of cleavage of UCG<u>UA</u>A is AA (underlined <u>UA</u> showing the location of the labile phosphodiester bond or the place where cleavage

[11] R. Kierzek, *Nucleic Acids Res.* **20,** 5077 (1992).
[12] R. Kierzek, *Nucleic Acids Res.* **20,** 5084 (1992).
[13] R. Markham and J. D. Smith, *Biochem. J.* **52,** 552 (1952).

of the oligoribonucleotide is expected). The identity of the 3′ product was confirmed by comparing its mobility with a commercially prepared dimer, AA, on a C-8 column (HPLC) and on a silica gel TLC plate (propanol-2/ammonia/water = 7 : 1 : 2 v/v/v).

Single-Stranded Character of Oligoribonucleotide Required for Cleavage

Cleavage only proceeds in single-stranded oligoribonucleotides. For example, GUCGUAGCC is cleaved to yield GUCGU > p and AGCC, but the presence of the complementary oligomer, GGCUACGAC, stops cleavage. UV-melting curves of GUCGUAGCC in the absence and presence of the complementary oligomer showed single- and double-stranded melting behavior, respectively. The UV-melting experiments were performed in standard conditions (1 *M* NaCl, 20 m*M* sodium cacodylate, 1 m*M* EDTA, pH 7) and in cleavage buffer (50 m*M* Tris-HCl, 1 m*M* EDTA, 1 m*M* spermidine) in presence and absence of 0.1% PVP.[11]

A similar behavior was observed with the self-complementary oligoribonucleotide UCGUACGA, which was stable at 37° at a 10^{-6} *M* concentration, but underwent cleavage at 50°. The calculated melting temperature (T_m) at this concentration is 43.8°.[11]

The experiments described above raise the question of how long a single-stranded RNA fragment has to be in order to undergo specific cleavage. To answer this question, the stability of GCUCGUAA was tested when paired with each of four different complementary oligoribonucleotides.[14] The first complementary oligomer, UUACGAGC, is able to form a perfect duplex with the studied octamer. The three following oligomers, UACGAGC, ACGAGC, and CGAGC, were respectively one, two, and three nucleotides shorter at the 5′ end. As a result, different fragments of the 3′ end of GCUCGUAA were exposed as single strands.

To confirm the hybridization between the tested oligomers during the cleavage reaction, the thermodynamic parameters were measured for the duplexes formed between GCUCGUAA and its four counterparts.[15] In addition, to check whether the reaction conditions were properly chosen, the stability of model duplexes in the presence of S1 nuclease was tested.

The cleavage reaction was performed in 50 m*M* Tris-HCl (pH 7.5), 50 m*M* NaCl, 1 m*M* spermidine, 1 m*M* EDTA, and 0.1% PVP at 15°. Reactions were performed with 0.5 m*M* cleaving oligomer, GCUCGUAA, and 1 m*M* complementary oligoribonucleotides. The reaction mixtures were analyzed by PAGE after 0, 24, 48, and 90 h of incubation. This demonstrated that the specific cleavage of GCUCGUAA took place only in the presence of the shortest complementary oligomer (CGAGC). Thus, for effective cleavage, both nucleotides participating in

[14] A. Bibillo, M. Figlerowicz, K. Ziomek, and R. Kierzek, *Nucleosides Nucleotides* **25,** 977 (2000).

[15] R. Kierzek, M. H. Caruthers, C. E. Longfellow, D. Swinton, D. H. Turner, and S. M. Freier, *Biochemistry* **25,** 7840 (1986).

the formation of the labile phosphodiester bond must be unpaired. Apparently, the lack of base pairing with complementary oligomer increases the flexibility of the labile phosphodiester bond. This allows an in-line orientation of the 2′-hydroxyl of uridine, the phosphorus atom, and the 5′-oxygen atom of adenosine, as required for cleavage of oligoribonucleotides.[16]

Important Factors for Cleavage

To determine which components of the standard reaction mixture are required for specific cleavage, the stability of a heptamer, UCGUAACp, was examined in various conditions. The results show that without the biogenic polyamine, there is no cleavage of UCGUAACp in the reaction buffer (50 m*M* Tris-HCl, 1 m*M* EDTA, pH 7.5), with or without 0.1% PVP. Specific cleavage is observed in the buffer containing 1 m*M* spermidine, but without PVP, its rate is about 7 times lower than when PVP is present.[12]

The observation that spermidine is capable of inducing selective cleavage raised the question whether two other biogenic polyamines, putrescine and spermine, would display the same activity. The cleavage of UCGUAACp in the presence of spermine or putrescine show that these polyamines also are able to induce the specific cleavage.[17,18] The rate of cleavage of UCGUAACp vs the polyamine concentration from 10 μM to 20 m*M* was measured. The three plots of rate vs polyamine concentration are all bell-shaped curves. However, a different concentration of each polyamine was required to reach the maximum rate of specific cleavage: 0.1 m*M* for spermine ($k_{cat} = 21.8 \times 10^{-5}$ min^{-1}), 1.0 m*M* for spermidine ($k_{cat} = 15.6 \times 10^{-5}$ min^{-1}), and 10 m*M* for putrescine ($k_{cat} = 7.8 \times 10^{-5}$ min^{-1}). This suggests that an active oligomer conformation is most abundant at a specific concentration of each polyamine.

It has also been observed that several organic polymers and proteins can accelerate specific oligoribonucleotide cleavage.[2] Complete cleavage of a U-A phosphodiester bond within AGAUGUAUUCU (Fig. 1) was observed after 4 h incubation with reverse transcriptase (700 U/ml), T7 RNA polymerase (5.000 U/ml), lysozyme (0.4 mg/ml), or trypsin (0.4 mg/ml), and after 8 h with EcoRI endonuclease (650 U/ml), pepsin (0.4 mg/ml), T4 polynucleotide kinase (330 U/ml), polyvinylpyrrolidone (0.1%, PVP, selected as a model cofactor for most experiments on the specific, nonenzymatic cleavage of oligoribonucleotides), dextrin (1%), and polyethylene glycol 1000 (1%). Less than 50% cleavage at U-A was observed after 24 h with ribonuclease inhibitors, human placental RNase inhibitor (RNasin) (170 U/ml), and Inhibit-ACE (38 U/ml), and with polyethylene glycol

[16] R. Kierzek, *Collect. Czech. Chem. Commun.* **61,** 253 (1996).
[17] A. Bibillo, K. Ziomek, M. Figlerowicz, and R. Kierzek, *Acta Biochimi. Polon.* **46,** 153 (1999).
[18] A. Bibillo, M. Figlerowicz, and R. Kierzek, *Nucleic Acids Res.* **27,** 3937 (1999).

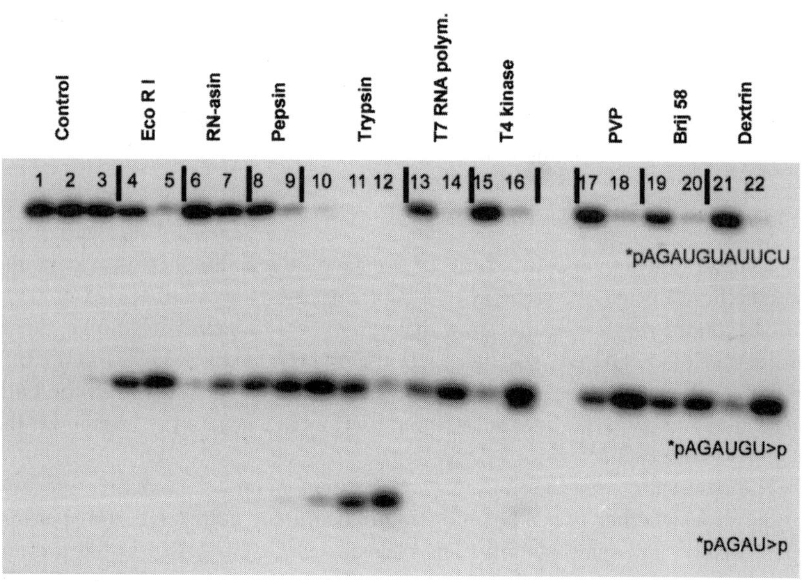

FIG. 1. Stability of *pAGAUGUAUUCU in the presence of different cofactors. Lanes 1–3: control in 50 mM Tris-HCl (pH 7.5), 1 mM spermidine, 1 mM EDTA after 0, 1, and 4 h, respectively. Lanes 4 and 5: endonuclease EcoRI (650 U/ml) after 1 and 4 h, respectively. Lanes 6 and 7: ribonuclease inhibitor RNasin (170 U/ml) after 1 and 4 h, respectively. Lanes 8 and 9: pepsin (0.4 mg/ml) after 1 and 4 h, respectively. Lanes 10, 11, and 12: trypsin (0.4 mg/ml) after 5 min, 1 h, and 4 h, respectively. Lanes 13 and 14: T7 RNA polymerase (5 000 U/ml) after 1 and 4 h, respectively. Lanes 15 and 16: T4 polynucleotide kinase (330 U/ml) after 1 and 4 h, respectively. Lanes 17 and 18: polyvinylpyrrolidone (PVP, 0.1%) after 1 and 4 h, respectively. Lanes 19 and 20: Brij 58 (0.5%) after 1 and 4 h, respectively. Lanes 21 and 22: dextrin (1%) after 1 and 4 h, respectively.

200 (1%) and 400 (1%). No cleavage of AGAUGUAUUCU was observed with polyglutamic acid, hexaglycine, L-arginine, RNA dinucleotide monophosphates, poly(A), poly(U), α-cyclodextrin, imidazole, sucrose, and proteinase K.

Stability of Chimeric DNA/RNA Oligomers

The identity of the cleavage products suggests that the 2′-hydroxyl group adjacent to the scissile diester bond is directly involved in cleavage and is necessary to produce it. However, as other 2′-hydroxyl groups of oligoribonucleotides may be indirectly involved in cleavage, the effect of substitution of individual ribonucleotides by 2′-deoxy analogs was studied. The stability of nine chimeric DNA/RNA oligonucleotides, sequential analogs of the oligoribonucleotide UCG(3)U(4)AA, were studied.[16] The chimeric oligomers can be divided into three groups. The first group contains oligomers tcguaa, UCGuAA, and UCGtAA (the capital letters denote ribonucleotides and lowercase letters denote deoxyribonucleotides), all of which have a deoxyribonucleotide at the fourth position. The

2′-hydroxyl of the nucleotide at that position is directly involved in the cleavage of oligoribonucleotides, as evidenced by the 2′,3′-cyclic phosphate product. The second group includes tcGUAa, tcGUaa, tcGUAA, and UCGUAA, with the common feature being the presence of guanosine immediately 5′ to uridine-4. The third group of oligomers, tcgUAa and tcgUaa, contains a deoxyguanosine 5′ to uridine-4.

The experiments demonstrate that the oligomers tcguaa, UCGuAA, and UCGtAA are stable to specific cleavage. This is consistent with the 5′ cleavage product having a 2′,3′-cyclic phosphate, and with an in-line mechanism of cleavage. The group of oligomers containing guanosine at position 3 are the best substrates and they cleave at similar rates. The third group of oligomers, containing deoxyguanosine at position 3, are substrates as well, but cleavage is 2- to 4-fold less efficient than for oligomers with guanosine at this position.

The results with DNA/RNA oligomers demonstrate that the presence of the ribonucleotide on the 5′-side of cleaved phosphodiester bond is sufficient to allow cleavage. However, the comparison of cleavage rate of the second and third groups of oligomers suggests that the type of the nucleotide at position 3 (guanosine vs deoxyguanosine) affects this process. The difference in stability of oligomers of group 2 and 3 is presumably due to the difference in stacking behavior of deoxy- and ribonucleotides.[19]

Nonenzymatic Nature of Cleavage

The oligoribonucleotide sequence preferences observed for the cleavage discussed here are similar to that reported for many ribonucleases, including ribonuclease A.[20–22] Many experiments were performed to prove that such contaminating ribonucleases are not responsible for the specific, nonenzymatic cleavage discussed here. Special attention was given to the purity of components (buffers, polyamines, and cofactors) used in cleavage reactions.

It was observed that oligoribonucleotide cleavage occurs even after preincubation of PVP and dextrin solution with proteinase K (50 mg/ml) for 12 h at 37°. However, no cleavage was observed in another experiment when lysozyme and pepsin (as alternatives to PVP and dextrin cofactors) were preincubated with proteinase K under similar conditions. Complete hydrolysis of both proteins by proteinase K under these conditions was observed by analyzing the mixtures on 15% polyacrylamide gel and Coomassie blue staining.

Especially strong evidence that cleavage is not due to contaminating RNase comes from experiments with ribonuclease inhibitors. Two of them, RNasin and

[19] S.-H. Chou, P. Flynn, and B. Reid, *Biochemistry* **28,** 2422 (1989).

[20] F. M. Richards and H. W. Wickoff, *in* "The Enzymes" (P. D. Boyer, ed.), p. 647. Academic Press, New York, 1971.

[21] H. Witzel, *in* "Progress in Nucleic Acids Research" (J. N. Davidson and W. E. Cohen, eds.), p. 647. Academic Press, New York, 1963.

[22] B. D. McLennan and B. G. Lane, *Can, J. Biochem.* **46,** 93 (1968).

Inhibit-ACE, stimulate cleavage of UCGUAA.[23] As with lysozyme and pepsin, preincubation of both inhibitors with proteinase K (50 μg/ml) for 12 h at 37° eliminates cleavage.

The rate of cleavage of UCGUAA and AGAUGUAUUCU over a wide range of PVP and polyamine concentrations shows a bell shape, not a linear shape, as a function of the concentrations of these compounds. This is inconsistent with the proposal that PVP and polyamines are a source of ribonuclease contamination.[22]

Moreover, it was demonstrated that silica gel column chromatography of PVP and purification by chloroform extraction does not change the ability of PVP to induce cleavage of the U-A phosphodiester bond. Additionally, experiments that examined cleavage of oligoribonucleotides in the presence of polyethylene glycol demonstrates that distillation of polyethylene glycol 200, or heating a 50% aqueous solution of this glycol for 14 h at 150° does not eliminate cleavage. Treatment of a 10% solution of polyethylene glycol with diethyl pyrocarbonate (DEPC) for 16 h at 37° followed by autoclaving also does not change the ability of polyethylene glycol to stimulate cleavage of oligoribonucleotides.

Effects of Polyamines on Cleavage of Oligoribonucleotides

As described earlier, the presence of polyamines is crucial for efficient nonenzymatic cleavage of oligoribonucleotides. For putrescine [$H_3N^+(CH_2)_4NH^+{}_3$], spermidine [$H_3N^+(CH_2)_3NH^+{}_2(CH_2)_4NH^+{}_3$], and spermine [$H_3N^+(CH_2)_3NH^+{}_2(CH_2)_4NH^+{}_2(CH_2)_4NH^+{}_3$], a bell-shaped curve representing correlation between rate of cleavage and polyamine concentration was observed and the maximum rate of cleavage was at 0.1, 1.0, and 10 m*M*, respectively.[18]

The results imply that rate of specific cleavage is influenced by the number of the protonated amino groups present in the polyamine cations. However, even at twice the concentration of putrescine, which has only terminal amines, the oligoribonucleotide cleavage is less efficient than in the presence of spermine, with two terminal and two internal ammonium groups. This observation suggests that, in addition to the number of positive charges, other elements of the polyamine structure affect the cleavage. The rate of cleavage in the presence of diamines such as 1,2-diaminoethane $H_3N^+(CH_2)_2NH^+{}_3$, Put2] and 1,3-diaminopropane [$H_3N^+(CH_2)_3NH^+{}_3$, Put3] and amino alcohols such as 2-aminoethanol [$H_3N^+(CH_2)_2OH$, Put2OH] and 3-aminopropanol [$H_3N^+(CH_2)_3OH$, Put3OH] were compared. In contrast to the diamines, amino alcohols could not induce oligoribonucleotide cleavage. This suggests that at least two linked, protonated amino groups are necessary to mediate cleavage.[18]

The next factor analyzed was the influence of the length of the methylene chain that links the protonated amino groups on the rate of cleavage. In this experiment,

[23] G. Scheel and P. Blackburn, *Proc. Natl. Acad. Sci. U.S.A.* **76,** 4898 (1979).

putrescine [1,4-diaminobutane; $[H_3N^+(CH_2)_4NH^+{}_3$, Put4] and its four homologs 1,2-diaminoethane (Put2), 1,3-diaminopropane (Put3), 1,5-diaminopentane $[H_3N^+(CH_2)_5NH^+{}_3$, Put5], and 1,6-diaminohexane $[H_3N^+(CH_2)_6NH^+{}_3$, Put6] were used. The cleavage rate was maximal with Put3 and slower for Put2 and Put4. For diamines with longer carbon chains, Put5 and Put6, specific oligoribonucleotide cleavage was remarkably reduced.[18]

Earlier observations indicate that the ability of the polyamine to induce RNA cleavage depends strongly on polyamine–oligoribonucleotide interactions.[24] These interactions can be achieved by hydrogen bond formation. To test this hypothesis, 1,3-diaminopropane (Put3) derivatives with reduced abilities to form hydrogen bonds were tested. In this series of experiments Put3 and the following methylated Put3 derivatives were tested: *N*-methyl-1,3-diaminopropane $[CH_3NH^+{}_2(CH_2)_3NH^+{}_3]$, *N,N*-dimethyl-1,3-diaminopropane $[(CH_3)_2NH^+(CH_2)_3NH^+{}_3]$, *N,N,N′,N′*-tetramethyl-1,3-diaminopropane $[(CH_3)_2NH^+(CH_2)_3NH^+(CH_3)_2]$, 2,2-dimethyl-1,3-diaminopropane $[NH^+{}_3CH_2C(CH_3)_2CH_2NH^+{}_3]$, and *N,N,N,N′,N′,N′*-hexamethyl-1,3-diammonium propane diiodide $[(CH_3)_3N^+(CH_2)_3N^+(CH_3)_3 \cdot 2I^-]$. The results demonstrated that progressive methylation of the diamine decreased its ability to induce oligoribonucleotide cleavage.[18] The methylation of just one ammonium group slightly reduces the effect of Put3 on cleavage. However, a large decrease in the rate of cleavage was observed when both amino groups were methylated. This could indicate that the amino groups perform different functions during the cleavage. For example, one may bind the oligomer and the other may activate the scissile phosphodiester bond. Permethylation of Put3 almost completely abolished oligoribonucleotide cleavage, although the hexamethyl derivative still possesses two positively charged groups linked with a trimethylene chain. Both Me_3N^+ groups can participate in electrostatic interactions with the phosphate anions, but they cannot form hydrogen bonds.

Effect of Oligoribonucleotide Structure on Cleavage

Sequence and Position of Scissile Diester Bonds within Oligomers

The stability of the phosphodiester bond is dependent on the flanking nucleotides. Analysis of cleavage in more than 20 different hexamers demonstrated that different phosphodiester bonds are cleaved at different rates.[11] The initially reported instability of the UA phosphodiester bond can be generalized to YA and YC (Y is a pyrimidine) phosphodiester bonds, which are cleaved with appreciable rates over 24 h at 37°. The phosphodiester bonds in YA cleave 3- to 5-fold faster than YC. The UA linkage cleaves 1.5-2 fold faster than CA. The phosphodiester bonds in YG and YU are 20- to 50-fold more stable than YA and YC analogs.

[24] C. W. Tabor and H. Tabor, *Ann. Rev. Biochem.* **53,** 749 (1984).

Phosphodiester bonds in RR and RY (R, purine) are stable under the standard cleavage conditions. Generally, the phosphodiester bond instabilities are in the following order, with least stable linkages first: UA > CA > YC > YG > YU. Moreover, the ability to cleavage of a given phosphodiester bond depends on its position in the oligoribonucleotide. For example, UA in UCGUAA is cleaved 5- to 15-fold faster than in UCGUUA. Also, CA in UCAUAA is cleaved 5- to 15-fold faster than in UCGUCA. In general, the presence of a second nucleoside 3′ to the scissile phosphodiester bond enhances cleavage.

Oligoribonucleotide Length

As described above, the reactivity of the YR phosphodiester bond depends on its position within the oligoribonucleotide. The presence of extra nucleotides 5′or 3′ to the scissile YR or YY phosphodiester bond increases its rate of cleavage. Quantitative stability analysis of two sets (UA, GUA, CGUA, UCGUA and UAA, GUAA, CGUAA, UCGUAA) confirmed the preliminary hypothesis. In both series the shortest oligomers (UA and UAA) were resistant to cleavage and the initial rate of cleavage was faster with longer oligoribonucleotides. The relative cleavage rates of GUA, CGUA, and UCGUA are 1 : 6 : 16, whereas the relative rates for GUAA, CGUAA, and UCGUAA are 1 : 4 : 6, respectively. Comparing these two series shows that an additional A at the 3′ end enhances cleavage. For GUA vs GUAA, CGUA vs CGUAA, and UCGUA, vs UCGUAA, the addition of the 3′-A enhances cleavage rates by factors of 8–10, 6–7, and 3.5, respectively.[11]

The influence of adjacent nucleotides 5′ and 3′ to a UA on the cleavage rate was also examined. The stability of UA diester bonds within two groups of oligoribonucleotides, NUAA and GUAN (where N is any nucleotide) were analyzed. In the NUAA series, the rate of cleavage was similar for any of the four nucleotides at the 5′ position. In the GUAN series, cleavage of the UA phosphodiester bonds depends on N, in the order from fastest to slowest: G > A > U > C. In general, cleavage of the UA diester is 2–3 fold faster when N is a purine than when N is a pyrimidine.[11]

Influence of Nucleobase Functional Groups at the Scissile Diester Bond

The cleavage of different YR diester bonds within oligoribonucleotides demonstrates that the nature of nucleotides flanking the cleaved bond affects the rate of this process. For example, the UA within an oligoribonucleotide is cleaved 1.5- to 2-fold faster than the CA analog. Similarly, the phosphodiester bond in UA is 20- to 50-fold more labile than in UG.[11] These observations suggest that structure of the nucleotides surrounding the cleaved diester bond affects this process and perhaps some functional groups of nucleobases are involved in interactions important for cleavage.

A study of the stability of two series of tetramers, GYAA and GURA, sheds light on the importance of adjacent nucleobases.[11] For the first group, GYAA, the following pyrimidine analogs were used: uridine, cytidine, 5-bromouridine, 5-methyluridine (U^{5Me}), 3-*N*-methyluridine (U^{3NMe}), and 2-thiouridine. Later experiments on stability of UCGYAA were extended to 6-methyluridine (U^{6Me}) and isocytidine (iC).[14] The results show that GUAA and GU^{5Me}AA hydrolyze the most rapidly. The three tetramers containing cytidine, 5-bromouridine and 2-thiouridine hydrolyze with similar rates, ca 0.5–0.7 of the GUAA cleavage rate. The tetramer GU^{3NMe}AA was resistant to cleavage in standard condition. The experiments on UCGYAA demonstrate stability of oligomers when the pyrimidine is 6-methyluridine and isocytidine.

The influence of pyrimidine substituents on the stability of YA phosphodiester bonds within the oligoribonucleotide apparently stems from several factors. The most interesting result is seen by comparing the effects of 5- and 6-methyluridine in the oligomer. Whereas GU^{5Me}AA is cleaved, UCGU^{6Me}AA is resistant to cleavage. The most reasonable explanation of this phenomenon is a change in the glycosidic bond orientation of the 6- methyluridine. With uridine and 5-methyluridine the base adopts the *anti* conformation, placing oxygen-2 of uracil near to the 2′-hydroxyl group. The presence of the methyl substituent at C-6 of uracil changes the glycosidic bond orientation to *syn,* and as a result the oxygen-2 is far away from the 2′-hydroxyl.[25] Thus, oxygen-2 and the 2′-hydroxyl group may interact via a hydrogen bond to increase the nucleophilicity of the hydroxyl. The importance of the interaction of the oxygen-2 is also strongly supported by the experiment where cytidine-4 in UCGCAA was replaced by isocytidine.[14] No cleavage was observed in UCGUiCAA, although the amino group at C-2 of isocytidine can serve as a proton donor in a hydrogen bond. Thus, the substituent at C-2 of the pyrimidine ring is crucial for oligoribonucleotide cleavage and must be a proton acceptor in a hydrogen bond. The similar behavior of oligoribonucleotides (GYAA) containing uridine, cytidine, 5-bromouridine, 5-methyluridine, 2-thiouridine could mean that substituents of the pyrimidine ring do not significantly affect the postulated hydrogen bond between the C-2 substituent and 2-hydroxyl group. The stability of GU^{3NMe}AA to cleavage may be due to stacking properties of the 3-*N*-methyl substituent. Model studies demonstrate that, as a 3′-dangling end, 3-*N*-methyluridine stabilizes duplexes more than uridine by 0.7 kcal/mol in free energy. The stronger stacking interaction in U^{3NMe}A, relative to UA, decreases the flexibility of the phosphodiester bond and thus inhibits cleavage.

To test the correlation between the ability to cleavage a phosphodiester bond and stacking interactions of the flanking nucleobases the following series of oligomers were employed: UCGU^{5R}AACp, where R = F, Cl, Br, I, H, Me, Et, *n*-Pr.[14] Analysis of the alkylated derivatives of UCGU^{5R}AACp demonstrate that more electrodonating 5-substituents (which tend to increase the stacking interaction of U^{5R} and A) reduce the rate of phosphodiester cleavage. The rate of U^{5R}-A

[25] D. Suck and W. Saenger, *J. Am. Chem. Soc.* **94,** 6520 (1972).

internucleotide bond cleavage is, from fastest to slowest, in the following order: methyl > ethyl > *n*-propyl. As with alkylation, halogenation at position 5 of uridine also affects stacking, but the electronegativity and the size of the halogen also affect cleavage. In UCGU^{5Hal}AACp, the rate of the cleavage of the U^{5Hal}-A phosphodiester bond, from fastest to slowest, is in the order iodo > bromo > chloro uridine derivatives. Halogens with lower electronegativity reduce the stacking interaction between U^{5Hal} and A(5), and cleavage is enhanced. The size of larger halogens also disturb the parallel orientation of U^{5Hal} and A5, reducing stacking interactions and thus increasing the rate of cleavage.[26]

The influence of the purine nucleotide on cleavage of the YR phosphodiester bond was also studied.[11] An analysis of several oligoribonucleotides indicated that NYAN and NYCN are susceptible to cleavage but NYGN and NYUN sequences are resistant. This suggests that cleavage requires an amino group in position 6 of a purine, or analogously, position 4 of a pyrimidine of the base 3′ to the cleavage site. To determine the importance of these functional groups, several purine analogs were tested in the sequence GURA. These analogs were adenosine, 6-methyladenosine (A^{6Me}), guanosine, 2-aminopurine riboside, purine riboside, and inosine. This was studied later with the sequence UCGURA, where the purine analogs were 6-chloropurine riboside, 2-amino-6-chloropurine riboside, 2,6-diaminopurine riboside (^{DA}P), isoguanosine (iG), and 8-bromoadenosine (A^{8Br}).[14] The results suggest that efficient cleavage requires the purine N1 to be unprotonated and available as a hydrogen bond acceptor. Furthermore, a hydrogen-bond donor or acceptor is required at C-6. Finally, efficient cleavage requires the absence of a functional group at position 2; although adenosine, isoguanosine, and 2,6-diaminopurine riboside have the same amino substituent at C6, only the oligomer with adenosine is cleaved. Most likely, the active oligomer conformation is sterically hindered by the carbonyl and amino groups at position 2 of iG and ^{DA}P, respectively.

The importance of the functional group at C-6 having a hydrogen bonding ability is confirmed by cleavage of oligomers UCGURA, where R is 6-chloropurine riboside or 2-amino-6-chloropurine riboside. The sequence GUA^{6Me}A is not cleaved under standard conditions. Presumably, the methyl substituent decreases the ability of the 6-amino group to hydrogen bond with the other part of the scissile internucleotide bond, perhaps the phosphorus center. This is consistent with the observation that the U-t^6A is stable in the anticodon loop of $tRNA_i^{Met}$ from *Lupinus luteus* whereas U-A is unstable in the anticodon loop of $tRNA_i^{Met}$ from *Escherichia coli*.[17] Moreover, experiments have demonstrated that the 6-*N*-alkyl substituents of adenosine destabilize RNA duplexes. This additionally explains the effect of 6-*N*-methyl on the stability of U-A^{6Me}.

[26] K. Ziomek, Ph.D. Thesis, Institute of Bioorganic Chemistry, Poznan, Poland, 1998.

A bulky bromine atom at position 8 of adenosine in UCGUA$^{8\text{Br}}$A does not affect cleavage rate, although a bromine at this position changes the glycosidic bond from *anti* to *syn*.[27] This suggests that orientation of the purine base (R) is not important for the cleavage. This contrasts with the importance of an *anti* orientation for the 5′-pyrimidine for cleavage in a YR bond.

Mechanism of Cleavage, and External Factors Affecting Cleavage

Cleavage via In-Line Transesterification

Determination of the cleavage mechanism of phosphodiester bonds is important for understanding the factors that control this process. Oligoribonucleotides containing a phosphorothioate diester bond at the scissile position were used to study the reaction mechanism. The oligoribonucleotide UCGU*p*AA (where the italicized *p* stands for a phosphorothioate diester bond) was chemically synthesized and, after deprotection, the R_p and S_p isomers were separated by polystyrene reversed-phase (PRP-1) HPLC.[18] The configurations of the phosphorothioate centers were determined by enzymatic cleavage (nuclease P1 and snake venom phosphodiesterase) and by ^{31}P NMR spectroscopy.

The R_p and S_p isomers of UCGU*p*AA were cleaved separately under standard conditions. The products of the cleavage were the dimer AA and the tetramer UCGU > *p* containing a 2′,3′-cyclic phosphorothioate. Next, the tetramer UCGU > *p* was digested by T1 nuclease to UCG > *p* and U > *p*. The configuration of U > *p* was identified by HPLC using a PRP-1 column. Cleavage of the R_p isomer of UCGU*p*AA generated the *endo* isomer of uridine-2′,3′-cyclic phosphorothioate, whereas the S_p isomer formed an *exo* isomer of U > *p*.[16]

The results of cleavage of UCGU*p*AA prove an S_N2 (in-line) mechanism, with inversion of configuration at phosphorus. Model studies of UCGU(4)*p*A(5)A indicate that to achieve an in-line orientation of the 2′-hydroxyl of uridine-4, the phosphorus atom, and the 5′-oxygen atom of adenosine-5, it is necessary to have an unstacked conformation of uridine-4 and adenosine-5. This arrangement is incompatible with the A-form conformation present in stacked or Watson–Crick-paired RNA, in which it is necessary to change phosphodiester torsion angles to achieve the active configuration. Such a conformation would break the stacking interaction of the nucleobases flanking the scissile phosphodiester bond.

The present data suggest the importance of a hydrogen bond between the C-2 substituent (oxygen or sulfur) of the pyrimidine and its 2′-hydroxyl for efficient cleavage of a YR linkage. Such a hydrogen bond can increase the nucleophilic character of the 2′-hydroxyl and accelerate cleavage. This is analogous to the role of divalent metal ions in many ribozymes.[5] The data also suggest that the purine

[27] S. S. Tavale and H. M. Sobell, *J. Mol. Biol.* **48,** 109 (1970).

nucleobase is involved in hydrogen bonding: N1 must be a proton acceptor and the substituent at C-6 either a hydrogen bond donor or acceptor. However, these are not sufficient for cleavage if a substituent at position C-2 of the purine is present. According to the model, hydrogen bonding between the purine base with the 5′ phosphate (either directly or bridged by water) is likely. This interaction would increase the electrophilic character of the scissile internucleotide bond.

Previous experiments have shown that a polyamine is necessary for cleavage.[12,18] Direct interactions of polyamines with RNA and DNA have been demonstrated.[24] In particular, the secondary ammonium groups of polyamines interact electrostatically with phosphate groups. The results suggest that secondary ammonium groups of polyamines interact with a phosphate anion while the primary ammonium groups form hydrogen bonds with the 5′-oxygen to make it a better leaving group.

Another factor important for RNA cleavage is the presence of cofactors such as organic polymers or nonribonuclease proteins.[12] The wide variety of molecules that can act as cofactors suggest nonspecific interactions with RNA, or with the RNA hydration shell. This interaction can change the hydrogen bond network of water molecules surrounding the RNA, thereby affecting reactivity of the RNA functional groups. Alternatively, the cofactor can exclude water from the RNA molecule, because some known cofactors, including polyethylene glycol and its derivatives, are useful for crystallization of nucleic acids and proteins.

Effect of Denaturants

It has been reported by several research groups that denaturing reagents affect cleavage specificity of the ribozymes from hepatitis delta virus (HDV)[28] and the Group I Intron from *Tetrahymena thermophila*.[29]

In studies of nonenzymatic cleavage of oligoribonucleotides, the effect of formamide, urea, ethanol, and sodium chloride on this process was examined.[14] The cleavage reaction of UCGUAA was performed under standard conditions at 37° with various concentrations of the denaturants.

The stability of UCGUAA was examined at seven concentrations of formamide, ranging from 5 m*M* to 3.2 *M*. The maximum rate of cleavage was observed at 20 m*M* formamide. As concentration increases from 20 to 200 m*M* formamide, the rate of cleavage decreases modestly. For 1 *M* and higher concentrations of formamide, the cleavage loses specificity and is observed at every position. With urea, the optimum cleavage occurs at 50 m*M* concentration. Above 200 m*M* urea, the cleavage rate is reduced significantly and is completely inhibited in 1 *M* urea. Ethanol had an effect similar to that of formamide and urea. The cleavage of

[28] J. B. Smith, P. A. Gottlieb, and G. Dinter-Gottlieb, *Biochemistry* **31,** 9629 (1992).

[29] A. J. Zaug, C. A. Grosshans, and T. R. Cech, *Biochemistry* **27,** 8924 (1988).

UCGUAA was most rapidly cleaved at 5% v/v (ca 0.85 *M*) ethanol. Finally, the stability of UCGUAA in seven different concentrations of sodium chloride (1 m*M* to 0.5 *M*) was analyzed. The maximum rate of cleavage of UCGUAA was observed at 5 m*M* sodium chloride, and the cleavage was significantly inhibited above 50 m*M* sodium chloride.

All the studied denaturants showed a similar influence on specific, nonenzymatic cleavage: a bell-shaped curve related the correlation between denaturant concentration and the rate of cleavage. The denaturants presumably affect the hydration network of the oligoribonucleotide and change the nucleophilic and electrophilic characters of functional groups directly involved in cleavage of the phosphodiester bond.

Stability of Biologically Active RNA Molecules

Stability of tRNA Molecules

The data presented heretofore concerned only chemically synthesized model RNA oligomers. To establish biological relevance, the stability of tRNA was tested under conditions similar to the nonenzymatic cleavage conditions for oligoribonucleotides. A tRNA was chosen because it is relatively small and well characterized; the secondary structure is well defined.[30] For experiments, the initiator $tRNA_i^{Met}$ from *Lupinus luteus* was used (see Fig. 2). The tRNA was labeled with ^{32}P at the 5′ terminus, or with [^{32}P]pCp at the 3′ terminus. The cleavage reactions were performed at 37° in 50 m*M* Tris-HCl (pH 7.5), 1 m*M* EDTA, and 0.1% PVP containing 50 m*M* NaCl. Sodium chloride was included to stabilize native secondary and tertiary structure of tRNA. Varying amounts of spermidine and spermine were used (0.1 μM to 10 m*M*) to induce cleavage. After 6 h the reaction products were analyzed by 14% polyacrylamide gel electrophoresis. The dominating cleavage position was the phosphodiester bond between C75 and A76, within the aminoacylation stem of the tRNA. Therefore 3′-labeling did not allow observation of the cleavage of other phosphodiester bonds, most of which are 5′ to this position, and consequently 5′-labeled $tRNA_i^{Met}$ was used for most experiments.

Analysis of the $tRNA_i^{Met}$ primary and secondary structures for single-stranded cleavage sites suggested that, besides the 3′-end degradation (cleavage between C75 and A76), there is cleavage within the anticodon loop between C34 and A35 and between U36 and t^6A37 [t^6A-N^6-(*N*-threonylcarbonyl)adenosine]. As expected, 5′-labeled $tRNA_i^{Met}$ from *Lupinus luteus* was cleaved in the anticodon loop between C34-A35 but the U36-t^6A37 linkage was resistant to cleavage. The maximum rates of tRNA cleavage were observed at 0.1 m*M* spermine or 1.0 m*M*

[30] J. G. Arnez and D. Moras, *in* "Oxford Handbook of Nucleic Acid Structure" (S. Neidle, ed.), p. 603. Oxford University Press, Oxford, 1999.

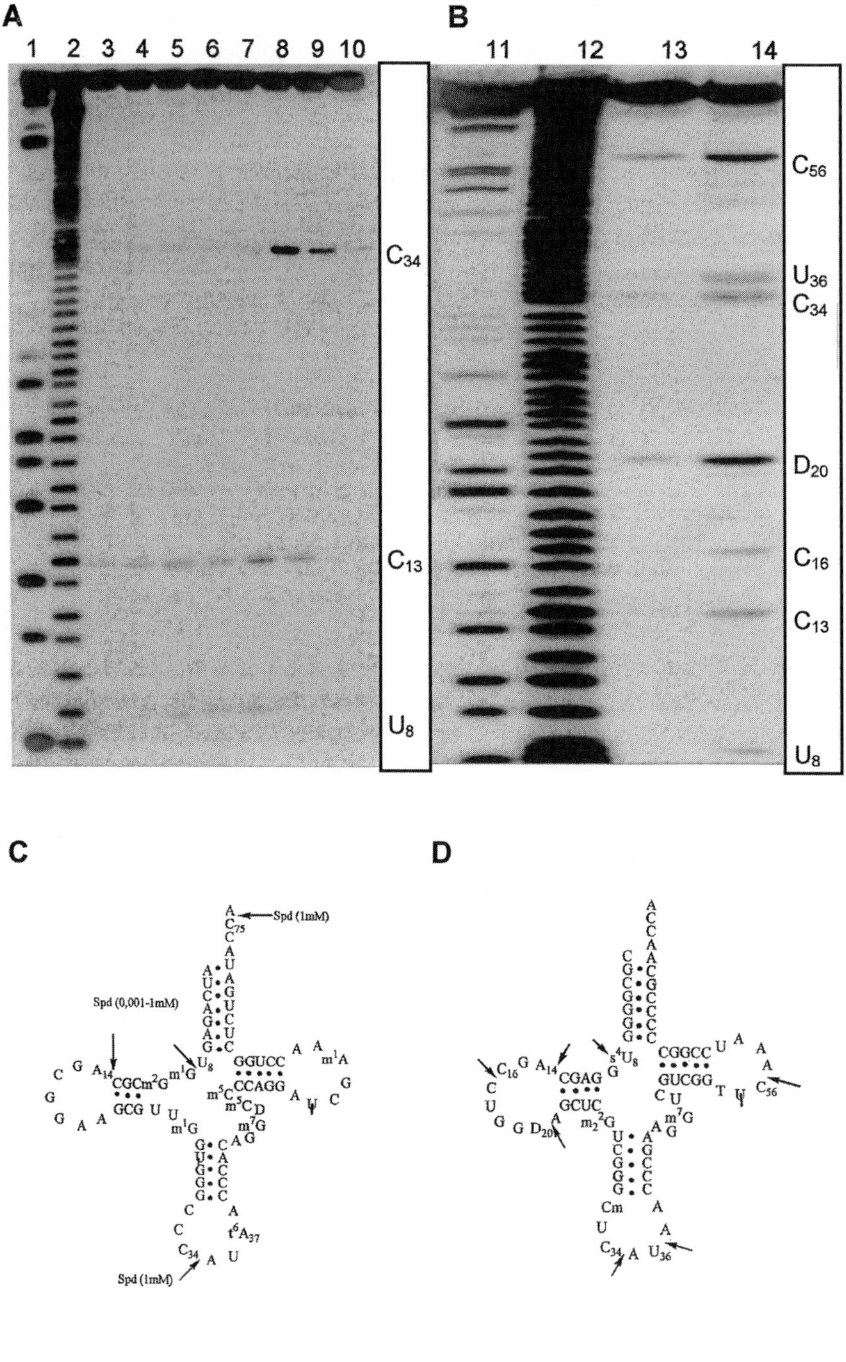
A
1 2 3 4 5 6 7 8 9 10
C34
C13
U8
B
11 12 13 14
C56
U36
C34
D20
C16
C13
U8
C
Spd (1mM)
Spd (0,001-1mM)
Spd (1mM)
D

spermidine. The $tRNA_i^{Met}$ from *Lupinus luteus* was stable in buffer without polyamine and in buffer with high polyamine concentrations (5 m*M* spermine and 10 m*M* spermidine). Other cleavage sites were observed between C13-A14 and U8-m^1G9. For the C13-A14 phosphodiester bond, cleavage was most efficient in the presence of 0.001 m*M* spermine or 0.1 m*M* spermidine. Also, 0.001 m*M* spermine was optimal for mediating cleavage of the U8-m^1G9 phosphodiester bond.

To determinate whether the stability of the U36-t^6A37 phosphodiester is due to adenosine modification, cleavage of the $tRNA_i^{Met}$ from *E. coli* was performed. It has an anticodon loop sequence very similar to $tRNA_i^{Met}$ from *L. luteus,* except C33 is replaced by U and t^6A37 by A. The data clearly indicate that $tRNA_i^{Met}$ from *E. coli* is efficiently cleaved between C34-A35, as well as between U36-A37. This confirmed previous observations that the 6-amino group of adenosine is crucial for cleavage (for example, stability of UCG$\underline{U^{6NMe}A}$A[11]) and explains why the anticodon loop of $tRNA_i^{Met}$ from *L. luteus* is cleaved in one site only. Analysis of the reaction mixtures of $tRNA_i^{Met}$ from *E. coli* revealed the presence of three additional cleavage sites. The first is located between C13 and A14, the second between dihydrouridine (D) at position 20 and A21, and the third between C56 and A57. This second cleavage is particularly fast, presumably because the nonaromatic D20 stacks poorly with A21.

Stability of Group I Intron from Tetrahymena thermophila

To explore the stability of a large RNA under standard conditions for nonenzymatic cleavage, the group I intron from *Tetrahymena thermophila* was tested.[5] The substrate was L-21 *Sca*I ribozyme derived from the *Tetrahymena thermophila* self-splicing large subunit (LSU) ribosomal RNA intron.[31,32] Since the L-21 *Sca*I ribozyme contains 409 nucleotides, it was difficult to precisely determine the position of every phosphodiester bond that was cleaved. Thus, the experiments

[31] T. R. Cech and B. L. Bass, *Ann. Rev. Biochem.* **55,** 596 (1986).
[32] T. R. Cech and S. L. Brehm, *Biochemistry* **22,** 2390 (1983).

FIG. 2. (A) The stability of $tRNA_i^{Met}$ from *Lupinus luteus* incubated for 6 h at 37° in the presence of 50 m*M* Tris-HCl (pH 7.5), 50 m*M* NaCl, 0.1% PVP, 1 m*M* EDTA and various concentrations of spermidine. Lane 1: incubated in the presence of T1 nuclease. Lane 2: alkaline ladder. Lanes 3–10: incubated in the presence of 0, 0.0001 m*M*, 0.001 m*M*, 0.01 m*M*, 1 m*M*, 5 m*M*, and 10 m*M* spermidine, respectively. (B) The stability of $tRNA_i^{fMet}$ from *E. coli* incubated for 6 h at 37° in the presence of 50 m*M* Tris-HCl (pH 7.5), 50 m*M* NaCl, 0.1% PVP, 1 m*M* EDTA, and various concentrations of spermidine. Lane 11: incubated in the presence of T1 nuclease. Lane 12: alkaline ladder. Lanes 13 and 14: incubated in the presence of 0 and 0.01 m*M* spermidine, respectively. (C) The secondary structure of $tRNA_i^{Met}$ from *Lupinus luteus*. The arrows indicate the cleavage site and optimal concentration of spermidine. (D) The secondary structure of $tRNA_i^{fMet}$ from *E. coli*. The arrows indicate the spermidine-dependent cleavage sites.

displayed more qualitative than quantitative character. However, they allowed an evaluation of the stability of the entire molecule.

The L-21 *Sca*I ribozyme was incubated in following buffers: 50 m*M* Tris-HCl (pH 7.5), 1 m*M* spermidine, 1 m*M* EDTA (buffer A); 50 m*M* Tris-HCl (pH 7.5), 1 m*M* spermidine, 1 m*M* EDTA, and 0.1% PVP (buffer B); 50 m*M* Tris-HCl (pH 7.5), 1 m*M* spermidine, 1 m*M* EDTA, 0.1% PVP, and 5 m*M* $MgCl_2$ (buffer C); and 50 m*M* Tris-HCl (pH 8.3), 75 m*M* KCl, 3 m*M* $MgCl_2$, and 10 m*M* dithiothreitol (DTT)(buffer D). After the reactions, aliquots of oligomers were analyzed by 10% polyacrylamide gel electrophoresis.

The conclusions from the gel analysis are as follows: (1) L-21 *Sca*I ribozyme is unstable in buffer B, and during the 4 h incubation it is completely cleaved to shorter fragments. The cleavage is selective and only certain phosphodiester bonds are cleaved. However, some internucleotide bonds are cleaved faster than others. In the electrophoretic pattern of the cleavage, both primary and secondary cleavage products are present. (2) The presence of 5 m*M* $MgCl_2$ in buffer C decreased the cleavage rate and even after 24 h a significant amount of L-21 *Sca*I ribozyme can be observed. Moreover, cleavage of phosphodiester bonds other than those appearing in the absence of $MgCl_2$ is observed. The cleavage is less extensive than with buffer B, and 3–4 predominant cleavage products are present. (3) In buffer A, the cleavage is slower than in the presence of PVP. However, the products of cleavage are the same as with the other buffers. (4) In the presence of T7 RNA polymerase in transcription buffer (buffer D) L-21 *Sca*I is cleaved to the same fragments as with buffer C, which contains 5 m*M* $MgCl_2$. This is presumably due to the presence of 3 m*M* $MgCl_2$ in the transcription buffer. The presence of Mg^{2+} helps maintain the structure of the ribozyme in a folded form, and in this way inhibits cleavage of single stranded fragments.

Biological Implication of Nonenzymatic RNA Cleavage

The results described above suggest that an internal single stranded YR sequence is unusually susceptible to cleavage. There have been previous observations of cleavage of large RNAs in the presence of detergents (Brij 58, Nonidet P-40 and Triton X-100) during transcription with RNA polymerase. These may reflect a similar phenomenon.[1–6]

The pattern of nonenzymatic cleavage of oligoribonucleotides is similar to that of ribonuclease A.[20–22] This similarity between the observed cleavage activity of oligoribonucleotides, especially the effects of cofactors and polyamines, and ribonuclease A could, however, reflect an evolutionary process. If RNA originally cleaved at sites intrinsically susceptible to cleavage, then it is likely that this same cleavage would be facilitated as proteins evolved. Literature data does not report any requirements for interactions of the pyrimidine and purine bases for cleavage by ribonuclease A. Thus it may be that ribonuclease A recognizes the most labile phosphodiester bond of oligoribonucleotides in the same manner as polyamines and cofactors assist in the nonenzymatic cleavage reaction.

The phenomenon of RNA-encoded cleavage could be important for regulation the stability of RNA *in vivo*. For example, the circular RNA formed from the LSU intron has a lifetime of only 6 sec in *Tetrahymena thermophila*.[33] The cleavage sites important for degradation of the polycistronic *lac* mRNA in *E. coli* have been reported to be pyrimidine-A.[1] In this case, $t_{1/2}$ for degradation is between 0.5 and 8 min at 37°. Cannistraro and co-workers observed the following distribution of the phosphodiester bonds cleavage during decay of polycistronic *lac* mRNA: 10 UA, 5 CA, 2 UU and 1 AA, and 1 GA.[34] Thus, 17 of 19 cleaved phosphodiester bonds are the ones preferred in nonenzymatic cleavage of oligoribonucleotides. Spontaneous cleavage of nine different tRNAs occurs predominantly in YA phosphodiester bonds. Dock-Bregeon and Moras reported that among 49 cuts observed, 41 precede an adenosine and 37 of them followed a pyrimidine.[3] Also, alkaline cleavage of several dinucleoside monophosphates showed that cleavage of phosphodiester bonds is preferential for YA and YC dimers.[33]

Conclusions

The specific, nonenzymatic cleavage of RNA is a structure-dependent process,[35] and certain conditions are necessary. The most important is the presence of a labile phosphodiester bond, such as YA or YC, in a single-stranded fragment. However, this single-stranded requirement is limited to only the two nucleotides flanking the phosphodiester bond. A polyamine is also necessary for efficient nonenzymatic RNA cleavage. The plot of correlation between concentration of polyamine and efficiency of cleavage is bell-shaped, and the concentration for optimal cleavage is specific to each polyamine. Other structural elements of the polyamine are also important for cleavage; at least two ammonium groups are necessary and the optimal distance between them is 0.5–0.6 nm (thus a 2–3 carbon methylene chain). Moreover, methylation of a primary ammonium group reduces the cleavage, particularly when both terminal ammonium groups are methylated. As discussed above, the correlation between structure, concentration of polyamine, efficiency of cleavage, and instability of only certain phosphodiester bonds shows that the polyamine does not merely promote alkaline cleavage. Moreover, interactions of functional groups of nucleobases adjacent to cleaved internucleotide bonds affect cleavage. Cofactors such as organic polymers and nonribonuclease protein enzymes can stimulate nonenzymatic cleavage via nonspecific interactions with the RNA.

[33] H. Witzel, *Justus Liebiegs Ann.* **635,** 182 (1960).
[34] V. J. Cannistraro, M. N. Subbarao, and D. Kennell, *J. Mol. Biol.* **192,** 257 (1986).
[35] G. A. Soukup and R. R. Breaker, *RNA* **5,** 1308 (1999).

Author Index

A

B

C

D

G

I

L

M

N

O

P

Q

R

S

T

X

Y

Z

Subject Index

A

I

K

L

M

N

Y

Z

ISBN 0-12-182242-7